Analysis of Variance, Design and Regression

Analysis of Variance, Design and Regression

Applied statistical methods

Ronald Christensen

Department of Mathematics and Statistics
University of New Mexico
Albuquerque, New Mexico
USA

CHAPMAN & HALL/CRC

Boca Raton London New York Washington, D.C.

Library of Congress Cataloging-in-Publication Data

Catalog record is available from the Library of Congress.

First edition 1996
First CRC Press reprint 1998
Originally published by Chapman & Hall

International Standard Book Number 0-412-06291-7
Library of Congress Card Number 96-84915
Printed in the United States of America 1 2 3 4 5 6 7 8 9 0
Printed on acid-free paper

To Mark, Karl, and John

It's been great fun.

Contents

Preface

This book examines the application of basic statistical methods: primarily analysis of variance and regression but with some discussion of count data. It is directed primarily towards Masters degree students in statistics studying analysis of variance, design of experiments, and regression analysis. I have found that the Masters level regression course is often popular with students outside of statistics. These students are often weaker mathematically and the book caters to that fact while continuing to give a complete matrix formulation of regression.

The book is complete enough to be used as a second course for upper division and beginning graduate students in statistics and for graduate students in other disciplines. To do this, one must be selective in the material covered, but the more theoretical material appropriate only for Statistics Masters students is generally isolated in separate subsections and, less often, in separate sections.

For a Masters level course in analysis of variance and design, I have the students review Chapter 2, I present Chapter 3 while simultaneously presenting the examples of Section 4.2, I present Chapters 5 and 6, very briefly review the first five sections of Chapter 7, present Sections 7.11 and 7.12 in detail and then I cover Chapters 9, 10, 11, 12, and 17. Depending on time constraints, I will delete material or add material from Chapter 16.

For a Masters level course in regression analysis, I again have the students review Chapter 2 and I review Chapter 3 with examples from Section 4.2. I then present Chapters 7, 13, and 14, Appendix A, Chapter 15, Sections 16.1.2, 16.3, 16.5 (along with analysis of covariance), Section 8.7 and finally Chapter 18 . All of this is done in complete detail. If any time remains I like to supplement the course with discussion of response surface methods.

As a second course for upper division and beginning graduate students in statistics and graduate students in other disciplines, I cover the first eight chapters with omission of the more technical material. A follow up course covers the less technical aspects of Chapters 9 through 15 and Appendix A.

I think the book is reasonably encyclopedic. It really contains everything I would like my students to know about applied statistics prior to them taking courses in linear model theory or log-linear models.

I believe that beginning students (even Statistics Masters students) often find statistical procedures to be a morass of vaguely related special techniques. As a result, this book focuses on four connecting themes.

1. Most inferential procedures are based on identifying a (scalar) parameter of interest, estimating that parameter, obtaining the standard error of the estimate, and identifying the appropriate reference distribution. Given these items, the inferential procedures are identical for various parameters.

2. Balanced one-way analysis of variance has a simple, intuitive interpretation in terms of comparing the sample variance of the group means with the mean of the sample variances for each group. All balanced analysis of variance problems are considered in terms of computing sample variances for various group means.

3. Comparing different models provides a structure for examining both balanced and unbalanced analysis of variance problems and for examining regression problems. In some problems the most reasonable analysis is simply to find a succinct model that fits the data well.

4. Checking assumptions is a crucial part of every statistical analysis.

The object of statistical data analysis is to reveal useful structure within the data. In a model-based setting, I know of two ways to do this. One way is to find a *succinct* model for the data. In such a case, the structure revealed is simply the model. The model selection approach is particularly appropriate when the ultimate goal of the analysis is making predictions. This book uses the model selection approach for multiple regression and for general unbalanced multifactor analysis of variance. The other approach to revealing structure is to start with a general model, identify interesting one-dimensional parameters, and perform statistical inferences on these parameters. This parametric approach requires that the general model involve parameters that are easily interpretable. We use the parametric approach for one-way analysis of variance, balanced multifactor analysis of variance, and simple linear regression. In particular, the parametric approach to analysis of variance presented here involves a strong emphasis on examining contrasts, including interaction contrasts. In analyzing two-way tables of counts, we use a partitioning method that is analogous to looking at contrasts.

All statistical models involve assumptions. Checking the validity of these assumptions is crucial because *the models we use are never correct. We hope that our models are good approximations to the true condition of the data and experience indicates that our models often work very well.* Nonetheless, to have faith in our analyses, we need to check the modeling assumptions as best we can. Some assumptions are very difficult to evaluate, e.g., the assumption that observations are statistically independent. For checking other assumptions, a variety of standard tools has been developed. Using these tools is as integral to a proper statistical analysis as is performing an appropriate confidence interval or test. For the most part, using model-checking tools without the aid of a computer is more trouble than most people are willing to tolerate.

My experience indicates that students gain a great deal of insight into balanced analysis of variance by actually doing the computations. The computation of the mean square for treatments in a balanced one-way analysis of variance is trivial on any hand calculator with a variance or standard deviation key. More importantly, the calculation reinforces the fundamental and intuitive idea behind the balanced analysis of variance test, i.e., that a mean square for treatments is just a multiple of the sample variance of the corresponding treatment means. I believe that as long as students find the balanced analysis of variance computations challenging, they should continue to do them by hand (calculator). I think that automated computation should be motivated by boredom rather than bafflement.

In addition to the four primary themes discussed above, there are several other characteristics that I have tried to incorporate into this book.

I have tried to use examples to motivate theory rather than to illustrate theory. Most chapters begin with data and an initial analysis of that data. After illustrating results for the particular data, we go back and examine general models and procedures. I have done this to make the book more palatable to two groups of people: those who only care about theory after seeing that it is useful and those unfortunates who can never bring themselves to care about theory. (The older I get, the more I identify with the first group. As for the other group, I find myself agreeing with W. Edwards Deming that experience without theory teaches nothing.) As mentioned earlier, the theoretical material is generally confined to separate subsections or, less often, separate sections, so it is easy to ignore.

I believe that the *ultimate* goal of all statistical analysis is prediction of observable quantities. I have incorporated predictive inferential procedures where they seemed natural.

The object of most statistics books is to illustrate techniques rather than to analyze data; this book is no exception. Nonetheless, I think we do students a disservice by not showing them a substantial portion of the work necessary to analyze even 'nice' data. To this end, I have tried to consistently examine residual plots, to present alternative analyses using different transformations and case deletions, and to give some final answers in plain English. I have also tried to introduce such material as early as possible. I have included reasonably detailed examinations of a three-factor analysis of variance and of a split plot design with four factors. I have included some examples in which, like real life, the final answers are not 'neat.' While I have tried to introduce statistical ideas as soon as possible, I have tried to keep the mathematics as simple as possible for as long as possible. For example, matrix formulations are postponed to the last chapter on multiple regression and the last section on unbalanced analysis of variance.

I never use side conditions or normal equations in analysis of variance.

In multiple comparison methods, (weakly) controlling the experimentwise error rate is discussed in terms of first performing an omnibus test for no treatment effects and then choosing a criterion for evaluating individual hypotheses. Most methods considered divide into those that use the omnibus F test, those that use the Studentized range test, and the Bonferroni method, which does not use any omnibus test.

I have tried to be very clear about the fact that experimental designs are set up for arbitrary groups of treatments and that factorial treatment structures are simply an efficient way of defining the treatments in some problems. Thus, the nature of a randomized complete block design does not depend on how the treatments happen to be defined. The analysis always begins with a breakdown of the sum of squares into treatments, blocks, and error. Further analysis of the treatments then focuses on whatever structure happens to be present.

The analysis of covariance chapter includes an extensive discussion of how the covariates must be chosen to maintain a valid experiment. Tukey's one degree of freedom test for nonadditivity is presented as an analysis of covariance test for the need to perform a power transformation rather than as a test for a particular type of interaction.

The chapter on confounding and fractional replication has more discussion of analyzing such data than many other books contain.

Minitab commands are presented for most analyses. Minitab was chosen because I find it the easiest of the common packages to use. However, the real point of including computer commands is to illustrate the kinds of things that one needs to specify for any computer program and the various auxiliary computations that may be necessary for the analysis. The other statistical packages used in creating the book were BMDP, GLIM, and MSUSTAT.

Acknowledgements

Many people provided comments that helped in writing this book. My colleagues Ed Bedrick, Aparna Huzurbazar, Wes Johnson, Bert Koopmans, Frank Martin, Tim O'Brien, and Cliff Qualls helped a lot. I got numerous valuable comments from my students at the University of New Mexico. Marjorie Bond, Matt Cooney, Jeff S. Davis, Barbara Evans, Mike Fugate, Jan Mines, and Jim Shields stand out in this regard. The book had several anonymous reviewers, some of whom made excellent suggestions.

I would like to thank Martin Gilchrist and Springer-Verlag for permission to reproduce Example 7.6.1 from *Plane Answers to Complex Questions: The Theory of Linear Models.* I also thank the Biometrika Trustees for permission to use the tables in Appendix B.5. Professor John Deely and the University of Canterbury in New Zealand were kind enough to support completion of the book during my sabbatical there.

Now my only question is what to do with the chapters on quality control, p^n factorials, and response surfaces that ended up on the cutting room floor.

Ronald Christensen
Albuquerque, New Mexico
February 1996

BMDP Statistical Software is located at 1440 Sepulveda Boulevard, Los Angeles, CA 90025, telephone: (213) 479-7799

MINITAB is a registered trademark of Minitab, Inc., 3081 Enterprise Drive, State College, PA 16801, telephone: (814) 238-3280, telex: 881612.

MSUSTAT is marketed by the Research and Development Institute Inc., Montana State University, Bozeman, MT 59717-0002, Attn: R.E. Lund.

Chapter 1

Introduction

In this chapter we introduce basic ideas of probability and some related mathematical concepts that are used in statistics. Values to be analyzed statistically are generally thought of as random variables; these are numbers that result from random events. The mean (average) value of a population is defined in terms of the expected value of a random variable. The variance is introduced as a measure of the variability in a random variable (population). We also introduce some special distributions (populations) that are useful in modeling statistical data. The purpose of this chapter is to introduce these ideas, so they can be used in analyzing data and in discussing statistical models.

In writing statistical models, we often use symbols from the Greek alphabet. A table of these symbols is provided in Appendix B.6.

Rumor has it that there are some students studying statistics who have an aversion to mathematics. Such people might be wise to focus on the concepts of this chapter and not let themselves get bogged down in the details. The details are given to provide a more complete introduction for those students who are not math averse.

1.1 Probability

Probabilities are numbers between zero and one that are used to explain random phenomena. We are all familiar with simple probability models. Flip a standard coin; the probability of heads is 1/2. Roll a die; the probability of getting a three is 1/6. Select a card from a well-shuffled deck; the probability of getting the queen of spades is 1/52 (assuming there are no jokers). One way to view probability models that many people find intuitive is in terms of random sampling from a fixed population. For example, the 52 cards form a fixed population and picking a card from a well-shuffled deck is a means of randomly selecting one element of the population. While we will exploit this idea of sampling from fixed populations, we should also note its limitations. For example, blood pressure is a very useful medical indicator, but even with a fixed population of people it would be very difficult to define a useful population of blood pressures. Blood pressure depends on the time of day, recent diet, current emotional state, the technique of the person taking the reading, and many other factors. Thinking about populations is very useful, but the concept can be very limiting both practically and mathematically. For measurements such as blood pressures and heights, there are difficulties in even specifying populations mathematically.

For mathematical reasons, probabilities are defined not on particular outcomes but on sets of outcomes (events). This is done so that continuous measurements can be dealt with. It seems much more natural to define probabilities on outcomes as we did in the previous paragraph, but consider some of the problems with doing that. For example, consider the problem of measuring the height of a corpse being kept in a morgue under controlled conditions. The only reason for getting morbid here is to have some hope of defining what the height is. Living people, to some extent, stretch and contract, so a height is a nebulous thing. But even given that someone has a fixed height, we can never know what it is. When someone's height is measured as 177.8 centimeters (5 feet 10 inches), their height is not really 177.8 centimeters, but (hopefully) somewhere between 177.75 and 177.85 centimeters. There is really no chance that anyone's height is *exactly* 177.8 cm, or exactly 177.8001 cm, or exactly 177.800000001 cm, or exactly 56.5955π cm, or exactly $(76\sqrt{5} + 4.5\sqrt{3})$ cm. In any neighborhood of 177.8, there are more numerical values than one could even imagine counting. The height should be somewhere in the neighborhood, but it won't be the particular value 177.8. The point is simply that trying to specify all the possible heights and their probabilities is a hopeless exercise. It simply cannot be done.

Even though individual heights cannot be measured exactly, when looking at a population of heights they follow certain patterns. There are not too many people over 8 feet (244 cm) tall. There are lots of males between 175.3 cm and 177.8 cm (5′9″ and 5′10″). With continuous values, each possible outcome has no chance of occurring, but outcomes do occur and occur with regularity. If probabilities are defined for sets instead of outcomes, these regularities can be reproduced mathematically. Nonetheless, initially the best way to learn about probabilities is to think about outcomes and their probabilities.

There are five key facts about probabilities:

1. Probabilities are between 0 and 1.
2. Something that happens with probability 1 is a sure thing.
3. If something has no chance of occurring, it has probability 0.
4. If something occurs with probability, say, .25, the probability that it will not occur is $1 - .25 = .75$.
5. If two events are mutually exclusive, i.e., if they cannot possibly happen at the same time, then the probability that either of them occurs is just the sum of their individual probabilities.

Individual outcomes are always mutually exclusive, e.g., you cannot flip a coin and get both heads and tails, so probabilities for outcomes can always be added together. Just to be totally correct, I should mention one other point. It may sound silly, but we need to assume that *something* occurring is always a sure thing. If we flip a coin, we must get either heads or tails with probability 1. We could even allow for the coin landing on its edge as long as the probabilities for all the outcomes add up to 1.

EXAMPLE 1.1.1. Consider the nine outcomes that are all combinations of three heights, tall (T), medium (M), short (S) and three eye colors, blue (Bl), brown (Br) and green (G). The combinations are displayed below.

Height–eye color combinations

		Eye color		
		Blue	Brown	Green
	Tall	T, Bl	T, Br	T, G
Height	Medium	M, Bl	M, Br	M, G
	Short	S, Bl	S, Br	S, G

The set of all outcomes is

$$\{(\mathrm{T},\mathrm{Bl}),(\mathrm{T},\mathrm{Br}),(\mathrm{T},\mathrm{G}),(\mathrm{M},\mathrm{Bl}),(\mathrm{M},\mathrm{Br}),(\mathrm{M},\mathrm{G}),(\mathrm{S},\mathrm{Bl}),(\mathrm{S},\mathrm{Br}),(\mathrm{S},\mathrm{G})\}\,.$$

The event that someone is tall consists of the three pairs in the first row of the table, i.e.,

$$\{\mathrm{T}\} = \{(\mathrm{T},\mathrm{Bl}),(\mathrm{T},\mathrm{Br}),(\mathrm{T},\mathrm{G})\}\,.$$

This is the union of the three outcomes (T, Bl), (T, Br), and (T, G). Similarly, the set of people with blue eyes is obtained from the first column of the table; it is the union of (T, Bl), (M, Bl), and (S, Bl) and can be written

$$\{\mathrm{Bl}\} = \{(\mathrm{T},\mathrm{Bl}),(\mathrm{M},\mathrm{Bl}),(\mathrm{S},\mathrm{Bl})\}\,.$$

If we know that $\{\mathrm{T}\}$ *and* $\{\mathrm{Bl}\}$ both occur, there is only one possible outcome, (T, Bl).

The event that $\{\mathrm{T}\}$ *or* $\{\mathrm{Bl}\}$ occurs consists of all outcomes in either the first row or the first column of the table, i.e.,

$$\{(\mathrm{T},\mathrm{Bl}),(\mathrm{T},\mathrm{Br}),(\mathrm{T},\mathrm{G}),(\mathrm{M},\mathrm{Bl}),(\mathrm{S},\mathrm{Bl})\}\,.$$ □

EXAMPLE 1.1.2. Table 1.1 contains probabilities for the nine outcomes that are combinations of height and eye color from Example 1.1.1.

TABLE 1.1. Height–eye color probabilities

		Eye color		
		Blue	Brown	Green
	Tall	.12	.15	.03
Height	Medium	.22	.34	.04
	Short	.06	.01	.03

Note that each of the nine numbers is between 0 and 1 and that the sum of all nine equals 1. The probability of blue eyes is

$$\begin{aligned}
\Pr(\mathrm{Bl}) &= \Pr[(\mathrm{T},\mathrm{Bl}),(\mathrm{M},\mathrm{Bl}),(\mathrm{S},\mathrm{Bl})] \\
&= \Pr(\mathrm{T},\mathrm{Bl}) + \Pr(\mathrm{M},\mathrm{Bl}) + \Pr(\mathrm{S},\mathrm{Bl}) \\
&= .12 + .22 + .06 \\
&= .4\,.
\end{aligned}$$

Similarly, $\Pr(\mathrm{Br}) = .5$ and $\Pr(\mathrm{G}) = .1$. The probability of not having blue eyes is

$$\begin{aligned}
\Pr(\text{not Bl}) &= 1 - \Pr(\mathrm{Bl}) \\
&= 1 - .4 \\
&= .6\,.
\end{aligned}$$

Note also that $\Pr(\text{not Bl}) = \Pr(\mathrm{Br}) + \Pr(\mathrm{G})$.

The (*marginal*) probabilities for the various heights are:

$$\Pr(\mathrm{T}) = .3, \quad \Pr(\mathrm{M}) = .6, \quad \Pr(\mathrm{S}) = .1\,.$$ □

Even if there are a countable (but infinite) number of possible outcomes, one can still define a probability by defining the probabilities for each outcome. It is only for measurement data that one really needs to define probabilities on sets.

Two random events are said to be independent if knowing that one of them occurs provides no information about the probability that the other event will occur. Formally, two events A and B are *independent* if

$$\Pr(A \text{ and } B) = \Pr(A)\Pr(B).$$

Thus the probability that *both* events A and B occur is just the product of the individual probabilities that A occurs and that B occurs. As we will begin to see in the next section, independence plays an important role is statistics.

EXAMPLE 1.1.3. Using the probabilities of Table 1.1 and the computations of Example 1.1.2, the events tall and brown eyes are independent because

$$\Pr(\text{tall and brown}) = \Pr(\text{T}, \text{Br}) = .15 = (.3)(.5) = \Pr(\text{T}) \times \Pr(\text{Br}).$$

On the other hand, medium height and blue eyes are *not* independent because

$$\Pr(\text{medium and blue}) = \Pr(\text{M}, \text{Bl}) = .22 \neq (.6)(.4) = \Pr(\text{M}) \times \Pr(\text{Bl}).$$ □

1.2 Random variables and expectations

A *random variable* is simply a function that relates outcomes with numbers. The key point is that any probability associated with the outcomes induces a probability on the numbers. The numbers and their associated probabilities can then be manipulated mathematically. Perhaps the most common and intuitive example of a random variable is rolling a die. The outcome is that a face of the die with a certain number of spots ends up on top. These can be pictured as

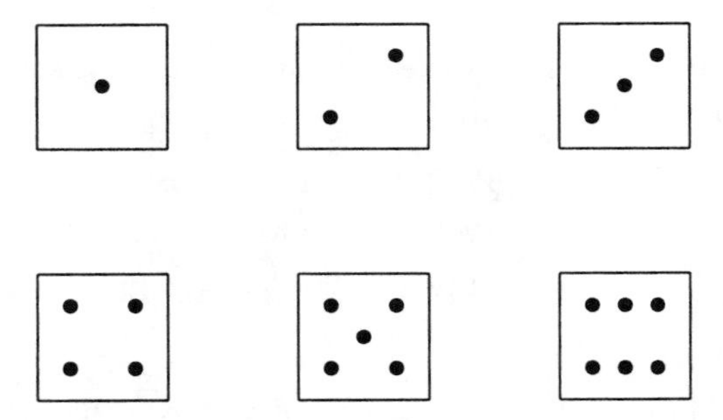

Without even thinking about it, we define a random variable that transforms these six faces into the numbers 1, 2, 3, 4, 5, 6.

In statistics we think of observations as random variables. These are often some number associated with a randomly selected member of a population. For example, one random variable is the height of a person who is to be randomly selected from among University of New Mexico students. (A random selection gives the same probability to every individual in the population. This random variable presumes that we have well-defined methods for measuring height and defining UNM students.) Rather than measuring height, we could define a different random variable by giving the person a score of 1 if that person is female and 0 if the person is male. We can also perform mathematical operations on random variables to yield new random variables. Suppose we plan to select a random sample of 10 students, then we would have 10 random variables with female and male scores. The sum of these random variables is another random variable that tells us the (random) number of females in the sample. Similarly, we would have 10 random variables for heights and we can define a new random variable consisting of the average of the 10 individual height random variables. Some random variables are related in obvious ways. In our example we measure both a height and a sex score on each person. If the sex score variable is a 1 (telling us that the person is female), it suggests that the height may be smaller than we would otherwise suspect. Obviously some female students are taller than some male students, but knowing a person's sex definitely changes our knowledge about their probable height.

We do similar things in tossing a coin.

EXAMPLE 1.2.1. Consider tossing a coin twice. The four outcomes are ordered pairs of heads (H) and tails (T). The outcomes can be denoted as

$$(H,H) \qquad (H,T) \qquad (T,H) \qquad (T,T)$$

where the outcome of the first toss is the first element of the ordered pair.

The standard probability model has the four outcomes equally probable, i.e., $1/4 = \Pr(H,H) = \Pr(H,T) = \Pr(T,H) = \Pr(T,T)$. Equivalently

		Second toss		
		Heads	Tails	Total
First toss	Heads	1/4	1/4	1/2
	Tails	1/4	1/4	1/2
	Total	1/2	1/2	1

The probability of heads on each toss is 1/2. The probability of tails is 1/2. We will define two random variables:

$$y_1(r,s) = \begin{cases} 1 & \text{if } r = H \\ 0 & \text{if } r = T \end{cases}$$

$$y_2(r,s) = \begin{cases} 1 & \text{if } s = H \\ 0 & \text{if } s = T \end{cases} .$$

Thus, y_1 is 1 if the first toss is heads and 0 otherwise. Similarly, y_2 is 1 if the second toss is heads and 0 otherwise.

The event $y_1 = 1$ occurs if and only if we get heads on the first toss. We get heads on the first toss by getting either of the outcome pairs (H,H) or (H,T). In other words, the event $y_1 = 1$ is equivalent to the event $\{(H,H),(H,T)\}$. The probability of $y_1 = 1$ is just the sum of the probabilities of the outcomes in $\{(H,H),(H,T)\}$.

$$\begin{aligned} \Pr(y_1 = 1) &= \Pr(H,H) + \Pr(H,T) \\ &= 1/4 + 1/4 = 1/2. \end{aligned}$$

Similarly,

$$\begin{aligned} \Pr(y_1 = 0) &= \Pr(T,H) + \Pr(T,T) \\ &= 1/2 \\ \Pr(y_2 = 1) &= 1/2 \\ \Pr(y_2 = 0) &= 1/2. \end{aligned}$$

Now define another random variable,

$$W(r,s) = y_1(r,s) + y_2(r,s)\,.$$

The random variable W is the total number of heads in two tosses:

$$\begin{aligned} W(H,H) &= 2 \\ W(H,T) &= W(T,H) = 1 \\ W(T,T) &= 0\,. \end{aligned}$$

Moreover,

$$\begin{aligned} \Pr(W=2) &= \Pr(H,H) = 1/4 \\ \Pr(W=1) &= \Pr(H,T) + \Pr(T,H) = 1/2 \\ \Pr(W=0) &= \Pr(T,T) = 1/4\,. \end{aligned}$$

These three equalities define a probability on the outcomes 0, 1, 2. In working with W, we can ignore the original outcomes of head-tail pairs and work only with the new outcomes 0, 1, 2 and their associated probabilities. We can do the same thing for y_1 and y_2. The probability table given earlier can be rewritten in terms of y_1 and y_2.

		y_2		
		1	0	y_1 totals
y_1	1	1/4	1/4	1/2
	0	1/4	1/4	1/2
	y_2 totals	1/2	1/2	1

Note that, for example, $\Pr[(y_1, y_2) = (1,0)] = 1/4$ and $\Pr(y_1 = 1) = 1/2$. This table shows the *distribution* of the probabilities for y_1 and y_2 both separately (marginally) and jointly. □

For any random variable, *a statement of the possible outcomes and their associated probabilities is referred to as the (marginal)* probability distribution *of the random variable. For two or more random variables, a table or other statement of the possible joint outcomes and their associated probabilities is referred to as the* joint probability distribution *of the random variables.*

All of the entries in the center of the distribution table given above for y_1 and y_2 are independent. For example,

$$\Pr[(y_1, y_2) = (1,0)] \equiv \Pr(y_1 = 1 \text{ and } y_2 = 0) = \Pr(y_1 = 1)\Pr(y_2 = 0).$$

We therefore say that y_1 and y_2 are independent. In general, *two random variables y_1 and y_2 are independent if any event involving only y_1 is independent of any event involving only y_2.*

Independence is an extremely important concept in statistics. Observations to be analyzed are commonly assumed to be independent. This means that *the random aspect of one observation contains no information about the random aspect of any other observation.* (However, every observation tells us about fixed aspects of the underlying population such as the population center.) *For most purposes in applied statistics, just this intuitive understanding of independence is sufficient.*

1.2.1 Expected Values and Variances

The *expected value (population mean)* of a random variable is a number characterizing the middle of the distribution. For a random variable y with a discrete distribution (i.e., one having a finite or countable number of outcomes), the expected value is

$$E(y) \equiv \sum_{\text{all } r} r \Pr(y = r) .$$

EXAMPLE 1.2.2. Let y be the result of picking one of the numbers 2, 4, 6, 8 at random. Because the numbers are chosen at random,

$$1/4 = \Pr(y = 2) = \Pr(y = 4) = \Pr(y = 6) = \Pr(y = 8) .$$

The expected value in this simple example is just the mean (average) of the four possible outcomes.

$$\begin{aligned} E(y) &= 2\left(\frac{1}{4}\right) + 4\left(\frac{1}{4}\right) + 6\left(\frac{1}{4}\right) + 8\left(\frac{1}{4}\right) \\ &= (2 + 4 + 6 + 8)/4 \\ &= 5 . \end{aligned}$$

□

EXAMPLE 1.2.3. Five pieces of paper are placed in a hat. The papers have the numbers 2, 4, 6, 6, and 8 written on them. A piece of paper is picked at random. The expected value of the number drawn is the mean of the numbers on the five pieces of paper. Let y be the random variable that relates a piece of paper to the number on that paper. Each piece of paper has the same probability of being chosen, so, because the number 6 appears twice, the distribution of the random variable y is

$$\begin{aligned} \frac{1}{5} &= \Pr(y = 2) = \Pr(y = 4) = \Pr(y = 8) \\ \frac{2}{5} &= \Pr(y = 6) . \end{aligned}$$

The expected value is

$$\begin{aligned} E(y) &= 2\left(\frac{1}{5}\right) + 4\left(\frac{1}{5}\right) + 6\left(\frac{2}{5}\right) + 8\left(\frac{1}{5}\right) \\ &= (2 + 4 + 6 + 6 + 8)/5 \\ &= 5.2 . \end{aligned}$$

□

EXAMPLE 1.2.4. Consider the coin tossing random variables y_1, y_2, and W from Example 1.2.1. Recalling that y_1 and y_2 have the same distribution,

$$\begin{aligned} E(y_1) &= 1\left(\frac{1}{2}\right) + 0\left(\frac{1}{2}\right) = \frac{1}{2} \\ E(y_2) &= \frac{1}{2} \\ E(W) &= 2\left(\frac{1}{4}\right) + 1\left(\frac{1}{2}\right) + 0\left(\frac{1}{4}\right) = 1 . \end{aligned}$$

The variable y_1 is the number of heads in the first toss of the coin. The two possible values 0 and 1 are equally probable, so the middle of the distribution is $1/2$. W is the number of heads in two tosses; the expected number of heads in two tosses is 1.

□

The expected value indicates the middle of a distribution, but does not indicate how spread out (dispersed) a distribution is.

EXAMPLE 1.2.5. Consider three gambles that I will allow you to take. In game z_1 you have equal chances of winning 12, 14, 16, or 18 dollars. In game z_2 you can again win 12, 14, 16, or 18 dollars, but now the probabilities are .1 that you will win either $14 or $16 and .4 that you will win $12 or $18. The third game I call z_3 and you can win 5, 10, 20, or 25 dollars with equal chances. Being no fool, I require you to pay me $16 for the privilege of playing any of these games. We can write each game as a random variable.

z_1	outcome	12	14	16	18
	probability	.25	.25	.25	.25

z_2	outcome	12	14	16	18
	probability	.4	.1	.1	.4

z_3	outcome	5	10	20	25
	probability	.25	.25	.25	.25

I try to be a good casino operator, so none of these games is fair. You have to pay $16 to play, but you only expect to win $15. It is easy to see that

$$E(z_1) = E(z_2) = E(z_3) = 15\,.$$

But don't forget that I'm taking a loss on the ice-water I serve to players and I also have to pay for the pictures of my extended family that I've decorated my office with.

Although the games z_1, z_2, and z_3 have the same expected value, the games (random variables) are very different. Game z_2 has the same outcomes as z_1, but much more of its probability is placed farther from the middle value 15. The extreme observations 12 and 18 are much more probable under z_2 than z_1. If you currently have $16, need $18 for your grandmother's bunion removal, and anything less than $18 has no value to you, then z_2 is obviously a better game for you than z_1.

Both z_1 and z_2 are much more tightly packed around 15 than is z_3. If you needed $25 for the bunion removal, z_3 is the game to play because you can win it all in one play with probability .25. In either of the other games you would have to win at least five times to get $25, a much less likely occurrence. Of course you should realize that the most probable result is that Grandma will have to live with her bunion. You are unlikely to win either $18 or $25. While the ethical moral of this example is that a fool and his money are soon parted, the statistical point is that there is more to a random variable than its mean. The variability of random variables is also important. □

The *(population) variance* is a measure of how spread out a distribution is from its expected value. Let y be a random variable having a discrete distribution with $E(y) = \mu$, then the variance of y is

$$\mathrm{Var}(y) \equiv \sum_{all\ r} (r - \mu)^2 \Pr(y = r)\,.$$

This is the average squared distance of the outcomes from the center of the population. More technically, it is the expected squared distance between the outcomes and the mean of the distribution.

EXAMPLE 1.2.6. Using the random variables of Example 1.2.5,

$$\begin{aligned}
\text{Var}(z_1) &= (12-15)^2(.25)+(14-15)^2(.25) \\
&\quad +(16-15)^2(.25)+(18-15)^2(.25) \\
&= 5 \\
\text{Var}(z_2) &= (12-15)^2(.4)+(14-15)^2(.1) \\
&\quad +(16-15)^2(.1)+(18-15)^2(.4) \\
&= 7.4 \\
\text{Var}(z_3) &= (5-15)^2(.25)+(10-15)^2(.25) \\
&\quad +(20-15)^2(.25)+(25-15)^2(.25) \\
&= 62.5
\end{aligned}$$

The increasing variances from z_1 through z_3 indicate that the random variables are increasingly spread out. However, the value $\text{Var}(z_3) = 62.5$ seems too large to measure the relative variabilities of the three random variables. More on this later. □

EXAMPLE 1.2.7. Consider the coin tossing random variables of Examples 1.2.1 and 1.2.4.

$$\begin{aligned}
\text{Var}(y_1) &= \left(1-\frac{1}{2}\right)^2\frac{1}{2}+\left(0-\frac{1}{2}\right)^2\frac{1}{2}=\frac{1}{4} \\
\text{Var}(y_2) &= \frac{1}{4} \\
\text{Var}(W) &= (2-1)^2\left(\frac{1}{4}\right)+(1-1)^2\left(\frac{1}{2}\right)+(0-1)^2\left(\frac{1}{4}\right)=\frac{1}{2}.
\end{aligned}$$ □

A problem with the variance is that it is measured on the wrong scale. If y is measured in meters, $\text{Var}(y)$ involves the terms $(r-\mu)^2$; hence it is measured in meters squared. To get things back on the original scale, we consider the *standard deviation* of y

$$\text{Std. dev. }(y) \equiv \sqrt{\text{Var}(y)}.$$

EXAMPLE 1.2.8. Consider the random variables of Examples 1.2.5 and 1.2.6.

$$\begin{aligned}
\text{Std. dev. }(z_1) &= \sqrt{5} &\doteq 2.236 \\
\text{Std. dev. }(z_2) &= \sqrt{7.4} &\doteq 2.720 \\
\text{Std. dev. }(z_3) &\equiv \sqrt{62.5} &\doteq 7.906
\end{aligned}$$

The standard deviation of z_3 is 3 to 4 times larger than the others. From examining the distributions, the standard deviations seem to be more intuitive measures of relative variability than the variances. The variance of z_3 is 8.5 to 12.5 times larger than the other variances; these values seem unreasonably inflated. □

Standard deviations and variances are useful as measures of the relative dispersions of different random variables. The actual numbers themselves do not mean much. Moreover, there are other equally good measures of dispersion that can give results that are somewhat inconsistent with these. One reason standard deviations and variances are so widely used is because they are convenient mathematically. In addition, normal (Gaussian) distributions are widely used in applied statistics and are completely characterized by their expected values (means) and variances (or standard deviations). Knowing these two numbers, the mean and variance, one knows everything about a normal distribution.

1.2.2 CHEBYSHEV'S INEQUALITY

Another place in which the numerical values of standard deviations are useful is in applications of Chebyshev's inequality. Chebyshev's inequality gives a lower bound on the probability that a random variable is within an interval. Chebyshev's inequality is important in quality control work (control charts) and in evaluating prediction intervals.

Let y be a random variable with $E(y) = \mu$ and $\text{Var}(y) = \sigma^2$. Chebyshev's inequality states that for any number $k > 1$,

$$\Pr[\mu - k\sigma < y < \mu + k\sigma] \geq 1 - \frac{1}{k^2}.$$

Thus the probability that y will fall within k standard deviations of μ is at least $1 - (1/k^2)$.

The beauty of Chebyshev's inequality is that it holds for absolutely any random variable y. Thus we can always make some statement about the probability that y is in a symmetric interval about μ. In many cases, for particular choices of y, the probability of being in the interval can be much greater than $1 - k^{-2}$. For example, if $k = 3$ and y has a normal distribution as discussed in the next section, the probability of being in the interval is actually .997, whereas Chebyshev's inequality only assures us that the probability is no less than $1 - 3^{-2} \doteq .889$. However, we know the lower bound of .889 applies regardless of whether y has a normal distribution.

1.2.3 COVARIANCES AND CORRELATIONS

Often we take two (or more) observations on the same member of a population. We might observe the height and weight of a person. We might observe the IQs of a wife and husband. (Here the population consists of married couples.) In such cases we may want a numerical measure of the relationship between the pairs of observations. Data analysis related to these concepts is known as regression analysis and is discussed in Chapters 7, 13, 14, and 15. These ideas are also briefly used for testing normality in Section 2.4.

The *covariance* is a measure of the linear relationship between two random variables. Suppose y_1 and y_2 are discrete random variables. Let $E(y_1) = \mu_1$ and $E(y_2) = \mu_2$. The covariance between y_1 and y_2 is

$$Cov(y_1, y_2) \equiv \sum_{\text{all } (r,s)} (r - \mu_1)(s - \mu_2) \Pr(y_1 = r, y_2 = s).$$

Positive covariances arise when relatively large values of y_1 tend to occur with relatively large values y_2 and small values of y_1 tend to occur with small values of y_2. On the other hand, negative covariances arise when relatively large values of y_1 tend to occur with relatively small values y_2 and small values of y_1 tend to occur with large values of y_2. It is simple to see from the definition that, for example,

$$\text{Var}(y_1) = Cov(y_1, y_1).$$

In an attempt to get a handle on what the numerical value of the covariance means, it is often rescaled into a *correlation coefficient*.

$$\text{Corr}(y_1, y_2) \equiv Cov(y_1, y_2) / \sqrt{\text{Var}(y_1)\text{Var}(y_2)}.$$

Positive values of the correlation have the same qualitative meaning as positive values of the covariance, but now a *perfect* increasing linear relationship is indicated by a correlation of 1. Similarly, negative correlations and covariances mean similar things, but a perfect decreasing linear relationship gives a correlation of -1. The absence of any linear relationship is indicated by a value of 0.

A perfect linear relationship between y_1 and y_2 means that an increase of one unit in, say, y_1 dictates an exactly proportional change in y_2. For example, if we make a series of very accurate temperature measurements on something and simultaneously use one device calibrated in Fahrenheit and one calibrated in Celsius, the pairs of numbers should have an essentially perfect linear relationship.

EXAMPLE 1.2.9. Let z_1 and z_2 be two random variables defined by the following probability table:

		z_2 0	1	2	z_1 totals
	6	0	1/3	0	1/3
z_1	4	1/3	0	0	1/3
	2	0	0	1/3	1/3
	z_2 totals	1/3	1/3	1/3	1

Then

$$\begin{aligned} E(z_1) &= 6\left(\frac{1}{3}\right) + 4\left(\frac{1}{3}\right) + 2\left(\frac{1}{3}\right) = 4, \\ E(z_2) &= 0\left(\frac{1}{3}\right) + 1\left(\frac{1}{3}\right) + 2\left(\frac{1}{3}\right) = 1, \end{aligned}$$

$$\begin{aligned} \text{Var}(z_1) &= (2-4)^2\left(\frac{1}{3}\right) + (4-4)^2\left(\frac{1}{3}\right) + (6-4)^2\left(\frac{1}{3}\right) \\ &= 8/3, \\ \text{Var}(z_2) &= (0-1)^2\left(\frac{1}{3}\right) + (1-1)^2\left(\frac{1}{3}\right) + (2-1)^2\left(\frac{1}{3}\right) \\ &= 2/3, \end{aligned}$$

$$\begin{aligned} Cov(z_1, z_2) &= (2-4)(0-1)(0) + (2-4)(1-1)(0) + (2-4)(2-1)\left(\frac{1}{3}\right) \\ &\quad + (4-4)(0-1)\left(\frac{1}{3}\right) + (4-4)(1-1)(0) + (4-4)(2-1)(0) \\ &\quad + (6-4)(0-1)(0) + (6-4)(1-1)\left(\frac{1}{3}\right) + (6-4)(2-1)(0) \\ &= -2/3, \end{aligned}$$

$$\begin{aligned} \text{Corr}(z_1, z_2) &= (-2/3)\Big/\sqrt{(8/3)(2/3)} \\ &= -1/2. \end{aligned}$$

This correlation indicates that relatively large z_1 values tend to occur with relatively small z_2 values. However, the correlation is considerably greater than -1, so the linear relationship is less than perfect. Moreover, the correlation measures the linear relationship and *fails to identify the perfect nonlinear relationship* between z_1 and z_2. If $z_1 = 2$, then $z_2 = 2$. If $z_1 = 4$, then $z_2 = 0$. If $z_1 = 6$, then $z_2 = 1$. If you know one random variable, you know the other, but because the relationship is nonlinear, the correlation is not ± 1. □

EXAMPLE 1.2.10. Consider the coin toss random variables y_1 and y_2 from Example 1.2.1. We earlier observed that these two random variables are independent. If so, there should be no relationship between them (linear or otherwise). We now show that their covariance is 0.

$$\begin{aligned} Cov(y_1, y_2) &= \left(0-\frac{1}{2}\right)\left(0-\frac{1}{2}\right)\frac{1}{4}+\left(0-\frac{1}{2}\right)\left(1-\frac{1}{2}\right)\frac{1}{4} \\ &\quad +\left(1-\frac{1}{2}\right)\left(0-\frac{1}{2}\right)\frac{1}{4}+\left(1-\frac{1}{2}\right)\left(1-\frac{1}{2}\right)\frac{1}{4} \\ &= \frac{1}{16}-\frac{1}{16}-\frac{1}{16}+\frac{1}{16}=0. \end{aligned}$$

□

In general, whenever two random variables are independent, their covariance (and thus their correlation) is 0. However, just because two random variables have 0 covariance does not *imply that they are independent. Independence has to do with not having any kind of relationship; covariance examines only linear relationships. Random variables with nonlinear relationships can have zero covariance but not be independent.*

1.2.4 RULES FOR EXPECTED VALUES AND VARIANCES

We now present some extremely useful results that allow us to show that statistical estimates are reasonable and to establish the variability associated with statistical estimates. These results relate to the expected values, variances, and covariances of linear combinations of random variables. A linear combination of random variables is something that only involves multiplying random variables by fixed constants, adding such terms together, and adding a constant.

Proposition 1.2.11. Let y_1, y_2, y_3, and y_4 be random variables and let a_1, a_2, a_3, and a_4 be real numbers.

1. $E(a_1y_1 + a_2y_2 + a_3) = a_1E(y_1) + a_2E(y_2) + a_3$.
2. If y_1 and y_2 are independent, $\text{Var}(a_1y_1 + a_2y_2 + a_3) = a_1^2\text{Var}(y_1) + a_2^2\text{Var}(y_2)$.
3. $\text{Var}(a_1y_1 + a_2y_2 + a_3) = a_1^2\text{Var}(y_1) + 2a_1a_2Cov(y_1, y_2) + a_2^2\text{Var}(y_2)$.
4. $Cov(a_1y_1 + a_2y_2, a_3y_3 + a_4y_4) = a_1a_3Cov(y_1, y_3) + a_1a_4Cov(y_1, y_4) + a_2a_3Cov(y_2, y_3) + a_2a_4Cov(y_2, y_4)$.

All of these results generalize to linear combinations involving more than two random variables.

EXAMPLE 1.2.12. Recall that when independently tossing a coin twice, the total number of heads, W, is the sum of y_1 and y_2, the number of heads on the first and second tosses respectively. We have already seen that $E(y_1) = E(y_2) = .5$ and that $E(W) = 1$. We now illustrate item 1 of the proposition by finding $E(W)$ again. Since $W = y_1 + y_2$,

$$E(W) = E(y_1 + y_2) = E(y_1) + E(y_2) = .5 + .5 = 1.$$

We have also seen that $\text{Var}(y_1) = \text{Var}(y_2) = .25$ and that $\text{Var}(W) = .5$. Since the coin tosses are independent, item 2 above gives

$$\text{Var}(W) = \text{Var}(y_1 + y_2) = \text{Var}(y_1) + \text{Var}(y_2) = .25 + .25 = .5\,.$$

The key point is that this is an easier way of finding the expected value and variance of W than using the original definitions.

□

We now illustrate the generalizations referred to in Proposition 1.2.11. We begin by looking at the problem of estimating the mean of a population.

EXAMPLE 1.2.13. Let y_1, y_2, y_3, and y_4 be four random variables each with the same (population) mean μ, i.e., $E(y_i) = \mu$ for $i = 1, 2, 3, 4$. We can compute the *sample mean* (average) of these, defining

$$\begin{aligned} \bar{y}_{\bullet} &\equiv \frac{y_1 + y_2 + y_3 + y_4}{4} \\ &= \frac{1}{4}y_1 + \frac{1}{4}y_2 + \frac{1}{4}y_3 + \frac{1}{4}y_4. \end{aligned}$$

The $\bullet$ in the subscript of $\bar{y}_{\bullet}$ indicates that the sample mean is obtained by summing over the subscripts of the y_is. The $\bullet$ notation is not necessary for this problem but becomes useful in dealing with the analysis of variance problems treated later in the book.

Using item 1 of Proposition 1.2.11 we find that

$$\begin{aligned} E(\bar{y}_{\bullet}) &= E\left(\frac{1}{4}y_1 + \frac{1}{4}y_2 + \frac{1}{4}y_3 + \frac{1}{4}y_4\right) \\ &= \frac{1}{4}E(y_1) + \frac{1}{4}E(y_2) + \frac{1}{4}E(y_3) + \frac{1}{4}E(y_4) \\ &= \frac{1}{4}\mu + \frac{1}{4}\mu + \frac{1}{4}\mu + \frac{1}{4}\mu \\ &= \mu. \end{aligned}$$

Thus one observation on $\bar{y}_{\bullet}$ would make a reasonable estimate of μ.

If we also assume that the y_is are independent with the same variance, say, σ^2, then from item 2 of Proposition 1.2.11

$$\begin{aligned} \text{Var}(\bar{y}_{\bullet}) &= \text{Var}\left(\frac{1}{4}y_1 + \frac{1}{4}y_2 + \frac{1}{4}y_3 + \frac{1}{4}y_4\right) \\ &= \left(\frac{1}{4}\right)^2 \text{Var}(y_1) + \left(\frac{1}{4}\right)^2 \text{Var}(y_2) \\ &\quad + \left(\frac{1}{4}\right)^2 \text{Var}(y_3) + \left(\frac{1}{4}\right)^2 \text{Var}(y_4) \\ &= \left(\frac{1}{4}\right)^2 \sigma^2 + \left(\frac{1}{4}\right)^2 \sigma^2 + \left(\frac{1}{4}\right)^2 \sigma^2 + \left(\frac{1}{4}\right)^2 \sigma^2 \\ &= \frac{\sigma^2}{4}. \end{aligned}$$

The variance of $\bar{y}_{\bullet}$ is only one fourth of the variance of an individual observation. Thus the $\bar{y}_{\bullet}$ observations are more tightly packed around their mean μ than the y_is are. This indicates that one observation on $\bar{y}_{\bullet}$ is more likely to be close to μ than an individual y_i. □

These results for $\bar{y}_{\bullet}$ hold quite generally; they are not restricted to the average of four random variables. *If $\bar{y}_{\bullet} = (1/n)(y_1 + \cdots + y_n) = \sum_{i=1}^{n} y_i/n$ is the sample mean of n independent random variables all with the same population mean μ and population variance σ^2,*

$$E(\bar{y}_{\bullet}) = \mu$$

and

$$\text{Var}(\bar{y}_{\bullet}) = \frac{\sigma^2}{n}.$$

In fact, proving these general results uses exactly the same ideas as the proofs for a sample of size 4.

As with a sample of size 4, the general results on $\bar{y}_\cdot$ are very important in statistical inference. If we are interested in determining the population mean μ from future data, the obvious estimate is the average of the individual observations, $\bar{y}_\cdot$. The observations are random, so the estimate $\bar{y}_\cdot$ is also a random variable and the middle of its distribution is $E(\bar{y}_\cdot) = \mu$, the original population mean. Thus $\bar{y}_\cdot$ is a reasonable estimate of μ. Moreover, $\bar{y}_\cdot$ is a better estimate than any particular observation y_i because $\bar{y}_\cdot$ has a smaller variance, σ^2/n as opposed to σ^2 for y_i. With less variability in the estimate, any one observation of $\bar{y}_\cdot$ is more likely to be near its mean μ than a single observation y_i. In practice, we obtain data and compute a sample mean. This constitutes one observation on the random variable $\bar{y}_\cdot$. If our sample mean is to be a good estimate of μ, our one look at $\bar{y}_\cdot$ had better have a good chance of being close to μ. This occurs when the variance of $\bar{y}_\cdot$ is small. Note that the larger the sample size n, the smaller is σ^2/n, the variance of $\bar{y}_\cdot$. We will return to these ideas later.

Generally, we will use item 1 of Proposition 1.2.11 to show that estimates are *unbiased*. In other words, we will show that the expected value of an estimate is what we are trying to estimate. In estimating μ, we have $E(\bar{y}_\cdot) = \mu$, so $\bar{y}_\cdot$ is an unbiased estimate of μ. All this really does is show that $\bar{y}_\cdot$ is a reasonable estimate of μ. More important than showing unbiasedness is using item 2 to find variances of estimates. Statistical inference depends crucially on having some idea of the variability of an estimate. Item 2 is the primary tool in finding the appropriate variance for different estimates.

1.3 Continuous distributions

As discussed in Section 1.1, many things that we would like to measure are, in the strictest sense, not measurable. We cannot find a building's exact height even though we can approximate it *extremely* accurately. This theoretical inability to measure things exactly has little impact on our practical world, but it has a substantial impact on the theory of statistics.

The data in most statistical applications can be viewed as counts of how often some event has occurred or as measurements. Probabilities associated with count data are easy to describe. We discuss some probability models for count data in Sections 1.4 and 1.5. With measurement data, we can never obtain an exact value, so we don't even try. With measurement data, we assign probabilities to intervals. Thus we do not discuss the probability that a person has the height 177.8 cm or 177.8001 cm or 56.5955π cm, but we do discuss the probability that someone has a height *between* 177.75 cm and 177.85 cm. Typically, we think of doing this in terms of pictures. We associate probabilities with areas under curves. (Mathematically, this involves integral calculus and is discussed in a brief appendix at the end of the chapter.) Figure 1.1 contains a picture of a continuous probability distribution (*a density*). Probabilities must be between 0 and 1, so the curve must always be nonnegative (to make all areas nonnegative) and the area under the entire curve must be 1.

Figure 1.1 also shows a point $K(1-\alpha)$. This point divides the area under the curve into two parts. The probability of obtaining a number less than $K(1-\alpha)$ is $1-\alpha$, i.e., the area under the curve to the left of $K(1-\alpha)$ is $1-\alpha$. The probability of obtaining a number greater than $K(1-\alpha)$ is α, i.e., the area under the curve to the right of $K(1-\alpha)$. $K(1-\alpha)$ is a particular number, so the probability is 0 that $K(1-\alpha)$ will actually occur. There is no area under a curve associated with any particular point.

Pictures such as Figure 1.1 are often used as models for populations of measurements. With a fixed population of measurements, it is natural to form a histogram, i.e., a bar chart that plots intervals for the measurement against the proportion of individuals that fall into a particular interval. Pictures such as Figure 1.1 can be viewed as approximations to such histograms. The probabilities described by pictures such as Figure 1.1 are those associated with randomly picking an individual from the population. Thus, randomly picking an individual from the population modeled by Figure 1.1 yields a measurement less than $K(1-\alpha)$ with probability $1-\alpha$.

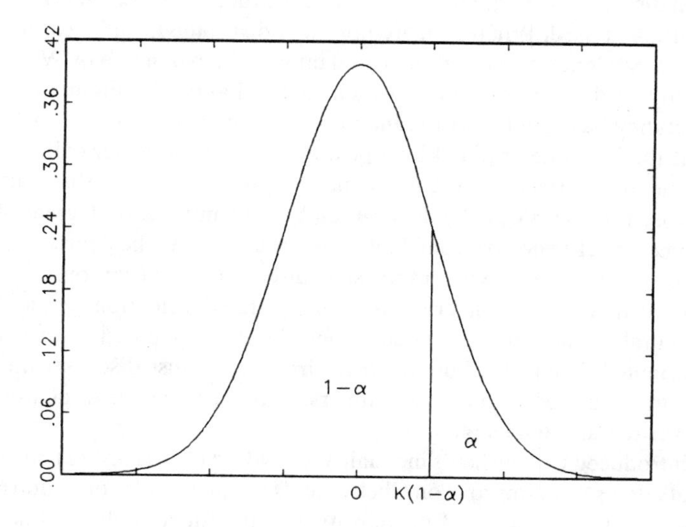

FIGURE 1.1. A continuous probability density.

Ideas similar to those discussed in Section 1.2 can be used to define expected values, variances, and covariances for continuous distributions. These extensions involve integral calculus and are discussed in the appendix. In any case, Proposition 1.2.11 continues to apply.

The most commonly used distributional model for measurement data is the *normal* distribution (also called the *Gaussian* distribution). The bell shaped curve in Figure 1.1 is referred to as the standard normal curve. The formula for writing the curve is not too ugly, it is

$$f(x) = \frac{1}{\sqrt{2\pi}} e^{-x^2/2}.$$

Here e is the base of natural logarithms. Unfortunately, even with calculus it is very difficult to compute areas under this curve. Finding standard normal probabilities requires a table.

By itself, the standard normal curve has little value in modeling measurements. For one thing, the curve is centered about 0. I don't take many measurements where I think the central value should be 0. To make the normal distribution a useful model, we need to expand the standard normal into a family of distributions with different centers (expected values) μ and different spreads (standard deviations) σ. By appropriate recentering and rescaling of the plot, all of these curves will have the same shape as Figure 1.1.

The standard normal distribution is the special case of a normal with $\mu = 0$ and $\sigma = 1$. The standard normal plays an important role because it is the only normal distribution that we need tabled. (Obviously, we could not table normal distributions for every possible value of μ and σ.) Suppose a measurement y has a normal distribution with mean μ, standard deviation σ, and variance σ^2. We write this as

$$y \sim N(\mu, \sigma^2).$$

Normal distributions have the property that

$$\frac{y-\mu}{\sigma} \sim N(0, 1),$$

cf. Exercise 1.6.2. This standardization process allows us to get by with only the standard normal table for finding probabilities for all normal distributions.

The standard normal distribution is sometimes used in constructing statistical inferences but more often a similar distribution is used. When data are normally distributed, statistical inferences often require something called Student's t distribution. (Student was the pen name of W. S. Gosset.) The t distribution is a family of distributions all of which look roughly like Figure 1.1. They are all symmetric about 0, but they have slightly different amounts of dispersion (spread). The amount of variability in each distribution is determined by a positive integer parameter called the *degrees of freedom*. With only 1 degree of freedom, the mathematical properties of a t distribution are fairly bizarre. (This special case is called a Cauchy distribution.) As the number of degrees of freedom get larger, the t distributions get better behaved and have less variability. As the degrees of freedom gets arbitrarily large, the t distribution approximates the standard normal distribution.

Two other distributions that come up later are the chi-squared distribution (χ^2) and the F distribution. These arise naturally when drawing conclusions about the population variance from data that are normally distributed. Both distributions differ from those just discussed in that both are asymmetric and both are restricted to positive numbers. However, the basic idea of probabilities being areas under curves remains unchanged.

In Section 1.2, we introduced Chebyshev's inequality. Shewhart (1931, p. 177) discusses work by Camp and Meidell that allows us to improve on Chebyshev's inequality for continuous distributions. Once again let $E(y) = \mu$ and $\text{Var}(y) = \sigma^2$. If the density, i.e., the function that defines the curve, is symmetric, unimodal (has only one peak), and always decreases as one moves farther away from the mode, then the inequality can be sharpened to

$$\Pr[\mu - k\sigma < y < \mu + k\sigma] \geq 1 - \frac{1}{(2.25)k^2}.$$

As discussed in the previous section, with y normal and $k = 3$, the true probability is .997, Chebyshev's inequality gives a lower bound of .889, and the new improved Chebyshev inequality gives a lower bound of .951. By making some relatively innocuous assumptions, we get a substantial improvement in the lower bound.

1.4 The binomial distribution

There are a few distributions that are used in the vast majority of statistical applications. The reason for this is that they tend to occur naturally. The normal distribution is one. As discussed in the next chapter, the normal distribution occurs in practice because a result called The central limit theorem dictates that many distributions can be approximated by the normal. Two other distributions, the binomial and the multinomial, occur in practice because they are very simple. In this section we discuss the binomial. The next section introduces the multinomial distribution. The results of this section are only used in Chapter 8 and in discussions of transformations.

If you have independent identical random trials and count how often something (anything) occurs, the appropriate distribution is the binomial. What could be simpler?

EXAMPLE 1.4.1. Being somewhat lonely in my misspent youth, I decided to go to a dating service. The service was to provide me with five dates. Being a very open-minded soul, I convinced myself that the results of one date would not influence my opinion about other dates. From my limited experience with the opposite sex, I have found that I enjoy about 40% of such brief encounters. I decided that my money would be well spent if I enjoyed two or more of the five dates. Unfortunately, my loan shark repossessed my 1954 Studebaker before I could indulge in this taste of nirvana. Back in those days, we chauvinists believed: no wheels – no women. Nevertheless, let us compute the probability that I would have been satisfied with the dating service. Let W be the number of dates I would have enjoyed. The simplest way to find the probability of satisfaction is

$$\begin{aligned} \Pr(W \geq 2) &= 1 - \Pr(W < 2) \\ &= 1 - \Pr(W = 0) - \Pr(W = 1), \end{aligned}$$

but that is much too easy. Let's compute

$$\Pr(W \geq 2) = \Pr(W = 2) + \Pr(W = 3) + \Pr(W = 4) + \Pr(W = 5).$$

In particular, we compute each term on the right-hand side.

Write the outcome of the five dates as an ordered collection of Ls and Ds. For example, (L, D, L, D, D) indicates that I like the first and third dates, but dislike the second, fourth, and fifth.

To like five dates, I must like everyone of them.

$$\Pr(W = 5) = \Pr(L, L, L, L, L).$$

Remember, I assumed that the dates were independent and that the probability of my liking any one is .4. Thus,

$$\begin{aligned} \Pr(W = 5) &= \Pr(L)\Pr(L)\Pr(L)\Pr(L)\Pr(L) \\ &= (.4)^5. \end{aligned}$$

The probability of liking four dates is a bit more complicated. I could only dislike one date, but there are five different choices for the date that I could dislike. It could be the fifth, the fourth, the third, the second, or the first. Any pattern of 4 Ls and a D excludes the other patterns from occurring, e.g., if the only date I dislike is the fourth, then the only date I dislike cannot be the second. Since the patterns are mutually exclusive (disjoint), the probability of disliking one date is the sum of the probabilities of the individual patterns.

$$\begin{aligned} \Pr(W = 4) &= \Pr(L, L, L, L, D) \\ &\quad + \Pr(L, L, L, D, L) \\ &\quad + \Pr(L, L, D, L, L) \\ &\quad + \Pr(L, D, L, L, L) \\ &\quad + \Pr(D, L, L, L, L). \end{aligned} \tag{1.4.1}$$

By assumption $\Pr(L) = .4$, so $\Pr(D) = 1 - \Pr(L) = 1 - .4 = .6$. The dates are independent, so

$$\begin{aligned} \Pr(L, L, L, L, D) &= \Pr(L)\Pr(L)\Pr(L)\Pr(L)\Pr(D) \\ &= (.4)^4 .6. \end{aligned}$$

Similarly,

$$\begin{aligned} \Pr(L, L, L, D, L) &= \Pr(L, L, D, L, L) \\ &= \Pr(L, D, L, L, L) \\ &= \Pr(D, L, L, L, L) \\ &= (.4)^4 .6. \end{aligned}$$

Summing up the values in equation (1.4.1),

$$\Pr(W = 4) = 5(.4)^4(.6).$$

Computing the probability of liking three dates is even worse.

$$\begin{aligned}\Pr(W=3) &= \Pr(L,L,L,D,D)\\ &\quad+\Pr(L,L,D,L,D)\\ &\quad+\Pr(L,D,L,L,D)\\ &\quad+\Pr(D,L,L,L,D)\\ &\quad+\Pr(L,L,D,D,L)\\ &\quad+\Pr(L,D,L,D,L)\\ &\quad+\Pr(D,L,L,D,L)\\ &\quad+\Pr(L,D,D,L,L)\\ &\quad+\Pr(D,L,D,L,L)\\ &\quad+\Pr(D,D,L,L,L)\end{aligned}$$

Again all of these patterns have exactly the same probability. For example, using independence

$$\Pr(D,L,D,L,L)=(.4)^3(.6)^2 .$$

Adding up all of the patterns

$$\Pr(W=3)=10(.4)^3(.6)^2 .$$

By now it should be clear that

$$\Pr(W=2)=(\text{no. of patterns with 2 Ls and 3 Ds})(.4)^2(.6)^3 .$$

The number of patterns can be computed as

$$\binom{5}{2}\equiv\frac{5!}{2!(5-2)!}\equiv\frac{5\bullet4\bullet3\bullet2\bullet1}{(2\bullet1)(3\bullet2\bullet1)}=10 .$$

The probability that I would be satisfied with the dating service is

$$\begin{aligned}\Pr(W\ge2) &= 10(.4)^2(.6)^3+10(.4)^3(.6)^2+5(.4)^4.6+(.4)^5\\ &= .663 .\end{aligned}$$

□

Binomial random variables can also be generated by sampling from a fixed population. If we were going to make 20 random selections from the UNM student body, the number of females would have a binomial distribution. Given a set of procedures for defining and sampling the student body, there would be some fixed number of students of which a given number would be females. Under random sampling, the probability of selecting a female on any of the 20 trials would be simply the proportion of females in the population. Although it is very unlikely to occur in this example, the sampling scheme must allow the possibility of students being selected more than once in the sample. If people were not allowed to be chosen more than once, each successive selection would change the proportion of females available for the subsequent selection. Of course, when making 20 selections out of a population of over 20,000 UNM students, even if you did not allow people to be reselected, the changes in the proportions of females are insubstantial and the binomial distribution makes a good approximation to the true distribution. On the other hand, if the entire student population was 40 rather than 20,000+, it might not be wise to use the binomial approximation when people are not allowed to be reselected.

Typically, the outcome of interest in a binomial is referred to as a success. If the probability of a success is p for each of N independent identical trials, then the number of successes y has a binomial distribution with parameters N and p. Write

$$y \sim Bin(N,p) .$$

The distribution of y is

$$\Pr(y = r) = \binom{N}{r} p^r (1 - p)^{N-r}$$

for $r = 0, 1, \ldots, N$. Here

$$\binom{N}{r} \equiv \frac{N!}{r!(N - r)!}$$

where for any positive integer m, $m! \equiv m(m - 1)(m - 2)\cdots(2)(1)$ and $0! \equiv 1$. The notation $\binom{N}{r}$ is read "N choose r" because it is the number of distinct ways of choosing r individuals out of a collection containing N individuals.

EXAMPLE 1.4.2. The random variables in Example 1.2.1 were y_1, the number of heads on the first toss of a coin, y_2, the number of heads on the second toss of a coin, and W, the combined number of heads from the two tosses. These have the following distributions:

$$\begin{aligned} y_1 &\sim Bin\left(1, \frac{1}{2}\right) \\ y_2 &\sim Bin\left(1, \frac{1}{2}\right) \\ W &\sim Bin\left(2, \frac{1}{2}\right). \end{aligned}$$

Note that W, the $Bin\left(2, \frac{1}{2}\right)$, was obtained by adding together the two independent $Bin\left(1, \frac{1}{2}\right)$ random variables y_1 and y_2. This result is quite general. Any $Bin(N, p)$ random variable can be written as the sum of N independent $Bin(1, p)$ random variables. □

Given the probability distribution of a binomial, we can find the mean (expected value) and variance. By definition, if $y \sim Bin(N, p)$, the mean is

$$E(y) = \sum_{r=0}^{N} r \binom{N}{r} p^r (1 - p)^{N-r}.$$

This is difficult to evaluate directly, but by writing y as the sum of N independent $Bin(1, p)$ random variables and using Exercise 1.6.1 and Proposition 1.2.11, it is easily seen that

$$E(y) = Np.$$

Similarly, the variance of y is

$$\text{Var}(y) = \sum_{r=0}^{N} (r - Np)^2 \binom{N}{r} p^r (1 - p)^{N-r}$$

but by again writing y as the sum of N independent $Bin(1, p)$ random variables and using Exercise 1.6.1 and Proposition 1.2.11, it is easily seen that

$$\text{Var}(y) = Np(1 - p).$$

Exercise 1.6.8 consists of proving these mean and variance formulae.

On occasion we will need to look at both the number of successes from a group of N trials and the number of failures at the same time. If the number of successes is y_1 and the number of failures is y_2, then

$$\begin{aligned} y_2 &= N - y_1 \\ y_1 &\sim Bin(N,p) \end{aligned}$$

and

$$y_2 \sim Bin(N, 1-p)\,.$$

The last result holds because, with independent identical trials, the number of outcomes that we call failures must also have a binomial distribution. If p is the probability of success, the probability of failure is $1-p$. Of course,

$$\begin{aligned} E(y_2) &= N(1-p) \\ \text{Var}(y_2) &= N(1-p)p\,. \end{aligned}$$

Note that $\text{Var}(y_1) = \text{Var}(y_2)$ regardless of the value of p. Finally,

$$Cov(y_1, y_2) = -Np(1-p)$$

and

$$\text{Corr}(y_1, y_2) = -1\,.$$

There is a perfect linear relationship between y_1 and y_2. If y_1 goes up one count, y_2 goes down one count. When we look at both successes and failures write

$$(y_1, y_2) \sim Bin\big(N, p, (1-p)\big)\,.$$

This is the simplest case of the multinomial distribution discussed in the next section.

1.5 The multinomial distribution

The multinomial distribution is a generalization of the binomial allowing more than two categories. The results in this section are only used in Chapter 8.

EXAMPLE 1.5.1. Consider the probabilities for the nine height and eye color categories given in Example 1.1.2. The probabilities are repeated below.

Height–eye color probabilities

		Eye color		
		Blue	Brown	Green
	Tall	.12	.15	.03
Height	Medium	.22	.34	.04
	Short	.06	.01	.03

Suppose a random sample of 50 individuals was obtained with these probabilities. For example, one might have a population of 100 people in which 12 were tall with blue eyes, 15 were tall with brown eyes, 3 were short with green eyes, etc. We could randomly select one of the 100 people as the first individual in the sample. Then, returning that individual to the population, take another random selection from the 100 to be the second individual. We are to proceed in this way until 50 people are selected. Note that with a population of 100 and a sample of 50 there is a substantial chance that some people would be selected more than once. The numbers of selections falling into each of the nine categories has a multinomial distribution with $N = 50$ and these probabilities.

It is unlikely that one would actually perform sampling from a population of 100 people as described above. Typically, one would not allow the same person to be chosen more than once. However, if we had a population of 10,000 people where 1200 were tall with blue eyes, 1500 were tall with brown eyes, 300 were short with green eyes, etc., with a sample size of 50 we might be willing to allow the possibility of selecting the same person more than once simply because it is extremely unlikely to happen. Technically, to obtain the multinomial distribution with $N = 50$ and these probabilities, when sampling from a fixed population we need to allow individuals to appear more than once. However, when taking a small sample from a large population, it does not matter much whether or not you allow people to be chosen more than once, so the multinomial often provides a good approximation even when individuals are excluded from reappearing in the sample. □

Consider a group of N independent identical trials in which each trial results in the occurrence of one of q events. Let y_i, $i = 1, \ldots, q$ be the number of times that the ith event occurs and let p_i be the probability that the ith event occurs on any trial. The p_is must satisfy $p_1 + p_2 + \cdots + p_q = 1$. We say that $(y_1, \ldots, y_q)$ has a multinomial distribution with parameters $N, p_1, \ldots, p_q$. Write

$$(y_1, \ldots, y_q) \sim Mult(N, p_1, \ldots, p_q).$$

The distribution is given by the probabilities

$$\begin{aligned}\Pr(y_1 = r_1, \ldots, y_q = r_q) &= \frac{N!}{r_1! \cdots r_q!} p_1^{r_1} \cdots p_q^{r_q} \\ &= \left(N! \Big/ \prod_{i=1}^{q} r_i!\right) \prod_{i=1}^{q} p_i^{r_i}.\end{aligned}$$

Here the r_is are allowed to be any whole numbers with each $r_i \geq 0$ and $r_1 + \cdots + r_q = N$. Note that if $q = 2$, this is just a binomial distribution. In general, each individual component y_i of a multinomial consists of N trials in which category i either occurs or does not occur, so individual components have the marginal distributions

$$y_i \sim Bin(N, p_i).$$

It follows that

$$E(y_i) = Np_i$$

and

$$\mathrm{Var}(y_i) = Np_i(1 - p_i).$$

It can also be shown that

$$Cov(y_i, y_j) = -Np_ip_j \qquad \text{for} \qquad i \neq j.$$

EXAMPLE 1.5.2. Suppose that the 50 individuals from Example 1.5.1 fall into the categories as listed below.

Height–eye color observations

		Eye color		
		Blue	Brown	Green
	Tall	5	8	2
Height	Medium	10	18	2
	Short	3	1	1

The probability of getting this particular table is

$$\frac{50!}{5!8!2!10!18!2!3!1!1!}(.12)^5(.15)^8(.03)^2(.22)^{10}(.34)^{18}(.04)^2(.06)^3(.01)^1(.03)^1.$$

This number is zero to over 5 decimal places. The fact that this is a very small number is not surprising. There are a lot of possible tables, so the probability of getting any particular table is very small. In fact, many of the possible tables are *much* less likely to occur than this table.

Let's return to thinking about the observations as random. The expected number of observations for each category is given by Np_i. It is easily seen that the expected counts for the cells are as given below.

Height–eye color expected values

		Eye color		
		Blue	Brown	Green
	Tall	6.0	7.5	1.5
Height	Medium	11.0	17.0	2.0
	Short	3.0	0.5	1.5

Note that the expected counts need not be integers.

The variance for, say, the number of tall blue-eyed people in this population is $50(.12)(1-.12) = 5.28$. The variance of the number of short green-eyed people is $50(.03)(1-.03) = 1.455$. The covariance between the number of tall blue-eyed people and the number of short green-eyed people is $-50(.12)(.03) = -.18$. The correlation between the numbers of tall blue-eyed people and short green-eyed people is $-.18/\sqrt{(5.28)(1.455)} = -0.065$. □

APPENDIX: PROBABILITY FOR CONTINUOUS DISTRIBUTIONS

As stated in Section 1.3, probabilities are sometimes defined as areas under a curve. The curve, called a probability density function or just a density, must be defined by some nonnegative function $f(\bullet)$. (Nonnegative to ensure that probabilities are always positive.) Thus the probability that a random observation y is between two numbers, say a and b, is the area under the curve measured between a and b. Using calculus, this is

$$\Pr[a < y < b] = \int_a^b f(y)\,dy.$$

Because we are measuring areas under curves, there is no area associated with any one point, so $\Pr[a < y < b] = \Pr[a \le y < b] = \Pr[a < y \le b] = \Pr[a \le y \le b]$. The area under the entire curve must be 1, i.e.,

$$1 = \Pr[-\infty < y < \infty] = \int_{-\infty}^{\infty} f(y)\,dy.$$

Figure 1.1 indicates that the probability below $K(1-\alpha)$ is $1-\alpha$, i.e.,

$$1-\alpha = \Pr[y < K(1-\alpha)] = \int_{-\infty}^{K(1-\alpha)} f(y)\,dy$$

and that the probability above $K(1-\alpha)$ is α, i.e.,

$$\alpha = \Pr[y > K(1-\alpha)] = \int_{K(1-\alpha)}^{\infty} f(y)\,dy.$$

The expected value of y is defined as

$$E(y) = \int_{-\infty}^{\infty} y f(y)\, dy.$$

For any function $g(y)$, the expected value is

$$E[g(y)] = \int_{-\infty}^{\infty} g(y) f(y)\, dy.$$

In particular, if we let $E(y) = \mu$ and $g(y) = (y - \mu)^2$, we define the variance as

$$\text{Var}(y) = E[(y - \mu)^2] = \int_{-\infty}^{\infty} (y - \mu)^2 f(y)\, dy.$$

To define the covariance between two random variables, say y_1 and y_2, we need a joint density $f(y_1, y_2)$. We can find the density for y_1 alone as

$$f_1(y_1) = \int_{-\infty}^{\infty} f(y_1, y_2)\, dy_2$$

and we can write $E(y_1)$ in two equivalent ways

$$E(y_1) = \int_{-\infty}^{\infty} \int_{-\infty}^{\infty} y_1 f(y_1, y_2)\, dy_1\, dy_2 = \int_{-\infty}^{\infty} y_1 f_1(y_1)\, dy_1.$$

Writing $E(y_1) = \mu_1$ and $E(y_2) = \mu_2$ we can now define the covariance between y_1 and y_2 as

$$Cov(y_1, y_2) = \int_{-\infty}^{\infty} \int_{-\infty}^{\infty} (y_1 - \mu_1)(y_2 - \mu_2) f(y_1, y_2)\, dy_1\, dy_2.$$

1.6 Exercises

EXERCISE 1.6.1. Use the definitions to find the expected value and variance of a $Bin(1, p)$ distribution.

EXERCISE 1.6.2. Let y be a random variable with $E(y) = \mu$ and $\text{Var}(y) = \sigma^2$. Show that

$$E\left(\frac{y - \mu}{\sigma}\right) = 0$$

and

$$\text{Var}\left(\frac{y - \mu}{\sigma}\right) = 1.$$

Let $\bar{y}_\cdot$ be the sample mean of n independent observations y_i with $E(y_i) = \mu$ and $\text{Var}(y_i) = \sigma^2$. What is the expected value and variance of

$$\frac{\bar{y}_\cdot - \mu}{\sigma/\sqrt{n}}?$$

Hint: For the first part, write

$$\frac{y - \mu}{\sigma} \quad \text{as} \quad \frac{1}{\sigma} y - \frac{\mu}{\sigma}$$

and use Proposition 1.2.11.

EXERCISE 1.6.3. Let y be the random variable consisting of the number of spots that face up upon rolling a die. Give the distribution of y. Find the expected value, variance, and standard deviation of y.

EXERCISE 1.6.4. Consider your letter grade for this course. Obviously, it is a random phenomenon. Define the 'grade point' random variable: $y(A) = 4$, $y(B) = 3$, $y(C) = 2$, $y(D) = 1$, $y(F) = 0$. If you were lucky enough to be taking the course from me, you would find that I am an easy grader. I give 5% As, 10% Bs, 35% Cs, 30% Ds, and 20% Fs. I also assign grades at random, that is to say, my tests generate random scores. Give the distribution of y. Find the expected value, variance, and standard deviation of the grade points a student would earn in my class. (Just in case you hadn't noticed, I'm being sarcastic.)

EXERCISE 1.6.5. Referring to Exercise 1.6.4, suppose I have a class of 40 students, what is the joint distribution for the numbers of students who get each of the five grades? Note that we are no longer looking at how many grade points an individual student might get, we are now counting how many occurrences we observe of various events. What is the distribution for the number of students who get Bs? What is the expected value of the number of students who get Cs? What is the variance and standard deviation of the number of students who get Cs? What is the probability that in a class of 5 students, 1 gets an A, 2 get Cs, 1 gets a D, and 1 fails?

EXERCISE 1.6.6. Graph the function $f(x) = 1$ if $0 < x < 1$ and $f(x) = 0$ otherwise. This is known as the uniform density on $(0, 1)$. If we use this curve to define a probability function, what is the probability of getting an observation larger than $1/4$? Smaller than $2/3$? Between $1/3$ and $7/9$?

EXERCISE 1.6.7. Arthritic ex-football players prefer their laudanum made with Old Pain-Killer Scotch by two to one. If we take a random sample of 5 arthritic ex-football players, what is the distribution of the number who will prefer Old Pain-Killer? What is the probability that only 2 of the ex-players will prefer Old Pain-Killer? What is the expected number who will prefer Old Pain-Killer? What are the variance and standard deviation of the number who will prefer Old Pain-Killer?

EXERCISE 1.6.8. Let $W \sim Bin(N,p)$ and for $i = 1, \ldots, N$ take independent y_is that are $Bin(1,p)$. Argue that W has the same distribution as $y_1 + \cdots + y_N$. Use this fact, along with Exercise 1.6.1 and Proposition 1.2.11, to find $E(W)$ and $\text{Var}(W)$.

EXERCISE 1.6.9. Appendix B.1 gives probabilities for a family of distributions that all look roughly like Figure 1.1. All members of the family are symmetric about zero and the members are distinguished by having different numbers of degrees of freedom (df). They are called t distributions. For $0 \leq \alpha \leq 1$, the α percentile of a t distribution with df degrees of freedom is the point x such that $\Pr[t(df) \leq x] = \alpha$. For example, from Table B.1 the row corresponding to $df = 10$ and the column for the .90 percentile tells us that $\Pr[t(10) \leq 1.372] = .90$.

(a) Find the .99 percentile of a $t(7)$ distribution.

(b) Find the .975 percentile of a $t(50)$ distribution.

(c) Find the probability that a $t(25)$ is less than or equal to 3.450.

(d) Find the probability that a $t(100)$ is less than or equal to 2.626.

(e) Find the probability that a $t(16)$ is greater than 2.92.

(f) Find the probability that a $t(40)$ is greater than 1.684.

(g) Recalling that t distributions are symmetric about zero, what is the probability that a $t(40)$ distribution is less than -1.684?

(h) What is the probability that a $t(40)$ distribution is between -1.684 and 1.684?

(i) What is the probability that a $t(25)$ distribution is less than -3.450?

(j) What is the probability that a $t(25)$ distribution is between -3.450 and 3.450?

EXERCISE 1.6.10. Consider a random variable that takes on the values 25, 30, 45, and 50 with probabilities .15, .25, .35, and .25, respectively. Find the expected value, variance, and standard deviation of this random variable.

EXERCISE 1.6.11. Consider three independent random variables X, Y, and Z. Suppose $E(X) = 25$, $E(Y) = 40$, and $E(Z) = 55$ with $\text{Var}(X) = 4$, $\text{Var}(Y) = 9$, and $\text{Var}(Z) = 25$.

(a) Find $E(2X + 3Y + 10)$ and $\text{Var}(2X + 3Y + 10)$.

(b) Find $E(2X + 3Y + Z + 10)$ and $\text{Var}(2X + 3Y + Z + 10)$.

EXERCISE 1.6.12. As of 1994, Duke University had been in the final four of the NCAA's national basketball championship tournament seven times in nine years. Suppose their appearances were independent and that they had a probability of .25 for winning the tournament in each of those years.

(a) What is the probability that Duke would win two national championships in those seven appearances?

(b) What is the probability that Duke would win three national championships in those seven appearances?

(c) What is the expected number of Duke championships in those seven appearances?

(d) What is the variance of the number of Duke championships in those seven appearances?

EXERCISE 1.6.13. Graph the function $f(x) = 2x$ if $0 < x < 1$ and $f(x) = 0$ otherwise. If we use this curve to define a probability function, what is the probability of getting an observation larger than $1/4$? Smaller than $2/3$? Between $1/3$ and $7/9$?

EXERCISE 1.6.14. A pizza parlor makes small, medium, and large pizzas. Over the years they make 20% small pizzas, 35% medium pizzas, and 45% large pizzas. On a given Tuesday night they were asked to make only 10 pizzas. If the orders were independent and representative of the long-term percentages, what is the probability that the orders would be for four small, three medium, and three large pizzas. On such a night, what is the expected number of large pizzas to be ordered and what is the expected number of small pizzas to be ordered? What is the variance of the number of large pizzas to be ordered and what is the variance of the number of medium pizzas to be ordered?

EXERCISE 1.6.15. When I order a limo, 65% of the time the driver is male. Assuming independence, what is the probability that 6 of my next 8 drivers are male? What is the expected number of male drivers among my next eight? What is the variance of the number of male drivers among my next eight?

EXERCISE 1.6.16. When I order a limo, 65% of the time the driver is clearly male, 30% of the time the driver is clearly female, and 5% of the time the gender of the driver is indeterminant. Assuming independence, what is the probability that among my next 8 drivers 5 are clearly male and 3 are clearly female? What is the expected number of indeterminant drivers among my next eight? What is the variance of the number of clearly female drivers among my next eight?

Chapter 2

One sample

In this chapter we examine the analysis of a single *random* sample consisting of n independent observations from some population.

2.1 Example and introduction

EXAMPLE 2.1.1. Consider the dropout rate from a sample of math classes at the University of New Mexico in the 1984–85 school year as reported by Koopmans (1987). The data are

$$5, 22, 10, 12, 8, 17, 2, 25, 10, 10, 7, 7, 40, 7, 9, 17, 12, 12, 1,$$

$$13, 10, 13, 16, 3, 14, 17, 10, 10, 13, 59, 11, 13, 5, 12, 14, 3, 14, 15.$$

This list of $n = 38$ observations is not very illuminating. A graphical display of the numbers is more informative. Figure 2.1 plots the data above a single axis. This is often called a *dot plot*. From Figure 2.1, we see that most of the observations are between 0 and 18. There are two conspicuously large observations. Going back to the original data we identify these as the values 40 and 59. In particular, these two *outlying* values strongly suggest that the data do not follow a bell shaped curve and thus that the data do not follow a normal distribution. □

Typically, for one sample of data we assume that the n observations are

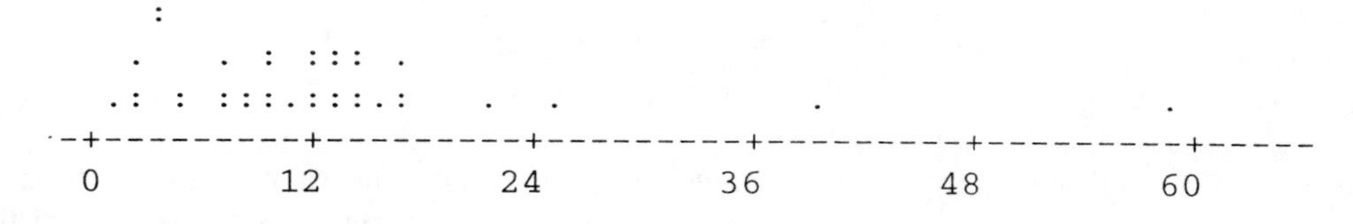

FIGURE 2.1. Dot plot for drop rate percentage data.

Data		Distribution
$y_1, y_2, \ldots, y_n$	independent	$N(\mu, \sigma^2)$

The key assumptions are that the observations are independent and have the same distribution. In particular, we assume they have the same (unknown) mean μ and the same (unknown) variance σ^2.

These assumptions of independence and a constant distribution should be viewed as only useful approximations to actual conditions. Often the most valuable approach to evaluating these assumptions is simply to think hard about whether they are reasonable. In any case, the conclusions we reach are only as good as the assumptions we have made. The only way to be positive that these assumptions are true is if we arrange for them to be true. If we have a fixed finite population and take a random sample from the population allowing elements of the population to be observed more than once, then the assumptions (other than normality) are true. In Example 2.1.1, if we had the dropout rates for all math classes in the year and randomly selected these 38 while allowing for classes to appear more than once in the sample, the assumptions of independence with the same distribution are satisfied.

The ideal conditions of independent sampling from a fixed population are difficult to achieve. Many populations refuse to hold still while we sample them. For example, the population of students at a large university changes almost continuously (during working hours). To my way of thinking, the populations associated with most interesting data are virtually impossible to define unambiguously. Who really cares about the dropout rates for 1984–85? As such, they can only be used to fix blame. Our real interest is in what the data can tell us about current and future dropout rates. If the data are representative of current or future conditions, the data can be used to fix problems. For example, one might find out whether certain instructors generate huge dropout rates and avoid taking classes from them. It is difficult to decide whether these or any data are representative of current or future conditions because we cannot possibly know the future population and we cannot practically know the current population. As mentioned earlier, often our best hope is to think hard about whether these data approximate independent observations from the population of interest.

Even when sampling from a fixed population, we use approximations. In practice we rarely allow elements of a fixed population to be observed more than once in a sample. This invalidates the assumptions. If the first sampled element is eliminated, the second element is actually being sampled from a different population than the first. (One element has been eliminated.) Fortunately, when the sample contains a small proportion of the fixed population, the standard assumptions make a good approximation. Moreover, the normal distribution is never more than an approximation to a fixed population. The normal distribution has an infinite number of possible outcomes, while fixed populations are finite. Often, the normal distribution makes a good approximation, especially if we do our best to validate it. In addition, the assumption of a normal distribution is only used when drawing conclusions from small samples. For large samples we can get by without the assumption of normality.

Our primary objective is to draw conclusions about the mean μ. We condense the data into summary statistics. These are the sample mean, the sample variance, and the sample standard deviation. The sample mean has the algebraic formula

$$\bar{y}_{\bullet} \equiv \frac{1}{n} \sum_{i=1}^{n} y_i = \frac{1}{n} [y_1 + y_2 + \cdots + y_n]$$

where the $\bullet$ in $\bar{y}_{\bullet}$ indicates that the mean is obtained by averaging the y_is over the subscript i. The sample mean $\bar{y}_{\bullet}$ estimates the population mean μ. The sample variance is an estimate of the population variance σ^2. The sample variance is *essentially* the average squared distance of the observations from the sample mean,

$$s^2 \equiv \frac{1}{n-1}\sum_{i=1}^{n}(y_i - \bar{y}_\cdot)^2 = \frac{1}{n-1}\left[(y_1 - \bar{y}_\cdot)^2 + (y_2 - \bar{y}_\cdot)^2 + \cdots + (y_n - \bar{y}_\cdot)^2\right]. \quad (2.1.1)$$

The sample standard deviation is just the square root of the sample variance,

$$s \equiv \sqrt{s^2}.$$

EXAMPLE 2.1.2. The sample mean of the dropout rate data is

$$\bar{y}_\cdot = \frac{5 + 22 + 10 + 12 + 8 + \cdots + 3 + 14 + 15}{38} = 13.11.$$

If we think of these data as a sample from the fixed population of math dropout rates in 1984–85, $\bar{y}_\cdot$ is obviously an estimate of the simple average of all the dropout rates of all the classes in that academic year. Equivalently, $\bar{y}_\cdot$ is an estimate of the expected value for the random variable defined as the dropout rate obtained when we randomly select one class from the fixed population. Alternatively, we may interpret $\bar{y}_\cdot$ as an estimate of the mean of some population that is more interesting but less well defined than the fixed population of math dropout rates for 1984–85.

The sample variance is

$$s^2 = \frac{\left[(5 - 13.11)^2 + (22 - 13.11)^2 + \cdots + (14 - 13.11)^2 + (15 - 13.11)^2\right]}{38 - 1} = 106.5.$$

This estimates the variance of the random variable obtained when randomly selecting one class from the fixed population. The sample standard deviation is

$$s = \sqrt{106.5} = 10.32. \qquad \square$$

The only reason s^2 is *not* the average squared distance of the observations from the sample mean is that the denominator in (2.1.1) is $n - 1$ instead of n. If μ were known, a better estimate of the population variance σ^2 would be $\hat{\sigma}^2 \equiv \sum_{i=1}^{n}(y_i - \mu)^2 / n$. In s^2, we have used $\bar{y}_\cdot$ to estimate μ. Not knowing μ, we know less about the population, so s^2 cannot be as good an estimate as $\hat{\sigma}^2$. The quality of a variance estimate can be measured by the number of observations on which it is based; $\hat{\sigma}^2$ makes full use of all n observations for estimating σ^2. In using s^2, we lose the functional equivalent of one observation for having estimated the parameter μ. Thus s^2 has $n - 1$ in the denominator of (2.1.1) and is said to have $n - 1$ *degrees of freedom*. In nearly all problems that we will discuss, there is one degree of freedom available for every observation. The degrees of freedom are assigned to various estimates and we will need to keep track of them.

The statistics $\bar{y}_\cdot$ and s^2 are estimates of μ and σ^2 respectively. The *law of large numbers* is a mathematical result implying that for large sample sizes n, $\bar{y}_\cdot$ gets arbitrarily close to μ and s^2 gets arbitrarily close to σ^2.

Both $\bar{y}_\cdot$ and s^2 are computed from the random observations y_i. The summary statistics are functions of random variables, so they must also be random. Each has a distribution and to draw conclusions about the unknown parameters μ and σ^2 we need to know the distributions. In particular, if the original data are normally distributed, the sample mean has the distribution

$$\bar{y}_{\cdot} \sim N\left(\mu, \frac{\sigma^2}{n}\right)$$

or equivalently,

$$\frac{\bar{y}_{\cdot} - \mu}{\sqrt{\sigma^2 / n}} \sim N(0, 1), \tag{2.1.2}$$

see Exercise 1.6.2. In Subsection 1.2.4 we established that $E(\bar{y}_{\cdot}) = \mu$ and $\text{Var}(\bar{y}_{\cdot}) = \sigma^2/n$, so the only new claim made here is that the sample mean computed from *independent, identically distributed (iid)* normal random variables is again normally distributed. Moreover, the *central limit theorem* is a mathematical result stating that these distributions are approximately true for 'large' samples n, regardless of whether the original data are normally distributed.

As we will see below, the distributions given above are only useful in drawing conclusions about data when σ^2 is known. Generally, we will need to estimate σ^2 with s^2 and proceed as best we can. By the law of large numbers, s^2 becomes arbitrarily close to σ^2, so for large samples we can substitute s^2 for σ^2 in the distributions above. In other words, for large samples the *approximation*

$$\frac{\bar{y}_{\cdot} - \mu}{\sqrt{s^2 / n}} \sim N(0, 1) \tag{2.1.3}$$

holds regardless of whether the data were originally normal.

For small samples we cannot rely on s^2 being close to σ^2, so we fall back on the assumption that the original data are normally distributed. For normally distributed data, the appropriate distribution is called a t distribution with $n - 1$ degrees of freedom. In particular,

$$\frac{\bar{y}_{\cdot} - \mu}{\sqrt{s^2 / n}} \sim t(n - 1). \tag{2.1.4}$$

The t distribution is similar to the standard normal but more spread out, see Figure 2.2. It only makes sense that if we need to estimate σ^2 rather than knowing it, our conclusions will be less exact. This is reflected in the fact that the t distribution is more spread out than the $N(0, 1)$. In the previous paragraph we argued that for large n the appropriate distribution is

$$\frac{\bar{y}_{\cdot} - \mu}{\sqrt{s^2 / n}} \sim N(0, 1).$$

We are now arguing that for normal data the appropriate distribution is $t(n - 1)$. It better be the case (and is) that for large n the $N(0, 1)$ distribution is approximately the same as the $t(n - 1)$ distribution. In fact, we define $t(\infty)$ to be a $N(0, 1)$ distribution where ∞ indicates an infinitely large number.

FORMAL DISTRIBUTION THEORY

By definition, the t distribution is obtained as the ratio of two things related to the sample mean and variance. We now present this general definition.

First, for normally distributed data, the sample variance s^2 has a known distribution that depends on σ^2. It is related to a distribution called the *chi-squared* (χ^2) distribution with $n - 1$ degrees of freedom. In particular,

$$\frac{(n - 1)s^2}{\sigma^2} \sim \chi^2(n - 1). \tag{2.1.5}$$

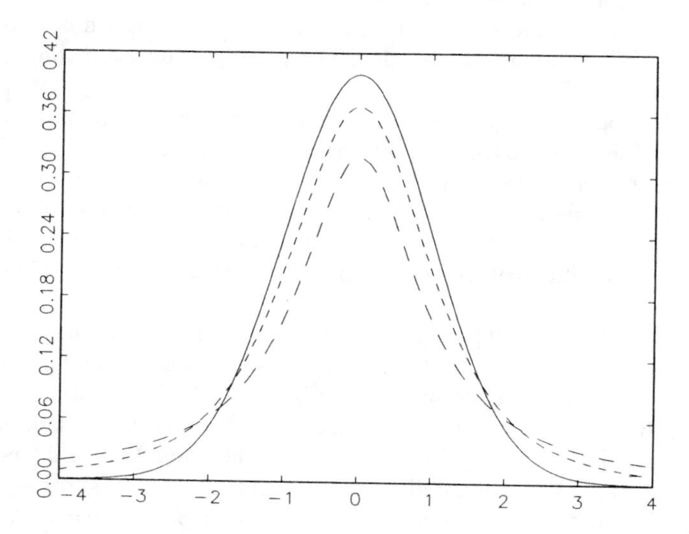

FIGURE 2.2. Three distributions: solid, $N(0, 1)$; long dashes, $t(1)$; short dashes, $t(3)$.

Moreover, for normal data, $\bar{y}_\cdot$ and s^2 are independent.

Definition 2.1.3. A t distribution is the distribution obtained when a random variable with a $N(0, 1)$ distribution is divided by an independent random variable that is the square root of a χ^2 random variable over its degrees of freedom. The t distribution has the same degrees of freedom as the chi-square.

In particular, $[\bar{y}_\cdot - \mu] / \sqrt{\sigma^2/n}$ is $N(0, 1)$, $\sqrt{[(n-1)s^2/\sigma^2]/(n-1)}$ is the square root of a chi-squared random variable over its degrees of freedom, and the two are independent because $\bar{y}_\cdot$ and s^2 are independent, so

$$\frac{\bar{y}_\cdot - \mu}{\sqrt{s^2/n}} = \frac{[\bar{y}_\cdot - \mu] / \sqrt{\sigma^2/n}}{\sqrt{[(n-1)s^2/\sigma^2]/(n-1)}} \sim t(n-1).$$

The t distribution has the same degrees of freedom as the estimate of σ^2; this is typically the case in other applications.

2.2 Inference about μ

Most statistical tests and confidence intervals are applications of a single theory. (Tests and confidence intervals for variances are exceptions.) To use this theory we need to know four things. In the one-sample problem the four things are

1. the parameter of interest, μ,
2. the estimate of the parameter, $\bar{y}_\cdot$,
3. the standard error of the estimate, $SE(\bar{y}_\cdot) \equiv \sqrt{s^2/n} = s/\sqrt{n}$, and
4. the appropriate distribution for $[\bar{y}_\cdot - \mu] / \sqrt{s^2/n}$.

Specifically, we need a known (tabled) distribution for $[\bar{y}_{\bullet} - \mu] / \sqrt{s^2/n}$ that is symmetric about zero and continuous. The standard error, $SE(\bar{y}_{\bullet})$, is the estimated standard deviation of $\bar{y}_{\bullet}$. Recall that the variance of $\bar{y}_{\bullet}$ is σ^2/n, so its standard deviation is $\sqrt{\sigma^2/n}$ and estimating σ^2 by s^2 gives the standard error $\sqrt{s^2/n}$.

The appropriate distribution for $[\bar{y}_{\bullet} - \mu] / \sqrt{s^2/n}$ when the data are normally distributed is the $t(n-1)$ as in (2.1.4). For large samples, the appropriate distribution is the $N(0, 1)$ as in (2.1.3). Recall that for large samples from a normal population, it is irrelevant whether we use the standard normal or the t distribution because they are essentially the same. In the unrealistic case where σ^2 is known we do not need to estimate it, so we use $\sqrt{\sigma^2/n}$ instead of $\sqrt{s^2/n}$ for the standard error. In this case, the appropriate distribution is (2.1.2) if either the original data are normal or the sample size is large.

We need notation for the percentage points of the known distribution and we need a name for the point that cuts off the top α of the distribution. Typically, we need to find points that cut off the top 5%, 2.5%, 1%, or 0.5% of the distribution, so α is .05, .025, .01, or .005. As discussed in the previous paragraph, the appropriate distribution depends on various circumstances of the problem, so we begin by discussing percentage points with a generic notation. We use the notation $K(1-\alpha)$ for the point that cuts off the top α of the distribution. Figure 2.3 displays this idea graphically for a value of α between 0 and .5. The distribution is described by the curve, which is symmetric about 0. $K(1-\alpha)$ is indicated along with the fact that the area under the curve to the right of $K(1-\alpha)$ is α. Formally the point that cuts off the top α of the distribution is $K(1-\alpha)$ where

$$\Pr\left[\frac{\bar{y}_{\bullet} - \mu}{SE(\bar{y}_{\bullet})} > K(1-\alpha)\right] = \alpha.$$

Note that the same point $K(1-\alpha)$ also cuts off the bottom $1-\alpha$ of the distribution, i.e.,

$$\Pr\left[\frac{\bar{y}_{\bullet} - \mu}{SE(\bar{y}_{\bullet})} < K(1-\alpha)\right] = 1 - \alpha.$$

This is illustrated in Figure 2.3 by the fact that the area under the curve to the left of $K(1-\alpha)$ is $1-\alpha$. The reason the point is labeled $K(1-\alpha)$ is because it cuts off the bottom $1-\alpha$ of the distribution. The labeling depends on the percentage to the left even though our interest is in the percentage to the right.

There are at least three different ways to label these percentage points; I have simply used the one I feel is most consistent with general usage in probability and statistics. The key point however is to be familiar with Figure 2.3. We need to find points that cut off a fixed percentage of the area under the curve. As long as we can find such points, what we call them is irrelevant. Ultimately, anyone doing statistics will need to be familiar with all three methods of labeling. One method of labeling is in terms of the area to the left of the point; this is the one we will use. A second method is labeling in terms of the area to the right of the point; thus the point we call $K(1-\alpha)$ could be labeled, say, $Q(\alpha)$. The third method is to call this number, say, $W(2\alpha)$, where the area to the right of the point is doubled in the label. For example, if the distribution is a $N(0, 1)$, the point that cuts off the bottom 97.5% of the distribution is 1.96. This point also cuts off the top 2.5% of the area. It makes no difference if we refer to 1.96 as the number that cuts off the bottom 97.5%, $K(.975)$, or as the number that cuts off the top 2.5%, $Q(.025)$, or as the number $W(.05)$ where the label involves $2 \times .025$; the important point is being able to identify 1.96 as the appropriate number. Henceforth, we will always refer to points in terms of $K(1-\alpha)$, the point that cuts off the bottom $1-\alpha$ of the distributions. No further reference to the alternative labelings will be made but all three labels are used in Appendix B.1. There $K(1-\alpha)$s are labeled as percentiles and, for reasons related to statistical tests, $Q(\alpha)$s and $W(2\alpha)$s are labeled as *one-sided* and *two-sided* α levels respectively.

A fundamental assumption in inference about μ is that the distribution of $[\bar{y}_{\bullet} - \mu] / SE(\bar{y}_{\bullet})$ is symmetric about 0. By the symmetry around zero, if $K(1-\alpha)$ cuts off the top α of the distribution,

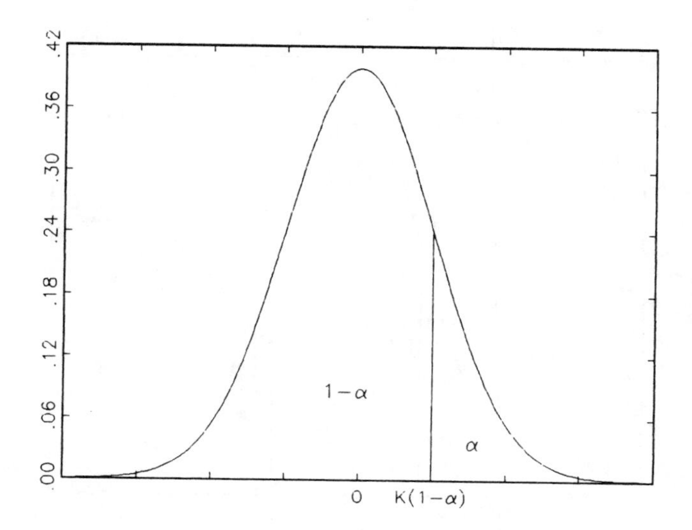

FIGURE 2.3. $1 - \alpha$ percentile of the distribution of $[\bar{y}_{\bullet} - \mu]/SE(\bar{y}_{\bullet})$.

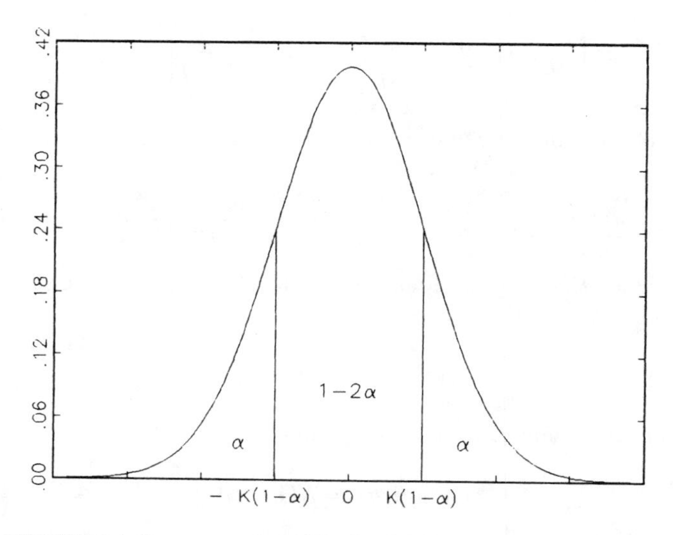

FIGURE 2.4. Symmetry about 0 in the distribution of $[\bar{y}_{\bullet} - \mu]/SE(\bar{y}_{\bullet})$.

$-K(1 - \alpha)$ must cut off the bottom α of the distribution. Thus for distributions that are symmetric about 0 we have $K(\alpha)$, the point that cuts off the bottom α of the distribution, equal to $-K(1 - \alpha)$. This fact is illustrated in Figure 2.4. Algebraically, we write

$$\Pr\left[\frac{\bar{y}_{\bullet} - \mu}{SE(\bar{y}_{\bullet})} < -K(1 - \alpha)\right] = \Pr\left[\frac{\bar{y}_{\bullet} - \mu}{SE(\bar{y}_{\bullet})} < K(\alpha)\right] = \alpha.$$

Frequently, we want to create a central interval that contains a specified probability, say $1 - \alpha$. Figure 2.5 illustrates the construction of such an interval. Algebraically, the middle interval with

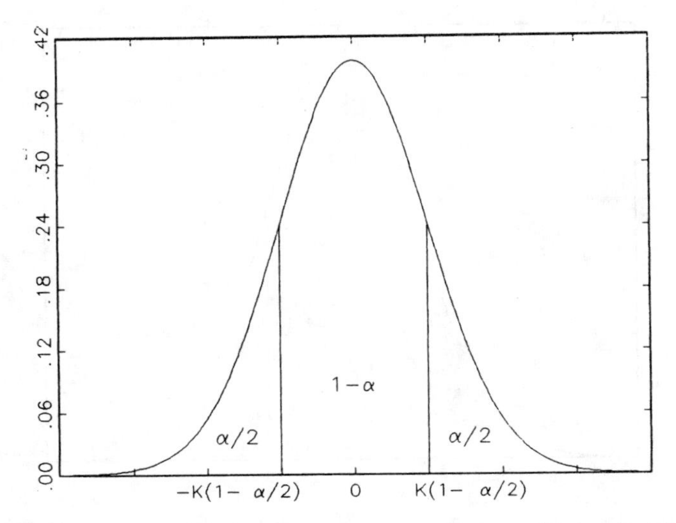

FIGURE 2.5. $1-\alpha$ central interval for the distribution of $[\bar{y}_\cdot - \mu]/SE(\bar{y}_\cdot)$.

probability $1-\alpha$ is obtained by

$$\Pr\left[-K\left(1-\frac{\alpha}{2}\right) < \frac{\bar{y}_\cdot - \mu}{SE(\bar{y}_\cdot)} < K\left(1-\frac{\alpha}{2}\right)\right] = 1-\alpha.$$

The probability of getting something outside of this interval is

$$\alpha = \frac{\alpha}{2} + \frac{\alpha}{2} = \Pr\left[\frac{\bar{y}_\cdot - \mu}{SE(\bar{y}_\cdot)} < -K\left(1-\frac{\alpha}{2}\right)\right] + \Pr\left[\frac{\bar{y}_\cdot - \mu}{SE(\bar{y}_\cdot)} > K\left(1-\frac{\alpha}{2}\right)\right].$$

In practice, the values $K(1-\alpha)$ are found from either a normal table or a t table. For normal percentage points, we use the notation

$$z(1-\alpha) = K(1-\alpha).$$

For percentage points of a t with df degrees of freedom, use

$$t(1-\alpha, df) = K(1-\alpha).$$

Recall that as df gets large, the $t(df)$ distribution converges to a $N(0,1)$, so

$$z(1-\alpha) = t(1-\alpha, \infty).$$

Percentiles of the t distribution are given in Appendix B.1 with the ∞ row giving percentiles of the $N(0,1)$ distribution.

2.2.1 CONFIDENCE INTERVALS

A confidence interval is an interval of possible μ values in which we are 'confident' that the true value of μ lies. Moreover, a numerical level of confidence is specified for the interval. Confidence intervals are commonly viewed as the most useful single procedure in statistical inference. A 95% confidence interval for μ is based on the following probability statements:

$$\begin{aligned} .95 &= \Pr\left[-K(.975) < \frac{\bar{y}_{\bullet} - \mu}{SE(\bar{y}_{\bullet})} < K(.975)\right] \\ &= \Pr[\bar{y}_{\bullet} - K(.975)SE(\bar{y}_{\bullet}) < \mu < \bar{y}_{\bullet} + K(.975)SE(\bar{y}_{\bullet})] \end{aligned}$$

The first equality given above holds simply by the definition of $K(.975)$ and the symmetry of the distribution; it expresses Figure 2.5 algebraically for $\alpha = .05$. The second equality follows from the fact that the statements within the two sets of square brackets can be shown to be algebraically equivalent.

More generally, a $(1-\alpha)100\%$ confidence interval for μ is based on the following probability statements:

$$\begin{aligned} 1-\alpha &= \Pr\left[-K\left(1-\frac{\alpha}{2}\right) < \frac{\bar{y}_{\bullet} - \mu}{SE(\bar{y}_{\bullet})} < K\left(1-\frac{\alpha}{2}\right)\right] \\ &= \Pr\left[\bar{y}_{\bullet} - K\left(1-\frac{\alpha}{2}\right)SE(\bar{y}_{\bullet}) < \mu < \bar{y}_{\bullet} + K\left(1-\frac{\alpha}{2}\right)SE(\bar{y}_{\bullet})\right] \end{aligned}$$

The first equality given above holds simply by the definition of $K\left(1-\frac{\alpha}{2}\right)$ and the symmetry of the distribution. Again, it is just an algebraic statement of Figure 2.5. The second equality follows from the fact that the statements within the square brackets are algebraically equivalent. A proof of the equivalence is given in the appendix to the next chapter.

The probability statement

$$1-\alpha = \Pr\left[\bar{y}_{\bullet} - K\left(1-\frac{\alpha}{2}\right)SE(\bar{y}_{\bullet}) < \mu < \bar{y}_{\bullet} + K\left(1-\frac{\alpha}{2}\right)SE(\bar{y}_{\bullet})\right]$$

is the basis of the confidence interval for μ. The $(1-\alpha)100\%$ confidence interval for μ is simply the interval within the square brackets, i.e., the points between $\bar{y}_{\bullet} - K\left(1-\frac{\alpha}{2}\right)SE(\bar{y}_{\bullet})$ and $\bar{y}_{\bullet} + K\left(1-\frac{\alpha}{2}\right)SE(\bar{y}_{\bullet})$ *with observed values substituted for $\bar{y}_{\bullet}$ and $SE(\bar{y}_{\bullet})$*. The endpoints can be written

$$\bar{y}_{\bullet} \pm K\left(1-\frac{\alpha}{2}\right)SE(\bar{y}_{\bullet}),$$

or, substituting the form of the standard error,

$$\bar{y}_{\bullet} \pm K\left(1-\frac{\alpha}{2}\right)\frac{s}{\sqrt{n}}.$$

Note that increasing the sample size n decreases the standard error and thus makes the confidence interval narrower. Narrower confidence intervals give more precise information about μ. In fact, by taking n large enough, we can make the confidence interval arbitrarily narrow.

EXAMPLE 2.2.1. For the dropout rate data presented at the beginning of the chapter, the parameter is the mean dropout rate for math classes, the estimate is $\bar{y}_{\bullet} = 13.11$, and the standard error is $s/\sqrt{n} = 10.32/\sqrt{38} = 1.67$. As seen in the dot plot, the original data are not normally distributed. The plot looks nothing at all like the bell shaped curve in Figure 1.1, which is a picture of a normal distribution. Thus we hope that a sample of size 38 is sufficiently large to justify use of the $N(0,1)$ distribution via the central limit theorem and the law of large numbers. For a 95% confidence interval, $95 = (1-\alpha)100$, $.95 = (1-\alpha)$, $\alpha = 1 - .95 = .05$, and $1 - \alpha/2 = .975$, so the number we need from the t table is $z(.975) = t(.975, \infty) = 1.96$. The endpoints of the confidence interval are

$$13.11 \pm 1.96(1.67)$$

giving an interval of

$$(9.8, 16.4).$$

Rounding to simple numbers, we are 95% confident that the true dropout rate is between 10% and 16.5% □

The confidence interval has probability $1 - \alpha$ that we are *going to get* a confidence interval that covers what we are trying to estimate, i.e., μ. However, once the data are observed and the interval computed, this is no longer true. The particular interval that we get either covers μ or it does not. There is no probability associated with the coverage; nothing is random, neither μ nor the endpoints of the interval. For this reason we say that, 'We are $(1 - \alpha)100\%$ *confident* that the true value of μ is in the interval.' I doubt that anybody has a good definition of what the word 'confident' means in that sentence. Having done my duty to explain the correct meaning of confidence intervals, you can (and will) go back to thinking that the probability is $1 - \alpha$ that your interval covers μ. It does not do any real harm and it can be justified using arguments from Bayesian statistics. This issue of interpretation is discussed in much more detail in the next chapter.

2.2.2 HYPOTHESIS TESTS

An hypothesis test is a procedure for checking the validity of a claim. Someone makes a claim which becomes the *null hypothesis*. We wish to test whether or not the claim is true. If relevant data are available, we can test the claim, but we cannot really test whether it is true or false, we can merely test whether the data are consistent or inconsistent with the claim. Data that are inconsistent with the claim suggest that the claim is false. Data that are consistent with the claim are just that, consistent with the claim; they do not imply that the claim is true because other circumstances could equally well have generated the data.

In a one sample problem, for some fixed known number m we may want to test the *null hypothesis*

$$H_0 : \mu = m$$

versus the *alternative hypothesis*

$$H_A : \mu \neq m.$$

The number m must be known; it is some number that is of interest for the specific data being analyzed. It is not just an unspecified symbol.

EXAMPLE 2.2.2. For the dropout rate data, we might be interested in the hypothesis that the true dropout rate is 10%. Thus the null hypothesis is $H_0 : \mu = 10$ and the alternative hypothesis is $H_A : \mu \neq 10$. □

The test is based on the assumption that H_0 is true and we check to see if the data are inconsistent with that assumption. The idea is much like the idea of a proof by contradiction. We make an assumption H_0. If the data contradict that assumption, we can conclude that the assumption H_0 is false. If the data do not contradict H_0, we can only conclude that the data are consistent with the assumption; *we cannot conclude that the assumption is true.*

Unfortunately, there are two complicating factors in a statistical test. First, data almost never yield an absolute contradiction to the assumption. We need to quantify the extent to which the data are inconsistent with the assumption. Second, while we wish to test a specific assumption H_0, there are other assumptions involved in any statistical procedure. A contradiction only invalidates H_0 if the other assumptions are valid. These other assumptions were discussed at the beginning of the chapter. They include such things as independence, normality, and all observations having the same mean and variance. While we can never confirm that these other assumptions are absolutely valid, it is a key aspect of modern statistical practice to validate the assumptions as far as is reasonably possible. When we are convinced that the other assumptions are reasonably valid, data that contradict the assumptions can be reasonably interpreted as contradicting the specific assumption H_0.

We need to be able to identify data that are inconsistent with the assumption that $\mu = m$. Note that, regardless of any hypotheses, $\bar{y}_\cdot$ is an estimate μ. For example, suppose $m = 10$. If $\bar{y}_\cdot = 10.1$, $\bar{y}_\cdot$ is an estimate of μ, so the data seem to be consistent with the idea that $\mu = 10$. On the other hand, if $\bar{y}_\cdot = 10,000$, we expect that μ will be near 10,000 and the observed $\bar{y}_\cdot$ seems to be inconsistent with $H_0 : \mu = 10$. The trick is in determining which values of $\bar{y}_\cdot$ are far enough away from 10 for us to be reasonably sure that $\mu \neq 10$. As a matter of fact, in the absence of information about the variability of $\bar{y}_\cdot$, we cannot really say that $\bar{y}_\cdot = 10.1$ is consistent with $\mu = 10$ or that $\bar{y}_\cdot = 10,000$ is inconsistent with $\mu = 10$. If the variability associated with $\bar{y}_\cdot$ is extremely small, $\bar{y}_\cdot = 10.1$ may be highly inconsistent with $\mu = 10$. On the other hand, if the variability associated with $\bar{y}_\cdot$ is extremely large, $\bar{y}_\cdot = 10,000$ may be perfectly consistent with $\mu = 10$. Obviously, the standard error of $\bar{y}_\cdot$, which is our measure of variability, must play a major role in the analysis.

Generally, since $\bar{y}_\cdot$ estimates μ, if $\mu > m$, then $\bar{y}_\cdot$ tends to be greater than m so that $\bar{y}_\cdot - m$ and thus $[\bar{y}_\cdot - m]/SE(\bar{y}_\cdot)$ tend to be large positive numbers (larger than they would be if $H_0 : \mu = m$ were true). On the other hand, if $\mu < m$, then $\bar{y}_\cdot - m$ and $[\bar{y}_\cdot - m]/SE(\bar{y}_\cdot)$ will tend to be a large negative numbers. Data that are inconsistent with the null hypothesis $\mu = m$ are large positive and large negative values of the *test statistic* $[\bar{y}_\cdot - m]/SE(\bar{y}_\cdot)$. The problem is in specifying what we mean by 'large'.

We reject the null hypothesis (disbelieve $\mu = m$) if the *test statistic*

$$\frac{\bar{y}_\cdot - m}{SE(\bar{y}_\cdot)}$$

is greater than some positive cutoff value or less than some negative cutoff value. Very large and very small (large negative) values of the test statistic are those that are most inconsistent with $\mu = m$. The problem is in specifying the cutoff values. For example, we do not want to reject $\mu = 10$ if the data are consistent with $\mu = 10$. One of our basic assumptions is that we know the distribution of $[\bar{y}_\cdot - \mu]/SE(\bar{y}_\cdot)$. Thus if $H_0 : \mu = 10$ is true, we know the distribution of the test statistic $[\bar{y}_\cdot - 10]/SE(\bar{y}_\cdot)$, so we know what kind of data are consistent with $\mu = 10$. For instance, when $\mu = 10$, 95% of the possible values of $[\bar{y}_\cdot - 10]/SE(\bar{y}_\cdot)$ are between $-K(.975)$ and $K(.975)$. Any values of $[\bar{y}_\cdot - 10]/SE(\bar{y}_\cdot)$ that fall between these numbers are reasonably consistent with $\mu = 10$ and values outside the interval are defined as being inconsistent with $\mu = 10$. Thus values of $[\bar{y}_\cdot - 10]/SE(\bar{y}_\cdot)$ greater than $K(.975)$ or less than $-K(.975)$ cause us to reject the null hypothesis. Note that we arbitrarily specified the central 95% of the distribution as being consistent with $\mu = 10$. That leaves a 5% chance of getting outside the central interval, so a 5% chance that we will reject $\mu = 10$ even when it is true. In other words, even when $\mu = 10$, 5% of the time $[\bar{y}_\cdot - 10]/SE(\bar{y}_\cdot)$ will be outside the limits. We could reduce this chance of error by specifying the central 99% of the distribution as consistent with $\mu = 10$. This reduces the chance of error to 1%, but then if $\mu \neq 10$, we are less likely to reject $\mu = 10$. Thus there are two types of possible errors that we need to play off against each other. *Type I error is rejecting H_0 when it is true. Type II error is not rejecting H_0 when it is not true. The probability of type I error is known as the α level of the test.*

EXAMPLE 2.2.3. For the dropout rate data, consider the null hypothesis $H_0 : \mu = 10$, i.e., that the mean dropout rate is 10%. The alternative hypothesis is $H_A : \mu \neq 10$. As discussed in the example on confidence intervals, these data are not normal, so we must hope that the sample size is large enough to justify use of the $N(0, 1)$ distribution. If we choose a central 90% interval and thus a type I error rate of $\alpha = .10$, the upper cutoff value is $K\left(1 - \frac{\alpha}{2}\right) = z\left(1 - \frac{\alpha}{2}\right) = z(1 - .05) = t(.95, \infty) = 1.645$.

The $\alpha = .10$ level test for $H_0 : \mu = 10$ versus $H_A : \mu \neq 10$ is to reject H_0 if

$$\frac{\bar{y}_{\cdot} - 10}{s/\sqrt{38}} > 1.645.$$

or if

$$\frac{\bar{y}_{\cdot} - 10}{s/\sqrt{38}} < -1.645.$$

The estimate of μ is $\bar{y}_{\cdot} = 13.11$ and the observed standard error is $s/\sqrt{n} = 10.32/\sqrt{38} = 1.67$, so *the observed value of the test statistic* is

$$\frac{13.11 - 10}{1.67} = 1.86.$$

Comparing this to the cutoff value of 1.645 we have $1.86 > 1.645$, so the null hypothesis is rejected. There is evidence at the $\alpha = .10$ level that the mean dropout rate is not 10%. In fact, since $\bar{y}_{\cdot} = 13.11 > 10$ there is the suggestion that the dropout rate is greater than 10%.

This conclusion depends on the choice of the α level. If we choose $\alpha = .05$, then the appropriate cutoff value is $z(.975) = 1.96$. Since the observed value of the test statistic is 1.86, which is neither greater than 1.96 nor less than -1.96, we do not reject the null hypothesis. When we do not reject H_0, we cannot say that the true mean dropout rate is 10%, but we can say that, at the $\alpha = .05$ level, the data are consistent with the (null) hypothesis that the true mean dropout rate is 10%. □

Generally, a test of hypothesis is based on controlling the probability of making an error when the null hypothesis is true. The α level of the test (the probability a type I error) is the probability of rejecting the null hypothesis (saying that it is false) when the null hypothesis is in fact true. The α level test for $H_0 : \mu = m$ versus $H_A : \mu \neq m$ is to reject H_0 if

$$\frac{\bar{y}_{\cdot} - m}{SE(\bar{y}_{\cdot})} > K\left(1 - \frac{\alpha}{2}\right)$$

or if

$$\frac{\bar{y}_{\cdot} - m}{SE(\bar{y}_{\cdot})} < -K\left(1 - \frac{\alpha}{2}\right).$$

This is equivalent to saying, reject H_0 if

$$\frac{|\bar{y}_{\cdot} - m|}{SE(\bar{y}_{\cdot})} > K\left(1 - \frac{\alpha}{2}\right).$$

Note that if H_0 is true, the probability that we will reject H_0 is

$$\Pr\left[\frac{\bar{y}_{\cdot} - m}{SE(\bar{y}_{\cdot})} > K\left(1 - \frac{\alpha}{2}\right)\right] + \Pr\left[\frac{\bar{y}_{\cdot} - m}{SE(\bar{y}_{\cdot})} < -K\left(1 - \frac{\alpha}{2}\right)\right] = \alpha/2 + \alpha/2 = \alpha.$$

Also note that we are rejecting H_0 for those values of $[\bar{y}_{\cdot} - m]/SE(\bar{y}_{\cdot})$ that are most inconsistent with H_0, these being the values of the test statistic with large absolute values.

A null hypothesis should never be accepted; it is either rejected or not rejected. A better way to think of a test is that one concludes that the data are either consistent or inconsistent with the null hypothesis. The statement that the data are inconsistent with H_0 is a strong statement. It disproves H_0 in some specified degree. The statement that the data are consistent with H_0 is not a strong statement; it does not prove H_0. For example, the dropout data happen to be consistent with $H_0 : \mu = 12$; the test statistic

$$\frac{\bar{y}_{\cdot} - 12}{SE(\bar{y}_{\cdot})} = \frac{13.11 - 12}{1.67} = .66$$

is very small. However, the data are equally consistent with $\mu = 12.00001$. These data cannot possibly indicate that $\mu = 12$ rather than $\mu = 12.00001$. However, when the null hypothesis is $H_0 : \mu = 12$, the value $\mu = 12.00001$ is part of the alternative hypothesis $H_A : \mu \neq 12$, so clearly data that are consistent with H_0 are also consistent with some elements of the alternative. In fact, we established earlier that based on an $\alpha = .05$ test, these data are even consistent with $\mu = 10$. *Data that are consistent with H_0 do not imply that the alternative is false.*

With these data there is very little hope of distinguishing between $\mu = 12$ and $\mu = 12.00001$. The probability of getting data that lead to rejecting $H_0 : \mu = 12$ when $\mu = 12.00001$ is only just slightly more than the probability of getting data that lead to rejecting H_0 when $\mu = 12$. The probability of getting data that lead to rejecting $H_0 : \mu = 12$ when $\mu = 12.00001$ is called the *power* of the test when $\mu = 12.00001$. *The power is the probability of appropriately rejecting H_0 and depends on the particular value of μ ($\neq 12$).* The fact that the power is very small for detecting $\mu = 12.00001$ is not much of a problem because no one would really care about the difference between a dropout rate of 12 and a dropout rate of 12.00001. However, a small power for a difference that one cares about is a major concern. The power is directly related to the standard error and can be increased by reducing the standard error. One natural way to reduce the standard error $s/\sqrt{n}$ is by increasing the sample size n.

One of the difficulties in a general discussion of hypothesis testing is that the actual null hypothesis is always context specific. You cannot give general rules for what to use as a null hypothesis because the null hypothesis needs to be some interesting claim about the population mean μ. When you sample different populations, the population mean differs, and interesting claims about the population mean depend on the exact nature of the population. The best practice for setting up null hypotheses is simply to look at lots of problems and ask yourself what claims about the population mean are of interest to you. As we examine more sophisticated data structures, some interesting hypotheses will arise from the structures themselves. For example, if we have two samples of similar measurements we might be interested in testing the null hypothesis that they have the same population means. Note that there are lots of ways in which the means could be different, but only one way in which they can be the same. Of course if the specific context suggests that one mean should be, say, 25 units greater than the other, we can use that as the null hypothesis. Similarly, if we have a sample of objects and two different measurements on each object, we might be interested in whether or not the measurements are related. In that case, an interesting null hypothesis is that the measurements are *not* related. Again, there is only one way in which measurements can be unrelated, but there are many ways for measurements to display a relationship.

We will see in the next chapter that there is a duality between testing and confidence intervals. Tests are used to examine whether a difference can be shown to exist between the hypothesized mean and the mean of the population being sampled. Confidence intervals are used to quantify what is known about the population mean. In particular, confidence intervals can be used to quantify how much difference exists between some hypothesized mean and the sampled population's mean. Of course, one must consider not only how much of a difference exists but also whether such a difference is meaningful in the context of the problem.

ONE-SIDED TESTS

Unless math classes were intentionally being used to weed out students (something I do not believe was true) high dropout rates are typically considered unfortunate. Math instructors might claim that dropout rates are 10% *or less* and students may want to test that claim. In such a case the claim is only contradicted by dropout rates greater than 10%

We can do one-sided tests in a similar manner to the two-sided testing discussed previously. The

α level test for $H_0 : \mu \leq m$ versus $H_A : \mu > m$ is to reject H_0 if

$$\frac{\bar{y}_{\bullet} - m}{SE(\bar{y}_{\bullet})} > K(1 - \alpha).$$

Again, the value m must be known; either someone tells it to you or you determine it from the subject being investigated. The alternative hypothesis is that μ is greater than something and the null hypothesis is rejected when the test statistic is greater than some cutoff value. *We reject the null hypothesis for those values of the test statistic that are most inconsistent with the null hypothesis and most consistent with the alternative hypothesis.* If the alternative is true, $\bar{y}_{\bullet}$ should be near μ, which is greater than m, so large positive values of $\bar{y}_{\bullet} - m$ or, equivalently, large positive values of $[\bar{y}_{\bullet} - m]/SE(\bar{y}_{\bullet})$ are consistent with the alternative and inconsistent with the null hypothesis. Note that if $\mu = m$ is true, the probability of rejecting the test is

$$\Pr\left[\frac{\bar{y}_{\bullet} - m}{SE(\bar{y}_{\bullet})} > K(1 - \alpha)\right] = \alpha.$$

Moreover, it is easily seen that if $\mu < m$,

$$\Pr\left[\frac{\bar{y}_{\bullet} - m}{SE(\bar{y}_{\bullet})} > K(1 - \alpha)\right] < \alpha.$$

Thus when H_0 is true, i.e., when $\mu \leq m$, the probability of rejecting the null hypothesis is *at most* α. As with the two-sided tests, we have controlled the probability of making an error when the null hypothesis is true.

EXAMPLE 2.2.4. The null hypothesis is that the dropout rate is 10% or less, i.e., $H_0 : \mu \leq 10$. The alternative is that the dropout rate is greater than 10%, i.e., $H_A : \mu > 10$. The $\alpha = .05$ level test rejects H_0 if

$$\frac{\bar{y}_{\bullet} - 10}{SE(\bar{y}_{\bullet})} > z(1-.05) = 1.645.$$

As seen earlier, the observed value of the test statistic is $1.86 > 1.645$, so the null hypothesis is rejected. Based on a one-sided $\alpha = .05$ test, we have evidence to reject the (null) hypothesis that the true dropout rate is 10% or less. In other words, we have evidence that the dropout rate is greater than 10%.

Students who are math averse might be interested in the claim that the dropout rate is *at least* 10%, i.e., $\mu \geq 10$. Setting this up as the null hypothesis is much less informative than the approach just demonstrated. In this case, the value of $\bar{y}_{\bullet} = 13.11$ is obviously consistent with μ being at least 10%. The question is whether $\bar{y}_{\bullet}$ is also inconsistent with $\mu \leq 10$. For $H_0 : \mu \geq 10$ a test will not be rejected. If you do not reject a test, α means very little. However, when you reject a test, α measures your chance of making an error. Setting up the test as we did allowed us to reject $H_0 : \mu \leq 10$ at $\alpha = .05$, which quantifies our chance for error. Accepting $H_0 : \mu \geq 10$ tells us nothing about the chance for error, so it is less informative. □

As we argued earlier, with a two-sided test you can *never* be sure that your H_0 claim is true. With a one-sided test, this is not the case. If the data are extreme enough, one hypothesis or the other is clearly indicated. In the dropout rate data example, with a standard error of 1.67, it is pretty clear that $\bar{y}_{\bullet} = 4$ indicates $\mu \leq 10$ and $\bar{y}_{\bullet} = 16$ indicates $\mu \geq 10$, *assuming that all other assumptions are valid.* The problem occurs with $\bar{y}_{\bullet}$ values close to 10, say $\bar{y}_{\bullet} = 9$ or $\bar{y}_{\bullet} = 11$. If $\bar{y}_{\bullet} = 9$, we cannot be sure that $\mu \leq 10$ because μ could be 10 or a little larger and we would still have a reasonable chance of observing $\bar{y}_{\bullet} = 9$. Similarly, $\bar{y}_{\bullet} = 11$ is reasonably consistent with μ values of 10 or a little smaller. The only really hard problem is whether we are sure $\mu \neq 10$. If μ is different from 10, it is obvious whether $\mu < 10$ or $\mu > 10$. And if you are bothering to run this test at all, $\mu = 10$ must have some special significance and it should be of interest to establish which way μ might differ from 10. This is one of several reasons I have for preferring two-sided tests.

The α level test for $H_0 : \mu \geq m$ versus $H_A : \mu < m$ is to reject H_0 if

$$\frac{\bar{y}_\cdot - m}{SE(\bar{y}_\cdot)} < -K(1-\alpha).$$

The alternative hypothesis is that μ is less than something and the null hypothesis is rejected when the test statistic is less than some cutoff value. Note that the form of the alternative determines the form of the rejection region. *In all cases we reject H_0 for the data that are most inconsistent with H_0.*

The one-sided null hypotheses involve inequalities, but $\mu = m$ is always part of the null hypothesis. The tests are set up assuming that $\mu = m$ and this needs to be part of the null hypothesis. In all cases, the test is set up so that if $\mu = m$, then the probability of making a mistake is α.

P VALUES

Rather than having formal rules for when to reject the null hypothesis, one can report the evidence against the null hypothesis. This is done by reporting the *significance level* of the test, also known as the *P value*. The P value is computed under the assumption that $\mu = m$ and *is the probability of seeing data that are as extreme or more extreme than those that were actually observed.* In other words, it is the α level at which the test would just barely be rejected.

EXAMPLE 2.2.5. For $H_0 : \mu = 10$ versus $H_A : \mu \neq 10$ the observed value of the test statistic is 1.86. Clearly, data that give values of the test statistic that are greater than 1.86 are more extreme than the actual data. Also, by symmetry, data that give a test statistic of -1.86 are just as extreme as data that yield a 1.86. Finally, data that give values smaller than -1.86 are more extreme than data yielding a 1.86. As before, we use the standard normal distribution z. From a standard normal table or an appropriate computer program,

$$\begin{aligned} P &= \Pr[z \geq 1.86] + \Pr[z \leq -1.86] \\ &= .0314 + .0314 \\ &= .0628. \end{aligned}$$

Thus the approximate P value is .06. The P value is approximate because the use of the standard normal distribution is an approximation based on large samples. Note that

$$P = \Pr[z \geq 1.86] + \Pr[z \leq -1.86] = \Pr\left[|z| \geq |1.86|\right].$$

In the t tables of Appendix B.1, the standard normal distribution corresponds to $t(\infty)$. Comparing $|1.86|$ to the tables, we see that

$$t(.95, \infty) = 1.645 < |1.86| < 1.96 = t(.975, \infty),$$

so for a two-sided test the P value satisfies

$$2(1-.95) = .10 > P > .05 = 2(1-.975).$$

In other words, $t(.95, \infty)$ is the cutoff value for an $\alpha = .10$ test and $t(.975, \infty)$ is the cutoff value for an $\alpha = .05$ test; $|1.86|$ falls between these values, so the P value is between .10 and .05. When only a t table is available, P values are most simply specified in terms of bounds such as these. □

The P value is a measure of the evidence against the null hypothesis in which the smaller the P value the more evidence against H_0. The P value can be used to perform various α level tests. In the example, the P value is .06. This is less than .10, so an $\alpha = .10$ level test of $H_0 : \mu = 10$ versus $H_A : \mu \neq 10$ will reject H_0. On the other hand, .06 is greater than .05, so an $\alpha = .05$ test does not reject $H_0 : \mu = 10$. Note that these are exactly the conclusions we reached in the earlier example on testing $H_0 : \mu = 10$ versus $H_A : \mu \neq 10$.

The P value for a one-sided test, say, $H_0 : \mu \geq m$ versus $H_A : \mu < m$, is one half of the P value from the test of $H_0 : \mu = m$ versus $H_A : \mu \neq m$ provided that $\bar{y}_{\cdot} < m$. If $\bar{y}_{\cdot} \geq m$, the P value is at least .5.

2.3 Prediction intervals

In many situations, rather than trying to learn about μ, it is more important to obtain information about future observations from the same process. With independent observations, the natural point prediction for a future observation is just the estimate of μ, but a prediction interval with, say, 99% confidence of containing a future observation differs from a 99% confidence interval for μ. Our ideas about where future observations will lie involves two sources of variability. First, there is the variability that a new observation y displays about its mean value μ. Second, we need to deal with the fact that we do not know μ, so there is variability associated with $\bar{y}_{\cdot}$, our estimate of μ. In the dropout rate example, $\bar{y}_{\cdot} = 13.11$ and $s^2 = 106.5$. If we could assume that the observations are normally distributed (which is a poor assumption), we could create a 99% prediction interval, i.e., an interval that contains a future observation with 99% confidence. The interval for the new observation is centered about $\bar{y}_{\cdot}$, our best point predictor, and is similar to a confidence interval but uses a standard error that is appropriate for prediction. The actual interval has endpoints

$$\bar{y}_{\cdot} \pm t(.995, n-1)\sqrt{s^2 + \frac{s^2}{n}}.$$

In our example, $n = 38$ and $t(.995, 37) = 2.71$, so this becomes

$$13.11 \pm 2.71\sqrt{106.5 + \frac{106.5}{38}}$$

or

$$13.11 \pm 28.33$$

for an interval of $(-15.22, 41.44)$. In practice, dropout percentages cannot be less than 0, so a more practical interval is $(0, 41.44)$. To the limits of our assumptions, we can be 99% confident that the dropout rate for a new, similar math class will be between 0 and 41.5%. It is impossible to validate assumptions about future observations (as long as they remain in the future), thus the exact confidence levels of prediction intervals are always suspect.

The key difference between the 99% prediction interval and a 99% confidence interval is the standard error. In a confidence interval, the standard error is $\sqrt{s^2/n}$. In a prediction interval, we mentioned the need to account for two sources of variability and the corresponding standard error is $\sqrt{s^2 + s^2/n}$. The first term in the square root estimates the variance of the new observation, while the second term in the square root estimates the variance of $\bar{y}_{\cdot}$, the point predictor.

As mentioned earlier and as will be shown in the next section, the assumption of normality is pretty poor for the 38 observations on dropout rates. Even without the assumption of normality we can get an approximate evaluation of the interval. The interval uses the value $t(.995, 37) = 2.71$, and we will see below that even without the assumption of normality, the approximate confidence level of this prediction interval is at least

$$100\left(1-\frac{1}{(2.71)^2}\right)\% = 86\%.$$

THEORY

In this chapter we assume that the observations y_i are independent from a population with mean μ and variance σ^2. We have assumed that all our previous observations on the process have been independent, so it is reasonable to assume that the future observation y is independent of the previous observations with the same mean and variance. The prediction interval is actually based on the difference $y - \bar{y}_\cdot$, i.e., we examine how far a new observation may reasonably be from our point predictor. Note that

$$E(y - \bar{y}_\cdot) = \mu - \mu = 0.$$

To proceed we need a standard error for $y - \bar{y}_\cdot$ and a distribution that is symmetric about 0. The standard error of $y - \bar{y}_\cdot$ is just the standard deviation of $y - \bar{y}_\cdot$ when available or, more often, an estimate of the standard deviation. First we need to find the variance. As $\bar{y}_\cdot$ is computed from the previous observations, it is independent of y and, using Proposition 1.2.11,

$$\mathrm{Var}(y - \bar{y}_\cdot) = \mathrm{Var}(y) + \mathrm{Var}(\bar{y}_\cdot) = \sigma^2 + \frac{\sigma^2}{n} = \sigma^2\left[1 + \frac{1}{n}\right].$$

The standard deviation is the square root of the variance. Typically, σ^2 is unknown, so we estimate it with s^2 and our standard error becomes

$$SE(y - \bar{y}_\cdot) = \sqrt{s^2 + \frac{s^2}{n}} = \sqrt{s^2\left[1 + \frac{1}{n}\right]} = s\sqrt{1 + \frac{1}{n}}.$$

To get an appropriate distribution, we assume that all the observations are normally distributed. In this case,

$$\frac{y - \bar{y}_\cdot}{SE(y - \bar{y}_\cdot)} \sim t(n-1).$$

The validity of the $t(n-1)$ distribution is established in Exercise 2.7.10. When the observations are not normally distributed, if we have a large sample we can use the law of large numbers and Chebyshev's inequality to approximate the worst case scenario.

Using the distribution based on normal observations, a 99% prediction interval is obtained from the following probability equalities:

$$\begin{aligned} .99 &= \Pr\left[-t(.995, n-1) < \frac{y - \bar{y}_\cdot}{SE(y - \bar{y}_\cdot)} < t(.995, n-1)\right] \\ &= \Pr[\bar{y}_\cdot - t(.995, n-1)SE(y - \bar{y}_\cdot) < y < \bar{y}_\cdot + t(.995, n-1)SE(y - \bar{y}_\cdot)]. \end{aligned}$$

The key point is that the two sets of inequalities within the square brackets are algebraically equivalent. Based on the last equality, the 99% prediction interval consists of all y values between $\bar{y}_\cdot - t(.995, n-1)SE(y - \bar{y}_\cdot)$ and $\bar{y}_\cdot + t(.995, n-1)SE(y - \bar{y}_\cdot)$. In other words, the 99% prediction interval has endpoints

$$\bar{y}_\cdot \pm t(.995, n-1)SE(y - \bar{y}_\cdot).$$

This looks similar to a 99% confidence interval for μ but the standard error is very different. In the prediction interval, the endpoints are actually

$$\bar{y}_{\cdot} \pm t(.995, n-1)\, s\sqrt{\left[1 + \frac{1}{n}\right]},$$

while in a confidence interval the endpoints are

$$\bar{y}_{\cdot} \pm t(.995, n-1)\, s\sqrt{\frac{1}{n}}.$$

The standard error for the prediction interval is typically much larger than the standard error for the confidence interval. Moreover, unlike the confidence interval, the prediction interval cannot be made arbitrarily small by taking larger and larger sample sizes n. Of course to compute an arbitrary $(1-\alpha)100\%$ prediction interval, simply replace the value $t(.995, n-1)$ with $t(1-\alpha/2, n-1)$.

As mentioned above, even when the data are not normally distributed, we can obtain an approximate worst case result for large samples. The approximation comes from using the law of large numbers to justify treating s as if it were the actual population standard deviation σ. With this approximation, Chebyshev's inequality states that

$$\begin{aligned} 1 &- \frac{1}{t(.995, n-1)^2} \\ &\leq \Pr\left[-t(.995, n-1) < \frac{y - \bar{y}_{\cdot}}{SE(y - \bar{y}_{\cdot})} < t(.995, n-1)\right] \\ &= \Pr[\bar{y}_{\cdot} - t(.995, n-1)SE(y - \bar{y}_{\cdot}) < y < \bar{y}_{\cdot} + t(.995, n-1)SE(y - \bar{y}_{\cdot})], \end{aligned}$$

cf. Subsection 1.2.2. As mentioned above, the 99% prediction interval based on 38 normal observations has a confidence level of at least

$$\left(1 - \frac{1}{(2.71)^2}\right) 100\% = 86\%.$$

This assumes that the past observations and the future observation form a random sample from the same population and assumes that 38 observations is large enough to justify using the law of large numbers. Similarly, if we can apply the improved version of Chebyshev's inequality from Section 1.3, we get a lower bound of $1 - [1/2.25(2.71)^2] = 93.9\%$ on the confidence coefficient.

Throughout, we have assumed that the process of generating the data yields independent observations from some population. In quality control circles this is referred to as having a process that is under *statistical control*.

2.4 Checking normality

From Figure 2.1, we identified two *outliers* in the dropout rate data, the 40% and the 59% dropout rates. If we delete these two points from the data, the remaining data may have a more nearly normal distribution. The dot plot with the two cases deleted is given in Figure 2.6. This is much more nearly normally distributed, i.e., looks much more like a bell shaped curve, than the complete data.

Dot plots and other versions of histograms are not effective in evaluating normality. Very large amounts of data are needed before one can evaluate normality from a histogram. A more useful technique for evaluating the normality of small and moderate size samples is the construction of a *normal probability plot*, also known as a *normal plot* or a *rankit plot*. The idea is to order the

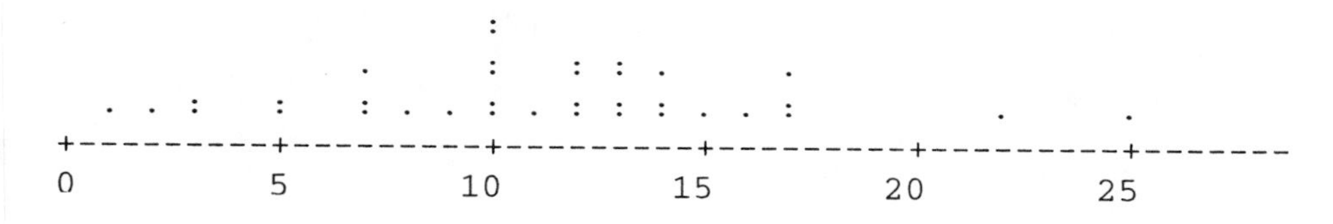

FIGURE 2.6. Dot plot for drop rate percentage data: outliers deleted.

data from smallest to largest and then to compare the ordered values to what one would expect the ordered values to be if they were truly a random sample from a normal distribution. These pairs of values should be roughly equal, so if we plot the pairs we would expect to see a line with a slope of about 1 that goes through the origin.

The problem with this procedure is that finding the expected ordered values requires us to know the mean μ and standard deviation σ of the appropriate population. These are generally not available. To avoid this problem, the expectations of the ordered values are computed assuming $\mu = 0$ and $\sigma = 1$. The expected ordered values from this standard normal distribution are called *normal scores* or *rankits*. Computing the expected values this way, we no longer anticipate a line with slope 1 and intercept 0. We now anticipate a line with slope σ and intercept μ. While it is possible to obtain estimates of the mean and standard deviation from a normal plot, our primary interest is in whether the plot looks like a line. A linear plot is consistent with normal data; a nonlinear plot is inconsistent with normal data. Christensen (1987, section XIII.2) gives a more detailed motivation for normal plots.

The normal scores are difficult to compute, so we generally get a computer program to do the work. In fact, just creating a plot is considerable work without a computer.

EXAMPLE 2.4.1. Consider the dropout rate data. Figure 2.7 contains the normal plot for the complete data. The two outliers cause the plot to be severely nonlinear. Figure 2.8 contains the normal plot for the dropout rate data with the two outliers deleted. It is certainly not horribly nonlinear. There is a little shoulder at the bottom end and some wiggling in the middle.

We can eliminate the shoulder in this plot by transforming the original data. Figure 2.9 contains a normal plot for the square roots of the data with the outliers deleted. While the plot no longer has a shoulder on the lower end, it seems to be a bit less well behaved in the middle.

We might now repeat our tests and confidence intervals for the 36 observations left when the outliers are deleted. We can do this for either the original data or the square roots of the original data. In either case, it now seems reasonable to treat the data as normal, so we can use a $t(36 - 1)$ distribution instead of hoping that the sample is large enough to justify use of the standard normal distribution. We will consider these tests and confidence intervals in the next chapter.

It is important to remember that *if outliers are deleted, the conclusions reached are not valid for data containing outliers*. For example, a confidence interval will be for the mean dropout rate excluding the occasional classes with extremely large dropout rates. If we are confident that any deleted outliers are not really part of the population of interest, this causes no problem. Thus, if we were sure that the large dropout rates were the result of clerical errors and did not provide any information about true dropout rates, our conclusions about the population should be based on the data excluding the outliers. More often though, we do not know that outliers are simple mistakes. *Often, outliers are true observations and often they are the most interesting and useful observations in the data.* If the outliers are true observations, systematically deleting them changes both the sample and the population of interest. In this case, the confidence interval is for the mean of a population implicitly defined by the process of deleting outliers. Admittedly, the idea of the mean dropout rate excluding the occasional outliers is not very clearly defined, but remember that the real population of interest is not too clearly defined either. We do not really want to learn about the

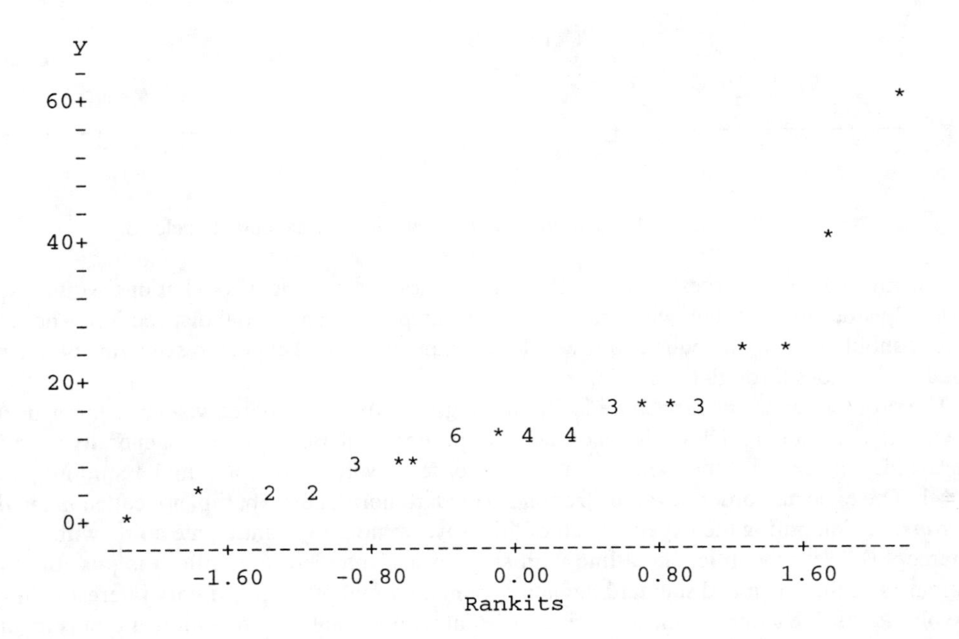

FIGURE 2.7. Normal plot for drop rate percentage data: full data.

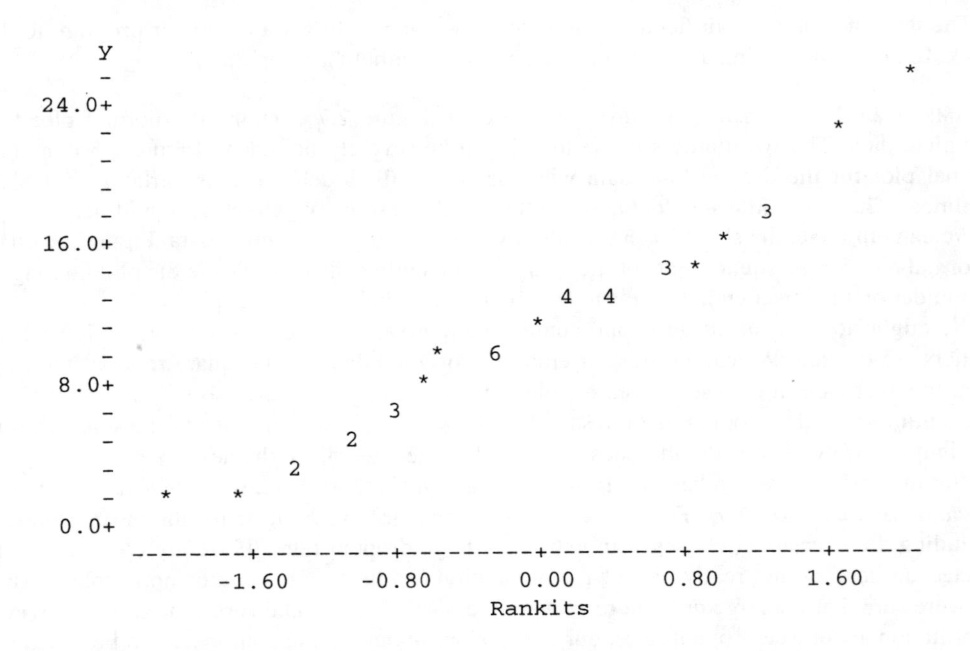

FIGURE 2.8. Normal plot for drop rate percentage data: outliers deleted.

clearly defined population of 1984–85 dropout rates, we really want to treat the dropout rate data as a sample from a population that allows us to draw useful inferences about current and future dropout rates. If we really cared about the fixed population, we could specify exactly what kinds of observations we would exclude and what we meant by the population mean of the observations that would be included. Given the nature of the true population of interest, I think that such technicalities

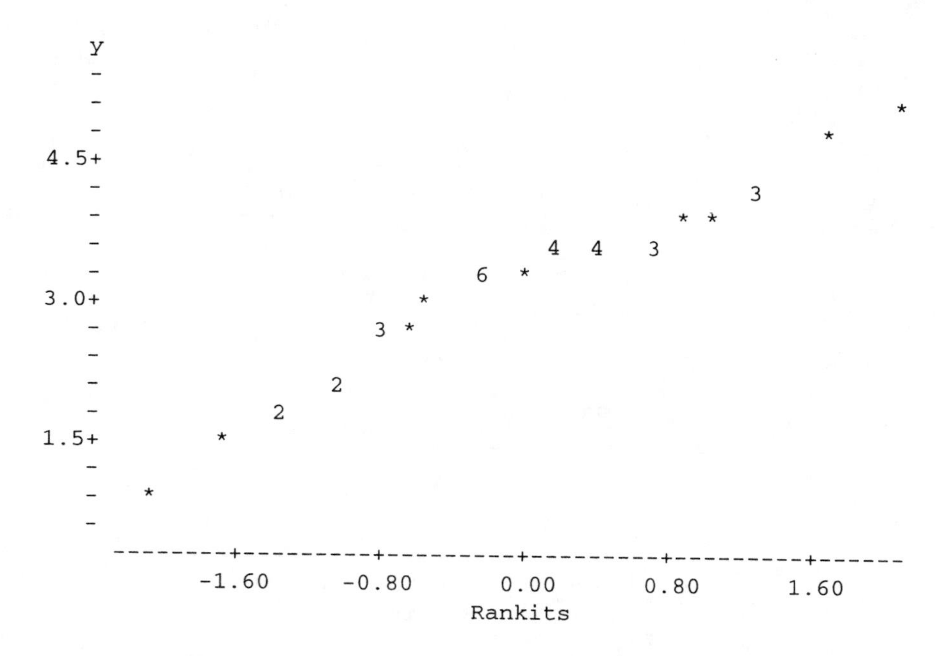

FIGURE 2.9. Normal plot for square roots of drop rate percentage data: outliers deleted.

are more trouble than they are worth at this point. □

Normal plots are subject to random variation because the data used in them are subject to random variation. Typically, normal plots are not perfectly straight. Figures 2.10 through 2.15 present six normal plots for which the data are in fact normally distributed. By comparison to these, Figures 2.8 and 2.9, the normal plots for the dropout rate data and the square root of the dropout rates both with outliers deleted, look reasonably normal. Of course, if the dropout rate data are truly normal, the square root of these data cannot be truly normal and vice versa. However, both are reasonably close to normal distributions.

Figures 2.10 through 2.15 contain normal plots based on 25 observations each. Normal plots based on larger normal samples tend to appear straighter than these. Normal plots based on smaller normal samples can look much more crooked.

TESTING NORMALITY

In an attempt to quantify the straightness of a normal plot, Shapiro and Francia (1972) proposed the summary statistic W', which is the squared sample correlation between the pairs of points in the plots. The population correlation coefficient was introduced in Subsection 1.2.3. The sample correlation coefficient is introduced in Chapter 7. At this point, it is sufficient to know that sample correlation coefficients near 0 indicate very little linear relationship between two variables and sample correlation coefficients near 1 or -1 indicate a very strong linear relationship. Since you need a computer to get the normal scores (rankits) anyway, just rely on the computer to give you the squared sample correlation coefficient.

A sample correlation coefficient near 1 indicates a strong tendency of one variable to increase (linearly) as the other variable increases and sample correlation coefficients near -1 indicate a strong tendency for one variable to decrease (linearly) as the other variable increases. In normal plots we are looking for a strong tendency for one variable, the ordered data, to increase as the other variable,

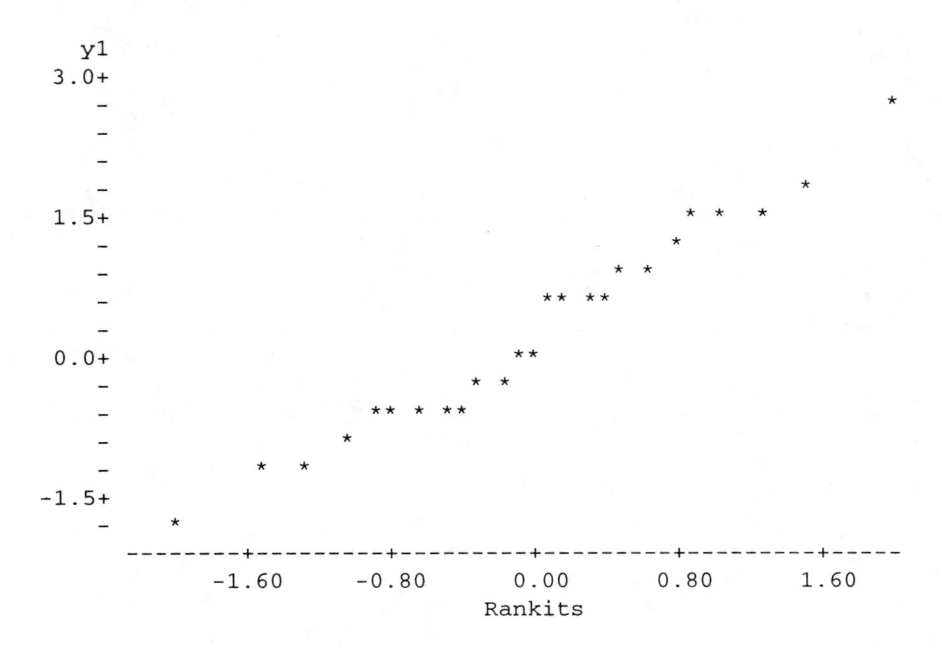

FIGURE 2.10. Normal plot.

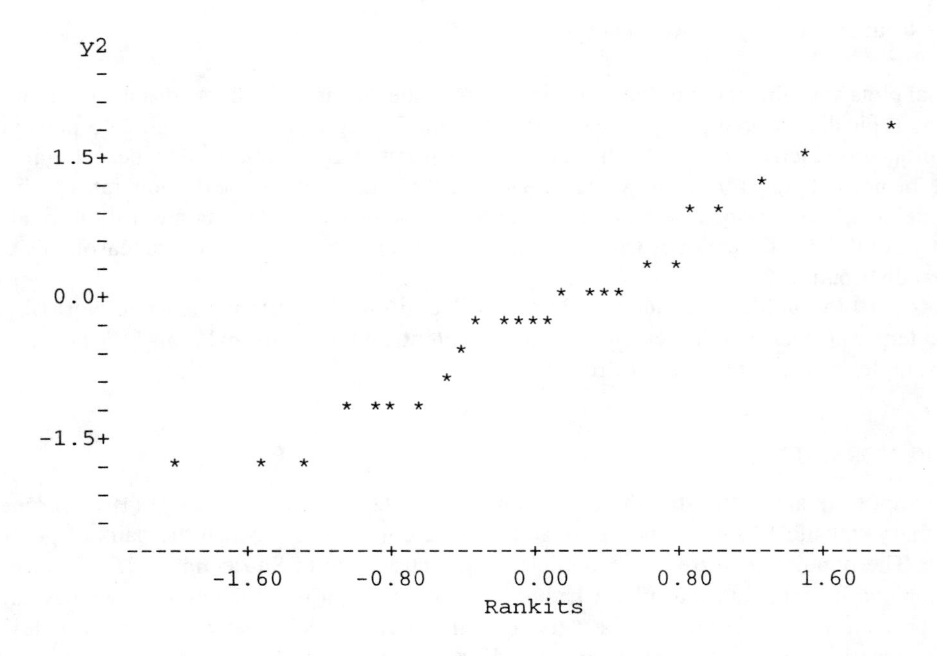

FIGURE 2.11. Normal plot.

the rankits, increases, so normal data should display a sample correlation coefficient near 1 and thus the square of the sample correlation, W', should be near 1. If W' is too small, it indicates that the data are inconsistent with the assumption of normality. If W' is smaller than, say, 95% of the values one would see from normally distributed data, it is substantial evidence that the data are not normally

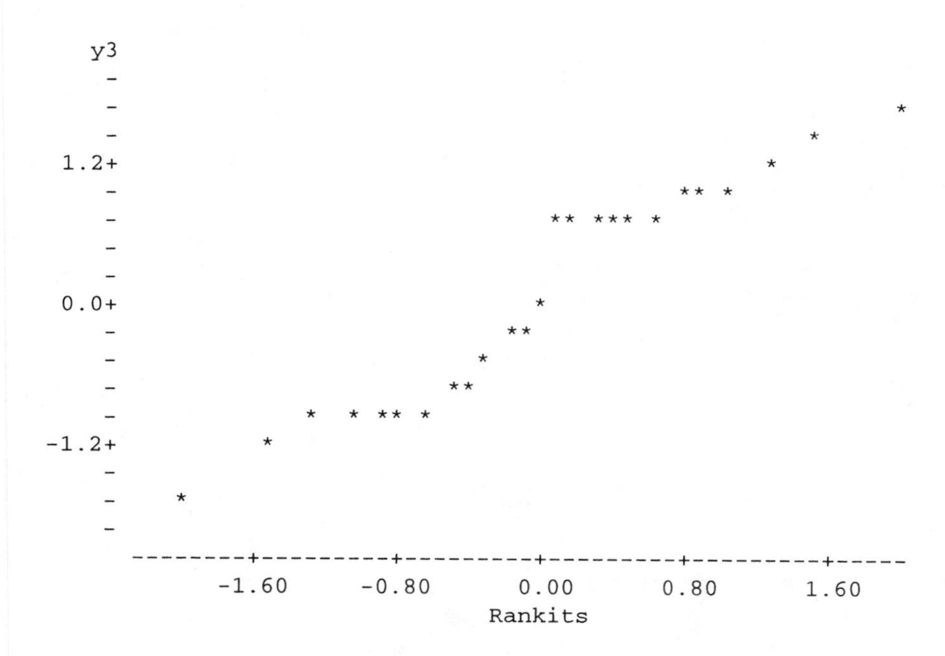

FIGURE 2.12. Normal plot.

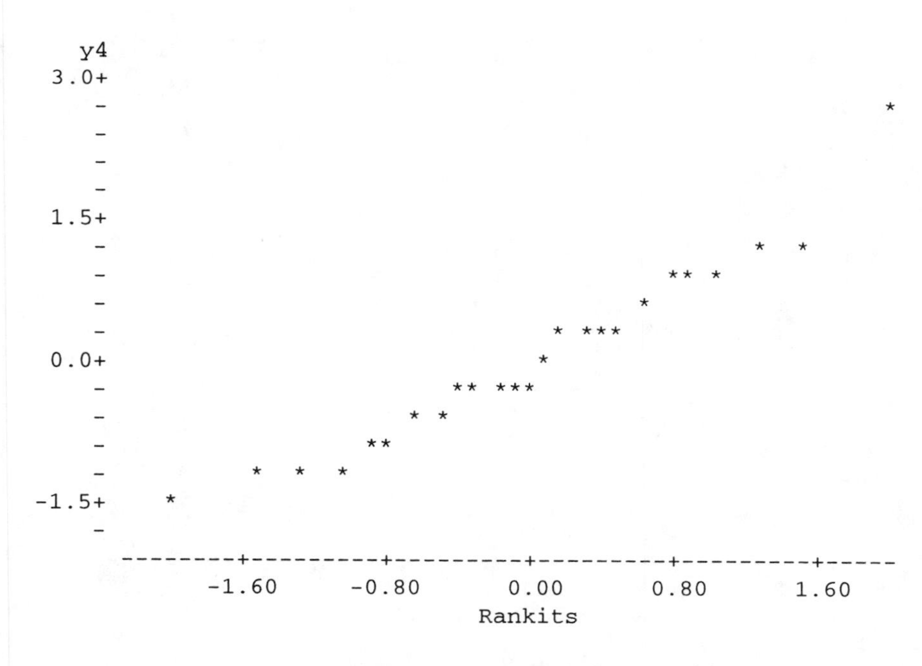

FIGURE 2.13. Normal plot.

distributed. If W' is smaller than, say, 99% of the values one would see from normally distributed data, it is strong evidence that the data are not normally distributed. Appendix B.3 presents tables of the values $W'(.05, n)$ and $W'(.01, n)$. These are the points above which fall, respectively, 95% and 99% of the W' values one would see from normally distributed data. Of course the W' percentiles are

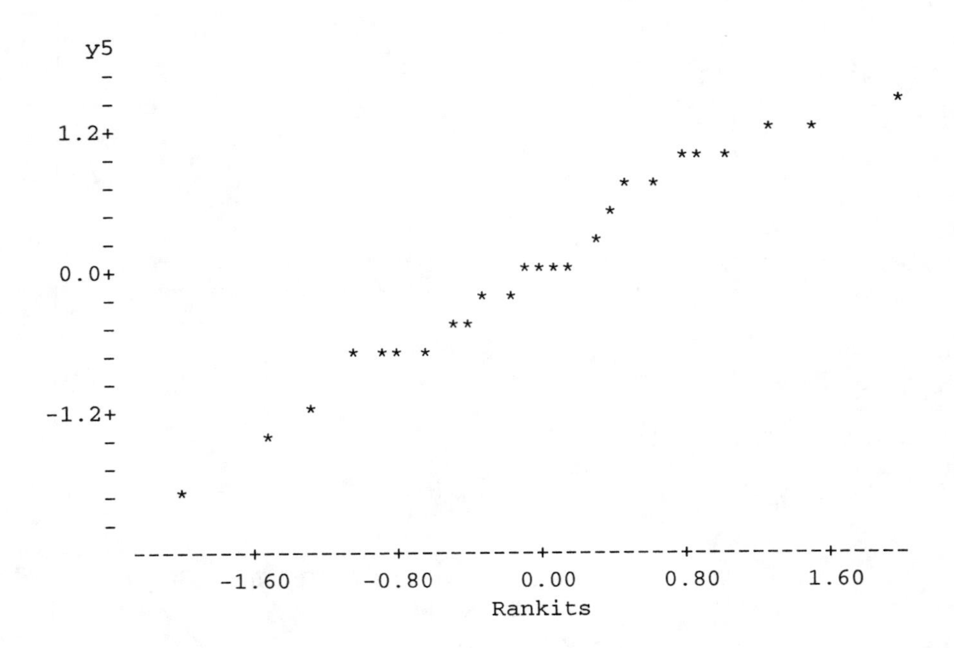

FIGURE 2.14. Normal plot.

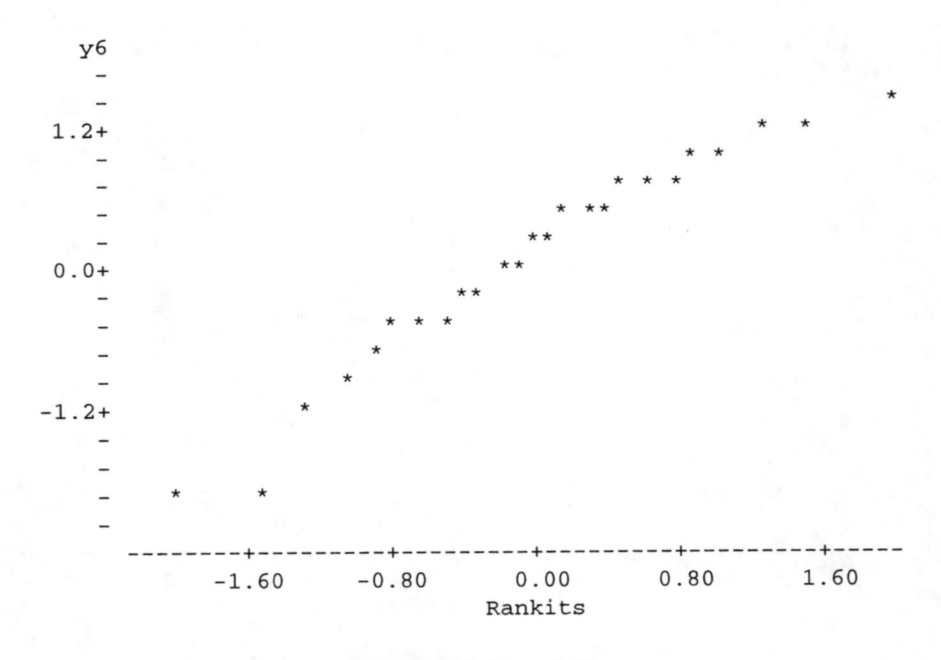

FIGURE 2.15. Normal plot.

computed using not only the assumption of normality, but also the assumptions that the observations are independent with the same mean and variance. Note also that the values of these percentiles depend on the sample size n. The tabled values are consistent with our earlier observation that the plots are more crooked for smaller numbers of observations and straighter for larger numbers of observations in that the tabled values get larger with n. For comparison, we give the observed W' values for the data used in Figures 2.10 through 2.15.

Shapiro–Francia statistics	
Figure	W'
2.10	0.966
2.11	0.974
2.12	0.937
2.13	0.956
2.14	0.958
2.15	0.978

These should be compared to $W'(.05, 25) \doteq .918$ and $W'(.01, 25) \doteq .88$ from Appendix B.3. None of these six values is below the 5% point.

EXAMPLE 2.4.2. For the dropout rate data we have three normal plots. The complete, untransformed data yield a W' value of .697. This value is inconsistent with the assumption that the dropout rate data has a normal distribution. Deleting the two outliers, W' is .978 for the untransformed data and .960 for the square roots of the data. The tabled percentiles are $W'(.05, 36) = .940$ and $W'(.01, 36) = .91$, so the untransformed data and the square root data look alright. In addition, W' was computed for the square roots of the complete data. Its value, .887, is still significantly low, but is a vast improvement over the untransformed complete data. The outliers are not nearly as strange when the square roots of the data are considered. Sometimes it is possible to find a transformation that eliminates outliers. □

MINITAB COMMANDS

A computer program is necessary for finding the normal scores and convenient for plotting the data and computing W'. The following Minitab commands provide a normal plot and the W' statistic for a variable in c1.

```
MTB > name c1 'y'
MTB > nscores c1 c2
MTB > plot c1 c2
MTB > corr c1 c2
MTB > note    The correlation is printed out, e.g., .987.
MTB > note    This correlation is used in the next command.
MTB > let k1=.987**2
MTB > note    k1 is W'
MTB > print k1
```

2.5 Transformations

In analyzing a collection of numbers, we assume that the observations are a random sample from some population. Often, the population from which the observations come is not as well defined as we might like. For example, if our observations are the yields of corn on 30 one acre plots of ground grown in the summer of 1990, what is the larger population from which this is a sample? Typically, we do not have a large number of one acre plots from which we randomly select 30. Even if we had a large collection of plots, these plots are subject to different weather conditions, have different fertilities, etc. Most importantly, we are rarely interested in corn grown in 1990 for its own sake. If we are studying corn grown in 1990, we are probably interested in predicting how that same type of corn would behave if we planted it at some time in the future. No population that currently exists could be completely appropriate for drawing conclusions about plant growths in a future year. Thus *the assumption that the observations are a random sample from some population is often only a useful approximation.*

When making approximations, it is often necessary to adjust things to make the approximations more accurate. In statistics, *two approximations we frequently make are that all the data have the same variance and that the data are normally distributed. Making numerical transformations of the data is a primary tool for improving the accuracy of these approximations.* When sampling from a fixed population, we are typically interested in transformations that improve the normality assumption because having different variances is not a problem associated with sampling from a fixed population. With a fixed population, the variance of an object is the variance of randomly choosing an object from the population. This is a constant regardless of which object we end up choosing. But data are rarely as simple as random samples from a fixed population. Once we have an object from the population, we have to obtain an observation (measurement or count) from the object. These observations on a given object are also subject to random error and the error may well depend on the specific object being observed.

We now examine the fact that observations often have different variances, depending on the object being observed. First consider taking length measurements using a 30 centimeter ruler that has millimeters marked on it. For measuring objects that are less than 30 centimeters long, like this book, we can make very accurate measurements. We should be able to measure things within half a millimeter. Now consider trying to measure the height of a dog house that is approximately 3.5 feet tall. Using the 30 cm ruler, we measure up from the base, mark 30 cm, measure from the mark up another 30 cm, make another mark, measure from the new mark up another 30 cm, mark again, and finally we measure from the last mark to the top of the house. With all the marking and moving of the ruler, we have much more opportunity for error than we have in measuring the length of the book. Obviously, if we try to measure the height of a house containing two floors, we will have much more error. If we try to measure the height of the Sears tower in Chicago using a 30 cm ruler, we will not only have a lot of error, but large psychiatric expenses as well. The moral of this tale is that, when making measurements, larger objects tend to have more variability. If the objects are about the same size, this causes little or no problem. One can probably measure female heights with approximately the same accuracy for all women in a sample. One probably cannot measure the weights of a large sample of marine animals with constant variability, especially if the sample includes both shrimp and blue whales. *When the observations are the measured* amounts *of something, often the standard deviation of an observation is proportional to its mean. When the standard deviation is proportional to the mean, analyzing the logarithms of the observations is more appropriate than analyzing the original data.*

Now consider the problem of counting up the net financial worth of a sample of people. For simplicity, let's think of just three people, me, my 10 year old son (at least he was 10 when I started writing this), and my rich uncle, Scrooge. In fact, let's just think of having a stack of one dollar bills in front of each person. My pile is of a decent size, my son's is small, and my uncle's is huge. When I count my pile, it is large enough that I could miscount somewhere and make a significant, but not major, error. When I count my son's pile, it is small enough that I should get it about right. When I count my uncle's pile, it is large enough that I will, almost inevitably, make several significant errors. As with measuring amounts of things, the larger the observation, the larger the potential error. However, the process of making these errors is very different than that described for measuring amounts. In such cases, the variance of the observations is often proportional to the mean of the observations. The standard corrective measure for counts is different from the standard corrective measure for amounts. *When the observations are* counts *of something, often the variance of the count is proportional to its mean. In this case, analyzing the square roots of the observations is more appropriate than analyzing the original data.*

Suppose we are looking at yearly sales for a sample of corporations. The sample may include both the corner gas (petrol) station and Exxon. It is difficult to argue that one can really *count* sales for a huge company such as Exxon. In fact, it may be difficult to count even yearly sales for a gas station. Although in theory one should be able to count sales, it may be better to think of yearly sales as measured amounts. It is not clear how to transform such data. Another example is age. We usually think of counting the years a person has been alive, but one could also argue that we are measuring the amount of time a person has been alive. *In practice, we often try both logarithmic and square root transformations and use the transformation that seems to work best*, even when the type of observation (count or amount) seems clear.

Finally, consider the proportion of times people drink a particular brand of soda pop, say, Dr. Pepper. The idea is simply that we ask a group of people what proportion of the time they drink Dr. Pepper. People who always drink Dr. Pepper are aware of that fact and should give a quite accurate proportion. Similarly, people who never drink Dr. Pepper should be able to give an accurate proportion. Moreover, people who drink Dr. Pepper about 90% of the time or about 10% of the time, can probably give a fairly accurate proportion. The people who will have a lot of variability in their replies are those who drink Dr. Pepper about half the time. They will have little idea whether they drink it 50% of the time, or 60%, or 40%, or just what. With observations that are counts or amounts, larger observations have larger variances. With observations that are proportions, observations near 0 and 1 have small variability and observations near .5 have large variability. Proportion data call for a completely different type of transformation. *The standard transformation for proportion data is the inverse sine (arcsine) of the square root of the proportion. When the observations are proportions, often the variance of the proportion is a constant times $\mu(1-\mu)/N$, where μ is the mean and N is the number of trials. In this case, analyzing the inverse sine (arcsine) of the square root of the proportion is more appropriate than analyzing the original data.*

In practice, the square root transformation is sometimes used with proportion data. After all, many proportions are obtained as a count divided by the total number of trials. For example, the best data we could get in the Dr. Pepper drinking example would be the count of the number of Dr. Peppers consumed divided by the total number of sodas devoured.

There is a subtle but important point that was glossed over in the previous paragraphs. If we take multiple measurements on a house, the variance depends on the true height, but the true height is the same for all observations. Such a dependence of the variance on the mean causes no problems. The problem arises when we measure a random sample of buildings each with a variance depending on its true height.

EXAMPLE 2.5.1. For the dropout rate data, we earlier considered the complete, untransformed data and after deleting two outliers, we looked at the untransformed data and the square roots of the data. In Examples 2.4.1 and 2.4.2 we saw that the untransformed data with the outliers deleted and the square roots of the data with the outliers deleted had approximate normal distributions. Based on the W' statistic, the untransformed data seemed to be more nearly normal. The data are proportions of people who drop from a class, so our discussion in this section suggests transforming by the inverse sine of the square roots of the proportions. Recall that proportions are values between 0 and 1, while the dropout rates were reported as values between 0 and 100, so the reported rates need to be divided by 100. For the complete data, this transformation yields a W' value of .85, which is much better than the untransformed value of .70, but worse than the value .89 obtained with the square root transformation. With the two outliers deleted, the inverse sine of the square roots of the proportions yields the respectable value $W' = .96$, but the square root transformation is simpler and gives almost the same value, while the untransformed data give a much better value of .98. Examination of the six normal plots (only three of which have been presented here) reinforce the conclusions given above.

With the outliers deleted, it seems reasonable to analyze the untransformed data and, to a lesser extent, the data after either transformation. *Other things being equal*, we prefer using the simplest transformation that seems to work. Simple transformations are easier to explain, justify, and interpret. The square root transformation is simpler, and thus better, than the inverse sine of the square roots of the proportions. Of course, not making a transformation seems to work best and not transforming is always the simplest transformation. Actually some people would point out, and it is undeniably true, that the act of deleting outliers is really a transformation of the data. However, we will not refer to it as such. □

MINITAB COMMANDS

Minitab commands for the three transformations discussed here and for the cubed root power transformation are given below. The cubed root is just to illustrate a general power transformation.

```
MTB > name c1 'y'
MTB > let c2 = loge(c1)
MTB > let c3 = sqrt(c1)
MTB > let c4 = asin(sqrt(c1))
MTB > let c5 = c1**(1/3)
```

THEORY

The standard transformations given above are referred to as *variance stabilizing transformations.* The idea is that each observation is a look at something with a different mean and variance, where the variance depends on the mean. For example, when we measure the height of a house, the house has some 'true' height and we simply take a measurement of it. The variability of the measurement depends on the true height of the house. Variance stabilizing transformations are designed to eliminate the dependence of the variance on the mean. Although variance stabilizing transformations are used quite generally for counts, amounts, and proportions, they are derived for certain assumptions about the relationship between the mean and the variance. These relationships are tied to theoretical distributions that are appropriate for some counts, amounts, and proportions. Rao (1973, section 6g) gives a nice discussion of the mathematical theory behind variance stabilizing transformations.

Proportions are related to the binomial distribution for the numbers of successes. We have a fixed number of trials; the proportion is the number of successes divided by the number of trials. The mean of a $Bin(N, p)$ distribution is Np and the variance is $Np(1-p)$. This relationship between the mean and variance of a binomial leads to the inverse sine of the square root transformation.

Counts are related to the Poisson distribution. The Poisson distribution is an approximation used for binomials with a very large number of trials, each having a very small probability of success. Poisson data has the property that the variance equals the mean of the observation. This relationship leads to the square root as the variance stabilizing transformation.

For amounts, the log transformation comes from having the standard deviation proportional to the mean. The standard deviation divided by the mean is called the *coefficient of variation*, so the log transformation is appropriate for observations that have a constant coefficient of variation. (The square root transformation comes from having the variance, rather than the standard deviation, proportional to the mean.) A family of continuous distributions called the gamma distributions has constant coefficient of variation.

The variance stabilizing transformations are given below. In each case we assume $E(y_i) = \mu_i$ and $\text{Var}(y_i) = \sigma_i^2$. The symbol $\propto$ means 'proportional to.'

Variance stabilizing transformations

Data	Distribution	Mean, variance relationship	Transformation
Count	Poisson	$\mu_i \propto \sigma_i^2$	$\sqrt{y_i}$
Amount	Gamma	$\mu_i \propto \sigma_i$	$\log(y_i)$
Proportion	Binomial/ N	$\frac{\mu_i(1-\mu_i)}{N} \propto \sigma_i^2$	$\sin^{-1}(\sqrt{y_i})$

I cannot honestly recommend using variance stabilizing transformations to analyze either binomial or Poisson data. In the past 20 years, a large body of statistical techniques has been developed specifically for analyzing binomial and Poisson data, see, for example, Christensen (1990b). I would recommend using these alternative methods. Many people would make a similar recommendation for gamma distributed data citing the applicability of generalized linear model theory, cf. McCullagh and Nelder (1989) or Christensen (1990b, chapter X). When applied to binomial, Poisson, or gamma distributed data, variance stabilizing transformations provide a way to force the methods developed for normally distributed data into giving a reasonable analysis for data that are not normally distributed. If you have a clear idea about the true distribution of the data, you should use methods developed specifically for that distribution. The problem is that we often have little idea of the appropriate distribution for a set of data. For example, if we simply ask people the proportion of times they drink Dr. Pepper, we have proportion data that is not binomial. In such cases, we seek a transformation that will make a normal theory analysis approximately correct. We often pick transformations by trial and error. *The variance stabilizing transformations provide little more than a place to start when considering transformations.*

At the beginning of this section, we mentioned two key approximations that we frequently make. These are that all the data have the same variance and that the data are normally distributed. While the rationale given above for picking transformations was based on stabilizing variances, in practice we typically choose a transformation for a single sample to attain approximate normality. To evaluate whether a transformation really stabilizes the variance, we need more information than is contained in a single sample. Control chart methods can be used to evaluate variance stabilization for a single sample, cf. Shewhart (1931). Those methods require formation of rational subgroups and that requires additional information. We could also plot the sample against appropriately chosen variables to check variance stabilization, but finding appropriate variables can be quite difficult and would depend on properties of the particular sampling process. Variance stabilizing transformations are probably best suited to problems that compare samples from several populations, where the variance in each population depends on the mean of the population.

On the other hand, we already have examined methods for evaluating the normality of a single sample. Thus, since we cannot (actually, do not) evaluate variance stabilization in a single sample, if we think that the variance of observations should increase with their mean, we might try both the log and square root transformations and pick the one for which the transformed data best approximate normality.

2.6 Inference about σ^2

If the data are normally distributed, we can also perform confidence intervals and tests for the population variance σ^2. While these are not typically of primary importance, they can be useful. They also tend to be sensitive to the assumption of normality. The procedures do not follow the same pattern used for most inferences that involve 1) a parameter of interest, 2) an estimate of the parameter, 3) the standard error of the estimate, and 4) a known distribution symmetric about zero; however, there are similarities. Procedures for variances typically require a parameter, an estimate, and a known distribution.

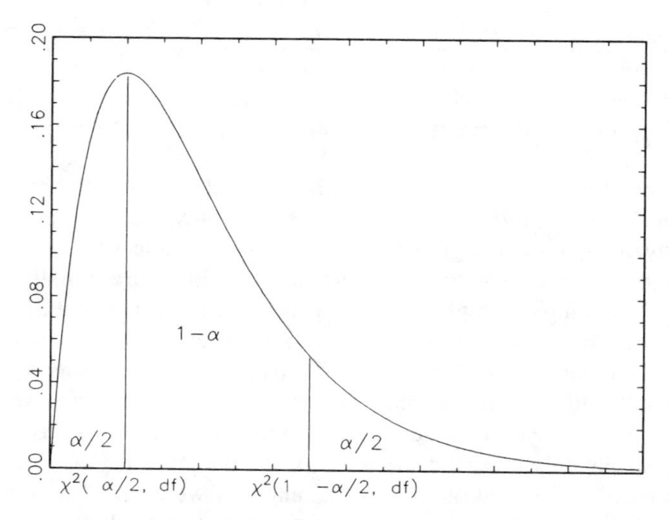

FIGURE 2.16. Central χ^2 interval with probability $1 - \alpha$.

The procedures discussed in this section actually apply to all the problems in this book that involve a single variance parameter σ^2. One need only substitute the relevant estimate of σ^2 and use its degrees of freedom. Applications to the data and models considered in Chapter 12 are not quite as straightforward because there the models involve more than one variance.

In the one-sample problem, the parameter is σ^2, the estimate is s^2, and the distribution, as discussed in equation (2.1.5), is

$$\frac{(n-1)s^2}{\sigma^2} \sim \chi^2(n-1).$$

The notation $\chi^2(1-\alpha, n-1)$ is used to denote the point that cuts off the bottom $1-\alpha$ (top α) of the χ^2 distribution with $n-1$ degrees of freedom. Note that $(n-1)s^2/\sigma^2$ is nonnegative, so the curve in Figure 2.16 illustrating the χ^2 distribution is also nonnegative. Figure 2.16 shows a central interval with probability $1-\alpha$ for a χ^2 distribution.

A $(1-\alpha)100\%$ confidence interval for σ^2 is based on the following equality:

$$\begin{aligned} 1-\alpha &= \Pr\left[\chi^2\left(\frac{\alpha}{2}, n-1\right) < \frac{(n-1)s^2}{\sigma^2} < \chi^2\left(1-\frac{\alpha}{2}, n-1\right)\right] \\ &= \Pr\left[\frac{(n-1)s^2}{\chi^2\left(1-\frac{\alpha}{2}, n-1\right)} < \sigma^2 < \frac{(n-1)s^2}{\chi^2\left(\frac{\alpha}{2}, n-1\right)}\right]. \end{aligned} \tag{2.6.1}$$

The first equality corresponds to Figure 2.16 and is just the definition of the percentage points $\chi^2\left(\frac{\alpha}{2}, n-1\right)$ and $\chi^2\left(1-\frac{\alpha}{2}, n-1\right)$. These are defined to be the points that cut out the middle $1-\alpha$ of the chi-squared distribution and are tabled in Appendix B.2. The second equality in (2.6.1) is based on algebraic manipulation of the terms in the square brackets. The actual derivation is given later in this section. The second equality gives an interval that contains σ^2. There is a probability of $1-\alpha$ that σ^2 is *going to be* in the interval

$$\left(\frac{(n-1)s^2}{\chi^2\left(1-\frac{\alpha}{2}, n-1\right)}, \frac{(n-1)s^2}{\chi^2\left(\frac{\alpha}{2}, n-1\right)}\right). \tag{2.6.2}$$

The derivation of the confidence interval for σ^2 requires the data to be normally distributed. This assumption is more vital for inferences about σ^2 than it is for inferences about μ. For inferences about μ, the central limit theorem indicates that the sample means are approximately normal even when the data are not normal. There is no similar result indicating that the sample variance is approximately χ^2 even when the data are not normal.

EXAMPLE 2.6.1. Consider again the dropout rate data. We have seen that the complete data are not normal, but that after deleting the two outliers, the remaining data are reasonably normal. We find a 95% confidence interval for σ^2 from the deleted data. The deleted data contain 36 observations and s^2 for the deleted data is 27.45. The percentage points for the $\chi^2(36-1)$ distribution are $\chi^2(.025, 35) = 20.57$ and $\chi^2(.975, 35) = 53.20$. Applying (2.6.2), the 95% confidence interval is

$$\left(\frac{35(27.45)}{53.20}, \frac{35(27.45)}{20.57},\right)$$

or equivalently $(18.1, 46.7)$. We are 95% confident that the true variance is between 18.1 and 46.7, but remember that this is the true variance *after the deletion of outliers.* Again, when we delete outliers we are a little fuzzy about the exact definition of our parameter, but we are also being fuzzy about the exact population of interest. The exception to this is when we believe that the only outliers that exist are observations that are not really part of the population. □

It is the endpoints of the interval (2.6.2) that are random. To use the interval, we replace the random variable s^2 with the observed value of s^2 and replace the term 'probability $(1-\alpha)$' with '$(1-\alpha)100\%$ confidence.' *Once the observed value of s^2 is substituted into the interval, nothing about the interval is random any longer*, the fixed unknown value of σ^2 is either in the interval or it is not; there is no probability associated with it. The probability statement about random variables is mystically transformed into a 'confidence' statement. This is not unreasonable, but the rationale is, to say the least, murky.

The α level test of $H_0 : \sigma^2 = \sigma_0^2$ versus $H_A : \sigma^2 \neq \sigma_0^2$ is again based on the first equality in equation (2.6.1). To actually perform a test, σ_0^2 *must be a known value*. As usual, we assume that the null hypothesis is true, i.e., $\sigma^2 = \sigma_0^2$, so under this assumption

$$1-\alpha = \Pr\left[\chi^2\left(\frac{\alpha}{2}, n-1\right) < \frac{(n-1)s^2}{\sigma_0^2} < \chi^2\left(1-\frac{\alpha}{2}, n-1\right)\right].$$

If we observe data yielding an s^2 such that $(n-1)s^2/\sigma_0^2$ is between the values $\chi^2\left(\frac{\alpha}{2}, n-1\right)$ and $\chi^2\left(1-\frac{\alpha}{2}, n-1\right)$, the data are consistent with the assumption that $\sigma^2 = \sigma_0^2$ at level α. Conversely, we reject $H_0 : \sigma^2 = \sigma_0^2$ with a two-sided α level test if

$$\frac{(n-1)s^2}{\sigma_0^2} > \chi^2\left(1-\frac{\alpha}{2}, n-1\right)$$

or if

$$\frac{(n-1)s^2}{\sigma_0^2} < \chi^2\left(\frac{\alpha}{2}, n-1\right).$$

A clear definition of 'confidence' can be given in terms of testing the hypothesis $H_0 : \sigma^2 = \sigma_0^2$ versus the alternative $H_A : \sigma^2 \neq \sigma_0^2$. The same algebraic manipulations that lead to equation (2.6.1) can be used to show that the $(1-\alpha)100\%$ confidence interval contains precisely those values of σ_0^2 that are consistent with the data when testing $H_0 : \sigma^2 = \sigma_0^2$ at level α. This idea is discussed in more detail in Section 3.4.

EXAMPLE 2.6.2. For the dropout rate data consider testing $H_0 : \sigma^2 = 50$ versus $H_A : \sigma^2 \neq 50$ with $\alpha = .01$. Again, we use the data with the two outliers deleted, so our concept of the population variance σ^2 must account for our deletion of weird cases. The test statistic is

$$\frac{(n-1)s^2}{\sigma_0^2} = \frac{35(27.45)}{50} = 19.215.$$

The *critical region*, the region for which we reject H_0, contains all values greater than $\chi^2(.995, 35) = 60.275$ and all values less than $\chi^2(.005, 35) = 17.19$. The test statistic is certainly not greater than 60.275 and it is also not less than 17.19, so we have no basis for rejecting the null hypothesis at the $\alpha = .01$ level. At the .01 level, the data are consistent with the claim that $\sigma^2 = 50$.

The 95% confidence interval (18.1, 46.7) from Example 2.6.1 contains all values of σ^2 that are consistent with the data as determined by a two-sided $\alpha = .05$ level test. The interval does not contain 50, so we do have evidence against $H_0 : \sigma^2 = 50$ at the $\alpha = .05$ level. □

While methods for drawing inferences about variances *do not* fit our standard pattern based on 1) a parameter of interest, 2) an estimate of the parameter, 3) the standard error of the estimate, and 4) a known distribution symmetric about zero, it should be noted that the basic logic behind these confidence intervals and tests is the same. Confidence intervals are based on a random interval that contains the parameter of interest with some specified probability. The unusable random interval is changed into a usable nonrandom interval by substituting the observed value of the random variable into the endpoints of the interval. The probability is then miraculously, if intuitively, turned into 'confidence.' Tests of hypotheses are based on evaluating whether the data are consistent with the null hypothesis. Consistency is defined in terms of a known distribution that applies when the null hypothesis is true. If the data are inconsistent with the null hypothesis, the null hypothesis is rejected as being inconsistent with the observed data.

Below is a series of equalities that justify equation (2.6.1).

$$\begin{aligned} 1-\alpha &= \Pr\left[\chi^2\left(\frac{\alpha}{2}, n-1\right) < \frac{(n-1)s^2}{\sigma^2} < \chi^2\left(1-\frac{\alpha}{2}, n-1\right)\right] \\ &= \Pr\left[\frac{1}{\chi^2(\frac{\alpha}{2}, n-1)} > \frac{\sigma^2}{(n-1)s^2} > \frac{1}{\chi^2(1-\frac{\alpha}{2}, n-1)}\right] \\ &= \Pr\left[\frac{1}{\chi^2(1-\frac{\alpha}{2}, n-1)} < \frac{\sigma^2}{(n-1)s^2} < \frac{1}{\chi^2(\frac{\alpha}{2}, n-1)}\right] \\ &= \Pr\left[\frac{(n-1)s^2}{\chi^2(1-\frac{\alpha}{2}, n-1)} < \sigma^2 < \frac{(n-1)s^2}{\chi^2(\frac{\alpha}{2}, n-1)}\right]. \end{aligned}$$

2.7 Exercises

EXERCISE 2.7.1. Mulrow et al. (1988) presented data on the melting temperature of biphenyl as measured on a differential scanning calorimeter. The data are given below; they are the observed melting temperatures in Kelvin less 340.

3.02, 2.36, 3.35, 3.13, 3.33, 3.67, 3.54, 3.11, 3.31, 3.41, 3.84, 3.27, 3.28, 3.30

Compute the sample mean, variance, and standard deviation. Give a 99% confidence interval for the population mean melting temperature of biphenyl as measured by this machine. (Note that we don't know whether the calorimeter is accurately calibrated.)

EXERCISE 2.7.2. Box (1950) gave data on the weights of rats that were about to be used in an experiment. The data are repeated in Table 2.1. Assuming that these are a random sample from a broader population of rats, give a 95% confidence interval for the population mean weight. Test the null hypothesis that the population mean weight is 60 using a .01 level test.

TABLE 2.1. Weights of rats

59	54	56	59	57	52	52	61	59
53	59	51	51	56	58	46	53	57
60	52	49	56	46	51	63	49	57

EXERCISE 2.7.3. Fuchs and Kenett (1987) presented data on citrus juice for fruits grown during a specific season at a specific location. The sample size was 80 but many variables were measured on each sample. Sample statistics for some of these variables are given below

Variable	BX	AC	SUG	K	FORM	PECT
Mean	10.4	1.3	7.7	1180.0	22.2	451.0
Variance	0.38	0.036	0.260	43590.364	6.529	16553.996

The variables are BX – total soluble solids produced at 20°C, AC – acidity as citric acid unhydrons, SUG – total sugars after inversion, K – potassium, FORM – formol number, PECT – total pectin. Give a 99% confidence interval for the population mean of each variable. Give a 99% prediction interval for each variable. Test whether the mean of BX equals 10. Test whether the mean of SUG is less than or equal to 7.5. Use $\alpha = .01$ for each test.

EXERCISE 2.7.4. Jolicoeur and Mosimann (1960) gave data on female painted turtle shell lengths. The data are presented in Table 2.2. Give a 95% confidence interval for the population mean length. Give a 99% prediction interval for the shell length of a new female.

TABLE 2.2. Female painted turtle shell lengths

98	138	123	155	105	147	133	159
103	138	133	155	109	149	134	162
103	141	133	158	123	153	136	177

EXERCISE 2.7.5. Mosteller and Tukey (1977) extracted data from the *Coleman Report*. Among the variables considered was the percentage of sixth-graders who's fathers were employed in white collar jobs. Data for 20 New England schools are given in Table 2.3. Are the data reasonably normal? Do any of the standard transformations improve the normality? After finding an appropriate transformation (if necessary), test the null hypothesis that the percentage of white collar fathers is 50%. Use a .05 level test. Give a 99% confidence interval for the percentage of fathers with white collar jobs. If a transformation was needed, relate your conclusions back to the original measurement scale.

EXERCISE 2.7.6. Give a 95% confidence interval for the population variance associated with the data of Exercise 2.7.5. Remember that inferences about variances require the assumption of normality. Could the variance reasonably be 10?

TABLE 2.3. Percentage of fathers with white collar jobs

28.87	20.10	69.05	65.40	29.59
44.82	77.37	24.67	65.01	9.99
12.20	22.55	14.30	31.79	11.60
68.47	42.64	16.70	86.27	76.73

EXERCISE 2.7.7. Give a 95% confidence interval for the population variance associated with the data of Exercise 2.7.4. Remember that the inferences about variances require the assumption of normality.

EXERCISE 2.7.8. Give 99% confidence intervals for the population variances of all the variables in Exercise 2.7.3. Assume that the original data were normally distributed. Using $\alpha = .01$, test whether the potassium variance could reasonably be 45000. Could the formol number variance be 8?

EXERCISE 2.7.9. Shewhart (1931, p. 62) reproduces Millikan's data on the charge of an election. These are repeated in Table 2.4. Check for outliers and nonnormality. Adjust the data appropriately if there are any problems. Give a 98% confidence interval for the population mean value. Give a 98% prediction interval for a new measurement. (Millikan argued that some adjustments were needed before these data could be used in an optimal fashion but we will ignore his suggestions.)

TABLE 2.4. Observations on the charge of an electron

4.781	4.764	4.777	4.809	4.761	4.769	4.795	4.776
4.765	4.790	4.792	4.806	4.769	4.771	4.785	4.779
4.758	4.779	4.792	4.789	4.805	4.788	4.764	4.785
4.779	4.772	4.768	4.772	4.810	4.790	4.775	4.789
4.801	4.791	4.799	4.777	4.772	4.764	4.785	4.788
4.779	4.749	4.791	4.774	4.783	4.783	4.797	4.781
4.782	4.778	4.808	4.740	4.790	4.767	4.791	4.771
4.775	4.747						

EXERCISE 2.7.10. Show that if $y, y_1, \ldots, y_n$ are independent $N(\mu, \sigma^2)$ random variables, $(y - \bar{y}_\cdot)/\sqrt{\sigma^2 + \sigma^2/n} \sim N(0, 1)$. Recalling that y, $\bar{y}_\cdot$, and s^2 are independent and that $(n-1)s^2/\sigma^2 \sim \chi^2(n-1)$, use Definition 2.1.3 to show that $(y - \bar{y}_\cdot)/\sqrt{s^2 + s^2/n} \sim t(n-1)$.

Chapter 3

A general theory for testing and confidence intervals

The most commonly used statistical tests and confidence intervals derive from a single theory. (Tests and confidence intervals about variances are an exception.) The basic ideas of this theory were illustrated in Chapter 2. The point of the current chapter is to present the theory in its general form and to reemphasize fundamental techniques. The general theory will then be used throughout the book. Because the theory is stated in quite general terms, some prior familiarity with the ideas, e.g., reading Sections 2.2 and 2.3, is highly recommended.

To use the general theory you need to know four things:

1. the parameter of interest, *Par*,
2. the estimate of the parameter, *Est*,
3. the standard error of the estimate, *SE*(*Est*), and
4. the appropriate reference distribution.

Specifically, what you need to know about the distribution is that

$$\frac{Est - Par}{SE(Est)}$$

has a known (tabled) distribution that is symmetric about zero. The estimate *Est* is taken to be a random variable. The standard error, *SE*(*Est*), is the standard deviation of the estimate if that is known, but more commonly it is an estimate of the standard deviation. If the *SE*(*Est*) is estimated, the known distribution is usually the t distribution with some known number of degrees of freedom. If the *SE*(*Est*) is known, then the distribution is usually the standard normal distribution, i.e., mean 0, variance 1. In some problems, e.g., problems involving the binomial distribution, the central limit theorem is used to get an approximate distribution and inferences proceed as if that distribution is correct. When appealing to the central limit theorem, the known distribution is the standard normal.

Identifying a parameter of interest and an estimate of that parameter is relatively easy. The more complicated part of the procedure is obtaining the standard error. To do this, one typically derives the variance, estimates it (if necessary), and takes the square root. Obviously, rules for deriving variances play an important role in the process.

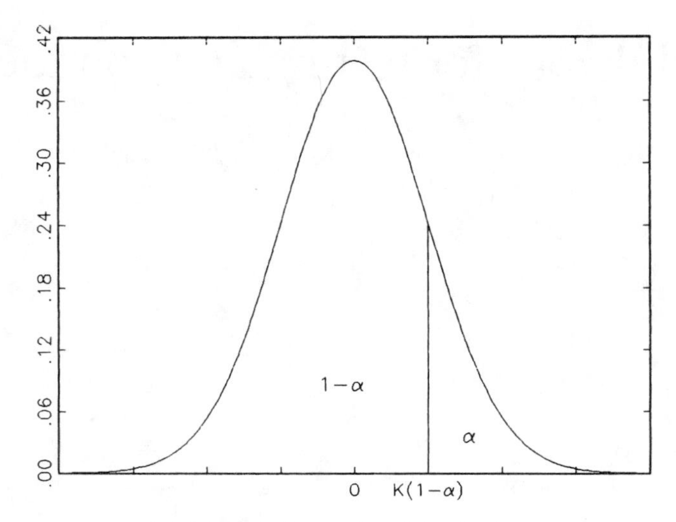

FIGURE 3.1. Percentiles of distributions.

We need notation for the percentage points of the known reference distribution. In particular, we need a name for the point that cuts off the top α of the distribution. The point that cuts off the top α of the distribution also cuts off the bottom $1 - \alpha$ of the distribution. These ideas are illustrated in Figure 3.1. The notation $K(1 - \alpha)$ is used for the point that cuts off the top α.

The illustration in Figure 3.1 is written formally as

$$\Pr\left[\frac{Est - Par}{SE(Est)} > K(1 - \alpha)\right] = \alpha.$$

By the symmetry about zero we also have

$$\Pr\left[\frac{Est - Par}{SE(Est)} < -K(1 - \alpha)\right] = \alpha.$$

The value $K(1 - \alpha)$ is called a percentile or percentage point; it is most often found from either a standard normal table or a t table. For t percentage points with df degrees of freedom, we use the notation

$$t(1 - \alpha, df) = K(1 - \alpha)$$

and for standard normal percentage points we use

$$z(1 - \alpha) = K(1 - \alpha).$$

As the degrees of freedom get arbitrarily large, the t distribution approximates the standard normal distribution. Thus we write

$$z(1 - \alpha) = t(1 - \alpha, \infty).$$

One can get a feeling for the quality of this approximation simply by examining the t tables in Appendix B.1 and noting how quickly the t percentiles approach the values given for infinite degrees of freedom.

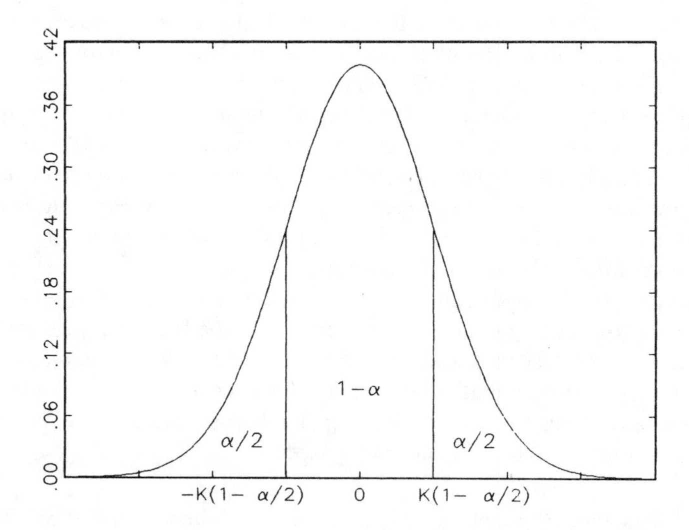

FIGURE 3.2. Symmetry about 0 in the distribution of $[Est - Par]/SE(Est)$.

3.1 Theory for confidence intervals

Confidence intervals are interval estimates of the parameter of interest. We have a specified 'confidence' that the parameter is in the interval. Confidence intervals are more valuable than simply reporting the estimate *Est* because confidence intervals provide an idea of the amount of error associated with the estimate.

A $(1 - \alpha)100\%$ confidence interval for *Par* is based on the following probability equalities

$$\begin{aligned} 1 - \alpha &= \Pr\left[-K\left(1 - \frac{\alpha}{2}\right) < \frac{Est - Par}{SE(Est)} < K\left(1 - \frac{\alpha}{2}\right)\right] \\ &= \Pr\left[Est - K\left(1 - \frac{\alpha}{2}\right) SE(Est) < Par < Est + K\left(1 - \frac{\alpha}{2}\right) SE(Est)\right] \end{aligned} \tag{3.1.1}$$

The first equality in (3.1.1) is simply a statement of the picture illustrated in Figure 3.2. It follows from the definition of $K\left(1 - \frac{\alpha}{2}\right)$ and the symmetry of the distribution. The second equality follows from the fact that the statements within the two sets of square brackets are algebraically equivalent. *A proof of the equivalence is given in the appendix at the end of the chapter.*

The probability statement

$$1 - \alpha = \Pr\left[Est - K\left(1 - \frac{\alpha}{2}\right) SE(Est) < Par < Est + K\left(1 - \frac{\alpha}{2}\right) SE(Est)\right].$$

is the basis for the confidence interval for *Par*. The $(1 - \alpha)100\%$ confidence interval for *Par* is simply the interval within the square brackets, i.e., the points between $Est - K\left(1 - \frac{\alpha}{2}\right) SE(Est)$ and $Est + K\left(1 - \frac{\alpha}{2}\right) SE(Est)$. However, the confidence interval is obtained by substituting observed values for *Est* and *SE*(*Est*). We are $(1 - \alpha)100\%$ 'confident' that *Par* is in this interval. The endpoints of the interval can be written succinctly as

$$Est \pm K\left(1 - \frac{\alpha}{2}\right) SE(Est).$$

I think everyone would agree with the statement 'The probability is $1 - \alpha$ that you are *going to get* a confidence interval that covers what you are trying to estimate, *Par*.' I did not indicate that the probability that your actual interval covers *Par* is $1 - \alpha$. The particular interval that you get uses the observed values of *Est* and *SE(Est)*, so it is a fixed interval and either covers *Par* or does not. There is no probability associated with *Par* being in the interval. For this reason the result of a confidence interval is described as, 'We are $(1 - \alpha)100\%$ *confident* that the true value of *Par* is in the interval.' I have no idea what this is supposed to mean, even though I find it intuitively appealing. I do, however, know of two acceptable interpretations for confidence intervals. As we will see in Section 3.4, a confidence interval contains all those parameter values that are consistent with the data. Consistency is measured by performing a statistical test with a specified error level α. The α in the test plays the same role as the α in a confidence interval. Since I think I understand the philosophical basis of hypothesis tests, I am comfortable with this interpretation.

The confidence intervals obtained from the theory presented in this chapter can frequently be obtained by another approach using 'Bayesian' arguments. In the Bayesian justification, the *correct* interpretation of a 95% confidence interval is that *the probability is 95% that the parameter is in the interval.* This is precisely the interpretation that most statistics students wish to adopt and that many statisticians strive so hard and so unsuccessfully to make their students reject. We will return to the issue of interpreting confidence intervals later in this section.

EXAMPLE 3.1.1. Years ago, 10 people were independently abducted by S.P.E.C.T.R.E after a Van Holland concert and forced to submit to psychological testing. Among the tests was a measure of audio acuity. From many past abductions in other circumstances, S.P.E.C.T.R.E knows that such observations form a normal population with variance 6. In this case, they found that $\bar{y}_{\bullet}$ was 17. They seek a 95% confidence interval for μ, the mean of the population.

1) $Par = \mu$,

2) $Est = \bar{y}_{\bullet}$,

3) $SE(Est) = \sqrt{6/10}$, in this case $SE(Est)$ is known and not estimated.

4) $[Est - Par]/SE(Est) = [\bar{y}_{\bullet} - \mu]/\sqrt{6/10}$ has a standard normal distribution.

To find the appropriate tabled values, observe that $(1 - \alpha)100 = 95$, so $1 - \alpha = .95$ and $\alpha = .05$. It follows that $K\left(1 - \frac{\alpha}{2}\right) = K(.975) = z(.975) = 1.96$.

The limits of the 95% confidence interval are

$$\bar{y}_{\bullet} \pm 1.96\sqrt{6/10}$$

or, since $\bar{y}_{\bullet} = 17$,

$$17 \pm 1.96\sqrt{6/10}.$$

S.P.E.C.T.R.E. was 95% confident that the mean hearing score for people at this concert (or at least for the population they were considering for abduction) was between 15.5 and 18.5. □

EXAMPLE 3.1.2. In Chapter 2 we considered data on dropout rates for math classes. We found that the 38 observations on dropout rates were not normally distributed; they contained two outliers. Our parameter for these data is μ, the population mean dropout rate for math classes, the estimate is the sample mean $\bar{y}_{\bullet}$, and the standard error is $\sqrt{s^2/38}$ where s^2 is the sample variance. Based on the central limit theorem and the law of large numbers, we used the *approximate* reference distribution

$$\frac{\bar{y}_{\bullet} - \mu}{\sqrt{s^2/38}} \sim N(0, 1).$$

From the 38 observations, we computed $\bar{y}_{\cdot} = 13.11$ and $s^2 = 106.421$ and found a 95% confidence interval for the dropout rate of (9.8, 16.4). The endpoints of the confidence interval are computed as

$$13.11 \pm 1.96(\sqrt{106.421/38}).$$

If we drop the two outliers, the remaining data seem to be normally distributed. Recomputing the sample mean and sample variance with the outliers deleted we get $\bar{y}_d = 11.083$ and $s_d^2 = 27.45$. Here the subscripts d are used as a reminder that the outliers have been deleted. Without the outliers, we can use the reference distribution

$$\frac{\bar{y}_d - \mu_d}{\sqrt{s_d^2/36}} \sim t(35).$$

The t distribution relies on the assumption of normality (which we have validated) rather than relying on the unvalidated large sample approximations from the central limit theorem and law of large numbers. The t distribution should give more accurate results. For a 95% confidence interval based on the data without the outliers, we need to find the appropriate tabled values. Observe once again that $(1 - \alpha)100 = 95$, so $1 - \alpha = .95$ and $\alpha = .05$. It follows that $K\left(1 - \frac{\alpha}{2}\right) = K(.975) = t(.975, 35) = 2.030$ and the confidence interval has endpoints

$$11.083 \pm 2.030(\sqrt{27.45/36}).$$

The actual interval is (9.3, 12.9). *Excluding the extremely high values that occasionally occur*, we are 95% confident that the population mean dropout rate is between 9.3 and 12.9 percent. Remember, this is a confidence interval for the mean of math classes; it *does not* indicate that you can be 95% confident that your next math class will have a dropout rate between 9.3 and 12.9 percent. Such an inference requires a prediction interval. The interval (9.3, 12.9) is much narrower than the one given in the previous paragraph, largely because our estimate of the variance is much smaller when the outliers have been deleted. Note also that with the outliers deleted, we are drawing inferences about a different parameter than when they are present. With the outliers deleted, our conclusions are only valid for the bulk of the observations. While occasional weird observations can be eliminated from our analysis, we cannot stop them from occurring.

In constructing the confidence interval we used the tabled value of 2.030 from the t distribution. This is larger than the 1.96 we obtained earlier from the standard normal distribution. Using the larger t value makes our confidence intervals wider. Other things being equal, we prefer narrower confidence intervals because they make more precise statements about the location of the mean. However, even though the value 1.96 is smaller than 2.030 and thus gives narrower intervals, we prefer to use the t distribution. The t distribution incorporates the fact that we do not know σ^2 and must estimate it. Thus an analysis using the $N(0, 1)$ distribution is much cruder in that it treats the estimate of σ^2 as if it were really σ^2. *Whenever we can establish that the data are reasonably normal, we will use the t distribution because it should give more accurate results.*

In the previous chapter we discussed the use of transformations. In particular, we looked at the square roots of the dropout rate data. *We now consider the effect on confidence intervals of transforming the data.* With the two outliers deleted and taking square roots of the observations, we found earlier that the data are reasonably normal. The sample mean and variance of the transformed, deleted data are $\bar{y}_{rd} = 3.218$ and $s_{rd}^2 = .749574$. Here the subscript r reminds us that square *roots* have been taken and the subscript d reminds us that outliers have been deleted. Using the reference distribution

$$\frac{\bar{y}_{rd} - \mu_{rd}}{\sqrt{s_{rd}^2/36}} \sim t(35),$$

we obtain a 95% confidence interval with endpoints

$$3.218 \pm 2.030 \left(\sqrt{\frac{.749574}{36}} \right).$$

The confidence interval reduces to $(2.925, 3.511)$. This is a 95% confidence interval for the population mean of the square roots of the dropout rate percentages with 'outliers' removed from the population.

The confidence interval $(2.925, 3.511)$ does not really address the issue that we set out to investigate. We wanted some idea of the value of the population mean dropout rate. We have obtained a 95% confidence interval for the population mean of the square roots of the dropout rate percentages (with outliers removed from the population). There is no simple, direct relationship between the population mean dropout rate and the population mean of the square roots of the dropout rate percentages, but a simple device can be used to draw conclusions about typical values for mean dropout rates when the analysis is performed on the square roots of the dropout rates. Since $(2.925, 3.511)$ provides a 95% confidence interval from the *square roots* of the dropout rate percentages, we simply *square* all the values in the interval to draw conclusions about the dropout rate percentages. Squaring the endpoints of the interval gives the new interval $(2.925^2, 3.511^2) = (8.6, 12.3)$. We are now 95% confident that the *central value* of the population of dropout rates is between 8.6 and 12.3. The central value referred to here is really the square of the population mean of the square roots of the dropout rate percentages. We are using this central value as a surrogate for the population mean of the (outlier deleted) dropout rate percentages; generally this central value will not equal the mean of the (deleted) dropout rates. For the most part we ignore the difference between the surrogate and the parameter that we set out to investigate. Interestingly, we will see in Section 3.5 that prediction intervals do not share these difficulties associated with transforming the data.

Note that the retransformed interval $(8.6, 12.3)$ obtained from the transformed, deleted data is similar to the interval $(9.3, 12.9)$ obtained earlier from the untransformed data with the outliers deleted. When, as in this case, two distinct analyses both seem reasonably valid, I would be very hesitant about drawing practical conclusions that could not be justified from *both* analyses. □

INTERPRETING CONFIDENCE INTERVALS

The interpretation of confidence intervals is actually a quite profound issue that statisticians have been arguing about for decades. This subsection presents the author's point of view in the context of some relatively simple problems. Although the problems are simple, the issues being discussed are not.

The disquieting thing about confidence intervals is the logic (or lack thereof) behind the leap from the probability of $1 - \alpha$ that a future interval will contain the parameter into a '$(1 - \alpha)100\%$ confidence' that the parameter is in a particular observed interval. The problem is in defining the meaning of confidence.

The standard interpretation of $(1-\alpha)100\%$ confidence intervals is that if you repeatedly performed many similar independent confidence intervals, about $(1-\alpha)100\%$ would contain the true parameter. The repeated sampling interpretation is exactly the same idea as saying that since a future coin toss has probability .5 of turning up heads, if you actually make many independent tosses of a coin, about 50% will be heads. This interpretation is not really saying anything new nor does it solve any problems because it still only relates to things that may be observed in the future. The fundamental problem of *inverting probabilities* for future observables into confidence about parameters remains. Moreover, the repeated sampling interpretation rarely applies to interesting problems. If you are obtaining a confidence interval for the height of corn plants grown outdoors, there is no way to perform independent replications of the experiment because there is no way to reproduce the exact growing conditions. In such cases, not only will the data behave differently but even the parameter of interest is likely to have a different meaning and value.

An alternative interpretation of confidence intervals based on statistical tests of hypotheses is presented in Section 3.4. I feel comfortable with the logic behind testing, so I like this interpretation. However, this interpretation makes no appeal whatsoever to the intuitive idea that 95% confidence means something similar to 95% probability.

I personally do not think it is possible to define confidence as anything other than probability. Two simple examples illustrate my point. I am going to flip a coin; we agree that the probability is .5 that it will land heads up. I flip the coin, look at it, but refuse to show it to you. Undoubtedly, you would feel comfortable saying that you are 50% confident that the coin is heads. I cannot imagine what that would mean except that you believe the probability is .5 that the coin is heads. Note that the 50% confidence is a statement about your beliefs and not a statement about the coin. The outcome of the coin toss is fixed (and known by someone other than you). This example has neither a fixed parameter nor any observable data but we can modify the example to make it more like a confidence interval problem. I place a coin either heads up or tails up and hide it from you, this is the parameter. You are going to flip a coin but I exercise my well known psychic powers. When I do this, the probability is .75 that the coin face I chose will be the face on your coin. When you toss your coin it lands either heads or tails and you observe this datum. The observed outcome of your toss is no longer random and it either matches mine or does not. Intuitively, you may reasonably feel that the probability is still .75 that the coins match, regardless of how I set my coin. But now the probability is no longer about what the outcome of your flip will be because you have seen your datum. The probability must now be about how you believe I set my coin. Such a probability can only exist in your head. (Of course, I have other ideas and probabilities because, having seen both coins, I know whether they match.) While the intuition behind this probability is appealing, the logic escapes me. Glossing over the problem by saying that you have confidence, but not probability, of .75 for matching my coin does nothing to clear up the real issue.

R. A. Fisher made an attempt to build a theory of inverting probabilities from future observables into probabilities for parameters using this sort of intuition that we all find appealing. While it was a noble effort, I do not know of anyone who thinks Fisher succeeded or anyone who thinks that such a theory can succeed. (Of course, I have a limited sphere of acquaintance.) For more information, see the discussion of fiducial probability in Fisher (1956).

Another method of inverting probabilities about future data into probabilities about parameters is the theory of Bayesian statistics. Let me briefly mention how a Bayesian could arrive at a probability of .75 for the second coin tossing example. The computations are illustrated in Exercise 3.7.1. I place my coin any way I want. To arrive at a probability, you need to decide on *your beliefs* about how I placed my coin. If you believe that I am equally likely to place it heads up or tails up, those are your prior beliefs. Your prior beliefs are then modified by any data. In this example, if your initial beliefs are that I was equally likely to place the coin heads up or tails up, using a result known as Bayes theorem the probability that your coin agrees with mine becomes .75, regardless of what face of the coin I chose to place upwards and regardless of what you actually saw on your flip.

Notice that there is a lot more structure here than the mere intuition referred to earlier. In the intuitive discussion, your personal probability of a 75 : 25 chance of matching exists regardless of how I set my coin. In this discussion, you need to specify your beliefs about how I set my coin and the final 75 : 25 chance is a result of your having chosen an initial 50 : 50 chance for how I set my coin. For example, if you thought I was four times more likely to select heads, the probability of matching would be $12/13$ if your coin turned up heads but only $3/7$ if it turned up tails. Note that these beliefs do not depend on how I *actually* set my coin because you cannot know that. These beliefs do depend on your knowledge of how the data relate to how I set my coin, i.e., what data are likely when I choose heads and what are likely when I choose tails.

Bayesian methods are often criticized for requiring you to specify your initial beliefs in terms of a probability distribution on the possible parameter values. The result of a Bayesian data analysis is then an updated version of *your beliefs*. Berger (1985), among many others, responds to such criticisms. Many of us think that Bayesian methods provide the only logically consistent (though I would *not* say the only useful) method for doing statistics.

As I see it, a person has three choices: one can ignore the problem of what confidence means, one can use the hypothesis testing interpretation of confidence intervals to be given later, or one can rely on Bayesian methods. As it turns out, the confidence intervals and prediction intervals used in this book can be obtained by reasonable Bayesian methods. In the Bayesian interpretation of these intervals, confidence simply means probability, as the data modify a particular set of prior beliefs that are chosen to have a minimum of influence on the results of the data analysis.

3.2 Theory for hypothesis tests

Hypothesis tests are used to check whether *Par* has some specified value. For some fixed known number m, we may want to test the *null hypothesis*

$$H_0 : Par = m$$

versus the *alternative hypothesis*

$$H_A : Par \neq m.$$

The number m must be known; it is some number that is of interest for the specific data being analyzed. It is impossible to give general rules for picking m because the choice must depend on the context of the data. As mentioned in the previous chapter, *the structure of the data (but not the actual values of the data) sometimes suggests interesting hypotheses* such as testing whether two populations have the same mean or testing whether there is a relationship between two variables, but ultimately the researcher must determine what hypotheses are of interest and these hypotheses determine m. In any case, m is never just an unspecified symbol; it must have meaning within the context of the problem. The test of $H_0 : Par = m$ versus $H_A : Par \neq m$ is based on the assumption that H_0 is true and consists of checking to see whether the data are inconsistent with that assumption.

To identify data that are inconsistent with the assumption that $Par = m$, we examine what happens when $Par \neq m$. Note that *Est* is always an estimate of *Par*; this has nothing to do with any hypothesis. With *Est* estimating *Par*, it follows that if $Par > m$ then *Est* tends to be larger than m. Equivalently, $Est - m$, and thus $[Est - m]/SE(Est)$, tend to be large positive numbers when $Par > m$ (larger than they would be if $H_0 : Par = m$ is true). On the other hand if $Par < m$, then $Est - m$ and $[Est - m]/SE(Est)$ tend to be large negative numbers. Data that are inconsistent with the null hypothesis $Par = m$ are large positive and large negative values of the test statistic $[Est - m]/SE(Est)$. The problem is in specifying what we mean by 'large.' In practice we conclude that the data contradict the null hypothesis $Par = m$ if we observe a value of $[Est - m]/SE(Est)$ that is further from 0 than some cutoff values. The problem is to make an intelligent choice for the cutoff values. The solution is based on the fact that if H_0 is true, the *test statistic*

$$\frac{Est - m}{SE(Est)}$$

has the known reference distribution that is symmetric about 0.

When we substitute the observed values of *Est* and *SE*(*Est*) into the test statistic we get one observation on the random test statistic. When H_0 is true, this observation comes from the reference distribution. The question is whether it is reasonable to believe that this one observation came from the reference distribution. If so, the data are consistent with H_0. If the observation could not reasonably have come from the reference distribution, the data contradict H_0. Contradicting H_0 is a strong inference; it implies that H_0 is false. On the other hand, inferring that the data are consistent with H_0 does not suggest that H_0 is true. Such data can also be consistent with some aspects of the alternative.

Before we can state the test formally, i.e., give intelligent cutoff values to determine the test, we need to consider the concept of error. Even if H_0 is true, it is usually possible (not probable but possible) to get any value at all for $[Est - m]/SE(Est)$. For that reason, no matter what we conclude about the null hypothesis, there is a possibility of error. A test of hypothesis is based on controlling the probability of making an error when the null hypothesis is true. We define the α level of the test as the probability of rejecting the null hypothesis (saying that it is false) when the null hypothesis is in fact true. *The α level is also called the probability of a type I error, with a type I error being the rejection of a true null hypothesis.*

The α level determines the cutoff values for testing. The α level test for $H_0 : Par = m$ versus $H_A : Par \neq m$ is to reject H_0 if

$$\frac{Est - m}{SE(Est)} > K\left(1 - \frac{\alpha}{2}\right)$$

or if

$$\frac{Est - m}{SE(Est)} < -K\left(1 - \frac{\alpha}{2}\right).$$

This is equivalent to saying, reject H_0 if

$$\frac{|Est - m|}{SE(Est)} > K\left(1 - \frac{\alpha}{2}\right).$$

To see that using $K\left(1 - \frac{\alpha}{2}\right)$ and $-K\left(1 - \frac{\alpha}{2}\right)$ as cutoff values gives an α level test, observe that if H_0 is true, the probability that we will reject H_0 is

$$\Pr\left[\frac{Est - m}{SE(Est)} > K\left(1 - \frac{\alpha}{2}\right)\right] + \Pr\left[\frac{Est - m}{SE(Est)} < -K\left(1 - \frac{\alpha}{2}\right)\right] = \alpha/2 + \alpha/2 = \alpha,$$

see Figure 3.2. Also note that we are rejecting H_0 for those values of $[Est - m]/SE(Est)$ that are most inconsistent with H_0, these being the values far from zero.

Actually, this test could be developed without any reference to the alternative hypothesis whatsoever. (In fact, I much prefer such a development since I believe that if you are willing to specify an alternative you should probably do a Bayesian analysis.) The only place where we used the alternative hypothesis was in determining which values of the test statistic were inconsistent with H_0. A different approach simply uses Figure 3.2 to decide which values of the test statistic are inconsistent. We can define the values that are most inconsistent as those that are the least likely to occur. The values that are least likely to occur are those where the density (i.e., the curve) is lowest. In Figure 3.2, the lowest values of the density are those corresponding to values of the test statistic that are far from 0. The density is symmetric, so our test should be symmetric. Thus an α level test has exactly the form given above. Of course this analysis relies on Figure 3.2 being an accurate portrayal of the distribution under H_0, but for all of our applications it is.

EXAMPLE 3.2.1. In Example 3.1.1 we considered past data on audio acuity in a post-rock environment. Those data were collected on fans of the group Van Holland in their Lee David Rothschild days. The nefarious organization responsible for this study found it necessary to update their findings after Rothschild was replaced by Slammy Hagar-Slacks. This time they abducted for themselves 16 independent observations and they were positive that the data would continue to follow a normal distribution. (Such arrogance is probably responsible for the failure of S.P.E.C.T.R.E.'s plans of world domination. In any case, their resident statistician was in no position to question this assumption.)

The observed values of $\bar{y}_\cdot$ and s^2 were 22 and .25 respectively for the audio acuity scores. Now the purpose of all this is that S.P.E.C.T.R.E. had a long standing plot that required the use of a loud rock band. They had been planning to use the group Audially Disadvantaged Leopard but Van Holland's fans offered certain properties they preferred, provided that those fans audio acuity scores were satisfactory. From extremely long experience with abducting Audially Disadvantaged Leopard fans, S.P.E.C.T.R.E. knows that they have a population mean of 20 on the audio acuity test. S.P.E.C.T.R.E. wishes to know whether Van Holland fans differ from this value. Naturally, they tested $H_0 : \mu = 20$ versus $H_A : \mu \neq 20$ and they chose an α level of .01.

1) $Par = \mu$

2) $Est = \bar{y}_\cdot$

3) $SE(Est) = s/\sqrt{16}$. In this case the $SE(Est)$ is estimated.

4) $[Est - Par]/SE(Est) = [\bar{y}_\cdot - \mu]/[s/\sqrt{16}]$ has a $t(15)$ distribution. This follows because the data are normally distributed and the standard error is estimated using s.

The $\alpha = .01$ test is to reject H_0 if

$$\frac{|\bar{y}_\cdot - 20|}{s/\sqrt{16}} > 2.947 = t(.995, 15).$$

Note that the sample size is $n = 16$ and $K(1 - \alpha/2) = K(1 - .005) = t(.995, 15)$. Since $\bar{y}_\cdot = 22$ and $s^2 = .25$ we reject H_0 if

$$\frac{|22 - 20|}{\sqrt{.25/16}} > 2.947.$$

Since $|22 - 20|/\sqrt{.25/16} = 16$ is greater than 2.947, we reject the null hypothesis at the $\alpha = .01$ level. There is clear (indeed, overwhelming) evidence that the Van Holland fans have higher scores. (Unfortunately, my masters will not let me inform you whether high scores mean better hearing or worse.) □

EXAMPLE 3.2.2. The National Association for the Abuse of Student Yahoos (also known as NAASTY) has established guidelines indicating that university dropout rates for math classes should be 15%. Based on an $\alpha = .05$ test, we wish to know if the University of New Mexico (UNM) meets these guidelines when treating the 1984–85 academic year data as a random sample. As is typical in such cases, NAASTY has specified that the central value of the distribution of dropout rates should be 15% but it has not stated a specific definition of the central value. We interpret the central value to be the population mean of the dropout rates and test the null hypothesis $H_0 : \mu = 15\%$ against the two-sided alternative $H_A : \mu \neq 15\%$.

The complete data consist of 38 observations from which we compute $\bar{y}_\cdot = 13.11$ and $s^2 = 106.421$. The data are nonnormal, so we have little choice but to hope that 38 observations constitute a sufficiently large sample to justify the use of

$$\frac{\bar{y}_\cdot - \mu}{\sqrt{s^2/38}} \sim N(0, 1)$$

as an approximate reference distribution. With an α level of .05 and the standard normal distribution, the two-sided test rejects H_0 if

$$\frac{\bar{y}_\cdot - 15}{\sqrt{s^2/38}} > 1.96 = z(.975) = z\left(1 - \frac{\alpha}{2}\right)$$

or if

$$\frac{\bar{y}_{\bullet} - 15}{\sqrt{s^2/38}} < -1.96.$$

Substituting the observed values for $\bar{y}_{\bullet}$ and s^2 gives the observed value of the test statistic

$$\frac{13.11 - 15}{\sqrt{106.421/38}} = -1.13.$$

The value of -1.13 is neither greater than 1.96 nor less than -1.96, so the null hypothesis cannot be rejected at the .05 level. The 1984–85 data provide no evidence that UNM violates the NAASTY guidelines.

If we delete the two outliers, the analysis changes somewhat. Without the outliers, the data are approximately normal and we can use the reference distribution

$$\frac{\bar{y}_d - \mu_d}{\sqrt{s_d^2/36}} \sim t(35).$$

For this reference distribution the two-sided $\alpha = .05$ test rejects $H_0 : \mu_d = 15$ if

$$\frac{\bar{y}_d - 15}{\sqrt{s_d^2/36}} > 2.030 = t(.975, 35)$$

or if

$$\frac{\bar{y}_d - 15}{\sqrt{s_d^2/36}} < -2.030 = -t(.975, 35).$$

With $\bar{y}_d = 11.083$ and $s_d^2 = 27.45$ from the data without the outliers, the observed value of the test statistic is

$$\frac{11.083 - 15}{\sqrt{27.45/36}} = -4.49.$$

The absolute value of -4.49 is greater than 2.030, i.e., $-4.49 < -2.030$, so we reject the null hypothesis of $H_0 : \mu_d = 15\%$ at the .05 level. When we exclude the two extremely high observations, we have evidence that the typical dropout rate was different from 15%. In particular, *since the test statistic is negative*, we have evidence that the population mean dropout rate with outliers deleted was actually *less than* 15%. Obviously, most of the UNM math faculty during 1984–85 were not sufficiently nasty.

Finally, we consider the role of transformations in testing. We again consider the square roots of the dropout rates with the two outliers deleted. As discussed earlier, NAASTY has specified that the central value of the distribution of dropout rates should be 15% but has not stated a specific definition of the central value. We are reasonably free to interpret their guideline and we now interpret it as though the population mean of the square roots of the dropout rates should be $\sqrt{15}$. This interpretation leads us to the null hypothesis $H_0 : \mu_{rd} = \sqrt{15}$ and the alternative $H_A : \mu_{rd} \neq \sqrt{15}$. As discussed earlier, a reasonably appropriate reference distribution is

$$\frac{\bar{y}_{rd} - \mu_{rd}}{\sqrt{s_{rd}^2/36}} \sim t(35),$$

so the test rejects H_0 if

$$\frac{|\bar{y}_{rd} - \sqrt{15}|}{\sqrt{s_{rd}^2/36}} > 2.030 = t(.975, 35).$$

The sample mean and variance of the transformed, deleted data are $\bar{y}_{rd} = 3.218$ and $s^2_{rd} = .749574$, so the observed value of the test statistic is

$$\frac{3.218 - 3.873}{\sqrt{.749574/36}} = -4.54.$$

The test statistic is similar to that in the previous paragraph. The null hypothesis is again rejected and all conclusions drawn from the rejection are essentially the same. As stated earlier, I believe that when two analyses both appear to be valid, either the practical conclusions agree or neither analysis should be trusted. □

ONE-SIDED TESTS

We can do one-sided tests in a similar manner. The α level test for $H_0 : Par \leq m$ versus $H_A : Par > m$ is to reject H_0 if

$$\frac{Est - m}{SE(Est)} > K(1 - \alpha).$$

The alternative hypothesis is that *Par* is greater than something and the null hypothesis is rejected when the test statistic is greater than some cutoff value. We reject the null hypothesis for the values of the test statistic that are most inconsistent with the null hypothesis and thus most consistent with the alternative hypothesis. If the alternative is true, *Est* should be near *Par*, which is greater than m, so large positive values of $Est - m$ or, equivalently, large positive values of $[Est - m]/SE(Est)$ are consistent with the alternative and inconsistent with the null hypothesis.

The α level test for $H_0 : Par \geq m$ versus $H_A : Par < m$ is to reject H_0 if

$$\frac{Est - m}{SE(Est)} < -K(1 - \alpha).$$

The alternative hypothesis is that *Par* is less than something and the null hypothesis is rejected when the test statistic is less than some cutoff value. *The form of the alternative determines the form of the rejection region.* In both cases we reject H_0 for the data that are most inconsistent with H_0

The null hypotheses involve inequalities but $Par = m$ is always part of the null hypotheses. The tests are set up assuming that $Par = m$ and this needs to be part of any null hypothesis. In both cases, if $Par = m$ then the probability of making a mistake is α and, more generally, if H_0 is true, the probability of making a mistake is no greater than α.

EXAMPLE 3.2.3. Again consider the Slammy Hagar-Slacks era Van Holland audio data. Recall that there are 16 independent observations taken from a normal population with observed statistics of $\bar{y}_\cdot = 22$ and $s^2 = .25$. This time I have been required to perform a one-sided test to see whether I can prove that the Van Holland mean audio acuity scores are lower than the Audially Disadvantaged Leopard mean. I now test $H_0 : \mu \geq 20$ versus $H_A : \mu < 20$ with $\alpha = .01$. Here I am claiming that the scores are not lower and check to see whether the data contradict this. If they do, then my claim must be false and I have proven that the scores must be lower. If I initially claimed that the scores were lower, I would not be able to prove it; I could only establish that the data were consistent with my claim. As before,

1) $Par = \mu$

2) $Est = \bar{y}_\cdot$

3) $SE(Est) = s/\sqrt{16}$. In this case the $SE(Est)$ is estimated.

4) $[Est - Par]/SE(Est) = [\bar{y}_\cdot - \mu]/[s/\sqrt{16}]$ has a $t(15)$ distribution.

The $\alpha = .01$ test is to reject $H_0 : \mu \geq 20$ if

$$\frac{\bar{y}_{\bullet} - 20}{s/\sqrt{16}} < -2.602 = -t(.99, 15).$$

Note that with a sample size of $n = 16$ we get $K(1 - \alpha) = K(1 - .01) = t(.99, 15)$. With $\bar{y}_{\bullet} = 22$ and $s^2 = .25$, we reject if

$$\frac{22 - 20}{\sqrt{.25/16}} < -2.602$$

Since $(22 - 20)/\sqrt{.25/16} = 16$ is greater than -2.602 we do not reject the null hypothesis at the $\alpha = .01$ level. There is no evidence that the Van Holland mean is lower than the Audially Disadvantaged Leopard mean. Observe that with the alternative $\mu < 20$, i.e., μ *less than* something, H_0 is only rejected when the test statistic is *less than* some cutoff value.

If you stop and think about it, we really did not have to go to all this trouble to discover the conclusion of this test. The null hypothesis is that $\mu \geq 20$. The observed $\bar{y}_{\bullet}$ value of 22 is obviously consistent with the hypothesis that the mean is greater than or equal to 20. Only $\bar{y}_{\bullet}$ values that are less than 20 could possibly contradict the null hypothesis. The only point at issue is how far $\bar{y}_{\bullet}$ must be below 20 before we can claim that $\bar{y}_{\bullet}$ contradicts the null hypothesis. As discussed in Example 2.2.4, given a choice it would be more informative to reverse the inequalities in H_0 and H_A for this problem.

□

EXAMPLE 3.2.4. A colleague of mine claims that, excluding classes with outrageous dropout rates, the math dropout rate at UNM was never more than 9% in any year during the 1980s. We now test this claim using the only data we have, that from the 1984–85 school year. My colleague excluded classes with outrageous dropout rates, so we use only the data with the outliers deleted. We again use $\alpha = .05$.

Based on the untransformed data, the null hypothesis is simply my colleague's claim, i.e., $H_0 : \mu \leq 9$. The alternative is $H_A : \mu > 9$. With $\alpha = .05$, the test is rejected if

$$\frac{\bar{y}_d - 9}{\sqrt{s_d^2/36}} > 1.690 = t(.95, 35).$$

With $\bar{y}_d = 11.083$ and $s_d^2 = 27.45$, the observed test statistic is

$$\frac{11.083 - 9}{\sqrt{27.45/36}} = 2.39,$$

so the test is easily, but not overwhelmingly, rejected.

Using the square roots of the data, the null hypothesis becomes $H_0 : \mu_{rd} \leq \sqrt{9}$. The alternative is $H_A : \mu_{rd} > \sqrt{9}$. With $\alpha = .05$, the test is rejected if

$$\frac{\bar{y}_{rd} - \sqrt{9}}{\sqrt{s_{rd}^2/36}} > 1.690 = t(.95, 35).$$

The sample mean and variance of the transformed, deleted data are $\bar{y}_{rd} = 3.218$ and $s_{rd}^2 = .749574$, so the observed value of the test statistic is

$$\frac{3.218 - 3}{\sqrt{.749574/36}} = 1.51.$$

The observed value is not greater than 1.690, so the test cannot be rejected at the .05 level.

In this case the two tests disagree. The untransformed data rejects the .05 level test easily. The transformed data does not quite achieve significance at the .05 level. To me, the data seem inconclusive. There is certainly some reason to suspect that the true dropout rate during 1984–85 was greater than 9%; one test rejected the null hypothesis and the other came somewhat close to being rejected. However, both analyses seem reasonable, so I cannot place great confidence in the rejection obtained using the untransformed data when the result is not fully corroborated by the transformed data. □

P VALUES

Rather than having formal rules for when to reject the null hypothesis, one can report the evidence against the null hypothesis. This is done by reporting the *P* value. The *P* value is computed under the assumption that $Par = m$. It is the probability of seeing data that are as extreme or more extreme than those that were actually observed. Formally, we write t_{obs} for the observed value of the test statistic, computed from the *observed* values of *Est* and *SE*(*Est*). Thus t_{obs} is our summary of the data that were actually observed. Recalling our earlier discussion of which values of *Est* would be most inconsistent with $Par = m$, the probability of seeing something as or more extreme than we actually saw is

$$P = \Pr\left[\left|\frac{Est - m}{SE(Est)}\right| \geq |t_{obs}|\right]$$

where *Est* (and usually *SE*(*Est*)) are viewed as random and it is assumed that $Par = m$. Under these conditions $(Est - m)/SE(Est)$ has the known reference distribution and t_{obs} is a known number, so we can actually compute *P*. The basic idea is that for, say, t_{obs} positive, any value of $(Est - m)/SE(Est)$ greater than t_{obs} is more extreme than t_{obs}. Any data that yield $(Est - m)/SE(Est) = -t_{obs}$ are just as extreme as t_{obs} and values of $(Est - m)/SE(Est)$ less than $-t_{obs}$ are more extreme than observing t_{obs}.

EXAMPLE 3.2.5. Again consider the Slammy Hagar-Slacks era Van Holland data. We have 16 observations taken from a normal population and we wish to test $H_0 : \mu = 20$ versus $H_A : \mu \neq 20$. As before, 1) $Par = \mu$, 2) $Est = \bar{y}_{\cdot}$, 3) $SE(Est) = s/\sqrt{16}$, and 4) $[Est - Par]/SE(Est) = [\bar{y}_{\cdot} - \mu]/[s/\sqrt{16}]$ has a $t(15)$ distribution. This time we take $\bar{y}_{\cdot} = 19.78$ and $s^2 = .25$, so the observed test statistic is

$$t_{obs} = \frac{19.78 - 20}{\sqrt{.25/16}} = -1.76.$$

From a *t* table, $t(.95, 15) = 1.75$, so

$$P = \Pr\left[|t(15)| \geq |-1.76|\right] \doteq \Pr\left[|t(15)| \geq 1.75\right] = .10.$$

Alternatively, $t(.95, 15) \doteq |1.76|$, so $P \doteq 2(1 - .95)$. □

Equivalently, the *P* value is the smallest α level for which the test would be rejected. With this definition, if we perform an α level test where α *is less than the P value, we can conclude immediately that the null hypothesis is not rejected.* If we perform an α level test where α *is greater than the P value, we know immediately that the null hypothesis is rejected.* Thus computing a *P* value eliminates the need to go through the formal testing procedures described above. Knowing the *P* value immediately gives the test results for any choice of α. The *P* value is a measure of how consistent the data are with H_0. Large values (near 1) indicate great consistency. Small values (near 0) indicate data that are inconsistent with H_0.

EXAMPLE 3.2.6. In Example 3.2.2 we considered two-sided tests for the drop rate data. Using the complete untransformed data, the null hypothesis $H_0 : \mu = 15$, and the alternative $H_A : \mu \neq 15$, we observed the test statistic

$$t_{obs} = \frac{13.11 - 15}{\sqrt{106.421/38}} = -1.13.$$

Using a standard normal table or a computer program, we can compute

$$P = \Pr\left[|z| \geq |-1.13|\right] = .26.$$

An $\alpha = .26$ test would be just barely rejected by these data. Any test with an α level smaller than .26 is more stringent (the cutoff values are farther from 0 than 1.13) and would not be rejected. Thus the standard $\alpha = .05$ and $\alpha = .01$ tests would not be rejected. Similarly, any test with an α level greater than .26 is less stringent and would be rejected. Of course, it is extremely rare that one would use a test with an α level greater than .26.

Using the untransformed data with outliers deleted, the null hypothesis $H_0 : \mu_d = 15$, and the alternative $H_A : \mu_d \neq 15$, we observed the test statistic

$$\frac{11.083 - 15}{\sqrt{27.45/36}} = -4.49.$$

We compute

$$P = \Pr\left[|t(35)| \geq |-4.49|\right] = .000.$$

This P value is not really zero; it is a number that is so small that when we round it off to three decimal places the number is zero. In any case, the test is rejected for any reasonable choice of α. In other words, the test is rejected for any choice of α that is greater than .000. (Actually for any α greater than .0005 because of the round off problem.)

Using the square roots of the data with outliers deleted, the null hypothesis $H_0 : \mu_{rd} = \sqrt{15}$, and the alternative $H_A : \mu_{rd} \neq \sqrt{15}$, the observed value of the test statistic is

$$\frac{3.218 - 3.873}{\sqrt{.749574/36}} = -4.54.$$

We compute

$$P = \Pr\left[|t(35)| \geq |-4.54|\right] = .000.$$

Once again, the test result is highly significant. □

EXAMPLE 3.2.7. In Example 3.2.4 we considered one-sided tests for the drop rate data. Using the deleted untransformed data, the null hypothesis $H_0 : \mu_d \leq 9$, and the alternative $H_A : \mu_d > 9$, we observed the test statistic

$$\frac{11.083 - 9}{\sqrt{27.45/36}} = 2.39.$$

Using Minitab, we compute

$$P = \Pr\left[t(35) \geq 2.39\right] = .011.$$

The probability is only for large positive values because negative values of the test statistic are consistent with H_0. The P value of .011 is just greater than .01, so we would not be able to reject an $\alpha = .01$ test. We can of course reject any test with α greater than .011. The P value for the one-sided test is exactly half of what the P value would be for testing $H_0 : \mu_d = 9$ versus $H_A : \mu_d \neq 9$.

Using the square roots of the data, the null hypothesis became $H_0 : \mu_{rd} \leq \sqrt{9}$ with the alternative $H_A : \mu_{rd} > \sqrt{9}$. The observed value of the test statistic was

$$\frac{3.218 - 3}{\sqrt{.749574/36}} = 1.51.$$

We compute

$$P = \Pr[t(35) \geq 1.51] = .07.$$

The P value here is small, .07, but not small enough to reject an $\alpha = .05$ test. There is some indication that the null hypothesis is not true but the indication is not very strong. To be precise, if we repeated this test procedure many times when the null hypothesis is true, 7% of the time we would expect to get results that are at least this suggestive of the incorrect conclusion that the null hypothesis is false. □

MINITAB COMMANDS

To find a P value using Minitab when the reference distribution is a t, start with the number $-|t_{obs}|$, where t_{obs} is the observed value of the test statistic. In other words, find the observed test statistic and make it a negative number. Then simply use this number with the 'cdf' command, specifying the t distribution and the degrees of freedom in the subcommand. The procedure for $t_{obs} = 1.51$ is illustrated below. The probability given by the cdf command is the appropriate P value for one-sided tests but *must be doubled if the test is two-sided.*

```
MTB > cdf -1.51;
SUBC> t 35.
```

CONCLUSION

To keep this discussion as simple as possible, the examples have been restricted to one-sample normal theory. However, the results of this section and Section 3.1 apply to more complicated problems such as two-sample problems, testing contrasts in analysis of variance, and testing coefficients in regression. All of these applications will be considered in later chapters.

3.3 Validity of tests and confidence intervals

In testing an hypothesis, we make an assumption, namely the null hypothesis, and check to see whether the data are consistent with the assumption or inconsistent with it. If the data are consistent with the null hypothesis, that is all that we can say. If the data are inconsistent with the null hypothesis, it suggests that our assumption was wrong. (This is very similar to the mathematical idea of a proof by contradiction.)

One of the problems with testing hypotheses is that we are really making a series of assumptions. The null hypothesis is one of these, but there are many others. Typically we assume that observations are independent. In most tests that we will consider, we assume that the data have normal distributions. As we consider more complicated data structures, we will need to make more assumptions. The proper conclusion from a test of hypothesis is that either the data are consistent with our assumptions or the data are inconsistent with our assumptions. If the data are inconsistent with the assumptions, it suggests that at least one of them is invalid. In particular, if the data are inconsistent with the assumptions, it does not necessarily imply that the particular assumption embodied in the null hypothesis is the one that is invalid. Before we can reasonably conclude that the null hypothesis is untrue, we need to ensure that the other assumptions are reasonable. Thus it is crucial to check our assumptions as fully as we can. Plotting the data plays a vital role in checking assumptions. Plots are used throughout the book, but special emphasis on plotting is given in Chapter 7.

Typically, it is quite easy to define parameters *Par* and estimates *Est*. The role of the assumptions is crucial in obtaining a valid $SE(Est)$ and an appropriate reference distribution. If our assumptions are reasonably valid, our $SE(Est)$ and reference distribution will be reasonably valid and the procedures outlined here for performing statistical inferences will be reasonably valid. This applies not only to testing but to confidence intervals as well. Of course the assumptions that need to be checked depend on the precise nature of the analysis being performed.

3.4 The relationship between confidence intervals and tests

The two most commonly used tools in statistical inference are tests and confidence intervals. Tests determine whether a difference can be established between an hypothesized parameter value and the true parameter for the data. Typically, one must consider not only whether a difference exists, but how much difference exists, and whether such a difference is important within the context of the problem. Confidence intervals are used to quantify what is known about the true parameter and thus can be used to quantify how much of a difference may exist. In particular, confidence intervals give all the possible parameter values that seem to be consistent with the data. Tests and confidence intervals are very closely related inferential tools and in this section we explore their relationship.

As discussed earlier, the term 'confidence' as used in confidence intervals is rather nebulously defined. Confidence intervals are based on the unusable probability statement

$$1-\alpha = \Pr\left[Est - K\left(1-\frac{\alpha}{2}\right) SE(Est) < Par < Est + K\left(1-\frac{\alpha}{2}\right) SE(Est)\right],$$

which is a statement about the unknown (unobserved) random variables Est and $SE(Est)$. It is a highly intuitive idea that this probability statement generates a usable interval for Par,

$$Est - K\left(1-\frac{\alpha}{2}\right) SE(Est) < Par < Est + K\left(1-\frac{\alpha}{2}\right) SE(Est),$$

in which the *observed* values of Est and $SE(Est)$ are used to define a known interval. However, the logic behind this intuitive idea is not clear and so we are left with an unclear definition of 'confidence.'

A clear definition of confidence can be made in terms of testing hypotheses. *The* $(1-\alpha)100\%$ *confidence interval for Par*,

$$Est - K\left(1-\frac{\alpha}{2}\right) SE(Est) < Par < Est + K\left(1-\frac{\alpha}{2}\right) SE(Est),$$

consists of all the values m that would not *be rejected by an* α *level test of* $H_0 : Par = m$ *versus* $H_A : Par \neq m$. To see this recall that the α level test is rejected when

$$\frac{Est - m}{SE(Est)} > K\left(1-\frac{\alpha}{2}\right)$$

or

$$\frac{Est - m}{SE(Est)} < -K\left(1-\frac{\alpha}{2}\right).$$

Conversely, the α level test is not rejected when

$$-K\left(1-\frac{\alpha}{2}\right) \le \frac{Est - m}{SE(Est)} \le K\left(1-\frac{\alpha}{2}\right).$$

Exactly the same algebraic manipulations that lead to equation (3.1.1) also lead to the conclusion that the test is not rejected when

$$Est - K\left(1-\frac{\alpha}{2}\right) SE(Est) < m < Est + K\left(1-\frac{\alpha}{2}\right) SE(Est).$$

Thus the confidence interval consists of all values of m for which the α level test of $H_0 : Par = m$ versus $H_A : Par \neq m$ is not rejected. In other words, *a* $(1-\alpha)100\%$ *confidence interval consists of all parameter values that are consistent with the data as judged by an* α *level test.*

We have now established that there is little point in performing the fixed α*, fixed m testing procedures discussed in Section 3.2. P values give the results of testing* $H_0 : Par = m$ *versus* $H_A : Par \neq m$ *for a fixed m but* every *choice of* α*. Confidence intervals give the results of testing* $H_0 : Par = m$ *versus* $H_A : Par \neq m$ *for a fixed* α *but* every *choice of m.*

EXAMPLE 3.4.1. In Example 3.2.1 we considered audio acuity data for Van Holland fans and tested whether their mean score differed from fans of Audially Disadvantaged Leopard. In this example we test whether their mean score differs from that of Tangled Female Sibling fans. Recall that the observed values of n, $\bar{y}_\cdot$, and s^2 for Van Holland fans were 16, 22, and .25, respectively and that the data were normal. Tangled Female Sibling fans have a population mean score of 22.325, so we test $H_0 : \mu = 22.325$ versus $H_A : \mu \neq 22.325$. The test statistic is $(22 - 22.325)/\sqrt{.25/16} = -2.6$. If we do an $\alpha = .05$ test, $|-2.6| > 2.13 = t(.975, 15)$, so we reject H_0, but if we do an $\alpha = .01$ test, $|-2.6| < 2.95 = t(.995, 15)$, so we do not reject H_0. In fact, $|-2.6| \doteq t(.99, 15)$, so the P value is essentially .02. The P value is larger than .01, so the .01 test does not reject H_0; the P value is less than .05, so the test rejects H_0 at the .05 level.

If we consider confidence intervals, the 99% interval has endpoints $22 \pm 2.95\sqrt{.25/16}$ for an interval of $(21.631, 22.369)$ and the 95% interval has endpoints $22 \pm 2.13\sqrt{.25/16}$ for an interval of $(21.734, 22.266)$. Notice that the hypothesized value of 22.325 is inside the 99% interval, so it is not rejected by a .01 level test, but 22.325 is outside the 95% interval, so a .05 two-sided test rejects $H_0 : \mu = 22.325$. The 98% interval has endpoints $22 \pm 2.60\sqrt{.25/16}$ for an interval of $(21.675, 22.325)$ and the hypothesized value is on the edge of the interval. □

3.5 Theory of prediction intervals

Some slight modifications of the general theory allow us to construct prediction intervals. Many of us would argue that the fundamental purpose of science is making accurate predictions of things that could be observed in the future. As with estimation, predicting the occurrence of a particular value (point prediction) is less valuable than interval prediction because a point prediction gives no idea of the variability associated with the prediction.

In constructing prediction intervals for a new observation y, we make a number of assumptions. The observations, including the new one, are assumed to be independent and normally distributed. Moreover, we take as our parameter $Par = E(y)$. $E(y)$ would be a reasonable point prediction for y but we do not know the value of $E(y)$. Est depends only on the observations other than y and it estimates $E(y)$, so Est makes a reasonable point prediction of y. We also assume that $\text{Var}(y) = \sigma^2$, that σ^2 has an estimate $\hat{\sigma}^2$, that $SE(Est) = \hat{\sigma}A$ for some known constant A, and that $(Est - Par)/SE(Est)$ has a t distribution with, say, r degrees of freedom. (Technically, we need Est to have a normal distribution, $r(\hat{\sigma}^2/\sigma^2)$ to have a $\chi^2(r)$ distribution, and independence of Est and $\hat{\sigma}^2$.) In some applications, these methods are used with the approximation $r \doteq \infty$, i.e., we act as if we know the variance and the appropriate distribution is taken to be a standard normal.

A prediction interval for y is based on the distribution of $y - Est$ because we need to evaluate how far y can reasonably be from our point prediction of y. The value of the future observation y is independent of the past observations and thus of Est. It follows that the variance of $y - Est$ is

$$\text{Var}(y - Est) = \text{Var}(y) + \text{Var}(Est) = \sigma^2 + \text{Var}(Est)$$

and that the standard error of $y - Est$ is

$$SE(y - Est) = \sqrt{\hat{\sigma}^2 + [SE(Est)]^2}. \tag{3.5.1}$$

One can then show that

$$\frac{y - Est}{SE(y - Est)} \sim t(r).$$

A $(1 - \alpha)100\%$ prediction interval is based on the probability equality,

$$1 - \alpha = \Pr\left[-t\left(1 - \frac{\alpha}{2}, r\right) < \frac{y - Est}{SE(y - Est)} < t\left(1 - \frac{\alpha}{2}, r\right)\right].$$

Rearranging the terms within the square brackets leads to the equality

$$1 - \alpha = \Pr\left[Est - t\left(1 - \frac{\alpha}{2}, r\right) SE(y - Est) < y < Est + t\left(1 - \frac{\alpha}{2}, r\right) SE(y - Est)\right].$$

The prediction interval consists of all y values that fall between the two observable limits in the probability statement. The endpoints of the interval are generally written

$$Est \pm t\left(1 - \frac{\alpha}{2}, r\right) SE(y - Est).$$

Of course, it is impossible to validate assumptions about observations to be taken in the future, so the confidence levels of prediction intervals are always suspect.

From the form of $SE(y - Est)$ given in (3.5.1), we see that

$$SE(y - Est) = \sqrt{\hat{\sigma}^2 + [SE(Est)]^2} \geq SE(Est).$$

Typically, the prediction standard error is much larger than the standard error of the estimate, so prediction intervals are much wider than confidence intervals. In particular, increasing the number of observations typically decreases the standard error of the estimate but has a *relatively* minor effect on the standard error of prediction. Increasing the sample size is not intended to make $\hat{\sigma}^2$ smaller, it only makes $\hat{\sigma}^2$ a more accurate estimate of σ^2.

EXAMPLE 3.5.1. As in Example 3.1.2, we eliminate the two outliers from the dropout rate data. The 36 remaining observations are approximately normal. A 95% confidence interval for the mean had endpoints

$$11.083 \pm 2.030\sqrt{27.45/36}.$$

A 95% prediction interval has endpoints

$$11.083 \pm 2.030\sqrt{27.45 + \frac{27.45}{36}}$$

or

$$11.083 \pm 10.782.$$

The prediction interval is $(.301, 21.865)$, which is much wider than the confidence interval of $(9.3, 12.9)$. We are 95% confident that the dropout rate for a new math class would be between .3% and 21.9%. We are 95% confident that the population mean dropout rate for math classes is between 9% and 13%. Of course the prediction interval assumes that the new class is from a population similar to the 1984–85 math classes with huge dropout rates deleted. Such assumptions are almost impossible to validate. Moreover, there is some chance that the new observation will be one with a huge dropout rate and this interval says nothing about such observations.

In Example 3.1.2 we also considered the square roots of the dropout rate data with the two outliers eliminated. To predict the square root of a new observation, we use the 95% interval

$$3.218 \pm 2.030\left(\sqrt{.749574 + \frac{.749574}{36}}\right),$$

which reduces to $(1.436, 5.000)$. This is a prediction interval for the square root of a new observation, so we are 95% confident that the actual value of the new observation will fall between $(1.436^2, 5.000^2)$, i.e., $(2.1, 25)$. Retransforming a prediction interval back into the original scale causes no problems of interpretation whatsoever. This prediction interval and the one in the previous paragraph are comparable. Both include values from near 0 up to the low to mid twenties. □

We have criticized commonly used definitions of the word 'confidence' but to this point the motivation for a prediction interval is exactly analogous to the motivation for confidence intervals. The endpoints of a prediction interval are obtained by taking a probability statement about random variables, substituting observed values for the random variables, and replacing 'probability' by 'confidence'. For some reason, explicitly stating that a 95% prediction interval gives 95% *confidence* that a future observation will fall within the interval seems to be a somewhat rare occurrence. Once again, a solution to the problem of defining confidence can be obtained by testing. If we wanted to test whether a new observation y was consistent with the old observations we could set up an α level test that would reject if $(y - Est)/SE(y - Est)$ was too far from zero, i.e., if its absolute value was greater than $K(1 - \alpha/2)$. Analogous to the relationship between tests of parameters and confidence intervals, this test of a new observation will not be rejected precisely when y is within the prediction interval. Thus the $(1 - \alpha)100\%$ prediction interval consists of all values of y that are consistent with the other data as determined by an α level test. Moreover, the testing approach gives some insight into why prediction intervals are based on the distribution of $y - Est$, i.e., because we are comparing the new observation y to the old data as summarized by Est.

LOWER BOUNDS ON PREDICTION CONFIDENCE

If the normal and χ^2 distributional assumptions stated at the beginning of the section break down, the prediction interval based on the t distribution is invalid. Relying primarily on the independence assumptions and there being sufficient data to use $\hat{\sigma}^2$ as an approximation to σ^2, we can find an approximate lower bound for the confidence that a new observation is in the prediction interval. Chebyshev's inequality from Subsection 1.2.2 gives

$$1 - t\left(1 - \frac{\alpha}{2}, r\right)^{-2} \leq \Pr\left[-t\left(1 - \frac{\alpha}{2}, r\right) < \frac{y - Est}{SE(y - Est)} < t\left(1 - \frac{\alpha}{2}, r\right)\right]$$

or equivalently

$$1 - t\left(1 - \frac{\alpha}{2}, r\right)^{-2} \leq \Pr\left[Est - t\left(1 - \frac{\alpha}{2}, r\right) SE(y - Est) < y < Est + t\left(1 - \frac{\alpha}{2}, r\right) SE(y - Est)\right].$$

This indicates that the confidence coefficient for the prediction interval given by

$$Est \pm t\left(1 - \frac{\alpha}{2}, r\right) SE(y - Est)$$

is (approximately) at least

$$\left[1 - t\left(1 - \frac{\alpha}{2}, r\right)^{-2}\right] 100\%.$$

If we can use the improved version of Chebyshev's inequality from Section 1.3, we can raise the confidence coefficient to

$$\left[1 - (2.25)^{-1} t\left(1 - \frac{\alpha}{2}, r\right)^{-2}\right] 100\%.$$

EXAMPLE 3.5.2. Assuming that a sample of 36 observations is enough to ensure that s^2 is essentially equal to σ^2, the nominal 95% prediction interval given in Example 3.5.1 for dropout rates has a confidence level, regardless of the distribution of the data, that is at least

$$\left(1 - \frac{1}{2.030^2}\right) = 76\% \text{ or even } \left(1 - \frac{1}{2.25(2.030)^2}\right) = 89\%.$$

3.6 Sample size determination and power

Suppose we wish to estimate the mean height of the men officially enrolled in statistics classes at the University of New Mexico on Thursday, February 4, 1993 at 3 pm. How many observations should we take? The answer to that question depends on how accurate our estimate needs to be and on our having some idea of the variability in the population.

To get a rough indication of the variability we argue as follows. Generally, men have a mean height of about 69 inches and I would guess that about 95% of them are between 63 inches and 75 inches. The probability that a $N(\mu, \sigma^2)$ random variable is between $\mu \pm 2\sigma$ is approximately .95, which suggests that $\sigma = [(\mu + 2\sigma) - (\mu - 2\sigma)]/4$ may be about $(75 - 63)/4 = 3$ for a typical population of men.

Before proceeding with sample size determination, observe that sample sizes have a real effect on the usefulness of confidence intervals. Suppose $\bar{y}_{\cdot} = 72$ and $n = 9$, so the 95% confidence interval for mean height has endpoints of roughly $72 \pm 2(3/\sqrt{9})$, or 72 ± 2, with an interval of $(70, 74)$. Here we use 3 as a rough indication of σ in the standard error and 2 as a rough indication of the tabled value for a 95% interval. If having an estimate that is off by 1 inch is a big deal, the confidence interval is totally inadequate. There is little point in collecting the data, because regardless of the value of $\bar{y}_{\cdot}$, we do not have enough accuracy to draw interesting conclusions. For example, if I claimed that the true mean height for this population was 71 inches and I cared whether my claim was off by an inch, the data are not only consistent with my claim but also with the claims that the true mean height is 70 inches and 72 inches and even 74 inches. The data are inadequate for my purposes. Now suppose $\bar{y}_{\cdot} = 72$ and $n = 3600$, the confidence interval has endpoints $72 \pm 2(3/\sqrt{3600})$ or $72 \pm .1$ with an interval of $(71.9, 72.1)$. We can tell that the population mean may be 72 inches but we are quite confident that it is not 72.11 inches. Would anyone really care about the difference between a mean height of 72 inches and a mean height of 72.11 inches? Three thousand six hundred observations gives us more information that we really need. We would like to find a middle ground.

Now suppose we wish to learn the mean height to within 1 inch with 95% confidence. From a sample of size n, a 95% confidence interval for the mean has endpoints that are roughly $\bar{y}_{\cdot} \pm 2(3/\sqrt{n})$. With 95% confidence, the mean height could be as high as $\bar{y}_{\cdot} + 2(3/\sqrt{n})$ or as low as $\bar{y}_{\cdot} - 2(3/\sqrt{n})$. We want the difference between these numbers to be no more than 1 inch. The difference between the two numbers is $12/\sqrt{n}$, so for the required difference of 1 inch set $1 = 12/\sqrt{n}$, so that $\sqrt{n} = 12/1$ or $n = 144$.

The semantics of these problems can be a bit tricky. We asked for an interval that would tell us the mean height to within 1 inch with 95% confidence. If instead we specified that we wanted our estimate to be off by no more than 1 inch, the estimate is in the middle of the interval, so the distance from the middle to the endpoint needs to be 1 inch. In other words, $1 = 2(3/\sqrt{n})$, so $\sqrt{n} = 6/1$ or

$n = 36$. Note that learning the parameter to within 1 inch is the same as having an estimate that is off by no more than $1/2$ inch.

The concepts illustrated above work quite generally. Typically an observation y has $\text{Var}(y) = \sigma^2$ and Est has $SE(Est) = \sigma A$. The constant A in $SE(Est)$ is a known function of the sample size (or sample sizes in situations involving more than one sample). In inference problems we replace σ in the standard error with an estimate of σ obtained from the data. In determining sample sizes, the data have not yet been observed, so σ has to be approximated from previous data or knowledge. The length of a $(1-\alpha)100\%$ confidence interval is

$$[Est + K(1-\alpha/2)SE(Est)] - [Est - K(1-\alpha/2)SE(Est)] = 2K(1-\alpha/2)SE(Est) = 2K(1-\alpha/2)\sigma A.$$

The tabled value $K(1-\alpha/2)$ can be approximated by $t(1-\alpha/2, \infty)$. If we specify that the confidence interval is to be w units wide, set

$$w = 2t(1-\alpha/2, \infty)\sigma A \tag{3.6.1}$$

and solve for the (approximate) appropriate sample size. In equation (3.6.1), w, $t(1-\alpha/2, \infty)$, and σ are all known and A is a known function of the sample size.

Unfortunately it is not possible to take equation (3.6.1) any further and show directly how it determines the sample size. The discussion given here is general and thus the ultimate solution depends on the type of data being examined. In the only case we have examined as yet, there is one-sample, $Par = \mu$, $Est = \bar{y}_{\cdot}$, and $SE(Est) = \sigma/\sqrt{n}$. Thus, $A = 1/\sqrt{n}$ and equation (3.6.1) becomes

$$w = 2t(1-\alpha/2, \infty)\sigma/\sqrt{n}.$$

Rearranging this gives

$$\sqrt{n} = 2t(1-\alpha/2, \infty)\sigma/w$$

and

$$n = (2t(1-\alpha/2, \infty)\sigma/w)^2.$$

But this formula only applies to one sample problems. For other problems considered in this book, e.g., comparing two independent samples, comparing more than two independent samples, and simple linear regression, equation (3.6.1) continues to apply but the constant A becomes more complicated. In cases where there is more than one sample involved, the various sample sizes are typically assumed to all be the same, and in general their relative sizes need to be specified, e.g., we could specify that the first sample will have 10 more observations than the second or that the first sample will have twice as many observations as the second.

Another approach to determining approximate sample sizes is based on the power of an α level test. Tests are set up assuming that, say, $H_0 : Par = m_0$ is true. Power is computed assuming that $Par \neq m_0$. Suppose that $Par = m_A \neq m_0$, then *the power when $Par = m_A$ is the probability that the $(1-\alpha)100\%$ confidence interval will not contain m_0.* Another way of saying that the confidence interval does not contain m_0 is saying that an α level two-sided test of $H_0 : Par = m_0$ rejects H_0. In determining sample sizes, you need to pick m_A as some value you care about. You need to care about it in the sense that if $Par = m_A$ rather than $Par = m_0$, you would like to have a reasonably good chance of rejecting $H_0 : Par = m_0$.

Cox (1958, p. 176) points out that it often works well to choose the sample size so that

$$|m_A - m_0| \doteq 3SE(Est). \tag{3.6.2}$$

Cox shows that this procedure gives reasonable powers for common choices of α. Here m_A and m_0 are known and $SE(Est) = \sigma A$, where σ is known and A is a known function of sample size. Also note that this suggestion does not depend on the α level of the test. As with equation (3.6.1), equation (3.6.2) can be solved to give n in particular cases, but a general solution for n is not possible because it depends on the exact nature of the value A.

Consider again the problem of determining the mean height. If my null hypothesis is $H_0 : \mu = 72$ and I want a reasonable chance of rejecting H_0 when $\mu = 73$, Cox's rule suggests that I should have $1 = |73 - 72| \doteq 3\left(3/\sqrt{n}\right)$ so that $\sqrt{n} \doteq 9$ or $n \doteq 81$.

It is important to remember that these are only rough guides for sample sizes. They involve several approximations, the most important of which is approximating σ. If there is more than one parameter of interest in a study, sample size computations can be performed for each and a compromise sample size can be selected.

For the past ten years I've been amazed at my own lack of interest in teaching students about statistical power. Cox (1958, p. 161) finally explained it for me. He points out that power is very important in planning investigations but it is not very important in analyzing them. I might even go so far as to say that once the data have been collected, a power analysis can at best tell you whether you have been wasting your time. In other words, a power analysis will only tell you how likely you were to find differences given the design of your experiment and the choice of test.

APPENDIX: DERIVATION OF CONFIDENCE INTERVALS

We wish to establish the validity of equation (3.1.1), i.e.,

$$\begin{aligned} 1-\alpha &= \Pr\left[-K\left(1-\frac{\alpha}{2}\right) < \frac{Est - Par}{SE(Est)} < K\left(1-\frac{\alpha}{2}\right)\right] \\ &= \Pr\left[Est - K\left(1-\frac{\alpha}{2}\right) SE(Est) < Par < Est + K\left(1-\frac{\alpha}{2}\right) SE(Est)\right] \end{aligned}$$

and in particular we wish to show that the expressions in the square brackets are equivalent. We do this by establishing a series of equivalences. The justifications for the equivalences are given at the end.

$$-K\left(1-\frac{\alpha}{2}\right) < \frac{Est - Par}{SE(Est)} < K\left(1-\frac{\alpha}{2}\right) \tag{1}$$

if and only if

$$-K\left(1-\frac{\alpha}{2}\right) SE(Est) < Est - Par < K\left(1-\frac{\alpha}{2}\right) SE(Est) \tag{2}$$

if and only if

$$K\left(1-\frac{\alpha}{2}\right) SE(Est) > -Est + Par > -K\left(1-\frac{\alpha}{2}\right) SE(Est) \tag{3}$$

if and only if

$$Est + K\left(1-\frac{\alpha}{2}\right) SE(Est) > Par > Est - K\left(1-\frac{\alpha}{2}\right) SE(Est) \tag{4}$$

if and only if

$$Est - K\left(1-\frac{\alpha}{2}\right) SE(Est) < Par < Est + K\left(1-\frac{\alpha}{2}\right) SE(Est). \tag{5}$$

JUSTIFICATION OF STEPS.

For (1) iff (2): if $c > 0$, then $a < b$ if and only if $ac < bc$.
For (2) iff (3): $a < b$ if and only if $-a > -b$.
For (3) iff (4): $a < b$ if and only if $a + c < b + c$.
For (4) iff (5): $a > b$ if and only if $b < a$.

3.7 Exercises

EXERCISE 3.7.1. This exercise illustrates the Bayesian computations discussed in the subsection of 3.1 on interpreting confidence intervals. I place a coin either heads up or tails up and hide it from you. Because of my psychic powers, when you subsequently flip a coin the probability is .75 that your coin face will be the same as mine. The four things of interest here are the outcomes that I have tails (IT), I have heads (IH), you have tails (YT), and you have heads (YH).

The computations involve ideas of conditional probability. For example, the probability that you get tails given that my coin was placed tails up is defined to be $\Pr(YT|IT) \equiv \Pr(YT \text{ and } IT)/\Pr(IT)$

Bayes' theorem relates different conditional probabilities. It states that

$$\Pr(IT|YT) = \frac{\Pr(YT|IT)\Pr(IT)}{\Pr(YT|IT)\Pr(IT) + \Pr(YT|IH)\Pr(IH)}.$$

Similarly,

$$\Pr(IH|YH) = \frac{\Pr(YH|IH)\Pr(IH)}{\Pr(YH|IH)\Pr(IH) + \Pr(YH|IT)\Pr(IT)}.$$

Clearly this problem is set up so that $\Pr(YT|IT) = \Pr(YH|IH) = .75$. Show that if your prior probability is $\Pr(IT) = \Pr(IH) = .5$, then $\Pr(IT|YT) = \Pr(IH|YH) = .75$ as claimed in the earlier discussion.

The earlier discussion also mentioned prior probabilities that were four times greater for me placing my coin heads up than tails up. In this case, $\Pr(IT) = 1/5$ and $\Pr(IH) = 4/5$. Find $\Pr(IT|YT)$ and $\Pr(IH|YH)$ and check whether these agree with the values given in Section 3.1.

Obviously, you should show all of your work.

EXERCISE 3.7.2. Identify the parameter, estimate, standard error of the estimate, and reference distribution for Exercise 2.7.1.

EXERCISE 3.7.3. Identify the parameter, estimate, standard error of the estimate, and reference distribution for Exercise 2.7.2.

EXERCISE 3.7.4. Identify the parameter, estimate, standard error of the estimate, and reference distribution for Exercise 2.7.4.

EXERCISE 3.7.5. Consider that I am collecting (normally distributed) data with a variance of 4 and I want to test a null hypothesis of $H_0 : \mu = 10$. What sample size should I take according to Cox's rule if I want a reasonable chance of rejecting H_0 when $\mu = 13$? What if I want a reasonable chance of rejecting H_0 when $\mu = 12$? What sample size should I take if I want a 95% confidence interval that is no more than 2 units long? What if I want a 99% confidence interval that is no more than 2 units long?

EXERCISE 3.7.6. The turtle shell data of Jolicoeur and Mosimann (1960) given in Exercise 2.7.4 has a standard deviation of about 21.25. If we were to collect a new sample, how large should the sample size be in order to have a 95% confidence interval with a length of (about) four units? According to Cox's rule, what sample size should I take if I want a reasonable chance of rejecting $H_0 : \mu = 130$ when $\mu = 140$?

EXERCISE 3.7.7. With reference to Exercise 2.7.3, give the approximate number of observations necessary to estimate the mean of BX to within .01 units with 99% confidence. How large a sample is needed to get a reasonable test of $H_0 : \mu = 10$ when $\mu = 11$ using Cox's rule?

EXERCISE 3.7.8. With reference to Exercise 2.7.3, give the approximate number of observations necessary to get a 99% confidence for the mean of K that has a length of 60. How large a sample is needed to get a reasonable test of $H_0 : \mu = 1200$ when $\mu = 1190$ using Cox's rule? What is the number when $\mu = 1150$?

EXERCISE 3.7.9. With reference to Exercise 2.7.3, give the approximate number of observations necessary to estimate the mean of FORM to within .5 units with 95% confidence. How large a sample is needed to get a reasonable test of $H_0 : \mu = 20$ when $\mu = 20.2$ using Cox's rule?

EXERCISE 3.7.10. With reference to Exercise 2.7.2, give the approximate number of observations necessary to estimate the mean rat weight to within 1 unit with 95% confidence. How large a sample is needed to get a reasonable test of $H_0 : \mu = 55$ when $\mu = 54$ using Cox's rule?

Chapter 4
Two sample problems

In this chapter we consider several situations where it is of interest to compare two samples. First we consider two samples of correlated data. These are data that consist of pairs of observations measuring comparable quantities. Next we consider two independent samples from populations with the same variance. We then examine two independent samples from populations with different variances. Finally we consider the problem of testing whether the variances of two populations are equal.

4.1 Two correlated samples: paired comparisons

Paired comparisons involve pairs of observations on similar variables. Often these are two observations taken on the same object under different circumstances or two observations taken on related objects. No new statistical methods are needed for analyzing such data.

EXAMPLE 4.1.1. Shewhart (1931, p. 324) presents data on the hardness of an item produced by welding two parts together. Table 4.1 gives the hardness measurements for each of the two parts. The hardness of part 1 is denoted y_1 and the hardness of part 2 is denoted y_2. For $i = 1, 2$, the data for part i are denoted $y_{ij}, j = 1, \ldots, 27$. The data are actually a subset of the data presented by Shewhart.

We are interested in the difference between μ_1, the population mean for part one, and μ_2, the population mean for part two. In other words, the parameter of interest is $Par = \mu_1 - \mu_2$. Note that if there is no difference between the population means, $\mu_1 - \mu_2 = 0$. The natural estimate of this parameter is the difference between the sample means, i.e., $Est = \bar{y}_{1\bullet} - \bar{y}_{2\bullet}$. Here we use the $\bullet$ subscript to indicate averaging over the second subscript in $\bar{y}_{i\bullet} = (y_{i1} + \cdots + y_{i27})/27$.

To perform statistical inferences, we need the standard error of the estimate, i.e., $SE(\bar{y}_{1\bullet} - \bar{y}_{2\bullet})$. As indicated earlier, finding an appropriate standard error is often the most difficult aspect of statistical inference. In problems such as this, where the data are paired, finding the standard error is complicated by the fact that the two observations in each pair are not independent. In data such as these, *different pairs are often independent but observations within a pair are not.*

In paired comparisons, we use a trick to reduce the problem to one sample. It is a simple algebraic fact that the difference of the sample means, $\bar{y}_{1\bullet} - \bar{y}_{2\bullet}$ is the same as the sample mean of the differences

TABLE 4.1. Shewhart's hardness data

Case	y_1	y_2	$d = y_1 - y_2$	Case	y_1	y_2	$d = y_1 - y_2$
1	50.9	44.3	6.6	15	46.6	31.5	15.1
2	44.8	25.7	19.1	16	50.4	38.1	12.3
3	51.6	39.5	12.1	17	45.9	35.2	10.7
4	43.8	19.3	24.5	18	47.3	33.4	13.9
5	49.0	43.2	5.8	19	46.6	30.7	15.9
6	45.4	26.9	18.5	20	47.3	36.8	10.5
7	44.9	34.5	10.4	21	48.7	36.8	11.9
8	49.0	37.4	11.6	22	44.9	36.7	8.2
9	53.4	38.1	15.3	23	46.8	37.1	9.7
10	48.5	33.0	15.5	24	49.6	37.8	11.8
11	46.0	32.6	13.4	25	51.4	33.5	17.9
12	49.0	35.4	13.6	26	45.8	37.5	8.3
13	43.4	36.2	7.2	27	48.5	38.3	10.2
14	44.4	32.5	11.9				

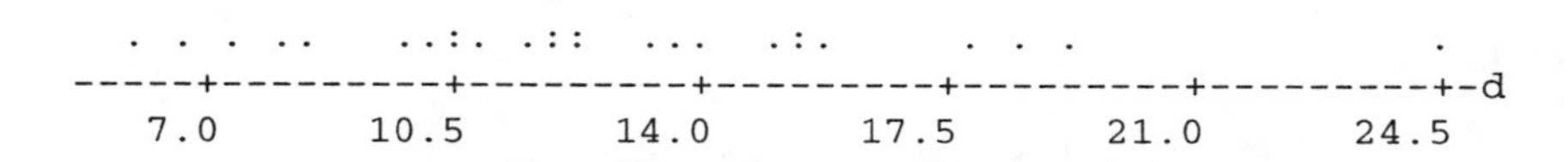

FIGURE 4.1. Dot plot of differences.

$d_j = y_{1j} - y_{2j}$, i.e., $\bar{d} = \bar{y}_{1\cdot} - \bar{y}_{2\cdot}$. Thus $\bar{d}$ is an estimate of the parameter of interest $\mu_1 - \mu_2$. The differences are given in Table 4.1 along with the data. Summary statistics are listed below for each variable and the differences. Note that for the hardness data, $\bar{d} = 12.663 = 47.552 - 34.889 = \bar{y}_{1\cdot} - \bar{y}_{2\cdot}$. In particular, if the positive value for $\bar{d}$ means anything (other than random variation), it indicates that part one is harder than part two.

Sample statistics

Variable	N_i	Mean	Variance	Std. dev.
y_1	27	47.552	6.79028	2.606
y_2	27	34.889	26.51641	5.149
$d = y_1 - y_2$	27	12.663	17.77165	4.216

Given that $\bar{d}$ is an estimate of $\mu_1 - \mu_2$, we can base the entire analysis on the differences. The differences constitute a single sample of data, so the standard error of $\bar{d}$ is simply the usual one-sample standard error,

$$SE(\bar{d}) = s_d / \sqrt{27},$$

where s_d is the sample standard deviation as computed from the 27 differences. The differences are plotted in Figure 4.1. Note that there is one potential outlier. We leave it as an exercise to reanalyze the data with the possible outlier removed.

We now have *Par*, *Est*, and *SE*(*Est*); it remains to find the appropriate distribution. Figure 4.2 gives a normal plot for the differences. While there is an upward curve at the top due to the possible outlier, the curve is otherwise reasonably straight. The Wilk–Francia statistic of $W' = 0.955$ is above the fifth percentile of the null distribution. With normal data we use the reference distribution

$$\frac{\bar{d} - (\mu_1 - \mu_2)}{s_d / \sqrt{27}} \sim t(27 - 1)$$

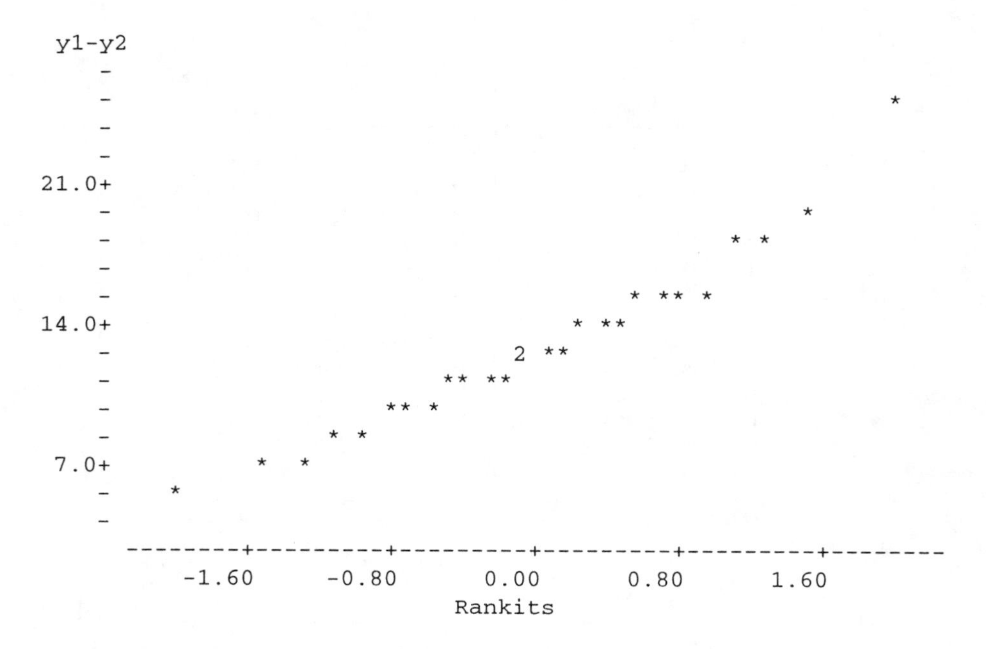

FIGURE 4.2. Normal plot of differences, $W' = .955$.

and we are now in a position to perform statistical inferences.

Our observed values of the mean and standard error are $\bar{d} = 12.663$ and $SE(\bar{d}) = 4.216/\sqrt{27} = 0.811$. From a $t(26)$ distribution, we find $t(.995, 26) = 2.78$. A 99% confidence interval for the difference in hardness has endpoints

$$12.663 \pm 2.78(.811),$$

which gives an interval of, roughly, $(10.41, 14.92)$. We are 99% confident that the population mean hardness for part 1 is between 10.41 and 14.92 units harder than that for part 2.

We can also get a 99% prediction interval for the difference in hardness to be observed on a new welded piece. The prediction interval has endpoints of

$$12.663 \pm 2.78\sqrt{4.216^2 + .811^2}$$

for an interval of $(.73, 24.60)$.

To test the hypothesis that the two parts have the same hardness, we set up the hypotheses $H_0 : \mu_1 = \mu_2$ versus $H_A : \mu_1 \neq \mu_2$, or equivalently, $H_0 : \mu_1 - \mu_2 = 0$ versus $H_A : \mu_1 - \mu_2 \neq 0$. The test statistic is

$$\frac{12.663 - 0}{.811} = 15.61.$$

This is far from zero, so the data are inconsistent with the null hypothesis. In other words, there is strong evidence that the hardness of part 1 is different than the hardness of part 2. Since the test statistic is positive, we conclude that $\mu_1 - \mu_2 > 0$ and that part 1 is harder than part 2. Note that this is consistent with our 99% confidence interval $(10.41, 14.92)$, which contains only positive values for $\mu_1 - \mu_2$.

Inferences and predictions for an individual population are made ignoring the other population, i.e., they are made using methods for one sample. For example, using the sample statistics for y_1

gives a 99% confidence interval for μ_1, the population mean hardness for part 1, with endpoints

$$47.552 \pm 2.78\sqrt{\frac{6.79028}{27}}$$

and a 99% prediction interval for the hardness of a new piece of part 1 has endpoints

$$47.552 \pm 2.78\sqrt{6.79028 + \frac{6.79028}{27}}$$

and interval $(40.175, 59.929)$. Of course, the use of the $t(26)$ distribution requires that we validate the assumption that the observations on part 1 are a random sample from a normal distribution.

When finding a prediction interval for y_1, we can typically improve the interval if we know the corresponding value of y_2. As we saw earlier, the 99% prediction interval for a new difference $d = y_1 - y_2$ has $.73 < y_1 - y_2 < 24.60$. If we happen to know that, say, $y_2 = 35$, the interval becomes $.73 < y_1 - 35 < 24.60$ or $35.73 < y_1 < 59.60$. As it turns out, with these data the new 99% prediction interval for y_1 is not an improvement over the interval in the previous paragraph. The new interval is noticeably wider. However, these data are somewhat atypical. Typically in paired data, the two measurements are highly correlated, so that the sample variance of the differences is substantially less than the sample variance of the individual measurements. In such situations, the new interval will be substantially narrower. In these data, the sample variance for the differences is 17.77165 and is actually much larger than the sample variance of 6.79028 for y_1. □

The trick of looking at differences between pairs is necessary because the two observations in a pair are not independent. While different pairs of welded parts are assumed to behave independently, it seems unreasonable to *assume* that two hardness measurements on a single item that has been welded together would behave independently. This lack of independence makes it difficult to find a standard error for comparing the sample means unless we look at the differences. In the remainder of this chapter, we consider two-sample problems in which all of the observations are assumed to be independent. The observations in each sample are independent of each other and independent of all the observations in the other sample. Paired comparison problems almost fit those assumptions but they break down at one key point. In a paired comparison, we assume that every observation is independent of the other observations in the same sample and that each observation is independent of all the observations in the other sample *except* for the observation in the other sample that it is paired with. When analyzing two samples, if we can find any reason to identify individuals as being part of a pair, that fact is sufficient to make us treat the data as a paired comparison.

The method of paired comparisons is also the name of a totally different statistical procedure. Suppose one wishes to compare five brands of chocolate chip cookies: A, B, C, D, E. It would be difficult to taste all five and order them appropriately. As an alternative, one can taste test pairs of cookies, e.g., (A,B), (A,C), (A,D), (A,E), (B,C), (B,D), etc. and identify the better of the two. The benefit of this procedure is that it is much easier to rate two cookies than to rate five. See David (1988) for a survey and discussion of procedures developed to analyze such data.

4.2 Two independent samples with equal variances

The most commonly used two-sample technique consists of comparing independent samples from two populations with the same variance. The sample sizes for the two groups are possibly different, say, N_1 and N_2, and we write the common variance as σ^2.

EXAMPLE 4.2.1. The data in Table 4.2 are final point totals for an introductory statistics class. The data are divided by the sex of the student. We investigate whether the data display sex differences.

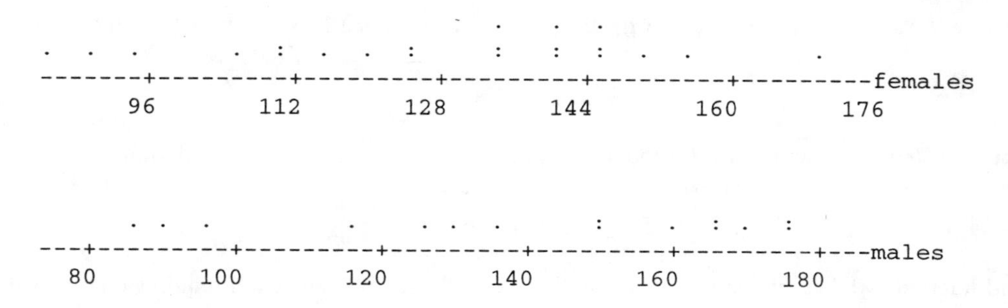

FIGURE 4.3. Dot plots for final point totals.

The data are plotted in Figure 4.3. Figures 4.4 and 4.5 contain normal plots for the two sets of data. Figure 4.4 is quite straight but Figure 4.5 looks curved. Our analysis is not particularly sensitive to nonnormality and the W' statistic for Figure 4.5 is .937, which is well above the fifth percentile, so we proceed under the assumption that both samples are normal. We also assume that all of the observations are independent. This assumption may be questionable because some students probably studied together, nonetheless, independence seems like a reasonable working assumption. □

TABLE 4.2. Final point totals for an introductory statistics class

Females					Males		
140	125	90	105	145	165	175	135
135	155	170	140	85	175	160	165
150	115	125	95		170	115	150
135	145	110	135		150	85	130
110	120	140	145		90	95	125

The methods in this section rely on the assumption that the two populations are normally distributed and have the same variance. In particular, we assume two independent samples

Sample	Data		Distribution
1	$y_{11}, y_{12}, \ldots, y_{1N_1}$	iid	$N(\mu_1, \sigma^2)$
2	$y_{21}, y_{22}, \ldots, y_{2N_2}$	iid	$N(\mu_2, \sigma^2)$

and compute summary statistics from the samples. The summary statistics are just the sample mean and the sample variance for each individual sample.

Sample statistics			
Sample	Size	Mean	Variance
1	N_1	$\bar{y}_{1\bullet}$	s_1^2
2	N_2	$\bar{y}_{2\bullet}$	s_2^2

Except for checking the validity of our assumptions, these summary statistics are more than sufficient for the entire analysis. Algebraically, the sample mean for population i, $i = 1, 2$, is

$$\bar{y}_{i\bullet} \equiv \frac{1}{N_i} \sum_{j=1}^{N_i} y_{ij} = \frac{1}{N_i} \left[y_{i1} + y_{i2} + \cdots + y_{iN_i} \right]$$

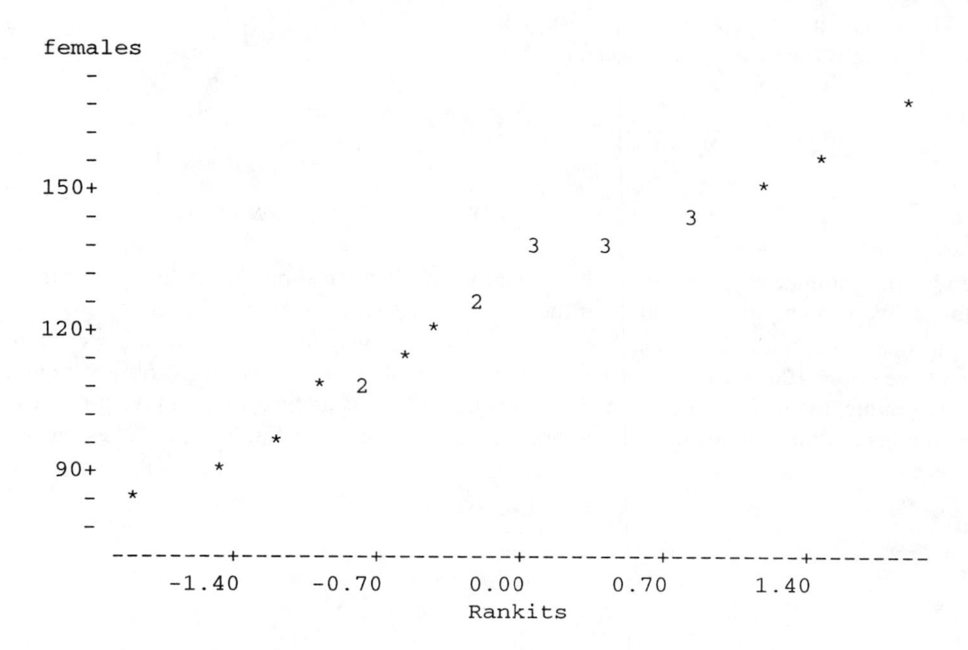

FIGURE 4.4. Normal plot for females, $W' = .974$.

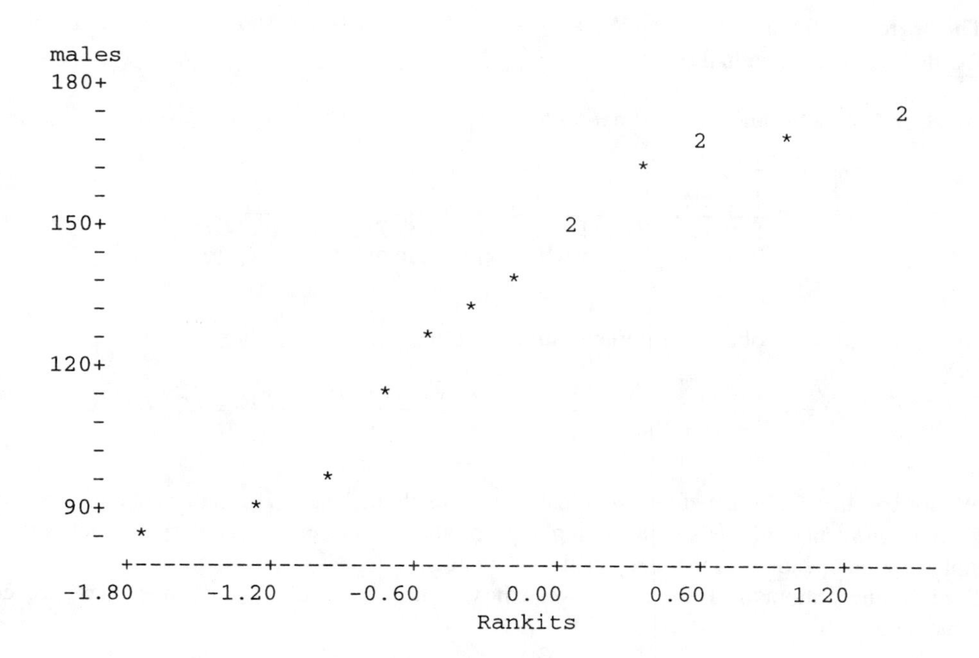

FIGURE 4.5. Normal plot for males, $W' = .937$.

where the $\bullet$ in $\bar{y}_{i\bullet}$ indicates that the mean is obtained by averaging over j, the second subscript in the y_{ij}s. The sample means, $\bar{y}_{1\bullet}$ and $\bar{y}_{2\bullet}$, are estimates of μ_1 and μ_2.

The sample variance for population i, $i = 1, 2$, is

$$\begin{aligned} s_i^2 &= \frac{1}{N_i - 1} \sum_{j=1}^{N_i} \left(y_{ij} - \bar{y}_{i\bullet}\right)^2 \\ &= \frac{1}{N_i - 1} \left[(y_{i1} - \bar{y}_{i\bullet})^2 + (y_{i2} - \bar{y}_{i\bullet})^2 + \cdots + \left(y_{iN_i} - \bar{y}_{i\bullet}\right)^2\right]. \end{aligned}$$

The s_i^2s both estimate σ^2. Combining the s_i^2s can yield a better estimate of σ^2 than either individual estimate. We form a pooled estimate of the variance, say s_p^2, by averaging s_1^2 and s_2^2. With unequal sample sizes an efficient pooled estimate of σ^2 must be a weighted average of the s_i^2s. Obviously, if we have $N_1 = 100\,000$ observations in the first sample and only $N_2 = 10$ observations in the second sample, the variance estimate s_1^2 is much better than s_2^2 and we want to give it more weight. The weights are the degrees of freedom associated with the estimates. The pooled estimate of the variance is

$$\begin{aligned} s_p^2 &\equiv \frac{(N_1 - 1)s_1^2 + (N_2 - 1)s_2^2}{(N_1 - 1) + (N_2 - 1)} \\ &= \frac{1}{N_1 + N_2 - 2} \left[\sum_{j=1}^{N_1} \left(\bar{y}_{1j} - \bar{y}_{1\bullet}\right)^2 + \sum_{j=1}^{N_2} \left(\bar{y}_{2j} - \bar{y}_{2\bullet}\right)^2\right] \\ &= \frac{1}{N_1 + N_2 - 2} \sum_{i=1}^{2} \sum_{j=1}^{N_i} \left(\bar{y}_{ij} - \bar{y}_{i\bullet}\right)^2. \end{aligned}$$

The degrees of freedom for s_p^2 are $N_1 + N_2 - 2 = (N_1 - 1) + (N_2 - 1)$, i.e., the sum of the degrees of freedom for the individual estimates s_i^2.

EXAMPLE 4.2.2. For the data on final point totals, the sample statistics are given below.

Sample Statistics

Sample	N_i	$\bar{y}_{i\bullet}$	s_i^2	s_i
females	22	127.954545	487.2835498	22.07
males	15	139.000000	979.2857143	31.29

From these values, we obtain the pooled estimate of the variance,

$$s_p^2 = \frac{(N_1 - 1)s_1^2 + (N_2 - 1)s_2^2}{N_1 + N_2 - 2} = \frac{(21)487.28 + (14)979.29}{35} = 684.08. \qquad \square$$

We are now in a position to draw statistical inferences about the μ_is. The main problem in obtaining tests and confidence intervals is in finding appropriate standard errors. The crucial fact is that the samples are independent so that the $\bar{y}_{i\bullet}$s are independent.

For inferences about the difference between the two means, say, $\mu_1 - \mu_2$, use the general procedure of Chapter 3 with

$$Par = \mu_1 - \mu_2$$

and

$$Est = \bar{y}_{1\bullet} - \bar{y}_{2\bullet}.$$

Note that $\bar{y}_{1\bullet} - \bar{y}_{2\bullet}$ is unbiased for estimating $\mu_1 - \mu_2$ because

$$E(\bar{y}_{1\bullet} - \bar{y}_{2\bullet}) = E(\bar{y}_{1\bullet}) - E(\bar{y}_{2\bullet}) = \mu_1 - \mu_2 .$$

The two means are independent, so the variance of $\bar{y}_{1\bullet} - \bar{y}_{2\bullet}$ is the variance of $\bar{y}_{1\bullet}$ plus the variance of $\bar{y}_{2\bullet}$, i.e.,

$$\text{Var}(\bar{y}_{1\bullet} - \bar{y}_{2\bullet}) = \text{Var}(\bar{y}_{1\bullet}) + \text{Var}(\bar{y}_{2\bullet}) = \frac{\sigma^2}{N_1} + \frac{\sigma^2}{N_2} = \sigma^2 \left[\frac{1}{N_1} + \frac{1}{N_2}\right].$$

The standard error of $\bar{y}_{1\bullet} - \bar{y}_{2\bullet}$ is the estimated standard deviation of $\bar{y}_{1\bullet} - \bar{y}_{2\bullet}$,

$$SE(\bar{y}_{1\bullet} - \bar{y}_{2\bullet}) = \sqrt{s_p^2 \left[\frac{1}{N_1} + \frac{1}{N_2}\right]}.$$

Under our assumption that the original data are normal, the reference distribution is

$$\frac{(\bar{y}_{1\bullet} - \bar{y}_{2\bullet}) - (\mu_1 - \mu_2)}{\sqrt{s_p^2 \left[\frac{1}{N_1} + \frac{1}{N_2}\right]}} \sim t(N_1 + N_2 - 2).$$

The degrees of freedom for the t distribution are the degrees of freedom for s_p^2. For nonnormal data with large sample sizes, the reference distribution is $N(0, 1)$.

Having identified the parameter, estimate, standard error, and distribution, inferences follow the usual pattern. A 95% confidence interval for $\mu_1 - \mu_2$ is

$$(\bar{y}_{1\bullet} - \bar{y}_{2\bullet}) \pm t(.975, N_1 + N_2 - 2)\sqrt{s_p^2 \left[\frac{1}{N_1} + \frac{1}{N_2}\right]}.$$

A test of hypothesis that the means are equal, say

$$H_0 : \mu_1 = \mu_2 \quad \text{versus} \quad H_A : \mu_1 \neq \mu_2$$

can be converted into the equivalent hypothesis involving $Par = \mu_1 - \mu_2$, namely

$$H_0 : \mu_1 - \mu_2 = 0 \quad \text{versus} \quad H_A : \mu_1 - \mu_2 \neq 0.$$

The test is handled in the usual way. An $\alpha = .01$ test rejects H_0 if

$$\frac{|(\bar{y}_{1\bullet} - \bar{y}_{2\bullet}) - 0|}{\sqrt{s_p^2 \left[\frac{1}{N_1} + \frac{1}{N_2}\right]}} > t(.995, N_1 + N_2 - 2).$$

In our discussion of comparing differences, we have defined the parameter as $\mu_1 - \mu_2$. We could just as well have defined the parameter as $\mu_2 - \mu_1$. This would have given an entirely equivalent analysis.

Inferences about a single mean, say, μ_2, use the general procedures with $Par = \mu_2$ and $Est = \bar{y}_{2\bullet}$. The variance of $\bar{y}_{2\bullet}$ is σ^2/N_2, so $SE(\bar{y}_{2\bullet}) = \sqrt{s_p^2/N_2}$. Note the use of s_p^2 rather than s_2^2. The reference distribution is $[\bar{y}_{2\bullet} - \mu_2]/SE(\bar{y}_{2\bullet}) \sim t(N_1 + N_2 - 2)$. A 95% confidence interval for μ_2 is

$$\bar{y}_{2\bullet} \pm t(.975, N_1 + N_2 - 2)\sqrt{s_p^2/N_2}.$$

A 95% prediction interval for a new observation on variable y_2 is

$$\bar{y}_{2\bullet} \pm t(.975, N_1 + N_2 - 2)\sqrt{s_p^2 + \frac{s_p^2}{N_2}}.$$

An $\alpha = .01$ test of the hypothesis, say

$$H_0 : \mu_2 = 5 \quad \text{versus} \quad H_A : \mu_2 \neq 5,$$

rejects H_0 if

$$\frac{|\bar{y}_{2\bullet} - 5|}{\sqrt{s_p^2 / N_2}} > t(.995, N_1 + N_2 - 2).$$

EXAMPLE 4.2.3. For comparing females and males on final point totals, the parameter of interest is

$$Par = \mu_1 - \mu_2$$

where μ_1 indicates the population mean final point total for females and μ_2 indicates the population mean final point total for males. The estimate of the parameter is

$$Est = \bar{y}_{1\bullet} - \bar{y}_{2\bullet} = 127.95 - 139.00 = -11.05\,.$$

The pooled estimate of the variance is $s_p^2 = 684.08$, so the standard error is

$$SE(\bar{y}_{1\bullet} - \bar{y}_{2\bullet}) = \sqrt{s_p^2\left(\frac{1}{N_1} + \frac{1}{N_2}\right)} = \sqrt{684.08\left(\frac{1}{22} + \frac{1}{15}\right)} = 8.7578\,.$$

The data have reasonably normal distributions and the variances are not too different (more on this later), so the reference distribution is taken as

$$\frac{(\bar{y}_{1\bullet} - \bar{y}_{2\bullet}) - (\mu_1 - \mu_2)}{\sqrt{s_p^2\left(\frac{1}{22} + \frac{1}{15}\right)}} \sim t(35)$$

where $35 = N_1 + N_2 - 2$. The tabled value for finding 95% confidence intervals and $\alpha = .05$ two-sided tests is

$$t(.975, 35) = 2.030\,.$$

A 95% confidence interval for $\mu_1 - \mu_2$ has endpoints

$$-11.05 \pm (2.030)8.7578$$

which yields an interval $(-28.8, 6.7)$. We are 95% confident that the population mean scores are between, roughly, 29 points *less* for females and 7 points *more* for females.

An $\alpha = .05$ two-sided test of $H_0 : \mu_1 - \mu_2 = 0$ versus $H_A : \mu_1 - \mu_2 \neq 0$ is not rejected because 0, the hypothesized value of $\mu_1 - \mu_2$, is contained in the 95% confidence interval for $\mu_1 - \mu_2$. The P value for the test is based on the observed value of the test statistic

$$t_{obs} = \frac{(\bar{y}_{1\bullet} - \bar{y}_{2\bullet}) - 0}{\sqrt{s_p^2\left(\frac{1}{22} + \frac{1}{15}\right)}} = \frac{-11.05 - 0}{8.7578} = -1.26\,.$$

The probability of obtaining an observation from a $t(35)$ distribution that is as extreme or more extreme than $|-1.26|$ is 0.216. There is very little evidence that the population mean final point total for females is different (smaller) than the population mean final point total for males. The P value is greater than .2, so, as we established earlier, neither an $\alpha = .05$ nor an $\alpha = .01$ test is rejected. If we were silly enough to do an $\alpha = .25$ test, we would then reject the null hypothesis.

If one claimed that, for whatever reason, females tend to do worse than males in statistics classes, a two-sided test would probably be inappropriate. To test $H_0 : \mu_1 - \mu_2 \leq 0$ versus $H_A : \mu_1 - \mu_2 > 0$, the test statistic is the same but the interpretation is very different. The negative value of the test statistic is consistent with the null hypothesis. The P value is the very large value $1 - .216/2 = .892$. Claiming that females do better would give the opposite one-sided test with a P value of $.216/2 = .108$.

A 95% confidence interval for μ_1, the mean of the females, has endpoints

$$127.95 \pm (2.030)\sqrt{684.08/22}$$

which gives the interval $(116.6, 139.3)$. We are 95% confident that the mean of the final point totals for females is between 117 and 139. A 95% prediction interval for a new observation on a female has endpoints

$$127.95 \pm (2.030)\sqrt{684.08 + \frac{684.08}{22}}$$

which gives the interval $(73.7, 182.2)$. We are 95% confident that a new observation on a female will be between 74 and 182. This assumes that the new observation is randomly sampled from the same population as the previous data.

A test of the assumption of equal variances is left for the final section but we will see in the next section that the results for these data do not depend substantially on the equality of the variances. □

4.3 Two independent samples with unequal variances

We now consider two independent samples with unequal variances σ_1^2 and σ_2^2. In this section we examine inferences about the means of the two populations. While inferences about means are important, some care is required when drawing practical conclusions about populations with unequal variances. For example, if you want to produce gasoline with an octane of at least 87, you may have a choice between two processes. One process y_1 gives octanes distributed as $N(89, 4)$ and the other y_2 gives $N(90, 4)$. The two processes have the same variance, so the process with the higher mean gives more gas with an octane of at least 87. On the other hand, if y_1 gives $N(89, 4)$ and y_2 gives $N(90, 16)$, the y_1 process with mean 89 has a higher probability of achieving an octane of 87 than the y_2 process with mean 90, see Exercise 4.5.10. This is a direct result of the y_2 process having more variability. Having given this warning, we proceed with our discussion on drawing statistical inferences for the means.

EXAMPLE 4.3.1. Jolicoeur and Mosimann (1960) present data on the sizes of turtle shells (carapaces). Table 4.3 presents data on the shell heights for 24 females and 24 males. These data are not paired; it is simply a caprice that 24 carapaces were measured for each sex. Our interest centers on estimating the population means for female and male heights, estimating the difference between the heights, and testing whether the difference is zero.

Following Christensen (1990a) and others, we take natural logarithms of the data, i.e.,

$$y_1 = log(\text{female height}) \qquad y_2 = log(\text{male height}).$$

(*All logarithms in this book are natural logarithms.*) The log data are plotted in Figure 4.6. The female heights give the impression of being both larger and more spread out. Figures 4.7 and 4.8 contain normal plots for the females and males respectively. Neither is exceptionally straight but

TABLE 4.3. Turtle shell heights

Female				Male			
51	38	63	46	39	42	37	43
51	38	60	51	39	45	35	41
53	42	62	51	38	45	35	41
57	42	63	51	40	45	39	41
55	44	61	48	40	46	38	40
56	50	67	49	40	47	37	44

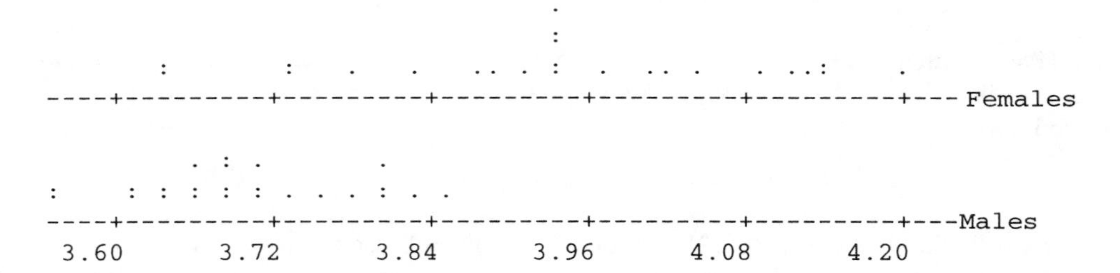

FIGURE 4.6. Plot of turtle shell log heights.

they do not seem too bad. Summary statistics are given below; they are consistent with the visual impressions given by Figure 4.6. The summary statistics will be used in later examples as the basis for our statistical inferences.

Group	Size	Mean	Variance	Standard deviation
Females	24	3.9403	0.02493979	0.1579
Males	24	3.7032	0.00677276	0.0823

□

In general we assume two independent samples

Sample	Data		Distribution
1	$y_{11}, y_{12}, \ldots, y_{1N_1}$	iid	$N(\mu_1, \sigma_1^2)$
2	$y_{21}, y_{22}, \ldots, y_{2N_2}$	iid	$N(\mu_2, \sigma_2^2)$

and compute summary statistics from the samples.

Sample	Size	Mean	Variance
1	N_1	$\bar{y}_{1\cdot}$	s_1^2
2	N_2	$\bar{y}_{2\cdot}$	s_2^2

Again, the sample means, $\bar{y}_{1\cdot}$ and $\bar{y}_{2\cdot}$, are estimates of μ_1 and μ_2, but now s_1^2 and s_2^2 estimate σ_1^2 and σ_2^2. We have two different variances, so it is inappropriate to pool the variance estimates. Once again, the crucial fact in obtaining a standard error is that the samples are independent.

For inferences about the difference between the two means, say, $\mu_1 - \mu_2$, again use the general procedure with

$$Par = \mu_1 - \mu_2$$

and

$$Est = \bar{y}_{1\cdot} - \bar{y}_{2\cdot}.$$

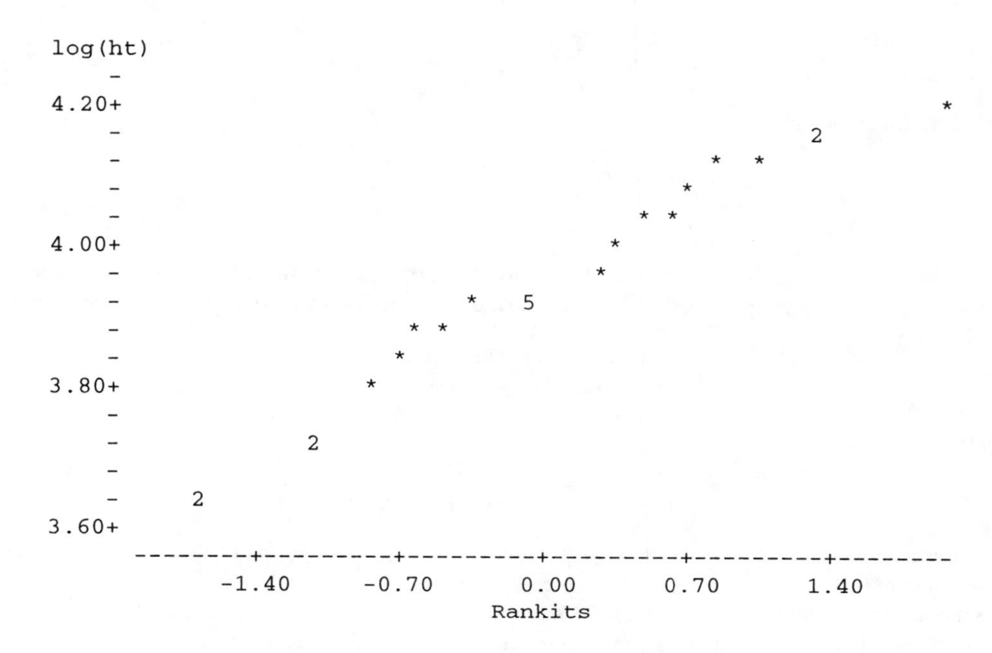

FIGURE 4.7. Normal plot for female turtle shell log heights.

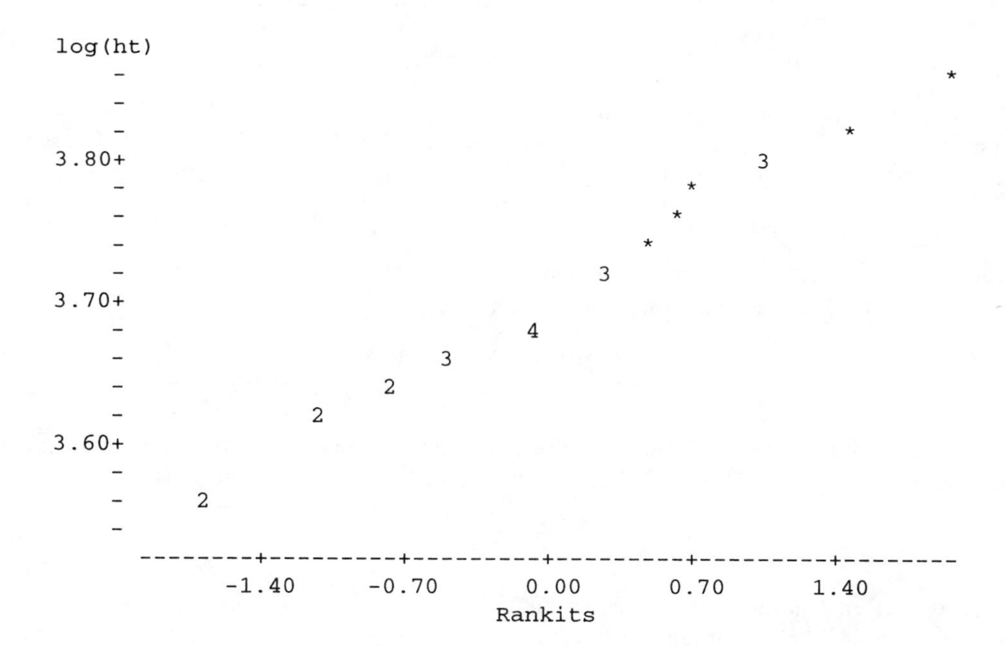

FIGURE 4.8. Normal plot for male turtle shell log heights.

Just as before, $\bar{y}_{1\bullet} - \bar{y}_{2\bullet}$ is unbiased for estimating $\mu_1 - \mu_2$. The two sample means are independent so

$$\text{Var}(\bar{y}_{1\bullet} - \bar{y}_{2\bullet}) = \text{Var}(\bar{y}_{1\bullet}) + \text{Var}(\bar{y}_{2\bullet}) = \frac{\sigma_1^2}{N_1} + \frac{\sigma_2^2}{N_2}.$$

The standard error of $\bar{y}_{1\bullet} - \bar{y}_{2\bullet}$ is

$$SE(\bar{y}_{1\bullet} - \bar{y}_{2\bullet}) = \sqrt{\frac{s_1^2}{N_1} + \frac{s_2^2}{N_2}}.$$

Even when the original data are normal, the appropriate reference distribution is not a t distribution. As a matter of fact, the appropriate reference distribution is not known. However, a good approximate distribution is

$$\frac{(\bar{y}_{1\bullet} - \bar{y}_{2\bullet}) - (\mu_1 - \mu_2)}{\sqrt{s_1^2/N_1 + s_2^2/N_2}} \sim t(\nu)$$

where

$$\nu \equiv \frac{\left(s_1^2/N_1 + s_2^2/N_2\right)^2}{\left(s_1^2/N_1\right)^2/(N_1 - 1) + \left(s_2^2/N_2\right)^2/(N_2 - 1)} \tag{4.3.1}$$

is an approximate number of degrees of freedom. This approximate distribution was proposed by Satterthwaite (1946) and is discussed by Snedecor and Cochran (1980).

For nonnormal data with large sample sizes, the reference distribution can be taken as $N(0, 1)$. Having identified the parameter, estimate, standard error and reference distribution, inferences follow the usual pattern.

EXAMPLE 4.3.2. Consider the turtle data. Recall that

Group	Size	Mean	Variance	Standard deviation
Females	24	3.9403	0.02493979	0.1579
Males	24	3.7032	0.00677276	0.0823

We begin by considering a test of $H_0 : \mu_1 = \mu_2$ versus $H_A : \mu_1 \neq \mu_2$ or equivalently $H_0 : \mu_1 - \mu_2 = 0$ versus $H_A : \mu_1 - \mu_2 \neq 0$. As before, $Par = \mu_1 - \mu_2$ and $Est = 3.9403 - 3.7032 = .2371$. The standard error is now

$$SE(\bar{y}_{1\bullet} - \bar{y}_{2\bullet}) = \sqrt{\frac{0.02493979}{24} + \frac{0.00677276}{24}} = .03635.$$

Using $s_1^2/N_1 = 0.02493979/24 = .001039158$ and $s_2^2/N_2 = 0.00677276/24 = .000282198$ in equation (4.3.1), the approximate degrees of freedom are

$$\nu = \frac{(.001039158 + .000282198)^2}{(.001039158)^2/23 + (.000282198)^2/23} = 34.6.$$

An $\alpha = .01$ test is rejected if the observed value of the test statistic is farther from zero than the cutoff value $t(.995, 34.6) \doteq t(.995, 35) = 2.72$. The observed value of the test statistic is

$$t_{obs} = \frac{.2371 - 0}{.03635} = 6.523$$

which is greater than the cutoff value, so the test is rejected. There is evidence at the .01 level that the mean shell height for females is different from the mean shell height for males. Obviously, since $\bar{y}_{1\bullet} - \bar{y}_{2\bullet} = .2371$ is positive, there is evidence that the females have shells of greater height. Actually, the conclusion is that the means of the *log*(heights) are different, but if these are different we conclude that the mean heights are different.

The 95% confidence interval for the difference between mean log shell heights for females and males, i.e., $\mu_1 - \mu_2$, uses $t(.975, 34.6) \doteq t(.975, 35) = 2.03$. The endpoints are

$$.2371 \pm 2.03\,(.03635),$$

and the interval is $(.163, .311)$. We took logs of the data, so if we transform back to the original scale the interval is $(e^{.163}, e^{.311})$ or $(1.18, 1.36)$. We are 95% confident that the population center for females is, roughly, between one and a sixth and one and a third *times* the shell heights for males. Note that a difference between .163 and .311 on the log scale transforms into a *multiplicative effect* between 1.18 and 1.36 on the original scale. This idea is discussed in more detail in Example 5.1.1.

It is inappropriate to pool the variance estimates, so inferences about μ_1 and μ_2 are performed just as for one sample. The 95% confidence interval for the mean shell height for females, μ_1, uses the estimate $\bar{y}_{1\bullet}$, the standard error $s_1/\sqrt{24}$, and the tabled value $t(.975, 24-1) = 2.069$. It has endpoints

$$3.9403 \pm 2.069\left(0.1579/\sqrt{24}\right)$$

which gives the interval $(3.87, 4.01)$. Transforming to the original scale gives the interval $(47.9, 55.1)$. We are 95% confident that the 'average' height for females' shells is between, roughly, 48 and 55 millimeters. Males also have 24 observations, so the interval for μ_2 also uses $t(.975, 24-1)$, has endpoints

$$3.7032 \pm 2.069\left(0.0823/\sqrt{24}\right),$$

and an interval $(3.67, 3.74)$. Transforming the interval back to the original scale gives $(39.3, 42.1)$. We are 95% confident that the 'average' height for males's shells is between, roughly, 39 and 42 millimeters. The 95% prediction interval for the transformed shell height of a future male has endpoints

$$3.7032 \pm 2.069\left(0.0823\sqrt{1+\frac{1}{24}}\right),$$

which gives the interval $(3.529, 3.877)$. Transforming the prediction interval back to the original scale gives $(34.1, 48.3)$. Transforming a prediction interval back to the original scale creates no problems of interpretation. □

EXAMPLE 4.3.3. Reconsider the final point totals data of Section 4.2. Without the assumption of equal variances, the standard error is

$$SE(\bar{y}_{1\bullet} - \bar{y}_{2\bullet}) = \sqrt{\frac{487.28}{22} + \frac{979.29}{15}} = 9.3507.$$

From equation (4.3.1), the degrees of freedom for the approximate t distribution are 23. A 95% confidence interval for the difference is $(-30.4, 8.3)$ and the observed value of the statistic for testing equal means is $t_{obs} = -1.18$. This gives a P value for a two-sided test of 0.22. These values are all quite close to those obtained using the equal variance assumption. □

It is an algebraic fact that if $N_1 = N_2$, the observed value of the test statistic for $H_0 : \mu_1 = \mu_2$ based on unequal variances is the same as that based on equal variances. In the turtle example, the sample sizes are both 24 and the test statistic of 6.523 is the same as the equal variances test statistic. The algebraic equivalence occurs because with equal sample sizes, the standard errors from the two procedures are the same. With equal sample sizes, the only practical difference between the two procedures for examining $Par = \mu_1 - \mu_2$ is in the choice of degrees of freedom for the t distribution.

In the turtle example above, the unequal variances procedure had approximately 35 degrees of freedom, while the equal variance procedure has 46 degrees of freedom. The degrees of freedom are sufficiently close that the substantive results of the turtle analysis are essentially the same, regardless of method. The other fact that should be recalled is that the reference distribution associated with $\mu_1 - \mu_2$ for the equal variance method is exactly correct for data that satisfy the assumptions. Even for data that satisfy the unequal variance method assumptions, the reference distribution is just an approximation.

4.4 Testing equality of the variances

Throughout this section we assume that the original data are normally distributed and that the two samples are independent. Our goal is to test the hypothesis that the variances are equal, i.e.,

$$H_0 : \sigma_2^2 = \sigma_1^2 \quad \text{versus} \quad H_A : \sigma_2^2 \neq \sigma_1^2.$$

The hypotheses can be converted into equivalent hypotheses,

$$H_0 : \frac{\sigma_2^2}{\sigma_1^2} = 1 \quad \text{versus} \quad H_A : \frac{\sigma_2^2}{\sigma_1^2} \neq 1.$$

An obvious test statistic is

$$\frac{s_2^2}{s_1^2}.$$

We will reject the hypothesis of equal variances if the test statistic is too much greater than 1 or too much less than 1. As always, the problem is in identifying a precise meaning for 'too much'. To do this, we need to know the distribution of the test statistic when the variances are equal. The distribution is known as an F distribution, i.e., if H_0 is true

$$\frac{s_2^2}{s_1^2} \sim F(N_2 - 1, N_1 - 1).$$

The distribution depends on the degrees of freedom for the two estimates. The first parameter in $F(N_2 - 1, N_1 - 1)$ is $N_2 - 1$, the degrees of freedom for the variance estimate in the numerator of s_2^2/s_1^2, and the second parameter is $N_1 - 1$, the degrees of freedom for the variance estimate in the denominator. The test statistic s_2^2/s_1^2 is nonnegative, so our reference distribution $F(N_2 - 1, N_1 - 1)$ is nonnegative. Tables are given in Appendix B.

In some sense, the F distribution is 'centered' around one and we reject H_0 if s_2^2/s_1^2 is too large or too small to have reasonably come from an $F(N_2 - 1, N_1 - 1)$ distribution. An $\alpha = .01$ level test is rejected, i.e., we conclude that $\sigma_2^2 \neq \sigma_1^2$, if

$$\frac{s_2^2}{s_1^2} > F(.995, N_2 - 1, N_1 - 1)$$

or if

$$\frac{s_2^2}{s_1^2} < F(.005, N_2 - 1, N_1 - 1)$$

where $F(.995, N_2 - 1, N_1 - 1)$ cuts off the top .005 of the distribution and $F(.005, N_2 - 1, N_1 - 1)$ cuts off the bottom .005 of the distribution. It is rare that one finds the bottom percentiles of an F distribution tabled but they can be obtained from the top percentiles. In particular,

$$F(.005, N_2 - 1, N_1 - 1) = \frac{1}{F(.995, N_1 - 1, N_2 - 1)}.$$

Note that the degrees of freedom have been reversed in the right-hand side of the equality.

The procedure for this test does not fit within the general procedures outlined in Chapter 3. It has been indicated all along that results for variances do not fit the general pattern. Although we have a parameter, σ_2^2/σ_1^2, and an estimate of the parameter, s_2^2/s_1^2, we do not have a standard error or a reference distribution that is symmetric about zero. In fact, the F distribution is not symmetric though we rely on it being 'centered' about 1.

EXAMPLE 4.4.1. We again consider the log turtle height data. The sample variance of log female heights is $s_1^2 = 0.02493979$ and the sample variance of log male heights is $s_2^2 = 0.00677276$. An $\alpha = .01$ level test is rejected, i.e., we conclude that $\sigma_2^2 \neq \sigma_1^2$, if

$$.2716 = \frac{0.00677276}{0.02493979} = \frac{s_2^2}{s_1^2} > F(.995, 23, 23) = 3.04$$

or if

$$.2716 < F(.005, 23, 23) = \frac{1}{F(.995, 23, 23)} = \frac{1}{3.04} = .33.$$

The second of these inequalities is true, so the null hypothesis of equal variances is rejected at the .01 level. We have evidence that $\sigma_2^2 \neq \sigma_1^2$ and, since the statistic is less than one, evidence that $\sigma_2^2 < \sigma_1^2$.

□

EXAMPLE 4.4.2. Consider again the final point total data. The sample variance for females is $s_1^2 = 487.28$ and the sample variance for males is $s_2^2 = 979.29$. The test statistic is

$$\frac{s_1^2}{s_2^2} = \frac{487.28}{979.29} = 0.498\,.$$

Naturally, it does not matter which variance estimate we put in the numerator as long as we keep the degrees of freedom straight. The observed test statistic is not less than $1/F(.95, 14, 21) = 1/2.197 = .455$ nor greater than $F(.95, 21, 14) = 2.377$, so the test cannot be rejected at the $\alpha = .10$ level. □

In practice, *tests for the equality of variances are rarely performed.* Typically, the main emphasis is on drawing conclusions about the μ_is; the motivation for testing equality of variances is frequently to justify the use of the pooled estimate of the variance. The test assumes that the null hypothesis of equal variances is true and data that are inconsistent with the assumptions indicate that the assumptions are false. We generally take this to indicate that the assumption about the null hypothesis is false, but, in fact, unusual data may be obtained if any of the assumptions are invalid. The equal variances test assumes that the data are independent and normal and that the variances are equal. Minor deviations from normality may cause the test to be rejected. While procedures for comparing μ_is based on the pooled estimate of the variance are sensitive to unequal variances, they are not particularly sensitive to nonnormality. The test for equality of variances is so sensitive to nonnormality that when rejecting this test one has little idea if the problem is really unequal variances or if it is nonnormality. Thus one has little idea whether there is a problem with the pooled estimate procedures or not. Since the test is not very informative, it is rarely performed. However, *studying this test prepares one for examining the important analysis of variance F test that is treated in the next chapter.*

MINITAB COMMANDS

Minitab can be used to get the F percentiles reported in Example 4.4.1.

```
MTB > invcdf .995
SUBC> f 23 23.
MTB > invcdf .005
SUBC> f 23 23.
```

THEORY

The F distribution used here is related to the fact that for normal data

$$\frac{(N_i - 1)s_i^2}{\sigma_i^2} \sim \chi^2(N_i - 1).$$

Definition 4.4.3. An F distribution is the ratio of two independent chi-squared random variables divided by their degrees of freedom. The numerator and denominator degrees of freedom for the F distribution are the degrees of freedom for the respective chi-squares.

In this problem, the two chi-squared random variables divided by their degrees of freedom are

$$\frac{(N_i - 1)s_i^2 / \sigma_i^2}{N_i - 1} = \frac{s_i^2}{\sigma_i^2}$$

$i = 1, 2$. They are independent because they are taken from independent samples and their ratio is

$$\frac{s_2^2}{\sigma_2^2} \Big/ \frac{s_1^2}{\sigma_1^2} = \frac{s_2^2}{s_1^2}\frac{\sigma_1^2}{\sigma_2^2}.$$

When the null hypothesis is true, i.e., $\sigma_2^2 / \sigma_1^2 = 1$, by definition, we get

$$\frac{s_2^2}{s_1^2} \sim F(N_2 - 1, N_1 - 1),$$

so the test statistic has an F distribution under the null hypothesis.

Note that we could equally well have reversed the roles of the two groups and set the test up as

$$H_0 : \frac{\sigma_1^2}{\sigma_2^2} = 1 \quad \text{versus} \quad H_A : \frac{\sigma_1^2}{\sigma_2^2} \neq 1$$

with the test statistic

$$\frac{s_1^2}{s_2^2}.$$

An α level test is rejected if

$$\frac{s_1^2}{s_2^2} > F\left(1 - \frac{\alpha}{2}, N_1 - 1, N_2 - 1\right)$$

or if

$$\frac{s_1^2}{s_2^2} < F\left(\frac{\alpha}{2}, N_1 - 1, N_2 - 1\right).$$

Using the fact that for any α between zero and one and any degrees of freedom r and s,

$$F(\alpha, r, s) = \frac{1}{F(1 - \alpha, s, r)}, \tag{4.4.1}$$

it is easily seen that this test is equivalent to the one we constructed. Relation (4.4.1) is a result of the fact that with equal variances both s_2^2 / s_1^2 and s_1^2 / s_2^2 have F distributions. Clearly, the smallest, say, 5% of values from s_2^2 / s_1^2 are also the largest 5% of the values of s_1^2 / s_2^2.

4.5 Exercises

EXERCISE 4.5.1. Box (1950) gave data on the weights of rats that were given the drug Thiouracil. The rats were measured at the start of the experiment and at the end of the experiment. The data are given in Table 4.4. Give a 99% confidence interval for the difference in weights between the finish and the start. Test the null hypothesis that the population mean weight gain was less than or equal to 50 with $\alpha = .02$.

TABLE 4.4. Weights of rats on thiouracil

Rat	Start	Finish	Rat	Start	Finish
1	61	129	6	51	119
2	59	122	7	56	108
3	53	133	8	58	138
4	59	122	9	46	107
5	51	140	10	53	122

EXERCISE 4.5.2. Box (1950) also considered data on rats given Thyroxin and a control group of rats. The weight *gains* are given in Table 4.5. Give a 95% confidence interval for the difference in weight gains between the Thyroxin group and the control group. Give an $\alpha = .05$ test of whether the control group has weight gains no greater than the Thyroxin group.

TABLE 4.5. Weight gain comparison

Control		Thyroxin	
115	107	132	88
117	90	84	119
133	91	133	
115	91	118	
95	112	87	

EXERCISE 4.5.3. Conover (1971, p. 226) considered data on the physical fitness of male seniors in a particular high school. The seniors were divided into two groups based on whether they lived on a farm or in town. The results in Table 4.6 are from a physical fitness test administered to the students. High scores indicate that an individual is physically fit. Give a 95% confidence interval for the difference in mean fitness scores between the town and farm students. Test the hypothesis of no difference at the $\alpha = .10$ level. Give a 99% confidence interval for the mean fitness of town boys. Give a 99% prediction interval for a future fitness score for a farm boy.

EXERCISE 4.5.4. Use the data of Exercise 4.5.3 to test whether the fitness scores for farm boys are more or less variable than fitness scores for town boys.

EXERCISE 4.5.5. Jolicoeur and Mosimann (1960) gave data on turtle shell *lengths*. The data for females and males are given in Table 4.7. Explore the need for a transformation. Test whether there is a difference in lengths using $\alpha = .01$. Give a 95% confidence interval for the difference in lengths.

TABLE 4.6. Physical fitness of male high school seniors

Town	12.7	16.9	7.6	2.4	6.2	9.9
Boys	14.2	7.9	11.3	6.4	6.1	10.6
	12.6	16.0	8.3	9.1	15.3	14.8
	2.1	10.6	6.7	6.7	10.6	5.0
	17.7	5.6	3.6	18.6	1.8	2.6
	11.8	5.6	1.0	3.2	5.9	4.0
Farm	14.8	7.3	5.6	6.3	9.0	4.2
Boys	10.6	12.5	12.9	16.1	11.4	2.7

TABLE 4.7. Turtle lengths

Females				Males			
98	138	123	155	121	104	116	93
103	138	133	155	125	106	117	94
103	141	133	158	127	107	117	96
105	147	133	159	128	112	119	101
109	149	134	162	131	113	120	102
123	153	136	177	135	114	120	103

EXERCISE 4.5.6. Koopmans (1987) gave the data in Table 4.8 on verbal ability test scores for 8 year-olds and 10 year-olds. Test whether the two groups have the same mean with $\alpha = .01$ and give a 95% confidence interval for the difference in means. Give a 95% prediction interval for a new 10 year old. Check your assumptions.

TABLE 4.8. Verbal ability test scores

8 yr. olds			10 yr. olds		
324	344	448	428	399	414
366	390	372	366	412	396
322	434	364	386	436	
398	350		404	452	

EXERCISE 4.5.7. Burt (1966) and Weisberg (1985) presented data on IQ scores for identical twins that were raised apart, one by foster parents and one by the genetic parents. Variable y_1 is the IQ score for a twin raised by foster parents, while y_2 is the corresponding IQ score for the twin raised by the genetic parents. The data are given in Table 4.9.

We are interested in the difference between μ_1, the population mean for twins raised by foster parents, and μ_2, the population mean for twins raised by genetic parents. Analyze the data. Check your assumptions.

EXERCISE 4.5.8. Table 4.10 presents data given by Shewhart (1939, p. 118) on various atomic weights as reported in 1931 and again in 1936. Analyze the data. Check your assumptions.

EXERCISE 4.5.9. Reanalyze the data of Example 4.1.1 after deleting the one possible outlier. Does the analysis change much? If so, how?

EXERCISE 4.5.10. Let $y_1 \sim N(89, 4)$ and $y_2 \sim N(90, 16)$. Show that $\Pr[y_1 \geq 87] > \Pr[y_2 \geq 87]$, so that the population with the lower mean has a higher probability of exceeding 87. Recall that

TABLE 4.9. Burt's IQ data

Case	y_1	y_2	Case	y_1	y_2	Case	y_1	y_2
1	82	82	10	93	82	19	97	87
2	80	90	11	95	97	20	87	93
3	88	91	12	88	100	21	94	94
4	108	115	13	111	107	22	96	95
5	116	115	14	63	68	23	112	97
6	117	129	15	77	73	24	113	97
7	132	131	16	86	81	25	106	103
8	71	78	17	83	85	26	107	106
9	75	79	18	93	87	27	98	111

TABLE 4.10. Atomic weights in 1931 and 1936

Compound	1931	1936	Compound	1931	1936
Arsenic	74.93	74.91	Lanthanum	138.90	138.92
Caesium	132.81	132.91	Osmium	190.8	191.5
Columbium	93.3	92.91	Potassium	39.10	39.096
Iodine	126.932	126.92	Radium	225.97	226.05
Krypton	82.9	83.7	Ytterbium	173.5	173.04

$(y_1 - 89)/\sqrt{4} \sim N(0, 1)$ with a similar result for y_2 so that both probabilities can be rewritten in terms of a $N(0, 1)$.

EXERCISE 4.5.11. Mandel (1972) reported stress test data on elongation for a certain type of rubber. Four pieces of rubber sent to one laboratory yielded a sample mean and variance of 56.50 and 5.66, respectively. Four different pieces of rubber sent to another laboratory yielded a sample mean and variance of 52.50 and 6.33, respectively. Are the data two independent samples or a paired comparison? Is the assumption of equal variances reasonable? Give a 99% confidence interval for the difference in population means and give an approximate P value for testing that there is no difference between population means.

EXERCISE 4.5.12. Bethea et al. (1985) reported data on the peel-strengths of adhesives. Some of the data are presented in Table 4.11. Give an approximate P value for testing no difference between adhesives, a 95% confidence interval for the difference between mean peel-strengths, and a 95% prediction interval for a new observation on Adhesive A.

TABLE 4.11. Peel-strengths

Adhesive	Observations					
A	60	63	57	53	56	57
B	52	53	44	48	48	53

EXERCISE 4.5.13. Garner (1956) presented data on the tensile strength of fabrics. Here we consider a subset of the data. The complete data and a more extensive discussion of the experimental procedure are given in Exercise 11.5.2. The experiment involved testing fabric strengths on different machines. Eight homogeneous strips of cloth were divided into samples and each machine was used on a sample from each strip. The data are given in Table 4.12. Are the data two independent samples or a

paired comparison? Give a 98% confidence interval for the difference in population means. Give an approximate P value for testing that there is no difference between population means. What is the result of an $\alpha = .05$ test?

TABLE 4.12. Tensile strength

Strip	1	2	3	4	5	6	7	8
m_1	18	9	7	6	10	7	13	1
m_2	7	11	11	4	8	12	5	11

EXERCISE 4.5.14. Snedecor and Cochran (1967) presented data on the number of acres planted in corn for two sizes of farms. Size was measured in acres. Some of the data are given in Table 4.13. Are the data two independent samples or a paired comparison? Is the assumption of equal variances reasonable? Test for differences between the farms of different sizes. Clearly state your α level. Give a 98% confidence interval for the mean difference between different farms.

TABLE 4.13. Acreage in corn for different farm acreages

Size	Corn acreage				
240	65	80	65	85	30
400	75	35	140	90	110

EXERCISE 4.5.15. Snedecor and Haber (1946) presented data on cutting dates of asparagus. On two plots of land, asparagus was grown every year from 1929 to 1938. On the first plot the asparagus was cut on June 1, while on the second plot the asparagus was cut on June 15. Note that growing conditions will vary considerably from year to year. Also note that the data presented have cutting dates confounded with the plots of land. If one plot of land is intrinsically better for growing asparagus than the other, there will be no way of separating that effect from the effect of cutting dates. Are the data two independent samples or a paired comparison? Give a 95% confidence interval for the difference in population means and give an approximate P value for testing that there is no difference between population means. Give a 95% prediction interval for the difference in a new year. The data are given in Table 4.14.

TABLE 4.14. Cutting dates

Year	29	30	31	32	33	34	35	36	37	38
June 1	201	230	324	512	399	891	449	595	632	527
June 15	301	296	543	778	644	1147	585	807	804	749

EXERCISE 4.5.16. Snedecor (1945b) presented data on a pesticide spray. The treatments were the number of units of active ingredient contained in the spray. Several different sources for breeding mediums were used and each spray was applied on each distinct breeding medium. The data consisted of numbers of dead adults flies found in cages that were set over the breeding medium containers. Some of the data are presented in Table 4.15. Give a 95% confidence interval for the difference in population means. Give an approximate P value for testing that there is no difference between

population means and an $\alpha = .05$ test. Give a 95% prediction interval for a new observation with 8 units. Give a 95% prediction interval for a new observation with 8 units when the corresponding 0 unit value is 300.

TABLE 4.15. Dead adult flies

Medium	A	B	C	D	E	F	G
0 units	423	326	246	141	208	303	256
8 units	414	127	206	78	172	45	103

EXERCISE 4.5.17. Using the data of Example 4.2.1 give a 95% prediction interval for the difference in total points between a new female and a new male. This was not discussed earlier so it requires a deeper understanding of Section 3.5.

Chapter 5

One-way analysis of variance

Analysis of variance (ANOVA) involves comparing random samples from several populations. Often the samples arise from observing experimental units with different treatments applied to them and we refer to the populations as treatment groups. The sample sizes for the treatment groups are possibly different, say, N_i and we assume that the samples are all independent. Moreover, we assume that each population has the same variance and is normally distributed.

5.1 Introduction and examples

EXAMPLE 5.1.1. Table 5.1 gives data from Koopmans (1987, p. 409) on the ages at which suicides were committed in Albuquerque during 1978. Ages are listed by ethnic group. The data are plotted in Figure 5.1. The assumption is that the observations in each group are a random sample from some population. While it is not clear what these populations would be, we proceed to examine the data. Note that there are fewer Native Americans in the study than either Hispanics or non-Hispanic Caucasians; moreover the ages for Native Americans seem to be both lower and less variable than for the other groups. The ages for Hispanics seem to be a bit lower than for non-Hispanic Caucasians.

Summary statistics are given below for the three groups.

Sample statistics: suicide ages

Group	N_i	$\bar{y}_{i\cdot}$	s_i^2	s_i
Caucasians	44	41.66	282.9	16.82
Hispanics	34	35.06	268.3	16.38
Native Am.	15	25.07	74.4	8.51

The sample standard deviation for the Native Americans is about half the size of the others. To evaluate the combined normality of the data, we subtracted the appropriate group mean from each observation, i.e., we computed *residuals*

$$\hat{\varepsilon}_{ij} = y_{ij} - \bar{y}_{i\cdot},$$

TABLE 5.1. Suicide ages

Non-Hispanic Caucasians				Hispanics			Native Americans	
21	31	28	52	50	27	45	26	23
55	31	24	27	31	22	57	17	25
42	32	53	76	29	20	22	24	23
25	43	66	44	21	51	48	22	22
48	57	90	35	27	60	48	16	
22	42	27	32	34	15	14	21	
42	34	48	26	76	19	52	36	
53	39	47	51	35	24	29	18	
21	24	49	19	55	24	21	48	
21	79	53	27	24	18	28	20	
31	46	62	58	68	43	17	35	
				38				

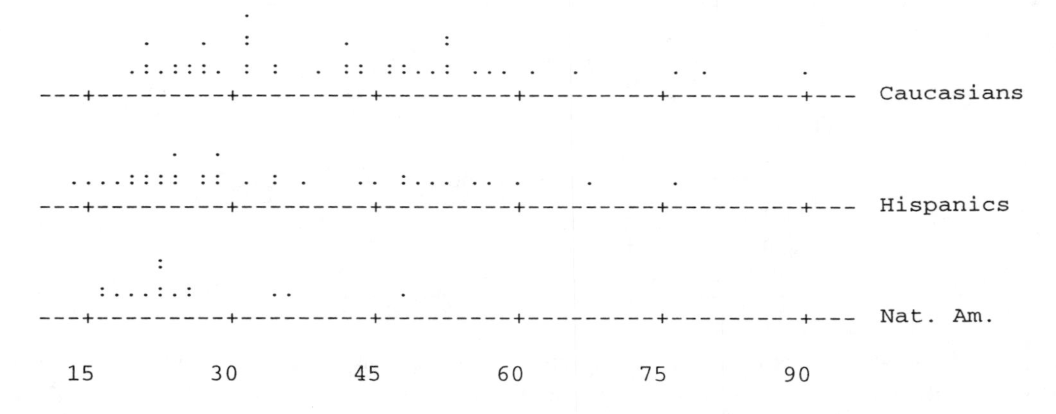

FIGURE 5.1. Dot plots of suicide age data.

where y_{ij} is the jth observation in the ith group and $\bar{y}_{i\cdot}$ is the sample mean from the ith group. We then did a normal plot of the residuals. One normal plot for all of the y_{ij}s would not be appropriate because they have different means, μ_i. The residuals adjust for the different means. Of course with the reasonably large samples available here for each group, it would be permissible to do three separate normal plots, but in other situations with small samples for each group, individual normal plots would not contain enough observations to be of any value. The normal plot for the residuals is given in Figure 5.2. The plot is based on $n = 44 + 34 + 15 = 93$ observations. This is quite a large number, so if the data are normal the plot should be quite straight. In fact, the plot seems reasonably curved.

In order to improve the quality of the assumptions of equal variances and normality, we consider transformations of the data. In particular, consider transforming to $\log(y_{ij})$. Figure 5.3 contains the plot of the transformed data. The variability in the groups seems more nearly the same. This is confirmed by the sample statistics given below.

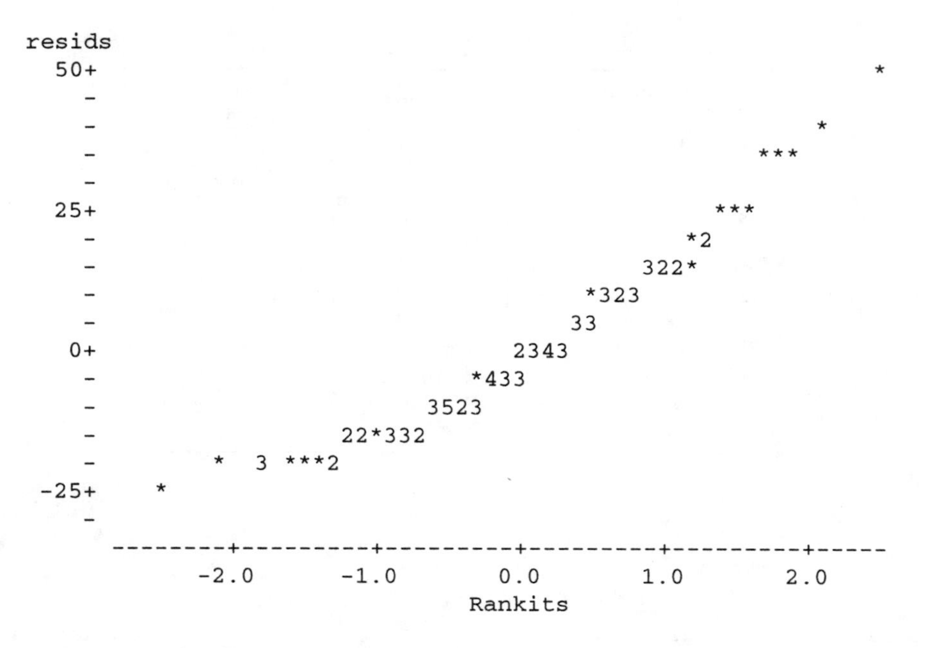

FIGURE 5.2. Normal plot of suicide residuals, $W' = .945$.

Sample statistics: log of suicide ages

Group	N_i	$\bar{y}_{i\bullet}$	s_i^2	s_i
Caucasians	44	3.6521	0.1590	0.3987
Hispanics	34	3.4538	0.2127	0.4612
Native Am.	15	3.1770	0.0879	0.2965

The largest sample standard deviation is only about 1.5 times the smallest. The normal plot of residuals for the transformed data is given in Figure 5.4; it seems considerably straighter than the normal plot for the untransformed data.

All in all, the logs of the original data seem to satisfy the assumptions reasonably well and considerably better than the untransformed data. The square roots of the data were also examined as a possible transformation. While the square roots seem to be an improvement over the original scale, they do not seem to satisfy the assumptions nearly as well as the log transformed data.

A basic assumption in analysis of variance is that the variance is the same for all populations. As we did for two independent samples with the same variance, we can compute a pooled estimate of the variance. Again, this is a weighted average of the variance estimates from the individual groups with weights that are the individual degrees of freedom. In analysis of variance, the pooled estimate of the variance is called the *mean squared error* (*MSE*). For the logs of the suicide age data, the mean squared error is

$$MSE = \frac{(44-1)(.1590) + (34-1)(.2127) + (15-1)(.0879)}{(44-1) + (34-1) + (15-1)} = .168.$$

The degrees of freedom for this estimate are the sum of the degrees of freedom for the individual estimates; the degrees of freedom for error (dfE) are

$$dfE = (44-1) + (34-1) + (15-1) = 44 + 34 + 15 - 3 = 90.$$

The data have an approximate normal distribution, so we can use $t(90)$ as the reference distribution for statistical inference.

```
        .     .    .         :    . :
     .  :.   :..:.  :: ..  . :...:.:. : . .    ..    .
-----+---------+---------+---------+---------+---------+- Caucasians

                   .
.  .    . .. .::   :   :.: .   .. .   . . ::.. ..    .  .
-----+---------+---------+---------+---------+---------+- Hispanics


    . .  .   ..: :...          :         .
-----+---------+---------+---------+---------+---------+- Nat. Am.

   2.80      3.15      3.50      3.85      4.20      4.55
```

FIGURE 5.3. Dotplots of log suicide age data.

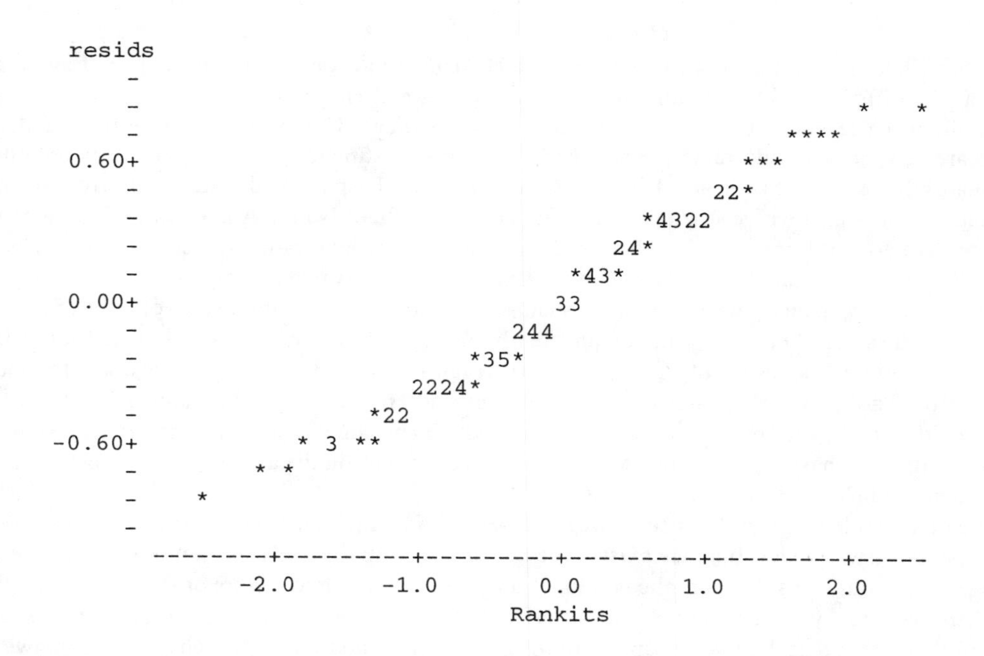

FIGURE 5.4. Normal plot of suicide residuals, log data, $W' = .986$.

We can now perform statistical inferences for a variety of parameters using our standard procedure involving a *Par*, an *Est*, a *SE*(*Est*), and a known distribution symmetric about 0 for $[Est-Par]/SE(Est)$. In this example, perhaps the most useful things to look at are simply whether there is evidence of any age differences in the three groups. Let μ_C, μ_H, and μ_N denote the population means for the log ages of the non-Hispanic Caucasian, Hispanic, and Native American groups respectively. Parameters of interest, with their estimates and the variances of the estimates, are given below.

Par	*Est*	Var(*Est*)
$\mu_C - \mu_H$	$3.6521 - 3.4538$	$\sigma^2\left(\frac{1}{44}+\frac{1}{34}\right)$
$\mu_C - \mu_N$	$3.6521 - 3.1770$	$\sigma^2\left(\frac{1}{44}+\frac{1}{15}\right)$
$\mu_H - \mu_N$	$3.4538 - 3.1770$	$\sigma^2\left(\frac{1}{34}+\frac{1}{15}\right)$

The estimates and variances are obtained exactly as in Section 4.2. The standard errors of the estimates are obtained by substituting MSE for σ^2 in the variance formula and taking the square root. Below are given the estimates, standard errors, the t_{obs} values for testing $H_0 : Par = 0$, the two-sided test P values, and the 99% confidence intervals for *Par*. The confidence intervals require the value $t(.995, 90) = 2.631$. This t table value appears repeatedly in our discussion.

Par	*Est*	*SE*(*Est*)	t_{obs}	*P*	99% CI
$\mu_C - \mu_H$	.1983	.0936	2.12	.037	$(-.04796, .44456)$
$\mu_C - \mu_N$	.4751	.1225	3.88	.000	$(.15280, .79740)$
$\mu_H - \mu_N$	.2768	.1270	2.18	.032	$(-.05734, .61094)$

Note that while the estimated difference between Hispanics and Native Americans is half again as large as the difference between non-Hispanic Caucasians and Hispanics, the t_{obs} values, and thus the significance levels of the differences, are almost identical. This occurs because the standard errors are substantially different. The standard error for the estimate of $\mu_C - \mu_H$ involves only the reasonably large samples for non-Hispanic Caucasians and Hispanics; the standard error for the estimate of $\mu_H - \mu_N$ involves the comparatively small sample of Native Americans, which is why this standard error is larger. On the other hand, the standards errors for the estimates of $\mu_C - \mu_N$ and $\mu_H - \mu_N$ are very similar. The difference in the standard error between having a sample of 34 or 44 is minor by comparison to the effect on the standard error of having a sample size of only 15.

The hypothesis $H_0 : \mu_C - \mu_H = 0$, or equivalently $H_0 : \mu_C = \mu_H$, is the only one rejected at the .01 level. Summarizing the results of the tests at the .01 level, we have no strong evidence of a difference between the ages at which non-Hispanic Caucasians and Hispanics commit suicide, we have no strong evidence of a difference between the ages at which Hispanics and Native Americans commit suicide, but we do have strong evidence that there is a difference in the ages at which non-Hispanic Caucasians and Native Americans commit suicide.

Note that establishing a difference between non-Hispanic Caucasians and Native Americans does little to explain why that difference exists. The reason that Native Americans committed suicide at younger ages could be some complicated function of socio-economic factors or it could be simply that there were many more young Native Americans than old ones in Albuquerque at the time. The test only indicates that the two groups were different, it says nothing about why the groups were different.

The confidence interval for the difference between non-Hispanic Caucasians and Native Americans was constructed on the log scale. Transforming the interval gives $(e^{.1528}, e^{.7974})$ or $(1.2, 2.2)$. We are 99% confident that the average age of suicides is between 1.2 and 2.2 times higher for non-Hispanic Caucasians than for Native Americans. Note that examining differences in log ages transforms to the original scale as a multiplicative factor between groups. The parameters μ_C and μ_N are means for the logs of the suicide ages. When we transform the interval $(.1528, .7974)$ for $\mu_C - \mu_N$ into the interval $(e^{.1528}, e^{.7974})$, we obtain a confidence interval for $e^{\mu_C - \mu_N}$ or equivalently for e^{μ_C}/e^{μ_N}. We can think of e^{μ_C} and e^{μ_N} as 'average' values for the age distributions of the non-Hispanic Caucasians and Native Americans although they are not the expected values of the distributions. Obviously, $e^{\mu_C} = (e^{\mu_C}/e^{\mu_N})\, e^{\mu_N}$, so e^{μ_C}/e^{μ_N} is the number of times greater the average suicide age is for non-Hispanic Caucasians. That is the basis for the interpretation of the interval $(e^{.1528}, e^{.7974})$.

With these data, the tests for differences in means do not depend crucially on the log transformation but interpretations of the confidence intervals do. For the untransformed data, the mean squared error is $MSE_u = 245$ and the observed value of the test statistic for comparing non-Hispanic Caucasians and Native Americans is

$$t_u = 3.54 = \frac{41.66 - 25.07}{\sqrt{245\left(\frac{1}{44} + \frac{1}{15}\right)}},$$

which is not far from the transformed value 3.88. However, the untransformed 99% confidence interval is $(4.3, 28.9)$, indicating a 4 to 29 year higher age for the mean non-Hispanic Caucasian suicide, rather than the transformed interval $(1.2, 2.2)$, indicating that typical non-Hispanic Caucasian suicide ages are 1.2 to 2.2 times greater than those for Native Americans.

The data do not strongly suggest that the means for Hispanics and Native Americans are different, so we *might* wish to compare the mean of the non-Hispanic Caucasians with the average of these groups. Typically, *averaging means will only be of interest if we feel comfortable treating the means as the same*. The parameter of interest is $Par = \mu_C - (\mu_H + \mu_N)/2$ or

$$Par = \mu_C - \frac{1}{2}\mu_H - \frac{1}{2}\mu_N$$

with

$$Est = \bar{y}_C - \frac{1}{2}\bar{y}_H - \frac{1}{2}\bar{y}_N = 3.6521 - \frac{1}{2}3.4538 - \frac{1}{2}3.1770 = .3367.$$

It is not appropriate to use our standard methods to test this *contrast* between the means because the contrast was suggested by the data. Nonetheless, we will illustrate the standard methods. From the independence of the data in the three groups and Proposition 1.2.11, the variance of the estimate is

$$\begin{aligned} &\text{Var}\left(\bar{y}_C - \frac{1}{2}\bar{y}_H - \frac{1}{2}\bar{y}_N\right) \\ &= \text{Var}(\bar{y}_C) + \left(\frac{-1}{2}\right)^2 \text{Var}(\bar{y}_H) + \left(\frac{-1}{2}\right)^2 \text{Var}(\bar{y}_N) \\ &= \frac{\sigma^2}{44} + \left(\frac{-1}{2}\right)^2 \frac{\sigma^2}{34} + \left(\frac{-1}{2}\right)^2 \frac{\sigma^2}{15} \\ &= \sigma^2 \left[\frac{1}{44} + \left(\frac{-1}{2}\right)^2 \frac{1}{34} + \left(\frac{-1}{2}\right)^2 \frac{1}{15}\right]. \end{aligned}$$

Substituting the *MSE* for σ^2 and taking the square root, the standard error is

$$.0886 = \sqrt{.168\left[\frac{1}{44} + \left(\frac{-1}{2}\right)^2 \frac{1}{34} + \left(\frac{-1}{2}\right)^2 \frac{1}{15}\right]}.$$

Note that the standard error happens to be smaller than any of those we have considered when comparing pairs of means. To test the null hypothesis that the mean for non-Hispanic Caucasians equals the average of the other groups, i.e., $H_0 : \mu_C - \frac{1}{2}\mu_H - \frac{1}{2}\mu_N = 0$, the test statistic is $[.3367 - 0]/.0886 = 3.80$, so the null hypothesis is easily rejected. This is an appropriate test statistic for evaluating H_0, but when letting the data suggest the contrast, the $t(90)$ distribution is no longer appropriate for quantifying the level of significance. Similarly, we could construct the 99% confidence interval

$$.3367 \pm 2.631(.0886)$$

but again, the confidence coefficient 99% is not really appropriate for a contrast suggested by the data.

While the parameter $\mu_C - \frac{1}{2}\mu_H - \frac{1}{2}\mu_N$ was suggested by the data, the theory of inference in Chapter 3 assumes that the parameter of interest does not depend on the data. In particular, the reference distributions we have used are invalid when the parameters depend on the data. Moreover, performing numerous inferential procedures complicates the analysis. Our standard tests are set up to check on one particular hypothesis. In the course of analyzing these data we have performed several tests. Thus we have had multiple opportunities to commit errors. In fact, the reason we have been discussing .01 level tests rather than .05 level tests is to help limit the number of errors made when all of the null hypotheses are true. In Chapter 6, we discuss methods of dealing with the problems that arise from making *multiple comparisons* among the means.

To this point, we have considered contrasts (comparisons) among the means. In constructing confidence intervals, prediction intervals, or tests for an individual mean, we continue to use the MSE and the $t(dfE)$ distribution. For example, the endpoints of a 99% confidence interval for μ_H, the mean of the log suicide age for this Hispanic population, are

$$3.4538 \pm 2.631\sqrt{\frac{.168}{34}}$$

for an interval of $(3.269, 3.639)$. Transforming the interval back to the original scale gives $(26.3, 38.1)$, i.e., we are 99% confident that the average age of suicides for this Hispanic population is between 26.3 years old and 38.1 years old. The word 'average' is used because this is not a confidence interval for the expected value of the suicide ages, it is a confidence interval for the exponential transformation of the expected value of the log suicide age. A 99% prediction interval for the age of a future suicide from this Hispanic population has endpoints

$$3.4538 \pm 2.631\sqrt{.168 + \frac{.168}{34}}$$

for an interval of $(2.360, 4.548)$. Transforming the interval back to the original scale gives $(10.6, 94.4)$, i.e., we are 99% confident that a future suicide from this Hispanic population would be between 10.6 years old and 94.4 years old. This interval happens to include all of the observed suicide ages for Hispanics in Table 5.1; that seems reasonable, if not terribly informative. □

5.1.1 THEORY

In analysis of variance, we assume that we have independent observations on, say, a different normal populations with the same variance. In particular, we assume the following data structure.

Sample	Data		Distribution
1	$y_{11}, y_{12}, \ldots, y_{1N_1}$	iid	$N(\mu_1, \sigma^2)$
2	$y_{21}, y_{22}, \ldots, y_{2N_2}$	iid	$N(\mu_2, \sigma^2)$
$\vdots$	$\vdots$	$\vdots$	$\vdots$
a	$y_{a1}, y_{a2}, \ldots, y_{aN_a}$	iid	$N(\mu_a, \sigma^2)$

Here each sample is independent of the other samples. These assumptions can be written more succinctly as the one-way analysis of variance model

$$y_{ij} = \mu_i + \varepsilon_{ij}, \quad \varepsilon_{ij}\text{s independent } N(0, \sigma^2) \tag{5.1.1}$$

$i = 1, \ldots, a$, $j = 1, \ldots, N_i$. The ε_{ij}s are unobservable random errors. We are writing each observation as its mean plus some random error. Alternatively, model (5.1.1) is often written as

$$y_{ij} = \mu + \alpha_i + \varepsilon_{ij}, \quad \varepsilon_{ij}\text{s independent } N(0, \sigma^2) \tag{5.1.2}$$

where $\mu_i = \mu + \alpha_i$. The parameter μ is viewed as a grand mean, while α_i is an effect for the ith treatment group. The μ and α_i parameters are not well defined. In model (5.1.2) they only occur as the sum $\mu + \alpha_i$, so for any choice of μ and α_i the choices, say, $\mu + 5$ and $\alpha_i - 5$ are equally valid. The 5 can be replaced by any number we choose. The parameters μ and α_i are not completely specified by the model. There would seem to be little point in messing around with model (5.1.2) except that it has useful relationships with other models that will be considered later.

To analyze the data, we compute summary statistics from each sample. These are the sample means and sample variances. For the ith group of observations, the sample mean is

$$\bar{y}_{i\cdot} = \frac{1}{N_i}\sum_{j=1}^{N_i} y_{ij}$$

and the sample variance is

$$s_i^2 = \frac{1}{N_i - 1}\sum_{j=1}^{N_i}\left(y_{ij} - \bar{y}_{i\cdot}\right)^2 .$$

With independent normal errors having the same variance, all *of the summary statistics are independent of one another.* Except for checking the validity of our assumptions, these summary statistics are more than sufficient for the entire analysis. Typically, we present the summary statistics in tabular form.

Sample statistics

Group	Size	Mean	Variance
1	N_1	$\bar{y}_{1\cdot}$	s_1^2
2	N_2	$\bar{y}_{2\cdot}$	s_2^2
$\vdots$	$\vdots$	$\vdots$	$\vdots$
a	N_a	$\bar{y}_{a\cdot}$	s_a^2

The sample means, the $\bar{y}_{i\cdot}$s, are estimates of the corresponding μ_is and the s_i^2s all estimate the common population variance σ^2. With unequal sample sizes an efficient pooled estimate of σ^2 must be a weighted average of the s_i^2s. The weights are the degrees of freedom associated with the various estimates. The pooled estimate of σ^2 is called the *mean squared error* (*MSE*),

$$\begin{aligned} MSE \equiv s_p^2 &\equiv \frac{(N_1 - 1)s_1^2 + (N_2 - 1)s_2^2 + \cdots + (N_a - 1)s_a^2}{\sum_{i=1}^{a}(N_i - 1)} \\ &= \frac{1}{(n-a)}\sum_{i=1}^{a}\sum_{j=1}^{N_i}\left(y_{ij} - \bar{y}_{i\cdot}\right)^2 \end{aligned}$$

where $n = \sum_{i=1}^{a} N_i$ is the total sample size. The degrees of freedom for the *MSE* are the *degrees of freedom for error*,

$$dfE \equiv n - a = \sum_{i=1}^{a}(N_i - 1).$$

This is the sum of the degrees of freedom for the individual variance estimates. Note that the *MSE* depends only on the sample variances, so, with independent normal errors having the same variance, *MSE is independent of the* $\bar{y}_{i\cdot}$*s.*

A simple average of the sample variances s_i^2 is not reasonable. If we had $N_1 = 1\,000\,000$ observations in the first sample and only $N_2 = 5$ observations in the second sample, obviously the variance estimate from the first sample is much better than that from the second and we want to give it more weight.

We need to check the validity of our assumptions. The errors in models (1) and (2) are assumed to be independent normals with mean 0 and variance σ^2, so we would like to use them to evaluate the distributional assumptions, e.g., equal variances and normality. Unfortunately, the errors are unobservable, we only see the y_{ij}s and we do not know the μ_is, so we cannot compute the ε_{ij}s. However, since $\varepsilon_{ij} = y_{ij} - \mu_i$ and we can estimate μ_i, we can estimate the errors with the *residuals*,

$$\hat{\varepsilon}_{ij} = y_{ij} - \bar{y}_{i\cdot}.$$

The residuals $y_{ij} - \bar{y}_{i\cdot}$ can be plotted against *predicted values* $\bar{y}_{i\cdot}$ to check whether the variance depends in some way on the means μ_i. They can also be plotted against rankits (normal scores) to check the normality assumption.

Using residuals to evaluate assumptions is a fundamental part of modern statistical data analysis. However, complications can arise. In later chapters we will discuss reasons for using standardized residuals rather than these raw residuals. Standardized residuals will be discussed in connection with regression analysis. In balanced analysis of variance, i.e., situations with equal numbers of observations on each group, the complications disappear. Thus, the unstandardized residuals are adequate for evaluating the assumptions in a balanced analysis of variance. In other analysis of variance situations, the problems are *relatively* minor.

If we are satisfied with the assumptions, we proceed to examine the parameters of interest. The basic parameters of interest in analysis of variance are the μ_is, which have natural estimates, the $\bar{y}_{i\cdot}$s. We also have an estimate of σ^2, so we are in a position to draw a variety of statistical inferences. The main problem in obtaining tests and confidence intervals is in finding appropriate standard errors. To do this we need to observe that each of the a samples are independent. The $\bar{y}_{i\cdot}$s are computed from different samples, so they are independent of each other. Moreover, $\bar{y}_{i\cdot}$ is the sample mean of N_i observations, so

$$\bar{y}_{i\cdot} \sim N\left(\mu_i, \frac{\sigma^2}{N_i}\right).$$

For inferences about a single mean, say, μ_2, use the general procedures with $Par = \mu_2$ and $Est = \bar{y}_{2\cdot}$. The variance of $\bar{y}_{2\cdot}$ is σ^2/N_2, so $SE(\bar{y}_{2\cdot}) = \sqrt{MSE/N_2}$. The reference distribution is $[\bar{y}_{2\cdot} - \mu_2]/SE(\bar{y}_{2\cdot}) \sim t(dfE)$. Note that the degrees of freedom for the t distribution are precisely the degrees of freedom for the *MSE*. The general procedures also provide prediction intervals using the *MSE* and $t(dfE)$ distribution.

For inferences about the difference between two means, say, $\mu_2 - \mu_1$, use the general procedures with $Par = \mu_2 - \mu_1$ and $Est = \bar{y}_{2\cdot} - \bar{y}_{1\cdot}$. The two means are independent, so the variance of $\bar{y}_{2\cdot} - \bar{y}_{1\cdot}$ is the variance of $\bar{y}_{2\cdot}$ plus the variance of $\bar{y}_{1\cdot}$, i.e., $\sigma^2/N_2 + \sigma^2/N_1$. The standard error of $\bar{y}_{2\cdot} - \bar{y}_{1\cdot}$ is

$$SE(\bar{y}_{2\cdot} - \bar{y}_{1\cdot}) = \sqrt{\frac{MSE}{N_2} + \frac{MSE}{N_1}} = \sqrt{MSE\left[\frac{1}{N_1} + \frac{1}{N_2}\right]}.$$

The reference distribution is

$$\frac{(\bar{y}_{2\cdot} - \bar{y}_{1\cdot}) - (\mu_2 - \mu_1)}{\sqrt{MSE\left[\frac{1}{N_1} + \frac{1}{N_2}\right]}} \sim t(dfE).$$

We might wish to compare one mean, μ_1, with the average of two other means, $(\mu_2 + \mu_3)/2$. In this case, the parameter can be taken as $Par = \mu_1 - (\mu_2 + \mu_3)/2 = \mu_1 - \frac{1}{2}\mu_2 - \frac{1}{2}\mu_3$. The estimate is $Est = \bar{y}_{1\cdot} - \frac{1}{2}\bar{y}_{2\cdot} - \frac{1}{2}\bar{y}_{3\cdot}$. By the independence of the sample means, the variance of the estimate is

$$\text{Var}\left(\bar{y}_{1\cdot} - \frac{1}{2}\bar{y}_{2\cdot} - \frac{1}{2}\bar{y}_{3\cdot}\right) = \text{Var}(\bar{y}_{1\cdot}) + \text{Var}\left(\frac{-1}{2}\bar{y}_{2\cdot}\right) + \text{Var}\left(\frac{-1}{2}\bar{y}_{3\cdot}\right)$$

$$= \frac{\sigma^2}{N_1} + \left(\frac{-1}{2}\right)^2 \frac{\sigma^2}{N_2} + \left(\frac{-1}{2}\right)^2 \frac{\sigma^2}{N_3}$$
$$= \sigma^2 \left[\frac{1}{N_1} + \frac{1}{4}\frac{1}{N_2} + \frac{1}{4}\frac{1}{N_3}\right].$$

The standard error is

$$SE\left(\bar{y}_{1\cdot} - \frac{1}{2}\bar{y}_{2\cdot} - \frac{1}{2}\bar{y}_{3\cdot}\right) = \sqrt{MSE\left[\frac{1}{N_1} + \frac{1}{4N_2} + \frac{1}{4N_3}\right]}.$$

The reference distribution is

$$\frac{\left(\bar{y}_{1\cdot} - \frac{1}{2}\bar{y}_{2\cdot} - \frac{1}{2}\bar{y}_{3\cdot}\right) - \left(\mu_1 - \frac{1}{2}\mu_2 - \frac{1}{2}\mu_3\right)}{\sqrt{MSE\left[\frac{1}{N_1} + \frac{1}{4N_2} + \frac{1}{4N_3}\right]}} \sim t(dfE).$$

Typically, in analysis of variance we are concerned with parameters that are *contrasts* (comparisons) among the μ_is. For *known* coefficients $\lambda_1, \ldots, \lambda_a$ with $\sum_{i=1}^a \lambda_i = 0$, a contrast is defined by $\sum_{i=1}^a \lambda_i \mu_i$. For example, $\mu_2 - \mu_1$ has $\lambda_1 = -1$, $\lambda_2 = 1$, and all other λ_is equal to 0. The contrast $\mu_1 - \frac{1}{2}\mu_2 - \frac{1}{2}\mu_3$ has $\lambda_1 = 1$, $\lambda_2 = -1/2$, $\lambda_3 = -1/2$, and all other λ_is equal to 0. The natural estimate of $\sum_{i=1}^a \lambda_i \mu_i$ substitutes the sample means for the population means, i.e., the natural estimate is $\sum_{i=1}^a \lambda_i \bar{y}_{i\cdot}$. In fact, Proposition 1.2.11 gives

$$E\left(\sum_{i=1}^a \lambda_i \bar{y}_{i\cdot}\right) = \sum_{i=1}^a \lambda_i E(\bar{y}_{i\cdot}) = \sum_{i=1}^a \lambda_i \mu_i,$$

so by definition this is an *unbiased* estimate of the contrast. Using the independence of the sample means and Proposition 1.2.11,

$$\begin{aligned} \mathrm{Var}\left(\sum_{i=1}^a \lambda_i \bar{y}_{i\cdot}\right) &= \sum_{i=1}^a \lambda_i^2 \,\mathrm{Var}(\bar{y}_{i\cdot}) \\ &= \sum_{i=1}^a \lambda_i^2 \frac{\sigma^2}{N_i} \\ &= \sigma^2 \sum_{i=1}^a \frac{\lambda_i^2}{N_i}. \end{aligned}$$

The standard error is

$$SE\left(\sum_{i=1}^a \lambda_i \bar{y}_{i\cdot}\right) = \sqrt{MSE \sum_{i=1}^a \frac{\lambda_i^2}{N_i}}$$

and the reference distribution is

$$\frac{\left(\sum_{i=1}^a \lambda_i \bar{y}_{i\cdot}\right) - \left(\sum_{i=1}^a \lambda_i \mu_i\right)}{\sqrt{MSE \sum_{i=1}^a \lambda_i^2 / N_i}} \sim t(dfE),$$

see Exercise 5.7.14. If the independence and equal variance assumptions hold, then the central limit theorem and law of large numbers can be used to justify a $N(0, 1)$ reference distribution even when the data are not normal. Moreover, in one-way ANOVA all of these results hold even when

$\sum_i \lambda_i \neq 0$, so they hold for linear combinations of the μ_is that are not contrasts. Nonetheless, our primary interest is in contrasts.

Having identified a parameter, an estimate, a standard error, and an appropriate reference distribution, inferences follow the usual pattern. A 95% confidence interval for $\sum_{i=1}^{a} \lambda_i \mu_i$ has endpoints

$$\sum_{i=1}^{a} \lambda_i \bar{y}_{i\cdot} \pm t(.975, dfE) \sqrt{MSE \sum_{i=1}^{a} \lambda_i^2 / N_i}.$$

An $\alpha = .05$ test of $H_0 : \sum_{i=1}^{a} \lambda_i \mu_i = 0$ versus $H_A : \sum_{i=1}^{a} \lambda_i \mu_i \neq 0$ rejects H_0 if

$$\frac{|\sum_{i=1}^{a} \lambda_i \bar{y}_{i\cdot} - 0|}{\sqrt{MSE \sum_{i=1}^{a} \lambda_i^2 / N_i}} > t(.975, dfE) \tag{5.1.3}$$

An equivalent procedure to the test in (5.1.3) is often useful. If we square both sides of (5.1.3), the test rejects if

$$\left(\frac{|\sum_{i=1}^{a} \lambda_i \bar{y}_{i\cdot} - 0|}{\sqrt{MSE \sum_{i=1}^{a} \lambda_i^2 / N_i}} \right)^2 > (t(.975, dfE))^2 .$$

The square of the test statistic leads to another statistic that will be useful later, the sum of squares for the contrast. Rewrite the test statistic as

$$\left(\frac{|\sum_{i=1}^{a} \lambda_i \bar{y}_{i\cdot} - 0|}{\sqrt{MSE \sum_{i=1}^{a} \lambda_i^2 / N_i}} \right)^2 = \frac{\left(\sum_{i=1}^{a} \lambda_i \bar{y}_{i\cdot} - 0\right)^2}{MSE \sum_{i=1}^{a} \lambda_i^2 / N_i}$$

$$= \frac{\left(\sum_{i=1}^{a} \lambda_i \bar{y}_{i\cdot}\right)^2 / \sum_{i=1}^{a} \lambda_i^2 / N_i}{MSE}$$

and define the *sum of squares for the contrast* as

$$SS\left(\sum_{i=1}^{a} \lambda_i \mu_i\right) \equiv \frac{\left(\sum_{i=1}^{a} \lambda_i \bar{y}_{i\cdot}\right)^2}{\sum_{i=1}^{a} \lambda_i^2 / N_i}. \tag{5.1.4}$$

The $\alpha = .05$ t test of $H_0 : \sum_{i=1}^{a} \lambda_i \mu_i = 0$ versus $H_A : \sum_{i=1}^{a} \lambda_i \mu_i \neq 0$ is equivalent to rejecting H_0 if

$$\frac{SS\left(\sum_{i=1}^{a} \lambda_i \mu_i\right)}{MSE} > [t(.975, dfE)]^2 .$$

It is a mathematical fact that for any α between 0 and 1 and any dfE,

$$\left[t\left(1 - \frac{\alpha}{2}, dfE\right)\right]^2 = F(1 - \alpha, 1, dfE) .$$

Thus the test based on the sum of squares for the contrast is an F test with 1 degree of freedom in the numerator. *Any contrast has 1 degree of freedom associated with it.*

A notational matter needs to be mentioned. Contrasts, by definition, have $\sum_{i=1}^{a} \lambda_i = 0$. If we use model (5.1.2) rather than model (5.1.1) we get

$$\sum_{i=1}^{a} \lambda_i \mu_i = \sum_{i=1}^{a} \lambda_i (\mu + \alpha_i) = \mu \sum_{i=1}^{a} \lambda_i + \sum_{i=1}^{a} \lambda_i \alpha_i = \sum_{i=1}^{a} \lambda_i \alpha_i.$$

Thus contrasts in model (5.1.2) involve only the treatment effects. This is of some importance later when dealing with more complicated models.

In our first example we transformed the suicide age data so that they better satisfy the assumptions of equal variances and normal distributions. In fact, analysis of variance tests and confidence intervals are frequently useful even when these assumptions are violated. Scheffé (1959, p. 345) concludes that (a) nonnormality is not a serious problem for inferences about means but it is a serious problem for inferences about variances, (b) unequal variances are not a serious problem for inferences about means from samples of the same size but are a serious problem for inferences about means from samples of unequal sizes, and (c) lack of independence can be a serious problem. Of course any such rules depend on just how bad the nonnormality is, how unequal the variances are, and how bad the lack of independence is. My own interpretation of these rules is that if you check the assumptions and they do not look too bad, you can probably proceed with a fair amount of assurance.

5.1.2 BALANCED ANOVA: INTRODUCTORY EXAMPLE

We now consider an example of a balanced one-way ANOVA. A balanced one-way ANOVA has equal numbers of observations in each group, say, $N = N_1 = \cdots = N_a$.

EXAMPLE 5.1.2. Ott (1949) presented data on an electrical characteristic associated with ceramic components for a phonograph. Ott and Schilling (1990) and Ryan (1989) have also considered these data. Ceramic pieces were cut from strips, each of which could provide 25 pieces. It was decided to take 7 pieces from each strip, manufacture the 7 ceramic phonograph components, and measure the electrical characteristic on each. The data from 4 strips are given below. (These are actually the third through sixth of the strips reported by Ott.)

Strip	Observations						
1	17.3	15.8	16.8	17.2	16.2	16.9	14.9
2	16.9	15.8	16.9	16.8	16.6	16.0	16.6
3	15.5	16.6	15.9	16.5	16.1	16.2	15.7
4	13.5	14.5	16.0	15.9	13.7	15.2	15.9

In the current analysis, we act as if the four strips are of intrinsic interest and investigate whether there are differences among them. In Subsection 13.4.2 we will consider an analysis in which we assume that the strips are themselves a random sample from some wider population. The data are displayed in Figure 5.5 and summary statistics follow.

Sample statistics: electrical characteristics

Strip	N	$\bar{y}_{i\cdot}$	s_i^2	s_i
1	7	16.4429	0.749524	0.866
2	7	16.5143	0.194762	0.441
3	7	16.0714	0.162381	0.403
4	7	14.9571	1.139524	1.067

The electrical characteristic appears to be lowest for strip 4 and highest for strips 1 and 2, but we need to use formal inferential procedures to establish whether the differences could be reasonably ascribed to random variation. The sample standard deviations, and thus the sample variances, are comparable. The ratio of the largest to the smallest standard deviation is just over 2.5, which is not small but which is also not large enough to cause major concern. As in Section 4.4, we could do F

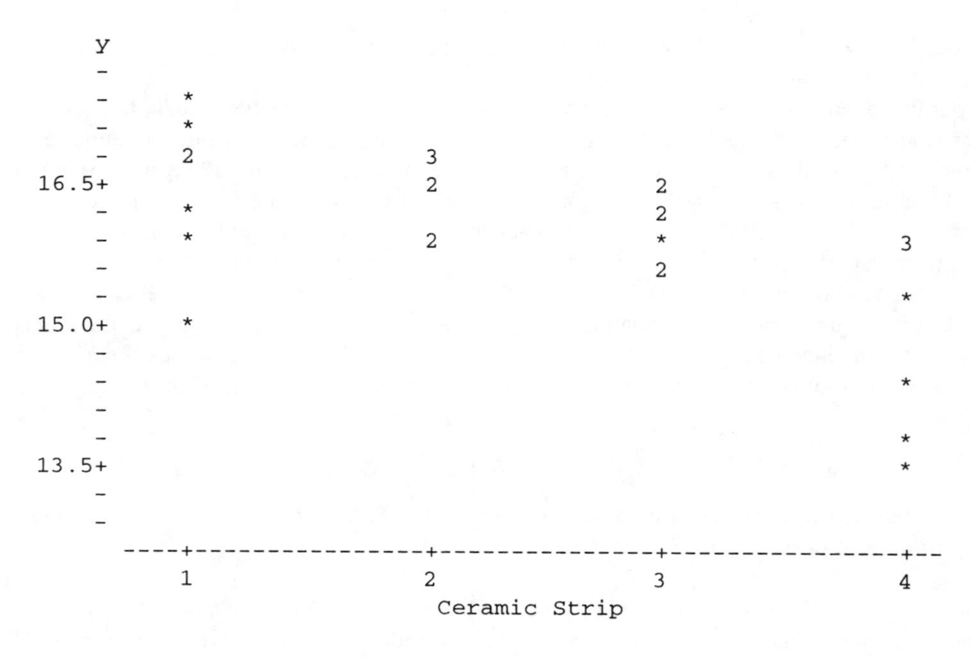

FIGURE 5.5. Plot of electrical characteristics data.

tests to determine whether any pairs of variances differ. The largest of these F tests is not significant at the .02 level and, after considering that there are six pairs to test, we conclude that there is no cause for major concern. Figure 5.5 is poorly suited to evaluate the variances visually because in Figure 5.5 the plot involves any differences in means as well as differences in variance. A better plot from which to evaluate the variances is given as Figure 5.6. Figure 5.6 is a plot of the residuals $\hat{\varepsilon}_{ij} \equiv y_{ij} - \bar{y}_{i\cdot}$ against the appropriate group. The residuals have been adjusted for their different means, so residuals, and thus residual plots, are centered at 0. Figure 5.6 is not wonderful in that we see differences in variability for the four groups, but it is also not outlandishly inconsistent with the assumption of equal variances. (Note that if one group had many more observations than another, the spread for that group would be greater even if the population variances were the same.) Figure 5.7 contains a normal plot of the residuals. The plot looks fairly reasonable, although it tails off at the top. The W' statistic of .956 gives a P value for the hypothesis of normality that is larger than .05 and in any case, analysis of variance procedures are not particularly sensitive to nonnormality.

With equal sample sizes in each group, the MSE reduces to the simple average of the sample variances.

$$\begin{aligned} MSE &= \frac{(7-1).74952 + (7-1).19476 + (7-1).16238 + (7-1)1.13952}{7+7+7+7-4} \\ &= \frac{.74952 + .19476 + .16238 + 1.13952}{4} \\ &= .56155 \end{aligned}$$

and has error degrees of freedom $dfE = 7+7+7+7-4 = 24$. Again, we compare all pairs of means. The value $t(.995, 24) = 2.797$ is required for constructing 99% confidence intervals. These intervals and two-sided tests of $H_0 : Par = 0$ are given below.

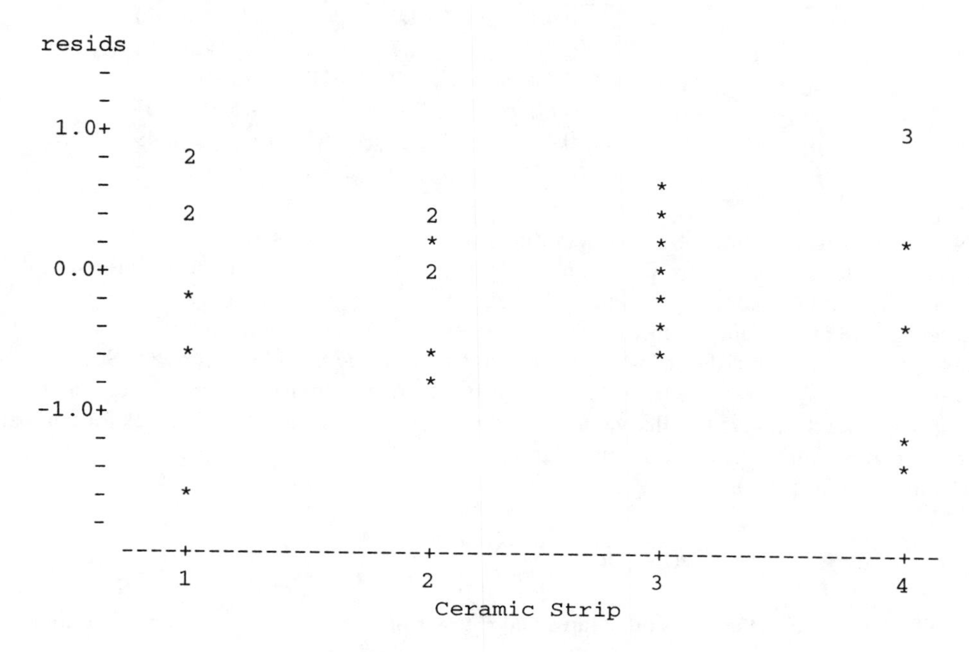

FIGURE 5.6. Residual plot.

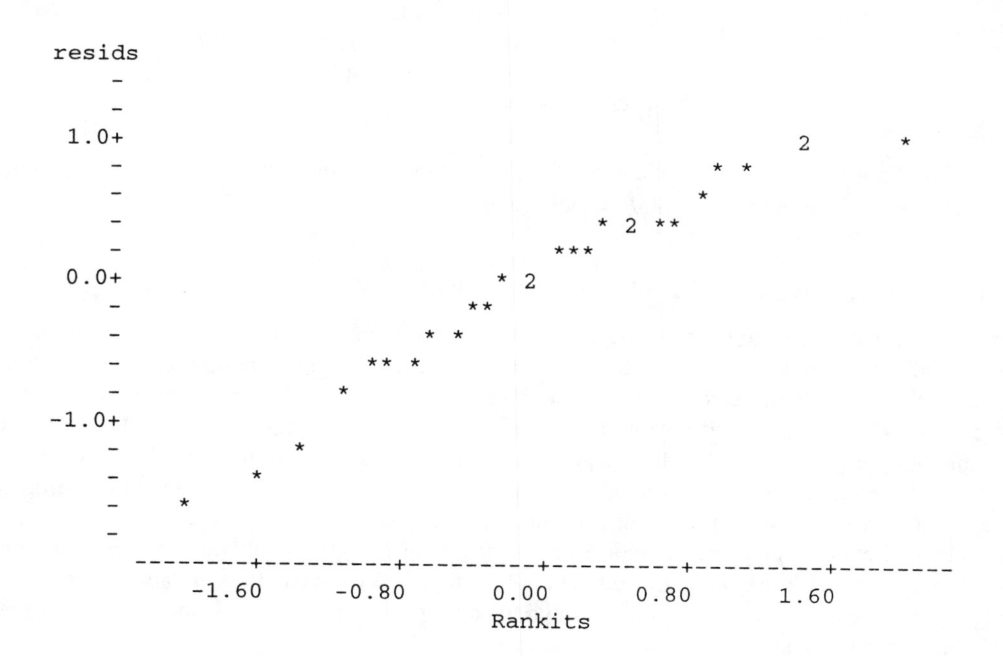

FIGURE 5.7. Normal plot of residuals, $W' = 0.956$.

Par	*Est*	*SE(Est)*	t_{obs}	*P*	99% CI
$\mu_1 - \mu_4$	1.4858	0.4006	3.709	.001	(0.37, 2.61)
$\mu_2 - \mu_4$	1.5572	0.4006	3.887	.001	(0.44, 2.68)
$\mu_3 - \mu_4$	1.1143	0.4006	2.782	.010	(−0.01, 2.23)
$\mu_1 - \mu_2$	−0.0714	0.4006	−0.178	.860	(−1.19, 1.05)
$\mu_1 - \mu_3$	0.3715	0.4006	0.927	.363	(−0.75, 1.49)
$\mu_2 - \mu_3$	0.4429	0.4006	1.106	.280	(−0.68, 1.56)

Note that with equal numbers of observations on each group, the standard errors are the same for each comparison of two means. Based on $\alpha = .01$ tests, the electrical characteristic for strip 4 differs significantly from those for strips 1 and 2, the decision for strip 3 is essentially a toss-up, and no other differences are significant. Even for strips 1 and 2, the 99% confidence intervals indicate that the data are consistent with differences from strip 4 as small as .37 and .44 respectively. Such differences may or may not be of practical importance. Clearly, the main source of differences among these data is that strip 4 tends to give smaller values than the other strips. In fact, the *P* values for comparisons among the other three strips are all quite large.

Using formula (5.1.4), the sum of squares for $\mu_1 - \mu_4$ is

$$SS(\mu_1 - \mu_4) = \frac{(1.4858)^2}{(1)^2/7 + (-1)^2/7} = 7.7266\,.$$

The table below gives the sums of squares and *F* tests for equality between all pairs of means.

Par	*SS*	F_{obs}	*P*
$\mu_1 - \mu_4$	7.727	13.76	.001
$\mu_2 - \mu_4$	8.487	15.11	.001
$\mu_3 - \mu_4$	4.346	7.74	.010
$\mu_1 - \mu_2$	0.018	0.03	.860
$\mu_1 - \mu_3$	0.483	0.86	.363
$\mu_2 - \mu_3$	0.687	1.22	.280

Note that the *F* statistics are just the sums of squares divided by the *MSE*. They equal the squares of the *t* statistics given earlier and the *P* values are identical. □

5.1.3 ANALYTIC AND ENUMERATIVE STUDIES

In one-sample, two-sample, and one-way ANOVA problems, we assume that we have random samples from various populations. In the more sophisticated models treated later, we continue to assume that at least the errors are a random sample from a $N(0, \sigma^2)$ population. The statistical inferences we draw are valid for the populations that were sampled. Often it is not clear what the sampled populations are. What are the populations from which the Albuquerque suicide ages were sampled? Presumably, our data were all of the suicides reported in 1978 for these ethnic groups. The electrical characteristic data has four ceramic strips divided into 25 pieces, of which seven pieces are taken. Are the seven pieces a random sample from the 25? They could be. Is the collection of 25 pieces the population that we really care about? Doubtful! What we really care about is whether the differences in ceramic strips are large enough to cause problems in the production of phonographs. (Not that anyone makes phonographs anymore.)

When we analyze data, we assume that the measurements are subject to errors and that the errors are consistent with our models. However, the populations from which these samples are taken may be nothing more than mental constructs. In such cases, it requires extrastatistical reasoning to justify applying the statistical conclusions to whatever issues we really wish to address. Moreover, the desire to predict the future underlies virtually all studies and, unfortunately, one can never be sure that data collected now will apply to the conditions of the future. So what can you do? Only your best. You can try to make your data as relevant as possible to your anticipation of future conditions. You can try to collect data for which the assumptions will be reasonably true. You can try to validate your assumptions. Studies in which it is not clear that the data are random samples from the population of immediate interest are often called *analytic studies*.

About the only time one can be really sure that statistical conclusions apply directly to the population of interest is when one has control of the population of interest. If we have a list of all the elements in the population, we can choose a random sample from the population. Of course, choosing a random sample is still very different from obtaining a random sample of observations. Without control or total cooperation, we may not be able to take measurements on the sample. (Even when you can find people that you want for a sample, many will not submit to a measurement process.) Studies in which one can arrange to have the assumptions met are often called *enumerative studies*. See Hahn and Meeker (1993) and Deming (1986) for additional discussion of these issues.

5.2 Balanced one-way analysis of variance: theory

We now examine in detail the important special case of one-way analysis of variance in which the numbers of observations for each sample are the same, cf. Subsection 5.1.2. In this case, the analysis of variance is referred to as *balanced*. Balanced one-way ANOVA is important because it is both understandable and extendable. The logic behind analysis of variance is much clearer when dealing with balanced samples and the standard methods for multifactor analysis of variance are extensions of the techniques developed for balanced one-way ANOVA. The standard methods for multifactor ANOVA also assume equal numbers of observations on all treatments.

For balanced analysis of variance, let $N \equiv N_1 = \cdots = N_a$ be the number of observations in each sample. In particular, we assume the data structure

Sample	Data		Distribution
1	$y_{11}, y_{12}, \ldots, y_{1N}$	iid	$N(\mu_1, \sigma^2)$
2	$y_{21}, y_{22}, \ldots, y_{2N}$	iid	$N(\mu_2, \sigma^2)$
$\vdots$	$\vdots$	$\vdots$	$\vdots$
a	$y_{a1}, y_{a2}, \ldots, y_{aN}$	iid	$N(\mu_a, \sigma^2)$

with all samples independent. The data structure can be rewritten as the balanced one-way ANOVA model

$$y_{ij} = \mu_i + \varepsilon_{ij}, \quad \varepsilon_{ij}\text{s independent } N(0, \sigma^2)$$

$i = 1, \ldots, a, j = 1, \ldots, N$. Again, we have assumed the same variance σ^2 for each sample.

In this section, we focus on testing the (null) hypothesis

$$H_0 : \mu_1 = \mu_2 = \cdots = \mu_a.$$

This is a test of whether there are *any* differences among the groups. If we use model (5.1.2), the null hypothesis can be written as $H_0 : \alpha_1 = \alpha_2 = \cdots = \alpha_a$. To perform the test, first compute summary statistics from the samples.

Sample statistics

Group	Size	Mean	Variance
1	N	$\bar{y}_{1\bullet}$	s_1^2
2	N	$\bar{y}_{2\bullet}$	s_2^2
$\vdots$	$\vdots$	$\vdots$	$\vdots$
a	N	$\bar{y}_{a\bullet}$	s_a^2

As before, the sample means, the $\bar{y}_{i\bullet}$s, are estimates of the μ_is and the s_i^2s all estimate σ^2.

The test of H_0 is based on estimating σ^2. We construct two estimates of the variance. The first estimate is always valid, assuming of course that our initial assumptions were correct. The second estimate is valid *only* when $\mu_1 = \mu_2 = \cdots = \mu_a$. If the μ_is are all equal, we have two estimates of σ^2, so they should be about the same. If the μ_is are not all equal, the second estimate tends to be bigger than σ^2, so it should be larger than the first estimate. We conclude that the data are consistent with $\mu_1 = \mu_2 = \cdots = \mu_a$ when the two estimates seem to be about the same and conclude that the μ_is are not all equal when the second estimate is substantially larger than the first. As usual, when the estimates are about the same we conclude that the data are consistent with the μ_is all being equal; *we do not conclude that the μ_is are really all equal.* If the μ_is are not quite equal but are very nearly so, we cannot expect to be able to detect the differences. On the other hand, two widely different variance estimates give substantial proof that the μ_is are not all the same.

The easy part of the process is creating the first estimate of the variance, the one that is always valid. From each sample, regardless of the value of μ_i, we have an estimate of σ^2, namely s_i^2. Obviously, the average of the s_i^2s must also be an estimate of σ^2. The average is the pooled estimate of the variance, i.e., the mean squared error is

$$\begin{aligned} MSE &\equiv \frac{s_1^2 + s_2^2 + \cdots + s_a^2}{a} \\ &= \frac{1}{a(N-1)} \sum_{i=1}^{a} \sum_{j=1}^{N} \left(y_{ij} - \bar{y}_{i\bullet}\right)^2 . \end{aligned}$$

As discussed earlier, a simple average such as this is not always appropriate. The simple average is only reasonable because we have the same number of observations in each sample.

Recall that each s_i^2 has $N-1$ degrees of freedom. Each s_i^2 is based on N observations but is functionally based on $N-1$ observations because of the need to estimate μ_i before estimating the variance. By pooling together the variance estimates, we also get to pool the degrees of freedom. We have combined a independent estimates of σ^2, each with $N-1$ degrees of freedom, so the pooled estimate has $a(N-1)$ degrees of freedom. In other words, the *MSE* is functionally based on $a(N-1)$ observations. The degrees of freedom associated with the *MSE* are the degrees of freedom for error (dfE), so we have

$$dfE = a(N-1).$$

The data, the y_{ij}s, are random, so the *MSE*, which is computed from them, must also be random. If we collected another set of similar data we would not expect to get exactly the same value for the *MSE*. If we are to evaluate whether this estimate of σ^2 is similar to another estimate, we need to have some idea of the variability in the *MSE*. Under the assumptions we have made, the distribution of the *MSE* depends only on dfE and σ^2. The distribution is related to the χ^2 family of distributions. In particular,

$$\frac{dfE \times MSE}{\sigma^2} \sim \chi^2(dfE)$$

where, on the right hand side, dfE indicates the particular member of the χ^2 family that is appropriate. A commonly used terminology in analysis of variance is the sum of squares for error (*SSE*). This is defined to be

$$SSE \equiv dfE \times MSE = \sum_{i=1}^{a} \sum_{j=1}^{N} \left(y_{ij} - \bar{y}_{i\bullet}\right)^2. \tag{5.2.1}$$

Note that $SSE/\sigma^2 \sim \chi^2(dfE)$. Note also that the *SSE is the sum of the squared residuals*, the residuals being

$$\hat{\varepsilon}_{ij} = y_{ij} - \bar{y}_{i\bullet}\,.$$

The second estimate of σ^2 is to be valid only when $\mu_1 = \mu_2 = \cdots = \mu_a$. We have already used the sample variances s_i^2 in constructing the *MSE*, so we use the rest of our summary statistics, the $\bar{y}_{i\bullet}$s, in constructing the second estimate of σ^2. In fact, the $\bar{y}_{i\bullet}$s are estimates of the μ_is, so it is only reasonable to use the $\bar{y}_{i\bullet}$s when trying to draw conclusions about the μ_is. Consider the distributions of the $\bar{y}_{i\bullet}$s. Each is the sample mean of N observations, so each has the distribution of a sample mean. In particular,

$$\begin{aligned}
\bar{y}_{1\bullet} &\sim N\left(\mu_1, \frac{\sigma^2}{N}\right) \\
\bar{y}_{2\bullet} &\sim N\left(\mu_2, \frac{\sigma^2}{N}\right) \\
&\vdots \\
\bar{y}_{a\bullet} &\sim N\left(\mu_a, \frac{\sigma^2}{N}\right)
\end{aligned}$$

The a different samples are independent of each other, so $\bar{y}_{1\bullet}, \bar{y}_{2\bullet}, \ldots, \bar{y}_{a\bullet}$ are all independent. They all have the same variance, σ^2/N, and they all have normal distributions. In fact, the only thing keeping them from having independent and identical distributions is that they have different means μ_i. If we assume that $\mu_1 = \mu_2 = \cdots = \mu_a$, they have independent and identical distributions and thus form a random sample from a population. Balanced analysis of variance is based on the fact that if the μ_is are the same, the $\bar{y}_{i\bullet}$s can be treated as a random sample. If the $\bar{y}_{i\bullet}$s are a random sample, we can compute their sample variance to get an estimate of the variance of the $\bar{y}_{i\bullet}$s. The variance of the $\bar{y}_{i\bullet}$s is σ^2/N and the sample variance of the $\bar{y}_{i\bullet}$s is

$$s_{\bar{y}}^2 = \frac{1}{a-1} \sum_{i=1}^{a} \left(\bar{y}_{i\bullet} - \bar{y}_{\bullet\bullet}\right)^2$$

where

$$\bar{y}_{\bullet\bullet} \equiv \frac{1}{a} \sum_{i=1}^{a} \bar{y}_{i\bullet}$$

is the sample mean of the $\bar{y}_{i\bullet}$s. We have $s_{\bar{y}}^2$ as an estimate of σ^2/N but we set out to find an estimate of σ^2. The obvious choice is

$$MSTrts \equiv N s_{\bar{y}}^2 = \frac{N}{a-1} \sum_{i=1}^{a} \left(\bar{y}_{i\bullet} - \bar{y}_{\bullet\bullet}\right)^2$$

where *MSTrts* abbreviates the commonly used term *mean squared treatments*. The estimate $s_{\bar{y}}^2$ is based on a sample of size a, so it, and thus *MSTrts*, has $a - 1$ degrees of freedom. These are referred to as the degrees of freedom for treatments (*dfTrts*). The sum of squares for treatments is defined as

$$SSTrts \equiv dfTrts \times MSTrts = N \sum_{i=1}^{a} \left(\bar{y}_{i\bullet} - \bar{y}_{\bullet\bullet}\right)^2. \tag{5.2.2}$$

Just as the *MSE* is random, the *MSTrts* is also random. The estimate $s_{\bar{y}}^2$ is the sample variance of a random sample of size a from a normal population with variance σ^2 / N, so

$$\frac{(a-1)s_{\bar{y}}^2}{\sigma^2 / N} = \frac{(a-1)MSTrts}{\sigma^2} \sim \chi^2(a-1).$$

The discussion above is based on the assumption that $\mu_1 = \mu_2 = \cdots = \mu_a$. If this is not true, the $\bar{y}_{i\cdot}$s do not form a random sample and $s_{\bar{y}}^2$ does not estimate σ^2 / N. Actually, it estimates σ^2 / N plus the 'variance' of the μ_is. Algebraically, $s_{\bar{y}}^2$ estimates

$$E(s_{\bar{y}}^2) = \frac{\sigma^2}{N} + \frac{1}{a-1}\sum_{i=1}^{a}(\mu_i - \bar{\mu}_\cdot)^2$$

where $\bar{\mu}_\cdot \equiv \sum_{i=1}^{a} \mu_i / a$ is the mean of the μ_is. Multiplying by N gives

$$E(MSTrts) = E(Ns_{\bar{y}}^2) = \sigma^2 + \frac{N}{a-1}\sum_{i=1}^{a}(\mu_i - \bar{\mu}_\cdot)^2, \tag{5.2.3}$$

so *MSTrts* is an estimate of σ^2 plus something that is always nonnegative. If $\mu_1 = \mu_2 = \cdots = \mu_a$, the μ_is are all equal to their average $\bar{\mu}_\cdot$, thus $(\mu_i - \bar{\mu}_\cdot)^2 = 0$ for all i, and

$$\frac{N}{a-1}\sum_{i=1}^{a}(\mu_i - \bar{\mu}_\cdot)^2 = 0.$$

As advertised earlier, if $\mu_1 = \cdots = \mu_a$, *MSTrts* is an estimate of σ^2. If the μ_is are not all the same, $[N/(a-1)]\sum_{i=1}^{a}(\mu_i - \bar{\mu}_\cdot)^2$ is positive. The larger this term is, the easier it is to conclude that the treatment means are different. The term increases when N, the number of observations in each group, increases and when the variability of the μ_is increases, i.e., when $\sum_{i=1}^{a}(\mu_i - \bar{\mu}_\cdot)^2/(a-1)$ increases.

A decision regarding the validity of the claim $\mu_1 = \mu_2 = \cdots = \mu_a$ is based on comparing *MSTrts* with *MSE*. If they are about the same, or equivalently if

$$F \equiv \frac{MSTrts}{MSE} \tag{5.2.4}$$

is about 1, the data are consistent with the idea that *MSTrts* and *MSE* both estimate the same (unknown) quantity σ^2 and thus are consistent with $\mu_1 = \mu_2 = \cdots = \mu_a$. The alternative is that the μ_is are not all equal, in which case *MSTrts* is estimating something larger than σ^2, while *MSE* continues to estimate σ^2. In this case, the ratio $F = MSTrts/MSE$ estimates something greater than 1. If F is much greater than 1, it provides clear evidence that the statistics are not estimating the same thing and thus that the μ_is are not all equal.

The nature of this evidence is probabilistic and one cannot eliminate the possibility of error. Although they are very unlikely to occur, F ratios much greater than 1 can arise even when the μ_is are all equal. Assuming that model (5.1.1) is appropriate, when the data yield a very large F ratio, the correct conclusion is either that the assumption of equal treatment means is violated or that the means are equal and a very rare event has occurred. The rarer the event, the stronger the suggestion of unequal treatment means. While we cannot directly quantify the strength of the suggestion of unequal treatment means, we can quantify it indirectly by evaluating how rarely large F ratios occur when the treatment means are equal. Under the assumption that $\mu_1 = \mu_2 = \cdots = \mu_a$, the F ratio is random and has an $F(a-1, a(N-1))$ distribution. (This distribution is called an F distribution in honor of the originator of analysis of variance, R. A. Fisher.)

The F distribution determines those values of the F ratio in (5.2.4) that commonly occur with equal treatment means. If the observed F ratio is so large as to be an uncommon occurrence when $\mu_1 = \mu_2 = \cdots = \mu_a$, we conclude that the μ_is are not all equal. To measure the strength of this conclusion, compute the probability of obtaining an F ratio as large or larger than that actually obtained from the data. This probability is called the *P value* or the *significance level* of the test. The smaller the P value, the more inconsistent the observed F ratio is with the assumption that the μ_is are all equal.

On occasion, it may be desired to have a fixed decision rule as to whether the data are inconsistent with the (null) hypothesis of equal means. One may decide that, with equal treatment means, common occurrences of the F ratio include 95% or 99% or more generally $(1-\alpha)100\%$ of the possible F values. Thus uncommon occurrences constitute 5% or 1% or $100\alpha\%$ of the observations. The hypothesis $\mu_1 = \mu_2 = \cdots = \mu_a$ is rejected at the α level if

$$\frac{MSTrts}{MSE} \geq F(1-\alpha, a-1, dfE).$$

Here $F(1-\alpha, a-1, dfE)$ is the number below which fall $(1-\alpha)100\%$ of the possible F ratios when the μ_is are all equal. There is a possibility that our data would yield an F ratio at least this large when the μ_is are all equal but it is pretty slim, α. We consider it more reasonable that the assumption of equal μ_is is violated. The number $F(1-\alpha, a-1, dfE)$ can be obtained from tables of the F distribution, see Appendix B.7. The number depends not only on the choice of α but also on the degrees of freedom for the estimate in the numerator of the ratio, $a-1$, and the degrees of freedom for the estimate in the denominator of the ratio, dfE. If we do not reject the hypothesis, the data are consistent with the hypothesis. Again, just because the data are consistent with the hypothesis does not mean that the hypothesis is true.

Fixed α level tests are easy to perform if the P value is available. To perform, say, an $\alpha = .05$ test, just compare the P value with .05. If the P value is greater than .05, a .05 level test does not reject the hypothesis of equal treatment means μ_i. If the P value is less than .05, a .05 test rejects the hypothesis.

5.2.1 THE ANALYSIS OF VARIANCE TABLE

The computations for the analysis of variance F test can be summarized in an analysis of variance table. The columns of the table are sources, degrees of freedom (df), sums of squares (SS), mean squares (MS), and F. There are rows for treatments, error, and total (corrected for the grand mean). The commonly used form for the analysis of variance table is given in Table 5.2. The sums of squares for error and treatments are just those given in equations (5.2.1) and (5.2.2). In each row, the mean square is the sum of squares divided by the degrees of freedom. The degrees of freedom and sums of squares for treatments and error can be added together to give the degrees of freedom and sum of squares total (corrected for the grand mean) respectively. Note that the sum of squares total divided by the degrees of freedom total is s_y^2, the sample variance of all aN observations computed without reference to any treatment groups. The degrees of freedom in the total line are just the degrees of freedom associated with the sample variance based on all aN observations. Traditionally, the total line does not include a mean square. The sample variance of all aN observations, and thus the total line, involves adjusting each observation for the grand mean. This can be accomplished as indicated in Table 5.2 or, alternatively, by the use of a *correction factor*. The correction factor is $C \equiv aN\bar{y}_{\cdot\cdot}^2$, so that $SSTot - C = \sum_{i=1}^{a}\sum_{j=1}^{N} y_{ij}^2 - C$, which is the sum of the squares of all the observations minus the correction factor.

A less commonly used form for the analysis of variance table, but one I prefer, is presented in Table 5.3. In this form, the total degrees of freedom consist of one degree of freedom for every observation, the sum of squares total is the sum of all of the squared observations, and an extra row

TABLE 5.2. Analysis of variance

Source	df	SS	MS	F
Treatments	$a-1$	$N\sum_{i=1}^{a}(\bar{y}_{i\cdot}-\bar{y}_{\cdot\cdot})^2$	$SSTrts/(a-1)$	$\frac{MSTrts}{MSE}$
Error	$aN-a$	$\sum_{i=1}^{a}\sum_{j=1}^{N}\left(y_{ij}-\bar{y}_{i\cdot}\right)^2$	$SSE/(n-a)$	
Total $-$ C	$aN-1$	$\sum_{i=1}^{a}\sum_{j=1}^{N}\left(y_{ij}-\bar{y}_{\cdot\cdot}\right)^2$		

has been added for the grand mean. The degrees of freedom and sums of squares for the grand mean, treatments, and error can be added together to obtain the degrees of freedom and sums of square total. In spite of my preference for Table 5.3, I will bow to tradition and generally use Table 5.2 with the $-C$ notation deleted from the Total line.

TABLE 5.3. Analysis of variance

Source	df	SS	MS	F
Grand mean	1	$aN\bar{y}_{\cdot\cdot}^2 \equiv C$	$aN\bar{y}_{\cdot\cdot}^2$	
Treatments	$a-1$	$N\sum_{i=1}^{a}(\bar{y}_{i\cdot}-\bar{y}_{\cdot\cdot})^2$	$SSTrts/(a-1)$	$\frac{MSTrts}{MSE}$
Error	$aN-a$	$\sum_{i=1}^{a}\sum_{j=1}^{N}\left(y_{ij}-\bar{y}_{i\cdot}\right)^2$	$SSE/(n-a)$	
Total	aN	$\sum_{i=1}^{a}\sum_{j=1}^{N}y_{ij}^2$		

EXAMPLE 5.2.1. We now examine the analysis of variance table for the electrical characteristic data of Example 5.1.2. The summary statistics for the four samples are repeated below.

Sample statistics: electrical characteristics

Strip	N	$\bar{y}_{i\cdot}$	s_i^2
1	7	16.4429	0.749524
2	7	16.5143	0.194762
3	7	16.0714	0.162381
4	7	14.9571	1.139524

The MSE for balanced data is the simple average of the s_i^2s,

$$MSE = \frac{.74952 + .19476 + .16238 + 1.13952}{4} = .56155.$$

The sample mean of the $\bar{y}_{i\cdot}$s is

$$\bar{y}_{\cdot\cdot} = \frac{16.4429 + 16.5143 + 16.0714 + 14.9571}{4} = 15.996425$$

and the sample variance of the $\bar{y}_{i\cdot}$s is

$$s_{\bar{y}}^2 = \frac{1}{4-1}\left[(16.4429 - 15.996425)^2 + (16.5143 - 15.996425)^2 + (16.0714 - 15.996425)^2 + (14.9571 - 15.996425)^2\right] = .517784\,.$$

The mean square treatments is the sample variance of the $\bar{y}_{i\cdot}$s times the number of observations in each $\bar{y}_{i\cdot}$,

$$MSTrts = Ns_{\bar{y}}^2 = 7(.517784) = 3.6245.$$

The analysis of variance table is given as Table 5.4. As discussed earlier in this section, all of the table entries are easily computed given the *MSE* and the *MSTrts*.

TABLE 5.4. Analysis of variance table: electrical characteristic data

Source	*df*	*SS*	*MS*	*F*	*P*
Treatments	3	10.873	3.624	6.45	0.002
Error	24	13.477	0.562		
Total	27	24.350			

The *F* statistic for these data is substantial and the *P* value is quite small. There is strong evidence that the treatments do not have the same mean. In other words, strips 1, 2, 3, and 4 do not have the same mean value for the electrical characteristic. The analysis of variance *F* test tells us that the means are not all equal but it does not tell us which particular means are unequal. Examining individual contrasts is required to answer more specific questions about the means. □

DISTRIBUTION THEORY

It has been stated that when there are no differences between the treatment means, the test statistic $F = MSTrts/MSE$ has an $F(a-1, dfE)$ distribution. We now briefly expand on that statement. By Definition 4.4.3, an F distribution is constructed from two independent χ^2 distributions. If $W_1 \sim \chi^2(r)$ and $W_2 \sim \chi^2(s)$ with W_1 and W_2 independent, then by definition

$$\frac{W_1/r}{W_2/s} \sim F(r, s).$$

In analysis of variance with the usual assumptions, the $\bar{y}_{i\cdot}$s and s_i^2s are all independent of each other. The *MSE* is computed from the s_i^2s and the *MSTrts* is computed from the $\bar{y}_{i\cdot}$s, so the *MSE* is independent of the *MSTrts*. We mentioned earlier that when the means are all equal

$$\frac{(a-1)MSTrts}{\sigma^2} \sim \chi^2(a-1)$$

and regardless of the mean structure

$$\frac{dfE \times MSE}{\sigma^2} \sim \chi^2(dfE),$$

so it follows from the definition of the F distribution that, when the means are all equal,

$$\frac{MSTrts}{MSE} = \frac{\left[(a-1)MSTrts/\sigma^2\right]/(a-1)}{\left[(dfE)MSE/\sigma^2\right]/dfE} \sim F(a-1, dfE).$$

When the treatment means are not all equal, the distribution of *MSTrts* depends on the value of

$$\frac{N}{(a-1)\sigma^2}\sum_{i=1}^{a}(\mu_i - \bar{\mu}_\cdot)^2.$$

Note the similarity of this number to the expected value of *MSTrts* given in (5.2.3).

5.3 Unbalanced analysis of variance

In unbalanced analysis of variance we allow different numbers N_i of observations on the groups. The analysis is slightly more difficult but it follows the same pattern as in Section 5.2. In particular, we assume that

Sample	Data		Distribution
1	$y_{11}, y_{12}, \ldots, y_{1N_1}$	iid	$N(\mu_1, \sigma^2)$
2	$y_{21}, y_{22}, \ldots, y_{2N_2}$	iid	$N(\mu_2, \sigma^2)$
$\vdots$	$\vdots$	$\vdots$	$\vdots$
a	$y_{a1}, y_{a2}, \ldots, y_{aN_a}$	iid	$N(\mu_a, \sigma^2)$

with independent samples and the same variance σ^2 for each sample. In other words, we assume

$$y_{ij} = \mu_i + \varepsilon_{ij}, \quad \varepsilon_{ij}\text{s independent } N(0, \sigma^2)$$

$i = 1, \ldots, a$ and $j = 1, \ldots, N_i$. The total number of observations is denoted $n = \sum_{i=1}^{a} N_i$. We wish to examine the (null) hypothesis

$$H_0 : \mu_1 = \mu_2 = \cdots = \mu_a.$$

Again we compute summary statistics from the samples.

Sample statistics			
Group	Size	Mean	Variance
1	N_1	$\bar{y}_{1\bullet}$	s_1^2
2	N_2	$\bar{y}_{2\bullet}$	s_2^2
$\vdots$	$\vdots$	$\vdots$	$\vdots$
a	N_a	$\bar{y}_{a\bullet}$	s_a^2

As before, the sample means, the $\bar{y}_{i\bullet}$s, are estimates of the corresponding μ_is and the s_i^2s all estimate σ^2. As discussed earlier, with unequal sample sizes an efficient pooled estimate of σ^2 must be a weighted average of the s_i^2s. The weights are the degrees of freedom associated with various estimates.

$$\begin{aligned} MSE &\equiv \frac{(N_1 - 1)s_1^2 + (N_2 - 1)s_2^2 + \cdots + (N_a - 1)s_a^2}{\sum_{i=1}^{a}(N_i - 1)} \\ &= \frac{1}{(n-a)} \sum_{i=1}^{a} \sum_{j=1}^{N_i} \left(y_{ij} - \bar{y}_{i\bullet}\right)^2. \end{aligned}$$

As before, $dfE = n - a$ and $SSE = (dfE)MSE$.

The second estimate of σ^2, the one based on the $\bar{y}_{i\bullet}$s, is not particularly intuitive. The $\bar{y}_{i\bullet}$s do not all have the same variance, so even when the μ_is are all equal, the $\bar{y}_{i\bullet}$s do not form a random sample. To get a variance estimate, the $\bar{y}_{i\bullet}$s must be weighted appropriately. It turns out that the appropriate estimate of σ^2 is

$$MSTrts = \frac{1}{a-1} \sum_{i=1}^{a} N_i \left(\bar{y}_{i\bullet} - \bar{y}_{\bullet\bullet}\right)^2$$

where

$$\bar{y}_{\bullet\bullet} = \frac{1}{n} \sum_{i=1}^{a} N_i \bar{y}_{i\bullet} = \frac{1}{n} \sum_{i=1}^{a} \sum_{j=1}^{N_i} y_{ij}.$$

Thus $\bar{y}_{\cdot\cdot}$ is the sample mean of all n observations, ignoring the treatment structure. As in the balanced case, the degrees of freedom are $a-1$ and $SSTrts = (a-1)MSTrts$. In general, $MSTrts$ is an estimate of

$$E(MSTrts) = \sigma^2 + \frac{1}{a-1}\sum_{i=1}^{a} N_i(\mu_i - \bar{\mu}_\cdot)^2$$

where

$$\bar{\mu}_\cdot \equiv \frac{1}{n}\sum_{i=1}^{a} N_i\mu_i$$

is the weighted mean of the μ_is. Once again, if the μ_is are all equal, $\mu_i = \bar{\mu}_\cdot$ for every i and $MSTrts$ is an estimate of σ^2. If the means are not all equal, $MSTrts$ is an estimate of something larger than σ^2. Values of $MSTrts/MSE$ that are much larger than 1 call in question the hypothesis of equal population means. Note that the computations for balanced data are just a special, simpler case of the computations for unbalanced data. In particular, the balanced case has $N_i = N$ and $n = aN$.

The computations are again summarized in an analysis of variance table. The commonly used form for the analysis of variance table is given below.

Analysis of variance

Source	*df*	*SS*	*MS*	*F*
Treatments	$a-1$	$\sum_{i=1}^{a} N_i(\bar{y}_{i\cdot} - \bar{y}_{\cdot\cdot})^2$	$SSTrts/(a-1)$	$\frac{MSTrts}{MSE}$
Error	$n-a$	$\sum_{i=1}^{a}\sum_{j=1}^{N_i}\left(y_{ij} - \bar{y}_{i\cdot}\right)^2$	$SSE/(n-a)$	
Total	$n-1$	$\sum_{i=1}^{a}\sum_{j=1}^{N_i}\left(y_{ij} - \bar{y}_{\cdot\cdot}\right)^2$		

The degrees of freedom and sums of squares for treatments and error can be added together to give the degrees of freedom and sum of squares total (corrected for the grand mean). Again,

$$SSE = \sum_{i=1}^{a}\sum_{j=1}^{N_i}\left(y_{ij} - \bar{y}_{i\cdot}\right)^2 = \sum_{i=1}^{a}\sum_{j=1}^{N_i}\hat{\varepsilon}_{ij}^2,$$

establishing that the sum of squares error is the sum of the squared residuals. Moreover, $SSTot/dfTot = s_y^2$, the sample variance of all n observations computed without reference to treatment groups. The degrees of freedom in the total line are the degrees of freedom associated with the sample variance based on all n observations. The total line is corrected for the grand mean, so that $SSTot = \sum_{i=1}^{a}\sum_{j=1}^{N_i} y_{ij}^2 - C$, which is the sum of the squares of all the observations minus the correction factor, $C \equiv n\bar{y}_{\cdot\cdot}^2$.

EXAMPLE 5.3.1. We now consider construction of the analysis of variance table for the logs of the suicide data. The sample statistics are repeated below.

Sample statistics: log of suicide ages

Group	N_i	$\bar{y}_{i\cdot}$	s_i^2
Caucasians	44	3.6521	0.1590
Hispanics	34	3.4538	0.2127
Native Am.	15	3.1770	0.0879

The mean squared error was computed earlier as .168. The sum of squares error is just the degrees of freedom error, 90, times the MSE. The sum of squares treatments is

$$SSTrts = 2.655 = 44(3.6521 - 3.5030)^2 + 34(3.4538 - 3.5030)^2 + 15(3.1770 - 3.5030)^2$$

where

$$3.5030 = \bar{y}_{\cdot\cdot} = \frac{44(3.6521) + 34(3.4538) + 15(3.1770)}{44 + 34 + 15}.$$

The ANOVA table is presented as Table 5.5.

TABLE 5.5. Analysis of variance, logs of suicide age data

Source	*df*	*SS*	*MS*	*F*	*P*
Groups	2	2.655	1.328	7.92	0.001
Error	90	15.088	0.168		
Total	92	17.743			

The extremely small P value for the analysis of variance F test establishes a clear difference between the mean log suicide ages. Again, more detailed comparisons are needed to identify which particular groups are different. We established earlier that at the .01 level, only non-Hispanic Caucasians and Native Americans display a pairwise difference. □

5.4 Choosing contrasts

You may be wondering why statisticians make a big fuss about analysis of variance. The procedures discussed in Sections 5.2 and 5.3 are not really of much use. The analysis of variance test involves only one hypothesis, that of equal treatment means μ_i. The more interesting issue of identifying which means are different is handled with a pooled estimate of the variance and the usual techniques involving a *Par*, an *Est*, a *SE*(*Est*), and a known distribution symmetric about zero for [*Est* − *Par*]/*SE*(*Est*). Actually, 'analysis of variance' is used as a name for the entire package of techniques used to compare more than two samples. The analysis of variance F test, from which the name devolves, is only one small part of the package. There are two reasons for examining the F test in detail. In more complicated situations than one-way ANOVA, the analysis of variance table becomes a very useful tool for identifying aspects of a complicated problem that deserve more attention. The other reason is that it introduces the *SSTrts* as a measure of treatment differences.

The *SSTrts* can be broken into components corresponding to the sums of squares for individual *orthogonal* contrasts. These components of *SSTrts* can then be used to explain the differences in the means. Recall that a contrast is a parameter $\sum_{i=1}^{a}\lambda_i\mu_i$ where the λ_is satisfy $\sum_{i=1}^{a}\lambda_i = 0$. The appropriate estimate and standard error were discussed earlier and the sum of squares for a contrast was given in (5.1.4) as

$$SS\left(\sum_{i=1}^{a}\lambda_i\mu_i\right) \equiv \frac{\left(\sum_{i=1}^{a}\lambda_i\bar{y}_{i\cdot}\right)^2}{\sum_{i=1}^{a}\lambda_i^2/N_i}.$$

In the balanced case with $N = N_i$ for all i,

$$SS\left(\sum_{i=1}^{a}\lambda_i\mu_i\right) = \frac{\left(\sum_{i=1}^{a}\lambda_i\bar{y}_{i\cdot}\right)^2}{\left(\sum_{i=1}^{a}\lambda_i^2\right)/N}.$$

The F test for $H_0 : \sum_{i=1}^{a}\lambda_i\mu_i = 0$ versus $H_A : \sum_{i=1}^{a}\lambda_i\mu_i \neq 0$ rejects H_0 for large values of $SS\left(\sum_{i=1}^{a}\lambda_i\mu_i\right)/MSE$.

Two contrasts $\sum_{i=1}^{a}\lambda_{i1}\mu_i$ and $\sum_{i=1}^{a}\lambda_{i2}\mu_i$ are defined to be *orthogonal* if

$$\sum_{i=1}^{a}\frac{\lambda_{i1}\lambda_{i2}}{N_i} = 0.$$

In balanced problems, $N_i = N$ for all i, so the condition of orthogonality becomes $\sum_{i=1}^{a} \lambda_{i1}\lambda_{i2}/N = 0$ or equivalently

$$\sum_{i=1}^{a} \lambda_{i1}\lambda_{i2} = 0.$$

Contrasts are only of interest when they define interesting functions of the μ_is. Orthogonal contrasts are most useful in balanced problems because a set of orthogonal contrasts can retain interesting interpretations. In unbalanced cases, orthogonality depends on the unequal N_is, so there is rarely more than one interpretable contrast in a set of orthogonal contrasts.

EXAMPLE 5.4.1. Consider again the electrical characteristic data. The sample statistics are

Sample statistics: electrical characteristics

Strip	N	$\bar{y}_{i\cdot}$	s_i^2
1	7	16.4429	0.749524
2	7	16.5143	0.194762
3	7	16.0714	0.162381
4	7	14.9571	1.139524

with $MSE = .56155$. We examine four contrasts

$$C_1 \equiv (1)\mu_1 + (-1)\mu_2 + (0)\mu_3 + (0)\mu_4 = \mu_1 - \mu_2,$$

$$C_2 \equiv (1/2)\mu_1 + (1/2)\mu_2 + (-1)\mu_3 + (0)\mu_4 = \frac{\mu_1 + \mu_2}{2} - \mu_3,$$

$$C_3 \equiv (1/3)\mu_1 + (1/3)\mu_2 + (1/3)\mu_3 + (-1)\mu_4 = \frac{\mu_1 + \mu_2 + \mu_3}{3} - \mu_4,$$

and

$$C_4 \equiv (-1)\mu_1 + (-1)\mu_2 + (2)\mu_3 + (0)\mu_4.$$

Contrasts C_1 and C_2 are orthogonal because

$$(1)(1/2) + (-1)(1/2) + (0)(-1) + (0)(0) = 0.$$

Similarly, C_1 and C_3 are orthogonal and C_2 and C_3 are orthogonal. We have previously examined the contrast C_1 and found the sum of squares to be

$$SS(C_1) = \frac{[(1)16.4429 + (-1)16.5143 + (0)16.0714 + (0)14.9571]^2}{\left[1^2 + (-1)^2 + 0^2 + 0^2\right]/7} = 0.0178\,.$$

The sum of squares for C_2 is

$$SS(C_2) = \frac{[(1/2)16.4429 + (1/2)16.5143 + (-1)16.0714 + (0)14.9571]^2}{\left[(1/2)^2 + (1/2)^2 + (-1)^2 + 0^2\right]/7} = 0.7738.$$

The sum of squares for C_3 is

$$\begin{aligned} SS(C_3) &= \frac{[(1/3)16.4429 + (1/3)16.5143 + (1/3)16.0714 + (-1)14.9571]^2}{\left[(1/3)^2 + (1/3)^2 + (1/3)^2 + (-1)^2\right]/7} \\ &= 10.0818. \end{aligned}$$

The decomposition referred to earlier follows from the fact that

$$10.873 = SSTrts = SS(C_1) + SS(C_2) + SS(C_3) = 0.0178 + 0.7738 + 10.0818\,.$$

SSTrts is a measure of the evidence for differences between means. Almost all of the *SSTrts* is accounted for by C_3. Thus, almost all of the differences between the means can be accounted for by the difference between μ_4 and the average of μ_1, μ_2, and μ_3. Almost none of the sum of squares for treatments is due to the difference between μ_1 and μ_2. A small amount is due to the difference between μ_3 and the average of μ_1 and μ_2. The data are consistent with the idea that the means for strips 1, 2, and 3 are the same.

The contrast C_4 was introduced to illustrate the fact that *multiplying a contrast by a constant has no real effect on the contrast.* Observe that

$$C_4 = -2C_2\,.$$

In particular, $C_4 = 0$ if and only if $C_2 = 0$. Note that

$$SS(C_4) = \frac{[(-1)16.4429 + (-1)16.5143 + (2)16.0714 + (0)14.9571]^2}{\left[(-1)^2 + (-1)^2 + 2^2 + 0^2\right]/7} = 0.7738,$$

so $SS(C_4) = SS(C_2)$ and the F test for $C_2 = 0$ is identical to the F test for $C_4 = 0$. It is also easily seen that a two-sided t test for $H_0 : C_4 = 0$ is identical to that for $H_0 : C_2 = 0$. The factor of -2 must be accounted for in estimation and in tests of C_2 and C_4 other than testing that they are zero, but, after suitable adjustment, estimation and testing are equivalent. The virtue of using C_4 rather than C_2 is that the λ_is in C_4 are all integers, so computations are simpler with C_4.

There are many ways to pick a set of orthogonal contrasts. We established that the data are consistent with the idea that ceramic strip 4 is different from the other strips and that there are no differences between the other strips. The data are even more consistent with another set of orthogonal contrasts. Consider the claim that the value for strip 4 is the average of the values for strips 1 and 2, i.e., $\mu_4 = (\mu_1 + \mu_2)/2$ or equivalently

$$C_5 \equiv (1)\mu_1 + (1)\mu_2 + (0)\mu_3 + (-2)\mu_4 = 0.$$

A contrast orthogonal to C_5 is C_1, considered earlier. A contrast orthogonal to both C_5 and C_1 is

$$C_6 \equiv (1)\mu_1 + (1)\mu_2 + (-3)\mu_3 + (1)\mu_4.$$

The sum of squares for C_5 is

$$SS(C_5) = \frac{[(1)16.4429 + (1)16.5143 + (0)16.0714 + (-2)14.9571]^2}{\left[1^2 + 1^2 + 0^2 + (-2)^2\right]/7} = 10.803\,.$$

The sum of squares for C_1 was given earlier, $SS(C_1) = .018$. The sum of squares for C_6 is

$$SS(C_6) = \frac{[(1)16.4429 + (1)16.5143 + (-3)16.0714 + (1)14.9571]^2}{\left[1^2 + 1^2 + (-3)^2 + 1^2\right]/7} = .052\,.$$

As before, with orthogonal contrasts

$$10.873 = SSTrts = SS(C_5) + SS(C_1) + SS(C_6) = 10.803 + .018 + .052\,.$$

For all practical purposes, these data are *totally* consistent with the claims $C_6 = 0$ and $C_1 = 0$ because $SS(C_6) \doteq 0 \doteq SS(C_1)$. Essentially, all the differences in means can be attributed to C_5 because $SS(C_5) \doteq SSTrts$. □

It is a mathematical fact that *there is always one contrast that accounts for all of SSTrts*, however, this contrast rarely has a simple interpretation because the coefficients of this contrast depend on the sample means. In a balanced one-way analysis of variance with, say, four treatments, the coefficients of the contrast that accounts for the entire *SSTrts* are $\lambda_1 = \bar{y}_{1\cdot} - \bar{y}_{\cdot\cdot}$, $\lambda_2 = \bar{y}_{2\cdot} - \bar{y}_{\cdot\cdot}$, $\lambda_3 = \bar{y}_{3\cdot} - \bar{y}_{\cdot\cdot}$, and $\lambda_4 = \bar{y}_{4\cdot} - \bar{y}_{\cdot\cdot}$. *Typically, a contrast with these coefficients will be difficult to interpret.* In Example 5.4.1, C_5 was constructed in this way, but to simplify the discussion we rounded the coefficients off. Rounding the coefficients helps to make the contrast more interpretable. In the case of C_5, the contrast became very simple. When rounding the coefficients, the contrast will not contain quite all of the sum of squares for treatments.

One reasonable approach to analysis of variance is to identify the contrast that accounts for all of the *SSTrts* and to try to interpret it. I prefer to look at the data and try to identify a contrast or a few orthogonal contrasts that are interpretable and account for most of *SSTrts*. Either of these approaches involves looking at the data to identify contrasts of interest. In such a situation, using the standard $F(1, dfE)$ or $t(dfE)$ distributions for statistical inference is inappropriate. Appropriate statistical methods are discussed in the next chapter.

In some situations, the structure of the treatments suggests orthogonal contrasts that are both interesting and interpretable. When the structure of the treatments, rather than the data, suggests the contrasts, standard methods of inference apply.

The key fact about orthogonal contrasts is that if $C_1, ..., C_{a-1}$ is any set of contrasts with each orthogonal to every other one, then

$$SSTrts = SS(C_1) + \cdots + SS(C_{a-1}).$$

In our example, $a = 4$, so there were sets of $a - 1 = 3$ orthogonal contrasts that decompose the *SSTrts*. We gave two such sets of contrasts. There are an infinite number of other ways to choose sets of orthogonal contrasts.

With a treatments, a set of orthogonal contrasts can contain no more than $a - 1$ elements. There can be at most $a - 1$ orthogonal contrasts but one can also choose sets of orthogonal contrasts with, say, $q < a - 1$ elements. In such a case,

$$SSTrts \geq SS(C_1) + \cdots + SS(C_q).$$

In particular, any one contrast C can be viewed as a set with $q = 1$, so

$$SSTrts \geq SS(C). \tag{5.4.1}$$

Interesting contrasts are determined by the structure of the treatments. We now illustrate this fact with an example.

EXAMPLE 5.4.2. Five diets were investigated to determine their effects on the growth of animals. If the diets do not have any recognizable structure, about the only interesting set of contrasts is to compare all pairs of population means. The collection of contrasts is $\mu_i - \mu_{i'}$ for $i, i' = 1, 2, 3, 4, 5$ with $i \neq i'$. Note that $\mu_1 = \mu_2 = \mu_3 = \mu_4 = \mu_5$ if and only if all 10 of these contrast are zero, i.e., if $\mu_i - \mu_{i'} = 0$ for all $i \neq i'$. These contrasts are not orthogonal. There can be at most $5 - 1 = 4$ members in a set of orthogonal contrasts; this collection of contrasts has 10 members. In fact, many of these 10 contrasts are redundant. For example, if $\mu_1 - \mu_2 = 0$ and $\mu_2 - \mu_3 = 0$, then, of course, $\mu_1 - \mu_3 = 0$. More generally, if you know the value of $\mu_i - \mu_j$ and the value of $\mu_j - \mu_k$, you also know the value of $\mu_i - \mu_k$.

Although these 10 contrasts may be redundant, statistical inferences about them may not be. For example, failing to reject $H_0 : \mu_1 - \mu_2 = 0$ and $H_0 : \mu_2 - \mu_3 = 0$ in no way implies the we will fail to reject $H_0 : \mu_1 - \mu_3 = 0$. Similarly, rejecting $H_0 : \mu_1 - \mu_2 = 0$ and $H_0 : \mu_2 - \mu_3 = 0$ does not imply that we will reject $H_0 : \mu_1 - \mu_3 = 0$.

Now suppose we are told that treatment 1 is the standard diet and that the other four treatments are new, experimental diets. In this case, the structure of the treatments suggests that we might examine only the contrasts $\mu_1 - \mu_i$ for $i = 2, 3, 4, 5$. These contrasts are not redundant. Knowing two or three of them will never tell you the values of any others. For example, if $\mu_1 - \mu_2 = 0$, $\mu_1 - \mu_3 = 0$, and $\mu_1 - \mu_4 = 0$, we still do not know the value of $\mu_1 - \mu_5$. On the other hand, if all 4 of the contrasts equal 0, we must have $\mu_1 = \mu_2 = \mu_3 = \mu_4 = \mu_5$, and if the treatment means are all equal, every contrast must be zero. These four contrasts are not orthogonal in any ANOVA.

Contrasts that are not redundant are said to be linearly independent. *With a treatments, one can have at most $a - 1$ linearly independent contrasts. Nontrivial orthogonal contrasts are always linearly independent. (The trivial contrast has $\lambda_i = 0$ for all i.) If any set of $a - 1$ linearly independent contrasts are all equal to 0, then $\mu_1 = \mu_2 = \cdots = \mu_a$.*

Additional structure on the treatments may suggest other contrasts. Suppose that the four new diets are, in order, two based on beef, one based on pork, and one based on soybeans. In this case contrasts with the following coefficients seem interesting.

	Diet treatments				
	Control	Beef	Beef	Pork	Beans
Contrast	λ_1	λ_2	λ_3	λ_4	λ_5
Ctrl vs others	4	−1	−1	−1	−1
Beef vs beef	0	1	−1	0	0
Beef vs pork	0	1	1	−2	0
Meat vs beans	0	1	1	1	−3

The first contrast, Ctrl vs others, compares the control (standard diet) to the average of the other four diets. This contrast would actually be

$$\mu_1 - \frac{\mu_2 + \mu_3 + \mu_4 + \mu_5}{4}$$

but multiplying the contrast by 4 gives the equivalent contrast

$$4\mu_1 - \mu_2 - \mu_3 - \mu_4 - \mu_5$$

which is the one tabled. The tabled contrast is simpler to work with because its contrast coefficients are all integers. The other three contrasts compare the two beef diets, the average of the beef diets with the pork diet, and the average of the meat diets with the soybean diet. In a balanced ANOVA, these four contrasts are all orthogonal to each other.

If the structure of the treatments was different, say, the first beef diet was instead a diet based on lima beans, the interesting orthogonal contrasts change.

	Diet treatments				
	Control	Lima	Beef	Pork	Soy
Contrast	λ_1	λ_2	λ_3	λ_4	λ_5
Ctrl vs others	4	−1	−1	−1	−1
Beef vs pork	0	0	1	−1	0
Lima vs soy	0	1	0	0	−1
Meat vs beans	0	−1	1	1	−1

These contrasts compare the control to the average of the other four diets, the two meat diets, the two bean diets, and the average of the meat diets with the average of the bean diets. Again, the contrasts are all orthogonal in a balanced ANOVA. □

5.5 Comparing models

The hypothesis

$$H_0 : \mu_1 = \mu_2 = \cdots = \mu_a$$

can be viewed as imposing a change in the analysis of variance model

$$y_{ij} = \mu_i + \varepsilon_{ij}, \tag{5.5.1}$$

$i = 1, \ldots, a, j = 1, \ldots, N_i$. If for some value μ, $\mu = \mu_1 = \mu_2 = \cdots = \mu_a$ the analysis of variance model can be rewritten as

$$y_{ij} = \mu + \varepsilon_{ij}, \tag{5.5.2}$$

which involves only a grand mean μ. This is just the special case of the analysis of variance model in which the μ_is do not really depend on the value of i. In (5.1.2) we wrote the analysis of variance model as $y_{ij} = \mu + \alpha_i + \varepsilon_{ij}$. Model (5.5.2) is the special case obtained by dropping the α_is. For simplicity, in this section models (5.5.1) and (5.5.2) will be referred to as models (1) and (2), respectively.

We wish to evaluate how well model (2) fits as compared to how well model (1) fits. A measure of how well any model fits is the sum of squared errors; a poor fitting model has much larger errors and thus a much larger SSE. In $y_{ij} = \mu_i + \varepsilon_{ij}$, the errors are $\varepsilon_{ij} = y_{ij} - \mu_i$ and the estimated errors (*residuals*) are $\hat{\varepsilon}_{ij} = y_{ij} - \bar{y}_{i\cdot}$. The sum of squares error in model (1) is the usual analysis of variance sum of squares error,

$$SSE(1) = \sum_{i=1}^{a}\sum_{j=1}^{N_i} \hat{\varepsilon}_{ij}^2 = \sum_{i=1}^{a}\sum_{j=1}^{N_i} \left(y_{ij} - \bar{y}_{i\cdot}\right)^2.$$

Recall that $MSE(1) = SSE(1)/(n-a)$ is an estimate of σ^2 and denote the error degrees of freedom

$$dfE(1) = n - a.$$

Model (2) treats all n observations as a random sample from one population with mean μ. Under model (2), an estimate of σ^2 is s_y^2, the sample variance of all n observations, so

$$MSE(2) \equiv s_y^2 = \frac{1}{n-1}\sum_{i=1}^{a}\sum_{j=1}^{N_i} \left(y_{ij} - \bar{y}_{\cdot\cdot}\right)^2$$

with error degrees of freedom

$$dfE(2) = n - 1.$$

We define the sum of squares error from model (2) to be

$$\begin{aligned} SSE(2) &= dfE(2) \times MSE(2) \\ &= \sum_{i=1}^{a}\sum_{j=1}^{N_i} \left(y_{ij} - \bar{y}_{\cdot\cdot}\right)^2. \end{aligned}$$

Since model (2) is a special case of model (1), the error from model (2) must be as large as the error from model (1), i.e., $SSE(2) \geq SSE(1)$. However, if $SSE(2)$ is much greater than $SSE(1)$, it suggests that the special case, model (2), is an inadequate substitute for the *full* model (1). In particular, large values of $SSE(2) - SSE(1)$ suggest that the *reduced* model (2) is inadequate to explain the data that, by assumption, were adequately explained using the full model (1). It can be established that, if the reduced model is true, the statistic

$$MSTest \equiv \frac{SSE(2) - SSE(1)}{dfE(2) - dfE(1)}$$

is an estimate of σ^2, which is independent of the estimate from the full model, $MSE(1)$. If the reduced model is not true, *MSTest* estimates σ^2 plus a positive number. A test of whether model (2) is an adequate substitute for model (1) is rejected if

$$F = \frac{[SSE(2) - SSE(1)] \,/\, [dfE(2) - dfE(1)]}{MSE(1)} \tag{5.5.3}$$

is too much larger than 1. In particular, an α level test rejects the adequacy of model (2) when

$$\frac{[SSE(2) - SSE(1)] \,/\, [dfE(2) - dfE(1)]}{MSE(1)} > F(1-\alpha, dfE(2) - dfE(1), dfE(1)). \tag{5.5.4}$$

To see that the numerator in (5.5.3) is a reasonable estimate of σ^2 when model (2) holds, write

$$\begin{aligned} MSE(2) &= \frac{1}{dfE(2)}[SSE(2) - SSE(1) + SSE(1)] \\ &= \frac{dfE(2) - dfE(1)}{dfE(2)} \left(\frac{SSE(2) - SSE(1)}{dfE(2) - dfE(1)} \right) + \frac{dfE(1)}{dfE(2)} MSE(1). \end{aligned}$$

$MSE(2)$ is a weighted average of *MSTest* and $MSE(1)$. $MSE(1)$ is certainly a reasonable estimate of σ^2 and, if the means are all equal, $MSE(2)$ is also a reasonable estimate of σ^2. Thus, if the means are all equal, $[SSE(2) - SSE(1)]/[dfE(2) - dfE(1)]$ must be a reasonable estimate of σ^2 because if it were not, a weighted average of it and $MSE(1)$ would not be a reasonable estimate of σ^2.

The F statistic in (5.5.3) is exactly the analysis of variance table F statistic. This follows because, relative to the analysis of variance table,

$$\begin{aligned} SSTot &= SSE(2) \\ dfTot &= dfE(2) \\ SSE &= SSE(1) \\ dfE &= dfE(1) \\ MSE &= SSE(1)/dfE(1) \\ SSTrts &= SSE(2) - SSE(1) \\ dfTrts &= dfE(2) - dfE(1) \\ MSTrts &= [SSE(2) - SSE(1)] \,/\, [dfE(2) - dfE(1)] \end{aligned}$$

This technique of testing the adequacy of a reduced (special case) model by comparing the error sum of squares for the full model and the reduced model is applicable very generally. In more sophisticated unbalanced analysis of variance situations and in regression analysis, this is a primary method used to test hypotheses. In particular, *the test in (5.5.4) applies for any ANOVA or regression model (2) that is a special case of any ANOVA or regression model (1) as long as the errors are independent* $N(0, \sigma^2)$.

5.6 The power of the analysis of variance F test

The power of a test is the probability of rejecting the null hypothesis when the null hypothesis is false. Thus, the power of the analysis of variance F test is the probability of correctly concluding that the μ_is are not all the same when they are in fact not all the same. In this section we give some intuition for the power of the analysis of variance F test. For simplicity we discuss only balanced analysis of variance.

As discussed in Section 5.2, whenever the analysis of variance model is correct, the *MSE* is an unbiased estimate of

$$E(MSE) = \sigma^2 \tag{5.6.1}$$

and *MSTrts* is an unbiased estimate of

$$E(MSTrts) = \sigma^2 + \frac{N}{a-1}\sum_{i=1}^{a}(\mu_i - \bar{\mu}_\bullet)^2. \tag{5.6.2}$$

Write

$$s_\mu^2 = \frac{1}{a-1}\sum_{i=1}^{a}(\mu_i - \bar{\mu}_\bullet)^2, \tag{5.6.3}$$

so s_μ^2 is the 'sample' variance of the μ_is. The word sample is in quotation marks because we do not really have a sample of μ_is, in fact we *never* get to observe the μ_is. s_μ^2 is a sample variance only in the sense that the computational formula (5.6.3) is identical to that for a sample variance. With the new notation, we can rewrite (5.6.2) as

$$E(MSTrts) = \sigma^2 + Ns_\mu^2. \tag{5.6.4}$$

The analysis of variance F statistic is defined as

$$F = \frac{MSTrts}{MSE}.$$

Since the *MSE* and the *MSTrts* estimate (5.6.1) and (5.6.4) respectively, by substitution we see that F is an estimate of

$$\frac{\sigma^2 + Ns_\mu^2}{\sigma^2} = 1 + \frac{N}{\sigma^2}s_\mu^2. \tag{5.6.5}$$

(F is *not* an unbiased estimate of this quantity but F is a reasonable estimate of it.)

The behavior of the F test depends crucially on the quantity that F estimates. First notice that if the μ_is are all equal, they have no variability and $s_\mu^2 = 0$. In fact *the μ_is are all equal if and only if* $s_\mu^2 = 0$. The statistic F always estimates the value in (5.6.5) and when $s_\mu^2 = 0$ that value is 1. Thus an F statistic that is too far above 1 suggests that $s_\mu^2 \neq 0$. Alternatively, when the μ_is are not all equal, they have positive variability and $s_\mu^2 > 0$. In this case, F is estimating a value in (5.6.5) that is greater than 1, so values of F substantially greater than 1 lead us to suspect that $s_\mu^2 > 0$ and hence that the μ_is are not all equal.

Remember that even when $s_\mu^2 = 0$, F is only an estimate of 1; it has a natural variability about 1. To reject the idea that $s_\mu^2 = 0$, an observed F value must be larger than would normally be experienced when $s_\mu^2 = 0$.

When $s_\mu^2 = 0$, the statistic F has an $F(dfTrts, dfE)$ distribution. This distribution specifies the values of F that would normally be experienced. Thus an α level test is rejected when the observed F value is larger than all but $100\alpha\%$ of the observations that normally occur, i.e., larger than $F(1-\alpha, dfTrts, dfE)$. When $s_\mu^2 = 0$ there is only a probability of α that the observed F value will exceed $F(1-\alpha, dfTrts, dfE)$.

Note that values of F that are much smaller than 1 *do not* suggest that $s_\mu^2 > 0$. Within the present discussion, values of F that are smaller than 1 are most consistent with $s_\mu^2 = 0$ and will not be considered further. It should be noted, however, that very small values of F are suggestive. In terms of modeling, they are suggestive of something fairly complicated, cf. Christensen (1989, 1991). Very small test statistics have also been known to occur when someone has manufactured data in order to justify a null hypothesis. For example, some data reported by Mendel that supported his theories of genetic inheritance were too good to be true.

We reject the hypothesis of equal μ_is for F values that are substantially greater than 1. It is natural to ask what causes F to take on values that are substantially greater than 1. In other words, what causes the test to have high power for detecting differences in the μ_is? Obviously, F will tend to be substantially greater than 1 when it is estimating something that is substantially greater than 1, i.e., when

$$1 + \frac{N}{\sigma^2} s_\mu^2$$

is substantially greater than 1. There are three items involved. To make $1 + Ns_\mu^2 / \sigma^2$ much larger than 1 we need some combination of N large, σ^2 small, and s_μ^2 large. The first two items are somewhat controllable. To increase the power of the F test we can increase N, the size of the various samples. The second item, σ^2, is a parameter, so we will never know it exactly, but improving one's experimental methods can make it smaller. For example, measuring the height of a house with a meter stick rather than a 30 centimeter ruler is likely to yield a much more accurate value for the height. In later chapters we discuss some general methods for designing experiments that enable us to reduce σ^2. The third item above, s_μ^2, we are simply stuck with. There is little we can do with the μ_is to cause a test to be powerful. If the differences among the μ_is are small, the μ_is have little variability and s_μ^2 is near zero. Other things being equal, it is unlikely that we will correctly reject the F test when s_μ^2 is near zero. More accurately, it is unlikely that we will correctly reject the F test when s_μ^2 is so small that Ns_μ^2/σ^2 is near zero. Even when s_μ^2 is small in absolute terms, if N is large or σ^2 is much smaller than s_μ^2, we have a good chance of correctly identifying that there are differences in the μ_is.

For specified values of Ns_μ^2 / σ^2 it is possible to compute the probability of rejecting the F test. To specify Ns_μ^2 / σ^2 one needs to know N and some approximation for σ^2; these are often available. The most difficult part of computing the power of an F test is in specifying a reasonable value for s_μ^2. In specifying a value for s_μ^2 we need both to specify a pattern for the differences in the μ_is and to quantify the extent of the differences. For example, our interest may be in detecting differences in the μ_is when all of the μ_is are equal except one, which is, say, d units larger than the others. We can compute the value for s_μ^2 by specifying d. Similarly, with an even number of treatments our interest may be in detecting differences in the μ_is in which half the μ_is equal one value and the other half equal a different value, with the two values d units apart. Again we can compute a value of s_μ^2 for any difference d but the value of s_μ^2 depends on d in a very different manner than in the first case.

5.7 Exercises

EXERCISE 5.7.1. In a study of stress at 600% elongation for a certain type of rubber, Mandel (1972) reported stress test data from five different laboratories. Summary statistics are given in Table 5.6. Compute the analysis of variance table and test for differences in means between all pairs of labs. Use $\alpha = .01$. Is there any reason to worry about the assumptions of the analysis of variance model?

EXERCISE 5.7.2. Snedecor and Cochran (1967, section 6.18) presented data obtained in 1942 from South Dakota on the relationship between the size of farms (in acres) and the number of acres planted in corn. Summary statistics are presented in Table 5.7. Note that the sample standard deviations rather than the sample variances are given. In addition, the pooled standard deviation is 0.4526.

(a) Give the one-way analysis of variance model with all of its assumptions. Can any problems with the assumptions be identified?

TABLE 5.6. Rubber stress at five laboratories

Lab.	Sample size	Sample mean	Sample variance
1	4	57.00	32.00
2	4	67.50	46.33
3	4	40.25	14.25
4	4	56.50	5.66
5	4	52.50	6.33

TABLE 5.7. Acreage in corn for different sized farms

Farm acres	Sample size	Sample mean	Sample std. dev.
80	5	2.9957	0.4333
160	5	3.6282	0.4056
240	5	4.1149	0.4169
320	5	4.0904	0.4688
400	5	4.4030	0.5277

(b) Give the analysis of variance table for these data. Test whether there are any differences in corn acreages due to the different size farms. Use $\alpha = .01$.

(c) Test for differences between all pairs of farm sizes using $\alpha = .01$ tests.

(d) Find the sum of squares for the following contrast:

Farm	80	160	240	320	400
Coeff.	-2	-1	0	1	2

What percentage is this of the treatment sum of squares?

e) Give 95% confidence and prediction intervals for the number of acres in corn for each farm size.

EXERCISE 5.7.3. Table 5.8 gives data on heights and weights of people. Give the analysis of variance table and test for differences among the four groups. Give a 99% confidence interval for the mean weight of people in the 72 inch height group.

TABLE 5.8. Weights (in pounds) for various heights (in inches)

Height	Sample size	Sample mean	Sample variance
63	3	$121.6\bar{6}$	$158.33\bar{3}$
65	4	131.25	$72.91\bar{3}$
66	2	142.50	112.500
72	3	$171.6\bar{6}$	$158.33\bar{3}$

EXERCISE 5.7.4. Conover (1971, p. 326) presented data on the amount of iron found in the livers of white rats. Fifty rats were randomly divided into five groups of ten and each group was given a different diet. We analyze the logs of the original data. The total sample variance of the 50 observations is 0.521767 and the means for each diet are given below.

Diet	A	B	C	D	E
Mean	1.6517	0.87413	0.89390	0.40557	0.025882

Compute the analysis of variance table and test whether there are differences due to diet.

If diets A and B emphasize beef and pork respectively, diet C emphasizes poultry, and diets D and E are based on dried beans and oats, the following contrasts may be of interest.

	Diet				
Contrast	A	B	C	D	E
Beef vs. pork	1	−1	0	0	0
Mammals vs. poultry	1	1	−2	0	0
Beans vs. oats	0	0	0	1	−1
Animal vs. vegetable	2	2	2	−3	−3

Show that the contrasts are orthogonal and compute sums of squares for each contrast. Interpret your results and draw conclusions about the data.

EXERCISE 5.7.5. In addition to the data discussed earlier, Mandel (1972) reported data from one laboratory on four different types of rubber. Four observations were taken on each type of rubber. The means are given below.

Material	A	B	C	D
Mean	26.4425	26.0225	23.5325	29.9600

The sample variance of the 16 observations is 14.730793. Compute the analysis of variance table, the overall F test, and test for differences between each pair of rubber types. Use $\alpha = .05$.

EXERCISE 5.7.6. In Exercise 5.7.5 on the stress of four types of rubber, the observations on material B were 22.96, 22.93, 22.49, and 35.71. Redo the analysis, eliminating the outlier. The sample variance of the 15 remaining observations is 9.3052838.

EXERCISE 5.7.7. Bethea et al. (1985) reported data on an experiment to determine the effectiveness of four adhesive systems for bonding insulation to a chamber. The data are a measure of the peel-strength of the adhesives and are presented in Table 5.9. A disturbing aspect of these data is that the values for adhesive system 3 are reported with an extra digit.

(a) Compute the sample means and variances for each group. Give the one-way analysis of variance model with all of its assumptions. Are there problems with the assumptions? If so, does an analysis on the square roots or logs of the data reduce these problems?

(b) Give the analysis of variance table for these (possibly transformed) data. Test whether there are any differences in adhesive systems. Use $\alpha = .01$.

(c) Test for differences between all pairs of adhesive systems using $\alpha = .01$ tests.

TABLE 5.9. Peel-strength of various adhesive systems

Adhesive system	Observations					
1	60	63	57	53	56	57
2	57	52	55	59	56	54
3	19.8	19.5	19.7	21.6	21.1	19.3
4	52	53	44	48	48	53

(d) Find the sums of squares i) for comparing system 1 with system 4 and ii) for comparing system 2 with system 3.

(e) Perform a .01 level F test for whether the mean peel-strength of systems 1 and 4 differs from the mean peel-strength of systems 2 and 3.

(f) What property is displayed by the sums of squares computed in (d) and (e)? Why do they have this property?

(g) Give a 99% confidence interval for the mean of every adhesive system.

(h) Give a 99% prediction interval for every adhesive system.

(i) Give a 95% confidence interval for the difference between systems 1 and 2.

EXERCISE 5.7.8. Table 5.10 contains weight gains of rats from Box (1950). The rats were given either Thyroxin or Thiouracil or were in a control group. Do a complete analysis of variance on the data. Give the model, check assumptions, make residual plots, give the ANOVA table, and examine appropriate contrasts.

TABLE 5.10. Weight gains of rats

Thyroxin	Thiouracil		Control	
132	68	68	107	115
84	63	52	90	117
133	80	80	91	133
118	63	61	91	115
87	89	69	112	95
88				
119				

EXERCISE 5.7.9. Aitchison and Dunsmore (1975) presented data on Cushing's syndrome. Cushing's syndrome is a condition in which the adrenal cortex overproduces cortisol. Patients are divided into one of three groups based on the cause of the syndrome: a – adenoma, b – bilateral hyperplasia, and c – carcinoma. The data are amounts of tetrahydrocortisone in the urine of the patients. The data are given in Table 5.11. Give a complete analysis.

EXERCISE 5.7.10. Draper and Smith (1966, p. 41) considered data on the relationship between the age of truck tractors (in years) and the cost (in dollars) of maintaining them over a six month period. The data are given in Table 5.12.

TABLE 5.11. Tetrahydrocortisone values for patients with Cushing's syndrome

a	b		c
3.1	8.3	15.4	10.2
3.0	3.8	7.7	9.2
1.9	3.9	6.5	9.6
3.8	7.8	5.7	53.8
4.1	9.1	13.6	15.8
1.9			

TABLE 5.12. Age and costs of maintenance for truck tractors

Age	Costs		
0.5	163	182	
1.0	978	466	549
4.0	495	723	681
4.5	619	1049	1033
5.0	890	1522	1194
5.5	987		
6.0	764	1373	

Note that there is only one observation at 5.5 years of age. This group does not yield an estimate of the variance and can be ignored for the purpose of computing the mean squared error. In the weighted average of variance estimates, the variance of this group is undefined but the variance gets 0 weight, so there is no problem.

Give the analysis of variance table for these data. Does cost differ with age? Is there a significant difference between the cost at 0.5 years as opposed to 1.0 year? Use several contrasts to determine whether there are any differences between costs at 4, 4.5, 5, 5.5, and 6 years. How much of the sum of squares for treatments is due to the following contrast?

Age	0.5	1.0	4.0	4.5	5.0	5.5	6.0
Coeff.	−5	−5	2	2	2	2	2

What is the sum of squares for the contrast that compares the average of 0.5 and 1.0 with the averages of 4, 4.5, 5, 5.5, and 6?

EXERCISE 5.7.11. George Snedecor (1945a) asked for the appropriate variance estimate in the following problem. One of six treatments was applied to the 10 hens contained in each of 12 cages. Each treatment was randomly assigned to two cages. The data were the number of eggs laid by each hen.

(a) What should you tell Snedecor? Were the treatments applied to the hens or to the cages? How will the analysis differ depending on the answer to this question?

(b) The mean of the 12 sample variances computed from the 10 hens in each cage was 297.8. The average of the 6 sample variances computed from the two-cage means for each treatment was 57.59. The sample variance of the 6 treatment means was 53.725. How should you construct an F test? Remember that the numbers reported above are not necessarily mean squares.

EXERCISE 5.7.12. Lehmann (1975), citing Heyl (1930) and Brownlee (1960), considered data on determining the gravitational constant of three elements: gold, platinum, and glass. The data Lehmann gives are the third and fourth decimal places in five determinations of the gravitational constant. They are presented below. Analyze the data.

Gold	Platinum	Glass
83	61	78
81	61	71
76	67	75
79	67	72
76	64	74

EXERCISE 5.7.13. Shewhart (1939, p. 69) also presented the gravitational constant data of Heyl (1930) that was considered in the previous problem, but Shewhart reports six observations for gold instead of five. Shewhart's data are given below. Analyze these data and compare your results to those of the previous exercise.

Gold	Platinum	Glass
83	61	78
81	61	71
76	67	75
79	67	72
78	64	74
72		

EXERCISE 5.7.14. Recall that if $Z \sim N(0, 1)$ and $W \sim \chi^2(r)$ with Z and W independent, then by Definition 2.1.3 $Z/\sqrt{W/r}$ has a $t(r)$ distribution. Also recall that in a one-way ANOVA with independent normal errors, a contrast has

$$\sum_{i=1}^{a} \lambda_i \bar{y}_{i\bullet} \sim N\left(\sum_{i=1}^{a} \lambda_i \mu_i, \sigma^2 \sum_{i=1}^{a} \frac{\lambda_i^2}{N_i}\right),$$

$$\frac{SSE}{\sigma^2} \sim \chi^2(dfE),$$

and MSE independent of all the $\bar{y}_{i\bullet}$s. Show that

$$\frac{\sum_{i=1}^{a} \lambda_i \bar{y}_{i\bullet} - \sum_{i=1}^{a} \lambda_i \mu_i}{\sqrt{MSE \sum_{i=1}^{a} \lambda_i^2 / N_i}} \sim t(dfE).$$

Chapter 6
Multiple comparison methods

As illustrated in Section 5.1, the most useful information from a one-way ANOVA is obtained through examining contrasts. The trick is in picking interesting contrasts to consider. Interesting contrasts are determined by the structure of the treatments or are suggested by the data.

The structure of the treatments often suggests a fixed group of contrasts that are of interest. For example, if one of the treatments is a standard treatment or a control, it is of interest to compare all of the other treatments to the standard. With a treatments, this leads to $a - 1$ contrasts. (These will not be orthogonal.) In Chapter 11 we will consider factorial treatment structures. These include cases such as four fertilizer treatments, say,

$$n_0p_0 \quad n_0p_1 \quad n_1p_0 \quad n_1p_1$$

where n_0p_0 is no fertilizer, n_0p_1 consists of no nitrogen fertilizer but application of a phosphorous fertilizer, n_1p_0 consists of a nitrogen fertilizer but no phosphorous fertilizer, and n_1p_1 indicates both types of fertilizer. Again the treatment structure suggests a fixed group of contrasts to examine. One interesting contrast compares the two treatments having nitrogen fertilizer against the two without nitrogen fertilizer, another compares the two treatments having phosphorous fertilizer against the two without phosphorous fertilizer, and a third contrast compares the effect of nitrogen fertilizer when phosphorous is not applied with the effect of nitrogen fertilizer when phosphorous is applied. Again, we have a treatments and $a - 1$ contrasts. In a balanced ANOVA, these $a - 1$ contrasts are orthogonal. Even when there is an apparent lack of structure in the treatments, the very lack of structure suggests a fixed group of contrasts. If there is no apparent structure, the obvious thing to do is compare all of the treatments with all of the other treatments. With three treatments, there are three distinct pairs of treatments to compare. With four treatments, there are six distinct pairs of treatments to compare. With five treatments, there are ten pairs. With seven treatments, there are 21 pairs. With 13 treatments, there are 78 pairs.

One problem is that, with a moderate number of treatment groups, there are many contrasts to look at. When we do tests or confidence intervals, there is a built in chance for error. The more statistical inferences we perform, the more likely we are to commit an error. The purpose of the multiple comparison methods examined in this chapter is to control the probability of making a specific type of error. When testing many contrasts, we have many null hypotheses. This chapter considers *multiple comparison methods that control (i.e., limit) the probability of making an error in any of the tests, when* all *of the null hypotheses are correct.* Limiting this probability is referred to as weak control of the *experimentwise error rate*. It is referred to as weak control because the control only applies under the very stringent assumption that all null hypotheses are correct. Some authors consider a different approach and define strong control of the experimentwise error rate as control of the probability of falsely rejecting any null hypothesis. Thus strong control limits the probability of false rejections even when some of the null hypotheses are false. Not everybody distinguishes between weak and strong control, so the definition of experimentwise error rate depends on whose work you are reading. One argument against weak control of the experimentwise error rate is that in designed experiments, you choose treatments that you expect to have different effects. In such cases, it makes little sense to concentrate on controlling the error under the assumption that all treatments have the same effect. On the other hand, strong control is more difficult to establish.

Our discussion of multiple comparisons focuses on testing whether contrasts are equal to 0. In all but one of the methods considered in this chapter, the experimentwise error rate is (weakly) controlled by first doing a test of the hypothesis $\mu_1 = \mu_2 = \cdots = \mu_a$. If this test is not rejected, we do not claim that any individual contrast is different from 0. In particular, if $\mu_1 = \mu_2 = \cdots = \mu_a$, any contrast among the means must equal 0, so all of the null hypotheses are correct. Since the error rate for the test of $\mu_1 = \mu_2 = \cdots = \mu_a$ is controlled, the weak experimentwise error rate for the contrasts is also controlled.

Many multiple testing procedures can be adjusted to provide multiple confidence intervals that have a guaranteed simultaneous coverage. Several such methods will be presented in this chapter.

Besides the treatment structure suggesting contrasts, the other source of interesting contrasts is having the data suggest them. If the data suggest a contrast, then the 'parameter' in our standard theory for statistical inferences is a function of the data and not a parameter in the usual sense of the word. When the data suggest the parameter, the standard theory for inferences does not apply. To handle such situations we can often include the contrasts suggested by the data in a broader class of contrasts and develop a procedure that applies to *all* contrasts in the class. In such cases we can ignore the fact that the data suggested particular contrasts of interest because these are still contrasts in the class and the method applies for all contrasts in the class. Of the methods considered in the current chapter, only Scheffé's method (discussed in Section 6.4) is generally considered appropriate for this kind of data dredging.

Recently, a number of books have been published on multiple comparison methods, e.g., Hochberg and Tamhane (1987). A classic discussion is Miller (1981), who also focuses on weak control of the experimentwise error rate, cf. Miller's section 1.2.

We present multiple comparison methods in the context of the one-way ANOVA model (5.1.1) but the methods extend easily to many other situations. We will use a single numerical example to illustrate most of the methods discussed in this chapter. The data are introduced in Example 6.0.1.

EXAMPLE 6.0.1. Mandel (1972) presented data on the stress at 600% elongation for natural rubber with a 40 minute cure at 140 ^{o}C. Stress was measured four times by each of 13 laboratories. The units for the data are kilograms per centimeter squared (kg/cm^2). The data are presented in Table 6.1. While an analysis of these data on the original scale is not unreasonable, the assumptions of equal variances and normality seem to be more nearly satisfied on the logarithmic scale. The standard summary statistics for computing the analysis of variance on the natural logs of the data are also given in Table 6.1.

This is a balanced one-way ANOVA, so the simple average of the 13 s_i^2s gives the *MSE*. There are three degrees of freedom for the variance estimate from each laboratory, so with 13 laboratories there are a total of 13(3) = 39 degrees of freedom for error. The mean squared error times the degrees of freedom for error gives the sum of squares for error. The sample variance of the 13 $\bar{y}_{i\cdot}$s is $s_{\bar{y}}^2 = .081806429$. Multiplying this by the number of observations in each group, 4, gives the *MSTrts*. The *MSTrts* times (13 − 1) gives the *SSTrts*. The sum of squares total is the sample variance of the

TABLE 6.1. Mandel's data on thirteen laboratories with summary statistics for the logs of the data

Lab	Observations				N	$\bar{y}_{i\cdot}$	s_i^2	s_i
1	133	129	123	156	4	4.9031	0.01061315	0.1030
2	129	125	136	127	4	4.8612	0.00134015	0.0366
3	121	125	109	128	4	4.7919	0.00502248	0.0709
4	57	58	59	67	4	4.0964	0.00540738	0.0735
5	122	98	107	110	4	4.6906	0.00814531	0.0903
6	109	120	112	107	4	4.7175	0.00252643	0.0503
7	80	72	76	64	4	4.2871	0.00915446	0.0957
8	135	151	143	142	4	4.9603	0.00210031	0.0458
9	69	69	73	70	4	4.2518	0.00071054	0.0267
10	132	129	141	137	4	4.9028	0.00155179	0.0394
11	118	109	115	106	4	4.7176	0.00239586	0.0489
12	133	133	129	128	4	4.8731	0.00040518	0.0201
13	86	84	96	81	4	4.4610	0.00535505	0.0732

logs of all 52 observations times $(52-1)$. The degrees of freedom total are $52-1$. These calculations are summarized in the analysis of variance table given in Table 6.2.

TABLE 6.2. Analysis of variance table for logs of Mandel's data

Source	df	SS	MS	F	P
Trts	12	3.92678	0.32723	77.73	0.000
Error	39	0.16418	0.00421		
Total	51	4.09097			

Figures 6.1, 6.2, and 6.3 give residual plots. Figure 6.1 is a plot of the residuals versus the predicted values. The group mean $\bar{y}_{i\cdot}$ is the predicted value for an observation from group i. Figure 6.1 shows no particular trend in the variabilities. Figure 6.2 is a plot of the residuals versus indicators of the 13 laboratories. Again, there are no obvious problems. Figure 6.3 gives a normal plot of the residuals; the plot looks quite straight.

For pedagogical purposes, on some occasions we consider only the first seven of the 13 treatment groups. We are not selecting these laboratories based on the data and we will continue to use the MSE and dfE from the full data. □

6.1 Fisher's least significant difference method

The easiest way to adjust for multiple comparisons is to use R. A. Fisher's least significant difference method. To put it as simply as possible, with this method you first look at the analysis of variance F test for whether there are differences between the groups. If this test provides no evidence of differences, you quit and go home. If the test is significant at, say, the $\alpha = .05$ level, you just ignore the multiple comparison problem and do all other tests in the usual way at the .05 level. *This method is generally considered inappropriate for use with contrasts suggested by the data.* While the theoretical basis for excluding contrasts suggested by the data is not clear (at least relative to weak control of the experimentwise error rate), experience indicates that the method rejects far too many individual null hypotheses if this exclusion is not applied. In addition, many people would not apply the method unless the number of comparisons to be made was quite small.

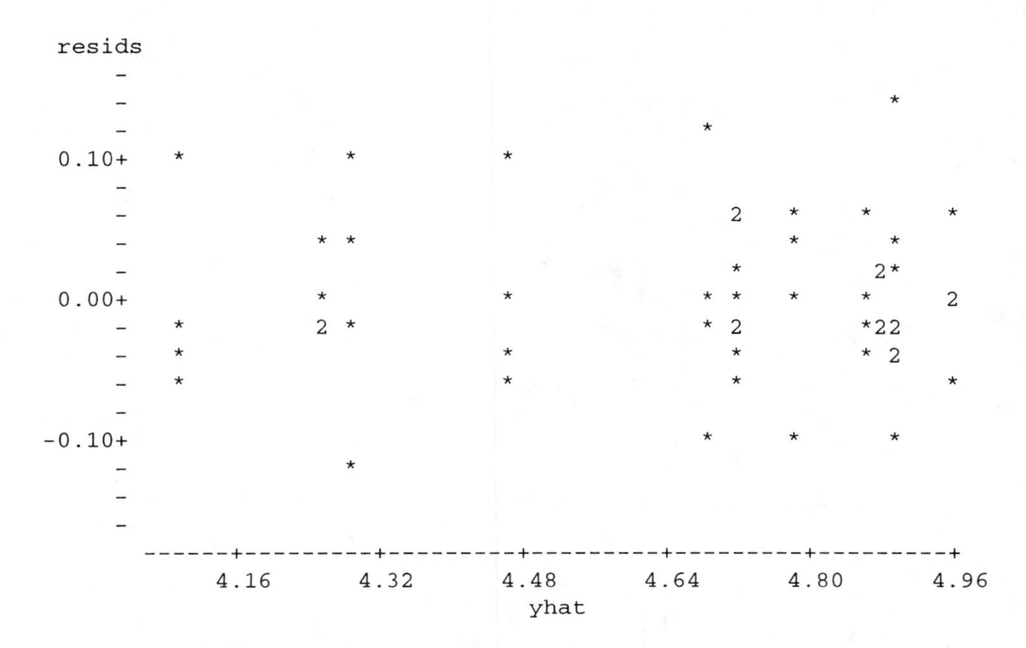

FIGURE 6.1. Plot of residuals versus predicted values.

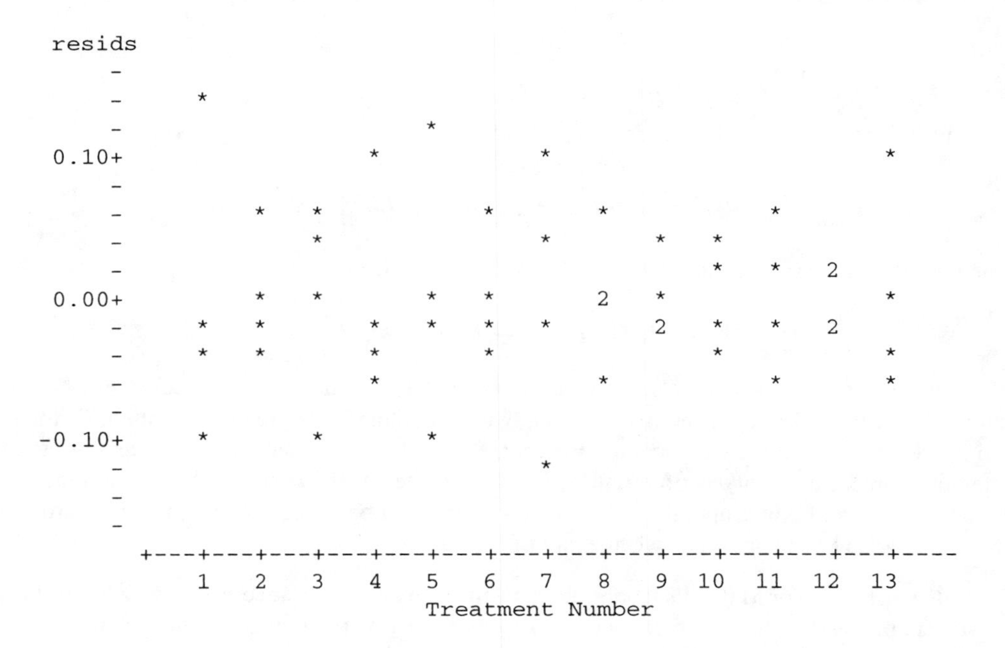

FIGURE 6.2. Plot of residuals versus treatment number.

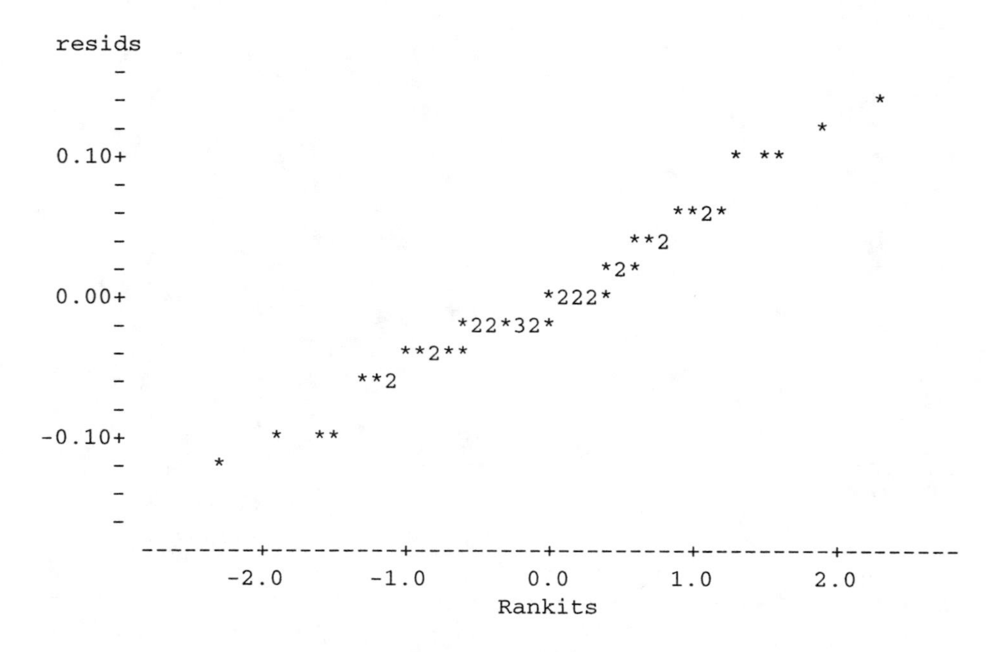

FIGURE 6.3. Normal plot of residuals, $W' = 0.976$.

The term 'least significant difference' comes from comparing pairs of means in a balanced ANOVA. There is a number, the least significant difference (*LSD*), such that the difference between two means must be greater than the *LSD* for the corresponding treatments to be considered significantly different. Generally, we have a significant difference between μ_i and μ_j if

$$\frac{|\bar{y}_{i\cdot} - \bar{y}_{j\cdot}|}{\sqrt{MSE\left[\frac{1}{N} + \frac{1}{N}\right]}} > t\left(1 - \frac{\alpha}{2}, dfE\right).$$

Multiplying both sides by the standard error leads to rejection if

$$|\bar{y}_{i\cdot} - \bar{y}_{j\cdot}| > t\left(1 - \frac{\alpha}{2}, dfE\right) \sqrt{MSE\left[\frac{1}{N} + \frac{1}{N}\right]}.$$

The number on the right is defined as the least significant difference,

$$LSD \equiv t\left(1 - \frac{\alpha}{2}, dfE\right) \sqrt{MSE\frac{2}{N}}.$$

Note that the *LSD* depends on the choice of α but does not depend on which means are being examined. If the absolute difference between two sample means is greater than the *LSD* the population means are declared significantly different. Recall, however, that these comparisons are never attempted unless the analysis of variance F test is rejected at the α level. The reason that a single number exists for comparing all pairs of means is that in a balanced ANOVA the standard error is the same for any comparison between a pair of means.

EXAMPLE 6.1.1. For Mandel's laboratory data, the analysis of variance F test is highly significant, so we can proceed to make individual comparisons among pairs of means. With $\alpha = .05$,

$$LSD = t(.975, 39)\sqrt{.00421\left[\frac{1}{4} + \frac{1}{4}\right]} = 2.023(.0459) = .093$$

Means that are greater than .093 apart are significantly different. Means that are less than .093 apart are not significantly different. We display the results visually. Order the sample means from smallest to largest and indicate groups of means that are not significantly different by underlining the group. Such a display is given below for comparing laboratories 1 through 7.

Lab.	4	7	5	6	3	2	1
Mean	4.0964	4.2871	4.6906	4.7175	4.7919	4.8612	4.9031

Laboratories 4 and 7 are distinct from all other laboratories. All the other consecutive pairs of labs are insignificantly different. Thus labs 5 and 6 cannot be distinguished. Similarly, labs 6 and 3 cannot be distinguished, 3 and 2 cannot be distinguished, and labs 2 and 1 cannot be distinguished. However, lab 5 is significantly different from labs 3, 2, and 1. Lab 6 is significantly different from labs 2 and 1. Also, lab 3 is different from lab 1.

To be completely correct, *when comparing just the first 7 laboratories the LSD method should be based on an F test for just those 7 laboratories* rather than the F test from Table 6.2 which is based on all 13 laboratories. This can be done by computing a *MSTrts* in the usual way from the sample means of just the first 7 labs. The resulting F test has 6 degrees of freedom in the numerator and is highly significant ($F = 89.84$). Unfortunately, this point is often ignored in practice.

We can also use the *LSD* to compare all 13 laboratories. Again, we use a visual display, but with more means we list the ordered means vertically and use letters, rather than lines, to indicate groups that are not significantly different.

Lab.	Mean			
4	4.0964	A		
9	4.2518	B		
7	4.2871	B		
13	4.4610	C		
5	4.6906	D		
6	4.7175	D	E	
11	4.7176	D	E	
3	4.7919	F	E	
2	4.8612	F	G	
12	4.8731	F	G	H
10	4.9028		G	H
1	4.9031		G	H
8	4.9603			H

For example, labs 12, 10, 1, and 8 all share the letter H, so there are no significant differences declared among those four labs. □

For testing a group of contrasts that are 1) not just comparisons between pairs of means or 2) not from a balanced ANOVA, first perform the analysis of variance F test at the α level and if it is rejected, test $H_0 : \sum_i \lambda_i \mu_i = 0$ by rejecting if

$$\frac{SS\left(\sum_i \lambda_i \mu_i\right)}{MSE} > F(1-\alpha, 1, dfE).$$

Alternatively, one can use the equivalent t tests for the contrasts.

EXAMPLE 6.1.2. Suppose that in Mandel's data the first two laboratories are in San Francisco, the second two are in Seattle, the fifth is in New York, and the sixth and seventh are in Boston. This structure to the treatments suggests some interesting orthogonal contrasts. We can compare the average of the labs on the West Coast with the average of the labs on the East Coast. On the West Coast we can compare the average of the San Francisco labs with the average of the Seattle labs, we can compare the San Francisco labs with each other and the Seattle labs with each other. On the East Coast we can compare the New York lab with the average of the Boston labs and the Boston labs with each other. The contrast coefficients along with estimates and sums of squares are given in Table 6.3. The contrasts involving averages have been multiplied by appropriate constants to get simple integer contrast coefficients.

TABLE 6.3. Orthogonal contrasts for the first seven laboratories in Mandel's data

	Contrast coefficients					
Lab.	C_1	C_2	C_3	C_4	C_5	C_6
1	3	1	1	0	0	0
2	3	1	-1	0	0	0
3	3	-1	0	1	0	0
4	3	-1	0	-1	0	0
5	-4	0	0	0	2	0
6	-4	0	0	0	-1	1
7	-4	0	0	0	-1	-1
Est	1.177	.8760	.0418	.6954	.3765	.4305
SS	.0660	.7674	.0035	.9673	.0945	.3706

Recalling that the overall F test is highly significant for the first 7 labs, to perform the $\alpha = .05$ level LSD method on the contrasts of Table 6.3, just divide each sum of squares by $MSE = .00421$ to get an F statistic and compare the F statistics to $F(.95, 1, 39) = 4.09$. The F statistics are given below.

Contrast	C_1	C_2	C_3	C_4	C_5	C_6
F	15.68	182.28	0.83	229.76	22.45	88.03

All of the contrasts are significantly different from zero except C_3, the comparison between the two labs in San Francisco. □

Apparently some people have taken to calling this method the Fisher significant difference (FSD) method. One suspects that this is a reaction to another meaning commonly associated with the letters LSD. I, for one, would *never* suggest that only people who are hallucinating would believe all differences declared by LSD are real.

6.2 Bonferroni adjustments

The Bonferroni method is the one method we consider that *does not* stem from a test of $\mu_1 = \mu_2 = \cdots = \mu_a$. Rather, it controls the experimentwise error rate by employing a simple adjustment to the significance level of each individual test. If you have planned to do s tests, you just perform each test at the α / s level rather than at the α level. This method is *absolutely not appropriate for contrasts that are suggested by the data.*

The justification for Bonferroni's method relies on a very simple result from probability: for two events, the probability that one or the other event occurs is no more than the sum of the probabilities for the individual events. Thus with two tests, say A and B, the probability that we reject A or reject B is less than or equal to the probability of rejecting A plus the probability of rejecting B. In particular, if we fix the probability of rejecting A at $\alpha/2$ and the probability of rejecting B at $\alpha/2$, then the probability of rejecting A or B is no more than $\alpha/2 + \alpha/2 = \alpha$. More generally, if we have s tests and control the probability of type I error for each test at α/s, then the probability of rejecting any of the tests when all s null hypotheses are true is no more than $\alpha/s + \cdots + \alpha/s = \alpha$.

To compare pairs of means in a balanced ANOVA, as with the least significant difference method, there is a single number to which we can compare the differences in means. For a fixed α, this number is called the *Bonferroni significant difference* and takes on the value

$$BSD \equiv t\left(1 - \frac{\alpha}{2s}, dfE\right) \sqrt{MSE\left[\frac{1}{N} + \frac{1}{N}\right]}.$$

Recall for comparison that with the least significant difference method, the necessary tabled value is $t(1 - \alpha/2, dfE)$, which is always smaller than the tabled value for the BSD. Thus the BSD is always larger than the LSD and the BSD tends to display fewer differences among the means than the LSD.

When testing a group of contrasts that are not just comparisons between pairs of means in a balanced ANOVA, reject a particular contrast hypothesis $H_0 : \sum_i \lambda_i\mu_i = 0$ if

$$\frac{SS\left(\sum_i \lambda_i\mu_i\right)}{MSE} > F\left(1 - \frac{\alpha}{s}, 1, dfE\right).$$

Equivalent adjustments can be made when performing t rather than F tests.

Bonferroni adjustments can also be used to obtain confidence intervals that have a simultaneous confidence of $(1 - \alpha)100\%$ for covering all of the contrasts. The endpoints of these intervals are

$$\sum_{i=1}^{a} \lambda_i\bar{y}_{i\cdot} \pm t\left(1 - \frac{\alpha}{2s}, dfE\right) SE\left(\sum_{i=1}^{a} \lambda_i\bar{y}_{i\cdot}\right).$$

Recall that for an unbalanced ANOVA,

$$SE\left(\sum_{i=1}^{a} \lambda_i\bar{y}_{i\cdot}\right) = \sqrt{MSE\sum_{i=1}^{a}\frac{\lambda_i^2}{N_i}}.$$

Only the tabled value distinguishes this interval from a standard confidence interval for $\sum_{i=1}^{a} \lambda_i\mu_i$. In the special case of comparing pairs of means in a balanced ANOVA, the Bonferroni confidence interval for, say, $\mu_i - \mu_j$ reduces to

$$\left(\bar{y}_{i\cdot} - \bar{y}_{j\cdot}\right) \pm BSD.$$

For these intervals, we are $(1 - \alpha)100\%$ confident that the collection of all such intervals simultaneously contain all of the corresponding differences between pairs of population means.

EXAMPLE 6.2.1. In comparing the first 7 laboratories, we have $\binom{7}{2} = 21$ pairs of laboratories to contrast. The Bonferroni significant difference for $\alpha = .05$ is

$$BSD = t\left(1 - \frac{.025}{21}, 39\right) \sqrt{.00421\left[\frac{1}{4} + \frac{1}{4}\right]} = t(.99881, 39).04588 = 3.2499(.04588) = .149\,.$$

Means that are greater than .149 apart are significantly different. Means that are less than .149 apart are not significantly different. Once again, we display the results visually. We order the sample means from smallest to largest and indicate groups of means that are not significantly different by underlining the group.

Lab.	4	7	5	6	3	2	1
Mean	4.0964	4.2871	4.6906	4.7175	4.7919	4.8612	4.9031
			———	———	———		
				———	———	———	
					———	———	———

Laboratories 4 and 7 are distinct from all other laboratories. Labs 5, 6, and 3 cannot be distinguished. Similarly, labs 6, 3, and 2 cannot be distinguished; however, lab 5 is significantly different from lab 2 and also lab 1. Labs 3, 2, and 1 cannot be distinguished, but lab 1 is significantly different from lab 6.

The Bonferroni simultaneous 95% confidence interval for, say, $\mu_2 - \mu_5$ has endpoints

$$(4.8612 - 4.6906) \pm .149$$

which gives the interval (.021,.320). Transforming back to the original scale from the logarithmic scale, we are 95% confident that values for lab 2 average being between $e^{.021} = 1.02$ and $e^{.320} = 1.38$ times greater than the values for lab 5. Similar conclusions are drawn for the other twenty comparisons between pairs of means.

If we examine all 13 means, we have $\binom{13}{2} = 78$ comparisons to make. The Bonferroni significant difference for $\alpha = .05$ is

$$BSD = t\left(1 - \frac{.025}{78}, 39\right)\sqrt{.00421\left[\frac{1}{4} + \frac{1}{4}\right]} = t(.9997, 39).04588 = 3.7125(.04588) = .170.$$

Unlike the *LSD*, with more means to consider the *BSD* is larger. Now, means that are greater than .170 apart are significantly different. Means that are less than .170 apart are not significantly different. Again, we use a visual display, but with more means we list the ordered means vertically and use letters, rather than lines, to indicate groups that are not significantly different.

Lab.	Mean			
4	4.0964	A		
9	4.2518	A	B	
7	4.2871		B	
13	4.4610	C		
5	4.6906	D		
6	4.7175	D	E	
11	4.7176	D	E	
3	4.7919	D	E	F
2	4.8612		E	F
12	4.8731		E	F
10	4.9028			F
1	4.9031			F
8	4.9603			F

Here, for example, labs 4 and 9 are not significantly different, nor are labs 9 and 7 but 4 and 7 are different. Lab 13 is significantly different from all other labs. □

EXAMPLE 6.2.2. Consider again the six contrasts from Example 6.1.2 and Table 6.3. To perform the $\alpha = .05$ level Bonferroni adjustments on these six contrasts, once again divide the sums of squares in Table 6.3 by the *MSE* to get *F* statistics but now compare the *F* statistics to $F(.991\bar{6}, 1, 39) = 7.73$, where $.991\bar{6} = 1 - .05/6$. As given in Example 6.1.2, the *F* statistics are

Contrast	C_1	C_2	C_3	C_4	C_5	C_6
F	15.68	182.28	0.83	229.76	22.45	88.03

Comparing these to 7.73 shows that once again all of the contrasts are significantly different from zero except C_3, the comparison between the two labs in San Francisco. □

MINITAB COMMANDS

Minitab can be used to obtain the *F* and *t* percentage points needed for Bonferroni's method. In this section we have used $t(.99881, 39)$, $t(.9997, 39)$, and $F(.991\bar{6}, 1, 39)$. To obtain these, use Minitab's inverse cumulative distribution function command.

```
MTB > invcdf .99881;
SUBC> t 39.
MTB > invcdf .9997;
SUBC> t 39.
MTB > invcdf .9916666;
SUBC> f 1 39.
```

6.3 Studentized range methods

Studentized range methods are generally used *only for comparing pairs of means in balanced analysis of variance problems.* They are not based on the analysis of variance *F* test but on an alternative test of $\mu_1 = \mu_2 = \cdots = \mu_a$.

The *range* of a random sample is the difference between the largest observation and the smallest observation. For a known variance σ^2, the *range* of a random sample from a normal population has a distribution that can be worked out. This distribution depends on σ^2 and the number of observations in the sample. It is only reasonable that the distribution depend on the number of observations because the difference between the largest and smallest observations ought to be larger in a sample of 75 observations than in a sample of 3 observations. Just by chance, we would expect the extreme observations to become more extreme in larger samples.

Knowing the distribution of the range is not very useful because the distribution depends on σ^2, which we do not know. To eliminate this problem, divide the range by an independent estimate of the standard deviation, say, $\hat{\sigma}$ having $r\hat{\sigma}^2/\sigma^2 \sim \chi^2(r)$. The distribution of this *studentized range* no longer depends on σ^2 but rather it depends on the degrees of freedom for the variance estimate. For a sample of *n* observations and a variance estimate with *r* degrees of freedom, the distribution of the studentized range is written as

$$Q(n, r).$$

Tables are given in Appendix B.5. The α percentile is denoted $Q(\alpha, n, r)$.

As discussed in Section 5.2, if $\mu_1 = \mu_2 = \cdots = \mu_a$ in a balanced ANOVA, the $\bar{y}_{i\cdot}$s form a random sample of size *a* from a $N(\mu_1, \sigma^2/N)$ population. Looking at the range of this sample and dividing by the natural independent chi-squared estimate of the standard deviation leads to the statistic

$$Q = \frac{\max \bar{y}_{i\cdot} - \min \bar{y}_{i\cdot}}{\sqrt{MSE/N}}$$

If the observed value of this studentized range statistic is consistent with its coming from a $Q(a, dfE)$ distribution, then the data are consistent with the null hypothesis of equal means μ_i. If the μ_is are not all equal, the studentized range Q tends to be larger than if the means were all equal; the difference between the largest and smallest observations will involve not only random variation but also the differences in the μ_is. Thus, for an $\alpha = .05$ level test, if the observed value of Q is larger than $Q(.95, a, dfE)$, we reject the claim that the means are all equal.

The studentized range multiple comparison methods discussed in this section begin with this studentized range test.

6.3.1 TUKEY'S HONEST SIGNIFICANT DIFFERENCE

John Tukey's honest significant difference method is to reject the equality of a pair of means, say, μ_i and μ_j at the $\alpha = .05$ level, if

$$\frac{|\bar{y}_{i\bullet} - \bar{y}_{j\bullet}|}{\sqrt{MSE/N}} > Q(.95, a, dfE).$$

Obviously, this test cannot be rejected for any pair of means unless the test based on the maximum and minimum sample means is also rejected. For an equivalent way of performing the test, reject equality of μ_i and μ_j if

$$|\bar{y}_{i\bullet} - \bar{y}_{j\bullet}| > Q(.95, a, dfE)\sqrt{MSE/N}.$$

With a fixed α, the honest significant difference is

$$HSD \equiv Q(1 - \alpha, a, dfE)\sqrt{MSE/N}.$$

For any pair of sample means with an absolute difference greater than the HSD, we conclude that the corresponding population means are significantly different. The HSD is the number that an observed difference must be greater than in order for the population means to have an 'honestly' significant difference. The use of the word 'honest' is a reflection of the view that the LSD method allows 'too many' rejections.

Tukey's method can be extended to provide simultaneous $(1 - \alpha)100\%$ confidence intervals for all differences between pairs of means. The interval for the difference $\mu_i - \mu_j$ has end points

$$\bar{y}_{i\bullet} - \bar{y}_{j\bullet} \pm HSD$$

where HSD depends on α. For $\alpha = .05$, we are 95% confident that the collection of all such intervals simultaneously contains all of the corresponding differences between pairs of population means.

EXAMPLE 6.3.1. For comparing the first 7 laboratories in Mandel's data with $\alpha = .05$, the honest significant difference is approximately

$$HSD = Q(.95, 7, 40)\sqrt{MSE/4} = 4.39\sqrt{.00421/4} = .142.$$

Here we have used $Q(.95, 7, 40)$ rather than the correct value $Q(.95, 7, 39)$ because the correct value was not available in the table used. Treatment means that are more than .142 apart are significantly different. Means that are less than .142 apart are not significantly different. Note that the HSD value is similar to the corresponding BSD value of .149; this frequently occurs. Once again, we display the results visually.

Lab.	4	7	5	6	3	2	1
Mean	4.0964	4.2871	4.6906	4.7175	4.7919	4.8612	4.9031

These results are nearly the same as for the *BSD* except that labs 6 and 2 are significantly different by the *HSD* criterion.

The HSD simultaneous 95% confidence interval for, say, $\mu_2 - \mu_5$ has endpoints

$$(4.8612 - 4.6906) \pm .142$$

which gives the interval (.029, .313). Transforming back to the original scale from the logarithmic scale, we are 95% confident that values for lab 2 average being between $e^{.029} = 1.03$ and $e^{.313} = 1.37$ times greater than values for lab 5. Again, there are 20 more intervals to examine.

If we consider all 13 means, the honest significant difference is approximately

$$HSD = Q(.95, 13, 40)\sqrt{MSE/4} = 4.98\sqrt{.00421/4} = .162$$

Unlike the *LSD*, but like the *BSD*, with more means to consider the *HSD* is larger. Now, means that are greater than .162 apart are significantly different. Means that are less than .162 apart are not significantly different. Again, we use a vertical display with letters to indicate groups that are not significantly different.

Lab.	Mean			
4	4.0964	A		
9	4.2518	A	B	
7	4.2871		B	
13	4.4610	C		
5	4.6906	D		
6	4.7175	D	E	
11	4.7176	D	E	
3	4.7919	D	E	F
2	4.8612	G	E	F
12	4.8731	G	E	F
10	4.9028	G		F
1	4.9031	G		F
8	4.9603	G		

The results are similar to those for the corresponding *BSD* of .170 except that labs 3 and 8 are now different. □

6.3.2 NEWMAN–KEULS MULTIPLE RANGE METHOD

The Newman–Keuls multiple range method involves repeated use of the honest significant difference method with some minor adjustments. Multiple range methods are difficult to describe in general, so we simply demonstrate how they work.

EXAMPLE 6.3.2. To use the Newman–Keuls method for comparing the first 7 laboratories in Mandel's data, we need the *HSD* value for comparing not only 7 laboratories, but also for comparing 6, 5, 4, 3, and 2 laboratories. Table 6.4 presents all the values needed, not only for comparing the first 7 labs, but also for comparing all 13 labs. Again, we approximate $Q(.95, r, 39)$ with $Q(.95, r, 40)$, so $HSD = Q(.95, r, 40)\sqrt{MSE/4}$ where $\sqrt{MSE/4} = \sqrt{.00421/4} = .0324423$.

As before, the seven means are ordered from smallest to largest. The smallest mean, 4.0964, and the largest mean, 4.9031, are compared using the $r = 7$ value of *HSD* from Table 6.4. These means are more than .142 apart so we go to the next stage.

At the second stage, the smallest mean, 4.0964, is compared with the second largest mean, 4.8612, and the second smallest mean, 4.2871, is compared to largest mean, 4.9031. These are groups of

TABLE 6.4. Comparison values for Newman–Keuls method as applied to Mandel's data

r	$Q(.95, r, 40)$	HSD	r	$Q(.95, r, 40)$	HSD
13	4.98	.162	7	4.39	.142
12	4.90	.159	6	4.23	.137
11	4.82	.156	5	4.04	.131
10	4.74	.154	4	3.79	.123
9	4.64	.151	3	3.44	.112
8	4.52	.147	2	2.86	.093

means that are 6 apart, so they are compared using the *HSD* value for $r = 6$. Both differences in means are greater than .137, so we progress to the third stage.

In the third stage, the smallest mean, 4.0964, is compared to the third largest mean, 4.7919, the second smallest mean, 4.2871, is compared to the second largest mean, 4.8612, and the third smallest mean, 4.6906, is compared to the largest mean, 4.9031. These are groups of means that are 5 apart, so they are compared using the *HSD* value for $r = 5$. All three differences in means are greater than .131, so we progress to the fourth stage and so on.

At any particular stage, means that are r apart get compared using the *HSD* value for comparing groups of r means. The only exception to this rule is that *if at any given stage we conclude that certain means are not significantly different, then at later stages we never reconsider the possibility that they may contain significant differences.* The standard visual display is given below.

Lab.	4	7	5	6	3	2	1
Mean	4.0964	4.2871	4.6906	4.7175	4.7919	4.8612	4.9031

All ordered means that were $r = 4$ apart were different. Of the means that were $r = 3$ apart, two groups were not significantly different. One of these consists of labs 5, 6, and 3, while the other group consists of labs 3, 2, and 1. For $r = 2$, we do not consider the possibility that there may be differences between labs 5, 6, and 3 or between labs 3, 2, and 1. We do consider possible differences between 4 and 7 and between 7 and 5.

If the mean for lab 6 was 4.7875, rather than its actual value 4.7175, the exception referred to in the previous paragraph would have come into play. In examining labs 5, 6, and 3, the difference between the largest and smallest of the three consecutive means 4.6906, 4.7875, and 4.7919 would still be less than the *HSD* for $r = 3$ which is .112. Thus the three labs would still be considered not significantly different. The rule is that, since the three are not significantly different, we no longer consider the possibility that any subset of the means could be different. If we allowed ourselves to compare the consecutive means 4.6906 and 4.7875 with $r = 2$, the appropriate *HSD* value is .093 and the means for labs 5 and 6 would be considered significantly different. However, because the triple 4.6906, 4.7875, and 4.7919 are not significantly different, we never compare 4.6906 and 4.7875 directly.

The visual display for all 13 laboratories is given below.

Lab.	Mean		
4	4.0964	A	
9	4.2518	B	
7	4.2871	B	
13	4.4610	C	
5	4.6906	D	
6	4.7175	D	
11	4.7176	D	
3	4.7919	D	E
2	4.8612	F	E
12	4.8731	F	E
10	4.9028	F	E
1	4.9031	F	E
8	4.9603	F	

□

6.4 Scheffé's method

Scheffé's method is valid for examining any and all contrasts simultaneously. *This method is primarily used with contrasts that were suggested by the data.* Scheffé's method should not be used for comparing pairs of means in a balanced ANOVA because the *HSD* method has properties comparable to Scheffé's but is better for comparing pairs of means.

Scheffé's method is closely related to the analysis of variance F test. Recalling the definition of the *MSTrts*, the analysis of variance F test is rejected when

$$\frac{SSTrts/(a-1)}{MSE} > F(1-\alpha, a-1, dfE). \tag{6.4.1}$$

Recall from Section 5.4 that for any contrast $\sum_i \lambda_i \mu_i$,

$$SS\left(\sum_i \lambda_i \mu_i\right) \leq SSTrts. \tag{6.4.2}$$

It follows immediately that

$$\frac{SS\left(\sum_i \lambda_i \mu_i\right)/(a-1)}{MSE} \leq \frac{SSTrts/(a-1)}{MSE}.$$

Scheffé's method is to replace *SSTrts* in (6.4.1) with $SS\left(\sum_i \lambda_i \mu_i\right)$ and to reject $H_0 : \sum_i \lambda_i \mu_i = 0$ if

$$\frac{SS\left(\sum_i \lambda_i \mu_i\right)/(a-1)}{MSE} > F(1-\alpha, a-1, dfE).$$

From (6.4.1) and (6.4.2), Scheffé's test cannot possibly be rejected unless the ANOVA test is rejected. This controls the experimentwise error rate for multiple tests. However, there always exists a contrast that contains all of the *SSTrts*, i.e., there is always a contrast that achieves equality in relation (6.4.2), so if the ANOVA test is rejected, there is always some contrast that can be rejected using Scheffé's method. This contrast may not be interesting but it exists, cf. Section 5.4.

Scheffé's method can be adapted to provide simultaneous $(1-\alpha)100\%$ confidence intervals for contrasts. These have the endpoints

$$\sum_{i=1}^{a} \lambda_i \bar{y}_{i\cdot} \pm \sqrt{(a-1)\,F(1-\alpha, a-1, dfE)}\; SE\left(\sum_{i=1}^{a} \lambda_i \bar{y}_{i\cdot}\right)$$

EXAMPLE 6.4.1. Just for a change, we reexamine the electrical characteristic data of Chapter 5 rather than illustrating the methods with Mandel's data. The electrical characteristic data has $MSE = .56155$ with $dfE = 24$. We examined the orthogonal contrasts

$$C_1 \equiv (1)\mu_1 + (-1)\mu_2 + (0)\mu_3 + (0)\mu_4 = \mu_1 - \mu_2,$$

$$C_2 \equiv (1/2)\mu_1 + (1/2)\mu_2 + (-1)\mu_3 + (0)\mu_4 = \frac{\mu_1 + \mu_2}{2} - \mu_3,$$

and

$$C_3 \equiv (1/3)\mu_1 + (1/3)\mu_2 + (1/3)\mu_3 + (-1)\mu_4 = \frac{\mu_1 + \mu_2 + \mu_3}{3} - \mu_4.$$

The sums of squares for C_1, C_2, and C_3 are

$$SS(C_1) = 0.0178, \ SS(C_2) = 0.7738, \text{ and } SS(C_3) = 10.0818.$$

These orthogonal contrasts were constructed because C_3 is easily interpretable and contains a large proportion of the available sum of squares for treatments; $SSTrts = 10.873$. The vast bulk of the treatment differences are due to the difference between sheet 4 and the average of the other sheets. The data suggest these contrasts, so it is not appropriate to ignore the selection process when testing whether the contrasts are 0. Scheffé's method compares

$$\frac{SS(C_3)/(a-1)}{MSE} = \frac{10.0818/3}{.56155} = 5.98$$

to an $F(3, 24)$ distribution. $F(.999, 3, 24) = 7.55$ and $F(.99, 3, 24) = 4.72$, so there is very substantial evidence that sheet 4 differs from the average of the other sheets as evaluated using Scheffé's method. The similar computation for C_1 gives an F of 0.01 and for C_2 an F of 0.46. Both are less than 1, so neither is significant.

In our earlier consideration of these data, we also examined the orthogonal contrasts

$$C_5 \equiv (1)\mu_1 + (1)\mu_2 + (0)\mu_3 + (-2)\mu_4,$$

C_1, and

$$C_6 \equiv (1)\mu_1 + (1)\mu_2 + (-3)\mu_3 + (1)\mu_4.$$

The sums of squares for C_5, C_1, and C_6 are

$$SS(C_5) = 10.803, \ SS(C_1) = 0.018, \text{ and } SS(C_6) = 0.052.$$

These contrasts were specifically constructed so that $S(C_5) \doteq SSTrts$. The only way to test a contrast that was constructed so as to contain all of the sums of squares treatments is to behave as if the contrast were the entire contribution from the treatments. Scheffé's method uses the test statistic

$$\frac{SS(C_5)/(a-1)}{MSE} = \frac{10.803/3}{.56155} = 6.41$$

and compares it to an $F(3, 24)$ distribution. This is essentially the analysis of variance F test.

The 95% Scheffé confidence interval for C_3 has endpoints

$$(1/3)16.4429 + (1/3)16.5143 + (1/3)16.0714 + (-1)14.9571$$
$$\pm \sqrt{3\,F(.95, 3, 24)} \sqrt{.56155 \frac{(1/3)^2 + (1/3)^2 + (1/3)^2 + (-1)^2}{7}}.$$

$F(.95, 3, 24) = 3.01$, so the endpoints reduce to $1.386 \pm .983$ and the interval is $(0.40, 2.37)$. □

As with the *LSD* method, the overall F test and thus Scheffé's method should be adapted to the contrasts of interest. For example, if we are considering only the first seven labs in Mandel's data, we would use an overall F test with only six degrees of freedom in the numerator and Scheffé's method for examining contrasts among the seven labs uses 6 in place of $a - 1 = 13$.

6.5 Other methods

Other multiple comparison methods have been developed that are similar in spirit to the studentized range methods. Just as studentized range methods were developed for comparing pairs of means in balanced analysis of variance problems, these other methods were developed for examining other sets of contrasts in balanced ANOVA. Again, the methods are not based on the analysis of variance F test but on alternative tests of $\mu_1 = \mu_2 = \cdots = \mu_a$. We will briefly discuss two of these methods: Ott's analysis of means method (AOM) and Dunnett's many-one t statistics. In addition, we mention another studentized range method proposed by Duncan that can also be modified for application with AOM and Dunnett's method.

6.5.1 OTT'S ANALYSIS OF MEANS METHOD

Ott (1967) introduced a graphical method called analysis of means for comparing each mean to the average of all the means. It is most often used in quality control work and the graphical method is closely related to control charts for means, cf. Shewhart (1931). Ott's work was founded upon earlier work that is referenced in his article. Nelson (1993) contains a brief, clear introduction, extensions, some tables, and references to other tables.

Balanced one-way ANOVA methods are founded on the fact that if $\mu_1 = \mu_2 = \cdots = \mu_a$, the $\bar{y}_{i\cdot}$s form a random sample of size a from a $N(\mu_1, \sigma^2 / N)$ population. We have already seen that the distribution of the studentized range is known when $\mu_1 = \mu_2 = \cdots = \mu_a$, so comparing the observed studentized range to the known distribution provides a test of $H_0 : \mu_1 = \mu_2 = \cdots = \mu_a$. This test was then modified to provide multiple comparison methods.

The AOM method is based on knowing the distribution of

$$\max_i \frac{|\bar{y}_{i\cdot} - \bar{y}_{\cdot\cdot}|}{SE(\bar{y}_{i\cdot} - \bar{y}_{\cdot\cdot})}$$

when the null hypothesis $H_0 : \mu_1 = \mu_2 = \cdots = \mu_a$ is true. The distribution depends on the number of treatments a and the dfE. The $1 - \alpha$ percentile of this distribution is often denoted $h(\alpha, a, dfE)$. An α level test of H_0 is rejected if

$$\max_i \frac{|\bar{y}_{i\cdot} - \bar{y}_{\cdot\cdot}|}{SE(\bar{y}_{i\cdot} - \bar{y}_{\cdot\cdot})} > h(\alpha, a, dfE).$$

When this test is rejected, we can do multiple comparisons to identify which individual means are different from the overall average. A particular mean μ_i is considered different if

$$\frac{|\bar{y}_{i\cdot} - \bar{y}_{\cdot\cdot}|}{SE(\bar{y}_{i\cdot} - \bar{y}_{\cdot\cdot})} > h(\alpha, a, dfE).$$

Clearly, if the overall test is not rejected, none of the individual means will be considered different.

The test given above is easily seen to be equivalent to the following: μ_i is considered different from the average of all the μs (or equivalently from the average of the other μs) if $\bar{y}_{i\cdot}$ is *not* between the values

$$\bar{y}_{\cdot\cdot} \pm h(\alpha, a, dfE) SE(\bar{y}_{i\cdot} - \bar{y}_{\cdot\cdot}).$$

This leads to a simple graphical procedure. Plot the pairs $(i, \bar{y}_{i\cdot})$. On this plot add horizontal lines at $\bar{y}_{\cdot\cdot} \pm h(\alpha, a, dfE) SE(\bar{y}_{i\cdot} - \bar{y}_{\cdot\cdot})$. Any $\bar{y}_{i\cdot}$ that lies outside the horizontal lines indicates a μ_i that is different from the others. While it is not crucial, traditionally the graphical display also includes a center line at $\bar{y}_{\cdot\cdot}$. This graphical display is very similar to a control chart for means. The AOM is focused on testing whether one particular mean is different from the rest of the means. This may be particularly appropriate for quality control problems.

The primary detail that we have not yet covered is the exact formula for $SE(\bar{y}_{i\cdot} - \bar{y}_{\cdot\cdot})$. To compute this, first note that

$$\begin{aligned} \bar{y}_{i\cdot} - \bar{y}_{\cdot\cdot} &= \bar{y}_{i\cdot} - \frac{\bar{y}_{1\cdot} + \cdots + \bar{y}_{a\cdot}}{a} \\ &= \frac{a-1}{a}\bar{y}_{i\cdot} - \frac{1}{a}\sum_{k \neq i} \bar{y}_{k\cdot}. \end{aligned}$$

It follows that the standard error is

$$\begin{aligned} SE(\bar{y}_{i\cdot} - \bar{y}_{\cdot\cdot}) &= \sqrt{MSE\left[\left(\frac{a-1}{a}\right)^2 + (a-1)\left(\frac{1}{a}\right)^2\right] \Big/ N} \\ &= \sqrt{MSE\left[\frac{a-1}{aN}\right]}. \end{aligned}$$

In fact, this argument explicates exactly what AOM is examining. AOM is simultaneously testing whether the contrasts $[(a-1)/a]\mu_i - (1/a)\sum_{k \neq i} \mu_k$, $i = 1, \ldots, a$ are all equal to 0. Equivalently, we can multiply the contrasts by a and think of the contrasts as being $(a-1)\mu_i - \sum_{k \neq i} \mu_k$, $i = 1, \ldots, a$. It is not difficult to see that these contrasts all equal 0 if and only if the μ_is are all equal.

A modification similar to the Newman–Keuls procedure can be used with AOM. The modification involves changing the value of a in $h(\alpha, a, dfE)$. Order the values of $|\bar{y}_{i\cdot} - \bar{y}_{\cdot\cdot}|$. When examining the largest value of $|\bar{y}_{i\cdot} - \bar{y}_{\cdot\cdot}|$ compare it to $h(\alpha, a, dfE)SE(\bar{y}_{i\cdot} - \bar{y}_{\cdot\cdot})$, when examining the second largest value of $|\bar{y}_{i\cdot} - \bar{y}_{\cdot\cdot}|$, compare it to $h(\alpha, a-1, dfE)SE(\bar{y}_{i\cdot} - \bar{y}_{\cdot\cdot})$, etc. To maintain consistency, if, say, the second largest value of $|\bar{y}_{i\cdot} - \bar{y}_{\cdot\cdot}|$ is not greater than $h(\alpha, a-1, dfE)SE(\bar{y}_{i\cdot} - \bar{y}_{\cdot\cdot})$, all of the smaller values of $|\bar{y}_{i\cdot} - \bar{y}_{\cdot\cdot}|$ should also be considered nonsignificant. Note that $h(\alpha, 1, dfE) = 0$ for any α, so that if all the other means are declared different, the mean with the smallest deviation from $\bar{y}_{\cdot\cdot}$ will also be declared different, assuming that the deviation is positive.

6.5.2 DUNNETT'S MANY-ONE t STATISTIC METHOD

Dunnett's method is designed for situations in which there is a standard treatment (or placebo or control) and where interest lies in comparing each of the other treatments to the standard. Miller (1981) contains a thorough discussion along with references to the early work by Dunnett and Paulson.

Suppose that the standard treatment is $i = 1$. Dunnett's method is based on knowing the distribution of

$$\max_i \frac{|\bar{y}_{i\cdot} - \bar{y}_{1\cdot}|}{SE(\bar{y}_{i\cdot} - \bar{y}_{1\cdot})}$$

when the null hypothesis $H_0 : \mu_1 = \mu_2 = \cdots = \mu_a$ is true. If we denote the $1 - \alpha$ percentile of the distribution as $d(1 - \alpha, a, dfE)$, an α level test of H_0 is rejected if

$$\max_i \frac{|\bar{y}_{i\cdot} - \bar{y}_{1\cdot}|}{SE(\bar{y}_{i\cdot} - \bar{y}_{1\cdot})} > d(1 - \alpha, a, dfE).$$

When the overall test is rejected, we can do multiple comparisons to identify which μ_is are different from μ_1. A particular mean μ_i is considered different if

$$\frac{|\bar{y}_{i\cdot} - \bar{y}_{1\cdot}|}{SE(\bar{y}_{i\cdot} - \bar{y}_{1\cdot})} > d(1 - \alpha, a, dfE).$$

Clearly, if the overall test is not rejected, none of the individual means will be considered different. It is also clear that $\mu_1 = \mu_2 = \cdots = \mu_a$ if and only if $\mu_i - \mu_1 = 0$ for all $i > 1$. The standard error is that for comparing two means, so

$$SE(\bar{y}_{i\cdot} - \bar{y}_{1\cdot}) = \sqrt{MSE\left[\frac{1}{N} + \frac{1}{N}\right]}.$$

Simultaneous $(1 - \alpha)100\%$ confidence intervals for the $\mu_i - \mu_1$s have endpoints

$$\bar{y}_{i\cdot} - \bar{y}_{1\cdot} \pm d(1 - \alpha, a, dfE) SE(\bar{y}_{i\cdot} - \bar{y}_{1\cdot}).$$

A modification similar to Newman–Keuls can be used with Dunnett's method. This modification orders the values of $|\bar{y}_{i\cdot} - \bar{y}_{1\cdot}|$ and adjusts the value of the a parameter in $d(1 - \alpha, a, dfE)$. When examining the largest value of $|\bar{y}_{i\cdot} - \bar{y}_{1\cdot}|$, use $d(1 - \alpha, a, dfE)$, when examining the second largest value of $|\bar{y}_{i\cdot} - \bar{y}_{1\cdot}|$, use $d(1 - \alpha, a - 1, dfE)$, etc. Of course to maintain consistency, when it is determined that, say, the second largest value of $|\bar{y}_{i\cdot} - \bar{y}_{1\cdot}|$ is not greater than $d(1 - \alpha, a - 1, dfE) SE(\bar{y}_{i\cdot} - \bar{y}_{1\cdot})$, all of the smaller values of $|\bar{y}_{i\cdot} - \bar{y}_{1\cdot}|$ should be considered as nonsignificant also.

6.5.3 DUNCAN'S MULTIPLE RANGE METHOD

Duncan has developed a multiple range procedure similar to that of Newman–Keuls. Newman–Keuls uses a series of tabled values $Q(1 - \alpha, a, dfE)$, $Q(1 - \alpha, a - 1, dfE)$, ..., $Q(1 - \alpha, 2, dfE)$. Duncan's method simply changes the tabled values. Duncan uses $Q\left([1 - \alpha]^{a-1}, a, dfE\right)$, $Q\left([1 - \alpha]^{a-2}, a - 1, dfE\right)$, ..., $Q(1 - \alpha, 2, dfE)$. See Miller (1981) for a discussion of the rationale behind these choices.

Using Duncan's value $Q\left([1 - \alpha]^{a-1}, a, dfE\right)$ to compare the largest and smallest means does not control the experimentwise error rate at α. (It controls it at $1 - [1 - \alpha]^{a-1}$.) As a result, Duncan suggests performing the analysis of variance F test first and proceeding only if the F test indicates that there are differences among the means at level α. Duncan's method is more likely to conclude that a pair of means is different than the Newman–Keuls method and less likely to establish a difference than the LSD method. Just as the Newman–Keuls approach can be used to modify the AOM and Dunnett's method, Duncan's idea can also be applied to the AOM and Dunnett's method.

6.6 Summary of multiple comparison procedures

In this section we review and compare the uses of the various multiple comparison procedures.

The most general procedures are the least significant difference, the Bonferroni, and the Scheffé methods. These can be used for arbitrary sets of preplanned contrasts. They are listed in order from least conservative (most likely to reject an individual null hypothesis) to most conservative (least likely to reject). Scheffé's method can also be used for examining contrasts suggested by the data. Bonferroni's method has the advantage that it can easily be applied to almost any multiple testing problem.

To compare all of the treatment groups in a balanced analysis of variance, we can use the least significant difference, the Duncan, the Newman–Keuls, the Bonferroni, and the Tukey methods. Again, these are (roughly) listed in the order from least conservative to most conservative. In some cases, for example when comparing Bonferroni and Tukey, an exact statement of which is more conservative is not possible.

To decide on a method, you need to decide on how conservative you want to be. If it is very important not to claim differences when there are none, you should be very conservative. If it is most important to identify differences that *may* exist, then you should choose less conservative methods.

Finally, we discussed two specialized methods for balanced ANOVA. The analysis of means provides for testing whether each group differs from all the other groups and Dunnett's method allows multiple testing of each group against a fixed standard group.

Many of the methods have corresponding methods for constructing multiple confidence intervals. Various computer programs execute these procedures. For example, newer versions of Minitab's 'oneway' command compute these for Fisher's LSD method, Tukey's method, and Dunnett's method.

6.7 Exercises

EXERCISE 6.7.1. Exercise 5.7.1 involved measurements from different laboratories on the stress at 600% elongation for a certain type of rubber. The summary statistics are repeated in Table 6.5. Ignoring any reservations you may have about the appropriateness of the analysis of variance model for these data, compare all pairs of laboratories using $\alpha = .10$ for the LSD, Bonferroni, Tukey, and Newman–Keuls methods. Give joint 95% confidence intervals using Tukey's method for all differences between pairs of labs.

TABLE 6.5. Rubber stress at five laboratories

Lab.	Sample size	Sample mean	Sample variance
1	4	57.00	32.00
2	4	67.50	46.33
3	4	40.25	14.25
4	4	56.50	5.66
5	4	52.50	6.33

EXERCISE 6.7.2. Use Scheffé's method with $\alpha = .01$ to test whether the contrast in Exercise 5.7.2d is zero.

EXERCISE 6.7.3. Use Bonferroni's method with an α near .01 to give simultaneous confidence intervals for the mean weight in each height group for Exercise 5.7.3.

EXERCISE 6.7.4. Use the LSD, Bonferroni, and Scheffé's methods to test whether the four orthogonal contrasts in Exercise 5.7.4 are zero. Use $\alpha = .05$.

EXERCISE 6.7.5. Exercise 5.7.5 contained data on stress measurements for four different types of rubber. Four observations were taken on each type of rubber; the means are repeated below

Material	A	B	C	D
Mean	26.4425	26.0225	23.5325	29.9600

and the sample variance of the 16 observations is 14.730793. Test for differences between all pairs of materials using $\alpha = .05$ for the LSD, Bonferroni, Tukey, and Newman–Keuls methods. Give 95% confidence intervals for the differences between all pairs of materials using the BSD method.

EXERCISE 6.7.6. In Exercise 5.7.6 on the stress of four types of rubber an outlier was noted in material B. Redo the multiple comparisons of the previous problem eliminating the outlier and using only the methods that are still applicable.

EXERCISE 6.7.7. In Exercise 5.7.7 on the peel-strength of different adhesive systems, parts (b) and (c) amount to doing LSD multiple comparisons for all pairs of systems. Compare the LSD results with the results obtained using the Tukey and Newman–Keuls methods with $\alpha = .01$.

EXERCISE 6.7.8. For the weight gain data of Exercise 5.7.8, use the LSD, Bonferroni, and Scheffé methods to test whether the following contrasts are zero: 1) the contrast that compares the two drugs and 2) the contrast that compares the control with the average of the two drugs. Pick an α level but clearly state the level chosen.

EXERCISE 6.7.9. For the Cushing's syndrome data of Exercise 5.7.9, use all appropriate methods to compare all pairwise differences among the three treatments. Pick an α level but clearly state the level chosen.

EXERCISE 6.7.10. Use Scheffé's method with $\alpha = .05$ and the data of Exercise 5.7.10 to test the significance of the contrast

Age	0.5	1.0	4.0	4.5	5.0	5.5	6.0
Coeff.	−5	−5	2	2	2	2	2

Chapter 7

Simple linear and polynomial regression

This chapter examines data that come as pairs of numbers, say (x, y), and the problem of fitting a line to them. More generally, it examines the problem of predicting one variable (y) from values of another variable (x). Consider for the moment the popular wisdom that people who read a lot tend to have large vocabularies and poor eyes. Thus reading causes both conditions: large vocabularies and poor eyes. If this is true, it may be possible to predict the size of someone's vocabulary from the condition of their eyes. Of course this does not mean that having poor eyes causes large vocabularies. Quite the contrary, if anything poor eyes probably keep people from reading and thus cause small vocabularies. Regression analysis is concerned with predictive ability, not with causation.

Section 7.1 of this chapter introduces an example along with many of the basic ideas and methods of simple linear regression. The next five sections go into the details of simple linear regression. Sections 7.7 and 7.8 deal with an idea closely related to simple linear regression: the correlation between two variables. Section 7.9 deals with methods for checking the assumptions made in simple linear regression. If the assumptions are violated, we need alternative methods of analysis. Section 7.10 presents methods for transforming the original data so that the assumptions become reasonable on the transformed data. *Sections 7.9 and 7.10 apply quite generally to analysis of variance and regression models*. They are not restricted to simple linear regression. Section 7.11 treats an alternative to transformations as a method for dealing with nonlinearity in the relationship between y and x, namely fitting polynomials (parabolas, etc.) to the data. Section 7.12 explores the relationship between one-way analysis of variance and fitting polynomials.

7.1 An example

Data from *The Coleman Report* were reproduced in Mosteller and Tukey (1977). The data were collected from schools in the New England and Mid-Atlantic states of the USA. In this chapter we consider only two variables: y – the mean verbal test score for sixth graders and x – a composite measure of socioeconomic status. The data are presented in Table 7.1.

Figure 7.1 contains a scatter plot of the data. Note that there is a rough linear relationship. The higher the composite socioeconomic status variable, the higher the mean verbal test score. However, there is a considerable amount of error in the relationship. By no means do the points lie exactly on

TABLE 7.1. Coleman Report data

School	y	x	School	y	x
1	37.01	7.20	11	23.30	−12.86
2	26.51	−11.71	12	35.20	0.92
3	36.51	12.32	13	34.90	4.77
4	40.70	14.28	14	33.10	−0.96
5	37.10	6.31	15	22.70	−16.04
6	33.90	6.16	16	39.70	10.62
7	41.80	12.70	17	31.80	2.66
8	33.40	−0.17	18	31.70	−10.99
9	41.01	9.85	19	43.10	15.03
10	37.20	−0.05	20	41.01	12.77

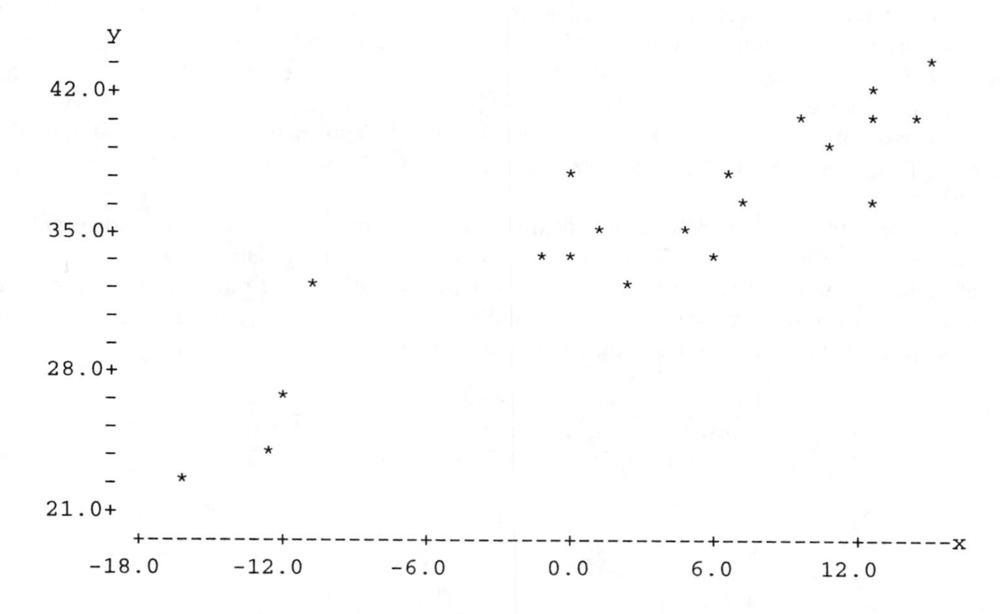

FIGURE 7.1. Plot of y versus x.

a straight line.

We assume a basic linear relationship between the ys and xs, something like $y = \beta_0 + \beta_1 x$. Here β_1 is the slope of the line and β_0 is the intercept. Unfortunately, the observed y values do not fit exactly on a line so $y = \beta_0 + \beta_1 x$ is only an approximation. We need to modify this equation to allow for the variability of the observations about the line. We do this by building a random error term into the linear relationship. Write the relationship as $y = \beta_0 + \beta_1 x + \varepsilon$, where ε indicates the random error. In this model for the behavior of the data, ε accounts for the deviations between the y values we actually observe and the line $\beta_0 + \beta_1 x$ where we expect to observe any y value that corresponds to x. As we are interested in predicting y from known x values, *we treat x as a known (nonrandom) variable.*

We assume that the relationship $y = \beta_0 + \beta_1 x + \varepsilon$ applies to all of our observations. For the current data, that means we assume this relationship holds for all of the 20 pairs of values in Table 7.1. This assumption is stated as *the simple linear regression model* for these data,

$$y_i = \beta_0 + \beta_1 x_i + \varepsilon_i, \tag{7.1.1}$$

$i = 1, \ldots, 20$. For this model to be useful, we need to make some assumptions about the errors, the ε_is. The standard assumption is that the

$$\varepsilon_i\text{s are independent } N(0, \sigma^2).$$

Given data for which these assumptions are reasonable, we can estimate the unknown parameters. Although we assume a linear relationship between the ys and xs, the model does not assume that we know the slope β_1 or the intercept β_0. Together these unknown parameters would tell us the exact nature of the linear relationship but both need to be estimated. We use the notation $\hat{\beta}_1$ and $\hat{\beta}_0$ to denote estimates of β_1 and β_0, respectively. To perform statistical inferences we also need to estimate the variance of the errors, σ^2. Note that σ^2 is also the variance of the y observations because none of β_0, β_1, and x are random.

Simple linear regression involves many assumptions. It assumes that the relationship between y and x is linear, it assumes that the errors are normally distributed, it assumes that the errors all have the same variance, it assumes that the errors are all independent, and it assumes that the errors all have mean 0. This last assumption is redundant. It turns out that the errors all have mean 0 if and only if the relationship between y and x is linear. As far as possible, we will want to verify (validate) that these assumptions are reasonable before we put much faith in the estimates and statistical inferences that can be obtained from simple linear regression. Section 7.9 deals with checking these assumptions.

Before getting into a detailed discussion of simple linear regression, we illustrate some highlights using the *Coleman Report* data. We need to fit model (7.1.1) to the data. A computer program typically yields parameter estimates, standard errors for the estimates, t ratios for testing whether the parameters are zero, P values for the tests, and an analysis of variance table. These results are often displayed in a fashion similar to that illustrated below.

Predictor	$\hat{\beta}_k$	$SE(\hat{\beta}_k)$	t	P
Constant	33.3228	0.5280	63.11	0.000
x	0.56033	0.05337	10.50	0.000

Analysis of Variance

Source	df	SS	MS	F	P
Regression	1	552.68	552.68	110.23	0.000
Error	18	90.25	5.01		
Total	19	642.92			

Much can be learned from these two tables of statistics. The estimated regression equation is

$$y = 33.3 + 0.560x.$$

This equation allows us to predict a value for y when the value of x is given. In particular, for these data an increase of one unit in socioeconomic status tends to increase mean verbal test scores by about .56 units. This is not to say that some program to increase socioeconomic statuses by one unit will increase mean verbal test scores by about .56 unit. The .56 *describes* the *current* data, *it does not imply a causal relationship*. If we want to predict the mean verbal test score for a school that is very similar to the ones in this study, this equation should give good predictions. If we want to predict the mean verbal test score for a school that is very different from the ones in this study, this equation is likely to give poor predictions. In fact, if we collect new data from schools with very different socioeconomic statuses, the data are not similar to these, so this fitted model would be highly questionable if applied to the new situation. Nevertheless, a simple linear regression model with a different intercept and slope might fit the new data well. Similarly, data collected after a successful program to raise socioeconomic statuses are unlikely to be similar to the data collected before such a program. The relationship between socioeconomic status and mean verbal test scores may be changed by such a program. In particular, the things causing both socioeconomic status and mean verbal test score may be changed in unknown ways by such a program. These are crucial points and bear repeating. *The regression equation describes an observed relationship between mean verbal test scores and socioeconomic status. It can be used to predict mean verbal test scores from socioeconomic status in similar situations. It does not imply that changing the socioeconomic status a fixed amount will cause the mean verbal test scores to change by a proportional amount.*

In simple linear regression, the reference distribution for statistical inferences is almost invariably $t(dfE)$ where dfE is the degrees of freedom for error from the analysis of variance table. For these data, $dfE = 18$. We now consider some illustrations of statistical inferences.

From our standard theory of Chapter 3, the 95% confidence interval for β_1 has endpoints

$$\hat{\beta}_1 \pm t(.975, dfE) SE(\hat{\beta}_1) .$$

From a t table, $t(.975, 18) = 2.101$, so, using the tabled statistics, the endpoints are

$$.56033 \pm 2.101(.05337).$$

The confidence interval is $(.448, .672)$, so we are 95% confident that the slope β_1 is between .448 and .672.

The t statistics for testing $H_0 : \beta_k = 0$ versus $H_A : \beta_k \neq 0$ are reported in the first table. For example, the test of $H_0 : \beta_1 = 0$ versus $H_A : \beta_1 \neq 0$ has

$$t_{obs} = \frac{0.56033}{.05337} = 10.50.$$

The significance level of the test is the P value,

$$P = \Pr[|t| > 10.50] = .000.$$

The value .000 indicates a large amount of evidence that $\beta_1 \neq 0$. Note that if $\beta_1 = 0$, the linear relationship becomes $y = \beta_0 + \varepsilon$, so there is no relationship between y and x, i.e., y does not depend on x. The small P value indicates that the slope is not zero and thus the variable x helps to explain the variable y.

The primary value of the analysis of variance table is that it gives the degrees of freedom, the sum of squares, and the mean square for error. The mean squared error is the estimate of σ^2 and the sum of squares error and degrees of freedom for error are vital for comparing different regression models that we may choose to consider. Note that the sums of squares for regression and error add up to the sum of squares total and that the degrees of freedom for regression and error also add up to the degrees of freedom total.

The analysis of variance table gives an alternative but equivalent test for whether the x variable helps to explain y. The alternative test of

$$H_0 : \beta_1 = 0 \quad \text{versus} \quad H_A : \beta_1 \neq 0$$

is based on

$$F = \frac{MSReg}{MSE} = \frac{552.68}{5.01} = 110.23.$$

Note that the value of this statistic is $110.23 = (10.50)^2$; the F statistic is just the square of the corresponding t statistic for testing $H_0 : \beta_1 = 0$ versus $H_A : \beta_1 \neq 0$. The F and t tests are equivalent.

In particular the P values are identical. In this case, both are infinitesimal, zero to three decimal places. Our conclusion that $\beta_1 \neq 0$ means that the x variable helps to explain the variation in the y variable. In other words, it is possible to predict the verbal test scores for a school's sixth grade class from the socioeconomic measure. Of course, the fact that some predictive ability exists does not mean that the predictive ability is sufficient to be useful.

The *coefficient of determination*, R^2, measures the percentage of the total variability in y that is explained by the x variable. If this number is large, it suggests a substantial predictive ability. In our example

$$R^2 \equiv \frac{SSReg}{SSTot} = \frac{552.68}{642.92} = 86.0\%,$$

so 86.0% of the total variability is explained by the regression model. This is a large percentage, so it appears that the x variable has substantial predictive power. However, *a large R^2 does not imply that the model is good in absolute terms.* It may be possible to show that this model does not fit the data adequately. In other words, while this model is explaining much of the variability, we may be able to establish that it is not explaining as much of the variability as it ought. (Example 7.9.2 involves a model with a high R^2 that is demonstrably inadequate.) Conversely, a model with a low R^2 value may be the perfect model but the data may simply have a great deal of variability. For example, if you have temperature measurements obtained by having someone walk outdoors and guess the Celsius temperature and then use the true Fahrenheit temperatures as a predictor, the exact linear relationship between Celsius and Fahrenheit temperatures may make a line the ideal model. Nonetheless, the obvious inaccuracy involved in people guessing Celsius temperatures may cause a low R^2. Moreover, even a high R^2 of 86% may provide inadequate predictions for the purposes of the study, while in other situations an R^2 of, say, 14% may be perfectly adequate. It depends on the purpose of the study. Finally, it must be recognized that a large R^2 may be an unrepeatable artifact of a particular data set. *The coefficient of determination is a useful tool but it must be used with care. In particular, it is a much better measure of the predictive ability of a model than of the correctness of a model.*

Consider the problem of estimating the value of the line at $x = -16.04$. This value of x is the minimum observed value for socioeconomic status, so it is somewhat dissimilar to the other x values in the data. Its dissimilarity causes there to be substantial variability in estimating the regression line (mean value of y) at this point. The point on the line is $\beta_0 + \beta_1(-16.04)$ and the estimator is

$$\hat{\beta}_0 + \hat{\beta}_1 x = 33.32 + .560(-16.04) = 24.34.$$

For constructing 95% t intervals, the percentile needed is $t(.975, 18) = 2.101$. The standard error for the estimate of the point on the line is usually available from computer programs; in this example it is 1.140. The 95% confidence interval for the point on the line $\beta_0 + \beta_1(-16.04)$ has endpoints

$$24.34 \pm 2.101(1.140)$$

which gives the interval (21.9, 26.7). We are 95% confident that the population mean of the school-wise mean verbal test scores for New England and Mid-Atlantic sixth graders with a school socioeconomic measure of -16.04 is between 21.9 and 26.7.

The prediction $\hat{y}$ for a new observation with $x = -16.04$ is simply the estimated point on the line

$$\hat{y} = \hat{\beta}_0 + \hat{\beta}_1(-16.04) = 24.34.$$

Prediction of a new observation is subject to more error than estimation of a point on the line. A new observation has the same variance as all other observations, so the prediction interval must account for this variance as well as for the variance of estimating the point on the line. The standard error for the prediction interval is computed as

$$SE(Prediction) = \sqrt{MSE + SE(Line)^2}. \qquad (7.1.2)$$

In this example,

$$SE(Prediction) = \sqrt{5.01 + (1.140)^2} = 2.512.$$

The prediction interval endpoints are

$$24.34 \pm 2.101(2.512).$$

and the 95% prediction interval is $(19.1, 29.6)$. We are 95% confident that sixth graders's mean verbal test scores would be between 19.1 and 29.6 for a *different* New England or Mid-Atlantic school with a socioeconomic measure of -16.04. Note that the prediction interval is considerably wider than the corresponding confidence interval. Note also that this is just another special case of the prediction theory in Section 3.5. As such, these results are analogous to those obtained for the one sample, two sample, and one-way ANOVA data structures.

Minitab commands

The Minitab commands given below generate the table of estimates and the analysis of variance table. Column c1 contains the test scores y and column c2 contains the composite socioeconomic statuses x. The primary command is to regress c1 on 1 predictor variable, c2. This same command allows for more predictor variables and we will use that capability in this chapter as well as in the chapters on multiple regression. In our example, the subcommand 'predict -16.04' was used; this subcommand gives the estimate of the line (prediction) when $x = -16.04$, the standard error for the estimate of the line, the 95% confidence interval for the value of the line at $x = -16.04$, and the 95% prediction interval when $x = -16.04$.

```
MTB > name c1 'test' c2 'socio'
MTB > regress c1 on 1 c2;
SUBC> predict -16.04.
```

7.2 The simple linear regression model

In general, simple linear regression seeks to fit a line to pairs of numbers (x, y) that are subject to error. These pairs of numbers may arise when there is a perfect linear relationship between x and a variable y_* but where y_* cannot be measured without error. Our actual observations y are then the sum of y_* and the measurement error. Alternatively, we may sample a population of objects and take two measurements on each object. In this case, both elements of the pair (x, y) are random. In simple linear regression we think of using the x measurement to predict the y measurement. While x is actually random in this scenario, we use it as if it were fixed because we cannot predict y until we have actually observed the x value. We want to use the particular observed value of x to predict y, so for our purposes x is a fixed number. In any case, *the xs are always treated as fixed numbers in simple linear regression.*

The model for simple linear regression is a line with the addition of errors

$$y_i = \beta_0 + \beta_1 x_i + \varepsilon_i, \quad i = 1, \ldots, n$$

where y is the variable of primary interest and x is the predictor variable. Both the y_is and the x_is are observable, the y_is are assumed to be random and the x_is are assumed to be known fixed constants. The unknown constants (regression parameters) β_0 and β_1 are the intercept and the slope of the line, respectively. The ε_is are unobservable errors that are assumed to be independent of each other with mean zero and the same variance, i.e.,

$$E(\varepsilon_i) = 0, \quad \text{Var}(\varepsilon_i) = \sigma^2.$$

Typically the errors are also assumed to have normal distributions, i.e.,

$$\varepsilon_i\text{s independent } N(0, \sigma^2).$$

Sometimes the assumption of independence is replaced by the assumption that $Cov(\varepsilon_i, \varepsilon_j) = 0$ for $i \neq j$.

Note that since β_0, β_1, and the x_is are all assumed to be fixed constants,

$$E(y_i) = E(\beta_0 + \beta_1 x_i + \varepsilon_i) = \beta_0 + \beta_1 x_i + E(\varepsilon_i) = \beta_0 + \beta_1 x_i,$$

$$\text{Var}(y_i) = \text{Var}(\varepsilon_i) = \sigma^2,$$

and if the ε_is are independent, the y_is are independent.

7.3 Estimation of parameters

The unknown parameters in the simple linear regression model are the slope, β_1, the intercept, β_0, and the variance, σ^2. All of the estimates $\hat{\beta}_1$, $\hat{\beta}_0$, and *MSE*, can be computed from just six summary statistics

$$n, \quad \bar{x}_\cdot, \quad s_x^2, \quad \bar{y}_\cdot, \quad s_y^2, \quad \sum_{i=1}^n x_i y_i,$$

i.e., the sample size, the sample mean and variance of the x_is, the sample mean and variance of the y_is, and $\sum_{i=1}^n x_i y_i$. The only one of these that is any real work to obtain on a decent hand calculator is $\sum_{i=1}^n x_i y_i$. The standard estimates of the parameters are, respectively,

$$\hat{\beta}_1 = \frac{\sum_{i=1}^n (x_i - \bar{x}_\cdot) y_i}{\sum_{i=1}^n (x_i - \bar{x}_\cdot)^2}$$

$$\hat{\beta}_0 = \bar{y}_\cdot - \hat{\beta}_1 \bar{x}_\cdot$$

and the *mean squared error*

$$\begin{aligned} MSE &= \frac{\sum_{i=1}^n \left(y_i - \hat{\beta}_0 - \hat{\beta}_1 x_i\right)^2}{n-2} \\ &= \frac{1}{n-2}\left[\sum_{i=1}^n (y_i - \bar{y}_\cdot)^2 - \hat{\beta}_1^2 \sum_{i=1}^n (x_i - \bar{x}_\cdot)^2\right] \\ &= \frac{1}{n-2}\left[(n-1)s_y^2 - \hat{\beta}_1^2 (n-1) s_x^2\right]. \end{aligned}$$

The slope estimate $\hat{\beta}_1$ given above is the form that is most convenient for deriving its statistical properties. In this form it is just a linear combination of the y_is. However, $\hat{\beta}_1$ is commonly written in a variety of ways to simplify various computations and, unfortunately for students, they are expected to recognize all of them. Observing that $0 = \sum_{i=1}^n (x_i - \bar{x}_\cdot)$ so that $0 = \sum_{i=1}^n (x_i - \bar{x}_\cdot)\bar{y}_\cdot$, we can also write

$$\hat{\beta}_1 = \frac{\sum_{i=1}^n (x_i - \bar{x}_\cdot)(y_i - \bar{y}_\cdot)}{\sum_{i=1}^n (x_i - \bar{x}_\cdot)^2} = \frac{s_{xy}}{s_x^2} = \frac{\left(\sum_{i=1}^n x_i y_i\right) - n\bar{x}_\cdot \bar{y}_\cdot}{(n-1)s_x^2}. \tag{7.3.1}$$

Here

$$s_{xy} = \frac{1}{n-1}\sum_{i=1}^n (x_i - \bar{x}_\cdot)(y_i - \bar{y}_\cdot)$$

is the sample covariance between x and y. The last equality on the right of equation (7.3.1) gives a form suitable for computing $\hat{\beta}_1$ from the summary statistics.

EXAMPLE 7.3.1. For the *Coleman Report* data,

$$n = 20, \quad \bar{x}_{\bullet} = 3.1405, \quad s_x^2 = 92.64798395,$$

$$\bar{y}_{\bullet} = 35.0825, \quad s_y^2 = 33.838125, \quad \sum_{i=1}^{n} x_i y_i = 3189.8793.$$

The estimates are

$$\hat{\beta}_1 = \frac{3189.8793 - 20(3.1405)(35.0825)}{(20-1)92.64798395} = .560325468,$$

$$\hat{\beta}_0 = 35.0825 - .560325468(3.1405) = 33.32279787$$

and

$$\begin{aligned} MSE & \\ &= \frac{1}{20-2}\left[(20-1)33.838125 - (.560325468)^2(20-1)92.64798395\right] \\ &= \frac{1}{18}[642.924375 - 552.6756109] \\ &= \frac{90.2487641}{18} = 5.01382. \end{aligned} \tag{7.3.2}$$

Up to round off error, these are the same results as tabled in Section 7.1. □

It is not clear that these estimates of β_0, β_1, and σ^2 are even reasonable. The estimate of the slope β_1 seems particularly unintuitive. However, from Proposition 7.3.2 below, the estimates are unbiased, so they are at least estimating what we claim that they estimate.

Proposition 7.3.2. $E(\hat{\beta}_1) = \beta_1$, $E(\hat{\beta}_0) = \beta_0$, and $E(MSE) = \sigma^2$.

Proofs of the unbiasedness of the slope and intercept are given in the appendix to this chapter.

The parameter estimates are unbiased but that alone does not ensure that they are good estimates. These estimates are the best estimates available in several senses. We briefly mention these optimality properties but for a detailed discussion see Christensen (1987, chapter II). Assuming that the errors have independent normal distributions, all of the estimates have the smallest variance of any unbiased estimates. The regression parameters are also maximum likelihood estimates. Maximum likelihood estimates are those values of the parameters that are most likely to generate the data that were actually observed. Without assuming that the errors are normally distributed, the regression parameters have the smallest variance of any unbiased estimates that are linear functions of the y observations. (Linear functions allow multiplying the y_is by constants and adding terms together. Remember, the x_is are constants, as are any functions of the x_is.) Note that with this weaker assumption, i.e., giving up normality, we get a weaker result, minimum variance among only linear unbiased estimates instead of all unbiased estimates. The regression parameter estimates are also least squares estimates. Least squares estimates are choices of β_0 and β_1 that minimize

$$\sum_{i=1}^{n} (y_i - \beta_0 - \beta_1 x_i)^2.$$

Under the standard assumptions, least squares estimates of the regression parameters are best (minimum variance) linear unbiased estimates (BLUEs), and for normally distributed data they are minimum variance unbiased estimates and maximum likelihood estimates.

To draw statistical inferences about the regression parameters, we need standard errors for the estimates. To find the standard errors we need to know the variance of each estimate.

Proposition 7.3.3.

$$\mathrm{Var}(\hat{\beta}_1) = \frac{\sigma^2}{\sum_{i=1}^{n}(x_i - \bar{x}_\cdot)^2} = \frac{\sigma^2}{(n-1)s_x^2}$$

and

$$\mathrm{Var}(\hat{\beta}_0) = \sigma^2\left[\frac{1}{n} + \frac{\bar{x}_\cdot^2}{\sum_{i=1}^{n}(x_i - \bar{x}_\cdot)^2}\right] = \sigma^2\left[\frac{1}{n} + \frac{\bar{x}_\cdot^2}{(n-1)s_x^2}\right].$$

The proof of this proposition is given in the appendix at the end of the chapter. Note that, except for the unknown parameter σ^2, the variances can be computed using the same six numbers we used to compute $\hat{\beta}_0$, $\hat{\beta}_1$, and MSE. Using MSE to estimate σ^2 and taking square roots, we get the standard errors.

$$SE(\hat{\beta}_1) = \sqrt{\frac{MSE}{(n-1)s_x^2}}$$

and

$$SE(\hat{\beta}_0) = \sqrt{MSE\left[\frac{1}{n} + \frac{\bar{x}_\cdot^2}{(n-1)s_x^2}\right]}.$$

EXAMPLE 7.3.4. For the *Coleman Report* data, using the numbers n, $\bar{x}_\cdot$, and s_x^2,

$$\mathrm{Var}(\hat{\beta}_1) = \frac{\sigma^2}{(20-1)92.64798395} = \frac{\sigma^2}{1760.311695}$$

and

$$\mathrm{Var}(\hat{\beta}_0) = \sigma^2\left[\frac{1}{20} + \frac{3.1405^2}{(20-1)92.64798395}\right] = \sigma^2\,[.055602837].$$

The MSE is 5.014, so the standard errors are

$$SE(\hat{\beta}_1) = \sqrt{\frac{5.014}{1760.311695}} = .05337$$

and

$$SE(\hat{\beta}_0) = \sqrt{5.014\,[.055602837]} = .5280.$$

□

We always like to have estimates with small variances. The forms of the variances show how to achieve this. For example, the variance of $\hat{\beta}_1$ gets smaller when n or s_x^2 gets larger. Thus, more observations (larger n) result in a smaller slope variance and more dispersed x_i values (larger s_x^2) also result in a smaller slope variance. Of course all of this assumes that the simple linear regression model is correct.

7.4 The analysis of variance table

A standard tool in regression analysis is the construction of an analysis of variance table. The best form is given in Table 7.2. In this form there is one degree of freedom for every observation, cf. the total line, and the sum of squares total is the sum of all of the squared observations. The degrees of freedom and sums of squares for intercept, regression, and error can be added to obtain the degrees of freedom and sums of squares total. We see that one degree of freedom is used to estimate the intercept, one is used for the slope, and the rest are used to estimate the variance.

TABLE 7.2. Analysis of variance

Source	df	SS	MS	F
Intercept(β_0)	1	$n\bar{y}_\cdot^2 \equiv C$	$n\bar{y}_\cdot^2$	
Regression(β_1)	1	$\hat{\beta}_1^2 \sum_{i=1}^n (x_i - \bar{x}_\cdot)^2$	$SSReg$	$\frac{MSReg}{MSE}$
Error	$n-2$	$\sum_{i=1}^n \left(y_i - \hat{\beta}_0 - \hat{\beta}_1 x_i\right)^2$	$SSE/(n-2)$	
Total	n	$\sum_{i=1}^n y_i^2$		

The more commonly used form for the analysis of variance table is given as Table 7.3. It eliminates the line for the intercept and corrects the total line so that the degrees of freedom and sums of squares still add up.

TABLE 7.3. Analysis of variance

Source	df	SS	MS	F
Regression(β_1)	1	$\hat{\beta}_1^2 \sum_{i=1}^n (x_i - \bar{x}_\cdot)^2$	$SSReg$	$\frac{MSReg}{MSE}$
Error	$n-2$	$\sum_{i=1}^n \left(y_i - \hat{\beta}_0 - \hat{\beta}_1 x_i\right)^2$	$SSE/(n-2)$	
Total	$n-1$	$\sum_{i=1}^n (y_i - \bar{y}_\cdot)^2$		

These two forms for the analysis of variance table are analogous to the two different forms discussed in Section 5.2 for the one-way ANOVA analysis of variance table.

EXAMPLE 7.4.1. Consider again the *Coleman Report* data. The analysis of variance table was given in Section 7.1; Table 7.4 illustrates the necessary computations. Most of the computations were made earlier in equation (7.3.2) during the process of obtaining the MSE and all are based on the usual six numbers, n, $\bar{x}_\cdot$, s_x^2, $\bar{y}_\cdot$, s_y^2, and $\sum x_i y_i$. More directly, the computations depend on n, $\hat{\beta}_1$, s_x^2, and s_y^2. The corrected version of $SSTot$ is $(n-1)s_y^2$. Note that the SSE is obtained as $SSTot - SSReg$. The correction factor C in Table 7.2 is $20(35.0825)^2$ but it is not used in these computations for Table 7.4. □

7.5 Inferential procedures

The general theory of Chapter 3 applies to inferences about regression parameters. The theory requires 1) a parameter (Par), 2) an estimate (Est) of the parameter, 3) the standard error of the

TABLE 7.4. Analysis of variance

Source	df	SS	MS	F
Regression(β_1)	1	$.560325^2(20-1)92.64798$	552.6756109	$\frac{552.68}{5.014}$
Error	$20-2$	90.2487641	$90.2487641/18$	
Total	$20-1$	$(20-1)33.838125$		

estimate ($SE(Est)$) and 4) a known (tabled) distribution for

$$\frac{Est - Par}{SE(Est)}$$

that is symmetric about 0. The computations for most of the applications considered in this section were illustrated in Section 7.1 for the *Coleman Report* data.

Consider inferences about the slope *parameter* β_1. The estimate $\hat{\beta}_1$ and the standard error of $\hat{\beta}_1$ are as given in Section 7.3. The appropriate reference distribution is

$$\frac{\hat{\beta}_1 - \beta_1}{SE(\hat{\beta}_1)} \sim t(n-2).$$

Using standard methods, the 99% confidence interval for β_1 has endpoints

$$\hat{\beta}_1 \pm t(.995, n-2)\, SE(\hat{\beta}_1)\,.$$

An $\alpha = .05$ test of, say, $H_0 : \beta_1 = 0$ versus $H_A : \beta_1 \neq 0$ rejects H_0 if

$$\frac{|\hat{\beta}_1 - 0|}{SE(\hat{\beta}_1)} > t(.975, n-2).$$

An $\alpha = .05$ test of $H_0 : \beta_1 \geq 1$ versus $H_A : \beta_1 < 1$ rejects H_0 if

$$\frac{\hat{\beta}_1 - 1}{SE(\hat{\beta}_1)} < -t(.95, n-2).$$

For inferences about the intercept parameter β_0, the estimate $\hat{\beta}_0$, and the standard error of $\hat{\beta}_0$ are as given in Section 7.3. The appropriate reference distribution is

$$\frac{\hat{\beta}_0 - \beta_0}{SE(\hat{\beta}_0)} \sim t(n-2).$$

A 95% confidence interval for β_0 has endpoints

$$\hat{\beta}_0 \pm t(.975, n-2)\, SE(\hat{\beta}_0)\,.$$

An $\alpha = .01$ test of $H_0 : \beta_0 = 0$ versus $H_A : \beta_0 \neq 0$ rejects H_0 if

$$\frac{|\hat{\beta}_0 - 0|}{SE(\hat{\beta}_0)} > t(.995, n-2).$$

An $\alpha = .05$ test of $H_0 : \beta_0 \leq 0$ versus $H_A : \beta_0 > 0$ rejects H_0 if

$$\frac{\hat{\beta}_0 - 0}{SE(\hat{\beta}_0)} > t(.95, n-2).$$

Typically inferences about β_0 are not of substantial interest. β_0 is the intercept, it is the value of the line when $x = 0$. Typically, the line is only an approximation to the behavior of the (x, y) pairs in the neighborhood of the observed data. This approximation is only valid in the neighborhood of the observed data. If we have not collected data near $x = 0$, the intercept is describing behavior of the line outside the range of valid approximation.

We can also draw inferences about a point on the line $y = \beta_0 + \beta_1 x$. For any fixed point x, $\beta_0 + \beta_1 x$ has an estimate

$$\hat{y} \equiv \hat{\beta}_0 + \hat{\beta}_1 x.$$

To get a standard error for $\hat{y}$, we first need its variance. As shown in the appendix to this chapter, the variance of $\hat{y}$ is

$$\mathrm{Var}\left(\hat{\beta}_0 + \hat{\beta}_1 x\right) = \sigma^2 \left[\frac{1}{n} + \frac{(x - \bar{x}_\bullet)^2}{(n-1)s_x^2}\right], \tag{7.5.1}$$

so the standard error of $\hat{y}$ is

$$SE\left(\hat{\beta}_0 + \hat{\beta}_1 x\right) = \sqrt{MSE\left[\frac{1}{n} + \frac{(x - \bar{x}_\bullet)^2}{(n-1)s_x^2}\right]}. \tag{7.5.2}$$

The appropriate distribution for inferences about the point $\beta_0 + \beta_1 x$ is

$$\frac{\left(\hat{\beta}_0 + \hat{\beta}_1 x\right) - \left(\beta_0 + \beta_1 x\right)}{SE\left(\hat{\beta}_0 + \hat{\beta}_1 x\right)} \sim t(n-2).$$

Using standard methods, the 99% confidence interval for $(\beta_0 + \beta_1 x)$ has endpoints

$$\left(\hat{\beta}_0 + \hat{\beta}_1 x\right) \pm t(.995, n-2)\, SE\left(\hat{\beta}_0 + \hat{\beta}_1 x\right).$$

We typically prefer to have small standard errors. Even when σ^2, and thus MSE, is large, from equation (7.5.2) we see that the standard error of $\hat{y}$ will be small when the number of observations n is large, when the x_i values are well spread out, i.e., s_x^2 is large, and when x is close to $\bar{x}_\bullet$. In other words, the line can be estimated efficiently in the neighborhood of $\bar{x}_\bullet$ by collecting a lot of data. Unfortunately, if we try to estimate the line far from where we collected the data, the standard error of the estimate gets large. The standard error gets larger as x gets farther away from the center of the data, $\bar{x}_\bullet$, because the term $(x - \bar{x}_\bullet)^2$ gets larger. This effect is standardized by the original observations; the term in question is $(x - \bar{x}_\bullet)^2 / (n-1)s_x^2$, so $(x - \bar{x}_\bullet)^2$ must be large relative to $(n-1)s_x^2$ before a problem develops. In other words, the distance between x and $\bar{x}_\bullet$ must be several times the standard deviation s_x before a problem develops. Nonetheless, large standard errors occur when we try to estimate the line far from where we collected the data. Moreover, the regression line is often just an approximation that holds in the neighborhood of where the data were collected. This approximation may be invalid for data points far from the original data. So, in addition to the problem of having large standard errors, estimates far from the neighborhood of the original data may be totally invalid.

Estimating a point on the line is distinct from prediction of a new observation for a given x value. Ideally, the prediction would be the true point on the line for the value x. However, the true line is an unknown quantity, so our prediction is the estimated point on the line at x. The distinction between prediction and estimating a point on the line arises because a new observation is subject to variability about the line. In making a prediction we must account for the variability of the new observation even when the line is known, as well as account for the variability associated with our need to estimate the line. The new observation is assumed to be independent of the past data, so the variance of the prediction is σ^2 (the variance of the new observation) plus the variance of the estimate of the line as given in (7.5.1). The standard error replaces σ^2 with MSE and takes the square root, i.e.,

$$SE(Prediction) = \sqrt{MSE\left[1 + \frac{1}{n} + \frac{(x - \bar{x}_{\cdot})^2}{(n-1)s_x^2}\right]}.$$

Note that this is the same as the formula given in equation (7.1.2). Prediction intervals follow in the usual way. For example, the 99% prediction interval associated with x has endpoints

$$(\hat{y}) \pm t(.995, n-2)\, SE(Prediction).$$

As discussed earlier, estimation of points on the line should be restricted to x values in the neighborhood of the original data. For similar reasons, predictions should also be made only in the neighborhood of the original data. While it is possible, by collecting a lot of data, to estimate the line well even when the variance σ^2 is large, it is not always possible to get good prediction intervals. Prediction intervals are subject to the variability of both the observations and the estimate of the line. The variability of the observations cannot be eliminated or reduced. If this variability is too large, we may get prediction intervals that are too large to be useful. If the simple linear regression model is the 'truth', there is nothing to be done, i.e., no way to improve the prediction intervals. If the simple linear regression model is only an approximation to the true process, a more sophisticated model may give a better approximation and produce better prediction intervals.

7.6 An alternative model

For some purposes, it is more convenient to work with an alternative to the model $y_i = \beta_0 + \beta_1 x_i + \varepsilon_i$. The alternative model is

$$y_i = \beta_{*0} + \beta_1 (x_i - \bar{x}_{\cdot}) + \varepsilon_i$$

where we have adjusted the predictor variable for its mean. The key difference between the parameters in the two models is that

$$\beta_0 = \beta_{*0} - \beta_1 \bar{x}_{\cdot}.$$

In fact, this is the basis for our formula for estimating β_0. The new parameter β_{*0} has a very simple estimate, $\hat{\beta}_{*0} \equiv \bar{y}_{\cdot}$. It then follows that

$$\hat{\beta}_0 = \bar{y}_{\cdot} - \hat{\beta}_1 \bar{x}_{\cdot}.$$

The reason that this model is useful is because the predictor variable $x_i - \bar{x}_{\cdot}$ has the property $\sum_{i=1}^{n} (x_i - \bar{x}_{\cdot}) = 0$. This property leads to the simple estimate of β_{*0} but also to the fact that $\bar{y}_{\cdot}$ and $\hat{\beta}_1$ are independent. Independence simplifies the computation of variances for regression line estimates. We will not go further into these claims at this point but the results follow trivially from the matrix approach to regression that will be treated in later chapters.

The key point about the alternative model is that it is equivalent to the original model. The β_1 parameters are the same, as are their estimates and standard errors. The models give the same predictions, the same ANOVA table F test, and the same R^2. Even the intercept parameters are equivalent, i.e., they are related in a precise fashion so that knowing about the intercept in either model yields equivalent information about the intercept in the other model.

7.7 Correlation

The correlation coefficient is a measure of the linear relationship between two variables. The population correlation coefficient, usually denoted ρ, was discussed in Chapter 1. The sample

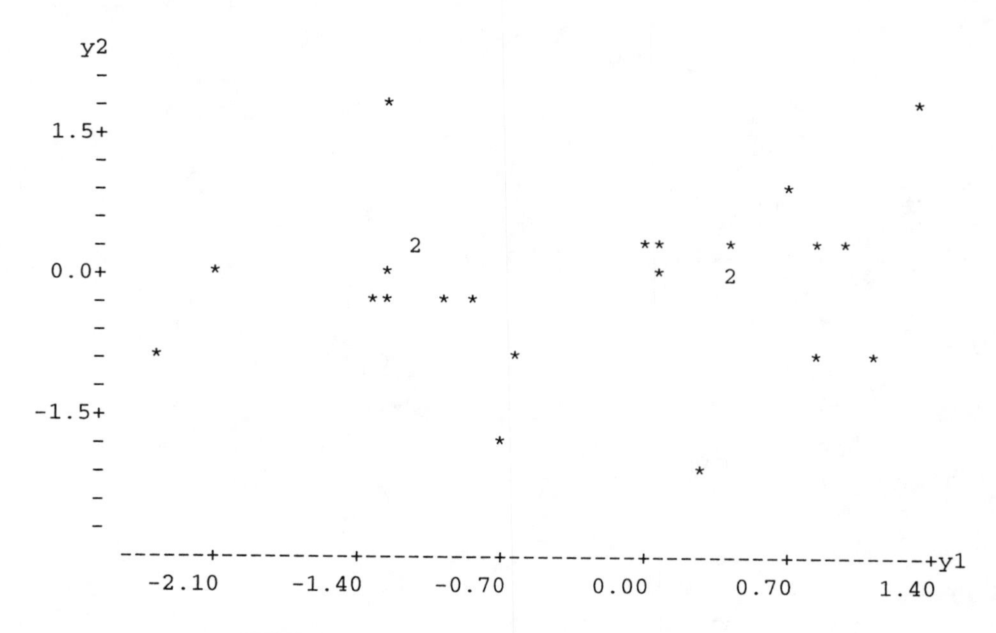

FIGURE 7.2. Correlation plot, $\rho = 0.000$, $r = 0.144$.

correlation is defined as

$$r = \frac{s_{xy}}{s_x s_y} = \frac{\sum_{i=1}^{n} (x_i - \bar{x}_\cdot)(y_i - \bar{y}_\cdot)}{\sqrt{\sum_{i=1}^{n} (x_i - \bar{x}_\cdot)^2 \sum_{i=1}^{n} (y_i - \bar{y}_\cdot)^2}}.$$

The sample correlation coefficient is related to the estimated slope. From equation (7.3.1) it is easily seen that

$$r = \hat{\beta}_1 \frac{s_x}{s_y}.$$

EXAMPLE 7.7.1. *Simulated data with various correlations*
Figures 7.2 through 7.5 contain plots of 25 correlated observations. These are presented so the reader can get some feeling for the meaning of various sample correlation values. The caption of each plot gives the sample correlation r and also the population correlation ρ. The population correlation is only useful in that it provides some feeling for the amount of sampling variation to be found in r based on samples of 25 from (jointly) normally distributed data. □

A commonly used statistic in regression analysis is the coefficient of determination,

$$R^2 \equiv \frac{SSReg}{SSTot}.$$

This is the percentage of the total variation in the dependent variable that is explained by the regression. For simple linear regression,

$$R^2 = \frac{\hat{\beta}_1^2 \sum_{i=1}^{n} (x_i - \bar{x}_\cdot)^2}{\sum_{i=1}^{n} (y_i - \bar{y}_\cdot)^2} = \hat{\beta}_1^2 \frac{s_x^2}{s_y^2} = r^2.$$

In later chapters we will consider regression problems with more than one predictor variable. For such problems R^2 does not equal r^2. In fact, with more than one predictor, there are several r^2s that one could compute. It is not clear which of these one would want to compare to R^2.

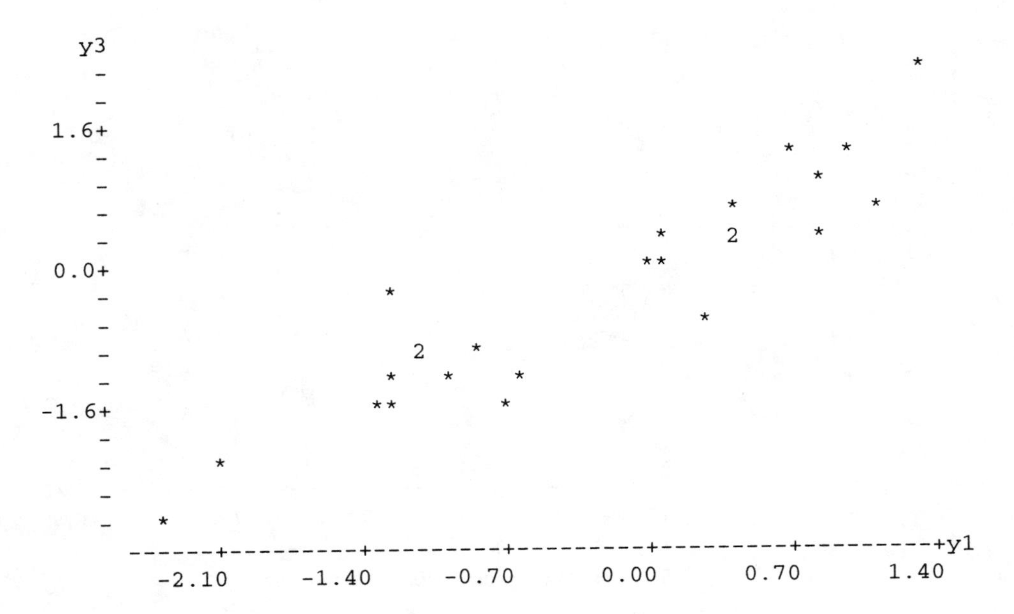

FIGURE 7.3. Correlation plot, $\rho = 0.894$, $r = 0.929$.

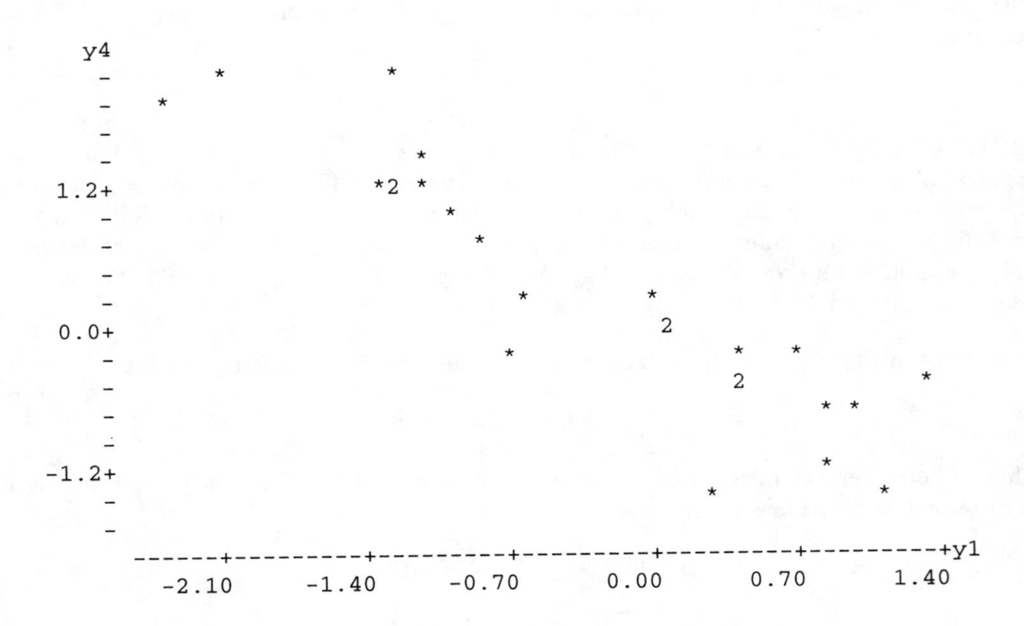

FIGURE 7.4. Correlation plot, $\rho = -0.894$, $r = -0.929$.

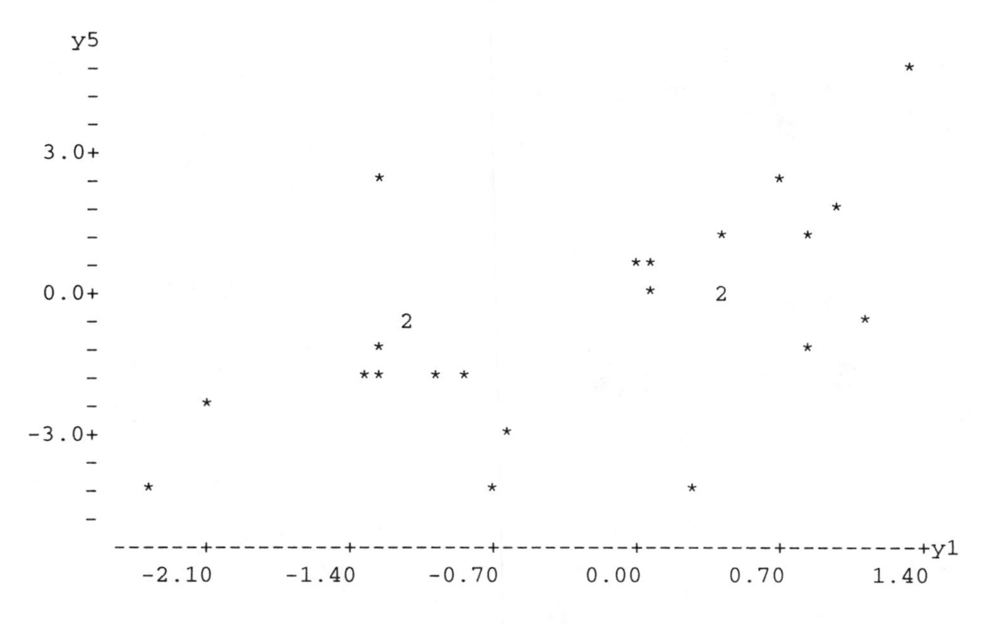

FIGURE 7.5. Correlation plot, $\rho = 0.447$, $r = 0.593$.

7.8 Recognizing randomness: simulated data with zero correlation

Just as it is important to be able to look at a plot and tell when the x and y variables are related, it is important to be able to look at a plot and tell that two variables are unrelated. In other words, we need to be able to identify plots that only display random variation. This skill is of particular importance in Section 7.9 where we use plots to evaluate the assumptions made in simple linear regression. To check the assumptions of the regression model, we use plots that should display only random variation when the assumptions are true. Any systematic pattern in the model checking plots indicates a problem with our assumed regression model.

EXAMPLE 7.8.1. *Simulated data with zero correlation*
We now examine data on six uncorrelated variables, C10 through C15. Figures 7.6 through 7.14 contain various plots of the variables. Since all the variable pairs have zero correlation, i.e., $\rho = 0$, any 'patterns' that are recognizable in these plots are due entirely to random variation. In particular, note that there is no real pattern in Figure 7.13.

The point of this example is to familiarize the reader with the appearance of random plots. The reader should try to identify systematic patterns in these plots, remembering that there are none. This suggests that in the model checking plots that appear later, any systematic pattern of interest should be more pronounced than anything that can be detected in Figures 7.6 through 7.15.

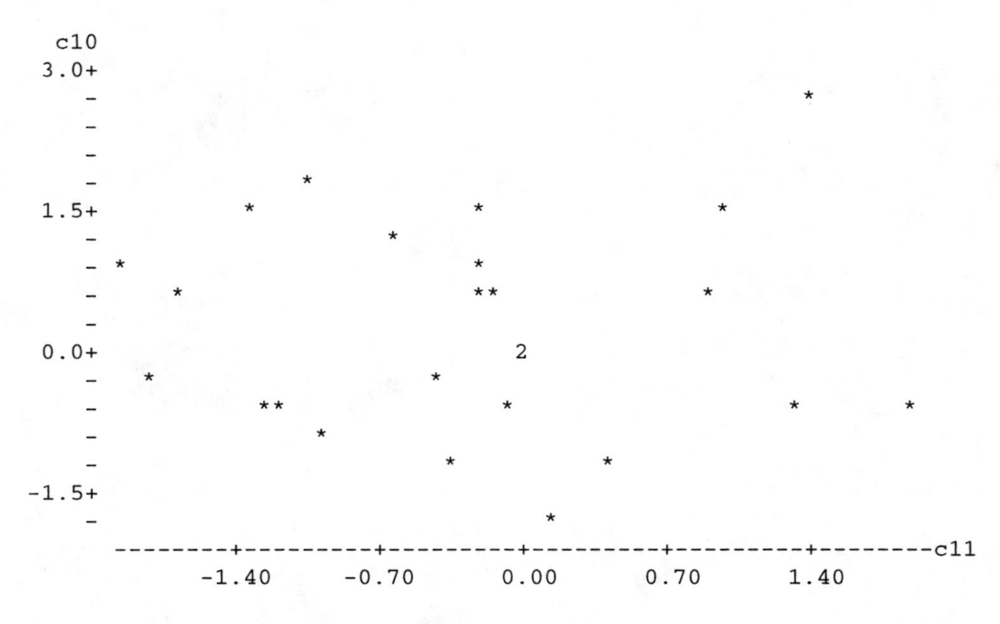

FIGURE 7.6. Plot of data with $\rho = 0$.

Below are the sample correlations r for each pair of variables. Although $\rho = 0$, none of the r values is zero and some of them are quite far from 0.

Sample correlations

	C10	C11	C12	C13	C14	C15
C10	1.000					
C11	0.005	1.000				
C12	−0.145	−0.209	1.000			
C13	−0.162	−0.416	0.488	1.000		
C14	−0.034	−0.038	−0.265	0.003	1.000	
C15	−0.218	−0.202	0.310	0.114	0.134	1.000

□

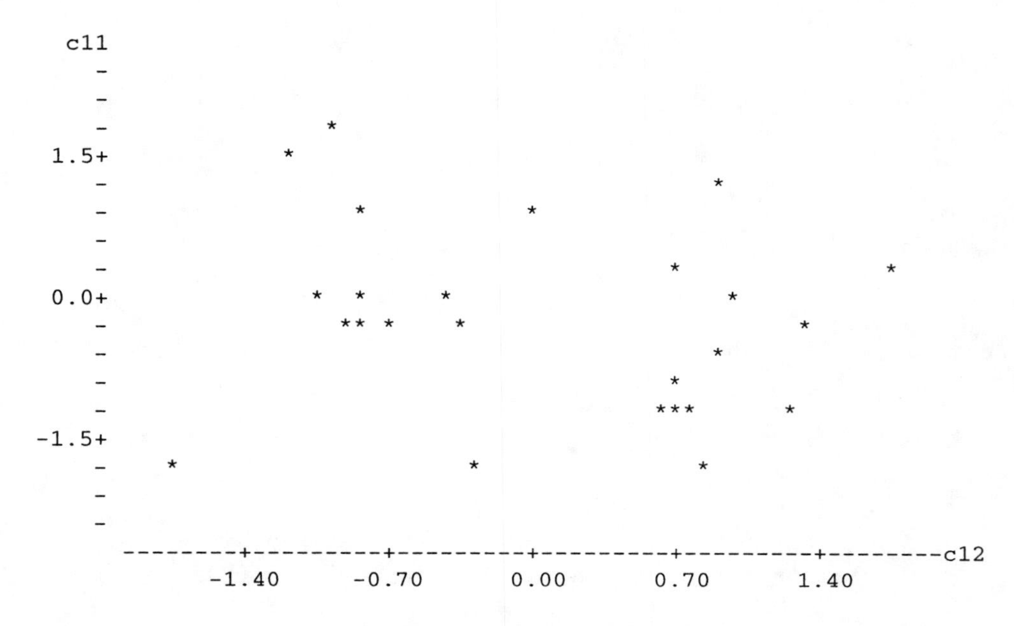

FIGURE 7.7. Plot of data with $\rho = 0$.

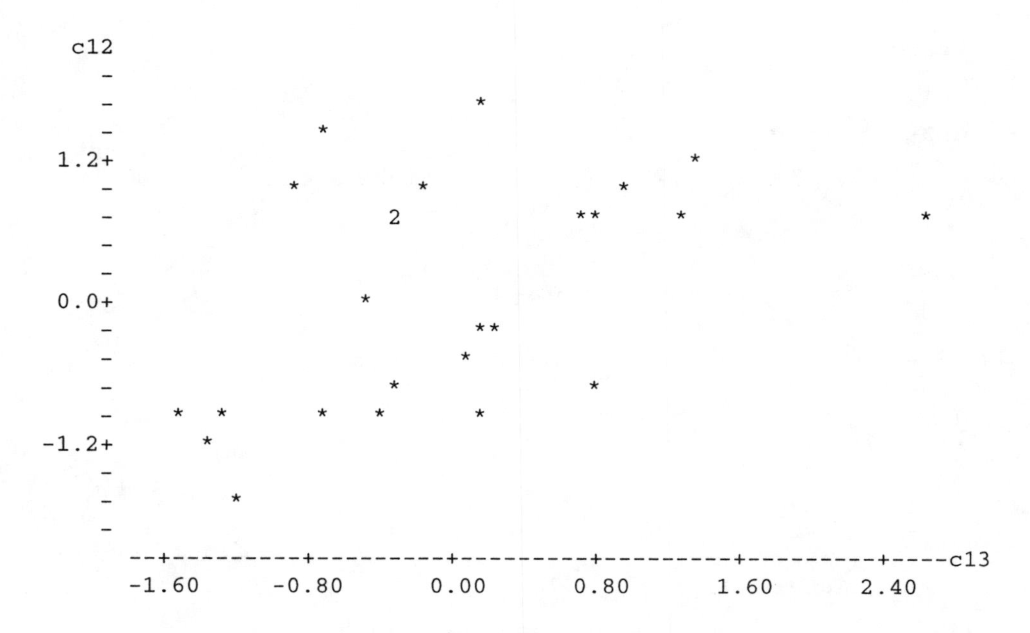

FIGURE 7.8. Plot of data with $\rho = 0$.

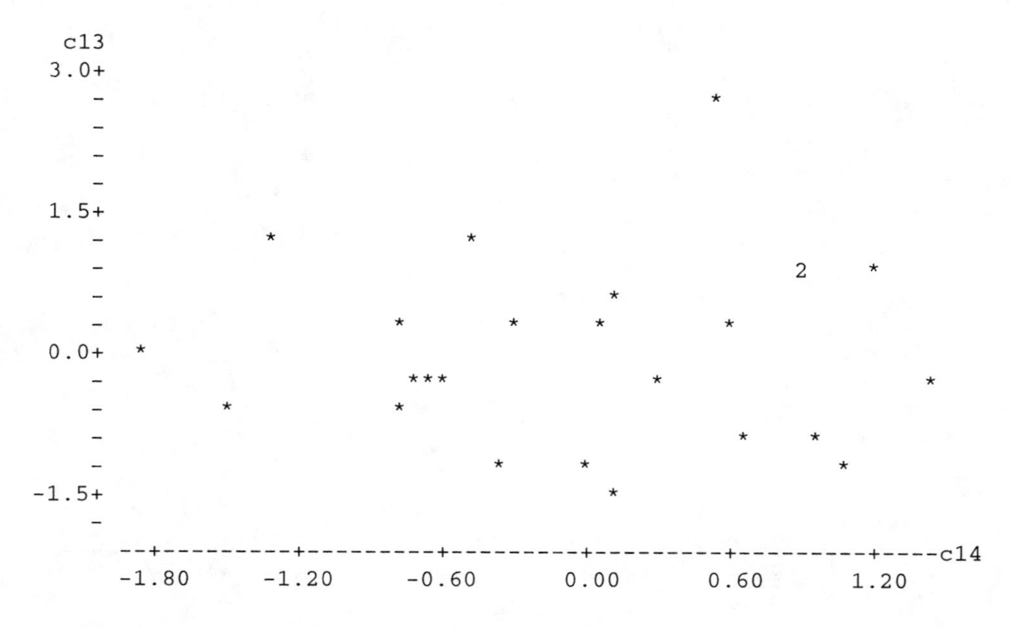

FIGURE 7.9. Plot of data with $\rho = 0$.

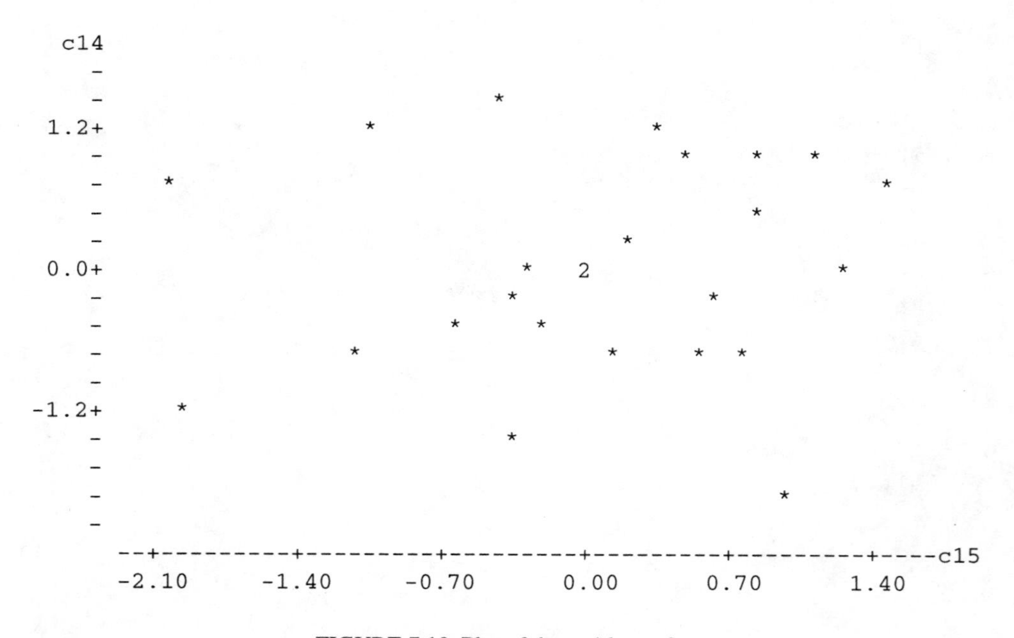

FIGURE 7.10. Plot of data with $\rho = 0$.

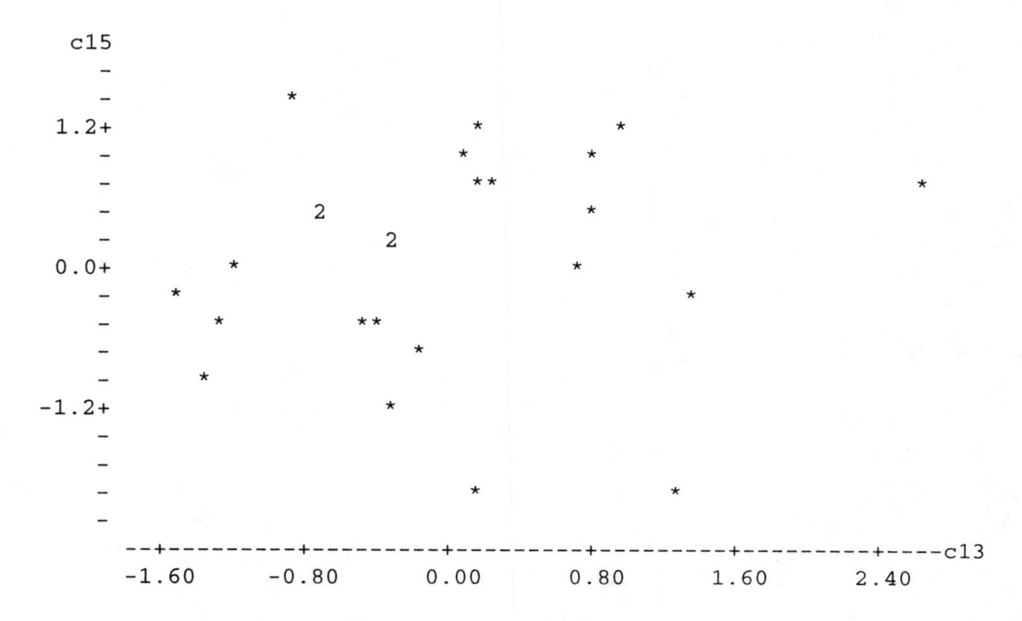

FIGURE 7.11. Plot of data with $\rho = 0$.

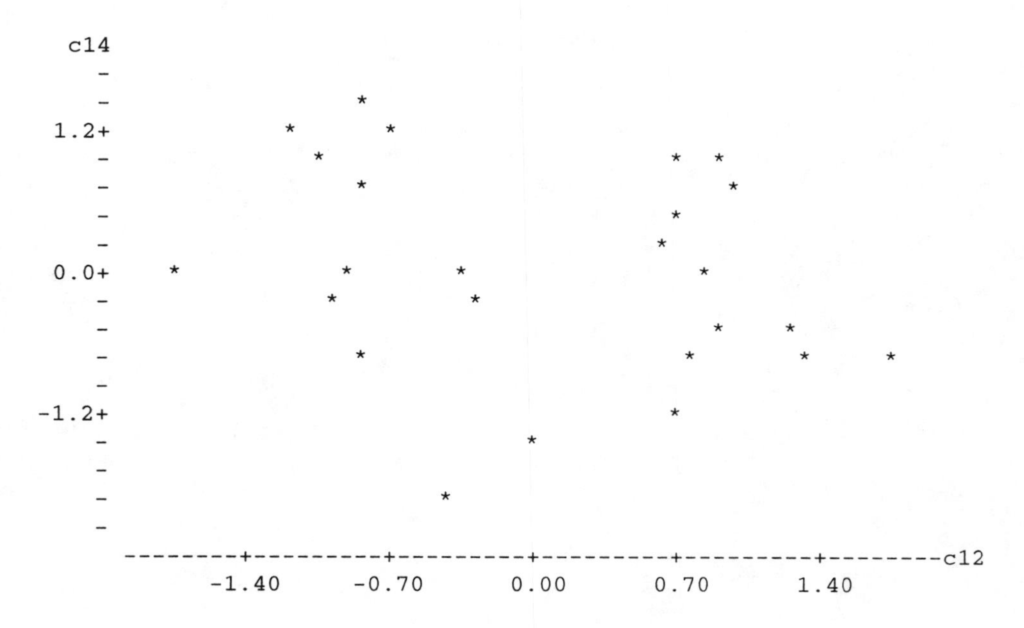

FIGURE 7.12. Plot of data with $\rho = 0$.

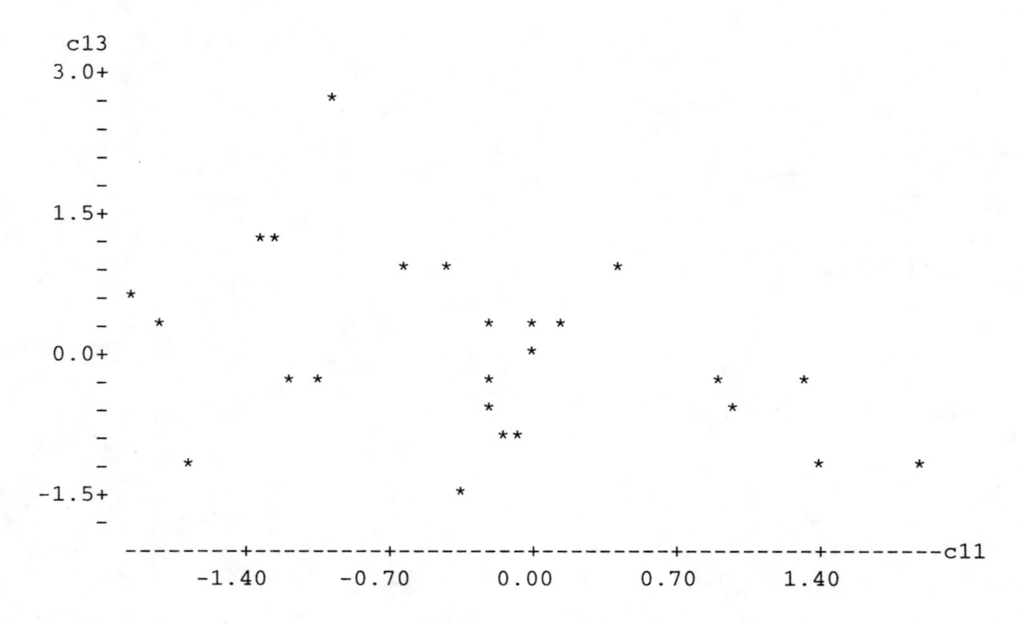

FIGURE 7.13. Plot of data with $\rho = 0$.

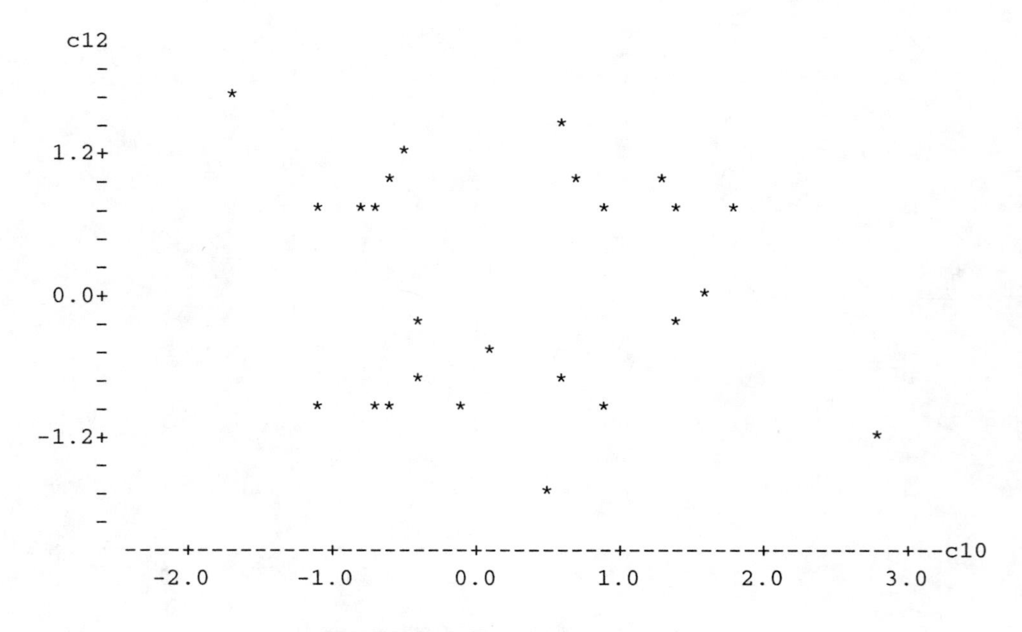

FIGURE 7.14. Plot of data with $\rho = 0$.

MINITAB COMMANDS

The plots and sample correlations in this section were obtained with the Minitab commands given below.

```
MTB > random 25 c10-c15;
SUBC> normal 0 1.
MTB > plot c10 c11
MTB > plot c11 c12
MTB > plot c12 c13
MTB > plot c13 c14
MTB > plot c14 c15
MTB > plot c15 c13
MTB > plot c14 c12
MTB > plot c13 c11
MTB > plot c12 c10
MTB > note      OBTAIN SAMPLE CORRELATION MATRIX
MTB > corr c10-c15
```

7.9 Checking assumptions: residual analysis

The assumptions involved in regression can all be thought of in terms of the errors. The assumptions are that

1. the ε_is are independent,
2. $E(\varepsilon_i) = 0$ for all i,
3. $\text{Var}(\varepsilon_i) = \sigma^2$ for all i,
4. the ε_is are normally distributed.

To have faith in our analysis, we need to validate these assumptions as far as possible. These are *assumptions* and cannot be validated completely, but we can try to detect gross violations of the assumptions.

The first assumption, that the ε_is are independent, is the most difficult to validate. If the observations are taken at regular time intervals, they may lack independence and standard time series methods may be useful in the analysis. We will not consider this further, the interested reader can consult the time series literature, e.g., Shumway (1988). In general, we rely on the data analyst to think hard about whether there are reasons for the data to lack independence.

The second assumption is that $E(\varepsilon_i) = 0$. This is violated when we have the wrong regression model. The simple linear regression model with $E(\varepsilon_i) = 0$ specifies that

$$E(y_i) = \beta_0 + \beta_1 x_i.$$

If we fit this model when it is incorrect, we will not have errors with $E(\varepsilon_i) = 0$. Having the wrong model is called *lack of fit*.

The last two assumptions are that the errors all have some common variance σ^2 and that they are normally distributed. The term *homoscedasticity* refers to having a constant (homogeneous) variance. The term *heteroscedasticity* refers to having nonconstant (heterogeneous) variances.

In checking the error assumptions, we are hampered by the fact that the errors are not observable; we must estimate them. The model involves

$$y_i = \beta_0 + \beta_1 x_i + \varepsilon_i$$

or equivalently,

$$y_i - \beta_0 - \beta_1 x_i = \varepsilon_i.$$

We can estimate ε_i with the *residual*

$$\hat{\varepsilon}_i = y_i - \hat{\beta}_0 - \hat{\beta}_1 x_i.$$

Actually, I prefer to call this predicting the error rather than estimating it. *One estimates fixed unknown parameters and predicts unobserved random variables.*

Previously, we used residuals to check assumptions in one-way analysis of variance. The discussion here is similar but more extensive. The methods presented here also apply to ANOVA models, but, especially in balanced ANOVA, many of the issues are not as crucial. Note also that, as in one-way ANOVA, the *SSE* for simple linear regression is precisely the sum of the squared residuals.

Two of the error assumptions are independence and homoscedasticity of the variances. Unfortunately, the residuals are neither independent nor do they have the same variance. The residuals all involve the random variables $\hat{\beta}_0$ and $\hat{\beta}_1$, so they are not independent. Moreover, the ith residual involves $\hat{\beta}_0 + \hat{\beta}_1 x_i$, the variance of which depends on $(x_i - \bar{x}_\bullet)$. Thus the variance of $\hat{\varepsilon}_i$ depends on x_i. There is little we can do about the lack of independence except hope that it does not cause severe problems. On the other hand, we can adjust for the differences in variances. The variance of a residual is

$$\text{Var}(\hat{\varepsilon}_i) = \sigma^2(1 - h_i)$$

where h_i is the *leverage* of the ith case. Leverages are discussed a bit later in this section and more extensively in relation to multiple regression.

Given the variance of a residual, we can obtain a standard error for it,

$$SE(\hat{\varepsilon}_i) = \sqrt{MSE(1 - h_i)}.$$

We can now adjust the residuals so they all have a variance of about 1; these *standardized residuals* are

$$r_i = \frac{\hat{\varepsilon}_i}{\sqrt{MSE(1 - h_i)}}.$$

The main tool used in checking assumptions is plotting the residuals or, more commonly, the standardized residuals. If the assumptions are correct, plots of the standardized residuals versus any variable should look random. If the variable plotted against the r_is is continuous with no major gaps, the plots should look similar to the plots given in the previous section. In analysis of variance problems, we often plot the residuals against indicators of the treatment groups, so the discrete nature of the number of groups keeps the plots from looking like those of the previous section. The single most popular diagnostic plot is probably the plot of the standardized residuals against the predicted values

$$\hat{y}_i = \hat{\beta}_0 + \hat{\beta}_1 x_i,$$

however the r_is can be plotted against any variable that provides a value associated with each case.

Violations of the error assumptions are indicated by any systematic pattern in the residuals. This could be, for example, a pattern of increased variability as the predicted values increase, or some curved pattern in the residuals, or any change in the variability of the residuals.

A residual plot that displays an increasing variance looks roughly like a horn opening to the right.

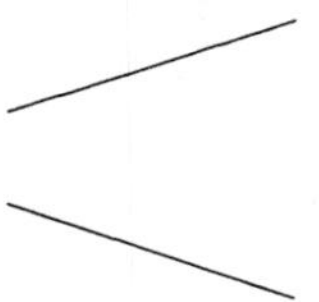

A residual plot indicating a decreasing variance is a horn opening to the left.

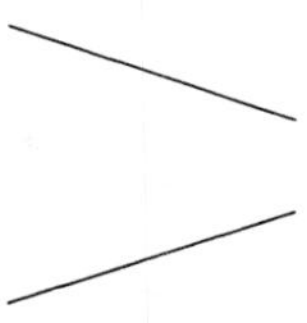

Plots that display curved shapes typically indicate lack of fit. One example of a curve is given below.

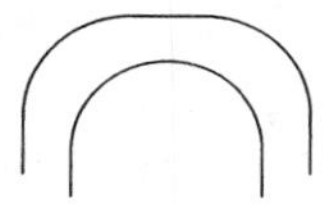

EXAMPLE 7.9.1. *Coleman Report data*
Figures 7.15 through 7.17 contain standardized residual plots for the *Coleman Report* Data. Figure 7.15 is a plot against the predicted values; Figure 7.16 is a plot against the sole predictor variable x. The shapes of these two plots are identical. This always occurs in simple linear regression because the predictions $\hat{y}$ are a linear function of the one predictor x. The one caveat to the claim of identical shapes is that the plots may be reversed. If the estimated slope is negative, the largest x values correspond to the smallest $\hat{y}$ values. Figures 7.15 and 7.16 look like random patterns but it should be noted that if the smallest standardized residual were dropped (the small one on the right), the plot might suggest decreasing variability. The normal plot of the standardized residuals in Figure 7.17 does not look too bad. □

MINITAB COMMANDS

We now illustrate the Minitab commands necessary for the analysis in Example 7.9.1. The subtlest thing going on here is in the command 'regress c1 on 1 c2 c11 c12'. The number 1 indicates that there is one predictor variable x and the first column (c2) after the number 1 is taken to be that predictor. The command recognizes columns c11 and c12 as not being predictors; in fact, the command puts the standardized residuals r_i in the first column (c11) listed after the predictor and the predicted values $\hat{y}_i$ in the second column (c12) listed after the predictor variable. Actually, Minitab will let you write the command as 'regress c1 on 1 c2 put standardized resids in c11 put predicted values in c12'. Minitab just ignores all the words in this command other than 'regress'.

```
MTB > names c1 'test' c2 'socio'
MTB > regress c1 on 1 c2 c11 c12
MTB > names c11 'r' c12 'yhat'
```

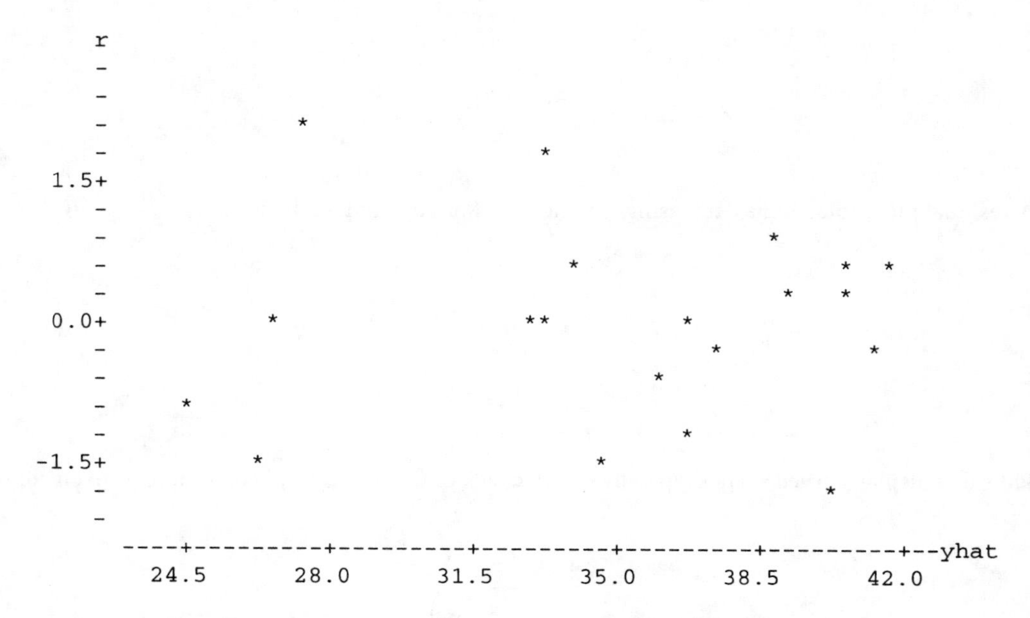

FIGURE 7.15. Plot of the standardized residuals r versus $\hat{y}$, *Coleman Report*.

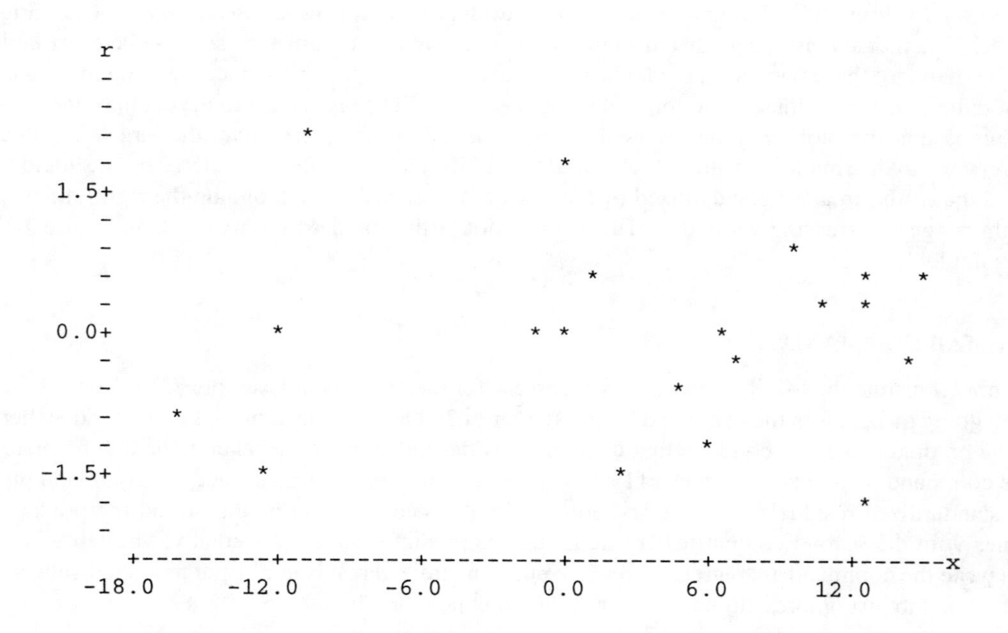

FIGURE 7.16. Plot of the standardized residuals r versus x, *Coleman Report*.

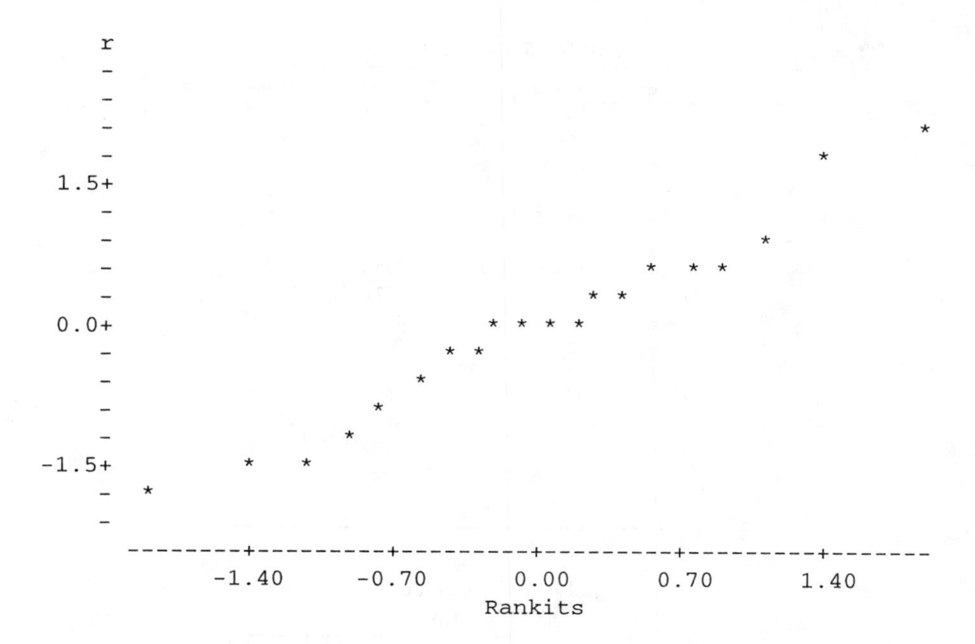

FIGURE 7.17. Normal plot, *Coleman Report*, $W' = .966$.

```
MTB > note      PLOT STD. RESIDS AGAINST PRED. VALUES
MTB > plot c11 c12
MTB > note      PLOT STD. RESIDS AGAINST x
MTB > plot c11 c2
MTB > note      COMPUTE NORMAL SCORES (RANKITS) FOR THE
MTB > note      STANDARDIZED RESIDUALS
MTB > nscores c11 c10
MTB > note      MAKE NORMAL PLOT
MTB > plot c11 c10
MTB > note      COMPUTE W' STATISTIC
MTB > corr c11 c10
MTB > note      CORR PRINTS OUT A NUMBER LIKE .978
MTB > let k1=.978**2
MTB > print k1
```

ANOTHER EXAMPLE

EXAMPLE 7.9.2. *Hooker data*
Forbes (1857) reported data on the relationship between atmospheric pressure and the boiling point of water that were collected in the Himalaya mountains by Joseph Hooker. Weisberg (1985, p. 28) presented a subset of 31 observations that are reproduced in Table 7.5.

A scatter plot of the data is given in Figure 7.18. The data appear to fit a line very closely. The usual summary tables are given below.

Predictor	$\hat{\beta}_k$	$SE(\hat{\beta}_k)$	t	P
Constant	−64.413	1.429	−45.07	0.000
Temperature	0.440282	0.007444	59.14	0.000

TABLE 7.5. Hooker data

Case	Temperature	Pressure	Case	Temperature	Pressure
1	180.6	15.376	17	191.1	19.490
2	181.0	15.919	18	191.4	19.758
3	181.9	16.106	19	193.4	20.480
4	181.9	15.928	20	193.6	20.212
5	182.4	16.235	21	195.6	21.605
6	183.2	16.385	22	196.3	21.654
7	184.1	16.959	23	196.4	21.928
8	184.1	16.817	24	197.0	21.892
9	184.6	16.881	25	199.5	23.030
10	185.6	17.062	26	200.1	23.369
11	185.7	17.267	27	200.6	23.726
12	186.0	17.221	28	202.5	24.697
13	188.5	18.507	29	208.4	27.972
14	188.8	18.356	30	210.2	28.559
15	189.5	18.869	31	210.8	29.211
16	190.6	19.386			

Analysis of variance

Source	df	SS	MS	F	P
Regression	1	444.17	444.17	3497.89	0.000
Error	29	3.68	0.13		
Total	30	447.85			

The coefficient of determination is an exceptionally large

$$R^2 = \frac{444.17}{447.85} = 99.2\%.$$

The plot of residuals versus predicted values is given in Figure 7.19. A pattern is very clear; the residuals form something like a parabola. In spite of a very large R^2 and a scatter plot that looks very linear, the residual plot shows that a lack of fit obviously exists. After seeing the residual plot, you can go back to the scatter plot and detect suggestions of nonlinearity. The simple linear regression model is clearly inadequate, so we do not bother presenting a normal plot. In the next two sections, we will examine ways of dealing with this lack of fit. □

OUTLIERS

Outliers are bizarre data points. They are points that do not seem to fit with the other observations in a data set. We can characterize bizarre points as having either bizarre x values or bizarre y values. There are two valuable tools for identifying outliers.

Leverages are values between 0 and 1 that measure how bizarre an x value is relative to the other x values in the data. *A leverage near 1 is a very bizarre point.* Leverages that are small are similar to the other data. The sum of all the leverages in a simple linear regression is always 2, thus the average leverage is $2/n$. Points with leverages larger that $4/n$ or $6/n$ are often considered high leverage points. The concept of leverage will be discussed in more detail when we discuss multiple regression.

Outliers in the y values can be detected from the *standardized deleted residuals*. Standardized deleted residuals are just standardized residuals, but the residual for the ith case is computed from a

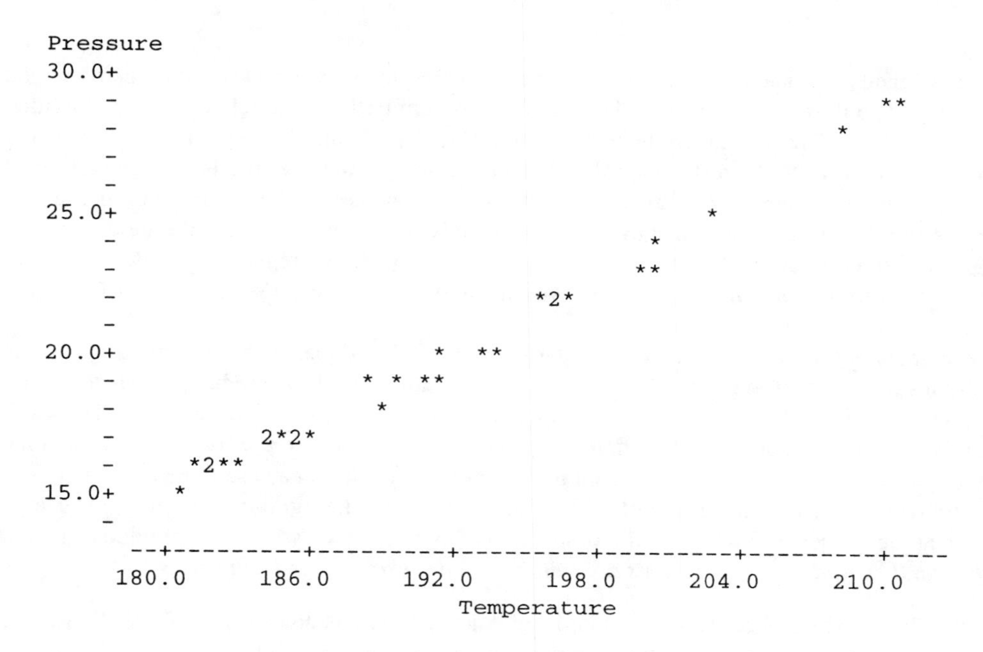

FIGURE 7.18. Scatter plot of Hooker data.

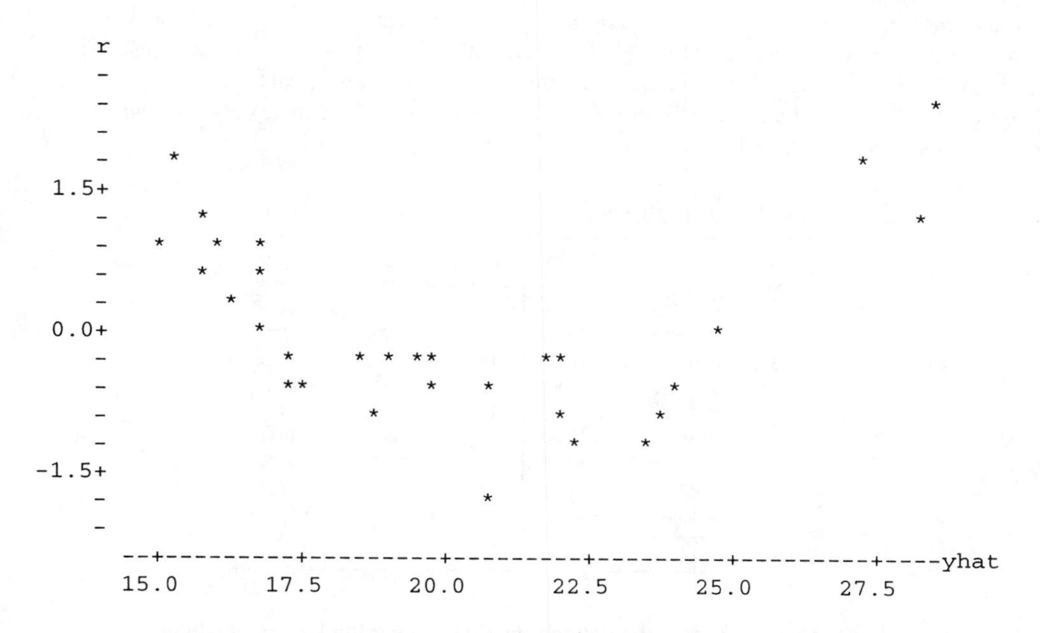

FIGURE 7.19. Standardized residuals versus predicted values for Hooker data.

regression that does not include the ith case. For example, the third deleted residual is

$$\hat{\varepsilon}_{[3]} = y_3 - \hat{\beta}_{0[3]} - \hat{\beta}_{1[3]}x_3$$

where the estimates $\hat{\beta}_{0[3]}$ and $\hat{\beta}_{1[3]}$ are computed from a regression in which case 3 has been dropped from the data. The third standardized deleted residual is simply the third deleted residual divided by its standard error. The standardized deleted residuals really contain the same information as the standardized residuals; the largest standardized deleted residuals are also the largest standardized residuals. The main virtue of the standardized deleted residuals is that they can be compared to a $t(n-3)$ distribution to test whether they could reasonably have occurred when the model is true. The degrees of freedom in the test are $n-3$ because the simple linear regression model was fitted without the ith case so there are only $n-1$ data points in the fit and $(n-1)-2$ degrees of freedom for error.

If one compares the *largest* absolute standardized deleted residual to a t distribution, one is essentially testing whether *every* case is an outlier. Thus a total of n tests are being performed and the overall error rate from an individual α level test may be as high as $n\alpha$. For $n = 20$ and $\alpha = .05$, $n\alpha = 1$, so we can reasonably expect to 'find an outlier' in the *Coleman Report* data even when none exists. Obviously, one needs a more stringent requirement for declaring a case to be an outlier. The criterion for declaring a case to be an outlier should be something like significance at the $.05/n$ level rather than at the $.05$ level. This is just the Bonferroni adjustment discussed in the previous chapter. *Often the standardized deleted residuals are simply called t residuals and denoted* t_i.

EXAMPLE 7.9.3. The leverages and standardized deleted residuals are given in Table 7.6 for the *Coleman Report* data with one predictor. Compared to the leverage rule of thumb $4/n = 4/20 = .2$, only case 15 has a noticeably large leverage. None of the cases is above the $6/n$ rule of thumb. In simple linear regression, one does not really need to evaluate the leverages directly because the necessary information about bizarre x values is readily available from the x, y plot. In multiple regression with three or more predictor variables, leverages are vital because no one scatter plot can give the information on bizarre x values. In the scatter plot of the *Coleman Report* data, Figure 7.1, there are no outrageous x values, although there is a noticeable gap between the smallest four values and the rest. From Table 7.1 we see that the cases with the smallest x values are 2, 11, 15, and 18. These cases also have the highest leverages reported in Table 7.6. The next two highest leverages are for cases 4 and 19; these have the largest x values.

TABLE 7.6. Outlier diagnostics for the *Coleman Report* data

Case	Leverages	Std. del. residuals	Case	Leverages	Std. del. residuals
1	0.059362	−0.15546	11	0.195438	−1.44426
2	0.175283	−0.12019	12	0.052801	0.61394
3	0.097868	−1.86339	13	0.051508	−0.49168
4	0.120492	−0.28961	14	0.059552	0.14111
5	0.055707	0.10792	15	0.258992	−0.84143
6	0.055179	−1.35054	16	0.081780	0.19341
7	0.101914	0.63059	17	0.050131	−1.41912
8	0.056226	0.07706	18	0.163429	2.52294
9	0.075574	1.00744	19	0.130304	0.63836
10	0.055783	1.92501	20	0.102677	0.24410

For an overall $\alpha = .05$ level test of the deleted residuals, the tabled value needed is

$$t\left(1 - \frac{.05}{2(20)}, 17\right) = 3.54\,.$$

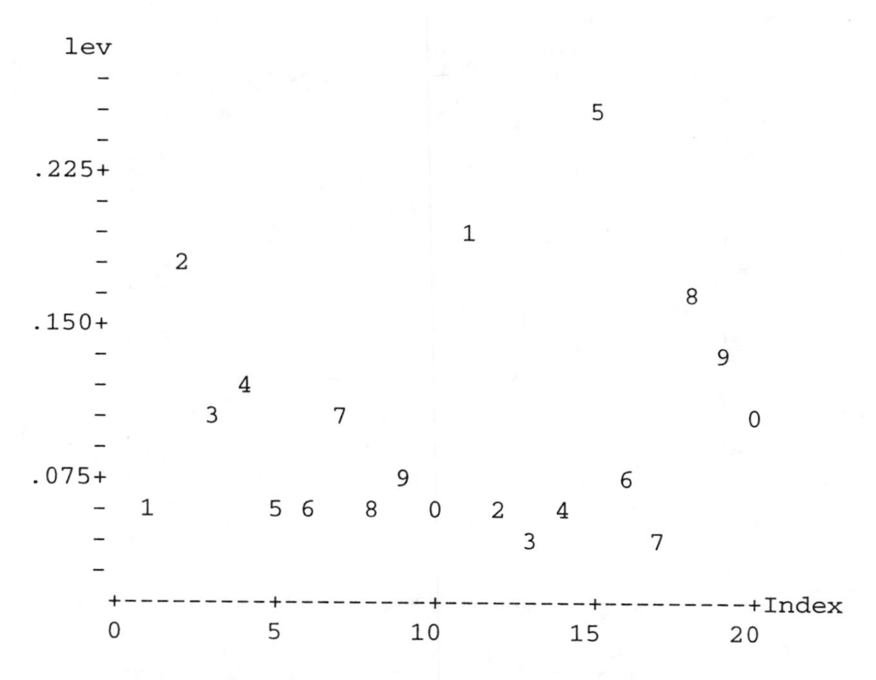

FIGURE 7.20. Index plot of leverages for the *Coleman Report* data.

None of the standardized deleted residuals approach this, so there is no evidence of any unaccountably bizarre y values.

A handy way to identify cases with large leverages, residuals, standardized residuals, or standardized deleted residuals is with an index plot. This is simply a plot of the value against the case number as in Figure 7.20 for leverages. In this version of the plot, the symbol plotted is the last digit of the case number. □

MINITAB COMMANDS

We now illustrate the Minitab commands necessary for obtaining the leverages and standardized deleted residuals for the *Coleman Report* data. Both sets of values are obtained by using subcommands of the regress command. The 'hi' subcommand gives leverages, while the 'tresid' subcommand gives standardized deleted residuals. The last command gives the index plot for leverages.

```
MTB > names c1 'test' c2 'socio'
MTB > regress c1 on 1 c2;
SUBC> hi c13;
SUBC> tresid c14.
MTB > tsplot c13
```

EFFECTS OF HIGH LEVERAGE

EXAMPLE 7.9.4. Figure 7.21 contains some data along with their least squares estimated line. The four points on the left form a perfect line with slope 1 and intercept 0. There is one high leverage point far away to the right. The actual data are given below along with their leverages.

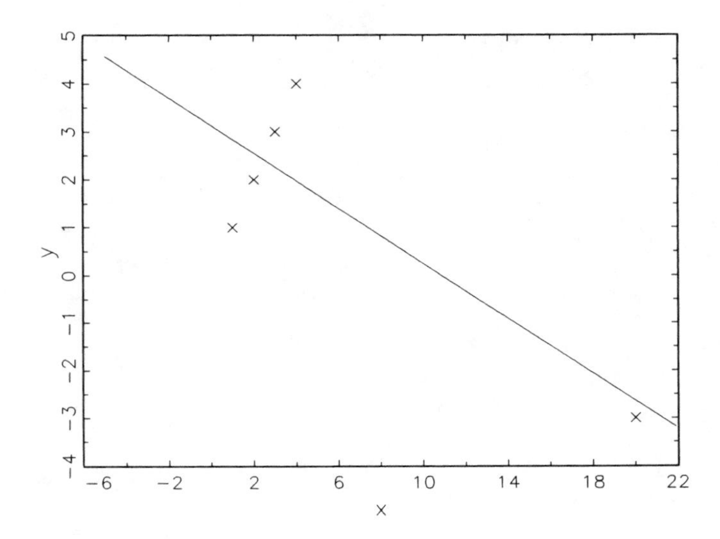

FIGURE 7.21. Plot of y versus x.

Case	1	2	3	4	5
y	1	2	3	4	−3
x	1	2	3	4	20
Leverage	.30	.26	.24	.22	.98

The case with $x = 20$ is an extremely high leverage point; it has a leverage of nearly 1. The estimated regression line is forced to go very nearly through this high leverage point. In fact, this plot has two clusters of points that are very far apart, so a rough approximation to the estimated line is the line that goes through the mean x and y values for each of the two clusters. This example has one cluster of four cases on the left of the plot and another cluster consisting solely of the one case on the right. The average values for the four cases on the left give the point $(\bar{x}, \bar{y}) = (2.5, 2.5)$. The one case on the right is $(20, -3)$. A little algebra shows the line through these two points to be $\hat{y} = 3.286 - 0.314x$. The estimated line using least squares turns out to be $\hat{y} = 3.128 - 0.288x$, which is not too different. The least squares line goes through the two points $(2.5, 2.408)$ and $(20, -2.632)$, so the least squares line is a little lower at $x = 2.5$ and a little higher at $x = 20$.

Obviously, the single point on the right of Figure 7.21 dominates the estimated straight line. For example, if the point on the right was $(20, 15)$, the estimated line would go roughly through this point and $(2.5, 2.5)$. Substantially changing the y value at $x = 20$ always gives an extremely different estimated line than the ones we just considered. Wherever the point on the right is, the estimated line follows it. This happens regardless of the fact that the four cases on the left follow a *perfect* straight line with slope 1 and intercept 0. The behavior of the four points on the left is almost irrelevant to the fitted line when there is a high leverage point on the right. They have an effect on the quality of the rough two-point approximation to the actual estimated line but their overall effect is small.

To summarize what can be learned from Figure 7.21, we have a reasonable idea about what happens to y for x values near the range 1 to 4 and we have some idea of what happens when x is 20 but, barring outside information, we have not the slightest idea what happens to y when x is between 4 and 20. Fitting a line to the complete data suggests that we know something about the behavior of y for any value of x between 1 and 20. This is just silly! We would be better off to analyze the two clusters of points separately and to admit that learning about y when x is between 4 and 20 requires us to obtain data on y when x is between 4 and 20. In this example, the two separate statistical

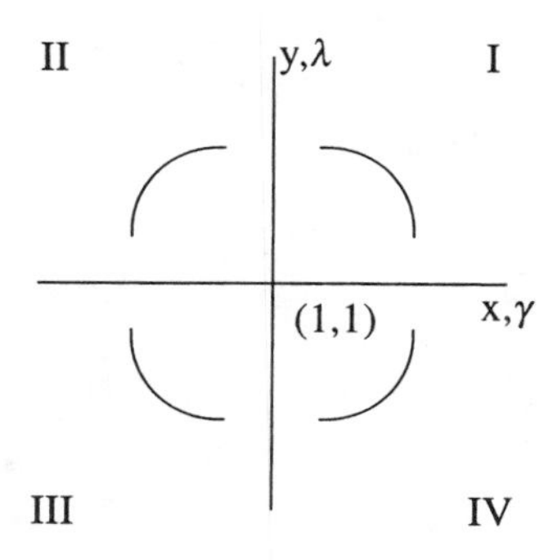

FIGURE 7.22. The circle of x, y transformations.

analyses are trivial. The cluster on the left follows a perfect line so we simply report that line. The cluster on the right is a single point so we report the point. □

7.10 Transformations

If the residuals show a problem with lack of fit, heteroscedasticity, or nonnormality, one way to deal with the problem is to try transforming the y_is. Typically, this only works well when y_{max}/y_{min} is reasonably large. The use of transformations is often a matter of trial and error. Various transformations are tried and the one that gives the best fitting model is used. In this context, the best fitting model should have residual plots indicating that the model assumptions are reasonably valid. The first approach to transforming the data should be to consider transformations that are suggested by any theory associated with the data collection. Another approach to choosing a transformation is to try a variance stabilizing transformation. These were discussed in Section 2.5 and are repeated below for data y_i with $E(y_i) = \mu_i$ and $\text{Var}(y_i) = \sigma_i^2$.

Variance stabilizing transformations

Data	Distribution	Mean, variance relationship	Transformation
Count	Poisson	$\mu_i \propto \sigma_i^2$	$\sqrt{y_i}$
Amount	Gamma	$\mu_i \propto \sigma_i$	$\log(y_i)$
Proportion	Binomial/ N	$\frac{\mu_i(1-\mu_i)}{N} \propto \sigma_i^2$	$\sin^{-1}\left(\sqrt{y_i}\right)$

Whenever the data have the indicated mean, variance relationship, the corresponding variance stabilizing transformation should work reasonably well.

The shape of an x, y plot can also suggest possible transformations to straighten it out. We consider power transformations of both y and x, thus y is transformed into, say, y^λ and x is transformed into x^γ. Note that $\lambda = 1$ and $\gamma = 1$ indicate no transformation. As we will justify later, we treat $\lambda = 0$ and $\gamma = 0$ as log transformations.

Figure 7.22 indicates the kinds of transformations appropriate for some different shapes of x, y curves. For example, if the x, y curve is similar to that in quadrant I, i.e., the y values decrease as x increases and the curve opens to the lower left, appropriate transformations involve increasing λ or

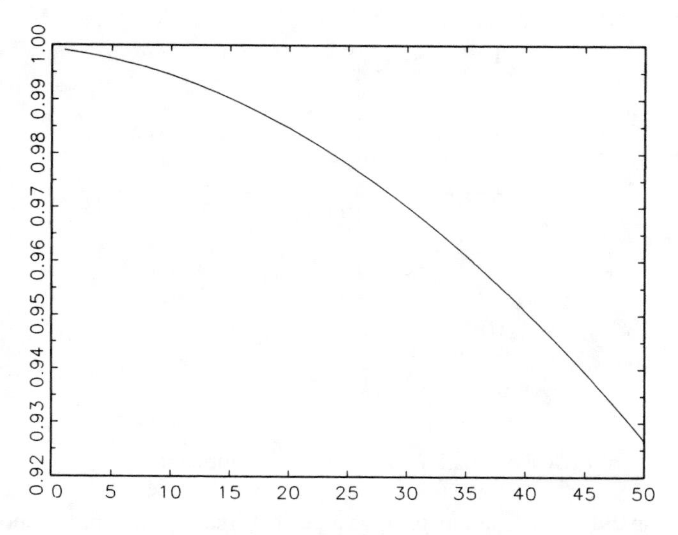

FIGURE 7.23. Curved x, y plot.

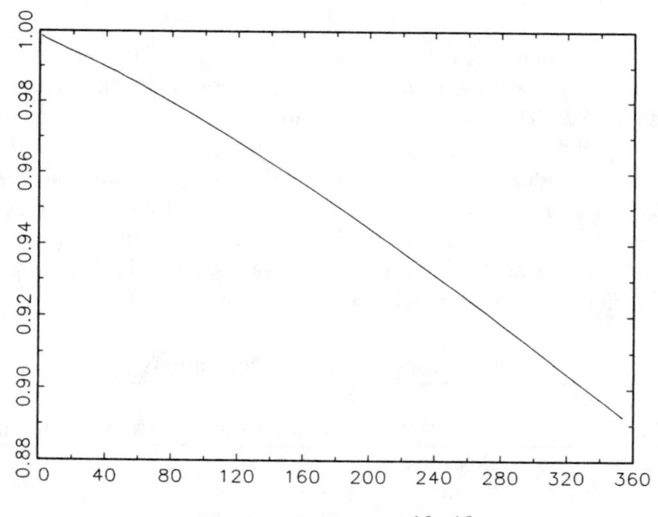

FIGURE 7.24. Plot of $x^{1.5}, y^{1.5}$.

increasing γ or both. Here we refer to increasing λ and γ relative to the no transformation values of $\lambda = 1$ and $\gamma = 1$. In particular, Figure 7.23 gives an x, y plot for part of a cosine curve that is shaped like the curve in quadrant I. Figure 7.24 is a plot of the numbers after x has been transformed into $x^{1.5}$ and y has been transformed into $y^{1.5}$. Note that the curve in Figure 7.24 is much straighter than the curve in Figure 7.23. If the x, y curve increases and opens to the lower right such as those in quadrant II, appropriate transformations involve increasing λ or decreasing γ or both. An x, y curve similar to that in quadrant III suggests decreasing λ or decreasing γ or both. The graph given in Figure 7.22 is often referred to as *the circle of x, y transformations*.

We established in the previous section that the Hooker data does not fit a straight line and that the scatter plot in Figure 7.18 increases with a slight tendency to open to the upper left. This is the same shaped curve as in quadrant IV of Figure 7.22. The circle of x, y transformations suggests that to straighten the curve, we should try transformations with decreased values of λ or increased values of γ or both. Thus we might try transforming y into $y^{1/2}$, $y^{1/4}$, $\log(y)$, or y^{-1}. Similarly, we might try transforming x into $x^{1.5}$ or x^2.

To get a preliminary idea of how well various transformations work, we should do a series of plots. We might begin by examining the four plots in which $y^{1/2}$, $y^{1/4}$, $\log(y)$, and y^{-1} are plotted against x. We might then plot y against both $x^{1.5}$ and x^2. We should also plot all possibilities involving one of $y^{1/2}$, $y^{1/4}$, $\log(y)$, and y^{-1} plotted against one of $x^{1.5}$ and x^2 and we may need to consider other choices of λ and γ. For the Hooker data, looking at these plots would probably only allow us to eliminate the worst transformations. Recall that Figure 7.18 looks remarkably straight and it is only after fitting a simple linear regression model and examining residuals that the lack of fit (the curvature of the x, y plot) becomes apparent. Evaluating the transformations would require fitting a simple linear regression for every pair of transformed variables that has a plot that looks reasonably straight.

Observe that many of the power transformations considered here break down with values of y that are negative. For example, it is difficult to take square roots and logs of negative numbers. Fortunately, data are often positive or at least nonnegative. Measured amounts, counts and proportions are almost always nonnegative. When problems arise, a small constant is often added to all cases so that they all become positive. Of course, it is unclear what constant should be added.

Obviously, the circle of transformations, just like the variance stabilizing transformations, provides only suggestions on how to transform the data. The process of choosing a particular transformation remains one of trial and error. We begin with reasonable candidates and examine how well these transformations agree with the simple linear regression model. When we find a transformation that agrees well with the assumptions of simple linear regression, we proceed to analyze the data. Obviously, an alternative to transforming the data is to change the model. In the next section we consider a new class of models that incorporate transformations of the x variable. In the remainder of this section, we focus on a systematic method for choosing a transformation of y.

7.10.1 Box–Cox transformations

We now consider a systematic method, introduced by Box and Cox (1964), for choosing a power transformation. Consider the family of power transformations, say, y_i^λ. This includes the square root transformation as the special case $\lambda = 1/2$ and other interesting transformations such as the reciprocal transformation y_i^{-1}. By making a minor adjustment, we can bring log transformations into the power family. Consider the transformations

$$y_i^{(\lambda)} = \begin{cases} (y_i^\lambda - 1)/\lambda & \lambda \neq 0 \\ \log(y_i) & \lambda = 0 \end{cases}.$$

For any fixed $\lambda \neq 0$, the transformation $y_i^{(\lambda)}$ is equivalent to y_i^λ, because the difference between the two transformations consists of subtracting a constant and dividing by a constant. In other words, fitting the model

$$y_i^\lambda = \beta_0 + \beta_1 x_i + \varepsilon_i$$

is equivalent to fitting the model

$$y_i^{(\lambda)} = \beta_0 + \beta_1 x_i + \varepsilon_i,$$

although the regression parameters in the two models have slightly different meanings. While the transformation $(y_i^\lambda - 1)/\lambda$ is undefined for $\lambda = 0$, as λ approaches 0, $(y_i^\lambda - 1)/\lambda$ approaches $\log(y_i)$, so the log transformation fits in naturally.

Unfortunately, the results of fitting models to $y_i^{(\lambda)}$ with different values of λ are not directly comparable. Thus it is difficult to decide which transformation in the family to use. This problem is easily evaded (cf. Cook and Weisberg, 1982) by further modifying the family of transformations so that the results of fitting with different λs are comparable. Let $\tilde{y}$ be the geometric mean of the y_is, i.e.,

$$\tilde{y} = \left[\prod_{i=1}^{n} y_i\right]^{1/n} = \exp\left[\frac{1}{n}\sum_{i=1}^{n}\log(y_i)\right]$$

and define the family of transformations

$$z_i^{(\lambda)} = \begin{cases} \left[y_i^{\lambda} - 1\right] / \left[\lambda\, \tilde{y}^{\lambda-1}\right] & \lambda \neq 0 \\ \tilde{y}\log(y_i) & \lambda = 0 \end{cases}.$$

The results of fitting the model

$$z_i^{(\lambda)} = \beta_0 + \beta_1 x_i + \varepsilon_i$$

can be summarized via $SSE(\lambda)$. These values are directly comparable for different values of λ. The choice of λ that yields the smallest $SSE(\lambda)$ is the best fitting model. (It maximizes the likelihood with respect to λ.) Actually, *this method of choosing a transformation works for any ANOVA or regression model.*

Box and Draper (1987, p. 290) discuss finding a confidence interval for the transformation parameter λ. An approximate $(1-\alpha)100\%$ confidence interval consists of all λ values that satisfy

$$\log SSE(\lambda) - \log SSE(\hat{\lambda}) \leq \chi^2(1-\alpha, 1)/\, dfE$$

where $\hat{\lambda}$ is the value of λ that minimizes $SSE(\lambda)$. When $y_{max}/\, y_{min}$ is not large, the interval tends to be wide.

EXAMPLE 7.10.1. *Hooker data*

In the previous section, we found that Hooker's data on atmospheric pressure and boiling points displayed a lack of fit when regressing pressure on temperature. We now consider using power transformations to eliminate the lack of fit.

TABLE 7.7. Choice of power transformation

λ	$1/2$	$1/3$	$1/4$	0	$-1/4$	$-1/2$
$SSE(\lambda)$	1.21	0.87	0.78	0.79	1.21	1.98

Table 7.7 contains $SSE(\lambda)$ values for some reasonable choices of λ. Assuming that $SSE(\lambda)$ is a very smooth (convex) function of λ, the best λ value is probably between 0 and $1/4$. If the curve being minimized is very flat between 0 and $1/4$, there is a possibility that the minimizing value is between $1/4$ and $1/3$. One could pick more λ values and compute more $SSE(\lambda)$s but I have a bias towards simple transformations. (They are easier to sell to clients.)

The log transformation of $\lambda = 0$ is simple (certainly simpler than the fourth root) and $\lambda = 0$ is near the optimum, so we will consider it further. We now use the simple log transformation, rather than adjusting for the geometric mean. The usual summary tables are given below.

Predictor	$\hat{\beta}_k$	$SE(\hat{\beta}_k)$	t	P
Constant	-1.02214	0.03365	-30.38	0.000
Temp.	0.0208698	0.0001753	119.08	0.000

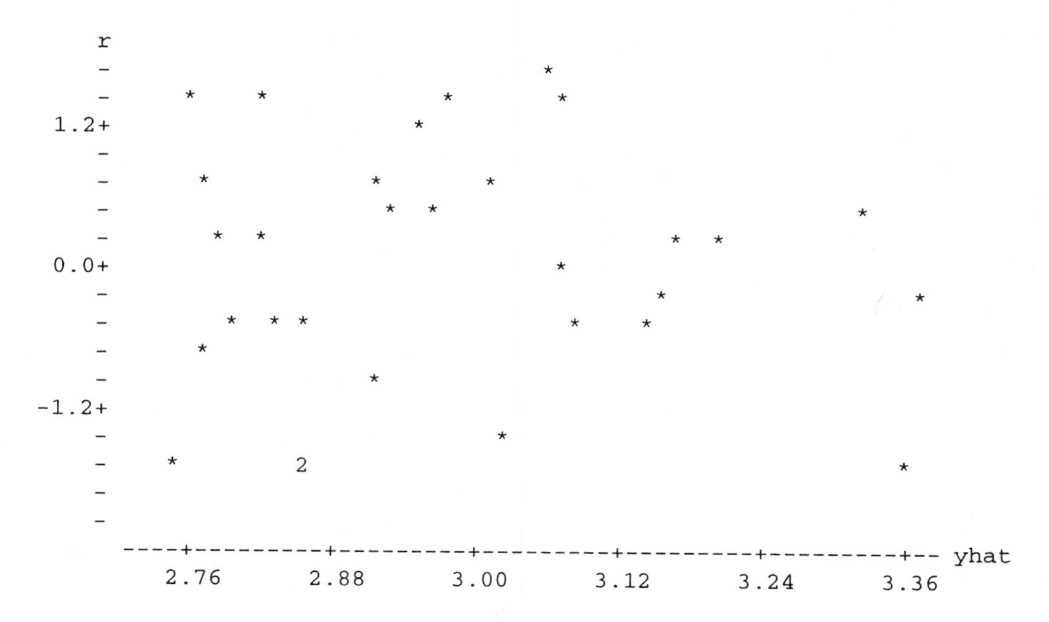

FIGURE 7.25. Standardized residuals versus predicted values, logs of Hooker data.

Analysis of variance: log Hooker data

Source	*df*	*SS*	*MS*	*F*	*P*
Regression	1	0.99798	0.99798	14180.91	0.000
Error	29	0.00204	0.00007		
Total	30	1.00002			

The coefficient of determination is again extremely high, $R^2 = 99.8\%$. The plot of the standardized residuals versus the predicted values is given in Figure 7.25. There is no obvious lack of fit or inconstancy of variances. Figure 7.26 contains a normal plot of the standardized residuals. The normal plot is not horrible but it is not wonderful either. There is a pronounced shoulder at the bottom and perhaps even an S shape.

If we are interested in the mean value of log pressure for a temperature of 205°F, the estimate is $3.2562 = -1.02214 + .0208698(205)$ with a standard error of 0.00276 and a 95% confidence interval of $(3.2505, 3.2618)$. In the original units, the estimate is $e^{3.2562} = 25.95$ and the confidence interval becomes $(e^{3.2505}, e^{3.2618})$ or $(25.80, 26.10)$. The point prediction for a new log observation at 205°F has the same value as the point estimate and has a 95% prediction interval of $(3.2381, 3.2742)$. In the original units, the prediction is again 25.95 and the prediction interval becomes $(e^{3.2381}, e^{3.2742})$ or $(25.49, 26.42)$. □

One way to test whether a transformation is needed is to use a *constructed variable* as introduced by Atkinson (1973). Let

$$w_i = y_i \left[\log(y_i/\tilde{y}) - 1\right]$$

and fit the multiple regression model

$$y_i = \beta_0 + \beta_1 x_i + \beta_2 w_i + \varepsilon_i.$$

Multiple regression gives results similar to those for simple linear regression; typical output includes estimates of the βs, standard errors, t statistics, and an ANOVA table. A test of $H_0 : \beta_2 = 0$ gives an approximate test that no transformation is needed. The test is performed using the standard methods

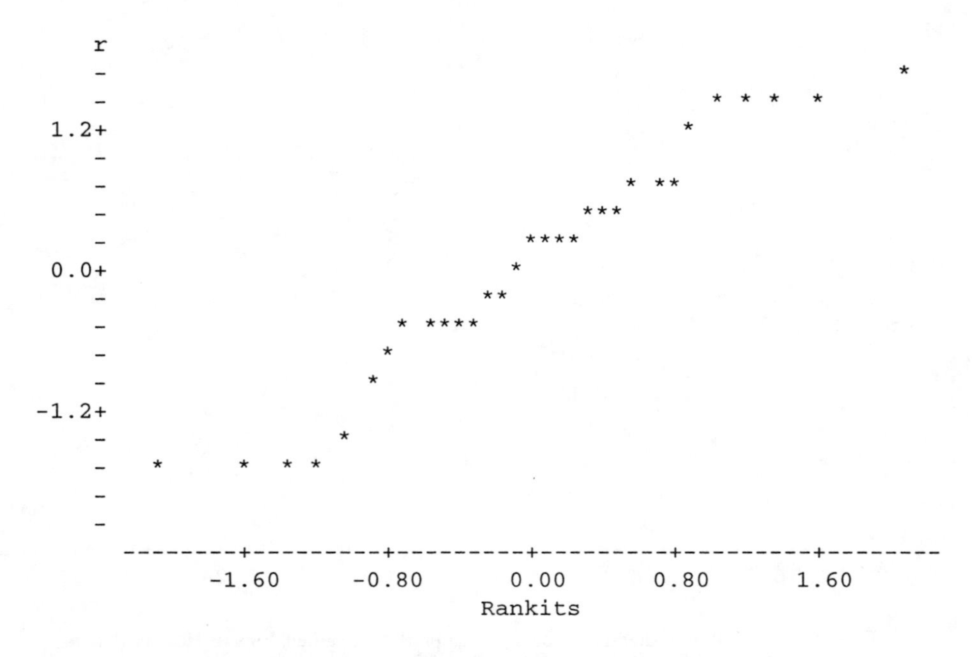

FIGURE 7.26. Normal plot for logs of Hooker data, $W' = 0.960$.

of Chapter 3. Details are illustrated in the example below and discussed in the chapters on multiple regression. In addition, the estimate $\hat{\beta}_2$ provides, indirectly, an estimate of λ,

$$\hat{\lambda} = 1 - \hat{\beta}_2.$$

Frequently, this is not a very good estimate of λ but it gives an idea of where to begin a search for good λs.

EXAMPLE 7.10.2. *Hooker data*
Performing the multiple regression of pressure on both temperature and the constructed variable w gives the following results.

Predictor	$\hat{\beta}_k$	$SE(\hat{\beta}_k)$	t	P
Constant	-43.426	2.074	-20.94	0.000
Temperature	0.411816	0.004301	95.75	0.000
w	0.80252	0.07534	10.65	0.000

The t statistic is $10.65 = .80252/.07534$ for testing that the regression coefficient of the constructed variable is 0. The P value is 0.000, which strongly indicates the need for a transformation. The estimate of λ is

$$\hat{\lambda} = 1 - \hat{\beta}_2 = 1 - 0.80 = .2,$$

which is consistent with what we learned from Table 7.7. From Table 7.7 we suspected that the best transformation would be between 0 and .25. Of course this estimate of λ is quite crude, finding the 'best' transformation requires a more extensive version of Table 7.7. I limited the choices of λ in Table 7.7 because I was unwilling to consider transformations that I did not consider simple. □

COMPUTATIONAL TECHNIQUES

Below are given Minitab commands for performing the Box–Cox transformations and the constructed variable test. To perform multiple regression using the 'regress' command, you need to specify the number of predictor variables, in this case 2.

```
MTB > name c1 'temp' c2 'press'
MTB > note      CONSTRUCT THE GEOMETRIC MEAN
MTB > let c9 = loge(c2)
MTB > mean c9 k1
   MEAN    =       2.9804
MTB > let k2 = expo(k1)
MTB > note      PRINT THE GEOMETRIC MEAN
MTB > print k2
K2        19.6960
MTB > note      CONSTRUCT THE z VARIABLES
MTB > note      FOR DIFFERENT LAMBDAS
MTB > let c20=(c2**.5-1)/(.5*k2**(.5-1))
MTB > let c21=(c2**.25-1)/(.25*k2**(.25-1))
MTB > let c22=(c2**.333333-1)/(.333333*k2**(.333333-1))
MTB > let c23=loge(c2)*k2
MTB > let c24=(c2**(-.5) - 1)/(-.5*k2**(-1.5))
MTB > let c25=(c2**(-.25) -1)/(-.25*k2**(-1.25))
MTB > note      REGRESS z FOR LAMBDA = .5 ON c1
MTB > regress c20 on 1 c1
MTB > note      4 MORE REGRESSIONS ARE NECESSARY
MTB > note
MTB > note      CONSTRUCT THE VARIABLE w
MTB > let c3=c2*(c9-k1-1)
MTB > note      PERFORM THE MULTIPLE REGRESSION ON x AND THE
MTB > note      CONSTRUCTED VARIABLE w
MTB > regress c2 on 2 c1 c3
```

TRANSFORMING THE PREDICTOR VARIABLE

Weisberg (1985, p. 156) suggests applying a log transformation to the predictor variable x whenever x_{max}/x_{min} is larger than 10 or so. There is also a procedure, originally due to Box and Tidwell (1962), that is akin to the constructed variable test but that is used for checking the need to transform x. As presented by Weisberg, this procedure consists of fitting the original model

$$y_i = \beta_0 + \beta_1 x_i + \varepsilon_i$$

to obtain $\hat{\beta}_1$ and then fitting the model

$$y_i = \eta_0 + \eta_1 x_i + \eta_2 x_i \log(x_i) + \varepsilon_i.$$

Here, $x_i \log(x_i)$ is just an additional predictor variable that we compute from the values of x_i. The test of $H_0 : \eta_2 = 0$ is a test for whether a transformation of x is needed. If $\eta_2 \neq 0$, transforming x into x^γ is suggested where a rough estimate of γ is

$$\hat{\gamma} = \frac{\hat{\eta}_2}{\hat{\beta}_1} + 1$$

and $\gamma = 0$ is viewed as the log transformation. Typically, only γ values between about -2 and 2 are considered useable. Of course none of this is going to make any sense if x takes on negative values, and if x_{max}/x_{min} is not large, computational problems may occur when trying to fit a model that contains both x_i and $x_i \log(x_i)$.

7.11 Polynomial regression

With Hooker's data, the simple linear regression of pressure on temperature showed a lack of fit. In the previous section, we used a power transformation in an attempt to eliminate the lack of fit. In this section we introduce an alternative method, a special case of multiple regression called *polynomial regression*. With a single predictor variable x, we can try to eliminate lack of fit by fitting larger models. In particular, we can fit the *quadratic* (parabolic) model

$$y_i = \beta_0 + \beta_1 x_i + \beta_2 x_i^2 + \varepsilon_i.$$

We could also try a *cubic* model

$$y_i = \beta_0 + \beta_1 x_i + \beta_2 x_i^2 + \beta_3 x_i^3 + \varepsilon_i,$$

the *quartic* model

$$y_i = \beta_0 + \beta_1 x_i + \beta_2 x_i^2 + \beta_3 x_i^3 + \beta_4 x_i^4 + \varepsilon_i,$$

or even higher degree polynomials. If we view our purpose as finding good, easily interpretable approximate models for the data, *high degree polynomials can behave poorly*. As we will see later, the process of fitting the observed data can cause high degree polynomials to give very erratic results in areas very near the observed data. A good approximate model should work well, not only at the observed data, but also near it. Thus, we should focus on low degree polynomials.

EXAMPLE 7.11.1. We again examine Hooker's data. Computer programs give output for polynomial regression that is very similar to that for simple linear regression. Typical summary tables for fitting the quadratic model are given below.

Predictor	$\hat{\beta}_k$	$SE(\hat{\beta}_k)$	t	P
Constant	88.02	13.93	6.32	0.000
Temp.	−1.1295	0.1434	−7.88	0.000
Temp squared	0.0040330	0.0003682	10.95	0.000

Analysis of variance

Source	df	SS	MS	F	P
Regression	2	447.15	223.58	8984.23	0.000
Error	28	0.70	0.02		
Total	30	447.85			

The MSE, regression parameter estimates, and standard errors are used in the usual way. The t statistics and P values are for the two-sided tests of whether the corresponding β parameters are 0. The t statistic for β_2 is -7.88, which is highly significant, so the quadratic model accounts for a significant amount of the lack of fit displayed by the simple linear regression model. (It is not clear yet that the quadratic accounts for all of the lack of fit.)

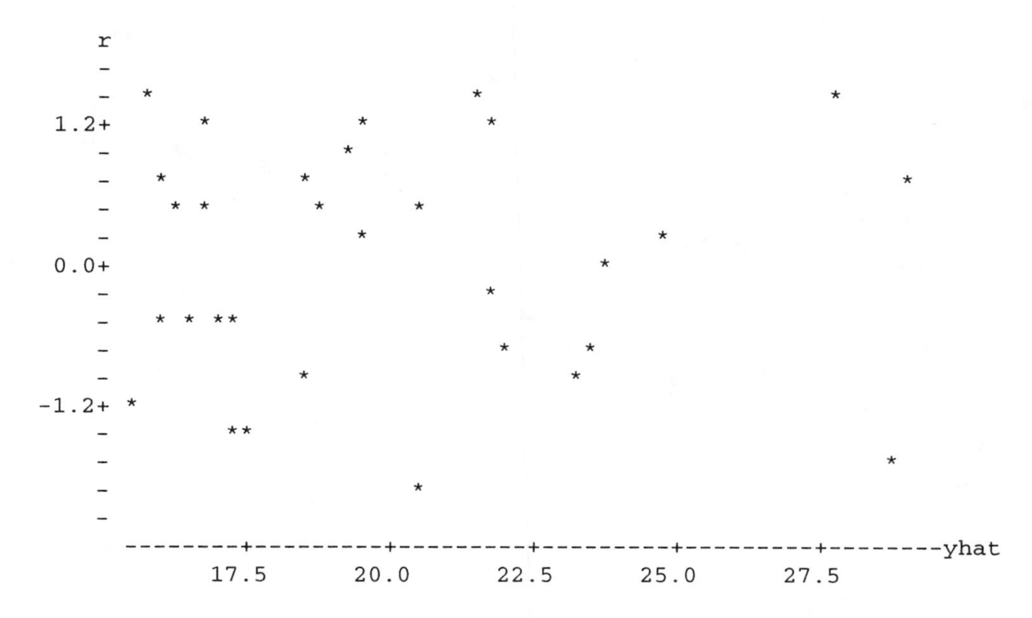

FIGURE 7.27. Standardized residuals versus predicted values, quadratic model.

Usually, *the only interesting test for a regression coefficient is the one for the highest term in the polynomial.* In particular, it usually makes little sense to have a quadratic (second degree) model that does not include a first degree term, so there is little point in testing $\beta_1 = 0$. One reason for this is that simple linear transformations of the predictor variable change the roles of lower order terms. For example, the Hooker data uses temperature measured in Fahrenheit as a predictor variable. While it is not actually the case, suppose the quadratic model for the Hooker data was consistent with $\beta_1 = 0$. If we then changed to measuring temperature in Celsius, we would be unlikely to have a new quadratic model that is still consistent with $\beta_1 = 0$. When there is a quadratic term in the model, a linear term based on Fahrenheit measurements has a completely different meaning than a linear term based on Celsius measurements. On the other hand, the Fahrenheit and Celsius quadratic models that include linear terms and intercepts are equivalent, just as the simple linear regressions based on Fahrenheit and Celsius are equivalent.

We will not discuss the ANOVA table in detail, but note that with two predictors, x and x^2, there are 2 degrees of freedom for regression. In general, if we fit a polynomial of degree a, there will be a degrees of freedom for regression, one degree of freedom for every term other than the intercept. Correspondingly, when fitting a polynomial of degree a, there are $n - a - 1$ degrees of freedom for error. *The ANOVA table F statistic provides a test of whether the quadratic model explains the data better than the model with only an intercept.*

The coefficient of determination is computed and interpreted as before. It is the *SSReg* divided by the *SSTot*, so it measures the amount of the total variability that is explained by the predictor variables temperature and temperature squared. For these data, $R^2 = 99.8\%$, which is an increase from 99.2% for the simple linear regression model. It is not appropriate to compare the R^2 for this model to the R^2 from the log transformed model of the previous section because they are computed from data that use different scales.

The standardized residual plots are given in Figures 7.27 and 7.28. The plot against the predicted values looks good, just as it did for the transformed data examined in the previous section. The normal plot for this model has a shoulder at the top but it looks much better than the normal plot for the simple linear regression on the log transformed data.

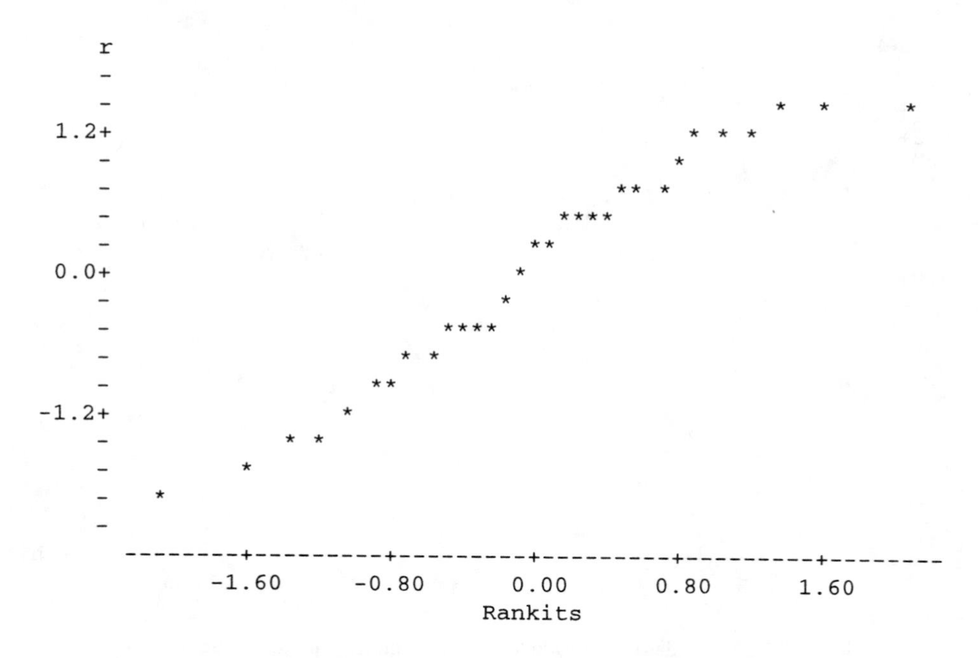

FIGURE 7.28. Normal plot for quadratic model, $W' = 0.966$.

If we are interested in the mean value of pressure for a temperature of 205°F, the quadratic model estimate is (up to a little of round off error)

$$\hat{y} = 25.95 = 88.02 - 1.1295\,(205) + .004033\,(205)^2.$$

The standard error (as reported by the computer program) is 0.0528 and a 95% confidence interval is (25.84, 26.06). This compares to a point estimate of 25.95 and a 95% confidence interval of (25.80, 26.10) obtained in the previous section from regressing the log of pressure on temperature. The quadratic model *prediction* for a new observation at 205°F is again 25.95 with a 95% prediction interval of (25.61, 26.29). The corresponding prediction interval from the log transformed data is (25.49, 26.42). In this example, the results of the two methods for dealing with lack of fit are qualitatively very similar, at least at 205°F.

Finally, we tried fitting a cubic model to these data. The cubic model suffers from substantial numerical instability. (Some computer programs object to fitting it.) This may be related to the fact that the R^2 is so high. The β_3 coefficient does not seem to be significantly different from 0, so considering the good residual plots, the quadratic model seems adequate. (One easy way to improve numerical stability is to adjust the predictor variables for their mean as in Section 7.6. In other words, one builds a polynomial using powers of the predictor variable $x_i - \bar{x}_{\cdot\cdot}$.) □

EXAMPLE 7.11.2. We now present a simple example that illustrates two points: that leverages depend on the model and that high order polynomials can fit the data in very strange ways. The data for the example are given below.

Case	1	2	3	4	5	6	7
y	0.445	1.206	0.100	−2.198	0.536	0.329	−0.689
x	0.0	0.5	1.0	10.0	19.0	19.5	20.0

I selected the x values. The y values are a sample of size 7 from a $N(0, 1)$ distribution. Note that with seven distinct x values, we can fit a polynomial of degree 6.

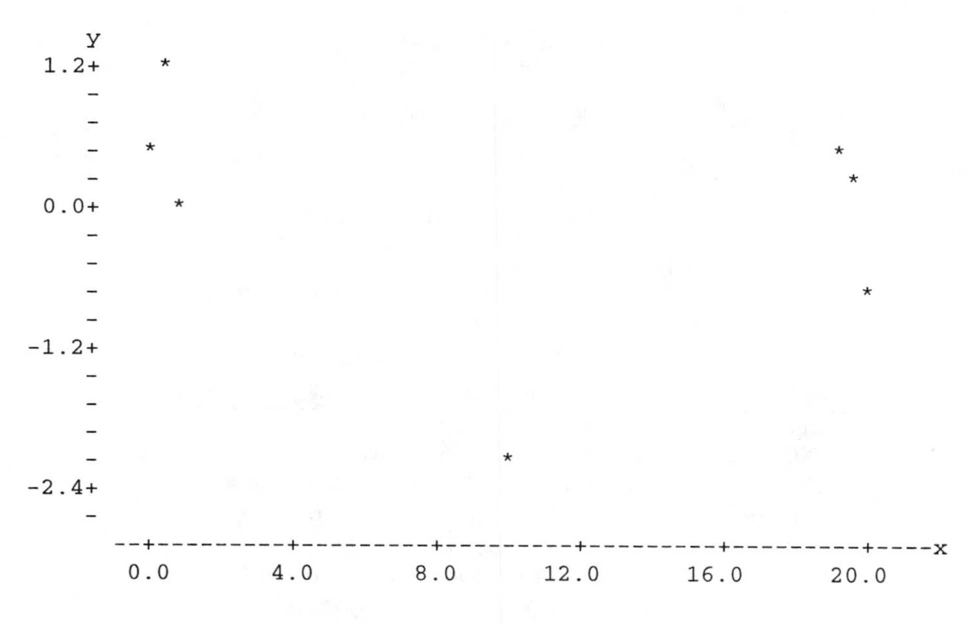

FIGURE 7.29. Plot of y versus x.

The data are plotted in Figure 7.29. Just by chance (honest folks), I observed a very small y value at $x = 10$, so the data appear to follow a parabola that opens up. The small y value at $x = 10$ totally dominates the impression given by Figure 7.29 If the y value at $x = 10$ had been near 3 rather than near -2, the data would appear to be a parabola that opens down. If the y value had been between 0 and 1, the data would appear to fit a line with a slightly negative slope. When thinking about fitting a parabola, the case with $x = 10$ is an extremely high leverage point.

Depending on the y value at $x = 10$, the data suggest a parabola opening up, a parabola opening down, or that we do not need a parabola to explain the data. Regardless of the y value observed at $x = 10$, the fitted parabola must go nearly through the point $(10, y)$. On the other hand, if we think only about fitting a line to these data, the small y value at $x = 10$ has much less effect. In fitting a line, the value $y = -2.198$ will look unusually small (it will have a very noticeable standardized residual), but it will not force the fitted line to go nearly through the point $(10, -2.198)$.

Table 7.8 gives the leverages for all of the polynomial models that can be fitted to these data. Note that there are no large leverages for the simple linear regression model (the linear polynomial). For the quadratic (parabolic) model, all of the leverages are reasonably small except the leverage of .96 at $x = 10$ which very nearly equals 1. Thus, in the quadratic model, the value of y at $x = 10$ dominates the fitted polynomial. The cubic model has extremely high leverage at $x = 10$, but the leverages are also beginning to get large at $x = 0, 1, 19, 20$. For the quartic model, the leverage at $x = 10$ is 1 to two decimal places; the leverages for $x = 0, 1, 19, 20$ are also nearly 1. The same pattern continues with the quintic model but the leverages at $x = 0.5, 19.5$ are also becoming large. Finally, with the sixth degree (hexic) polynomial, all of the leverages are exactly one. This indicates that the sixth degree polynomial has to go through every data point exactly and thus every data point is extremely influential on the estimate of the sixth degree polynomial.

As we fit larger polynomials, we get more high leverage cases (and more numerical instability). *The estimated polynomials must go very nearly through all high leverage cases. To accomplish this the estimated polynomials may get very strange*. Below we give all of the fitted polynomials for these data.

TABLE 7.8. Leverages

x	Model					
	Linear	Quadratic	Cubic	Quartic	Quintic	Hexic
0.0	0.33	0.40	0.64	0.87	0.94	1.00
0.5	0.31	0.33	0.33	0.34	0.67	1.00
1.0	0.29	0.29	0.55	0.80	0.89	1.00
10.0	0.14	0.96	0.96	1.00	1.00	1.00
19.0	0.29	0.29	0.55	0.80	0.89	1.00
19.5	0.31	0.33	0.33	0.34	0.67	1.00
20.0	0.33	0.40	0.64	0.87	0.94	1.00

Model	Estimated polynomial
Linear	$\hat{y} = 0.252 - 0.029x$
Quadratic	$\hat{y} = 0.822 - 0.536x + 0.0253x^2$
Cubic	$\hat{y} = 1.188 - 1.395x + 0.1487x^2 - 0.0041x^3$
Quartic	$\hat{y} = 0.713 - 0.141x - 0.1540x^2 + 0.0199x^3 - 0.00060x^4$
Quintic	$\hat{y} = 0.623 + 1.144x - 1.7196x^2 + 0.3011x^3 - 0.01778x^4 + 0.000344x^5$
Hexic	$\hat{y} = 0.445 + 3.936x - 5.4316x^2 + 1.2626x^3 - 0.11735x^4 + 0.004876x^5 - 0.00007554x^6$

Figures 7.30 and 7.31 contain graphs of these estimated polynomials.

Figure 7.30 contains the estimated linear, quadratic, and cubic polynomials. The linear and quadratic curves fit about as one would expect from looking at the scatter plot Figure 7.29. For x values near the range 0 to 20, we could use these curves to predict y values and get reasonable, if not necessarily good, results. One could not say the same for the estimated cubic polynomial. The cubic curve takes on $\hat{y}$ values near -3 for some x values that are near 6. The y values in the data are between about -2 and 1.2; nothing in the data suggests that y values near -3 are likely to occur. Such predicted values are entirely the product of fitting a cubic polynomial. If we really knew that a cubic polynomial was correct for these data, the estimated polynomial would be perfectly appropriate. But most often we use polynomials to approximate the behavior of the data and for these data the cubic polynomial gives a poor approximation.

Figure 7.31 gives the estimated quartic, quintic, and hexic polynomials. Note that the scale on the y axis has changed drastically from Figure 7.30. Qualitatively, the fitted polynomials behave like the cubic except their behavior is even worse. These polynomials do very strange things everywhere except near the observed data.

Another phenomenon that sometimes occurs when fitting large models to data is that the mean squared error gets unnaturally small. Table 7.9 gives the analysis of variance tables for all of the polynomial models. Our original data were a sample from a $N(0, 1)$ distribution. The data were constructed with no regression structure so the best estimate of the variance comes from the total line and is $7.353/6 = 1.2255$. This value is a reasonable estimate of the true value 1. The *MSE* from the simple linear regression model also provides a reasonable estimate of $\sigma^2 = 1$. The larger models do not work as well. Most have variance estimates near .5, while the hexic model does not even allow an estimate of σ^2 because it fits every data point perfectly. By fitting models that are too large one can often make the *MSE* artificially small. For example, the quartic model has a *MSE* of .306 and an F statistic of 5.51; if it were not for the small value of dfE, such an F value would be highly significant. *If you find a large model that has an unnaturally small MSE with a reasonable*

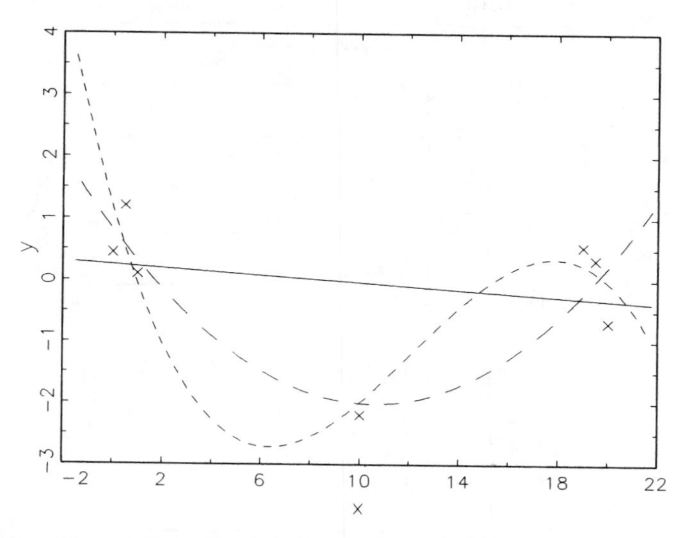

FIGURE 7.30. Plots of linear (solid), quadratic (long dashes), and cubic (short dashes) regression curves.

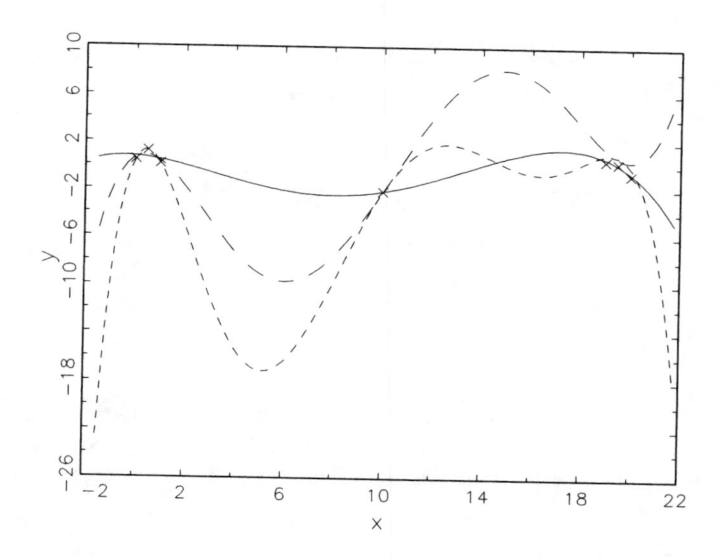

FIGURE 7.31. Plots of quartic (solid), quintic (long dashes), and hexic (short dashes) regression curves.

number of degrees of freedom, everything can appear to be significant even though nothing you look at is really significant.

Just as the mean squared error often gets unnaturally small when fitting large models, R^2 gets unnaturally large. As we have seen, there can be no possible reason to use a larger model than the quadratic with its R^2 of .71, but the cubic, quartic, quintic, and hexic models have R^2s of .78, .92, .93, and 1, respectively. □

TABLE 7.9. Analysis of variance tables

Source	df	SS	MS	F	P
Simple linear regression					
Regression	1	0.457	0.457	0.33	0.59
Error	5	6.896	1.379		
Total	6	7.353			
Quadratic model					
Regression	2	5.185	2.593	4.78	0.09
Error	4	2.168	0.542		
Total	6	7.353			
Cubic model					
Regression	3	5.735	1.912	3.55	0.16
Error	3	1.618	0.539		
Total	6	7.353			
Quartic model					
Regression	4	6.741	1.685	5.51	0.16
Error	2	0.612	0.306		
Total	6	7.353			
Quintic model					
Regression	5	6.856	1.371	2.76	0.43
Error	1	0.497	0.497		
Total	6	7.353			
Hexic model					
Regression	6	7.353	1.2255	—	—
Error	0	0.000	—		
Total	6	7.353			

MINITAB COMMANDS

Below we illustrate Minitab commands for fitting quadratic, cubic, and quartic models. These include the prediction subcommand used with the quadratic model for $x = 205$. Note that the prediction subcommand requires us to enter both the value of x and the value of x^2 when using the quadratic model.

```
MTB > names c1 'y' c2 'x'
MTB > note      FIT QUADRATIC MODEL
MTB > let c22=c2**2
MTB > regress c2 on 2 c2 c22;
SUBC> pred 205 42025.
MTB > note      FIT CUBIC MODEL
MTB > let c23=c2**3
MTB > regress c1 on 3 c2 c22 c23
MTB > note      FIT QUARTIC MODEL
MTB > let c24=c2**4
MTB > regress c1 on 4 c2 c22-c24
```

7.12 Polynomial regression and one-way ANOVA

The main reason for introducing polynomial regression at this point is to exploit its relationships with analysis of variance. In some analysis of variance problems, the treatment groups are determined by quantitative levels of a factor. For example, one might take observations on the depth of hole made by a drill press in a given amount of time with 20, 30, or 40 pounds of downward thrust applied. The groups are determined by the quantitative levels, 20, 30, and 40. In such a situation we could fit a one-way analysis of variance with three groups, or we could fit a simple linear regression model. Simple linear regression is appropriate because all the data come as pairs. The pairs are (x_i, y_{ij}), where x_i is the numerical level of thrust and y_{ij} is the depth of the hole on the jth trial with x_i pounds of downward thrust. Not only can we fit a simple linear regression, but we can fit polynomials to the data. In this example, we could fit no polynomial above second degree (quadratic), because three points determine a parabola and we only have three distinct x values. If we ran the experiment with 20, 25, 30, 35, and 40 pounds of thrust, we could fit at most a fourth degree (quartic) polynomial because five points determine a fourth degree polynomial and we would only have five x values.

In general, some number a of distinct x values allows fitting of an $a - 1$ degree polynomial. Moreover, *fitting the $a - 1$ degree polynomial is equivalent to fitting the one-way ANOVA with groups defined by the a different x values.* However, as discussed in the previous section, fitting high degree polynomials is often a very questionable procedure. The problem is not with how the model fits the observed data but with the suggestions that a high degree polynomial makes about the behavior of the process for x values other than those observed. In the example with 20, 25, 30, 35, and 40 pounds of thrust, the quartic polynomial will fit as well as the one-way ANOVA model but the quartic polynomial may have to do some very weird things in the areas between the observed x values. Of course, the ANOVA model gives no indications of behavior for x values other than those that were observed. When performing regression, we usually like to have some smooth fitting model giving predictions that, in some sense, interpolate between the observed data points. High degree polynomials often fail to achieve this goal.

EXAMPLE 7.12.1. Beineke and Suddarth (1979) and Devore (1991, p. 380) consider data on roof supports involving trusses that use light gauge metal connector plates. Their dependent variable is an axial stiffness index (ASI) measured in kips per inch. The predictor variable is the length of the light gauge metal connector plates. The data are given in Table 7.10 in a format consistent with performing a regression analysis on the data.

TABLE 7.10. Axial stiffness index data

Plate	ASI	Plate	ASI	Plate	ASI	Plate	ASI	Plate	ASI
4	309.2	6	402.1	8	392.4	10	346.7	12	407.4
4	409.5	6	347.2	8	366.2	10	452.9	12	441.8
4	311.0	6	361.0	8	351.0	10	461.4	12	419.9
4	326.5	6	404.5	8	357.1	10	433.1	12	410.7
4	316.8	6	331.0	8	409.9	10	410.6	12	473.4
4	349.8	6	348.9	8	367.3	10	384.2	12	441.2
4	309.7	6	381.7	8	382.0	10	362.6	12	465.8

The data could also be considered as an analysis of variance with plate lengths being different treatments and with seven observations on each treatment. Table 7.11 gives the usual summary statistics for a one-way ANOVA.

Viewed as regression data, we might think of fitting a simple linear regression model

$$y_h = \beta_0 + \beta_1 x_h + \varepsilon_h,$$

TABLE 7.11. ASI summary statistics

Plate	N	$\bar{y}_{i\cdot}$	s_i^2	s_i
4	7	333.2143	1338.6981	36.59
6	7	368.0571	816.3629	28.57
8	7	375.1286	433.7990	20.83
10	7	407.3571	1981.1229	44.51
12	7	437.1714	675.8557	26.00

$h = 1, \ldots, 35$. Note that while h varies from 1 to 35, there are only five distinct values of x_h that occur in the data. As an analysis of variance, we usually use two subscripts to identify an observation: one to identify the treatment group and one to identify the observation within the group. The ANOVA model would often be written as

$$y_{ij} = \mu_i + \varepsilon_{ij} \tag{7.12.1}$$

where $i = 1, 2, 3, 4, 5$ and $j = 1, \ldots, 7$. We can also rewrite the regression model using the two subscripts i and j in place of h,

$$y_{ij} = \beta_0 + \beta_1 x_i + \varepsilon_{ij},$$

where $i = 1, 2, 3, 4, 5$ and $j = 1, \ldots, 7$. Note that all of these models account for exactly 35 observations.

Figure 7.32 contains a scatter plot of the data. With multiple observations at each x value, the regression is really only fitted to the mean of the y values at each x value. The means of the ys are plotted against the x values in Figure 7.33. The overall trend of the data is easier to evaluate in this plot than in the full scatter plot. We see an overall increasing trend which is very nearly linear except for a slight anomaly with 6 inch plates. We need to establish if these visual effects are real or just random variation. We would also like to establish whether there is a simple regression model that is appropriate for any trend that may exist. With only five distinct x values, we can fit at most a quartic (fourth degree) polynomial, say,

$$y_{ij} = \beta_0 + \beta_1 x_i + \beta_2 x_i^2 + \beta_3 x_i^3 + \beta_4 x_i^4 + \varepsilon_{ij}, \tag{7.12.2}$$

so a simple model should be something smaller than a quartic, i.e., either a cubic, quadratic, or a linear polynomial.

Table 7.12 contains ANOVA tables for fitting the linear, quadratic, cubic, and quartic polynomial regressions and for fitting the one-way ANOVA model. From our earlier discussion, the F test in the simple linear regression ANOVA table strongly suggests that there is an overall trend in the data. From Figure 7.33 we see that this trend must be increasing, i.e., as lengths go up, by and large the ASI readings go up. ANOVA tables for higher degree polynomial models have been discussed briefly in the previous section but for now the key point to recognize is that *the ANOVA table for the quartic polynomial is identical to the ANOVA table for the one-way analysis of variance*. This occurs because models (7.12.1) and (7.12.2) are equivalent.

The first question of interest is whether a quartic polynomial is needed or whether a cubic model would be adequate. This is easily evaluated from the table of estimates and standard errors for the quartic fit. For computational reasons, the results reported are for a polynomial involving powers of $x - \bar{x}_\cdot$ rather than powers of x, cf. Section 7.6. This has *no* effect on our subsequent discussion, see Exercise 7.13.15.

Predictor	$\hat{\beta}_k$	$SE(\hat{\beta}_k)$	t	P
Constant	375.13	12.24	30.64	0.000
$(x - \bar{x}_\cdot)$	8.768	5.816	1.51	0.142
$(x - \bar{x}_\cdot)^2$	3.983	4.795	0.83	0.413
$(x - \bar{x}_\cdot)^3$	0.2641	0.4033	0.65	0.517
$(x - \bar{x}_\cdot)^4$	−0.2096	0.2667	−0.79	0.438

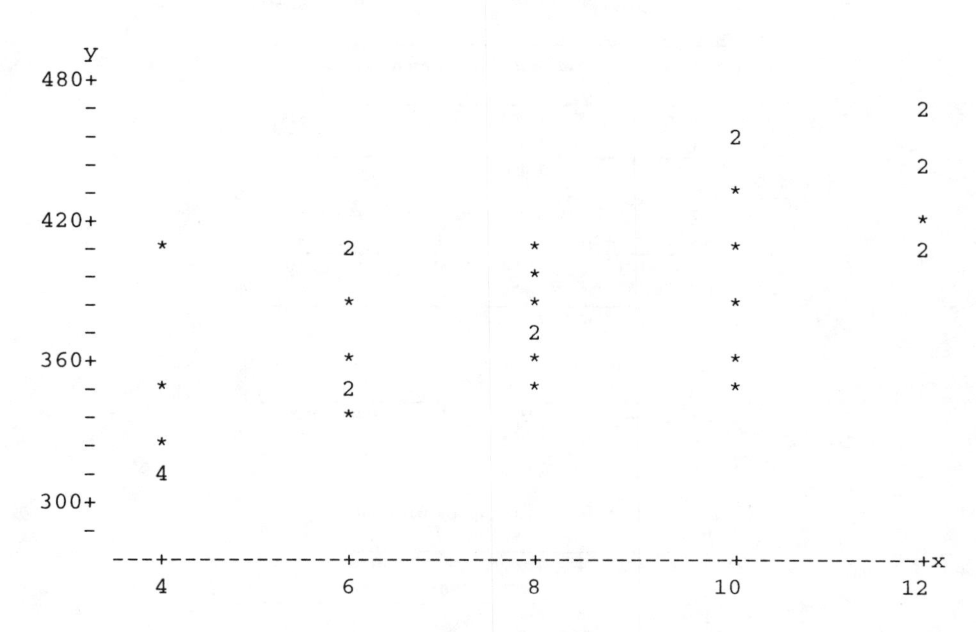

FIGURE 7.32. ASI data versus plate length.

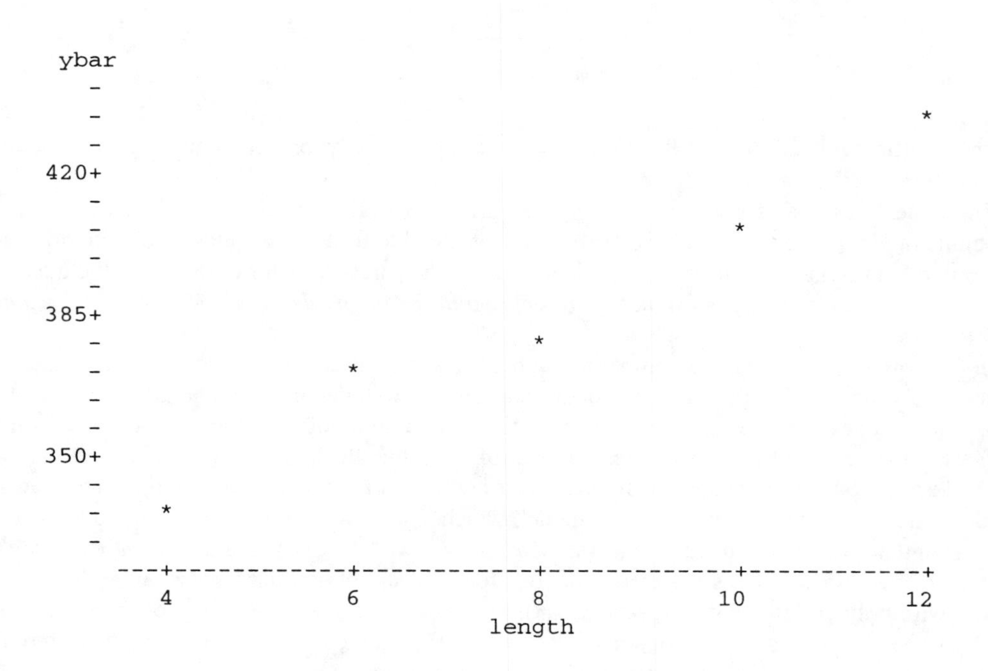

FIGURE 7.33. ASI means versus plate length.

TABLE 7.12. Analysis of variance tables for ASI data

Analysis of variance: simple linear regression

Source	df	SS	MS	F	P
Regression	1	42780	42780	43.19	0.000
Error	33	32687	991		
Total	34	75468			

Analysis of variance: quadratic polynomial

Source	df	SS	MS	F	P
Regression	2	42894	21447	21.07	0.000
Error	32	32573	1018		
Total	34	75468			

Analysis of variance: cubic polynomial

Source	df	SS	MS	F	P
Regression	3	43345	14448	13.94	0.000
Error	31	32123	1036		
Total	34	75468			

Analysis of variance: quartic polynomial

Source	df	SS	MS	F	P
Regression	4	43993	10998	10.48	0.000
Error	30	31475	1049		
Total	34	75468			

Analysis of variance: one-way ANOVA

Source	df	SS	MS	F	P
Trts(plates)	4	43993	10998	10.48	0.000
Error	30	31475	1049		
Total	34	75468			

There is little evidence ($P = .438$) that $\beta_4 \neq 0$, so a cubic polynomial seems to be an adequate explanation of the data.

The table of estimates given above is inappropriate for evaluating β_3 in the cubic model (even the cubic model based on $x - \bar{x}_{\bullet}$). To evaluate β_3, we need to fit the cubic model. If we then decide that a parabola is an adequate model, evaluating β_2 in the parabola requires one to fit the quadratic model. *In general, regression estimates are* only *valid for the model fitted. A new model requires new estimates.*

In Section 5.5, we discussed comparing models as a way of arriving at the F test in a one-way analysis of variance. Comparing a submodel against a larger model to determine the adequacy of the submodel is a key method in regression analysis. Recall that model comparisons are based on the difference between the sums of squares for error of the submodel and the sums of squares for error of the larger model. Obviously, the simple linear regression model is a submodel of the quadratic model which is a submodel of the cubic model, which is a submodel of the quartic model, and we have seen that the quartic model is equivalent to the one-way ANOVA model. Given below are the degrees of freedom and sums of squares for error for the four polynomial regression models and the model with only an intercept β_0 (grand mean). (See Section 5.5 for discussion of the grand mean model.) The differences in sums of squares error for adjacent models are also given; the differences in degrees of freedom error are just 1.

Model comparisons

Model	dfE	SSE	Difference
Intercept	34	75468	—
Linear	33	32687	42780
Quadratic	32	32573	114
Cubic	31	32123	450
Quartic	30	31475	648

Note that the dfE and SSE for the intercept model are those from the corrected Total lines in the ANOVAs of Table 7.12. The dfEs and SSEs for the other models also come from Table 7.12.

To test the quartic model against the cubic model we take

$$F = \frac{648/1}{31475/30} = .62.$$

This is just the square of the t statistic for testing $\beta_4 = 0$ in the quartic model. The reference distribution for the F statistic is $F(1,30)$ and the P value is .44, as it was for the t test.

If we decide that we do not need the quartic term, we can test whether we need the cubic term. We can test the quadratic model against the cubic model with

$$F = \frac{450/1}{32123/31} = 0.434.$$

The reference distribution is $F(1,31)$. This test is equivalent to the t test of $\beta_3 = 0$ *in the cubic model. The t test of* $\beta_3 = 0$ *in the quartic model is inappropriate.* An alternative to this F test can also be used. The denominator of this test is 32123/31, the mean squared error from the cubic model. If we accepted the cubic model only after testing the quartic model, the result of the quartic test is open to question and thus the estimate of σ^2 from the cubic model, i.e., the MSE from the cubic model, is open to question. It might be better just to use the estimate of σ^2 from the quartic model, which is the mean squared error from the one-way ANOVA. If we do this, the test statistic for the cubic term becomes

$$F = \frac{450/1}{31475/30} = 0.429.$$

The reference distribution for the alternative test is $F(1,30)$. In this example the two F tests give essentially the same answers. This should, by definition, almost always be the case. If, for example, one test were significant at .05 and the other were not, they are both likely to have P values near .05 and the fact that one is a bit larger than .05 and the other is a bit smaller than .05 should not be a cause for concern.

If we decide that neither the quartic nor the cubic terms are important, we can test whether we need the quadratic term. Testing the quadratic model against the simple linear model gives

$$F = \frac{114/1}{32573/32} = 0.112$$

which is compared to an $F(1,32)$ distribution. This test is equivalent to the t test of $\beta_2 = 0$ in the quadratic model. Again, an alternative test can also be used. The denominator of this test is 32573/32, the mean squared error from the quadratic model. If we accepted the quadratic model only after testing the cubic and quartic models, this estimate of σ^2 may be biased and it might be better to use the estimate of σ^2 from the quartic model, i.e., the one-way ANOVA model. If we do this, the test statistic for the quadratic term becomes

$$F = \frac{114/1}{31475/30} = 0.109$$

and the reference distribution is $F(1,30)$.

If we decide that we need none of the higher order terms, we can test whether we need the linear term. Testing the intercept model against the simple linear model gives

$$F = \frac{42780/1}{32687/33} = 43.190.$$

This is just the test for zero slope *in the simple linear model.* Again, the alternative test can be used. The denominator of this test is the mean squared error from the linear model, 32687/33. If we accepted the linear model only after testing the higher order models, it may be better to use the mean squared error from the one-way ANOVA model. The alternative F test for the linear term has

$$F = \frac{42780/1}{31475/30} = 40.775.$$

The model comparison tests just discussed can be reconstructed from contrasts in the one-way ANOVA. Below are given some simple contrasts that correspond to the differences in sums of squares error for the model comparisons.

	Orthogonal polynomial contrasts				
Plate	Linear	Quadratic	Cubic	Quartic	$\bar{y}_{i\cdot}$
4	−2	2	−1	1	333.2143
6	−1	−1	2	−4	368.0571
8	0	−2	0	6	375.1286
10	1	−1	−2	−4	407.3571
12	2	2	1	1	437.1714
Est	247.2142	15.1000	25.3571	−80.4995	
SS	42780.4	114.0	450.1	648.0	

Recall that the estimate of, say, the linear contrast is

$$(-2)(333.2143) + (-1)(368.0571) + (0)(375.1286) + (1)(407.3571) + (2)(437.1714) = 247.2142$$

and that with seven observations on each plate length, the sum of squares for the linear contrast is

$$SS(\text{linear}) = \frac{(247.2142)^2}{[(-2)^2 + (-1)^2 + 0^2 + 1^2 + 2^2]/7} = 42780.4.$$

This is precisely the difference in error sums of squares between the intercept and straight line models. Similar results hold for the other contrasts.

These contrasts are called *orthogonal polynomial contrasts* because they are orthogonal in balanced ANOVAs and reproduce the sums of squares for comparing different polynomial regression models. We leave it to the reader to verify that the contrasts are orthogonal, cf. Section 5.4, but recall that with orthogonal contrasts we have the identity

$$SSTrts = 43992.5 = 42780.4 + 114.0 + 450.1 + 648.0.$$

Table 7.13 contains an expanded analysis of variance table for the one-way ANOVA that incorporates the information from the contrasts. Using the orthogonal polynomial contrasts allows us to make all of the model comparisons by using simple analysis of variance computations rather than fitting polynomial regression models.

TABLE 7.13. Analysis of variance for ASI data

Source	*df*	*SS*	*MS*	*F*	*P*
Treatments	4	43993	10998	10.48	0.000
(linear)	(1)	(42780)	(42780)	(40.78)	
(quadratic)	(1)	(114)	(114)	(0.11)	
(cubic)	(1)	(450)	(450)	(0.43)	
(quartic)	(1)	(648)	(648)	(0.62)	
Error	30	31475	1049		
Total	34	75468			

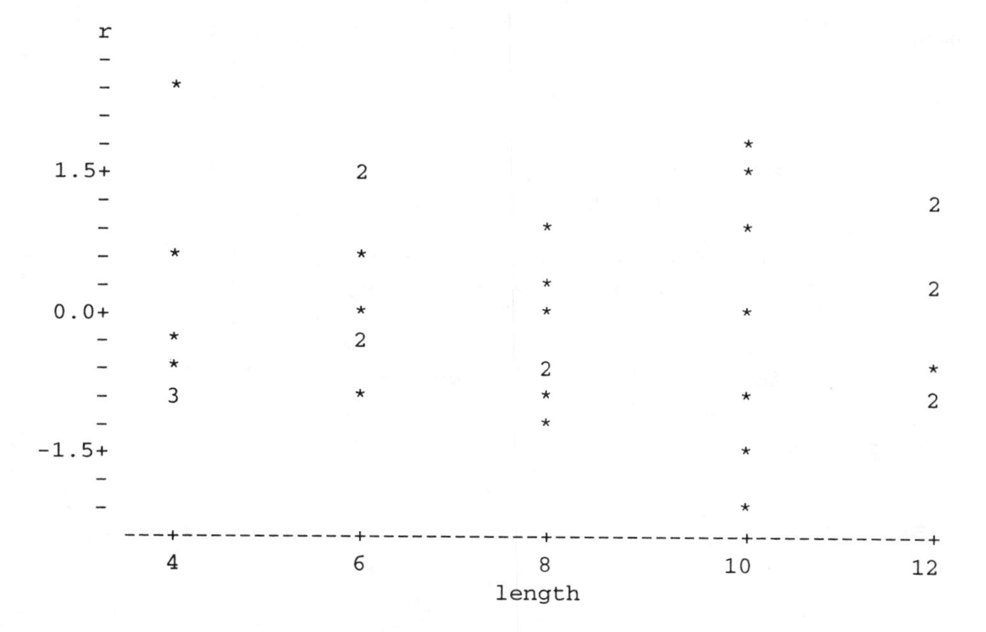

FIGURE 7.34. ASI SLR standardized residuals versus plate length.

From Table 7.13, the P value of .000 indicates strong evidence that the five groups are different, i.e., there is strong evidence for the quartic polynomial. The results from the contrasts are so clear that we did not bother to report P values for them. There is a huge effect for the linear contrast. The other three F statistics are all much less than 1, so there is no evidence of the need for a quartic, cubic, or quadratic polynomial. As far as we can tell, a line fits the data just fine. For completeness, some residual plots are presented as Figures 7.34 through 7.38. Note that the normal plot for the simple linear regression in Figure 7.35 is less than admirable, while the normal plot for the one-way ANOVA in Figure 7.38 is only slightly better. It appears that one should not put great faith in the normality assumption. □

The linear, quadratic, cubic, and quartic contrasts for the ASI data are simple only because the ANOVA is balanced and the treatment groups are equally spaced. The treatments occur at 4, 6, 8, 10, and 12 inches. Thus the treatments occur at intervals of 2 inches. If the treatments were at irregular intervals or if the group sample sizes were unequal, orthogonal linear, quadratic, cubic, and quartic contrasts still exist, but they are difficult to find. With either unequal spacings or unequal numbers, it is easier just to do the appropriate regressions. *With a balanced ANOVA and regularly spaced intervals, the orthogonal polynomial contrasts can be determined from the number of treatment*

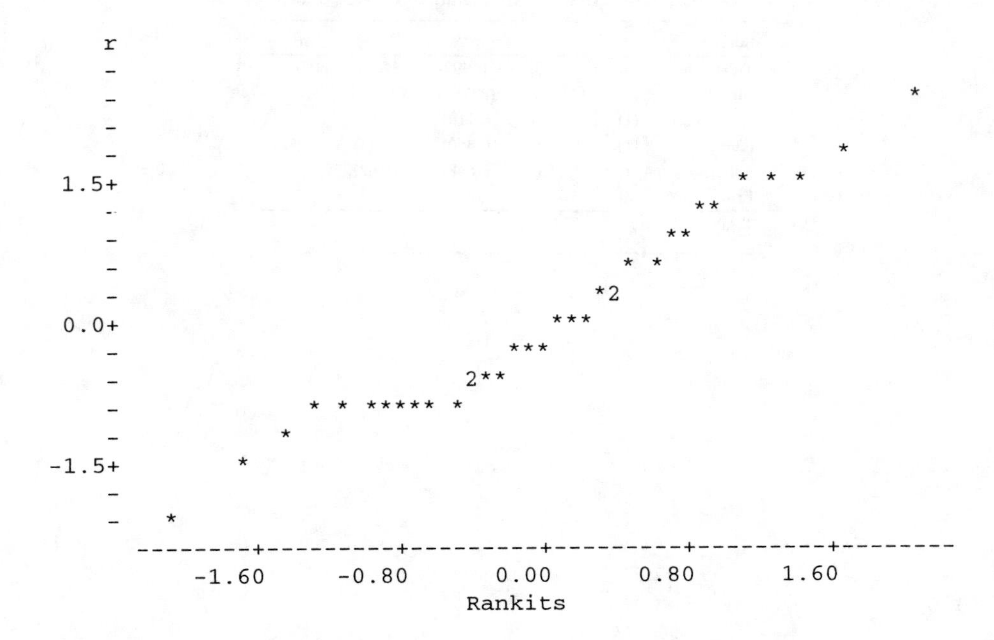

FIGURE 7.35. ASI SLR standardized residuals normal plot, $W' = .960$.

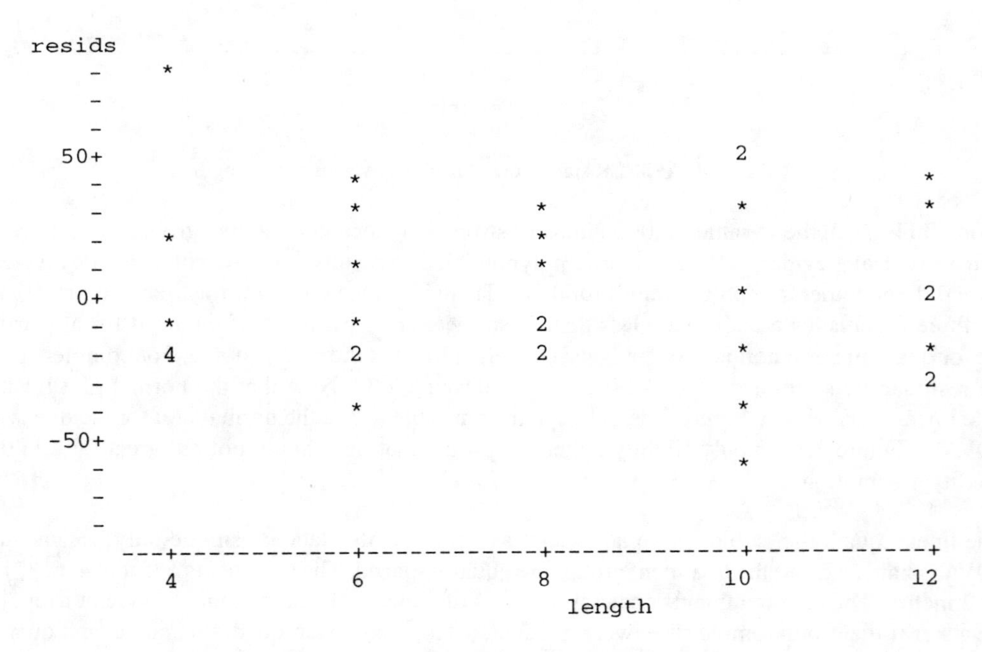

FIGURE 7.36. ASI ANOVA residuals versus plate length.

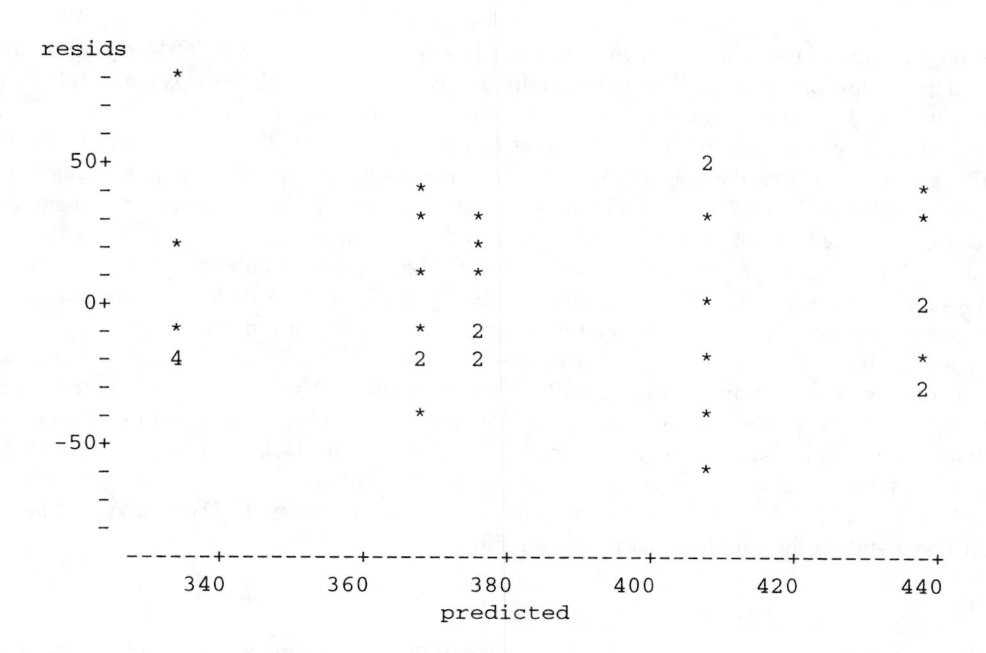

FIGURE 7.37. ASI ANOVA residuals versus predicted values.

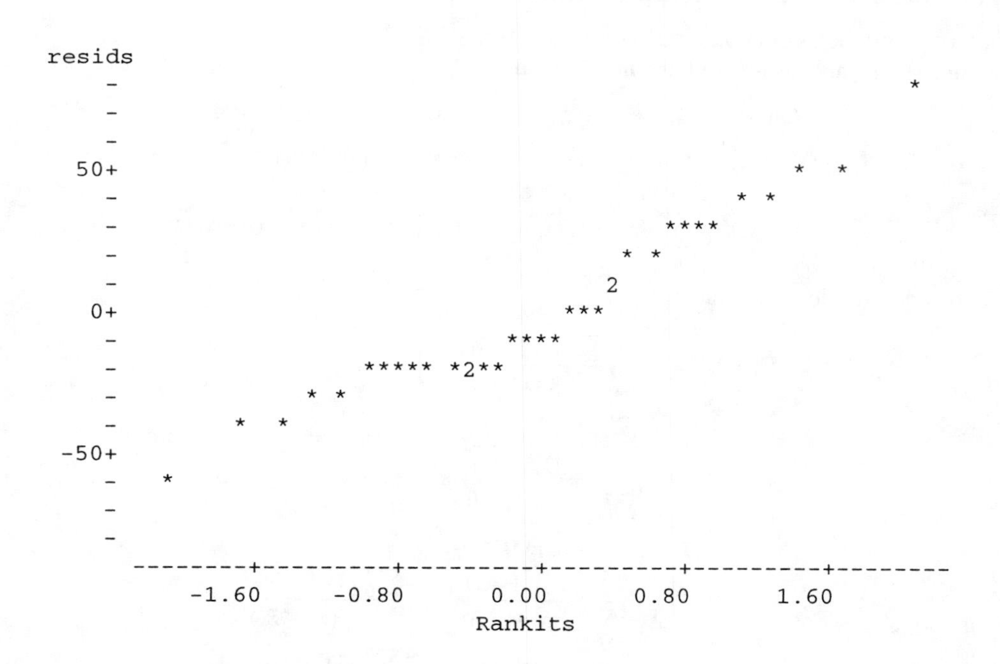

FIGURE 7.38. ASI ANOVA residuals normal plot, $W' = .966$.

groups and thus they can be tabled. Such a table is given in Appendix B.4 for linear, quadratic, and cubic contrasts.

Comparing one of the reduced polynomial models against the one-way ANOVA model is often referred to as a test of *lack of fit*. This is especially true when the reduced model is the simple linear regression model. In these tests, the degrees of freedom, sums of squares, and mean squares used in the numerator of the tests are all described as being for *lack of fit*. The denominator of the test is based on the error from the one-way ANOVA. The mean square, sum of squares, and degrees of freedom for error in the one-way ANOVA are often referred to as the mean square, sum of squares, and degrees of freedom for *pure error*. This lack of fit test can be performed *whenever* the data contain multiple observations at *any* x values. Often the appropriate unbalanced one-way ANOVA includes treatments with only one observation on them. These treatments do not provide an estimate of σ^2, so they simply play no role in obtaining the mean square for pure error.

For testing lack of fit in the simple linear regression model with the ASI data, the numerator sum of squares can be obtained either by differencing the sums of squares for error in the simple linear regression model and the one-way ANOVA model or by adding up the sums of squares for the quadratic, cubic, and quartic contrasts. Here the sum of squares for lack of fit is $32687 - 31475 = 1212 = 114 + 450 + 648$ and the degrees of freedom for lack of fit are $33 - 30 = 3$. The mean square for lack of fit is $1212/3 = 404$. The pure error comes from the one-way ANOVA table. The lack of fit test F statistic for the simple linear regression model is

$$F = \frac{404}{1049} = .39$$

which is less than 1, so there is no evidence of a lack of fit in the simple linear regression model. If an $\alpha = .05$ test were desired, the test statistic would be compared to $F(.95, 3, 30)$.

APPENDIX: SIMPLE LINEAR REGRESSION PROOFS

PROOF OF UNBIASEDNESS FOR THE REGRESSION ESTIMATES.
To begin, The βs and x_is are all fixed numbers so

$$E(y_i) = E(\beta_0 + \beta_1 x_i + \varepsilon_i) = \beta_0 + \beta_1 x_i + E(\varepsilon_i) = \beta_0 + \beta_1 x_i \,.$$

Also note that $\sum_{i=1}^n (x_i - \bar{x}_\cdot) = 0$, so $\sum_{i=1}^n (x_i - \bar{x}_\cdot)\bar{x}_\cdot = 0$. It follows that

$$\sum_{i=1}^n (x_i - \bar{x}_\cdot)^2 = \sum_{i=1}^n (x_i - \bar{x}_\cdot)x_i - \sum_{i=1}^n (x_i - \bar{x}_\cdot)\bar{x}_\cdot = \sum_{i=1}^n (x_i - \bar{x}_\cdot)x_i \,.$$

Now consider the slope estimate.

$$\begin{aligned}
E(\hat{\beta}_1) &= E\left(\frac{\sum_{i=1}^n (x_i - \bar{x}_\cdot)y_i}{\sum_{i=1}^n (x_i - \bar{x}_\cdot)^2}\right) \\
&= \frac{\sum_{i=1}^n (x_i - \bar{x}_\cdot)E(y_i)}{\sum_{i=1}^n (x_i - \bar{x}_\cdot)^2} \\
&= \frac{\sum_{i=1}^n (x_i - \bar{x}_\cdot)(\beta_0 + \beta_1 x_i)}{\sum_{i=1}^n (x_i - \bar{x}_\cdot)^2} \\
&= \beta_0 \frac{\sum_{i=1}^n (x_i - \bar{x}_\cdot)}{\sum_{i=1}^n (x_i - \bar{x}_\cdot)^2} + \beta_1 \frac{\sum_{i=1}^n (x_i - \bar{x}_\cdot)x_i}{\sum_{i=1}^n (x_i - \bar{x}_\cdot)^2} \\
&= \beta_0 \frac{0}{\sum_{i=1}^n (x_i - \bar{x}_\cdot)^2} + \beta_1 \frac{\sum_{i=1}^n (x_i - \bar{x}_\cdot)^2}{\sum_{i=1}^n (x_i - \bar{x}_\cdot)^2} \\
&= \beta_1
\end{aligned}$$

The proof for the intercept goes as follows:

$$
\begin{aligned}
E(\hat{\beta}_0) &= E(\bar{y}_\bullet - \hat{\beta}_1 \bar{x}_\bullet) \\
&= E\left(\frac{1}{n}\sum_{i=1}^{n} y_i\right) - E(\hat{\beta}_1)\,\bar{x}_\bullet \\
&= \frac{1}{n}\sum_{i=1}^{n} E(y_i) - \beta_1 \bar{x}_\bullet \\
&= \frac{1}{n}\sum_{i=1}^{n} (\beta_0 + \beta_1 x_i) - \beta_1 \bar{x}_\bullet \\
&= \beta_0 + \beta_1 \frac{1}{n}\sum_{i=1}^{n} (x_i) - \beta_1 \bar{x}_\bullet \\
&= \beta_0 + \beta_1 \bar{x}_\bullet - \beta_1 \bar{x}_\bullet \\
&= \beta_0 \,.
\end{aligned}
$$

PROOF OF VARIANCE FORMULAE.
To begin,

$$\text{Var}(y_i) = \text{Var}(\beta_0 + \beta_1 x_i + \varepsilon_i) = \text{Var}(\varepsilon_i) = \sigma^2.$$

Now consider the slope estimate. Recall that the y_is are independent.

$$
\begin{aligned}
\text{Var}(\hat{\beta}_1) &= \text{Var}\left(\frac{\sum_{i=1}^{n} (x_i - \bar{x}_\bullet)\, y_i}{\sum_{i=1}^{n} (x_i - \bar{x}_\bullet)^2}\right) \\
&= \frac{1}{\left[\sum_{i=1}^{n} (x_i - \bar{x}_\bullet)^2\right]^2} \text{Var}\left(\sum_{i=1}^{n} (x_i - \bar{x}_\bullet)\, y_i\right) \\
&= \frac{1}{\left[\sum_{i=1}^{n} (x_i - \bar{x}_\bullet)^2\right]^2} \sum_{i=1}^{n} (x_i - \bar{x}_\bullet)^2 \,\text{Var}(y_i) \\
&= \frac{1}{\left[\sum_{i=1}^{n} (x_i - \bar{x}_\bullet)^2\right]^2} \sum_{i=1}^{n} (x_i - \bar{x}_\bullet)^2 \,\sigma^2 \\
&= \frac{\sigma^2}{\sum_{i=1}^{n} (x_i - \bar{x}_\bullet)^2} \\
&= \frac{\sigma^2}{(n-1)s_x^2}.
\end{aligned}
$$

Rather than establishing the variance of $\hat{\beta}_0$ directly, we find $\text{Var}(\hat{\beta}_0 + \hat{\beta}_1 x)$ for an arbitrary value x. The variance of $\hat{\beta}_0$ is the special case with $x = 0$. A key result is that $\bar{y}_\bullet$ and $\hat{\beta}_1$ are independent. This was discussed in relation to the alternative regression model of Section 7.6. The independence of these estimates is based on the errors having independent normal distributions with the same variance. More generally, if the errors have the same variance and zero covariance, we still get $Cov(\bar{y}_\bullet, \hat{\beta}_1) = 0$, see Exercise 7.13.14.

$$
\begin{aligned}
\text{Var}(\hat{\beta}_0 + \hat{\beta}_1 x) &= \text{Var}(\bar{y}_\bullet - \hat{\beta}_1 \bar{x}_\bullet + \hat{\beta}_1 x) \\
&= \text{Var}(\bar{y}_\bullet + \hat{\beta}_1 (x - \bar{x}_\bullet))
\end{aligned}
$$

$$
\begin{aligned}
&= \text{Var}(\bar{y}_{\cdot}) + \text{Var}\left(\hat{\beta}_1\right)(x - \bar{x}_{\cdot})^2 - 2(x - \bar{x}_{\cdot})Cov\left(\bar{y}_{\cdot}, \hat{\beta}_1\right) \\
&= \frac{1}{n^2}\sum_{i=1}^{n} \text{Var}(y_i) + \text{Var}\left(\hat{\beta}_1\right)(x - \bar{x}_{\cdot})^2 \\
&= \frac{1}{n^2}\sum_{i=1}^{n} \sigma^2 + \frac{\sigma^2(x - \bar{x}_{\cdot})^2}{(n-1)s_x^2} \\
&= \sigma^2\left[\frac{1}{n} + \frac{(x - \bar{x}_{\cdot})^2}{(n-1)s_x^2}\right].
\end{aligned}
$$

In particular, when $x = 0$ we get

$$
\text{Var}\left(\hat{\beta}_0\right) = \sigma^2\left[\frac{1}{n} + \frac{\bar{x}_{\cdot}^2}{(n-1)s_x^2}\right].
$$

7.13 Exercises

EXERCISE 7.13.1. Draper and Smith (1966, p. 41) considered data on the relationship between the age of truck tractors (in years) and the cost (in dollars) of maintaining them over a six month period. The data are given in Table 7.14. Plot cost versus age and fit a regression of cost on age. Give 95% confidence intervals for the slope and intercept. Give a 99% confidence interval for the mean cost of maintaining tractors that are 2.5 years old. Give a 99% prediction interval for the cost of maintaining a particular tractor that is 2.5 years old.

TABLE 7.14. Age and maintenance costs of truck tractors

Age	Cost	Age	Cost	Age	Cost
0.5	163	4.0	495	5.0	890
0.5	182	4.0	723	5.0	1522
1.0	978	4.0	681	5.0	1194
1.0	466	4.5	619	5.5	987
1.0	549	4.5	1049	6.0	764
		4.5	1033	6.0	1373

Reviewing the plot of the data, how much faith should be placed in these estimates for tractors that are 2.5 years old?

EXERCISE 7.13.2. Stigler (1986, p. 6) reported data from Cassini (1740) on the angle between the plane of the equator and the plane of the earth's revolution about the sun. The data are given in Table 7.15. The years −229 and −139 indicate 230 B.C. and 140 B.C. respectively. The angles are listed as the minutes above 23 degrees.

Plot the data. Are there any obvious outliers? If outliers exist, compare the fit of the line with and without the outliers. In particular, compare the different 95% confidence intervals for the slope and intercept.

EXERCISE 7.13.3. Mulrow et al. (1988) presented data on the calibration of a differential scanning calorimeter. The melting temperatures of mercury and naphthalene are known to be 234.16 and 353.24 Kelvin, respectively. The data are given in Table 7.16. Plot the data. Fit a simple linear

TABLE 7.15. Angle between the plane of the equator and the plane of rotation about the sun

Year	Angle	Year	Angle	Year	Angle	Year	Angle
−229	$51.33\bar{3}$	880	35.000	1500	28.400	1600	31.000
−139	$51.33\bar{3}$	1070	34.000	1500	$29.26\bar{6}$	1656	$29.03\bar{3}$
140	$51.16\bar{6}$	1300	32.000	1570	$29.91\bar{6}$	1672	28.900
390	30.000	1460	30.000	1570	31.500	1738	$28.33\bar{3}$

regression $y = \beta_0 + \beta_1 x + \varepsilon$ to the data. Under ideal conditions, the simple linear regression should have $\beta_0 = 0$ and $\beta_1 = 1$; test whether these hypotheses are true using $\alpha = .05$. Give a 95% confidence interval for the population mean of observations taken on this calorimeter for which the true melting point is 250. Give a 95% prediction interval for a new observation taken on this calorimeter for which the true melting point is 250.

Is there any way to check whether it is appropriate to use a line in modeling the relationship between x and y? If so, do so.

TABLE 7.16. Melting temperatures

Chemical	x	y
Naphthalene	353.24	354.62
	353.24	354.26
	353.24	354.29
	353.24	354.38
Mercury	234.16	234.45
	234.16	234.06
	234.16	234.61
	234.16	234.48

EXERCISE 7.13.4. Using the complete data of Exercise 7.13.2, test the need for a transformation of the simple linear regression model. Repeat the test after eliminating any outliers. Compare the results.

EXERCISE 7.13.5. Dixon and Massey (1969) presented data on the relationship between IQ scores and results on an achievement test in a general science course. Table 7.17 contains a subset of the data. Fit the simple linear regression model of achievement on IQ and the quadratic model of achievement on IQ and IQ squared. Evaluate both models and decide which is the best.

EXERCISE 7.13.6. Snedecor and Cochran (1967, Section 6.18) presented data obtained in 1942 from South Dakota on the relationship between the size of farms (in acres) and the number of acres planted in corn. The data are given in Table 7.18.

Plot the data. Fit a simple linear regression to the data. Examine the residuals and discuss what you find. Test the need for a power transformation. Is it reasonable to examine the square root or log transformations? If so, do so.

EXERCISE 7.13.7. In Exercises 5.7.2 and 7.13.6 we considered data on the relationship between farm sizes and the acreage in corn. Fit the linear, quadratic, cubic, and quartic polynomial models to the logs of the acreages in corn. Find the model that fits best. Check the assumptions for this model.

TABLE 7.17. IQs and achievement scores

IQ	Achiev.	IQ	Achiev.	IQ	Achiev.	IQ	Achiev.	IQ	Achiev.
100	49	105	50	134	78	107	43	122	66
117	47	89	72	125	39	121	75	130	63
98	69	96	45	140	66	90	40	116	43
87	47	105	47	137	69	132	80	101	44
106	45	95	46	142	68	116	55	92	50
134	55	126	67	130	71	137	73	120	60
77	72	111	66	92	31	113	48	80	31
107	59	121	59	125	53	110	41	117	55
125	27	106	49	120	64	114	29	93	50

TABLE 7.18. Acreage in corn for different farm acreages

Farm x	Corn y	Farm x	Corn y	Farm x	Corn y
80	25	160	45	320	110
80	10	160	40	320	30
80	20	240	65	320	55
80	32	240	80	320	60
80	20	240	65	400	75
160	60	240	85	400	35
160	35	240	30	400	140
160	20	320	70	400	90
				400	110

Compute the sums of squares for the following contrasts using the means of the logs of the corn acreages:

	Farm acreages				
Contrast	80	160	240	320	400
Linear	−2	−1	0	1	2
Quadratic	2	−1	−2	−1	2
Cubic	−1	2	0	−2	1
Quartic	1	−4	6	−4	1
Means	2.9957	3.6282	4.1149	4.0904	4.4030

Compare the contrast sums of squares to the polynomial model fitting procedure.

EXERCISE 7.13.8. Repeat Exercise 7.13.6 but instead of using the number of acres of corn as the dependent variable, use the proportion of acreage in corn as the dependent variable. Compare the results to those given earlier.

EXERCISE 7.13.9. In Exercises 7.13.1 and 5.7.10, we performed a simple linear regression and a one-way ANOVA on the data of Table 7.14. Test for lack of fit, i.e., whether the simple linear regression is an adequate reduced model as compared to the one-way ANOVA model.

EXERCISE 7.13.10. The analysis of variance in Exercise 5.7.3 was based on the height and weight data given in Table 7.19. Fit a simple linear regression of weight on height for these data and check the assumptions. Give a 99% confidence interval for the mean weight of people with a 72 inch height and compare it to the interval from Exercise 5.7.3. Test the lack of fit of the simple linear regression model compared to the larger one-way ANOVA model.

TABLE 7.19. Weights for various heights

Ht.	Wt.	Ht.	Wt.
65	120	63	110
65	140	63	135
65	130	63	120
65	135	72	170
66	150	72	185
66	135	72	160

EXERCISE 7.13.11. Jensen (1977) and Weisberg (1985, p. 101) considered data on the outside diameter of crank pins that were produced in an industrial process. The diameters of batches of crank pins were measured on various days; if the industrial process is 'under control' the diameters should not depend on the day they were measured. A subset of the data is given in Table 7.20 in a format consistent with performing a regression analysis on the data. The diameters of the crank pins are actually $.742 + y_{ij}10^{-5}$ inches, where the y_{ij}s are reported in Table 7.20. Perform an analysis of variance and polynomial regressions on the data. Give the lack of fit test for the simple linear regression.

TABLE 7.20. Jensen's crank pin data

Days	Diameters	Days	Diameters	Days	Diameters	Days	Diameters
4	93	10	93	16	82	22	90
4	100	10	88	16	72	22	92
4	88	10	87	16	80	22	82
4	85	10	87	16	72	22	77
4	89	10	87	16	89	22	89

EXERCISE 7.13.12. Exercise 7.13.3 involves the calibration of a measuring instrument. Often, calibration curves are used in reverse, i.e., we would use the calorimeter to measure a melting point y and use the regression equation to give a point estimate of x. If a new substance has a measured melting point of 300 Kelvin, using the simple linear regression model what is the estimate of the true melting point? Use a prediction interval to determine whether the measured melting point of $y = 300$ is consistent with the true melting point being $x = 300$. Is an observed value of 300 consistent with a true value of 310?

EXERCISE 7.13.13. Working-Hotelling confidence bands are a method for getting confidence intervals for every point on a line with a guaranteed simultaneous coverage. The method is essentially the same as Scheffé's method for simultaneous confidence intervals discussed in Section 6.4. For estimating the point on the line at a value x, the endpoints of the $(1-\alpha)100\%$ simultaneous confidence intervals are

$$(\hat{\beta}_0 + \hat{\beta}_1 x) \pm \sqrt{2F(1-\alpha, 2, dfE)}\, SE(\hat{\beta}_0 + \hat{\beta}_1 x).$$

Using the *Coleman Report* data of Table 7.1, find 95% simultaneous confidence intervals for the values $x = -17, -6, 0, 6, 17$. Plot the estimated regression line and sketch the Working-Hotelling confidence bands. We are 95% confident that the entire line $\beta_0 + \beta_1 x$ lies between the confidence bands. Compute the regular confidence intervals for $x = -17, -6, 0, 6, 17$ and compare them to the results of the Working-Hotelling procedure.

EXERCISE 7.13.14. Use part (4) of Proposition 1.2.11 to show that $Cov(\bar{y}_{\cdot}, \hat{\beta}_1) = 0$ whenever $\text{Var}(\varepsilon_i) = \sigma^2$ for all i and $Cov(\varepsilon_i, \varepsilon_j) = 0$ for all $i \neq j$. Hint: write out $\bar{y}_{\cdot}$ and $\hat{\beta}_1$ in terms of the y_is.

EXERCISE 7.13.15. Using the axial stiffness index data of Table 7.10, fit linear, quadratic, cubic, and quartic polynomial regression models using powers of x, the plate length, and using powers of $x - \bar{x}_{\cdot}$, the plate length minus the average plate length. Compare the results of the two procedures. If your computer program will not fit some of the models, report on that in addition to comparing results for the models you could fit.

Chapter 8

The analysis of count data

For the most part, this book concerns itself with measurement data and the corresponding analyses based on normal distributions. In this chapter we consider data that consist of counts. We begin in Section 8.1 by examining a set of data on the number of females admitted into graduate school at the University of California, Berkeley. A key feature of these data is that only two outcomes are possible: admittance or rejection. Data with only two outcomes are referred to as *binary (or dichotomous) data*. Often the two outcomes are referred to generically as success and failure. In Section 8.2, we expand our discussion by comparing two sets of dichotomous data; we compare Berkeley graduate admission rates for females and males. Section 8.3 examines *polytomous data*, i.e., count data in which there are more than two possible outcomes. For example, numbers of Swedish females born in the various months of the year involve counts for 12 possible outcomes. Section 8.4 examines comparisons between two samples of polytomous data, e.g., comparing the numbers of females and males that are born in the different months of the year. Section 8.5 looks at comparisons among more than two samples of polytomous data. The penultimate section considers a method of reducing large tables of counts that involve several samples of polytomous data into smaller more interpretable tables. The final section deals with a count data analogue of simple linear regression.

Sections 8.1 and 8.2 involve analogues of Chapters 2 and 4 that are appropriate for dichotomous data. The basic analyses in these sections simply involve new applications of the ideas in Chapter 3. Analyzing polytomous data requires techniques that are different from the methods of Chapter 3. Sections 8.3, 8.4, and 8.5 are polytomous data analogues of Chapters 2, 4, and 5. Everitt (1977) and Fienberg (1980) give more detailed introductions to the analysis of count data. Sophisticated analyses of count data frequently use analogues of ANOVA and regression called log-linear models. Christensen (1990b) provides an intermediate level account of log-linear models.

8.1 One binomial sample

The few distributions that are most commonly used in statistics arise naturally. The normal distribution arises for measurement data because the variability in the data often results from the mean of a large number of small errors and the central limit theorem indicates that such means tend to be normally distributed.

The binomial distribution arises naturally with count data because of its simplicity. Consider a number of trials, say n, each a success or failure. If each trial is independent of the other trials and if the probability of obtaining a success is the same for every trial, then the random number of successes has a binomial distribution. *The beauty of discrete data is that the probability models can often be justified solely by how the data were collected. This does not happen with measurement data.* The binomial distribution depends on two parameters, n, the number of independent trials, and the constant probability of success, say p. Typically, we know the value of n, while p is the unknown parameter of interest. Binomial distributions were examined in Section 1.4.

Bickel et al. (1975) report data on admissions to graduate school at the University of California, Berkeley. The numbers of females that were admitted and rejected are given below along with the total number of applicants.

Graduate admissions at Berkeley

	Admitted	Rejected	Total
Female	557	1278	1835

It seems reasonable to view the 1835 females as a random sample from a population of potential female applicants. We are interested in the probability p that a female applicant is admitted to graduate school. A natural estimate of the parameter p is the proportion of females that were actually admitted, thus our estimate of the parameter is

$$\hat{p} = \frac{557}{1835} = .30354.$$

We have a parameter of interest, p, and an estimate of that parameter, $\hat{p}$. If we can identify a standard error and an appropriate distribution, we can use the methods of Chapter 3 to perform statistical inferences.

The key to finding a standard error is to find the variance of the estimate. As we will see later,

$$\text{Var}(\hat{p}) = \frac{p(1-p)}{n}. \tag{8.1.1}$$

To estimate the standard deviation of $\hat{p}$, we simply use $\hat{p}$ to estimate p in (8.1.1) and take the square root. Thus the standard error is

$$SE(\hat{p}) = \sqrt{\frac{\hat{p}(1-\hat{p})}{n}} = \sqrt{\frac{.30354(1-.30354)}{1835}} = .01073.$$

The final requirement for using the results of Chapter 3 is to find an appropriate reference distribution for

$$\frac{\hat{p}-p}{SE(\hat{p})}.$$

We can think of each trial as scoring either a 1, if the trial is a success, or a 0, if the trial is a failure. With this convention $\hat{p}$, the proportion of successes, is really the average of the 0–1 scores and since $\hat{p}$ is an average we can apply the central limit theorem. (In fact, $SE(\hat{p})$ is very nearly $s/\sqrt{n}$, where s is computed from the 0–1 scores.) The central limit theorem simply states that for a large number of trials n, the distribution of $\hat{p}$ is approximately normal with a population mean that is the population mean of $\hat{p}$ and a population variance that is the population variance of $\hat{p}$. We have already given the variance of $\hat{p}$ and we will see later that $E(\hat{p}) = p$. Thus for large n we have the approximation

$$\hat{p} \sim N\left(p, \frac{p(1-p)}{n}\right).$$

The variance is unknown but by the law of large numbers it is approximately equal to our estimate of it, $\hat{p}(1-\hat{p})/n$. Standardizing the normal distribution (cf. Exercise 1.6.2) gives the approximation

$$\frac{\hat{p}-p}{SE(\hat{p})} \sim N(0,1). \tag{8.1.2}$$

This distribution requires a sample size that is large enough for both the central limit theorem approximation and the law of large numbers approximation to be reasonably valid. For values of p that are not too close to 0 or 1, the approximation works reasonably well with sample sizes as small as 20.

We now have $Par = p$, $Est = \hat{p}$, $SE(\hat{p}) = \sqrt{\hat{p}(1-\hat{p})/n}$ and the distribution in (8.1.2). As in Chapter 3, a 95% confidence interval for p has limits

$$\hat{p} \pm 1.96\sqrt{\frac{\hat{p}(1-\hat{p})}{n}}.$$

Here $1.96 = z(.975) = t(.975, \infty)$. Recall that a $(1-\alpha)100\%$ confidence interval requires the $(1-\alpha/2)$ percentile of the distribution. For the female admissions data, the limits are

$$.30354 \pm 1.96(.01073)$$

which gives the interval $(.28, .32)$. We are 95% confident that the population proportion of females admitted to Berkeley's graduate school is between .28 and .32. (As is often the case, it is not exactly clear what population these data relate to.)

We can also perform, say, an $\alpha = .01$ test of the null hypothesis $H_0 : p = 1/3$ versus the alternative $H_A : p \neq 1/3$. The test rejects H_0 if

$$\frac{\hat{p}-1/3}{SE(\hat{p})} > 2.58$$

or if

$$\frac{\hat{p}-1/3}{SE(\hat{p})} < -2.58.$$

Here $2.58 = z(.995) = t(.975, \infty)$. An α level two-sided test requires the $(1-\frac{\alpha}{2})100\%$ point of the distribution. The Berkeley data yield the test statistic

$$\frac{.30354 - .33333}{.01073} = -2.78$$

which is smaller than -2.58, so we reject the null hypothesis of $p = 1/3$ with $\alpha = .01$. In other words, we can reject, with strong assurance, the claim that one third of female applicants are admitted to graduate school at Berkeley. Since the test statistic is negative, we have evidence that the true proportion is less than one third. The test as constructed here is equivalent to checking whether $p = 1/3$ is within a 99% confidence interval.

There is an alternative, slightly different, way of performing tests such as $H_0 : p = 1/3$ versus $H_A : p \neq 1/3$. The difference involves using a different standard error. The variance of the estimate $\hat{p}$ is $p(1-p)/n$. In obtaining a standard error, we estimated p with $\hat{p}$ and took the square root of the estimated variance. Recalling that tests are performed *assuming that the null hypothesis is true*, it makes sense in the testing problem to use the assumption $p = 1/3$ in computing a standard error for $\hat{p}$. Thus an alternative standard error for $\hat{p}$ in this testing problem is

$$\sqrt{\frac{1}{3}\left(1-\frac{1}{3}\right)\bigg/1835} = .01100.$$

The test statistic now becomes

$$\frac{.30354-.33333}{.01100} = -2.71.$$

Obviously, since the test statistic is slightly different, one could get slightly different answers for tests using the two different standard errors. Moreover, the results of this test will not always agree with a corresponding confidence interval for p because this test uses a different standard error than the confidence interval.

We should remember that the $N(0, 1)$ distribution being used for the test is only a large sample approximation. (In fact, all of our results are only approximations.) The difference between the two standard errors is often minor compared to the level of approximation inherent in using the standard normal as a reference distribution. In any case, whether we ascribe the differences to the standard errors or to the quality of the normal approximations, the exact behavior of the two test statistics can be quite different when the sample size is small. Moreover, *when p is near 0 or 1, the sample sizes must be quite large to get a good normal approximation.*

The main theoretical results for a single binomial sample are establishing that $\hat{p}$ is a reasonable estimate of p and that the variance formula given earlier is correct. The data are $y \sim Bin(n, p)$. As seen in Section 1.4, $E(y) = np$ and $\text{Var}(y) = np(1-p)$. The estimate of p is $\hat{p} = y/n$. The estimate is unbiased because

$$E(\hat{p}) = E(y/n) = E(y)/n = np/n = p.$$

The variance of the estimate is

$$\text{Var}(\hat{p}) = \text{Var}(y/n) = \text{Var}(y)/n^2 = np(1-p)/n^2 = p(1-p)/n.$$

8.1.1 THE SIGN TEST

We now consider an alternative analysis for paired comparisons based on the binomial distribution. Consider Burt's data on IQs of identical twins raised apart from Exercise 4.5.7 and Table 4.9. The earlier discussion of paired comparisons involved assuming and validating the normal distribution for the differences in IQs between twins. In the current discussion, we make the same assumptions as before except we replace the normality assumption with the weaker assumption that the distribution of the differences is symmetric. In the earlier discussion, we would test $H_0 : \mu_1 - \mu_2 = 0$. In the current discussion, we test whether there is a 50 : 50 chance that y_1, the IQ for the foster parent raised twin, is larger than y_2, the IQ for the genetic parent raised twin. In other words, we test whether $\Pr(y_1 - y_2 > 0) = .5$. We have a sample of $n = 27$ pairs of twins. If $\Pr(y_1 - y_2 > 0) = .5$, the number of pairs with $y_1 - y_2 > 0$ has a $Bin(27, .5)$ distribution. From Table 4.9, 13 of the 27 pairs have larger IQs for the foster parent raised child. (These are the differences with a positive sign, hence the name sign test.) The proportion is $\hat{p} = 13/27 = .481$. The test statistic is

$$\frac{.481-.5}{\sqrt{.5(1-.5)/27}} = -.20$$

which is nowhere near significant.

A similar method could be used to test, say, whether there is a 50 : 50 chance that y_1 is at least 3 IQ points greater than y_2. This hypothesis translates into $\Pr(y_1 - y_2 \geq 3) = .5$. The test is then based on the number of differences that are 3 or more.

The point of the sign test is the weakening of the assumption of normality. If the normality assumption is appropriate, the t test of Section 4.1 is more powerful. When the normality assumption is not appropriate, some modification like the sign test should be used. In this book, the usual approach is to check the normality assumption and, if necessary, to transform the data to make the normality assumption reasonable. For a more detailed introduction to *nonparametric* methods such as the sign test, see, for example, Conover (1971).

8.2 Two independent binomial samples

In this section we compare two independent binomial samples. Consider again the Berkeley admissions data. Table 8.1 contains data on admissions and rejections for the 1835 females considered in Section 8.1 along with data on 2691 males. We assume that the sample of females is independent of the sample of males. Throughout, we refer to the females as the first sample and the males as the second sample.

TABLE 8.1. Graduate admissions at Berkeley

	Admitted	Rejected	Total
Females	557	1278	1835
Males	1198	1493	2691

We consider being admitted to graduate school a 'success'. Assuming that the females are a binomial sample, they have a sample size of $n_1 = 1835$ and some probability of success, say, p_1. The observed proportion of female successes is

$$\hat{p}_1 = \frac{557}{1835} = .30354.$$

Treating the males as a binomial sample, the sample size is $n_2 = 2691$ and the probability of success is, say, p_2. The observed proportion of male successes is

$$\hat{p}_2 = \frac{1198}{2691} = .44519.$$

Our interest is in comparing the success rate of females and males. The appropriate parameter is the difference in proportions,

$$Par = p_1 - p_2.$$

The natural estimate of this parameter is

$$Est = \hat{p}_1 - \hat{p}_2 = .30354 - .44519 = -.14165.$$

With independent samples, we can find the variance of the estimate and thus the standard error. Since the females are independent of the males,

$$\text{Var}(\hat{p}_1 - \hat{p}_2) = \text{Var}(\hat{p}_1) + \text{Var}(\hat{p}_2).$$

Using the variance formula in equation (8.1.1),

$$\text{Var}(\hat{p}_1 - \hat{p}_2) = \frac{p_1(1-p_1)}{n_1} + \frac{p_2(1-p_2)}{n_2}. \tag{8.2.1}$$

Estimating p_1 and p_2 and taking the square root gives the standard error,

$$\begin{aligned} SE(\hat{p}_1 - \hat{p}_2) &= \sqrt{\frac{\hat{p}_1(1-\hat{p}_1)}{n_1} + \frac{\hat{p}_2(1-\hat{p}_2)}{n_2}} \\ &= \sqrt{\frac{.30354(1-.30354)}{1835} + \frac{.44519(1-.44519)}{2691}} \\ &= .01439. \end{aligned}$$

For large sample sizes n_1 and n_2, both $\hat{p}_1$ and $\hat{p}_2$ have approximate normal distributions and they are independent, so $\hat{p}_1 - \hat{p}_2$ has an approximate normal distribution and the appropriate reference distribution is approximately

$$\frac{(\hat{p}_1 - \hat{p}_2) - (p_1 - p_2)}{SE(\hat{p}_1 - \hat{p}_2)} \sim N(0, 1).$$

We now have all the requirements for applying the results of Chapter 3. A 95% confidence interval for $p_1 - p_2$ has endpoints

$$(\hat{p}_1 - \hat{p}_2) \pm 1.96\sqrt{\frac{\hat{p}_1(1 - \hat{p}_1)}{n_1} + \frac{\hat{p}_2(1 - \hat{p}_2)}{n_2}},$$

where the value $1.96 = z(.975)$ is obtained from the $N(0, 1)$ distribution. For comparing the female and male admissions, the 95% confidence interval for the population difference in proportions has endpoints

$$-.14165 \pm 1.96(.01439).$$

The interval is $(-.17, -.11)$. Thus we are 95% confident that the proportion of women being admitted to graduate school at Berkeley is between .11 and .17 less than that for men.

To test $H_0 : p_1 = p_2$, or equivalently $H_0 : p_1 - p_2 = 0$, against $H_A : p_1 - p_2 \neq 0$, reject an $\alpha = .10$ test if

$$\frac{(\hat{p}_1 - \hat{p}_2) - 0}{SE(\hat{p}_1 - \hat{p}_2)} > 1.645$$

or if

$$\frac{(\hat{p}_1 - \hat{p}_2) - 0}{SE(\hat{p}_1 - \hat{p}_2)} < -1.645.$$

Again, the value 1.645 is obtained from the $N(0, 1) \equiv t(\infty)$ distribution. With the Berkeley data, the observed value of the test statistic is

$$\frac{-.14165 - 0}{.01439} = -9.84.$$

This is far smaller than -1.645, so the test rejects the null hypothesis of equal proportions at the .10 level. The test statistic is negative, so there is evidence that the proportion of women admitted to graduate school is lower than the proportion of men.

Once again, an alternative standard error is often used in testing problems. The test assumes that the null hypothesis is true and under the null hypothesis $p_1 = p_2$, so in constructing a standard error for the test statistic it makes sense to pool the data into one estimate of this common proportion. The pooled estimate is a weighted average of the individual estimates,

$$\begin{aligned} \hat{p}_* &= \frac{n_1\hat{p}_1 + n_2\hat{p}_2}{n_1 + n_2} \\ &= \frac{1835(.30354) + 2691(.44519)}{1835 + 2691} \\ &= \frac{557 + 1198}{1835 + 2691} \\ &= .38776. \end{aligned}$$

Using $\hat{p}_*$ to estimate both p_1 and p_2 in equation (8.2.1) and taking the square root gives the alternative standard error

$$SE(\hat{p}_1 - \hat{p}_2) = \sqrt{\frac{\hat{p}_*(1 - \hat{p}_*)}{n_1} + \frac{\hat{p}_*(1 - \hat{p}_*)}{n_2}}$$

$$= \sqrt{\hat{p}_*(1-\hat{p}_*)\left[\frac{1}{n_1}+\frac{1}{n_2}\right]}$$

$$= \sqrt{.38776(1-.38776)\left[\frac{1}{1835}+\frac{1}{2691}\right]}$$

$$= .01475$$

The alternative test statistic is

$$\frac{-.14165-0}{.01475} = -9.60.$$

Again, the two test statistics are slightly different but the difference should be minor compared to the level of approximation involved in using the normal distribution.

A final note. Before you conclude that the data in Table 8.1 provide evidence of sex discrimination, you should realize that females tend to apply to different graduate programs than males. A more careful examination of the complete Berkeley data shows that the difference observed here results from females applying more frequently than males to highly restrictive programs, cf. Christensen (1990b, p. 96).

8.3 One multinomial sample

In this section we investigate the analysis a single polytomous variable, i.e., a count variable with more than two possible outcomes. In particular, we assume that the data are a sample from a *multinomial* distribution, cf. Section 1.5. The multinomial distribution is a generalization of the binomial that allows more than two outcomes. We assume that each trial gives one of, say, q possible outcomes. Each trial must be independent and the probability of each outcome must be the same for every trial. The multinomial distribution gives probabilities for the number of trials that fall into each of the possible outcome categories. The binomial distribution is a special case of the multinomial distribution in which $q = 2$.

The first two columns of Table 8.2 give months and numbers of Swedish females born in each month. The data are from Cramér (1946) who did not name the months. We assume that the data begin in January.

TABLE 8.2. Swedish female births by month

Month	Females	$\hat{p}$	Probability	E	$(O-E)/\sqrt{E}$
January	3537	.083	1/12	3549.25	−0.20562
February	3407	.080	1/12	3549.25	−2.38772
March	3866	.091	1/12	3549.25	5.31678
April	3711	.087	1/12	3549.25	2.71504
May	3775	.087	1/12	3549.25	3.78930
June	3665	.086	1/12	3549.25	1.94291
July	3621	.085	1/12	3549.25	1.20435
August	3596	.084	1/12	3549.25	0.78472
September	3491	.082	1/12	3549.25	−0.97775
October	3391	.080	1/12	3549.25	−2.65629
November	3160	.074	1/12	3549.25	−6.53372
December	3371	.079	1/12	3549.25	−2.99200
Total	42591	1	1	42591.00	

With polytomous data such as those listed in Table 8.2, there is no one parameter of primary interest. One might be concerned with the proportions of births in January, or December, or in any of the twelve months. With no one parameter of interest, the methods of Chapter 3 do not apply. Column 3 of Table 8.2 gives the observed proportions of births for each month. These are simply the monthly births divided by the total births for the year. Note that the proportion of births in March seems high and the proportion of births in November seems low.

A simplistic, yet interesting, hypothesis is that the proportion of births is the same for every month. To test this null hypothesis, we compare the number of observed births to the number of births we would expect to see if the hypothesis were true. The number of births we expect to see in any month is just the probability of having a birth in that month times the total number of births. The equal probabilities are given in column 4 of Table 8.2 and the expected values are given in column 5. The entries in column 5 are labeled E for expected value and are computed as $(1/12)42591 = 3549.25$. *It cannot be overemphasized that the expectations are computed under the assumption that the null hypothesis is true.*

Comparing observed values with expected values can be tricky. Suppose an observed value is 2145 and the expected value is 2149. The two numbers are off by 4; the observed value is pretty close to the expected. Now suppose the observed value is 1 and the expected value is 5. Again the two numbers are off by 4 but now the difference between observed and expected seems quite substantial. A difference of 4 means something very different depending on how large both numbers are. To account for this phenomenon, we standardized the difference between observed and expected counts. We do this by dividing the difference by the square root of the expected count. Thus, when we compare observed counts with expected counts we look at

$$\frac{O - E}{\sqrt{E}} \tag{8.3.1}$$

where O stands for the observed count and E stands for the expected count. The values in (8.3.1) are called *Pearson residuals*, after Karl Pearson.

The Pearson residuals for the Swedish female births are given in column 6 of Table 8.2. As noted earlier, the two largest deviations from the assumption of equal probabilities occur for March and November. Reasonably large deviations also occur for May and to a lesser extent December, April, October, and February. In general, *the Pearson residuals can be compared to observations from a* $N(0, 1)$ *distribution to evaluate whether a residual is large*. For example, the residuals for March and November are 5.3 and −6.5. These are not values one is likely to observe from a $N(0, 1)$ distribution; they provide strong evidence that birth rates in March are really larger than $1/12$ and that birth rates in November are really smaller than $1/12$.

Births seem to peak in March and they, more or less, gradually decline until November. After November, birth rates are still low but gradually increase until February. In March birth rates increase markedly. Birth rates are low in the fall and lower in the winter; they jump in March and remain relatively high, though decreasing, until September. This analysis could be performed using the monthly proportions of column 2 but the results are clearer using the residuals.

A statistic for testing whether the null hypothesis of equal proportions is reasonable can be obtained by squaring the residuals and adding them together. This statistic is known as *Pearson's* χ^2 (chi-squared) statistic and is computed as

$$X^2 = \sum_{\text{all cells}} \frac{(O - E)^2}{E}.$$

For the female Swedish births,

$$X^2 = 121.24.$$

Note that small values of X^2 indicate observed values that are similar to the expected values, so small values of X^2 are consistent with the null hypothesis. Large values of X^2 occur whenever one or more observed values are far from the expected values. To perform a test, we need some idea of how large X^2 could reasonably be when the null hypothesis is true. It can be shown that for a problem such as this with 1) a fixed number of cells q, here $q = 12$, with 2) a null hypothesis consisting of known probabilities such as those given in column 4 of Table 8.2, and with 3) large sample sizes for each cell, the null distribution of X^2 is approximately

$$X^2 \sim \chi^2(q-1).$$

The degrees of freedom are only $q-1$ because the $\hat{p}$s *must* add up to 1. Thus, if we know $q-1=11$ of the proportions, we can figure out the last one. Only $q-1$ of the cells are really free to vary. From Appendix B.2, the 99.5th percentile of a $\chi^2(11)$ distribution is $\chi^2(.995, 11) = 26.76$. The observed X^2 value of 121.24 is much larger than this, so the observed value of X^2 could not reasonably come from a $\chi^2(11)$ distribution. In particular, an $\alpha = .005$ test of the null hypothesis is rejected easily, so the P value for the test is 'much' less than .005. It follows that there is overwhelming evidence that the proportion of female Swedish births is not the same for all months.

In this example, our null hypothesis was that the probability of a female birth was the same in every month. A more reasonable hypothesis might be that the probability of a female birth is the same on every day. The months have different numbers of days so under this null hypothesis they have different probabilities. For example, assuming a 365 day year, the probability of a female birth in January is $31/365$ which is somewhat larger than $1/12$. Exercise 8.8.4 involves testing this alternative null hypothesis.

We can use results from Section 8.1 to help in the analysis of multinomial data. If we consider only the month of December, we can view each trial as a success if the birth is in December and a failure otherwise. Writing the probability of a birth in December as p_{12}, from Table 8.2 the estimate of p_{12} is

$$\hat{p}_{12} = \frac{3371}{42591} = .07915$$

with standard error

$$SE(\hat{p}_{12}) = \sqrt{\frac{.07915(1-.07915)}{42591}} = .00131$$

and a 95% confidence interval has endpoints

$$.07915 \pm 1.96(.00131).$$

The interval reduces to $(.077, .082)$. Tests for monthly proportions can be performed in a similar fashion. Bonferroni adjustments can be made to all tests and confidence intervals to control the experimentwise error rate for multiple tests or intervals, cf. Section 6.2.

8.4 Two independent multinomial samples

Table 8.3 gives monthly births for Swedish females and males along with various marginal totals. We wish to determine whether monthly birth rates differ for females and males. Denote the females as population 1 and the males as population 2. Thus we have a sample of 42591 females and, by assumption, an independent sample of 45682 males.

In fact, it is more likely that there is actually only one sample here, one consisting of 88273 births. It is more likely that the births have been divided into 24 categories depending on sex and birth month. Such data can be treated as two independent samples with (virtually) no loss of generality. The interpretation of results for two independent samples is considerably simpler than the interpretation necessary for one sample cross-classified by both sex and month. Thus we discuss such data as

TABLE 8.3. Swedish births: monthly observations (O_{ij}s) and monthly proportions by sex

Month	Observations Female	Male	Total	Proportions Female	Male
January	3537	3743	7280	.083	.082
February	3407	3550	6957	.080	.078
March	3866	4017	7883	.091	.088
April	3711	4173	7884	.087	.091
May	3775	4117	7892	.089	.090
June	3665	3944	7609	.086	.086
July	3621	3964	7585	.085	.087
August	3596	3797	7393	.084	.083
September	3491	3712	7203	.082	.081
October	3391	3512	6903	.080	.077
November	3160	3392	6552	.074	.074
December	3371	3761	7132	.079	.082
Total	42591	45682	88273	1.000	1.000

though they are independent samples. The alternative interpretation involves a multinomial sample with the probabilities for month and sex pairs all being independent.

The number of births in month i for sex j is denoted O_{ij}, where $i = 1, \ldots, 12$ and $j = 1, 2$. Thus, for example, the number of males born in December is $O_{12,2} = 3761$. Let $O_{i\bullet}$ be the total for month i, $O_{\bullet j}$ be the total for sex j, and $O_{\bullet\bullet}$ be the total over all months and sexes. For example, May has $O_{5\bullet} = 7892$, males have $O_{\bullet 2} = 45682$, and the grand total is $O_{\bullet\bullet} = 88273$.

Our interest now is in whether the population proportion of births for each month is the same for females as for males. We no longer make any assumption about what these proportions are, our null hypothesis is simply that the proportions are the same in each month. Again, we wish to compare the observed values, the O_{ij}s with expected values, but now, since we do not have hypothesized proportions for any month, we must estimate the expected values.

Under the null hypothesis that the proportions are the same for females and males, it makes sense to pool the male and female data to get an estimate of the proportion of births in each month. Using the column of monthly totals in Table 8.3, the estimated proportion for January is the January total divided by the total for the year, i.e.,

$$\hat{p}_1^0 = \frac{7280}{88273} = .0824714.$$

In general, for month i we have

$$\hat{p}_i^0 = \frac{O_{i\bullet}}{O_{\bullet\bullet}}$$

where the superscript of 0 is used to indicate that these proportions are estimated under the null hypothesis of identical monthly rates for males and females. The estimate of the expected number of females born in January is just the number of females born in the year times the estimated probability of a birth in January,

$$\hat{E}_{11} = 42591(.0824714) = 3512.54.$$

The expected number of males born in January is the number of males born in the year times the estimated probability of a birth in January,

$$\hat{E}_{12} = 45682(.0824714) = 3767.46.$$

In general,

$$\hat{E}_{ij} = O_{\bullet j}\hat{p}_i^0 = O_{\bullet j}\frac{O_{i\bullet}}{O_{\bullet\bullet}} = \frac{O_{i\bullet}O_{\bullet j}}{O_{\bullet\bullet}}.$$

Again, *the estimated expected values are computed assuming that the proportions of births are the same for females and males in every month, i.e., assuming that the null hypothesis is true.* The estimated expected values under the null hypothesis are given in Table 8.4. Note that the totals for each month and for each sex remain unchanged.

TABLE 8.4. Estimated expected Swedish births by month ($\hat{E}_{ij}$s) and pooled proportions

	Expectations			Pooled
Month	Female	Male	Total	proportions
January	3512.54	3767.46	7280	.082
February	3356.70	3600.30	6957	.079
March	3803.48	4079.52	7883	.089
April	3803.97	4080.03	7884	.089
May	3807.83	4084.17	7892	.089
June	3671.28	3937.72	7609	.086
July	3659.70	3925.30	7585	.086
August	3567.06	3825.94	7393	.084
September	3475.39	3727.61	7203	.082
October	3330.64	3572.36	6903	.078
November	3161.29	3390.71	6552	.074
December	3441.13	3690.87	7132	.081
Total	42591.00	45682.00	88273	1.000

The estimated expected values are compared to the observations using Pearson residuals, just as in Section 8.3. The Pearson residuals are

$$\tilde{r}_{ij} \equiv \frac{O_{ij} - \hat{E}_{ij}}{\sqrt{\hat{E}_{ij}}}.$$

A more apt name for the Pearson residuals in this context may be *crude standardized residuals*. It is the standardization here that is crude and not the residuals. The standardization in the Pearson residuals ignores the fact that $\hat{E}$ is itself an estimate. Better, but considerably more complicated, standardized residuals can be defined for count data, cf. Christensen (1990b, Section IV.9). For the Swedish birth data, the Pearson residuals are given in Table 8.5. Note that when compared to a $N(0, 1)$ distribution, none of the residuals is very large; all are smaller than 1.51 in absolute value.

As in Section 8.3, the sum of the squared Pearson residuals gives Pearson's χ^2 statistic for testing the null hypothesis of no differences between females and males. Pearson's test statistic is

$$X^2 = \sum_{ij} \frac{(O_{ij} - \hat{E}_{ij})^2}{\hat{E}_{ij}}.$$

For the Swedish birth data, computing the statistic from the 24 cells in Table 8.5 gives

$$X^2 = 14.9858.$$

For a formal test, X^2 is compared to a χ^2 distribution. The appropriate number of degrees of freedom for the χ^2 test is the number of cells in the table adjusted to account for all the parameters we have estimated as well as the constraint that the sex totals sum to the grand total. There are 12×2 cells but only $12 - 1$ free months and only $2 - 1$ free sex totals. The appropriate distribution is $\chi^2((12 - 1)(2 - 1)) = \chi^2(11)$. *The degrees of freedom are the number of data rows in Table 8.3 minus 1 times the number of data columns in Table 8.3 minus 1.* The 90th percentile of a $\chi^2(11)$

TABLE 8.5. Pearson residuals for Swedish birth months, ($\tilde{r}_{ij}$s)

Month	Female	Male
January	0.41271	−0.39849
February	0.86826	−0.83837
March	1.01369	−0.97880
April	−1.50731	1.45542
May	−0.53195	0.51364
June	−0.10365	0.10008
July	−0.63972	0.61770
August	0.48452	−0.46785
September	0.26481	−0.25570
October	1.04587	−1.00987
November	−0.02288	0.02209
December	−1.19554	1.15438

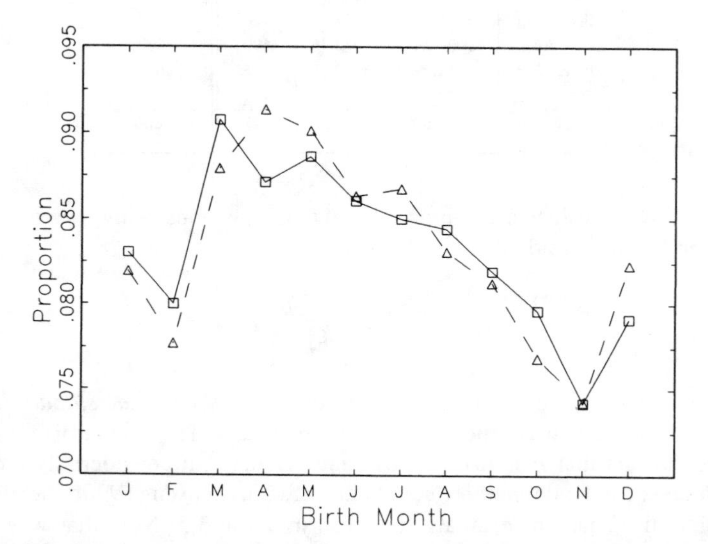

FIGURE 8.1. Monthly Swedish birth proportions by sex: solid line, female; dashed line, male.

distribution is $\chi^2(.9, 11) = 17.28$, so the observed test statistic $X^2 = 14.9858$ could reasonably come from a $\chi^2(11)$ distribution. In particular, the test is not significant at the .10 level. Moreover, $\chi^2(.75, 11) = 13.70$, so the test has a P value between .25 and .10. There is no evidence of any differences in the monthly birth rates for males and females.

Another way to evaluate the null hypothesis is by comparing the observed monthly birth proportions by sex. These observed proportions are given in Table 8.3. If the populations of females and males have the same proportions of births in each month, the observed proportions of births in each month should be similar (except for sampling variation). One can compare the numbers directly in Table 8.3 or one can make a visual display of the observed proportions as in Figure 8.1.

The methods just discussed apply equally well to the binomial data of Table 8.1. Applying the X^2 test given here to the data of Table 8.1 gives

$$X^2 = 92.2.$$

The statistic X^2 is equivalent to the test statistic given in Section 8.2 using the pooled estimate $\hat{p}_*$ to compute the standard error. The test statistic in Section 8.2 is -9.60, and if we square this we get

$$(-9.60)^2 = 92.2 = X^2.$$

The -9.60 is compared to a $N(0, 1)$, while the 92.2 is compared to a $\chi^2(1)$ because Table 8.1 has 2 rows and 2 columns. A $\chi^2(1)$ distribution is obtained by squaring a $N(0, 1)$ distribution, so P values are identical and critical values are equivalent.

MINITAB COMMANDS

Minitab commands for generating the analysis of Swedish birth rates are given below. Column c1 contains the observations, the O_{ij}s. Column c2 contains indices from 1 to 12 indicating the month of each observation and c3 contains indices for the two sexes. The subcommand 'colpercents' provides the proportions discussed in the analysis. The subcommand 'chisquare 3' gives the observations, estimated expected values, and Pearson residuals along with the Pearson test statistic.

```
MTB > read 'swede2.dat' c1 c2 c3
MTB > table c2 c3;
SUBC> frequencies c1;
SUBC> colpercents;
SUBC> chisquare 3.
```

8.5 Several independent multinomial samples

The methods of Section 8.4 extend easily to dealing with more than two samples. Consider the data in Table 8.6 that was extracted from Lazerwitz (1961). The data involve samples from three religious groups and consist of numbers of people in various occupational groups. The occupations are labeled A, professions; B, owners, managers, and officials; C, clerical and sales; and D, skilled. The three religious groups are Protestant, Roman Catholic, and Jewish. This is a subset of a larger collection of data that includes many more religious and occupational groups. The fact that we are restricting ourselves to a subset of a larger data set has no effect on the analysis. As discussed in Section 8.4, the analysis of these data is essentially identical regardless of whether the data come from one sample of 1926 individuals cross-classified by religion and occupation, or four independent samples of sizes 348, 477, 411, and 690 taken from the occupational groups, or three independent samples of sizes 1135, 648, and 143 taken from the religious groups. We choose to view the data as independent samples from the three religious groups. The data in Table 8.6 constitutes a 3×4 table because, excluding the totals, the table has 3 rows and 4 columns.

TABLE 8.6. Religion and occupations

	Occupation				
Religion	A	B	C	D	Total
Protestant	210	277	254	394	1135
Roman Catholic	102	140	127	279	648
Jewish	36	60	30	17	143
Total	348	477	411	690	1926

We again test whether the populations are the same. In other words, the null hypothesis is that the probability of falling into any occupational group is identical for members of the various religions.

Under this null hypothesis, it makes sense to pool the data from the three religions to obtain estimates of the common probabilities. For example, under the null hypothesis of identical populations, the estimate of the probability that a person is a professional is

$$\hat{p}_1^0 = \frac{348}{1926} = .180685.$$

For skilled workers the estimated probability is

$$\hat{p}_4^0 = \frac{690}{1926} = .358255.$$

Denote the observations as O_{ij} with i identifying a religious group and j indicating occupation. We use a dot to signify summing over a subscript. Thus the total for religious group i is

$$O_{i\bullet} = \sum_j O_{ij},$$

the total for occupational group j is

$$O_{\bullet j} = \sum_i O_{ij},$$

and

$$O_{\bullet\bullet} = \sum_{ij} O_{ij}$$

is the grand total. Recall that the null hypothesis is that the probability of being in an occupation group is the same for each of the three populations. Pooling information over religions, we have

$$\hat{p}_j^0 = \frac{O_{\bullet j}}{O_{\bullet\bullet}}$$

as the estimate of the probability that someone in the study is in occupational group j. *This estimate is only appropriate when the null hypothesis is true.*

The estimated expected count under the null hypothesis for a particular occupation and religion is obtained by multiplying the number of people sampled in that religion by the probability of the occupation. For example, the estimated expected count under the null hypothesis for Jewish professionals is

$$\hat{E}_{31} = 143(.180685) = 25.84.$$

Similarly, the estimated expected count for Roman Catholic skilled workers is

$$\hat{E}_{24} = 648(.358255) = 232.15.$$

In general,

$$\hat{E}_{ij} = O_{i\bullet}\hat{p}_j^0 = O_{i\bullet}\frac{O_{\bullet j}}{O_{\bullet\bullet}} = \frac{O_{i\bullet}O_{\bullet j}}{O_{\bullet\bullet}}.$$

Again, *the estimated expected values are computed assuming that the null hypothesis is true.* The expected values for all occupations and religions are given in Table 8.7.

The estimated expected values are compared to the observations using Pearson residuals. The Pearson residuals are

$$\tilde{r}_{ij} = \frac{O_{ij} - \hat{E}_{ij}}{\sqrt{\hat{E}_{ij}}}.$$

These crude standardized residuals are given in Table 8.8 for all occupations and religions. The largest negative residual is -4.78 for Jewish people with occupation D. This indicates that Jewish

TABLE 8.7. Estimated expected counts ($\hat{E}_{ij}$s)

Religion	A	B	C	D	Total
Protestant	205.08	281.10	242.20	406.62	1135
Roman Catholic	117.08	160.49	138.28	232.15	648
Jewish	25.84	35.42	30.52	51.23	143
Total	348.00	477.00	411.00	690.00	1926

people were substantially underrepresented among skilled workers relative to the other two religious groups. On the other hand, Roman Catholics were substantially overrepresented among skilled workers, with a positive residual of 3.07. The other large residual in the table is 4.13 for Jewish people in group B. Thus Jewish people were more highly represented among owners, managers, and officials than the other religious groups. Only one other residual is even moderately large, the 2.00 indicating a high level of Jewish people in the professions. The main feature of these data seems to be that the Jewish group was different from the other two. A substantial difference appears in every occupational group except clerical and sales.

TABLE 8.8. Residuals ($\tilde{r}_{ij}$s)

Religion	A	B	C	D
Protestant	0.34	−0.24	0.76	−0.63
Roman Catholic	−1.39	−1.62	−0.96	3.07
Jewish	2.00	4.13	−0.09	−4.78

As in Sections 8.3 and 8.4, the sum of the squared Pearson residuals gives Pearson's χ^2 statistic for testing the null hypothesis that the three populations are the same. Pearson's test statistic is

$$X^2 = \sum_{ij} \frac{(O_{ij} - \hat{E}_{ij})^2}{\hat{E}_{ij}}.$$

Summing the squares of the values in Table 8.8 gives

$$X^2 = 60.0.$$

The appropriate number of degrees of freedom for the χ^2 test is the number of data rows in Table 8.6 minus 1 times the number of data columns in Table 8.6 minus 1. Thus the appropriate reference distribution is $\chi^2((3-1)(4-1)) = \chi^2(6)$. The 99.5th percentile of a $\chi^2(6)$ distribution is $\chi^2(.995, 6) = 18.55$ so the observed statistic $X^2 = 60.0$ could not reasonably come from a $\chi^2(6)$ distribution. In particular, the test is significant at the .005 level, clearly indicating that the proportions of people in the different occupation groups differ with religious group.

As in the previous section, we can informally evaluate the null hypothesis by examining the observed proportions for each religious group. The observed proportions are given in Table 8.9. Under the null hypothesis, the observed proportions in each occupation category should be the same for all the religions (up to sampling variability). Figure 8.2 displays the observed proportions graphically. The Jewish group is obviously very different from the other two groups in occupations B and D and is very similar in occupation C. The Jewish proportion seems somewhat different for occupation A. The Protestant and Roman Catholic groups seem similar except that the Protestants are a bit underrepresented in occupation D and therefore are overrepresented in the other three categories. (Remember that the four proportions for each religion must add up to one, so being underrepresented in one category forces an overrepresentation in one or more other categories.)

TABLE 8.9. Observed proportions by religion

Religion	Occupation A	B	C	D	Total
Protestant	.185	.244	.224	.347	1.00
Roman Catholic	.157	.216	.196	.431	1.00
Jewish	.252	.420	.210	.119	1.00

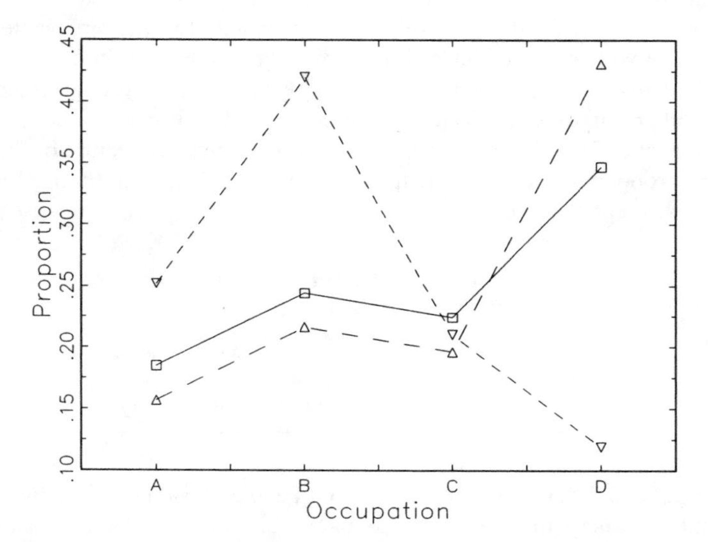

FIGURE 8.2. Occupational proportions by religion: solid – protestant, long dashes – catholic, short dashes – jewish.

8.6 Lancaster–Irwin partitioning

Lancaster–Irwin partitioning is a method for breaking a table of count data into smaller tables. When used to its maximum extent, partitioning is similar in spirit to looking at contrasts in analysis of variance. The basic idea is that a table of counts can be broken into two component tables, a reduced table and a collapsed table. Table 8.10 illustrates such a partition for the data of Table 8.6. In the reduced table, the row for the Jewish group has been eliminated, leaving a subset of the original table. In the collapsed table, the two rows in the reduced table, Protestant and Roman Catholic, have been collapsed into a single row.

In Lancaster–Irwin partitioning, we pick a group of either rows or columns, say rows. The reduced table involves all of the columns but only the chosen subgroup of rows. The collapsed table involves all of the columns and all of the rows *not* in the chosen subgroup, along with a row that combines (collapses) all of the subgroup rows into a single row. In Table 8.10 the chosen subgroup of rows contains the Protestants and Roman Catholics. The reduced table involves all occupational groups but only the Protestants and Roman Catholics. In the collapsed table the occupational groups are unaffected but the Protestants and Roman Catholics are combined into a single row. The other rows remain the same; in this case the other rows consist only of the Jewish row. As alluded to above, rather than picking a group of rows to form the partitioning, we could select a group of columns.

Lancaster–Irwin partitioning is by no means a unique process. There are as many ways to partition a table as there are ways to pick a group of rows or columns. In Table 8.10 we made a particular

TABLE 8.10. A Lancaster–Irwin partition of Table 8.6

Reduced table

Religion	A	B	C	D	Total
Protestant	210	277	254	394	1135
Roman Catholic	102	140	127	279	648
Total	312	417	381	673	1783

Collapsed table

Religion	A	B	C	D	Total
Prot. & R.C.	312	417	381	673	1783
Jewish	36	60	30	17	143
Total	348	477	411	690	1926

selection based on the residual analysis of these data from the previous section. The main feature we discovered in the residual analysis was that the Jewish group seemed to be different from the other two groups. Thus it seemed to be of interest to compare the Jewish group with a combination of the others and then to investigate what differences there might be among the other religious groups. The partitioning of Table 8.10 addresses precisely these questions.

Tables 8.11 and 8.12 provide statistics for the analysis of the reduced table and collapsed table. The reduced table simply reconfirms our previous conclusions. The X^2 value of 12.3 indicates substantial evidence of a difference between Protestants and Roman Catholics. The percentage point $\chi^2(.995, 3) = 12.84$ indicates that the P value for the test is a bit greater than .005. The residuals indicate that the difference was due almost entirely to the fact that Roman Catholics have relatively higher representation among skilled workers. (Or equivalently, that Protestants have relatively lower representation among skilled workers.) Overrepresentation of Roman Catholics among skilled workers forces their underrepresentation among other occupational groups but the level of underrepresentation in the other groups was approximately constant as indicated by the approximately equal residuals for Roman Catholics in the other three occupation groups. We will see later that for Roman Catholics in the other three occupation groups, their distribution among those groups was almost the same as those for Protestants. This reinforces the interpretation that the difference was due almost entirely to the difference in the skilled group.

The conclusions that can be reached from the collapsed table are also similar to those drawn in the previous section. The X^2 value of 47.5 on 3 degrees of freedom indicates overwhelming evidence that the Jewish group was different from the combined Protestant–Roman Catholic group. The residuals can be used to isolate the sources of the differences. The two groups differed in proportions of skilled workers and proportions of owners, managers, and officials. There was a substantial difference in the proportions of professionals. There was almost no difference in the proportion of clerical and sales workers between the Jewish group and the others.

The X^2 value computed for Table 8.6 was 60.0. The X^2 value for the collapsed table is 47.5 and the X^2 value for the reduced table is 12.3. Note that $60.0 \doteq 59.8 = 47.5 + 12.3$. It is not by chance that the sum of the X^2 values for the collapsed and reduced tables is approximately equal to the X^2 value for the original table. In fact, this relationship is a primary reason for using the Lancaster–Irwin partitioning method. The approximate equality $60.0 \doteq 59.8 = 47.5 + 12.3$ indicates that the vast bulk of the differences between the three religious groups is due to the collapsed table, i.e., the difference between the Jewish group and the other two. Roughly 80% $(47.5/60)$ of the original X^2 value is due to the difference between the Jewish group and the others. Of course the X^2 value 12.2 for the reduced table is still large enough to strongly suggest differences between Protestants and Roman Catholics.

TABLE 8.11. Reduced table

Religion	Observations A	B	C	D	Total
Protestant	210	277	254	394	1135
Roman Catholic	102	140	127	279	648
Total	312	417	381	673	1783

Religion	Estimated expected counts A	B	C	D	Total
Protestant	198.61	265.45	242.53	428.41	1135
Roman Catholic	113.39	151.55	138.47	244.59	648
Total	312.00	417.00	381.00	673.00	1783

Religion	Pearson residuals A	B	C	D
Protestant	0.81	0.71	0.74	−1.66
Roman Catholic	−1.07	−0.94	−0.97	2.20

$X^2 = 12.3$, $df = 3$

TABLE 8.12. Collapsed table

Religion	Observations A	B	C	D	Total
Prot. & R.C.	312	417	381	673	1783
Jewish	36	60	30	17	143
Total	348	477	411	690	1926

Religion	Estimated expected counts A	B	C	D	Total
Prot. & R.C.	322.16	441.58	380.48	638.77	1783
Jewish	25.84	35.42	30.52	51.23	143
Total	348.00	477.00	411.00	690.00	1926

Religion	Pearson residuals A	B	C	D
Prot. & R.C.	−0.57	−1.17	0.03	1.35
Jewish	2.00	4.13	−0.09	−4.78

$X^2 = 47.5$, $df = 3$

Not all data will yield an approximation as close as $60.0 \doteq 59.8 = 47.5 + 12.3$ for the partitioning. The fact that we have an approximate equality rather than an exact equality is due to our choice of the test statistic X^2. Pearson's statistic is simple and intuitive; it compares observed values with expected values and standardizes by the size of the expected value. An alternative test statistic also exists called the likelihood ratio test statistic. The motivation behind the likelihood ratio test statistic is not as transparent as that behind Pearson's statistic, so we will not discuss the likelihood ratio test statistic in any detail. However, one advantage of the likelihood ratio test statistic is that the sum of its values for the reduced table and collapsed table gives *exactly* the likelihood ratio test statistic for the original table. For more discussion of the likelihood ratio test statistic, see Christensen (1990b, chapter II).

FURTHER PARTITIONING

We began this section with the 3×4 data of Table 8.6 that has 6 degrees of freedom for its X^2 test. We partitioned the data into two 2×4 tables, each with 3 degrees of freedom. We can continue to use the Lancaster–Irwin method to partition the reduced and collapsed tables given in Table 8.10. The process of partitioning previously partitioned tables can be continued until the original table is broken into a collection of 2×2 tables. Each 2×2 table has one degree of freedom for its chi-squared test, so partitioning provides a way of breaking a large table into one degree of freedom components. This is similar in spirit to looking at contrasts in analysis of variance. Contrasts break the sum of squares for treatments into one degree of freedom components.

What we have been calling the reduced table involves all four occupational groups along with the two religious groups Protestant and Roman Catholic. The table was given in both Table 8.10 and Table 8.11. We now consider this table further. It was discussed earlier that the difference between Protestants and Roman Catholics can be ascribed almost entirely to the difference in the proportion of skilled workers in the two groups. To explore this we choose a new partition based on a group of *columns* that includes all occupations other than the skilled workers. Thus we get the 'reduced' table in Table 8.13 with occupations A, B, and C and the 'collapsed' table in Table 8.14 with occupation D compared to the accumulation of the other three.

TABLE 8.13.

	Observations			
Religion	A	B	C	Total
Protestant	210	277	254	741
Roman Catholic	102	140	127	369
Total	312	417	381	1110

	Estimated expected counts			
Religion	A	B	C	Total
Protestant	208.28	278.38	254.34	741
Roman Catholic	103.72	138.62	126.66	369
Total	312.00	417.00	381.00	1110

	Pearson residuals		
Religion	A	B	C
Protestant	0.12	−0.08	0.00
Roman Catholic	−0.17	0.12	0.03

$X^2 = .065$, $df = 2$

Table 8.13 allows us to examine the proportions of Protestants and Catholics in the occupational groups A, B, and C. We are not investigating whether Catholics were more or less likely than Protestants to enter these occupational groups; we are examining their distribution *within* the groups. The analysis is based only on those individuals *that were in this collection of three occupational groups*. The X^2 value is exceptionally small, only .065. There is no evidence of any difference between Protestants and Catholics for these three occupational groups.

Table 8.13 is a 2×3 table. We could partition it again into two 2×2 tables but there is little point in doing so. We have already established that there is no evidence of differences.

Table 8.14 has the three occupational groups A, B, and C collapsed into a single group. This table allows us to investigate whether Catholics were more or less likely than Protestants to enter this group of three occupations. The X^2 value is a substantial 12.2 on one degree of freedom, so we can tentatively conclude that there was a difference between Protestants and Catholics. From the residuals, we see that *among people in the four occupational groups*, Catholics were more likely

TABLE 8.14.

Religion	Observations A & B & C	D	Total
Protestant	741	394	1135
Roman Catholic	369	279	648
Total	1110	673	1783

Religion	Estimated expected counts A & B & C	D	Total
Protestant	706.59	428.41	1135
Roman Catholic	403.41	244.59	648
Total	1110.00	673.00	1783

Religion	Pearson residuals A & B & C	D
Protestant	1.29	−1.66
Roman Catholic	−1.71	2.20

$X^2 = 12.2, df = 1$

than Protestants to be in the skilled group and less likely to be in the other three.

Table 8.14 is a 2×2 table so no further partitioning is possible. Note again that the X^2 of 12.3 from Table 8.11 is approximately equal to the sum of the .065 from Table 8.13 and the 12.2 from Table 8.14.

Finally, we consider additional partitioning of the collapsed table given in Tables 8.10 and 8.12. It was noticed earlier that the Jewish group seemed to differ from Protestants and Catholics in every occupational group except C, clerical and sales. Thus we choose a partitioning that isolates group C. Table 8.15 gives a collapsed table that compares C to the combination of groups A, B, and D. Table 8.16 gives a reduced table that involves only occupational groups A, B, and D.

TABLE 8.15.

Religion	Observations A & B & D	C	Total
Prot. & R.C.	1402	381	1783
Jewish	113	30	143
Total	1515	411	1926

Religion	Estimated expected counts A & B & D	C	Total
Prot. & R.C.	1402.52	380.48	1783
Jewish	112.48	30.52	143
Total	1515.00	411.00	1926

Religion	Pearson residuals A & B & D	C
Prot. & R.C.	−0.00	0.03
Jewish	0.04	−0.09

$X^2 = .01, df = 1$

Table 8.15 demonstrates no difference between the Jewish group and the combined Protestant–Catholic group. Thus the proportion of people in clerical and sales was the same for the Jewish group as for the combined Protestant and Roman Catholic group. Any differences between the Jewish and Protestant–Catholic groups must be in the proportions of people *within* the three occupational groups A, B, and D.

TABLE 8.16.

	Observations			
Religion	A	B	D	Total
Prot. & R.C.	312	417	673	1402
Jewish	36	60	17	113
Total	348	477	690	1515

	Estimated expected counts			
Religion	A	B	D	Total
Prot. & R.C.	322.04	441.42	638.53	1402
Jewish	25.96	35.58	51.47	113
Total	348.00	477.00	690.00	1515

	Pearson residuals		
Religion	A	B	D
Prot. & R.C.	−0.59	−1.16	1.36
Jewish	1.97	4.09	−4.80

$X^2 = 47.2, df = 2$

Table 8.16 demonstrates major differences between occupations A, B, and D for the Jewish group and the combined Protestant–Catholic group. As seen earlier and reconfirmed here, skilled workers had much lower representation among the Jewish group, while professionals and especially owners, managers, and officials had much higher representation among the Jewish group.

Table 8.16 can be further partitioned into Tables 8.17 and 8.18. Table 8.17 is a reduced 2 × 2 table that considers the difference between the Jewish group and others with respect to occupational groups B and D. Table 8.18 is a 2 × 2 collapsed table that compares occupational group A with the combination of groups B and D.

TABLE 8.17.

	Observations		
Religion	B	D	Total
Prot. & R.C.	417	673	1090
Jewish	60	17	77
Total	477	690	1167

	Estimated expected counts		
Religion	B	D	Total
Prot. & R.C.	445.53	644.47	1090
Jewish	31.47	45.53	77
Total	477.00	690.00	1167

	Pearson residuals	
Religion	B	D
Prot. & R.C.	−1.35	1.12
Jewish	5.08	−4.23

$X^2 = 46.8, df = 1$

Table 8.17 shows a major difference between occupational groups B and D. Table 8.18 may or may not show a difference between group A and the combination of groups B and D. The X^2 values are 46.8 and 5.45 respectively. The question is whether an X^2 value of 5.45 is suggestive of a difference between religious groups when we have examined the data in order to choose the partitions of Table 8.6. Note that the two X^2 values sum to 52.25, whereas the X^2 value for Table 8.16, from which they were constructed, is only 47.2. The approximate equality is a very rough approximation. Nonetheless, we see from the relative sizes of the two X^2 values that the majority of the difference between the Jewish group and the other religious groups was in the proportion of owners, managers, and officials as compared to the proportion of skilled workers.

TABLE 8.18.

	Observations		
Religion	A	B & D	Total
Prot. & R.C.	312	1090	1402
Jewish	36	77	113
Total	348	1167	1515

	Estimated expected counts		
Religion	A	B & D	Total
Prot. & R.C.	322.04	1079.96	1402
Jewish	25.96	87.04	113
Total	348.00	1167.00	1515

	Pearson residuals	
Religion	A	B & D
Prot. & R.C.	-0.56	0.30
Jewish	1.97	-1.08

$X^2 = 5.45$, $df = 1$

Ultimately, we have partitioned Table 8.6 into Tables 8.13, 8.14, 8.15, 8.17, and 8.18. These are all 2×2 tables except for Table 8.13. We could also have partitioned Table 8.13 into two 2×2 tables but we chose to leave it because it showed so little evidence of any difference between Protestants and Roman Catholics for the three occupational groups considered. The X^2 value of 60.0 for Table 8.6 was approximately partitioned into X^2 values of .065, 12.2, .01, 46.8, and 5.45 respectively. Except for the .065 from Table 8.13, each of these values is computed from a 2×2 table, so each has 1 degree of freedom. The .065 is computed from a 2×3 table, so it has 2 degrees of freedom. The sum of the five X^2 values is 64.5 which is roughly equal to the 60.0 from Table 8.6.

The five X^2 values can all be used in testing. Not only does such testing involve the usual problems associated with multiple testing but we even let the data suggest the partitions. It is inappropriate to compare these X^2 values to their usual χ^2 percentage points to obtain tests. A simple way to adjust for both the multiple testing and the data dredging (letting the data suggest partitions) is to compare all X^2 values to the percentage points appropriate for Table 8.6. For example, the $\alpha = .05$ test for Table 8.6 uses the critical value $\chi^2(.95, 6) = 12.58$. By this standard, Table 8.17 with $X^2 = 46.8$ shows a significant difference between religious groups and Table 8.14 with $X^2 = 12.2$ nearly shows a significant difference between religious groups. The value of $X^2 = 5.45$ for Table 8.18 gives no evidence of a difference based on this criterion even though such a value would be highly suggestive if we could compare it to a $\chi^2(1)$ distribution. This method is similar in spirit to Scheffé's method from Section 6.4 and suffers from the same extreme conservatism.

8.7 Logistic regression

Logistic regression is a method of modeling the relationships between probabilities and predictor variables. We begin with an example.

EXAMPLE 8.7.1. Woodward et al. (1941) reported data on 120 mice divided into 12 groups of 10. The mice in each group were exposed to a specific dose of chloracetic acid and the observations consist of the number in each group that lived and died. Doses were measured in grams of acid per kilogram of body weight. The data are given in Table 8.19, along with the proportions of mice who died at each dose. We could analyze these data using the methods discussed earlier in this chapter; we have samples from twelve populations and we could test to see if the populations are the same. In addition though, we can try to model the relationship between dose level and the probability of dying. If we can model the probability of dying as a function of dose, we can make predictions about the probability of dying for any dose levels that are similar to those in the original data. □

TABLE 8.19. Lethality of chloracetic acid

Dose	Group	Died	Survived	Total	$\hat{p}_i$
.0794	1	1	9	10	.1
.1000	2	2	8	10	.2
.1259	3	1	9	10	.1
.1413	4	0	10	10	.0
.1500	5	1	9	10	.1
.1588	6	2	8	10	.2
.1778	7	4	6	10	.4
.1995	8	6	4	10	.6
.2239	9	4	6	10	.4
.2512	10	5	5	10	.5
.2818	11	5	5	10	.5
.3162	12	8	2	10	.8

Logistic regression as applied to this example is somewhat like fitting a simple linear regression to one-way ANOVA data as discussed in Section 7.12. In Section 7.12 we considered data on the ASI indices given in Table 7.10. These data have seven observations on each of five plate lengths. The data can be analyzed as either a one-way ANOVA or as a simple linear regression, and in Section 7.12 we examined relationships between the two approaches. In particular, we mentioned that the estimated regression line could be obtained by fitting a line to the sample means for the five groups. The analysis of the lethality data takes a similar approach. Instead of fitting a line to sample means, we perform a regression on the observed proportions. Unfortunately, a standard regression is inappropriate because the observed proportions do not have constant variance. For $i = 1, \ldots, q$, $\hat{p}_i$ is the observed proportion from N_i binomial trials, so as discussed in Section 8.1, $\text{Var}(\hat{p}_i) = p_i(1-p_i)/N_i$. One approach is to use the variance stabilizing transformation from Sections 2.3 and 7.10 on the $\hat{p}_i$s and then apply standard regression methods. As alluded to in Section 2.3, there are better methods available and this section briefly introduces some of them.

We begin with a reasonably simple analysis of the chloracetic acid data. This analysis involves not only a transformation of the $\hat{p}_i$s but incorporating weights into the simple linear regression procedure. Weighted regression is a method for dealing with nonconstant variances in the observations. If the variances are not constant, *observations with large variances should be given relatively little weight, while observations with small variances are given increased weight.* The details of weighted regression are discussed in Section 15.7. The discussion given there requires one to know the material in Chapter 13 and the first five sections of Chapter 15, but considerable insight can be obtained from Examples 15.7.1 and 15.7.2. These examples merely require the background from Section 7.12.

In weighted regression for binomial data we take the observations on the dependent variable as

$$\log[\hat{p}_i/(1-\hat{p}_i)].$$

We then fit the model

$$\log[\hat{p}_i/(1-\hat{p}_i)] = \beta_0 + \beta_1 x_i + \varepsilon_i$$

with weights

$$w_i = N_i \hat{p}_i(1-\hat{p}_i).$$

The regression estimates from this method minimize the weighted sum of squares

$$\sum_{i=1}^{q} w_i \left(\log[\hat{p}_i/(1-\hat{p}_i)] - \beta_0 - \beta_1 x_i\right)^2 .$$

There are a couple of serious drawbacks to this procedure. First, *the weights are really only appropriate if all the samples sizes N_i are large*. The weights rely on large sample variance formulae and the law of large numbers. Second, *the values* $\log[\hat{p}_i/(1-\hat{p}_i)]$ *are not always defined.* If we have an observed proportion with $\hat{p}_i = 0$ or 1, $\log[\hat{p}_i/(1-\hat{p}_i)]$ is undefined. Either we are trying to take the log of zero or we are trying to divide by zero. With $\hat{p}_i = y_i/N_i$, so that y_i is the number of 'successes,' this problem occurs whenever y_i equals 0 or N_i. The problem is often dealt with by adding or subtracting a small number to y_i. Generally, *the size of the small number should be chosen to be small in relation to the size of N_i*. In many applications, *all* of the N_is are 1. *In any case with $N_i = 1$, $\hat{p}_i$ is always either 0 or 1, so* $\log[\hat{p}_i/(1-\hat{p}_i)]$ *is always undefined.* These drawbacks are not as severe with another method of analysis that we will examine later.

EXAMPLE 8.7.1 CONTINUED. We now return to the chloracetic acid data. In this example $N_i = 10$ for all i, so the sample sizes are all reasonably large. For dose $x = .1413$, the number of deaths was 0, so the observed proportion was zero. We handle this problem by treating the observed count as .5, so the observed proportion is taken as .5/10 = .05. A computer program for regression analysis will typically give output such as the following tables.

Raw parameter table

Predictor	$\hat{\beta}_k$	$SE(\hat{\beta}_k)$	t	P
Constant	−3.1886	0.5914	−5.39	0.000
Dose	13.181	2.779	4.74	0.000

Analysis of variance: weighted simple linear regression

Source	df	SS	MS	F	P
Regression	1	15.282	15.282	22.50	0.000
Error	10	6.791	0.679		
Total	11	22.074			

The estimates of the regression parameters are appropriate but everything involving variances in these tables is wrong! The problem is that with binomial data the variance depends solely on the probability and we have already accounted for the variance in defining the weights. Thus there is no separate parameter σ^2 to deal with but standard regression output is designed to adjust for such a parameter. To obtain appropriate standard errors, we need to divide the reported standard errors by $\sqrt{MSE}$. The adjusted table is given below.

Adjusted parameter table

Predictor	$\hat{\beta}_k$	$SE(\hat{\beta}_k)$	t	P
Constant	−3.1886	0.7177	−4.44	0.000
Dose	13.181	3.373	3.91	0.000

The table provides clear evidence of the need for both parameters. To predict the probability of death for rats given a dose x, the predicted probability $\hat{p}$ satisfies

$$\log[\hat{p}/(1-\hat{p})] = \hat{\beta}_0 + \hat{\beta}_1 x = -3.1886 + 13.181x.$$

Solving for $\hat{p}$ gives

$$\hat{p} = \frac{\exp(\hat{\beta}_0 + \hat{\beta}_1 x)}{1+\exp(\hat{\beta}_0 + \hat{\beta}_1 x)} = \frac{\exp(-3.1886 + 13.181x)}{1+\exp(-3.1886+13.181x)}.$$

For example, if $x = .3$, $-3.1886 + 13.181(.3) = .7657$ and $\hat{p} = e^{.7657}/(1+e^{.7657}) = .68$.

The only interest in the ANOVA table is in the error line. As we have seen, $\sqrt{MSE}$ is needed to adjust the standard errors. In addition, the SSE provides a lack of fit test similar in spirit to that discussed in Section 7.12. To test for lack of fit compare SSE to a $\chi^2(dfE)$ distribution. Large values of SSE indicate lack of fit. In this example $SSE = 6.791$, which is smaller than $dfE = 10$, so the χ^2 test gives no evidence of lack of fit. A line seems to fit these data adequately. □

MINITAB COMMANDS

In Minitab let c1 contain the doses, c2 contain the number of deaths, and c3 contain the number of trials (10 in each case). The commands for this analysis are given below.

```
MTB > let c5=c2/c3
MTB > let c6=1-c5
MTB > let c7=loge(c5/c6)
MTB > let c8=c3*c5*c6
MTB > regress c7 on 1 c1;
SUBC> weights c8.
```

THE LOGISTIC MODEL AND MAXIMUM LIKELIHOOD

When we have a one-way ANOVA with treatments that are quantitative levels of some factor, we can fit either the one-way ANOVA model

$$y_{ij} = \mu_i + \varepsilon_{ij}$$

or the simple linear regression model

$$y_{ij} = \beta_0 + \beta_1 x_i + \varepsilon_{ij}.$$

We can think of the regression as a model for the μ_is, i.e.,

$$\mu_i = \beta_0 + \beta_1 x_i.$$

Logistic regression uses a very similar idea. The binomial situation here has 'observations' $\hat{p}_i = y_i/N_i$ where $y_i \sim Bin(N_i, p_i)$, $i = 1, \ldots, q$. In logistic regression, we model the parameters p_i. In particular, the model is

$$\log[p_i/(1-p_i)] = \beta_0 + \beta_1 x_i. \quad (8.7.1)$$

The question is then how to fit this model. The weighted regression approach was discussed earlier. The weighted regression estimates are the values of β_0 and β_1 that minimize the function

$$\sum_{i=1}^{q} w_i \left(\log[\hat{p}_i / (1 - \hat{p}_i)] - \beta_0 - \beta_1 x_i\right)^2 .$$

An alternative method for estimating the parameters is to maximize something called the likelihood function.

Recall from Section 1.4 that the probability function for an individual binomial, say, $y_i \sim Bin(N_i, p_i)$ is

$$\Pr(y_i = r_i) = \binom{N_i}{r_i} p_i^{r_i} (1 - p_i)^{N_i - r_i}.$$

We are dealing with q *independent* binomials, so probabilities for the entire collection of random variables are obtained by multiplying the probabilities for the individual events.

One of the things that students initially find confusing about statistical theory is that we often use the same symbols for random variables and for observations from those random variables. I am about to do the same thing. I want to write down the probability of the data that we actually saw. If we saw y_i, the probability of seeing that is

$$\binom{N_i}{y_i} p_i^{y_i} (1 - p_i)^{N_i - y_i}.$$

If all together we saw $y_1, \ldots, y_q$, the probability of obtaining all those values is the product of the individual probabilities, i.e.,

$$\prod_{i=1}^{q} \binom{N_i}{y_i} p_i^{y_i} (1 - p_i)^{N_i - y_i}. \tag{8.7.2}$$

This probability of getting the observed data is called the *likelihood function*. In the likelihood function we know all of the N_is and y_is but we do not know the p_is. Thus the likelihood is a function of the p_is. It is not too difficult to show that the maximum value of the likelihood function is obtained by taking $p_i = \hat{p}_i = y_i / N_i$ for all i. The observed proportions $\hat{p}_i$ are the values of the parameters that maximize the probability of getting the observed data. We say that such values are *maximum likelihood estimates (mles)* of the parameters p_i.

The model (8.7.1) specifies the p_is in terms of β_0 and β_1. We can solve (8.7.1) for p_i by writing

$$p_i = \frac{\exp(\beta_0 + \beta_1 x_i)}{1 + \exp(\beta_0 + \beta_1 x_i)}. \tag{8.7.3}$$

If we now substitute this formula for p_i into equation (8.7.2) we get the likelihood as a function of β_0 and β_1. The maximum likelihood estimates of β_0 and β_1 are simply the values of β_0 and β_1 that maximize the likelihood function. Equations (8.7.1) and (8.7.3) are equivalent ways of writing the model. Equation (8.7.3) is actually the logistic regression model and equation (8.7.1) is the corresponding *logit* model.

Computer programs are available for finding maximum likelihood estimates. Such programs typically give standard errors that are valid for large samples. If the large sample approximations are appropriate, the parameters, estimates, and standard errors can be used as in Chapter 3 with a $N(0, 1)$ reference distribution. For the approximations to be valid, it is typically enough that the total number of trials in the entire data be large; the individual sample sizes N_i need not be large.

Maximum likelihood theory also provides a test of lack of fit similar to the weighted regression χ^2 test using the *SSE*. In maximum likelihood theory the test examines the value of the likelihood (8.7.2) when using the mles of β_0 and β_1 in equation (8.7.3) to determine the p_is, and compares that value to the likelihood when using the observed proportions $\hat{p}_i$ as the p_is. Using the observed proportions involves less structure so the likelihood value will be greater using them. The lack of fit test statistic is -2 times the log of the ratio of the likelihood using the estimated β_ks to the likelihood using the $\hat{p}_i$s. This test statistic is properly called the (generalized) likelihood ratio test statistic but is often simply called the *deviance*. (The likelihood ratio test was also mentioned in the previous section.) The deviance is compared to a $\chi^2(q-2)$ distribution where q is the number of independent binomials and 2 is the number of regression parameters in the logistic model. Unlike the standard errors for the β_is, *all the sample sizes N_i must be large for the lack of fit test to be valid!*

EXAMPLE 8.7.2. Maximum likelihood for the chloracetic acid data gives the following results.

Predictor	$\hat{\beta}_k$	$SE(\hat{\beta}_k)$	t	P
Constant	-3.570	0.7040	-5.07	0.000
Dose	14.64	3.326	4.40	0.000

These are similar to the weighted regression results. The deviance of the maximum likelihood fit is 10.254 with $12 - 2 = 10$ degrees of freedom for the lack of fit test. The sample sizes are all reasonably large, so a χ^2 test is appropriate. The test statistic is approximately equal to the degrees of freedom, so a test would not be rejected. A simple line seems to fit the data adequately. The maximum likelihood results were obtained using the computer program GLIM. □

We will not analyze more sophisticated count data in this book but we should mention that *both the maximum likelihood methods and the weighted regression methods extend to much more general models*, such as those treated in the remainder of the book. Both methods work when there are many predictors, so we can perform multiple logistic regression which is similar in spirit to multiple regression as treated in Chapters 13, 14, and 15. By modifying the matrix approach to ANOVA problems discussed in Section 16.5, the methods introduced here can be applied to models that are structured like analysis of variance and even analysis of covariance. Christensen (1990b) contains a more complete discussion of logistic regression and logit models. It also contains references to additional work.

8.8 Exercises

EXERCISE 8.8.1. Reiss et al. (1975) and Fienberg (1980) reported that 29 of 52 virgin female undergraduate university students who used a birth control clinic thought that extramarital sex is not always wrong. Give a 99% confidence interval for the population proportion of virgin undergraduate university females who use a birth control clinic and think that extramarital sex is not always wrong.

In addition, 67 of 90 virgin females who did not use the clinic thought that extramarital sex is not always wrong. Give a 99% confidence interval for the difference in proportions between the two groups and give a .05 level test that there is no difference.

EXERCISE 8.8.2. Pauling (1971) reports data on the incidence of colds among French skiers who where given either ascorbic acid or a placebo. Of 139 people given ascorbic acid, 17 developed colds. Of 140 people given the placebo, 31 developed colds. Do these data suggest that the proportion of people who get colds differs depending on whether they are given ascorbic acid?

EXERCISE 8.8.3. Quetelet (1842) and Stigler (1986, p. 175) report data on conviction rates in the French Courts of Assize (Law Courts) from 1825 to 1830. The data are given in Table 8.20. Test whether the conviction rate is the same for each year. Use $\alpha = .05$. (Hint: Table 8.20 is written in a

TABLE 8.20. French convictions

Year	Convictions	Accusations
1825	4594	7234
1826	4348	6988
1827	4236	6929
1828	4551	7396
1829	4475	7373
1830	4130	6962

nonstandard form. You need to modify it before applying the methods of this chapter.) If there are differences in conviction rates, use residuals to explore these differences.

EXERCISE 8.8.4. Use the data in Table 8.2 to test whether the probability of a birth in each month is the number of days in the month divided by 365. Thus the null probability for January is $31/365$ and the null probability for February is $28/365$.

EXERCISE 8.8.5. Snedecor and Cochran (1967) report data from an unpublished report by E. W. Lindstrom. The data concern the results of cross-breeding two types of corn (maize). In 1301 crosses of two types of plants, 773 green, 231 golden, 238 green-golden, and 59 golden-green-striped plants were obtained. If the inheritance of these properties is particularly simple, Mendelian genetics suggests that the probabilities for the four types of corn may be $9/16$, $3/16$, $3/16$, and $1/16$, respectively. Test whether these probabilities are appropriate. If they are inappropriate, identify the problem.

EXERCISE 8.8.6. In France in 1827, 6929 people were accused in the courts of assize and 4236 were convicted. In 1828, 7396 people were accused and 4551 were convicted. Give a 95% confidence interval for the proportion of people convicted in 1827. At the .01 level, test the null hypothesis that the conviction rate in 1827 was greater than or equal to $2/3$. Does the result of the test depend on the choice of standard error? Give a 95% confidence interval for the difference in conviction rates between the two years. Test the hypothesis of no difference in conviction rates using $\alpha = .05$ and both standard errors.

EXERCISE 8.8.7. Table 8.21 contains additional data from Lazerwitz (1961). These consist of a breakdown of the Protestants in Table 8.6 but with the addition of four more occupational categories. The additional categories are E, semiskilled; F, unskilled; G, farmers; H, no occupation. Analyze the data with an emphasis on partitioning the table.

TABLE 8.21. Occupation and religion

Religion	A	B	C	D	E	F	G	H
White Baptist	43	78	64	135	135	57	86	114
Black Baptist	9	2	9	23	47	77	18	41
Methodist	73	80	80	117	102	58	66	153
Lutheran	23	36	43	59	46	26	49	46
Presbyterian	35	54	38	46	19	22	11	46
Episcopalian	27	27	20	14	7	5	2	15

EXERCISE 8.8.8. Stigler (1986, p. 208) reports data from the *Edinburgh Medical and Surgical Journal* (1817) on the relationship between heights and chest circumferences for Scottish militia

men. Measurements were made in inches. We concern ourselves with two groups of men, those with 39 inch chests and those with 40 inch chests. The data are given in Table 8.22. Test whether the distribution of heights is the same for these two groups.

TABLE 8.22. Heights and chest circumferences

	Heights					
Chest	64–65	66–67	68–69	70–71	71–73	Total
39	142	442	341	117	20	1062
40	118	337	436	153	38	1082
Total	260	779	777	270	58	2144

EXERCISE 8.8.9. Use weighted least squares to fit a logistic model to the data of Table 8.20 that relates probability of conviction to year. Is there evidence of a trend in the conviction rates over time? Is there evidence for a lack of fit?

EXERCISE 8.8.10. Is it reasonable to fit a logistic regression to the data of Table 8.22? Why or why not? Explain what such a model would be doing. Whether reasonable or not, fitting such a model can be done. Use weighted least squares to fit a logistic model and discuss the results. Is there evidence for a lack of fit?

Chapter 9

Basic experimental designs

In this chapter we examine the three most basic experimental designs: completely randomized designs (CRDs), randomized complete block (RCB) designs, and Latin square designs. Completely randomized designs are the simplest of these and have been used previously without having been named. Also, we have previously performed an analysis for a randomized complete block design.

The basic object of experimental design is to construct an experiment that allows for a valid estimate of σ^2, the variance of the observations. Obtaining a valid *estimate of error* requires appropriate replication of the experiment. Having one observation on each treatment is not sufficient. All three of the basic designs considered in this chapter allow for a valid estimate of the variance.

A second important consideration is to construct a design that yields a small variance. A smaller variance leads to sharper statistical inferences, i.e., narrower confidence intervals and more powerful tests. A fundamental tool for reducing variability is *blocking*. The basic idea is to examine the treatments on homogeneous experimental material. With four drug treatments and observations on eight animals, a valid estimate of the error can be obtained by randomly assigning each of the drugs to two animals. *If the treatments are assigned completely at random to the experimental units* (animals), *the design is a completely randomized design*. Generally, a smaller variance for treatment comparisons is obtained when the eight animals consist of two litters of four siblings and each treatment is applied to one randomly selected animal from each litter. With each treatment applied in every litter, all comparisons among treatments can be performed *within* each litter. Having at least two litters is necessary to get a valid estimate of the variance of the comparisons. *Randomized complete block designs: 1) identify blocks of homogeneous experimental material (units) and 2) randomly assign each treatment to an experimental unit within each block.* The blocks are complete in the sense that each block contains all of the treatments.

Latin squares use two forms of blocking at once. For example, if we suspect that birth order within the litter might also have an important effect on our results, we continue to take observations on each treatment within every litter, but we also want to have each treatment observed in every birth order. This is accomplished by having four litters with treatments arranged in a Latin square design.

Another fundamental concept in experimental design is the idea that the experimenter has the ability to randomly assign the treatments to the experimental material available. This is not always the case.

EXAMPLE 9.0.1. In Chapter 5, we considered two examples of one-way analysis of variance; neither were designed experiments. For the suicide ages, a designed experiment would require that we take a group of people who we know will commit suicide and randomly assign one of the ethnic groups, non-Hispanic Caucasian, Hispanic, or Native American, to the people. Obviously a difficult task. With the electrical characteristic data, rather than having ceramic sheets divided into strips, a designed experiment would require starting with different pieces of ceramic material and randomly assigning the pieces to have come from a particular ceramic strip. □

Random assignment of treatments to experimental units allows one to infer causation from a designed experiment. If treatments are *randomly* assigned to experimental units, then the only systematic differences between the units are the treatments. Barring an unfortuitous randomization, such differences must be caused by the treatments because they cannot be caused by anything else. However, as discussed below, the 'treatments' may be more involved than the experimenter realizes.

Random assignment of treatments does not mean haphazard assignment. Haphazard assignment is subject to the (unconscious) biases of the person making the assignments. Random assignment uses a reliable table of random numbers or a reliable computer program to generate random numbers. It then uses these numbers to assign treatments. For example, suppose we have four experimental units labeled u_1, u_2, u_3, and u_4 and four treatments labeled A, B, C, and D. Given a program or table that provides random numbers between 0 and 1 (i.e., random samples from a uniform(0,1) distribution), we associate numbers between 0 and .25 with treatment A, numbers between .25 and .50 with treatment B, numbers between .50 and .75 with treatment C, and numbers between .75 and 1 with treatment D. The first random number selected determines the treatment for u_1. If the first number is .6321, treatment C is assigned to u_1 because $.50 < .6321 < .75$. If the second random number is .4279, u_2 gets treatment B because $.25 < .4279 < .50$. If the third random number is .2714, u_3 would get treatment B, but we have already assigned treatment B to u_2, so we throw out the third number. If the fourth number is .9153, u_3 is assigned treatment D. Only one unit and one treatment are left, so u_4 gets treatment A. Any reasonable rule (decided ahead of time) can be used to make the assignment if a random number hits a boundary, e.g., if a random number comes up, say, .2500.

In cases such as those discussed previously, i.e., in *observational studies* where the treatments are not randomly assigned to experimental units, it is much more difficult to infer causation. If we find differences, there are differences in the corresponding populations, but it does not follow that the differences are caused by the labels given to the populations. If the average suicide age is lower for Native Americans, we know only that the phenomenon exists, we do not know what aspects of being Native American cause the phenomenon. Perhaps low socioeconomic status causes early suicides and Native Americans are over represented in the low socioeconomic strata. We don't know and it will be difficult to ever know using the only possible data on such matters, data that come from observational studies.

One also needs to realize that the treatments in an experiment may not be what the experimenter thinks they are. Suppose you want to test whether artificial sweeteners made with a new chemical cause cancer. You get some rats, randomly divide them into a treatment group and a control. You inject the treatment rats with a solution of the sweetener combined with another (supposedly benign) chemical. You leave the control rats alone. For simplicity you keep the treatment rats in one cage and the control rats in another cage. Eventually, you find an increased risk of cancer among the treatment rats as compared to the control rats. You can reasonably conclude that the treatments caused the increased cancer rate. Unfortunately, you do not really know whether the sweetener or the supposedly benign chemical or the combination of the two caused the cancer. In fact, you do not really know that it was the chemicals that caused the cancer. Perhaps the process of injecting the rats caused the cancer or perhaps something about the environment in the treatment rats's cage caused the cancer. A treatment consists of *all the ways* in which a group is treated differently from other groups. It is crucially important to *treat all experimental units as similarly as possible so that (as nearly as possible) the only differences between the units are the agents that were meant to be investigated.*

Ideas of blocking can also be useful in observational studies. While one cannot really create blocks in observational studies, one can adjust for important groupings.

EXAMPLE 9.0.2. If we wish to study whether cocaine users are more paranoid than other people, we may decide that it is important to block on socioeconomic status. This is appropriate if the underlying level of paranoia in the population differs by socioeconomic status. Conducting an experiment in this setting is difficult. Given groups of people of various socioeconomic statuses, it is a rare researcher who has the luxury of deciding which subjects will ingest cocaine and which will not. □

The seminal work on experimental design was written by Fisher (1935). It is still well worth reading. My favorite source on the ideas of experimentation is Cox (1958). The books by Cochran and Cox (1957) and Kempthorne (1952) are classics. Cochran and Cox is more applied. Kempthorne is more theoretical. There is a huge literature in both journal articles and books on the general subject of designing experiments. The article by Coleman and Montgomery (1993) is interesting in that it tries to formalize many aspects of planning experiments that are often poorly specified.

9.1 Completely randomized designs

In a completely randomized design, a group of experimental units are available and the experimenter randomly assigns treatments to the experimental units. The data consist of a group of observations on each treatment. These groups of observations are subjected to a one-way analysis of variance.

EXAMPLE 9.1.1. In Example 6.0.1, we considered data from Mandel (1972) on the elasticity measurements of natural rubber made by 13 laboratories. While Mandel did not discuss how the data were obtained, it could well have the result of a completely randomized design. For a CRD, we would need 52 pieces of the type of rubber involved. These should be randomly divided into 13 groups (using a table of random numbers or random numbers generated by a reliable computer program). The first group of samples is then sent to the first lab, the second group to the second lab, etc. For a CRD, it is important that a sample is not sent to a lab because the sample somehow seems appropriate for that particular lab.

Personally, I would also be inclined to send the four samples to a given lab at different times. If the four samples are sent at the same time, they might be analyzed by the same person, on the same machines, at the same time. Samples sent at different times might be treated differently. If samples are treated differently at different times, this additional source of variation should be included in any predictive conclusions we wish to make about the labs.

When samples sent at different times are treated differently, sending a batch of four samples at the same time constitutes *subsampling*. There are two sources of variation to deal with: variation from time to time and variation within a given time. The values from four samples at a given time help reduce the effect on treatment comparisons due to variability at a given time, but samples analyzed at different times are still *required* to obtain a valid estimate of the error. In fact, with subsampling, a perfectly valid analysis can be based on the means of the four subsamples. In our example, such an analysis gives only one 'observation' at each time, so the need for sending samples at more than one time is obvious. If the four samples were sent at the same time, there would be no replication, hence no estimate of error. Subsection 12.4.1 and Christensen (1987, section XI.4) discuss subsampling in more detail. □

9.2 Randomized complete block designs

In a randomized complete block design the experimenter obtains (constructs) blocks of homogeneous material that contain as many experimental units as there are treatments. The experimenter then randomly assigns a different treatment to each of the units in the block. The random assignments are performed independently for each block. The advantage of this procedure is that treatment comparisons are subject only to the variability within the blocks. Block to block variation is eliminated in the analysis. In a completely randomized design applied to the same experimental material, the treatment comparisons would be subject to both the within block and the between block variability.

The key to a good blocking design is in obtaining blocks that have little within block variability. Often this requires that the blocks be relatively small. A difficulty with RCB designs is that the blocks must be large enough to allow all the treatments to be applied within each block. This can be a serious problem if there is a substantial number of treatments or if maintaining homogeneity within blocks requires the blocks to be very small. If the treatments cannot all be fitted into each block, we need some sort of *incomplete block* design. Such designs will be considered in Chapters 16 and 17.

The analysis of a randomized complete block design is a two-way ANOVA without replication or interaction. The analysis is illustrated below and discussed in general in the following subsection.

EXAMPLE 9.2.1. Inman, Ledolter, Lenth, and Niemi (1992) studied the performance of an optical emission spectrometer. Table 9.1 gives some of their data on the percentage of manganese (Mn) in a sample. The data were collected using a sharp counterelectrode tip with the sample to be analyzed partially covered by a boron nitride disk. Data were collected under three temperature conditions. Upon fixing a temperature, the sample percentage of Mn was measured using 1) a new boron nitride disk with light passing through a clean window (new-clean), 2) a new boron nitride disk with light passing through a soiled window (new-soiled), 3) a used boron nitride disk with light passing through a clean window (used-clean), and 4) a used boron nitride disk with light passing through a soiled window (used-soiled). The four conditions, new-clean, new-soiled, used-clean, used-soiled are the treatments. The temperature was then changed and data were again collected for each of the four treatments. A block is always made up of experimental units that are homogeneous. The temperature conditions were held constant while observations were taken on the four treatments so the temperature levels identify blocks.

TABLE 9.1. Spectrometer data

		Block		Trt.
Treatment	1	2	3	means
New-clean	0.9331	0.8664	0.8711	$0.8902\bar{0}$
New-soiled	0.9214	0.8729	0.8627	$0.8856\bar{6}$
Used-clean	0.8472	0.7948	0.7810	$0.8076\bar{6}$
Used-soiled	0.8417	0.8035	0.8099	$0.8183\bar{6}$
Block means	0.885850	0.834400	0.831175	0.850475

In analyzing a one-way ANOVA, the analysis of variance table is of little direct importance. For a randomized complete block design the analysis of variance table is crucial. Before we can proceed with any analysis of the treatments, we need an estimate of the variance σ^2. In one-way ANOVA, the *MSE* is simply a pooled estimate obtained from group sample variances. In a RCB design, the replications of the experiment occur in different blocks and the effect of these blocks must be taken into account. In particular, the three observations on each treatment do not form a random sample from a population, so it is inappropriate to compute the sample variance within each treatment group and it is totally inappropriate to pool such variance estimates. Instead, we expand the analysis of variance table by accounting for both treatments and blocks and estimate σ^2 with the leftover sum of squares.

The sum of squares total (corrected for the grand mean) is computed just as in one-way ANOVA; it is the sample variance of all 12 observations multiplied by the degrees of freedom $12 - 1$, e.g.,

$$SSTot = (12 - 1)s_y^2 = (11).002277797 = .025055762.$$

The mean square and sum of squares for treatments are also computed as in one-way ANOVA. Using Table 9.1, the treatment means are averages of 3 observations and the sample variance of the treatment means is .001893343, so

$$MSTrts = 3(.001893343) = .005680028\,.$$

There are 4 treatments, so the sum of squares is the mean square multiplied by the degrees of freedom $(4 - 1)$,

$$SSTrts = (4 - 1).005680028 = .017040083\,.$$

The mean square and sum of squares for blocks are also computed as if they were treatments in a one-way ANOVA. The block means are averages of 4 observations and the sample variance of the block means is .000941143, so

$$MSBlocks = 4(.000941143) = .003764572\,.$$

There are 3 blocks, so the sum of squares is the mean square times the degrees of freedom $(3 - 1)$,

$$SSBlocks = (3 - 1).003764572 = .007529145\,.$$

The sum of squares error is obtained by subtraction,

$$\begin{aligned} SSE &= SSTot - SSTrts - SSBlocks \\ &= .025055762 - .017040083 - .007529145 \\ &= .000486534\,. \end{aligned}$$

Similarly,

$$\begin{aligned} dfE &= dfTot - dfTrts - dfBlocks \\ &= 11 - 3 - 2 \\ &= 6\,. \end{aligned}$$

The estimate of σ^2 is

$$MSE = \frac{SSE}{dfE} = \frac{.000486534}{6} = .000081089\,.$$

Given the definitions of SSE and dfE, it is tautological that the sums of squares for treatments, blocks, and error add up to the sum of squares total, and similarly for the degrees of freedom.

All of the calculations are summarized in the analysis of variance table, Table 9.2. Table 9.2 also gives the analysis of variance F test for the null hypothesis that the effects are the same for each treatment. By definition the F statistic is $MSTrts/MSE$ and in this example it is huge, 70.05. The P value is infinitesimal, so there is clear evidence that the 4 treatments do not behave the same.

Table 9.2 also contains an F test for blocks. In a true blocking experiment, there is not much interest in testing whether block means are different. After all, one *chooses* the blocks so that they have different means. Nonetheless, the F statistic $MSBlks/MSE$ is of some interest because it indicates how effective the blocking was, i.e., it indicates how much the variability was reduced

TABLE 9.2. Analysis of variance for spectrometer data

Source	*df*	*SS*	*MS*	*F*	*P*
Trts	3	0.0170401	0.0056800	70.05	0.000
Blocks	2	0.0075291	0.0037646	46.43	0.000
Error	6	0.0004865	0.0000811		
Total	11	0.0250558			

by blocking. For this example, *MSBlks* is 46 times larger than *MSE*, indicating that blocking was definitely worthwhile. In the model for RCB designs presented in the following subsection, there is no reason not to test for blocks, but some models used for RCBs do not allow a test for blocks. Regardless of the particular model, the analysis of treatments works in the same way.

Now that we have an estimate of the variance, we can proceed with the more interesting questions about how the treatments differ. We begin by examining pairwise differences. The multiple comparison methods of Chapter 6 all apply in the usual way after adjusting for the difference in dfE. For example, there are 4 treatment means, each based on 3 observations, and the *MSE* is 0.0000811 with 6 degrees of freedom, so for $\alpha = .05$ the honest significant difference is

$$HSD = Q(.95, 4, 6)\sqrt{0.0000811/3} = 4.90(.00519936) = .02548\,.$$

The differences between the treatments are illustrated below.

Treatment	Used-clean	Used-soiled	New-soiled	New-clean
Mean	$0.8076\bar{6}$	$0.8183\bar{6}$	$0.8856\bar{6}$	$0.8902\bar{0}$
	———	———		
			———	———

We have no evidence of an effect due to the condition of the window when considering used boron nitride disks. The higher yields occur for soiled windows, but they are not significantly different. We also have no evidence of an effect due to the condition of the window for new boron nitride disks. The higher yields occur for clean windows but again the difference is not significant. Evidence does exist that the two means for used disks are different (less than) the two means for new disks.

The structure of the treatments suggests particular orthogonal contrasts that are of interest. Contrast coefficients, estimated contrasts, and sums of squares for the contrasts are given in Table 9.3.

TABLE 9.3. Contrasts for the spectrometer data

	Contrast labels			Trt.
Treatment	D	W	DW	means
New-clean	1	1	1	$0.8902\bar{0}$
New-soiled	1	-1	-1	$0.8856\bar{6}$
Used-clean	-1	1	-1	$0.8076\bar{6}$
Used-soiled	-1	-1	1	$0.8183\bar{6}$
Est	.149833	$-.006167$	.015233	
SS	.0168375	.0000285	.0001740	

The contrast labeled D looks at the difference in disks by averaging over windows. This involves averaging the two means for new disks, say, μ_{NC} and μ_{NS}, and contrasting this average with the average of the two means for used disks, say, μ_{UC} and μ_{US}. The contrast examining the difference in disks averaging over windows is $(\mu_{NC} + \mu_{NS})/2 - (\mu_{UC} + \mu_{US})/2$ or

$$\frac{1}{2}\mu_{NC} + \frac{1}{2}\mu_{NS} - \frac{1}{2}\mu_{UC} - \frac{1}{2}\mu_{US}.$$

As discussed earlier, multiplying a contrast by a constant does not really change the contrast, so to eliminate the fractional multiplications and make the contrast a bit easier to work with, we multiply this contrast by 2 and make it

$$\mu_{NC} + \mu_{NS} - \mu_{UC} - \mu_{US}.$$

This is the contrast D reported in Table 9.3. Recall that the estimate of the D contrast is computed as

$$\hat{D} = (1)0.89020\bar{0} + (1)0.88566\bar{6} + (-1)0.80766\bar{6} + (-1)0.81833\bar{6} = .149833$$

and the sum of squares is computed as

$$SS(D) = \frac{(.149833)^2}{[1^2 + 1^2 + (-1)^2 + (-1)^2]/3} = .0168375.$$

We define the contrast W similarly; it looks at the difference in windows by averaging over disks. Again, we multiplied the averages by 2 to simplify the contrast.

Contrast DW looks at the *interaction* between disks and windows, i.e., how the difference between disks changes as we go from a clean window to a soiled window. The difference between new and used disks with a clean window is $(\mu_{NC} - \mu_{UC})$ and the difference between new and used disks with a soiled window is $(\mu_{NS} - \mu_{US})$. The change in the disk difference between clean and soiled windows is $(\mu_{NC} - \mu_{UC}) - (\mu_{NS} - \mu_{US})$, or equivalently

$$\mu_{NC} - \mu_{NS} - \mu_{UC} + \mu_{US}.$$

This is the DW contrast. Note that the DW contrast coefficients in Table 9.3 can be obtained by multiplying the corresponding D and W contrast coefficients. *This procedure for obtaining interaction contrast coefficients by multiplying main effect contrast coefficients works quite generally.* Note that the DW contrast can also be obtained by looking at $(\mu_{NC} - \mu_{NS}) - (\mu_{UC} - \mu_{US})$, which is the change in the difference between clean and soiled windows as we go from new disks to used disks.

The analysis of variance is balanced; there are three observations on each treatment and four observations on each block. Thus, using the definition of orthogonality for a balanced one-way ANOVA, the treatment contrasts are orthogonal. It follows, both numerically and theoretically, that

$$SS(D) + SS(W) + SS(DW) = SSTrts.$$

From Table 9.3 we see that the vast majority of the sum of squares for treatments is due to the difference between disks averaged over windows (the D contrast). In particular, $SS(D)/SSTrts = .0168375/.0170401$, so 99% of the sum of squares for treatments is due to the D contrast. We also see that there is relatively little effect due to windows averaged over disks (the W contrast) and little effect due to the change in the disk differences due to windows (the DW contrast). In particular, the unadjusted F statistic for testing whether the DW contrast is zero is

$$F = \frac{SS(DW)}{MSE} = \frac{.0001740}{0.0000811} = 2.15$$

which has a P value of .193. Recall that a contrast has one degree of freedom, so $SS(DW) = MS(DW)$ in constructing the F statistic. Similarly, the test for contrast W has $F = .35$ and $P = .575$.

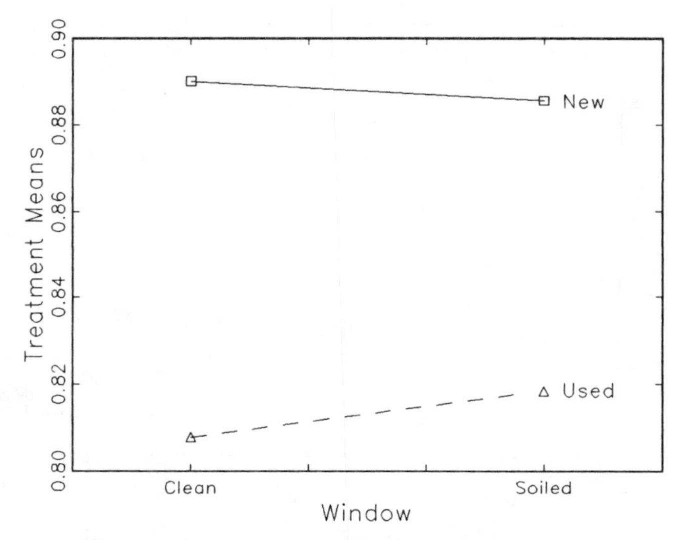

FIGURE 9.1. Disk–window interaction plot for spectrometer data.

While there is no statistical evidence for the existence of an interaction in the example, this does not prove that interaction does not exist. For example, if *MSE* had turned out to be one-third of its actual value, the F test for interaction would have been significant. When interactions exist, it is important to explore their nature. We now discuss some methods and ideas for examining interactions. This discussion is merely a precursor of the more extensive examination of interaction in Chapter 11. The difference between clean and soiled windows for new disks is $.004533$ and the difference for used disks is $-.010700$. These effects are actually in different directions! In one case, clean windows give higher readings and in the other case clean windows give lower readings. This seems to indicate that the effect of windows changes depending on the type of disk used, but in this example the *MSE* is large enough that the difference can reasonably be ascribed to random variation. In other words, this change in effect is not statistically significant because the interaction contrast is not statistically significant.

The contrast examining the different windows averaged over disks (the W contrast) was insignificant. However, if the DW interaction existed, the windows would still have a demonstrable effect on yields. The windows would have an effect because the disks behave differently for clean windows than for soiled windows. Additionally, the large effect for D would be of less interest if interaction were present because D is obtained by averaging over windows even though we would know that the disk effect depends on the window used.

Figure 9.1 contains a plot of the treatment means. There are two curves, one for new disks and another for used disks. The differences between disks is indicated by the separation between the two lines. The differences in the windows are indicated visually by the slopes of the new and used disk lines. If the effect of windows was the same regardless of disk condition, these slopes would be the same and the line segments would be parallel *up to sampling error*. With these data the lines are reasonably parallel. When interaction exists, the plot indicates the nature of the interaction.

Rather than describing the interaction as a difference in how the windows react to disks, we can describe the interaction in terms of how the effect of disk changes with type of window. As mentioned earlier, the two approaches are equivalent. Figure 9.2 contains another plot of the treatment means. There are two curves, one for clean windows and another for soiled windows. In this plot, the slopes indicate the differences due to disks and the separation of the lines indicates the differences due to

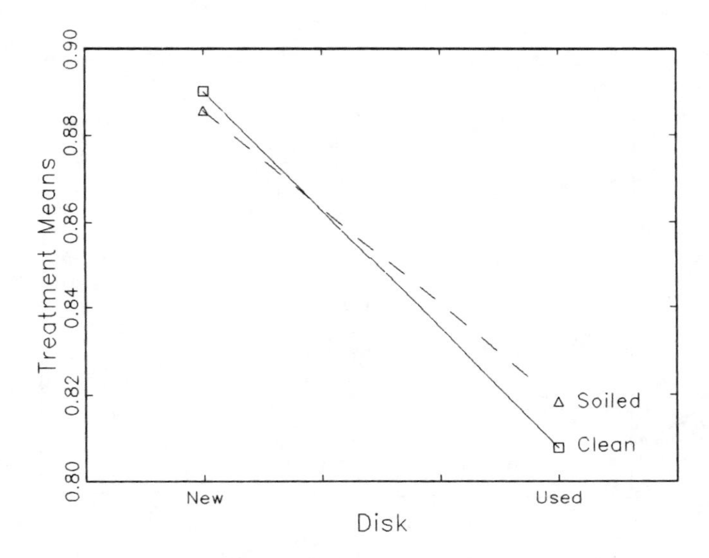

FIGURE 9.2. Disk–window interaction plot for spectrometer data.

windows. If there is no interaction, the curves should be parallel *up to sampling variation*. In this example, the curves intersect, suggesting that the curves are not parallel. However, after considering the level of sampling error, there is no evidence that the curves are not parallel.

Residual plots for the data are given in Figures 9.3 through 9.6. The residuals now must adjust for both the treatments and the blocks. The residual for an observation y_{ij} with treatment i in block j is defined as

$$\hat{\varepsilon}_{ij} = y_{ij} - \bar{y}_{i\cdot} - \bar{y}_{\cdot j} + \bar{y}_{\cdot\cdot},$$

where $\bar{y}_{i\cdot}$ is the ith treatment mean, $\bar{y}_{\cdot j}$ is the jth block mean, and $\bar{y}_{\cdot\cdot}$ is the grand mean of all 12 observations. By subtracting out both the treatment and block means, we have over adjusted for the overall level of the numbers, so the grand mean must be added back in. In balanced analysis of variance problems all the residuals have the same variance, so it is not necessary to standardize the residuals.

Figure 9.3 is a plot of the residuals versus the predicted values. The predicted values are

$$\hat{y}_{ij} = \bar{y}_{i\cdot} + \bar{y}_{\cdot j} - \bar{y}_{\cdot\cdot}\,.$$

Figure 9.4 plots the residuals versus indicators of the treatments. While the plot looks something like a bow tie, I am not overly concerned. Figure 9.5 contains a plot of the residuals versus indicators of blocks. The residuals look pretty good. From Figure 9.6, the residuals look reasonably normal. In the normal plot there are 12 residuals but the analysis has only 6 degrees of freedom for error. If you want to do a W' test for normality, you might use a sample size of 12 and compare the value $W' = .970$ to $W'(\alpha, 12)$, but it may be appropriate to use the dfE as the sample size for the test and use $W'(\alpha, 6)$. □

MINITAB COMMANDS

The following Minitab commands generate the analysis of variance. Column c1 contains the spectrometer data, while column c2 contains integers 1 through 4 indicating the appropriate treatment, and c3 contains integers 1 through 3 that indicate the block. The predicted values are given by the 'fits' subcommand.

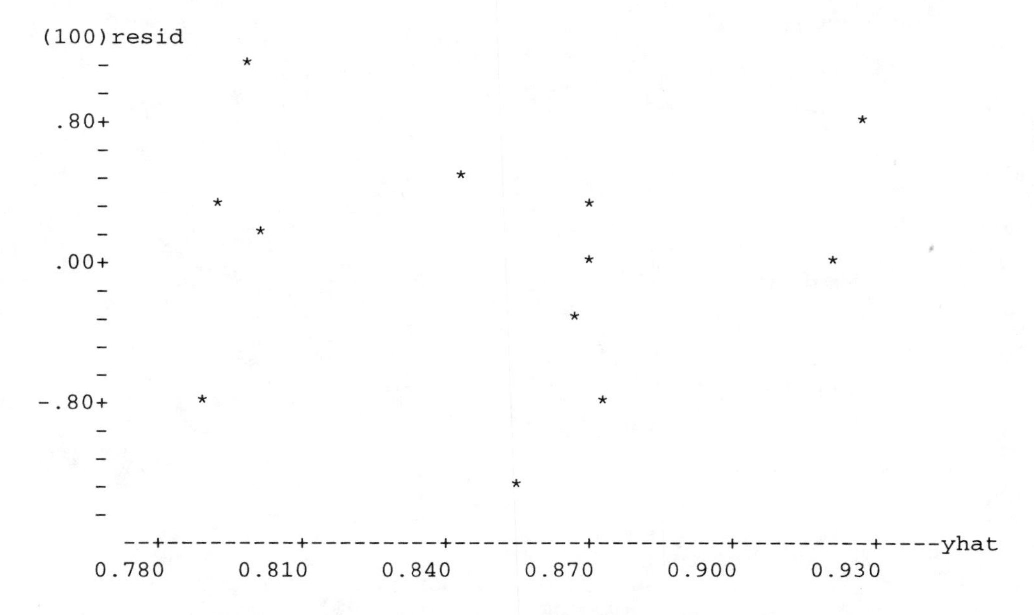

FIGURE 9.3. Plot of residuals versus predicted values, spectrometer data.

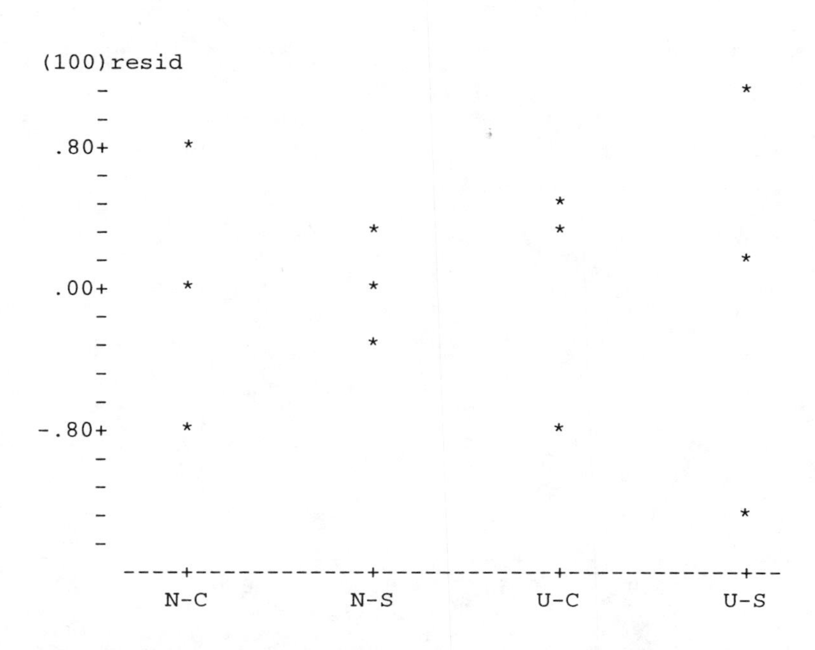

FIGURE 9.4. Plot of residuals versus treatment groups, spectrometer data.

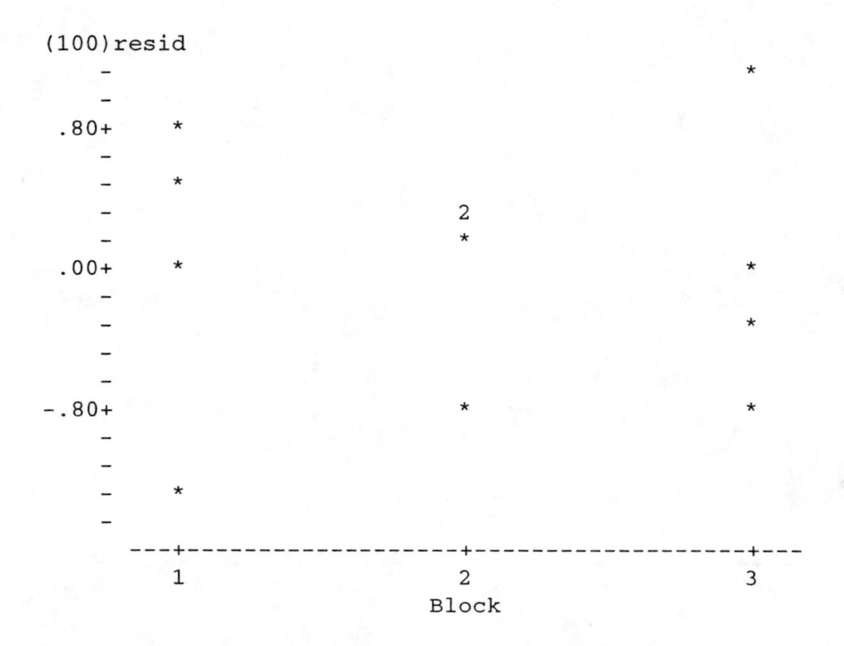

FIGURE 9.5. Plot of residuals versus blocks, spectrometer data.

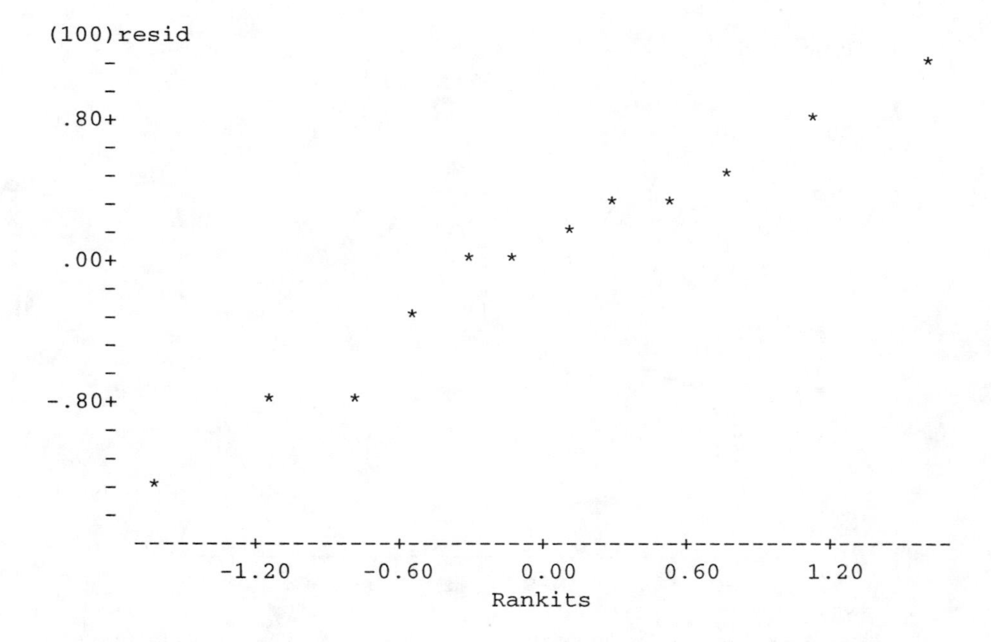

FIGURE 9.6. Normal plot of residuals, spectrometer data, $W' = 0.970$.

```
MTB > names c1 'y' c2 'Trts' c3 'Blks'
MTB > anova c1 = c2 c3;
SUBC> means c2 c3;
SUBC> resid c10;
SUBC> fits  c11.
```

BALANCED TWO-WAY ANALYSIS OF VARIANCE

The model for a randomized complete block design is a two-way analysis of variance,

$$y_{ij} = \mu + \alpha_i + \beta_j + \varepsilon_{ij}, \quad \varepsilon_{ij}\text{s independent } N(0, \sigma^2), \tag{9.2.1}$$

$i = 1, \ldots, a$, $j = 1, \ldots, b$. There are b blocks with a treatments observed within each block. The parameter μ is viewed as a grand mean, α_i is an unknown fixed effect for the ith treatment, and β_j is an unknown fixed effect for the jth block. The necessary summary statistics are the sample variance of all ab observations and the means for each treatment and block. It is frequently convenient to display the data as follows.

Treatment i	Block j: 1	2	$\cdots$	b	Trt. means
1	y_{11}	y_{12}	$\cdots$	y_{1b}	$\bar{y}_{1\bullet}$
2	y_{21}	y_{22}	$\cdots$	y_{2b}	$\bar{y}_{2\bullet}$
$\vdots$	$\vdots$	$\vdots$	$\ddots$	$\vdots$	$\vdots$
a	y_{a1}	y_{a2}	$\cdots$	y_{ab}	$\bar{y}_{a\bullet}$
Blk. means	$\bar{y}_{\bullet 1}$	$\bar{y}_{\bullet 2}$	$\cdots$	$\bar{y}_{\bullet b}$	$\bar{y}_{\bullet\bullet}$

The predicted values from this model are

$$\hat{y}_{ij} \equiv \bar{y}_{i\bullet} + \bar{y}_{\bullet j} - \bar{y}_{\bullet\bullet}$$

and the residuals are

$$\hat{\varepsilon}_{ij} \equiv y_{ij} - \hat{y}_{ij} = y_{ij} - \bar{y}_{i\bullet} - \bar{y}_{\bullet j} + \bar{y}_{\bullet\bullet}\,.$$

The computations involved in estimating σ^2 can be summarized in an analysis of variance table. The commonly used form for the analysis of variance table is given in Table 9.4. The degrees of freedom and sums of squares for treatments, blocks, and error add up to the degrees of freedom and sums of squares total (corrected for the grand mean). Note that the mean square for treatments is just the sample variance of the $\bar{y}_{i\bullet}$s times b and that the mean square for blocks is just the sample variance of the $\bar{y}_{\bullet j}$s times a.

The F statistic $MS(\alpha)/\,MSE$ is the ratio of the mean square treatments to the mean square error. It is used to test whether there are treatment effects, i.e., it is used to test

$$H_0 : \alpha_1 = \cdots = \alpha_a.$$

Note that if all the α_is are equal, we cannot *distinguish* between the effects of different treatments. In other words, we cannot isolate anything that can be identified as the effect of a treatment. H_0 does not imply that the treatments have no effect, it implies that they have the same effect. Generally, the effect of a treatment (an α_i) is impossible to isolate (or estimate) because we cannot distinguish it from the overall effect of running the experiment (μ) or indeed from any effect common to every

TABLE 9.4. Analysis of variance

Source	df	SS	MS	F
Trts(α)	$a-1$	$b\sum_{i=1}^{a}(\bar{y}_{i\cdot}-\bar{y}_{\cdot\cdot})^2$	$SS(\alpha)/(a-1)$	$\frac{MS(\alpha)}{MSE}$
Blks(β)	$b-1$	$a\sum_{j=1}^{b}\left(\bar{y}_{\cdot j}-\bar{y}_{\cdot\cdot}\right)^2$	$SS(\beta)/(b-1)$	$\frac{MS(\beta)}{MSE}$
Error	$(a-1)(b-1)$	$\sum_{i,j}\hat{\varepsilon}_{ij}^2$	SSE/dfE	
Total	$ab-1$	$\sum_{i=1}^{a}\sum_{j=1}^{b}\left(y_{ij}-\bar{y}_{\cdot\cdot}\right)^2$		

block. The same thing is true of the block effects β_j; they cannot be isolated from μ or common effects of the treatments. What we *can* isolate are comparative differences in the effects of treatments (and blocks). *The F statistic provides a test of whether there are differences in the treatment effects and not whether any treatment effects exist.* The only way you can test whether treatment effects exist is to redefine what you mean by treatment effects, so that they only exist when they are different.

The treatments are dealt with exactly as in a one-way ANOVA. For known λ_i*s that sum to zero*, a contrast in the treatment effects is

$$Par = \sum_{i=1}^{a} \lambda_i \alpha_i$$

with

$$Est = \sum_{i=1}^{a} \lambda_i \bar{y}_{i\cdot}\,.$$

Each treatment mean is the average of b observations, so

$$SE\left(\sum_{i=1}^{a} \lambda_i \bar{y}_{i\cdot}\right) = \sqrt{MSE \sum_{i=1}^{a} \lambda_i^2 / b}.$$

The reference distribution is $t(dfE)$. The sum of squares for the contrast is

$$SS\left(\sum_{i=1}^{a} \lambda_i \alpha_i\right) = \left[\sum_{i=1}^{a} \lambda_i \bar{y}_{i\cdot}\right]^2 \Big/ \left[\sum_{i=1}^{a} \lambda_i^2 / b\right]$$

If of interest, similar results hold for the block effects. For known ξ_js that add to zero,

$$Par = \sum_{j=1}^{b} \xi_j \beta_j$$

with

$$Est = \sum_{j=1}^{b} \xi_j \bar{y}_{\cdot j}\,.$$

The block means are the average of a observations, so

$$SE\left(\sum_{j=1}^{b} \xi_j \bar{y}_{\cdot j}\right) = \sqrt{MSE \sum_{j=1}^{b} \xi_j^2 / a}.$$

The reference distribution is $t(dfE)$. The sum of squares for the contrast is

$$SS\left(\sum_{j=1}^{b} \xi_j \beta_j\right) = \left[\sum_{j=1}^{b} \xi_j \bar{y}_{\cdot j}\right]^2 \Big/ \left[\sum_{j=1}^{b} \xi_j^2 / a\right].$$

The F statistic $MS(\beta)/MSE$ is the ratio of the mean square blocks to the mean square error. It is used to test whether there are block effects, i.e., it is used to test

$$H_0 : \beta_1 = \cdots = \beta_b.$$

Again, if all the β_is are equal we cannot distinguish between the effects of different blocks, so the F statistic provides a test of whether we can isolate comparative differences in the block effects. As discussed earlier, some models for RCB designs do not allow testing for block effects, but in any case, the ratio $MS(\beta)/MSE$ is of interest in that large values indicate that blocking was a worthwhile exercise.

The theoretical basis for this analysis of model (9.2.1) is precisely as in the balanced one-way ANOVA. Consider the analysis of treatment effects. (The analysis for block effects is similar.) The only thing random about a y_{ij} is the corresponding ε_{ij}. The ε_{ij}s are independent, so the y_{ij}s are independent. If follows that the $\bar{y}_{i\cdot}$s are independent because they are computed from distinct groups of observations. Since $y_{ij} = \mu + \alpha_i + \beta_j + \varepsilon_{ij}$, obviously $\bar{y}_{i\cdot} = \mu + \alpha_i + \bar{\beta}_\cdot + \bar{\varepsilon}_{i\cdot}$. Using Proposition 1.2.11,

$$E(\bar{y}_{i\cdot}) = \mu + \alpha_i + \bar{\beta}_\cdot + E(\bar{\varepsilon}_{i\cdot}) = \mu + \alpha_i + \bar{\beta}_\cdot$$

and

$$\mathrm{Var}(\bar{y}_{i\cdot}) = \mathrm{Var}(\bar{\varepsilon}_{i\cdot}) = \frac{\sigma^2}{b}.$$

More directly, the equalities follow because $\bar{\varepsilon}_{i\cdot}$ is the sample mean from a random sample of b variables each with population mean 0 and population variance σ^2. If the errors are normally distributed, the $\bar{y}_{i\cdot}$s are independent $N(\mu + \alpha_i + \bar{\beta}_\cdot, \sigma^2/b)$ random variables. In any case, if b is reasonably large, the normal distribution holds approximately because of the central limit theorem.

As in a balanced one-way ANOVA, the estimated contrast $\sum_{i=1}^{a} \lambda_i \bar{y}_{i\cdot}$ is an unbiased estimate of the parameter

$$E\left(\sum_{i=1}^{a} \lambda_i \bar{y}_{i\cdot}\right) = \sum_{i=1}^{a} \lambda_i (\mu + \alpha_i + \bar{\beta}_\cdot) = \sum_{i=1}^{a} \lambda_i \alpha_i\ .$$

This follows because $\sum_{i=1}^{a} \lambda_i = 0$. A derivation that is identical to the derivation used for a balanced one-way ANOVA gives

$$\mathrm{Var}\left(\sum_{i=1}^{a} \lambda_i \bar{y}_{i\cdot}\right) = \sigma^2 \frac{\sum_{i=1}^{a} \lambda_i^2}{b},$$

from which the standard error follows. The reference distribution for tests and confidence intervals relies on the fact that for normal errors,

$$\frac{SSE}{\sigma^2} \sim \chi^2(dfE)$$

with MSE independent of the $\bar{y}_{i\cdot}$s and the $\bar{y}_{\cdot j}$s.

Also as in a balanced one-way ANOVA, the $\bar{y}_{i\cdot}$s are a random sample from a normal population with variance σ^2/b if and only if the $\bar{y}_{i\cdot}$s have the same means, i.e., if and only if all the α_is are equal. When the α_is are all equal, the sample variance of the $\bar{y}_{i\cdot}$s is an estimate of σ^2/b and the *MSTrts* is an estimate of σ^2. In general, *MSTrts* estimates

$$E(MSTrts) = \sigma^2 + \frac{b}{a-1}\sum_{i-1}^{a}(\alpha_i - \bar{\alpha}_\cdot)^2$$

which is much larger than σ^2 if the treatment effects are very different relative to the size of σ^2 or if the number of blocks b is large. Moreover, the structure of the treatment means implies that all of the multiple comparison methods of Chapter 6 can continue to be applied.

Proposition 1.2.11 also shows that the predicted values have

$$E(\hat{y}_{ij}) = \mu + \alpha_i + \beta_j = E(y_{ij}),$$

that the residuals have

$$E(\hat{\varepsilon}_{ij}) = 0,$$

and that the *MSE* is unbiased for σ^2, i.e.,

$$E(MSE) = \sigma^2.$$

(Showing the last of these is much the most complicated.)

PAIRED COMPARISONS

An interesting special case of complete block data is paired comparison data as discussed in Section 4.1. In paired comparison data, there are two treatments to contrast and each pair constitutes a complete block.

EXAMPLE 9.2.2. *Shewhart's hardness data*
In Section 4.1, we examined Shewhart's data on hardness of two items that were welded together. In this case, it is impossible to group arbitrary formless pairs of parts and then randomly assign a part to be either part 1 or part 2, so the data do not actually come from an RCB experiment. Nonetheless, the two-way ANOVA model remains reasonable with pairs playing the role of blocks.

The data were given in Section 4.1 along with the means for each of the two parts. The two-way ANOVA analysis also requires the mean for each pair of parts. The analysis of variance table for the blocking analysis is given in Table 9.5. In comparing the blocking analysis to the paired comparison analysis given earlier, allowance for round-off errors must be made. The *MSE* is exactly half the value of $s_d^2 = 17.77165$ given in Section 4.1. The two-way ANOVA t test for differences between the two parts has

$$t_{obs} = \frac{47.552 - 34.889}{\sqrt{8.8858\,[(1/27) + (1/27)]}} = 15.61\,.$$

This is exactly the same t statistic as used in Section 4.1. The reference distribution is $t(26)$, again exactly the same. The analysis of variance F statistic is just the square of the t_{obs} and gives equivalent results for two-sided tests. Confidence intervals for the difference in means are also exactly the same in the blocking analysis and the paired comparison analysis. The one real difference between this analysis and the analysis of Section 4.1 is that this analysis provides an indication of whether pairing was worthwhile. □

TABLE 9.5. Analysis of variance for hardness data

Source	df	SS	MS	F	P
Pairs(Blocks)	26	634.94	24.42	2.75	0.006
Parts(Trts)	1	2164.73	2164.73	243.62	0.000
Error	26	231.03	8.89		
Total	53	3030.71			

9.3 Latin square designs

Latin square designs involve two simultaneous but distinct definitions of blocks. The treatments are arranged so that every treatment is observed in every block for both kinds of blocks.

EXAMPLE 9.3.1. Mercer and Hall (1911) and Fisher (1925, section 49) consider data on the weights of mangold roots. They used a Latin square design with 5 rows, columns, and treatments. The rectangular field on which the experiment was run was divided into five rows and five columns. This created 25 plots, arranged in a square, on which to apply the treatments A, B, C, D, and E. Each row of the square was viewed as a block, so every treatment was applied in every row. The unique feature of Latin square designs is that there is a second set of blocks. Every column was also considered a block, so every treatment was also applied in every column. The data are given in Table 9.6, arranged by rows and columns with the treatment given in the appropriate place and the observed root weight given in parentheses. The table also contains the means for rows, columns, and treatments. In each case, the mean is the average of 5 observations.

TABLE 9.6. Mangold root data

	Columns					Row
Rows	1	2	3	4	5	means
1	D(376)	E(371)	C(355)	B(356)	A(335)	358.6
2	B(316)	D(338)	E(336)	A(356)	C(332)	335.6
3	C(326)	A(326)	B(335)	D(343)	E(330)	332.0
4	E(317)	B(343)	A(330)	C(327)	D(336)	330.6
5	A(321)	C(332)	D(317)	E(318)	B(306)	318.8
Col. means	331.2	342.0	334.6	340.0	327.8	335.12

Treatments	A	B	C	D	E
Trt. means	333.6	331.2	334.4	342.0	334.4

The analysis of variance table is constructed like that for a randomized complete block design except that now both rows and columns play roles similar to blocks. The sum of squares total (corrected for the grand mean) is computed just as in one-way ANOVA, the sample variance of all 25 observations is computed and multiplied by 25 − 1, i.e.,

$$SSTot = (25 - 1)s_y^2 = (24)292.77\bar{6} = 7026.6.$$

The mean square and sum of squares for treatments are also computed as in one-way ANOVA. The treatment means are averages of 5 observations and the sample variance of the treatment means is 16.512, so

$$MSTrts = 5(16.512) = 82.56.$$

There are 5 treatments, so treatments have $5-1$ degrees of freedom and the sum of squares is the mean square times $(5-1)$,

$$SSTrts = (5-1)82.56 = 330.24\,.$$

The mean square and sum of squares for columns are also computed as if they were treatments in a one-way ANOVA. The column means are averages of 5 observations and the sample variance of the column means is 35.092, so

$$MSCols = 5(35.092) = 175.46\,.$$

There are 5 columns, so the sum of squares is the mean square times $(5-1)$,

$$SSCols = (5-1)175.46 = 701.84\,.$$

The mean square and sum of squares for rows are again computed as if they were treatments in a one-way ANOVA. The row means are the average of 5 observations and the sample variance of the row means is 212.012, so

$$MSRows = 5(212.012) = 1060.06\,.$$

There are 5 rows, so the sum of squares is the mean square times $(5-1)$,

$$SSRows = (5-1)1060.06 = 4240.24\,.$$

The sum of squares error is obtained by subtraction,

$$\begin{aligned} SSE &= SSTot - SSTrts - SSCols - SSRows \\ &= 7026.6 - 330.2 - 701.8 - 4240.2 \\ &= 1754.3\,. \end{aligned}$$

Similarly,

$$\begin{aligned} dfE &= dfTot - dfTrts - dfCols - dfRows \\ &= 24 - 4 - 4 - 4 \\ &= 12\,. \end{aligned}$$

The estimate of σ^2 is

$$MSE = \frac{SSE}{dfE} = \frac{1754.3}{12} = 146.2\,.$$

All of the calculations are summarized in the analysis of variance table, Table 9.7. Table 9.7 also gives the analysis of variance F test for the null hypothesis that the effects are the same for every treatment. The F statistic $MSTrts/MSE$ is very small, 0.56, so there is no evidence that the treatments behave differently. Blocking on columns was not very effective as evidenced by the F statistic of 1.20, but blocking on rows was very effective, $F = 7.25$.

Many experimenters are less than thrilled when told that there is no evidence for their treatments having any differential effects. Inspection of the treatment means given in Table 9.6 leads to the obvious conclusion that most of the differences are due to the fact that treatment D is much larger than the others, so we look at this a bit more. (Besides, this gives us an excuse to look at a contrast in a Latin square design.) If we construct the sum of squares for a contrast that compares D with the other means, say,

$$\mu_A + \mu_B + \mu_C - (4)\mu_D + \mu_E,$$

we get a sum of squares that contains the vast majority of the treatment sum of squares, i.e.,

$$SS(D \text{ vs. others}) = \frac{[333.6 + 331.2 + 334.4 - (4)342.0 + 334.40]^2}{[1 + 1 + 1 + (-4)^2 + 1]/5} = 295.84\,.$$

TABLE 9.7. Analysis of variance for mangold root data

Source	*df*	*SS*	*MS*	*F*	*P*
Trts	4	330.2	82.6	0.56	.696
Columns	4	701.8	175.5	1.20	.360
Rows	4	4240.2	1060.1	7.25	.003
Error	12	1754.3	146.2		
Total	24	7026.6			

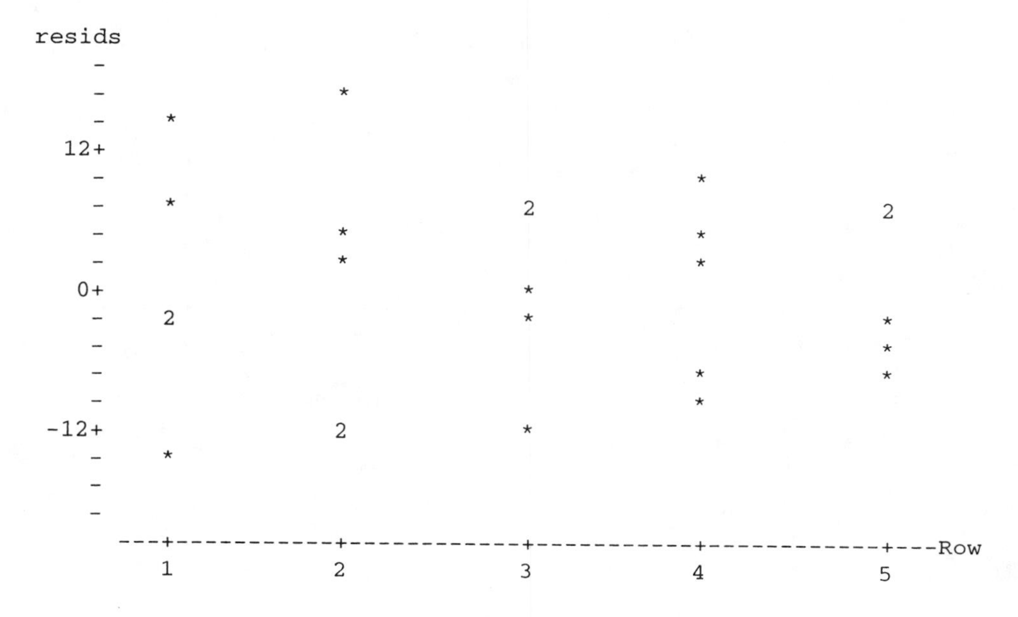

FIGURE 9.7. Plot of residuals versus rows.

However, the F ratio for the contrast is quite small,

$$F = \frac{295.84}{146.2} = 2.02 .$$

This is too small to provide any evidence for a difference between D and the average of the other treatments, *even if we had not let the data suggest the contrast*. If this F test cannot be rejected at even the unadjusted .05 level, there is no point in examining any multiple comparison methods to see if they will detect a difference, they will not.

Standard residual plots are given in Figures 9.7 through 9.10. They look quite good. □

COMPUTING TECHNIQUES

The following Minitab commands will give the sums of squares, means, and residuals necessary for the analysis. Here c1 is a column containing the mangold root yields, c2 has values from 1 to 5 indicating the row, c3 has values from 1 to 5 indicating the column, and c4 has values from 1 to 5 indicating the treatment.

```
MTB > names c1 'y' c2 'Rows' c3 'Cols' c4 'Trts'
MTB > ancova c1 = c2 c3 c4;
```

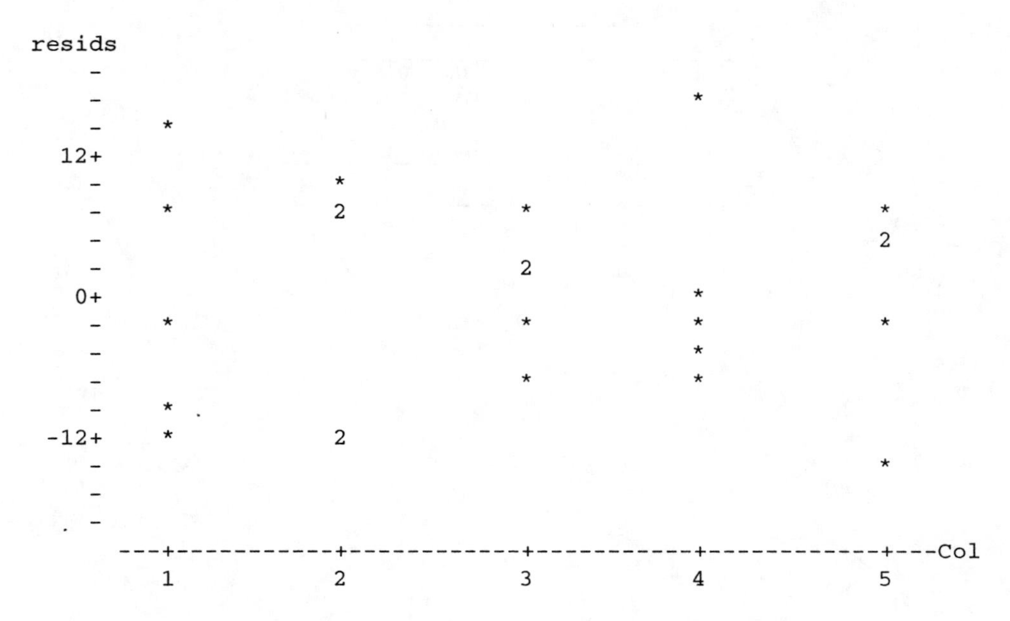

FIGURE 9.8. Plot of residuals versus columns.

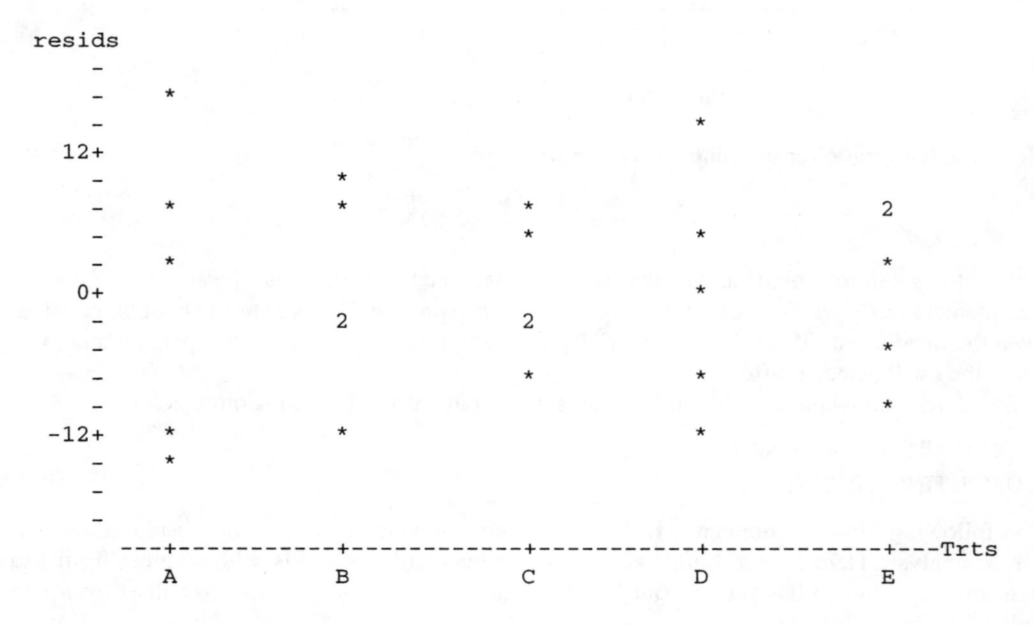

FIGURE 9.9. Plot of residuals versus treatment groups.

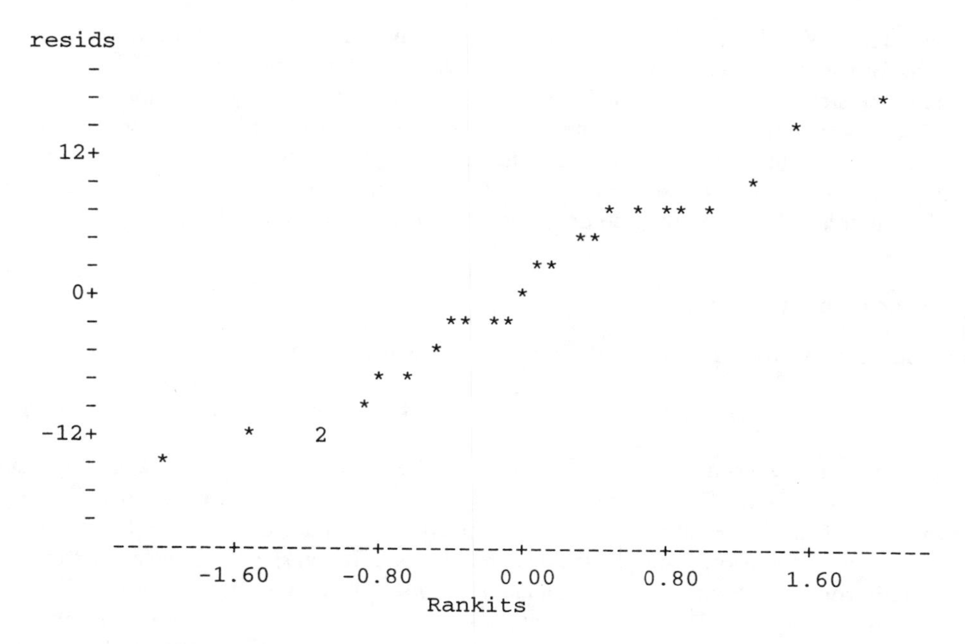

FIGURE 9.10. Normal plot of residuals for mangold root data, $W' = 0.978$.

```
SUBC> means c2 c3 c4;
SUBC> resid c11.
```

The 'glm' command can also be used in place of the 'ancova' command but gives more complicated output.

Computer programs for doing balanced analysis of variance are frequently incapable of dealing with Latin squares. For example, Minitab's 'anova' command will not give the analysis. In such cases, a simple trick can obtain the necessary results but the analysis must be constructed out of pieces of the output. The commands are given below. There are many simple ways to get the correct ANOVA table but a key aspect of these commands is that they give the correct residuals.

```
MTB > names c1 'y' c2 'Rows' c3 'Cols' c4 'Trts'
MTB > anova c1 = c2 c3;
SUBC> means c2 c3;
SUBC> resid c10.
MTB > anova c10 = c4;
SUBC> means c4;
SUBC> resid c11.
```

The degrees of freedom, sums of squares, and mean squares for rows, cols, and trts in the two ANOVA tables will be correct. The degrees of freedom total and the sum of squares total from the first ANOVA table (the one computed on c1) will be correct. The first ANOVA is a two-way with y as the dependent variable and rows and columns as the effects. The SSE from the second ANOVA table (on c10) will be correct but dfE and thus MSE will be incorrect. The second ANOVA is a one-way using the treatments as groups and the residuals from the first ANOVA as the dependent variable. From these pieces, a correct ANOVA table can be constructed. In particular, none of the F tests reported by these commands are appropriate. The residuals from the second ANOVA are the appropriate residuals for the Latin square analysis. These are in column c11. The means reported for rows and cols in the first ANOVA will be correct. The means for Trts in the second ANOVA are adjusted for the rows and columns, so they are not the actual treatment means. However, the means for treatments reported in the second ANOVA can be used for treatment comparisons (contrasts) just as if they were the original treatment means because the row and column adjustments cancel out when performing contrasts. Two additional points should be made. First, the residuals used as the dependent variable in the second ANOVA must be raw residuals, they cannot be standardized. Second, the roles played by rows, columns, and treatments can be interchanged.

LATIN SQUARE MODELS

The model for an $r \times r$ Latin square design is a three-way analysis of variance,

$$y_{ijk} = \mu + \alpha_i + \kappa_j + \rho_k + \varepsilon_{ijk}, \quad \varepsilon_{ijk}\text{s independent } N(0, \sigma^2). \tag{9.3.1}$$

The parameter μ is viewed as a grand mean, α_i is an effect for the ith treatment, κ_j is an effect for the jth column, and ρ_k is an effect for the kth row. The subscripting for this model is peculiar. All of the subscripts run from 1 to r but not freely. If you specify a row and a column, the design tells you the treatment. Thus, if you know k and j, the design tells you i. If you specify a row and a treatment, the design tells you the column, so k and i dictate j. In fact, if you know any two of the subscripts, the design tells you the third. The summary statistics necessary for the analysis are the sample variance of all r^2 observations and the means for each treatment, column, and row. The predicted values are

$$\hat{y}_{ijk} \equiv \bar{y}_{i\cdot\cdot} + \bar{y}_{\cdot j\cdot} + \bar{y}_{\cdot\cdot k} - 2\bar{y}_{\cdot\cdot\cdot}$$

and the residuals are

$$\hat{\varepsilon}_{ijk} \equiv y_{ijk} - \hat{y}_{ijk} = y_{ijk} - \bar{y}_{i\cdot\cdot} - \bar{y}_{\cdot j\cdot} - \bar{y}_{\cdot\cdot k} + 2\bar{y}_{\cdot\cdot\cdot}\,.$$

The computations for the *MSE* can be summarized in an analysis of variance table. The commonly used form for the analysis of variance table is given in Table 9.8. Notice that because of the peculiarity in the subscripting, the sums for error and total are taken over only i and j. The choice of these two subscripts is arbitrary; summing over any two subscripts sums over all of the observations in the Latin square. The degrees of freedom and sums of squares for treatments, columns, rows, and error add up to the degrees of freedom and sums of squares total (corrected for the grand mean).

TABLE 9.8. Analysis of variance

Source	df	SS	MS	F
Trts(α)	$r-1$	$r\sum_{i=1}^{r}(\bar{y}_{i\cdot\cdot} - \bar{y}_{\cdot\cdot\cdot})^2$	$SS(\alpha)/(r-1)$	$\frac{MS(\alpha)}{MSE}$
Columns(κ)	$r-1$	$r\sum_{j=1}^{r}\left(\bar{y}_{\cdot j\cdot} - \bar{y}_{\cdot\cdot\cdot}\right)^2$	$SS(\kappa)/(r-1)$	$\frac{MS(\kappa)}{MSE}$
Rows(ρ)	$r-1$	$r\sum_{k=1}^{r}(\bar{y}_{\cdot\cdot k} - \bar{y}_{\cdot\cdot\cdot})^2$	$SS(\rho)/(r-1)$	$\frac{MS(\rho)}{MSE}$
Error	$(r-1)(r-2)$	$\sum_{ij}\hat{\varepsilon}_{ijk}^2$	$SSE/(dfE)$	
Total	r^2-1	$\sum_i\sum_j\left(y_{ijk} - \bar{y}_{\cdot\cdot\cdot}\right)^2$		

The F statistic $MS(\alpha)/MSE$ is the ratio of the mean square treatments to the mean square error. It is used to test whether there are treatment effects, i.e., it is used to test

$$H_0 : \alpha_1 = \cdots = \alpha_r.$$

Again if all the α_is are equal, we cannot distinguish between the effects of different treatments, so the treatment F statistic provides a test of whether we can isolate comparative differences in the treatment effects.

The F statistic $MS(\kappa)/\,MSE$ is the ratio of the mean square columns to the mean square error. It is used to test whether there are column effects, i.e., it is used to test

$$H_0 : \kappa_1 = \cdots = \kappa_r.$$

The F statistic provides a test of whether we can isolate comparative differences in the column effects. The ratio of the mean square rows to the mean square error gives the F statistic $MS(\rho)/\,MSE$. This is used to test

$$H_0 : \rho_1 = \cdots = \rho_r.$$

The F statistic provides a test of whether we can isolate comparative differences in the row effects. Some models for Latin square designs do not allow testing for row and column effects but in any case the ratios $MS(\rho)/MSE$ and $MS(\kappa)/MSE$ are of interest in that large values indicate, respectively, that blocking on rows and columns was worthwhile.

The treatments are dealt with exactly as in a one-way ANOVA. A contrast in the treatment effects is, for known λ_is that sum to zero,

$$Par = \sum_{i=1}^{r} \lambda_i \alpha_i$$

with

$$Est = \sum_{i=1}^{r} \lambda_i \bar{y}_{i\bullet\bullet}\,.$$

The treatment means are the average of r observations, so

$$SE\left(\sum_{i=1}^{r} \lambda_i \bar{y}_{i\bullet\bullet}\right) = \sqrt{MSE \sum_{i=1}^{r} \lambda_i^2 / r}.$$

The reference distribution is $t(dfE)$. The sum of squares for the contrast is

$$SS\left(\sum_{i=1}^{r} \lambda_i \alpha_i\right) = \left[\sum_{i=1}^{r} \lambda_i \bar{y}_{i\bullet\bullet}\right]^2 \Big/ \left[\sum_{i=1}^{r} \lambda_i^2 / r\right].$$

If of interest, similar results hold for the row and column effects. For known ξ_js that add to zero, inferences for, say, the column effects can be based on

$$Par = \sum_{j=1}^{r} \xi_j \kappa_j$$

with

$$Est = \sum_{j=1}^{r} \xi_j \bar{y}_{\bullet j\bullet}\,.$$

The column means are averages of r observations so

$$SE\left(\sum_{j=1}^{r} \xi_j \bar{y}_{\bullet j\bullet}\right) = \sqrt{MSE \sum_{j=1}^{r} \xi_j^2 / r}.$$

The reference distribution is $t(dfE)$. The sum of squares for the contrast is

$$SS\left(\sum_{j=1}^{r} \xi_j \kappa_j\right) = \left[\sum_{j=1}^{r} \xi_j \bar{y}_{\cdot j \cdot}\right]^2 \Big/ \left[\sum_{j=1}^{r} \xi_j^2 / r\right].$$

The theoretical justification for the analysis is similar to that for a balanced one-way and a balanced two-way.

Discussion of Latin squares

The idea of simultaneously having two distinct sets of complete blocks is quite useful. For example, suppose you wish to compare the performance of four machines in producing something. Productivity is notorious for depending on the day of the week, with Mondays and Fridays often having low productivity; thus we may wish to block on days. The productivity of the machine is also likely to depend on who is operating the machine, so we may wish to block on operators. Thus we may decide to run the experiment on Monday through Thursday with four machine operators and using each operator on a different machine each day. One possible design is

	Operator			
Day	1	2	3	4
Mon	A	B	C	D
Tue	B	C	D	A
Wed	C	D	A	B
Thu	D	A	B	C

where the numbers 1 through 4 are randomly assigned to the four people who will operate the machines and the letters A through D are randomly assigned to the machines to be examined. Moreover, the days of the week should actually be randomly assigned to the rows of the Latin square. In general, the rows, columns, and treatments should all be randomized in a Latin square.

Another distinct Latin square design for this situation is

	Operator			
Day	1	2	3	4
Mon	A	B	C	D
Tue	B	A	D	C
Wed	C	D	B	A
Thu	D	C	A	B

This square cannot be obtained from the first one by any interchange of rows, columns, and treatments. Typically, one would randomly choose a possible Latin square design from a list of such squares (see, for example, Cochran and Cox, 1957) in addition to randomly assigning the numbers, letters, and rows to the operators, machines, and days.

The use of Latin square designs can be extended in numerous ways. One modification is the incorporation of a third kind of block; such designs are called *Graeco-Latin squares*. The use of Graeco-Latin squares is explored in the exercises for this chapter. A problem with Latin squares is that small squares give poor variance estimates because they provide few degrees of freedom for error, cf. Table 9.8. For example, a 3×3 Latin square gives only 2 degrees of freedom for error. In such cases, the Latin square experiment is often performed several times, giving additional replications that provide improved variance estimation. Section 11.4 presents an example in which several Latin squares are used.

9.4 Discussion of experimental design

Data are frequently collected with the intention of evaluating a change in the current system of doing things. If you really want to know the effect of a change in the system, you have to execute the change. It is not enough to look at conditions in the past that were similar to the proposed change because, along with the past similarities, there were dissimilarities. For example, suppose you think that instituting a good sex education program in schools will decrease teenage pregnancies. To evaluate this, it is not enough to compare schools that currently have such programs with schools that do not, because along with the differences in sex education programs there are other differences in the schools that affect teen pregnancy rates. Such differences may include parents' average socio-economic status and education. While adjustments can be made for any such differences that can be identified, there is no assurance that all important differences can be found. Moreover, initiating the proposed program involves making a change and the very act of change can affect the results. For example, current programs may exist and be effective because of the enthusiasm of the school staff that initiated them. Such enthusiasm is not likely to be duplicated when the new program is mandated from above.

To establish the effect of instituting a sex education program in a population of schools, you really need to (randomly) choose schools and actually institute the program. The schools at which the program is instituted should be chosen randomly, so no (unconscious) bias creeps in due to the selection of schools. For example, the people conducting the investigation are likely to favor or oppose the project. They could (perhaps unconsciously) choose the schools in such a way that makes the evaluation likely to reflect their prior attitudes. Unconscious bias occurs frequently and should *always* be assumed. Other schools without the program should be monitored to establish a base of comparison. These other schools should be treated as similarly as possible to the schools with the new program. For example, if the district school administration or the news media pay a lot of attention to the schools with the new program but ignore the other schools, we will be unable to distinguish the effect of the program from the effect of the attention. In addition, blocking similar schools together can improve the precision of the experimental results.

One of the great difficulties in learning about human populations is that obtaining the best data often requires morally unacceptable behavior. We object to having our lives randomly changed for the benefit of experimental science and typically the more important the issue under study, the more we object to such changes. Thus we find that in studying humans, the best data available are often historical. In our example we might have to accept that the best data available will be an historical record of schools with and without sex education programs. We must then try to identify and adjust for *all* differences in the schools that could potentially affect our conclusions. It is the extreme difficulty of doing this that leads to the relative unreliability of many studies in the social sciences. On the other hand, it would be foolish to give up the study of interesting and important phenomena just because they are difficult to study.

9.5 Exercises

EXERCISE 9.5.1. Garner (1956) presented data on the tensile strength of fabrics. Here we consider a subset of the data. The complete data and a more extensive discussion of the experimental procedure are given in Exercise 11.5.2. The experiment involved testing fabric strengths on four different machines. Eight homogeneous strips of cloth were divided into four samples. Each sample was tested on one of four machines. The data are given in Table 9.9.

(a) Identify the design for this experiment and give an appropriate model. List all of the assumptions made in the model.

TABLE 9.9. Tensile strength of uniform twill

Fabric strips	Machines m_1	m_2	m_3	m_4
s_1	18	7	5	9
s_2	9	11	12	3
s_3	7	11	11	1
s_4	6	4	10	8
s_5	10	8	6	10
s_6	7	12	3	15
s_7	13	5	15	16
s_8	1	11	8	12

(b) Analyze the data. Give an appropriate analysis of variance table. Examine appropriate contrasts using Tukey's method with $\alpha = .05$

(c) Check the assumptions of the model and adjust the analysis appropriately.

EXERCISE 9.5.2. Snedecor (1945b) presented data on a spray for killing adult flies as they emerged from a breeding medium. The data were numbers of adults found in cages that were set over the medium containers. The treatments were different levels of the spray's active ingredient, namely 0, 4, 8, and 16 units. (Actually, it is not clear whether a spray with 0 units was actually applied or whether no spray was applied. The former might be preferable.) Seven different sources for the breeding mediums were used and each spray was applied on each distinct breeding medium. The data are presented in Table 9.10.

TABLE 9.10. Dead adult flies

Medium	Units of active ingredient 0	4	8	16
A	423	445	414	247
B	326	113	127	147
C	246	122	206	138
D	141	227	78	148
E	208	132	172	356
F	303	31	45	29
G	256	177	103	63

(a) Identify the design for this experiment and give an appropriate model. List all the assumptions made in the model.

(b) Analyze the data. Give an appropriate analysis of variance table. Examine a contrast that compares the treatment with no active ingredient to the average of the three treatments that contain the active ingredient. Ignoring the treatment with no active ingredient, the other three treatments are quantitative levels of the active ingredient. On the log scale, these levels are equally spaced, so the tabled polynomial contrasts can be used to examine the polynomial regression of numbers killed on the log of the amount of active ingredient. The contrasts are given below.

Treatment	Contrasts Active vs. inactive	log(active) linear	log(active) quadratic
0	3	0	0
4	−1	−1	1
8	−1	0	−2
16	−1	1	1

Examine these contrasts. Compare the results given by the LSD, Bonferroni, and Scheffé methods. Use $\alpha = .10$ for LSD and Scheffé and something close to .05 for Bonferroni. Are the polynomial contrasts orthogonal to the first contrast?

(c) Check the assumptions of the model and adjust the analysis appropriately.

EXERCISE 9.5.3. Cornell (1988) considered data on scaled thickness values for five formulations of vinyl designed for use in automobile seat covers. Eight groups of material were prepared. The production process was then set up and the five formulations run with the first group. The production process was then reset and another group of five was run. In all, the production process was set eight times and a group of five formulations was run with each setting. The data are displayed in Table 9.11.

TABLE 9.11. Cornell's scaled vinyl thickness values

	Production setting							
Formulation	1	2	3	4	5	6	7	8
1	8	7	12	10	7	8	12	11
2	6	5	9	8	7	6	10	9
3	10	11	13	12	9	10	14	12
4	4	5	6	3	5	4	6	5
5	11	10	15	11	9	7	13	9

(a) From the information given, identify the design for this experiment and give an appropriate model. List all the assumptions made in the model.

(b) Analyze the data. Give an appropriate analysis of variance table. Examine appropriate contrasts using the Bonferroni method with an α of about .05.

(c) Check the assumptions of the model and adjust the analysis appropriately.

EXERCISE 9.5.4. In data related to that of the previous problem, Cornell (1988) has scaled thickness values for vinyl under four different process conditions. The process conditions were A, high rate of extrusion, low drying temperature; B, low rate of extrusion, high drying temperature; C, low rate of extrusion, low drying temperature; D, high rate of extrusion, high drying temperature. An initial set of data with these conditions was collected and later a second set was obtained. The data are given below.

	Treatments A	B	C	D
Rep 1	7.8	11.0	7.4	11.0
Rep 2	7.6	8.8	7.0	9.2

Identify the design, give the model, check the assumptions, give the analysis of variance table and interpret the F test for treatments.

The structure of the treatments suggest some interesting contrasts. These are given below.

	Treatments			
Contrast	A	B	C	D
Rate	1	−1	−1	1
Temp	−1	1	−1	1
RT	−1	−1	1	1

The rate contrast examines the difference between the two treatments with a high rate of extrusion and those with a low rate. The temp contrast examines the difference between the two treatments with a high drying temperature and those with a low temperature. The RT contrast is an interaction contrast that examines whether the effect of extrusion rate is the same for high drying temperatures as for low temperatures. Show that the contrasts are orthogonal and use the contrasts to analyze the data.

EXERCISE 9.5.5. Johnson (1978) and Mandel and Lashof (1987) present data on measurements of P_2O_5 (phosphorous pentoxide) in fertilizers. Table 9.12 presents data for five fertilizers, each analyzed in five labs. Our interest is in differences among the labs. Analyze the data.

TABLE 9.12. Phosphorous fertilizer data

	Laboratory				
Fertilizer	1	2	3	4	5
F	20.20	19.92	20.91	20.65	19.94
G	30.20	30.09	29.10	29.85	30.29
H	31.40	30.42	30.18	31.34	31.11
I	45.88	45.48	45.51	44.82	44.63
J	46.75	47.14	48.00	46.37	46.63

EXERCISE 9.5.6. Table 9.13 presents data on yields of cowpea hay. Four treatments are of interest, variety I of hay was planted 4 inches apart (I4), variety I of hay was planted 8 inches apart (I8), variety II of hay was planted 4 inches apart (II4), and variety II of hay was planted 8 inches apart (II8). Three blocks of land were each divided into four plots and one of the four treatments was randomly applied to each plot. These data are actually a subset of a larger data set given by Snedecor and Cochran (1980, p. 309) that involves three varieties and three spacings in four blocks. Analyze the data. Check your assumptions. Examine appropriate contrasts.

TABLE 9.13. Cowpea hay yields

	Block			Trt.
Treatment	1	2	3	means
I4	45	43	46	$44.\overline{666}$
I8	50	45	48	$47.66\bar{6}$
II4	61	60	63	$61.33\bar{3}$
II8	58	56	60	58.000
Block means	53.50	51.00	54.25	$52.91\bar{6}$

EXERCISE 9.5.7. In the study of the optical emission spectrometer discussed in Example 9.2.1 and Table 9.1, the target value for readings was .89. Subtract .89 from each observation and repeat the analysis. What new questions are of interest? Which aspects of the analysis have changed and which have not?

EXERCISE 9.5.8. An experiment was conducted to examine differences among operators of Suter hydrostatic testing machines. These machines are used to test the water repellency of squares of fabric. One large square of fabric was available but its water repellency was thought to vary along the length (warp) and width (fill) of the fabric. To adjust for this, the square was divided into four equal parts along the length of the fabric and four equal parts along the width of the fabric, yielding 16 smaller pieces. These pieces were used in a Latin square design to investigate differences among four operators: A, B, C, D. The data are given in Table 9.14. Construct an analysis of variance table. What, if any, differences can be established among the operators? Compare the results of using the Tukey, Newman–Keuls, and Bonferroni methods for comparing the operators.

TABLE 9.14. Hydrostatic pressure tests: operator, yield

A	B	C	D
40.0	43.5	39.0	44.0
B	A	D	C
40.0	42.0	40.5	38.0
C	D	A	B
42.0	40.5	38.0	40.0
D	C	B	A
40.0	36.5	39.0	38.5

EXERCISE 9.5.9. Table 9.15 contains data similar to that in the previous exercise except that in this Latin square differences among four machines: 1, 2, 3, 4, were investigated rather than differences among operators. Machines 1 and 2 were operated with a hand lever, while machines 3 and 4 were operated with a foot lever. Construct an analysis of variance table. What, if any, differences can be established among the machines? To this end, construct appropriate orthogonal contrasts.

TABLE 9.15. Hydrostatic pressure tests: machine, yield

2	4	3	1
39.0	39.0	41.0	41.0
1	3	4	2
36.5	42.5	40.5	38.5
4	2	1	3
40.0	39.0	41.5	41.5
3	1	2	4
41.5	39.5	39.0	44.0

EXERCISE 9.5.10. Table 9.15 is incomplete. The data were actually obtained from a Graeco-Latin square that incorporates four different operators as well as the four different machines. The correct design is given in Table 9.16. Note that this is a Latin square for machines when we ignore the operators and a Latin square for operators when we ignore the machines. Moreover, every operator works once with every machine. Using the four operator means, compute a sum of squares for operators and subtract this from the error computed in Exercise 9.5.9. Give the new analysis of variance table. How do the results on machines change? What evidence is there for differences among operators. Was the analysis for machines given earlier incorrect or merely inefficient?

TABLE 9.16. Hydrostatic pressure tests: operator, machine

B,2	A,4	D,3	C,1
A,1	B,3	C,4	D,2
D,4	C,2	B,1	A,3
C,3	D,1	A,2	B,4

Operators are A, B, C, D.
Machines are 1, 2, 3, 4.

EXERCISE 9.5.11. Table 9.17 presents data given by Nelson (1993) on disk drives from a Graeco-Latin square design (see Exercise 9.5.10). The experiment was planned to investigate the effect of four different substrates on the drives. The dependent variable is the amplitude of a signal read from the disk where the signal written onto the disk had a fixed amplitude. Blocks were constructed from machines, operators, and day of production. (In Table 9.17, Days are indicated by lower case Latin letters.) The substrata consist of A, aluminum; B, nickel plated aluminum; and two types of glass, C and D. Analyze the data. In particular, check for differences between aluminum and glass, between the two types of glass, and between the two types of aluminum. Check your assumptions.

TABLE 9.17. Amplitudes of disk drives

	Machine			
Operator	1	2	3	4
I	Aa 8	Cd 7	Db 3	Bc 4
II	Cc 11	Ab 5	Bd 9	Da 5
III	Dd 2	Ba 2	Ac 7	Cb 9
IV	Bb 8	Dc 4	Ca 9	Ad 3

Chapter 10

Analysis of covariance

Analysis of covariance incorporates one or more regression variables into an analysis of variance. The regression variables are referred to as covariates (relative to the dependent variable), hence the name analysis of covariance. Covariates are also known as supplementary or concomitant observations. Cox (1958, chapter 4) gives a particularly nice discussion of the ideas behind analysis of covariance and illustrates various useful plotting techniques. In 1957 and 1982, *Biometrics* devoted entire issues to the analysis of covariance. In this chapter, we only examine the use of a single covariate. We begin our discussion with an example that involves one-way analysis of variance and a covariate.

In Sections 1 and 4 of this chapter, we make extensive use of model comparisons. To simplify the discussions within these sections, we will often refer to a model such as (10.1.1) as simply model (1).

10.1 An example

Fisher (1947) gives data on the body weights (in kilograms) and heart weights (in grams) for domestic cats of both sexes that were given digitalis. A subset of the data is presented in Table 10.1. Our primary interest is to determine whether females' heart weights differ from males' heart weights when both have received digitalis.

As a first step, we might fit a one-way ANOVA model,

$$\begin{aligned} y_{ij} &= \mu_i + \varepsilon_{ij} \qquad (10.1.1) \\ &= \mu + \alpha_i + \varepsilon_{ij}, \end{aligned}$$

where the y_{ij}s are the heart weights, $i = 1, 2$, and $j = 1, \ldots, 24$. This model yields the analysis of variance given in Table 10.2. Note the overwhelming effect due to sexes.

Fisher provided both heart weights and body weights, so we can ask a more complex question, 'Is there a sex difference in the heart weights over and above the fact that male cats are naturally larger than female cats?' To examine this we add a regression term to model (1) and fit the traditional *analysis of covariance model*,

$$y_{ij} = \mu_i + \gamma z_{ij} + \varepsilon_{ij} \qquad (10.1.2)$$

TABLE 10.1. Fisher's data on body weights (kg) and heart weights (g) of domestic cats given digitalis

Females				Males			
Body	Heart	Body	Heart	Body	Heart	Body	Heart
2.3	9.6	2.0	7.4	2.8	10.0	2.9	9.4
3.0	10.6	2.3	7.3	3.1	12.1	2.4	9.3
2.9	9.9	2.2	7.1	3.0	13.8	2.2	7.2
2.4	8.7	2.3	9.0	2.7	12.0	2.9	11.3
2.3	10.1	2.1	7.6	2.8	12.0	2.5	8.8
2.0	7.0	2.0	9.5	2.1	10.1	3.1	9.9
2.2	11.0	2.9	10.1	3.3	11.5	3.0	13.3
2.1	8.2	2.7	10.2	3.4	12.2	2.5	12.7
2.3	9.0	2.6	10.1	2.8	13.5	3.4	14.4
2.1	7.3	2.3	9.5	2.7	10.4	3.0	10.0
2.1	8.5	2.6	8.7	3.2	11.6	2.6	10.5
2.2	9.7	2.1	7.2	3.0	10.6	2.5	8.6

TABLE 10.2. One-way analysis of variance on heart weights

Source	df	SS	MS	F	P
Sex	1	56.117	56.117	23.44	.0000
Error	46	110.11	2.3936		
Total	47	166.223			

$$= \mu + \alpha_i + \gamma z_{ij} + \varepsilon_{ij}.$$

Here the z_{ij}s are the body weights and γ is a slope parameter associated with body weights. Note that model (2) is an extension of the simple linear regression between the ys and the zs in which we allow a different intercept μ_i for each sex. An analysis of variance table for model (2) is given as Table 10.3. The interpretation of this table is different from the ANOVA tables examined earlier. For example, the sums of squares for body weights, sex, and error *do not* add up to the sum of squares total. The sums of squares in Table 10.3 are referred to as *adjusted sums of squares* (*Adj. SS*) because the body weight sum of squares is adjusted for sexes and the sex sum of squares is adjusted for body weights. In this section, we focus on the interpretation of Table 10.3; in Section 10.3 we discuss its computation.

TABLE 10.3. Analysis of variance for heart weights based on model (2)

Source	df	$Adj.\ SS$	MS	F	P
Body weights	1	37.828	37.828	23.55	0.000
Sex	1	4.499	4.499	2.80	0.101
Error	45	72.279	1.606		
Total	47	166.223			

The error line in Table 10.3 is simply the error from fitting model (2). The body weights line comes from comparing model (2) with the reduced model (1). Note that the only difference between models (1) and (2) is that (1) does not involve the regression on body weights, so by testing the two models we are testing whether there is a significant effect due to the regression on body weights. The standard way of comparing a full and a reduced model is by comparing their error terms. Model

(2) has one more parameter, γ, than model (1), so there is one more degree of freedom for error in model (1) than in model (2), hence one degree of freedom for body weights. The adjusted sum of squares for body weights is the difference between the sum of squares error in model (1) and the sum of squares error in model (2). Given the sum of squares and the mean square, the F statistic for body weights is constructed in the usual way, cf. Section 5.5. Examining Table 10.3, we see a major effect due to the regression on body weights.

The sex line in Table 10.3 provides a test of whether there are differences in sexes *after adjusting for the regression on body weights*. This comes from comparing model (2) to a similar model in which sex differences have been eliminated. In model (2), the sex differences are incorporated as μ_1 and μ_2 in the first version and as α_1 and α_2 in the second version. To eliminate sex differences in model (2), we simply eliminate the distinctions between the μs (the αs). Such a model can be written as

$$y_{ij} = \mu + \gamma z_{ij} + \varepsilon_{ij}.$$

In this example, the analysis of covariance model without treatment effects is just a simple linear regression of heart weight on body weight. We have reduced the two sex parameters to one overall parameter, so the difference in degrees of freedom between this model and model (2) is 1. The difference in the sums of squares error between this model and model (2) is the adjusted sum of squares for sex. Examining Table 10.3 we see that the evidence for a sex effect over and above the effect due to the regression on body weights is not great.

Our current data come from an observational study rather than a designed experiment. It is difficult to take a group of cats and randomly assign them to sex groups. As discussed in the next section, principles of experimental design focus attention on models such as (2). However, these data are from an observational study, so yet another model is of interest. There is little reason to assume that when regressing heart weight on body weight the relationships are the same for females and males. Model (2) allows different intercepts for these regressions but uses the same slope γ. We should test the assumption of a common slope by fitting the more general model that allows different slopes for females and males, i.e.,

$$\begin{aligned} y_{ij} &= \mu_i + \gamma_i z_{ij} + \varepsilon_{ij} \qquad (10.1.3) \\ &= \mu + \alpha_i + \gamma_i z_{ij} + \varepsilon_{ij}. \end{aligned}$$

In model (3) the γs depend on i and thus the slopes are allowed to differ between the sexes. While model (3) may look complicated, it consists of nothing more than fitting a simple linear regression to each group: one to the female data and a separate simple linear regression to the male data. The sum of squares error for model (3) comes from adding the error sums of squares for the two simple linear regressions. It is easily seen that for females the simple linear regression has an error sum of squares of 22.459 on 22 degrees of freedom and the males have an error sum of squares of 49.614 also on 22 degrees of freedom. Thus model (3) has an error sum of squares of $22.459 + 49.614 = 72.073$ on $22 + 22 = 44$ degrees of freedom. The mean squared error for model (3) is

$$MSE(3) = \frac{72.073}{44} = 1.638$$

and using results from Table 10.3, the test of model (3) against the reduced model (2) has

$$F = \frac{[72.279 - 72.073]/[45 - 44]}{1.638} = \frac{.206}{1.638} = .126.$$

The F statistic is very small; there is no evidence that we need to fit different slopes for the two sexes. We now return to the analysis of model (2).

Frequently, computer programs for fitting model (2) give information on the regression parameter γ. Often, this is presented in the same way the program gives information on parameters in pure regression problems, e.g.,

Covariate	$\hat{\gamma}$	$SE(\hat{\gamma})$	t	P
Body weight	2.7948	0.5759	4.853	0.000.

Note that the t statistic here is the square root of the F statistic for body weights in Table 10.3. The P values are identical. Again, we find clear evidence for the effect of body weights. A 95% confidence interval for γ has end points

$$2.7948 \pm 2.014(0.5759)$$

which yields the interval $(1.6, 4.0)$. We are 95% confident that, for data comparable to the data in this study, an increase in body weight of one kilogram corresponds to a mean increase in heart weight of between 1.6 g and 4.0 g.

In model (2), comparing treatments by comparing the treatment means $\bar{y}_{i\cdot}$ is inappropriate because of the complicating effect of the covariate. Adjusted means are often used to compare treatments. The formula and the actual values for the adjusted means are given below along with the raw means for body weights and heart rates.

$$\text{Adjusted means} \equiv \bar{y}_{i\cdot} - \hat{\gamma}(\bar{z}_{i\cdot} - \bar{z}_{\cdot\cdot})$$

Sex	N	Body	Heart	Adj. heart
Female	24	2.333	8.887	9.580
Male	24	2.829	11.050	10.357
Combined	48	2.581	9.969	

We have seen previously that there is little evidence of a differential effect on heart weights due to sexes after adjusting for body weights. Nonetheless, from the adjusted means what evidence exists suggests that, even after adjusting for body weights, a typical heart weight for males, 10.357, is larger than a typical heart weight for females, 9.580.

Figures 10.1 through 10.3 contain residual plots. The plot of residuals versus predicted values looks exceptionally good. The plot of residuals versus sexes shows slightly less variability for females than for males. The difference is probably not enough to worry about. The normal plot of the residuals is alright with W' above the appropriate percentile.

MINITAB COMMANDS

The following Minitab commands were used to generate the analysis of these data. The means given by the 'ancova' subcommand 'means' are the adjusted treatment means.

```
MTB > names c1 'body' c2 'heart' c3 'sex'
MTB > note   Fit model (1).
MTB > oneway c2 c3
MTB > note   Fit model (2).
MTB > ancova c2 = c3;
SUBC> covar c1;
SUBC> resid c10;
SUBC> fits c11;
SUBC> means c3.
MTB > plot c10 c11
MTB > plot c10 c3
MTB > note   Split the data into females and males and
MTB > note   perform two regressions to fit model (3).
MTB > copy c1 c2 to c11 c12;
SUBC> use c3=1.
```

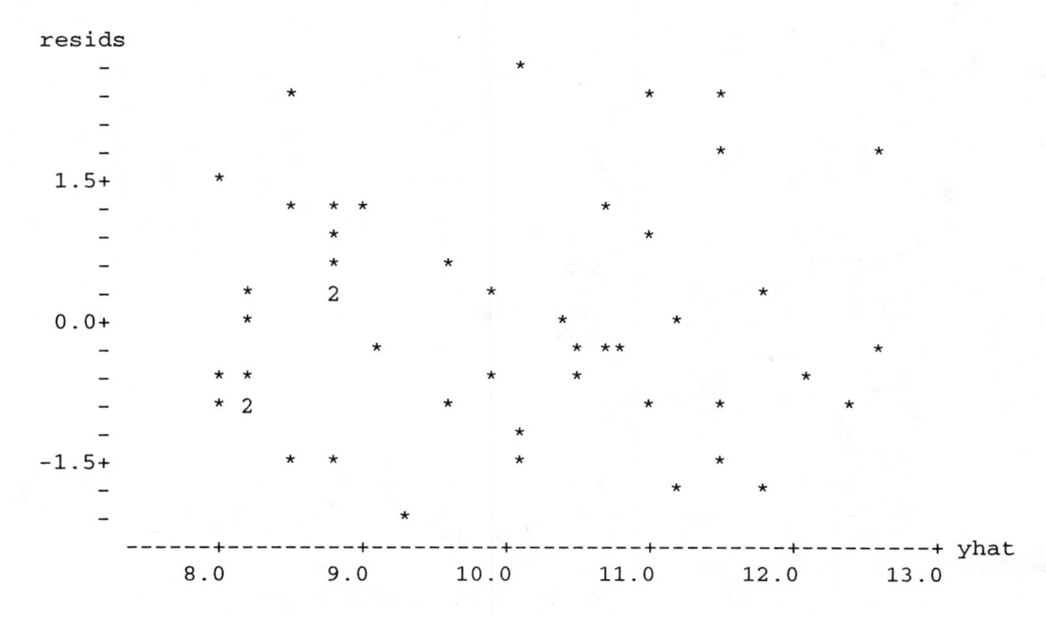

FIGURE 10.1. Residuals versus predicted values.

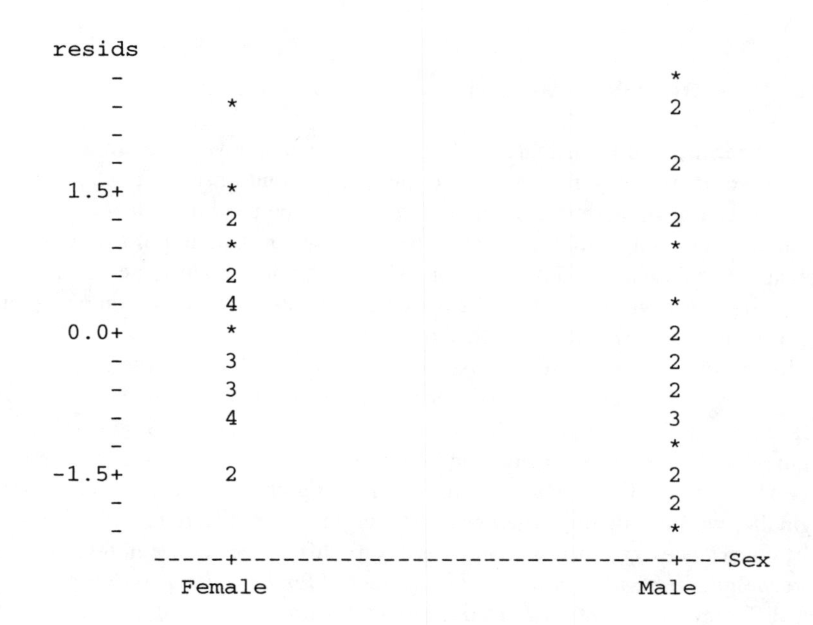

FIGURE 10.2. Residuals versus sex.

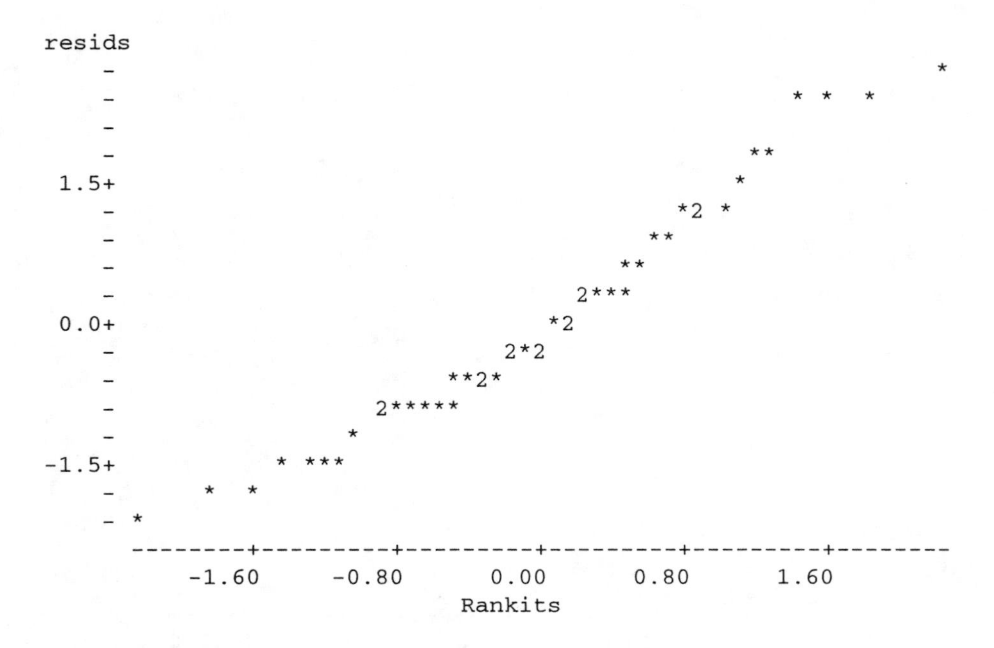

FIGURE 10.3. Normal plot of residuals, $W' = 0.968$.

```
MTB > regress c12 on 1 c11
MTB > copy c1 c2 to c21 c22;
SUBC> use c3=2.
MTB > regress c22 on 1 c21
```

10.2 Analysis of covariance in designed experiments

In designing an experiment to investigate a group of treatments, covariates are used to reduce the error of treatment comparisons. One way to use the concomitant observations is to define blocks based on them. For example, income, IQ, and heights can all be used to collect people into similar groups for a randomized complete block design. In fact, any construction of blocks must be based on information not otherwise incorporated into the ANOVA model, so any experiment with blocking uses concomitant information. In analysis of covariance we use the concomitant observations more directly, as regression variables in the statistical model.

Obviously, for a covariate to help our analysis it must be related to the dependent variable. Unfortunately, improper use of concomitant observations can invalidate, or at least alter, comparisons among the treatments. In the example of Section 10.1, the original ANOVA demonstrated an effect on heart weights due to sex but after adjusting for body weights, there was little evidence for a sex difference. The very nature of what we were comparing changed when we adjusted for body weights. Originally, we investigated whether heart weights were different for females and males. The analysis of covariance examined whether there were differences between female heart weights and male heart weights *beyond what could be accounted for by the regression on body weights*. These are very different interpretations. In a designed experiment, we want to investigate the effects of the treatments and not the treatments adjusted for some covariates. To this end, in a designed experiment we require that the covariates be logically independent of the treatments. In particular, we require that the concomitant observations be made before assigning the treatments to the experimental units,

the concomitant observations be made after assigning treatments to experimental units but before the effect of the treatments has developed, or

the concomitant observations be such that they are unaffected by treatment differences.

For example, suppose the treatments are five diets for cows and we wish to investigate milk production. Milk production is related to the size of the cow, so we might pick height of the cow as a covariate. For immature cows over a long period of time, diet may well affect both height and milk production. Thus to use height as a covariate we should measure heights before treatments begin or we could measure heights, say, two days after treatments begin. Two days on any reasonable diet should not affect a cow's height. Alternatively, if we use only mature cows their heights should be unaffected by diet and thus the heights of mature cows could be measured at any time during the experiment. Typically, *one should be very careful when claiming that a covariate measured near the end of an experiment is unaffected by treatments.*

The requirements listed above on the nature of covariates in a designed experiment are imposed so that the treatment effects do not depend on the presence or absence of covariates in the analysis. The treatment effects are logically identical regardless of whether covariates are actually measured or incorporated into the analysis. Recall that in the observational study of Section 10.1, the nature of the treatment (sex) effects changed depending on whether covariates were incorporated in the model. The role of the covariates in the analysis of a designed experiment is solely to reduce the error. In particular, using good covariates should reduce both the variance of the observations σ^2 and its estimate, the *MSE*. On the other hand, we will see in the next section that one pays a price for using covariates. Variances of treatment comparisons are σ^2 times a constant. With covariates in the model, the constant is larger than when they are not present. However, with well chosen covariates the appropriate value of σ^2 should be sufficiently smaller that the reduction in *MSE* overwhelms the increase in the multiplier. Nonetheless, in designing an experiment we need to play off these aspects against one another. We need covariates whose reduction in *MSE* more than makes up for the increase in the constant.

The requirements imposed on the nature of the covariates in a designed experiment have little affect on the analysis illustrated in the Section 10.1. The analysis focuses on a model such as (10.1.2). In Section 10.1, we also considered model (10.1.3) that has different slope parameters for the different treatments (sexes). The requirements on the covariates in a designed experiment imply that the relationship between the dependent variable y and the covariate z *cannot* depend on the treatments. Thus with covariates chosen for a designed experiment *it is inappropriate to have slope parameters that depend on the treatment.* There is one slope that is valid for the entire analysis and the treatment effects do not depend on the presence or absence of the covariates. If a model such as (10.1.3) fits better than (10.1.2) when the covariate has been chosen appropriately, it suggests that the effects of treatments may differ from experimental unit to experimental unit. In such cases a treatment cannot really be said to have *an* effect, it has a variety of effects depending on which units it is applied to. A suitable transformation of the dependent variable may alleviate the problem.

10.3 Computations and contrasts

Analysis of covariance begins with an analysis of variance model and adds a regressor to the model. The original idea in computing an analysis of covariance was to use the simple computations available for one-way ANOVAs and balanced higher-way ANOVAs to expedite the computations for the more complicated analysis of covariance model. With modern computing machines this is less crucial, but the original computational methods deserve consideration. In particular, they are extremely useful in statistical theory. To illustrate, consider a randomized complete block (RCB) experiment with a covariate z. The model is

$$y_{ij} = \mu + \alpha_i + \beta_j + \gamma z_{ij} + \varepsilon_{ij} \tag{10.3.1}$$

with $i = 1, \ldots, a$ indicating treatments, $j = 1, \ldots, b$ indicating blocks, and independent $N(0, \sigma^2)$ errors. The computational method involves performing RCB analyses on both the y_{ij}s and the z_{ij}s. In addition to computing the usual sums of *squares* for the ys and the zs, we need to compute sums of *cross products*. The formulae are given in Table 10.4. The entire analysis of covariance can be computed from the sums of squares and cross products in Table 10.4 along with the mean values needed to perform the two RCB analyses. In particular, the analysis focuses on the three error lines in Table 10.4. *For analysis of covariance models other than (10.3.1), similar methods applied to the error lines yield the appropriate analysis.* The analogous computations for model (10.1.2) are illustrated at the end of the section.

TABLE 10.4. RCB analysis of covariance, one covariate

Source	df	SS_{yy}
Trt	$a-1$	$b\sum_{i=1}^{a}(\bar{y}_{i\cdot} - \bar{y}_{\cdot\cdot})^2$
Blocks	$b-1$	$a\sum_{j=1}^{b}(\bar{y}_{\cdot j} - \bar{y}_{\cdot\cdot})^2$
Error	$(a-1)(b-1)$	subtraction
Total	$ab-1$	$\sum_{i=1}^{a}\sum_{j=1}^{b}(\bar{y}_{ij} - \bar{y}_{\cdot\cdot})^2$

Source	df	SS_{yz}
Trt	$a-1$	$b\sum_{i=1}^{a}(\bar{y}_{i\cdot} - \bar{y}_{\cdot\cdot})(\bar{z}_{i\cdot} - \bar{z}_{\cdot\cdot})$
Blocks	$b-1$	$a\sum_{j=1}^{b}(\bar{y}_{\cdot j} - \bar{y}_{\cdot\cdot})(\bar{z}_{\cdot j} - \bar{z}_{\cdot\cdot})$
Error	$(a-1)(b-1)$	subtraction
Total	$ab-1$	$\sum_{i=1}^{a}\sum_{j=1}^{b}(\bar{y}_{ij} - \bar{y}_{\cdot\cdot})(\bar{z}_{ij} - \bar{z}_{\cdot\cdot})$

Source	df	SS_{zz}
Trt	$a-1$	$b\sum_{i=1}^{a}(\bar{z}_{i\cdot} - \bar{z}_{\cdot\cdot})^2$
Blocks	$b-1$	$a\sum_{j=1}^{b}(\bar{z}_{\cdot j} - \bar{z}_{\cdot\cdot})^2$
Error	$(a-1)(b-1)$	subtraction
Total	$ab-1$	$\sum_{i=1}^{a}\sum_{j=1}^{b}(\bar{z}_{ij} - \bar{z}_{\cdot\cdot})^2$

From the error sums of squares and cross products in Table 10.4, compute the SSE for model (10.3.1) as

$$SSE = SSE_{yy} - \frac{(SSE_{yz})^2}{SSE_{zz}}.$$

Model (10.3.1) has one more parameter (γ) than the corresponding RCB model, so

$$dfE = (a-1)(b-1) - 1$$

and

$$MSE = \frac{SSE}{(a-1)(b-1)-1}.$$

The estimate of γ is computed as

$$\hat{\gamma} = \frac{SSE_{yz}}{SSE_{zz}}$$

and the standard error is

$$SE(\hat{\gamma}) = \sqrt{\frac{MSE}{SSE_{zz}}}.$$

The sum of squares for the covariate can be computed as

$$SS(\hat{\gamma}) = \hat{\gamma}^2 SSE_{zz}.$$

All of these formulae are very similar to formulae used in simple linear regression. In fact, if model (10.3.1) had no treatment or block effects, it would be a simple linear regression model and these formula would give the usual analysis for a simple linear regression.

The estimate of a contrast in the treatment effects, say $\sum_{i=1}^{a} \lambda_i \alpha_i$, is

$$\sum_{i=1}^{a} \lambda_i \left(\bar{y}_{i\cdot} - \hat{\gamma}\bar{z}_{i\cdot}\right).$$

The variance of the estimated contrast is

$$\mathrm{Var}\left(\sum_{i=1}^{a} \lambda_i \left(\bar{y}_{i\cdot} - \bar{z}_{i\cdot}\hat{\gamma}\right)\right) = \sigma^2 \left[\frac{\sum_{i=1}^{a} \lambda_i^2}{b} + \frac{\left(\sum_{i=1}^{a} \lambda_i \bar{z}_{i\cdot}\right)^2}{SSE_{zz}}\right].$$

Recall that in an RCB model without covariates, the variance of the estimate of $\sum_{i=1}^{a} \lambda_i \alpha_i$ is $\sigma^2 \left[\sum_{i=1}^{a} \lambda_i^2 / b\right]$ which is the variance given above without the term involving the z_{ij}s. The RCB variance appears to be strictly smaller than the variance from model (10.3.1). This illusion occurs because the variance parameters σ^2 are not the same in the covariate model (10.3.1) and the RCB model without covariates. With good covariates, the variance σ^2 in the covariate model should be much smaller than the corresponding variance in the model without covariates. In fact, the σ^2 in the covariate model should be sufficiently small to *more than make up for* the increase in the term that is multiplying σ^2.

The standard error of the estimated contrast is obtained immediately from the variance formula. It is

$$SE\left(\sum_{i=1}^{a} \lambda_i \left(\bar{y}_{i\cdot} - \bar{z}_{i\cdot}\hat{\gamma}\right)\right) = \sqrt{MSE\left[\frac{\sum_{i=1}^{a} \lambda_i^2}{b} + \frac{\left(\sum_{i=1}^{a} \lambda_i \bar{z}_{i\cdot}\right)^2}{SSE_{zz}}\right]}.$$

Because of the complications caused by having the regressor z_{ij} in the model, orthogonal contrasts are difficult to specify and of little interest.

For what they are worth, adjusted treatment means are often defined as

$$\bar{y}_{i\cdot} - \hat{\gamma}\left(\bar{z}_{i\cdot} - \bar{z}_{\cdot\cdot}\right).$$

These adjusted treatment means can be used in place of the values $\bar{y}_{i\cdot} - \hat{\gamma}\bar{z}_{i\cdot}$ when estimating contrasts. Using adjusted treatment means has no affect on the standard error of the estimated contrast.

To test for the existence of treatment effects, we test model (10.3.1) against the reduced model

$$y_{ij} = \mu + \beta_j + \gamma z_{ij} + e_{ij}. \tag{10.3.2}$$

In model (10.3.2) the treatment effects α_i have been eliminated, so the sum of squares for treatments is incorporated into the error term of model (10.3.2). To find the SSE for model (10.3.2) we combine the treatment and error lines in Table 10.4 and use the standard formula.

$$SSE(2) = \left[\left(SSTrt_{yy} + SSE_{yy} \right) - \frac{\left(SSTrt_{yz} + SSE_{yz} \right)^2}{SSTrt_{zz} + SSE_{zz}} \right].$$

The sum of squares used in the numerator of the F statistic for testing treatments is

$$\begin{aligned} SSTrt &= SSE(2) - SSE(1) \\ &= \left[\left(SSTrt_{yy} + SSE_{yy} \right) - \frac{\left(SSTrt_{yz} + SSE_{yz} \right)^2}{SSTrt_{zz} + SSE_{zz}} \right] - \left[SSE_{yy} - \frac{SSE_{yz}^2}{SSE_{zz}} \right]. \end{aligned}$$

This has the standard number of degrees of freedom, $a - 1$. The F statistic for treatments is

$$F = \frac{SSTrt/(a-1)}{MSE}$$

with MSE coming from model (10.3.1).

If it is of interest to test for block effects or investigate block contrasts, the methods given above apply with appropriate substitutions.

While the discussion in this section has been in terms of analyzing randomized complete block designs, analogous procedures work for any analysis of covariance model in which the corresponding analysis of variance computations are tractable. We illustrate the computations with the balanced one-way ANOVA data of Section 10.1.

EXAMPLE 10.3.1. Table 10.1 gave Fisher's heart and body weight data, Table 10.2 gave the one-way analysis of variance for heart weights, and Table 10.3 gave an analysis of variance table for the covariate model. Table 10.5 is analogous to Table 10.4; it gives the standard analysis of covariance table used to compute Table 10.3. Table 10.5 has the same degrees of freedom and sums of squares as given in Table 10.2 for heart weights along with a similar one-way ANOVA for body weights and cross product terms.

TABLE 10.5. Analysis of covariance for heart weights, body weights

Source	df	Heart SS_{yy}	Cross SS_{yz}	Body SS_{zz}
Sex	1	56.117	12.867	2.950
Error	46	110.106	13.535	4.843
Total	47	166.223	26.402	7.793

Compute the SSE as

$$SSE = 110.106 - \frac{(13.535)^2}{4.843} = 72.279.$$

The model has one more parameter (γ) than the corresponding one-way ANOVA model, so

$$dfE = 46 - 1 = 45$$

and

$$MSE = \frac{72.279}{45} = 1.606.$$

The estimate of γ is computed as

$$\hat{\gamma} = \frac{13.535}{4.843} = 2.7948$$

and the standard error is

$$SE(\hat{\gamma}) = \sqrt{\frac{1.606}{4.843}} = 0.5759.$$

The information on γ can be summarized as in Section 10.1.

Covariate	$\hat{\gamma}$	$SE(\hat{\gamma})$	t	P
Body weight	2.7948	0.5759	4.853	0.000

The sum of squares for body weight reported in Table 10.3 is computed as

$$SS(\hat{\gamma}) = (2.7948)^2 4.843 = 37.828.$$

Consider the contrast between females and males, say, $(1)\alpha_1 + (-1)\alpha_2$. To estimate this we need the body weight and heart weight means,

Sex	N	Body	Heart
Female	24	2.333	8.887
Male	24	2.829	11.050
Combined	48	2.581	9.969

The estimated contrast is

$$[8.887 - 2.7948(2.333)] - [11.050 - 2.7948(2.829)] = -0.777.$$

The standard error is

$$SE(\hat{\alpha}_1 - \hat{\alpha}_2) = \sqrt{1.606\left[\frac{1^2 + (-1)^2}{24} + \frac{(2.333 - 2.829)^2}{4.843}\right]} = .464.$$

With an estimate and a standard error, we can use the $t(45)$ distribution to form tests and confidence intervals for the difference between females and males after adjusting for body weights. Note that the t test is

$$t_{obs} = \frac{-.777}{.464} = -1.6746$$

and squaring this gives $(-1.6746)^2 = 2.80$, which is the F statistic for sexes in Table 10.3. Of course if there were more than 1 degree of freedom for treatments, the sum of squares for treatments would not typically equal the sum of squares for a contrast.

For what they are worth, the adjusted treatment mean for females is

$$8.887 - 2.7948(2.333 - 2.581) = 9.580$$

and for males is

$$11.050 - 2.7948(2.829 - 2.581) = 10.357.$$

These adjusted treatment means can be used in place of the values $\bar{y}_{i\cdot} - \hat{\gamma}\bar{z}_{i\cdot}$ when estimating contrasts. For example,

$$9.580 - 10.357 = -0.777$$

just as in the estimated contrast.

To test for the existence of treatment effects, the sum of squares used in the numerator of the F statistic is

$$\begin{aligned} SSTrt &= \left[(56.117 + 110.106) - \frac{(12.867 + 13.535)^2}{2.950 + 4.843}\right] - 72.279 \\ &= \left[166.223 - \frac{(26.402)^2}{7.793}\right] - 72.279 \\ &= 4.5 \end{aligned}$$

and the F statistic is

$$F = \frac{4.5/1}{1.606} = 2.80.$$

Note that, in general, computing the $SSTrt$ involves $SSTrt_{ab} + SSE_{ab}$, but in the one-way ANOVA application these equal $SSTot_{ab}$. □

10.4 Power transformations and Tukey's one degree of freedom

One application for analysis of covariance is in detecting nonadditivity in the results of a randomized complete block design. The model for a RCB is

$$y_{ij} = \mu + \alpha_i + \beta_j + \varepsilon_{ij} \tag{10.4.1}$$

in which the grand mean μ and the effects of treatments α and of blocks β are added together. It is possible that the effects could be something other than additive, say, multiplicative with a model

$$y_{ij} = MA_i B_j \epsilon_{ij}. \tag{10.4.2}$$

The multiplicative model can be transformed into an additive model using logarithms. Taking logs of both sides in equation (2) gives

$$\log(y_{ij}) = \log(M) + \log(A_i) + \log(B_j) + \log(\epsilon_{ij}).$$

If we redefine the parameters as $\mu = \log(M)$, $\alpha_i = \log(A_i)$, $\beta_j = \log(B_j)$, and $\varepsilon_{ij} = \log(\epsilon_{ij})$, we get the additive model

$$\log(y_{ij}) = \mu + \alpha_i + \beta_j + \varepsilon_{ij}.$$

Generally, when the additive model (1) is inappropriate, we want to find a transformation of the y_{ij}s that makes the additive analysis useful. Thus, we seek a value λ for which

$$y_{ij}^{(\lambda)} = \mu + \alpha_i + \beta_j + \varepsilon_{ij}$$

where, as in our discussion of transformations for simple linear regression, we use the family of modified power transformations

$$y_i^{(\lambda)} = \begin{cases} (y_i^\lambda - 1)/\lambda & \lambda \neq 0 \\ \log(y_i) & \lambda = 0 \end{cases}.$$

In Section 7.10, we used a constructed variable to obtain an approximate test of whether a power transformation was needed. Exactly the same method can be applied to the RCB model (1). The constructed variable is

$$w_{ij} = y_{ij}\left[\log(y_{ij}/\tilde{y}) - 1\right],$$

where $\tilde{y}$ is again the geometric mean of all of the observations. The only difference between this variable and the constructed variable in Section 7.10 is the difference between having one subscript and two on the ys. Incorporating the constructed variable into the RCB model (1) gives the analysis of covariance model

$$y_{ij} = \mu + \alpha_i + \beta_j + \gamma w_{ij} + \varepsilon_{ij}. \qquad (10.4.3)$$

As was mentioned above and in Section 7.10, fitting a model with a constructed variable and performing the usual test of $H_0 : \gamma = 0$ gives only an *approximate* test of whether a power transformation is needed. The usual t distribution is not really appropriate for this test. The problem is that the constructed variable w involves the ys, so the ys appear on both sides of the equality in model (3). This is enough to invalidate the theory behind the usual test. It turns out that this difficulty can be avoided by using the predicted values from model (1). We write these as $\hat{y}_{ij(1)}$s, where the subscript (1) is a reminder that the predicted values come from the additive model (1). We can now define a new constructed variable,

$$\tilde{w}_{ij} = \hat{y}_{ij(1)} \log(\hat{y}_{ij(1)})$$

and fit the analysis of covariance model

$$y_{ij} = \mu + \alpha_i + \beta_j + \gamma \tilde{w}_{ij} + \varepsilon_{ij} \qquad (10.4.4)$$

The new constructed variable $\tilde{w}_{ij}$ simply replaces y_{ij} with $\hat{y}_{ij(1)}$ in the definition of w_{ij} and deletes some terms made redundant by using the $\hat{y}_{ij(1)}$s. If model (1) is valid, the usual test of $H_0 : \gamma = 0$ from model (4) has the standard t distribution in spite of the fact that the $\tilde{w}_{ij}$s depend on the y_{ij}s. By basing the constructed variable on the $\hat{y}_{ij(1)}$s, we are able to get an exact t test for $\gamma = 0$ and restrict the weird behavior of the test statistic to situations in which $\gamma \neq 0$.

Tukey's (1949) one degree of freedom test for nonadditivity uses neither the constructed variable w_{ij} nor $\tilde{w}_{ij}$ but a third constructed variable that is an approximation to $\tilde{w}_{ij}$. Using a method from calculus known as Taylor's approximation (expanding about $\bar{y}_{\cdot\cdot}$) and simplifying the approximation by eliminating terms that have no effect on the test of $H_0 : \gamma = 0$, we get $\hat{y}^2_{ij(1)}$ as a new constructed variable. This leads to fitting the ACOVA model

$$y_{ij} = \mu + \alpha_i + \beta_j + \gamma \hat{y}^2_{ij(1)} + \varepsilon_{ij}. \qquad (10.4.5)$$

and testing the need for a transformation by testing $H_0 : \gamma = 0$. This is Tukey's one degree of freedom test for nonadditivity. Recall that t tests are equivalent to F tests with one degree of freedom in the numerator, hence the reference to one degree of freedom in the name of Tukey's test.

Models (3), (4), and (5) all provide *rough* estimates of the appropriate power transformation. From models (3) and (4), the appropriate power is estimated by $\hat{\lambda} = 1 - \hat{\gamma}$. In model (5), because of the simplification employed after the approximation, the estimate is $\hat{\lambda} = 1 - 2\bar{y}_{\cdot\cdot}\hat{\gamma}$.

Atkinson (1985, section 8.1) gives an extensive discussion of various constructed variables for testing power transformations. In particular, he suggests (on p. 158) that while the tests based on $\tilde{w}_{ij}$ and $\hat{y}^2_{ij(1)}$ have the advantage of giving exact t tests and being easier to compute, the test using w_{ij} may be more sensitive in detecting the need for a transformation, i.e., may be more powerful.

EXAMPLE 10.4.1. Consider the randomized complete block experiment of Example 9.2.1. The data are reproduced in Table 10.6. The predicted values from model (1) and the three constructed variables are given in Table 10.7. Note that the numerical values of the constructed variables are very different.

The results of testing $H_0 : \gamma = 0$ for models (3), (4), and (5) are given below.

TABLE 10.6. Spectrometer data

Treatment	Block 1	2	3	Trt. means
New-clean	0.9331	0.8664	0.8711	$0.8902\bar{0}$
New-soiled	0.9214	0.8729	0.8627	$0.8856\bar{6}$
Used-clean	0.8472	0.7948	0.7810	$0.8076\bar{6}$
Used-soiled	0.8417	0.8035	0.8099	$0.8183\bar{6}$
Block means	0.885850	0.834400	0.831175	0.850475

TABLE 10.7. Additive model predictions and constructed variables for the spectrometer data

Trt.	Block	$\hat{y}_{(1)}$	w	$\tilde{w}$	$\hat{y}^2_{(1)}$
1	1	0.925575	−0.845246	−0.071584	0.856689
1	2	0.874125	−0.849083	−0.117598	0.764095
1	3	0.870900	−0.848976	−0.120383	0.758467
2	1	0.921042	−0.846274	−0.075756	0.848318
2	2	0.869592	−0.848929	−0.121509	0.756190
2	3	0.866367	−0.849149	−0.124278	0.750591
3	1	0.843042	−0.849252	−0.143940	0.710719
3	2	0.791592	−0.847471	−0.185003	0.626617
3	3	0.788367	−0.846435	−0.187467	0.621522
4	1	0.853742	−0.849221	−0.134999	0.728875
4	2	0.802292	−0.848000	−0.176731	0.643672
4	3	0.799067	−0.848329	−0.179239	0.638508

Model	Constructed variable	$\hat{\gamma}$	$SE(\hat{\gamma})$	t	P
(3)	w	1.505	2.24	0.6717	0.532
(4)	$\tilde{w}$	2.165	2.40	0.9022	0.408
(5)	$\hat{y}^2_{(1)}$	1.275	1.40	0.9128	0.403

Models (3) and (4) involve distinct methods of defining the constructed variable, but the tests from the two methods are roughly similar for these data. Model (5) involves an approximation to and simplification of model (4). While the estimates and standard errors are very different in models (4) and (5), the t statistics are very similar. With these data, the t statistics for the constructed variables are all small, so there is no suggestion of the need for a power transformation. This level of agreement does not always occur. Atkinson (1985, p. 160) gives an example in which the various t statistics are not all numerically similar, though all suggest the need for a transformation.

If a power transformation were needed, model (3) suggests examining λ values in the neighborhood of $1 - (1.505) = -.505$. Model (4) suggests examining λ values in the neighborhood of $1 - (2.165) = -1.165$. Model (5) suggests examination of λ values in the neighborhood of $1 - 2(0.850475)(1.275) = -1.169$. Again, note the similarity between the results for models (4) and (5). □

So far, *the methods discussed have made no use of the special structure of model (1). Indeed, these methods can be used to test the need for a power transformation in any analysis of variance or regression model with independent normal errors having mean zero and the same variance*. In particular, they can be used with any of the models we have discussed so far and with the factorial treatment structure models of the next chapter and the multiple regression models examined later. (They should not be used with models that have more than one error (random) term, cf. Chapter 12.)

ANOTHER LOOK AT TUKEY'S TEST

Traditionally, Tukey's test is presented in a different context, one that relies on the special structure of model (1). To examine interaction in a RCB model, one might consider fitting the *nonlinear* models

$$y_{ij} = \mu + \alpha_i + \beta_j + \gamma\alpha_i\beta_j + \varepsilon_{ij}$$

or

$$y_{ij} = \mu + \alpha_i + \beta_j + \gamma\frac{\alpha_i\beta_j}{\mu} + \varepsilon_{ij}.$$

These differ from any other models that we have considered because they include parameters that are multiplied together, e.g., $\alpha_i\beta_j$. Such models are much more difficult to fit. We can finesse the fitting problem by fitting the models

$$y_{ij} = \mu + \alpha_i + \beta_j + \gamma\hat{\alpha}_{(1)i}\hat{\beta}_{(1)j} + \varepsilon_{ij} \tag{10.4.6}$$

or

$$y_{ij} = \mu + \alpha_i + \beta_j + \gamma\frac{\hat{\alpha}_{(1)i}\hat{\beta}_{(1)j}}{\hat{\mu}_{(1)}} + \varepsilon_{ij} \tag{10.4.7}$$

where the models incorporate some estimates of parameters that are appropriate in model (1), i.e.,

$$\hat{\mu}_{(1)} = \bar{y}_{\cdot\cdot} \quad \hat{\alpha}_{(1)i} = \bar{y}_{i\cdot} - \bar{y}_{\cdot\cdot} \quad \hat{\beta}_{(1)j} = \bar{y}_{\cdot j} - \bar{y}_{\cdot\cdot}.$$

Both models (6) and (7) give tests of $H_0 : \gamma = 0$ that are identical to the test from model (5). Note that the γ parameters in these three models are not the same. In particular, the estimate of γ in model (6) equals $\hat{\mu}_{(1)}$ times the estimate in model (7). Model (7) has one advantage in that for this model, just like model (3), $1 - \hat{\gamma}$ provides a rough estimate of the appropriate power transformation.

EXAMPLE 10.4.2. Fitting model (7) gives

Model	Constructed variable	$\hat{\gamma}$	$SE(\hat{\gamma})$	t	P
(10.4.7)	$\hat{\alpha}_{(1)i}\hat{\beta}_{(1)j}/\hat{\mu}_{(1)}$	2.168	2.38	0.9128	0.403

As mentioned, the t test is identical to that given earlier for model (5). The suggested power transformation from model (7) is $\hat{\lambda} = 1 - (2.168) = -1.168$, which is identical (up to round off error) to the value from model (5). With many computer programs, it is much easier to obtain the constructed variable for model (5) than for either of models (6) or (7) yet the results from models (6) and (7) are equivalent to the results for model (5). □

The tests used with models (4), (5), (6), and (7) are special cases of a general procedure introduced by Rao (1965) and Milliken and Graybill (1970). In addition, Atkinson (1985), Cook and Weisberg (1982), and Emerson (1983) contain useful discussions of constructed variables and methods related to Tukey's test.

MINITAB COMMANDS

Minitab commands are given below for this analysis of the data.

```
MTB > names c1 'y' c2 'Trts' c3 'Blk' c4 'yhat_{(1)}'
MTB > names c5 'w' c6 'wtilde' c8 'yhat^2'
MTB > note      FIT MODEL (10.4.1)
```

```
MTB > anova c1=c2 c3;
SUBC> fits c4.
MTB > note      CONSTRUCT w AND FIT MODEL (10.4.3)
MTB > let c10=loge(c1)
MTB > mean c9 k1
MTB > let c5=c1*(c10-k1-1)
MTB > ancova c1=c2 c3;
SUBC> covar c5.
MTB > note      CONSTRUCT wtilde AND FIT MODEL (10.4.4)
MTB > let c6=c4*loge(c4)
MTB > ancova c1=c2 c3;
SUBC> covar c6.
MTB > note      FIT MODEL (10.4.5)
MTB > let c8=c4**2
MTB > ancova c1=c2 c3;
SUBC> cova c8.
```

10.5 Exercises

EXERCISE 10.5.1. Table 10.8 contains data from Sulzberger (1953) and Williams (1959) on y, the maximum compressive strength parallel to the grain of wood from ten hoop pine trees. The data also include the temperature of the evaluation and a covariate z, the moisture content of the wood. Analyze the data. Examine (tabled) polynomial contrasts in the temperatures.

TABLE 10.8. Compressive strength of ten hoop pine trees (y) at different temperatures and with various moisture contents (z)

	Temperature									
	−20° C		0° C		20° C		40° C		60° C	
Tree	z	y	z	y	z	y	z	y	z	y
1	42.1	13.14	41.1	12.46	43.1	9.43	41.4	7.63	39.1	6.34
2	41.0	15.90	39.4	14.11	40.3	11.30	38.6	9.56	36.7	7.27
3	41.1	13.39	40.2	12.32	40.6	9.65	41.7	7.90	39.7	6.41
4	41.0	15.51	39.8	13.68	40.4	10.33	39.8	8.27	39.3	7.06
5	41.0	15.53	41.2	13.16	39.7	10.29	39.0	8.67	39.0	6.68
6	42.0	15.26	40.0	13.64	40.3	10.35	40.9	8.67	41.2	6.62
7	40.4	15.06	39.0	13.25	34.9	10.56	40.1	8.10	41.4	6.15
8	39.3	15.21	38.8	13.54	37.5	10.46	40.6	8.30	41.8	6.09
9	39.2	16.90	38.5	15.23	38.5	11.94	39.4	9.34	41.7	6.26
10	37.7	15.45	35.7	14.06	36.7	10.74	38.9	7.75	38.2	6.29

EXERCISE 10.5.2. Smith, Gnanadesikan, and Hughes (1962) gave data on urine characteristics of young men. The men were divided into four categories based on obesity. The data contain a covariate z that measures specific gravity. The dependent variable is y_1; it measures pigment creatinine. These variables are included in Table 10.9. Perform an analysis of covariance on y_1. How do the conclusions about obesity effects change between the ACOVA and the results of the ANOVA that ignores the covariate?

TABLE 10.9. Excretory characteristics from Smith et al. (1962)

Group I			Group II		
z	y_1	y_2	z	y_1	y_2
24	17.6	5.15	31	18.1	9.00
32	13.4	5.75	23	19.7	5.30
17	20.3	4.35	32	16.9	9.85
30	22.3	7.55	20	23.7	3.60
30	20.5	8.50	18	19.2	4.05
27	18.5	10.25	23	18.0	4.40
25	12.1	5.95	31	14.8	7.15
30	12.0	6.30	28	15.6	7.25
28	10.1	5.45	21	16.2	5.30
24	14.7	3.75	20	14.1	3.10
26	14.8	5.10	15	17.5	2.40
27	14.4	4.05	26	14.1	4.25
			24	19.1	5.80
			16	22.5	1.55

Group III			Group IV		
z	y_1	y_2	z	y_1	y_2
18	17.0	4.55	32	12.5	2.90
10	12.5	2.65	25	8.7	3.00
33	21.5	6.50	28	9.4	3.40
25	22.2	4.85	27	15.0	5.40
35	13.0	8.75	23	12.9	4.45
33	13.0	5.20	25	12.1	4.30
31	10.9	4.75	26	13.2	5.00
34	12.0	5.85	34	11.5	3.40
16	22.8	2.85			
31	16.5	6.55			
28	18.4	6.60			

EXERCISE 10.5.3. Smith, Gnanadesikan, and Hughes (1962) also give data on the variable y_2 that measures chloride in the urine of young men. These data are also reported in Table 10.9. As in the previous problem, the men were divided into four categories based on obesity. Perform an analysis of covariance on y_2 again using the specific gravity as the covariate z. Compare the results of the ACOVA to the results of the ANOVA that ignores the covariate.

EXERCISE 10.5.4. Test the need for a power transformation in each of the following problems from the previous chapter. Use all three constructed variables on each data set and compare results.

(a) Exercise 9.5.1.

(b) Exercise 9.5.2.

(c) Exercise 9.5.3.

(d) Exercise 9.5.4.

(e) Exercise 9.5.5.

(f) Exercise 9.5.6.

(g) Exercise 9.5.8.

EXERCISE 10.5.5. Consider the analysis of covariance for a completely randomized design with one covariate. Find the form for a 99% prediction interval for an observation, say, from the first treatment group with a given covariate value z.

EXERCISE 10.5.6. Assuming that in model (10.3.1) $Cov(\bar{y}_{i\cdot}, \hat{\gamma}) = 0$, show that

$$\text{Var}\left(\sum_{i=1}^{a} \lambda_i \left(\bar{y}_{i\cdot} - \bar{z}_{i\cdot}\hat{\gamma}\right)\right) = \sigma^2 \left[\frac{\sum_{i=1}^{a} \lambda_i^2}{b} + \frac{\left(\sum_{i=1}^{a} \lambda_i \bar{z}_{i\cdot}\right)^2}{SSE_{zz}}\right].$$

Chapter 11

Factorial treatment structures

Factorial treatment structures are simply an efficient way of defining the treatments used in an experiment. They can be used with any of the standard experimental designs discussed in Chapter 9. Factorial treatment structures have two great advantages, they give information that is not readily available from other methods and they use experimental material very efficiently. Section 11.1 introduces factorial treatment structures with an examination of treatments that involve two factors. Section 11.2 gives general results for two-factor treatment structures. Section 11.3 presents an example with three factors. Section 11.4 examines extensions of the Latin square designs that were discussed in Section 9.3.

11.1 Two factors

The effect of alcohol and sleeping pills taken together is much greater than one would suspect based on examining the effects of alcohol and sleeping pills separately. If we did one experiment with 20 subjects to establish the effect of a 'normal' dose of alcohol and a second experiment with 20 subjects to establish the effect of a 'normal' dose of sleeping pills, the temptation would be to conclude (incorrectly) that the effect of taking a normal dose of both alcohol and sleeping pills would be just the sum of the individual effects. Unfortunately, the two separate experiments provide no basis for either accepting or rejecting such a conclusion.

We can redesign the investigation to be both more efficient and more informative by using a *factorial treatment structure*. The alcohol experiment would involve 10 people getting no alcohol (a_0) and 10 people getting a normal dose of alcohol (a_1). Similarly, the sleeping pill experiment would have 10 people given no sleeping pills (s_0) and 10 people getting a normal dose of sleeping pills (s_1). The two *factors* in this investigation are alcohol (A) and sleeping pills (S). Each factor is at two *levels*, no drug (a_0 and s_0, respectively) and a normal dose (a_1 and s_1, respectively). A factorial treatment structure uses treatments that are all combinations of the different levels of the factors. Thus a factorial experiment to investigate alcohol and sleeping pills might have 5 people given no alcohol and no sleeping pills (a_0s_0), 5 people given no alcohol but a normal dose of sleeping pills (a_0s_1), 5 people given alcohol but no sleeping pills (a_1s_0), and 5 people given both alcohol and sleeping pills (a_1s_1).

Assigning the treatments in this way has two major advantages. First, it is more informative in that it provides direct evidence about the effect of taking alcohol and sleeping pills together. If the joint effect is different from the sum of the effect of alcohol plus the effect of sleeping pills, the factors are said to *interact*. If the factors interact, there does not exist a single effect for alcohol, the effect of alcohol depends on whether the person has taken sleeping pills or not. Similarly, there is no one effect for sleeping pills, the effect depends on whether a person has taken alcohol or not. Note that if the factors interact, the separate experiments described earlier have very limited value.

The second advantage of using factorial treatments is that *if the factors do not interact*, the factorial experiment is more efficient than performing the two separate experiments. The two separate experiments involve the use of 40 people, the factorial experiment involves the use of only 20 people, yet the factorial experiment contains just as much information about both alcohol effects and sleeping pill effects as the two separate experiments. The effect of alcohol can be studied by contrasting the 5 a_0s_0 people with the 5 a_1s_0 people and also by comparing the 5 a_0s_1 people with the 5 a_1s_1 people. Thus we have a total of 10 no alcohol people to compare with 10 alcohol people, just as we had in the separate experiment for alcohol. Recall that with no interaction, the effect of factor A is the same regardless of the dose of factor S, so we have 10 valid comparisons of the effect of alcohol. A similar analysis shows that we have 10 no sleeping pill people to compare with 10 people using sleeping pills, the same as in the separate experiment for sleeping pills. Thus, *when there is no interaction*, the 20 people in the factorial experiment are as informative about the effects of alcohol and sleeping pills as the 40 people in the two separate experiments. Moreover, the factorial experiment provides information about possible interactions between the factors that is unavailable from the separate experiments.

The factorial treatment concept involves only the definition of the treatments. Factorial treatment structure can be used in completely randomized designs, randomized complete block designs, and in Latin square designs. All of these designs allow for arbitrary treatments, so the treatments can be chosen to have factorial structure if the experimenter wishes.

Experiments involving factorial treatment structures are often referred to as *factorial experiments* or *factorial designs*. A useful notation for factorial experiments identifies the number of factors and the number of levels of each factor. For example, the alcohol–sleeping pill experiment has 4 treatments because there are 2 levels of alcohol times 2 levels of sleeping pills. This is described as a 2×2 factorial experiment. If we had 3 levels of alcohol and 4 doses (levels) of sleeping pills we would have a 3×4 experiment involving 12 treatments.

EXAMPLE 11.1.1. *A 2×2 factorial in 3 randomized complete blocks*
Consider again the spectroscopy data of Example 9.2.1. The treatments were all combinations of two disks (new, used) and two windows (clean, soiled), so the treatments have a 2×2 factorial structure. The full data are repeated in Table 11.1 along with treatment and block means. The analysis of variance table is given in Table 11.2.

TABLE 11.1. Spectrometer data

	Block			Trt.
Treatment	1	2	3	means
New-clean	0.9331	0.8664	0.8711	$0.8902\bar{0}$
New-soiled	0.9214	0.8729	0.8627	$0.8856\bar{6}$
Used-clean	0.8472	0.7948	0.7810	$0.8076\bar{6}$
Used-soiled	0.8417	0.8035	0.8099	$0.8183\bar{6}$
Block means	0.885850	0.834400	0.831175	0.850475

We now decompose the treatments line in the ANOVA table to account for the effects of the two

TABLE 11.2. Analysis of variance for spectrometer data

Source	df	SS	MS	F	P
Trts	3	0.0170401	0.0056800	70.05	0.000
Blocks	2	0.0075291	0.0037646	46.43	0.000
Error	6	0.0004865	0.0000811		
Total	11	0.0250558			

factors, disks and windows, and the possible interaction between them. The four treatment means can be rearranged to provide disk means and windows means.

	Treatment means		
	Window		Disk
Disk	Clean	Soiled	means
New	$.8902\bar{0}$	$.8856\bar{6}$	$.88793\bar{3}$
Used	$.8076\bar{6}$	$.8183\bar{6}$	$.81301\bar{6}$
Window means	$.84893\bar{3}$	$.85201\bar{6}$	.850475

The *main effect* for disks comes from treating them as the treatments in a one-way ANOVA. The disk means were reported above, so a mean square and sum of squares for disks can be computed as in a one-way ANOVA. The treatment means are averages of 3 observations, so the disk means are averages of 6 observations. The sample variance of the disk means is .002806253, so

$$MS(Disk) = 6(.002806253) = .0168375\,.$$

With 2 disk types, the sum of squares is the mean square times $(2-1)$,

$$SS(Disk) = (2-1).0168375 = .0168375\,.$$

To examine the main effect for windows, the mean square and sum of squares for windows are also computed as if they were treatments in a one-way ANOVA. The window means are the average of 6 observations and the sample variance of the window means is .000004753, so

$$MS(Wind) = 6(.000004753) = .0000285\,.$$

There are 2 windows, so the sum of squares is the mean square times $(2-1)$,

$$SS(Wind) = (2-1).0000285 = .0000285\,.$$

The sum of squares for interaction between disks and windows is simply what is left of the *SSTrts* after explaining the main effects for disks and windows. It is obtained by subtraction from the full sum of squares treatments,

$$\begin{aligned} SS(Disk * Wind) &= SSTrts - SS(Disk) - SS(Wind) \\ &= .0170401 - .0168375 - .0000285 \\ &= .0001741\,. \end{aligned}$$

This has some round off error in the last decimal place. Similarly,

$$\begin{aligned} df(Disk * Wind) &= dfTrts - df(Disk) - df(Wind) \\ &= 3 - 1 - 1 \\ &= 1\,. \end{aligned}$$

Alternatively, the degrees of freedom can be computed as $df(Disk * Wind) = [df(Disk)][df(Wind)]$. The mean square for interaction is

$$MS(Disk * Wind) = \frac{30.08}{1} = 30.08\,.$$

Having decomposed the treatments line, the analysis of variance table becomes

Analysis of variance for spectrometer data

Source	*df*	*SS*	*MS*	*F*	*P*
Disks	1	.0168375	.0168375	207.64	0.000
Windows	1	.0000285	.0000285	0.35	0.575
Disk*Wind.	1	.0001740	.0001740	2.15	0.193
Blocks	2	.0075291	.0037646	46.43	0.000
Error	6	.0004865	.0000811		
Total	11	.0250558			

The F statistic for disk–window interaction is not significant. This indicates a lack of evidence that disks behave differently with clean windows than with soiled windows. In examining the four disk–window means, the difference between clean and soiled windows differs in sign between new and used disks, but the effect is not significantly different from 0. This is the same conclusion we reached in Chapter 9 using an interaction contrast and interaction plots. The F statistic for disks indicates that disk types have different effects when averaging over the two window types. From the disk means, it is clear that new disks give greater yields than used disks. The F statistic for windows shows no evidence that the window types affect yield when averaged over disks. If interaction existed, this would be merely an artifact of averaging over disks, and the windows would be important because the interaction would imply that disks behave differently with the different types of windows. However, we possess no evidence of interaction.

Normally we would need to investigate contrasts in the interaction, disks, and windows, but for these data each source has only one degree of freedom, hence only one contrast. Thus the analysis of variance table provides F statistics for all the interesting contrasts and the analysis given in the previous paragraph is complete. When we considered this example earlier as a randomized complete block with four treatments, we had three degrees of freedom for treatments and examined three orthogonal contrasts. The contrasts, estimates, and sums of squares are given again in Table 11.3. Contrast D looks at the difference in disks averaging over windows. Contrast W looks at the difference in windows averaging over disks. Contrast DW looks at how the difference between disks changes from clean to soiled windows. Note that the sums of squares for the D, W, and DW contrasts are exactly the same as the sums of squares for disks, windows, and interaction in the ANOVA table. This is a phenomenon that occurs quite generally in factorial experiments. *The sums of squares for the individual factors (factorial main effects) and interactions can be constructed from either orthogonal contrasts in the original treatments or one-way ANOVA computations.*

The factorial treatment structure also suggests two residual plots that were not examined earlier. These are plots of the residuals versus disk and the residuals versus window. The plots are given in Figures 11.1 and 11.2. They give no particular cause for alarm. □

MINITAB COMMANDS

The following Minitab commands generate this analysis of variance. Here c1 is a column containing the spectrometer data, c2 has values from 1 to 4 indicating the treatment, c3 has values from 1 to 3 indicating the block, c4 has values 1 and 2 indicating the disk type, and c5 has values 1 and 2 indicating the window type.

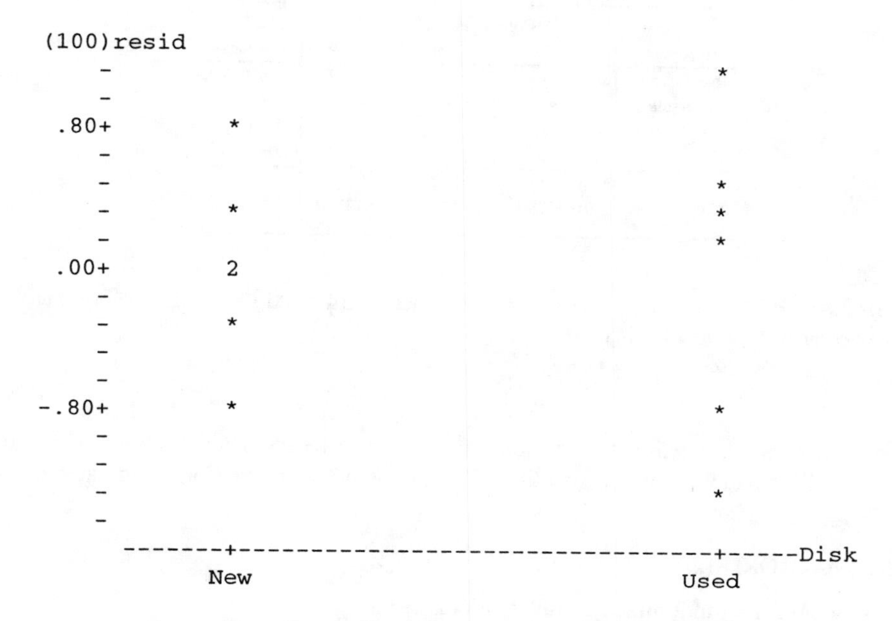

FIGURE 11.1. Plot of residuals versus disk, spectrometer data.

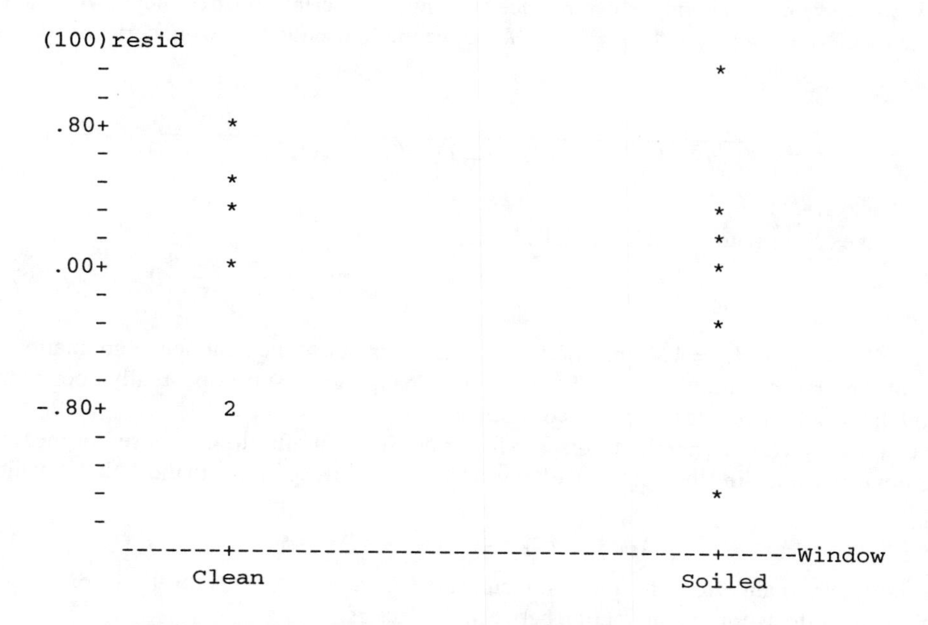

FIGURE 11.2. Plot of residuals versus window, spectrometer data.

TABLE 11.3. Contrasts for the spectrometer data

	Contrast coefficients			Trt.
Treatment	D	W	DW	means
New-clean	1	1	1	$0.8902\bar{0}$
New-soiled	1	-1	-1	$0.8856\bar{6}$
Used-clean	-1	1	-1	$0.8076\bar{6}$
Used-soiled	-1	-1	1	$0.8183\bar{6}$
Est	.149833	$-.006167$	.015233	
SS	.0168375	.0000285	.0001740	

```
MTB > names c1 'y' c2 'Trts' c3 'Blks' c4 'Disk' c5 'Window'
MTB > anova c1 = c4|c5 c3;
SUBC> means c3 c4 c5 c4*c5;
SUBC> resid c10.
```

In the 'anova' command, $c4|c5$ could be replaced by $c4$ $c5$ $c4 * c5$. The $c4$ and $c5$ terms indicate main effects for disks and windows, respectively. Disk by window interaction is indicated by $c4 * c5$.

MODELING FACTORIALS

The general model for a randomized complete block design is

$$y_{ij} = \mu + \alpha_i + \beta_j + \varepsilon_{ij}, \quad \varepsilon_{ij}\text{s independent } N(0, \sigma^2),$$

$i = 1, \ldots, a, j = 1, \ldots, b$, where i denotes the treatments and j denotes blocks. With factorial treatment structure involving two factors, one factor with levels $g = 1, \ldots, G$ and the other with levels $h = 1, \ldots, H$, we must have $a = GH$ and we can replace the single subscript i for treatments with the pair of subscripts gh. For example, with $G = 3$ and $H = 2$ we might use the following correspondence.

i	1	2	3	4	5	6
(g,h)	$(1,1)$	$(1,2)$	$(2,1)$	$(2,2)$	$(3,1)$	$(3,2)$

Moreover, we can rewrite the RCB model as

$$y_{ghj} = \mu + \alpha_{gh} + \beta_j + \varepsilon_{ghj}, \quad \varepsilon_{ghj}\text{s independent } N(0, \sigma^2), \tag{11.1.1}$$

$g = 1, \ldots, G, h = 1, \ldots, H, j = 1, \ldots, b$, where α_{gh} is the effect due to the treatment combination having level g of the first factor and level h of the second. Changing the subscripts really does nothing to the model; the subscripting is merely a convenience.

We can also rewrite the model to display factorial effects similar to those used in the analysis. This is done by expanding the treatment effects into effects corresponding to the ANOVA table lines. Write

$$y_{ghj} = \mu + \gamma_g + \xi_h + (\gamma\xi)_{gh} + \beta_j + \varepsilon_{ghj}, \tag{11.1.2}$$

where the γ_gs are main effects for the first factor, the ξ_hs are main effects for the second factor, and the $(\gamma\xi)_{gh}$s are effects for the interaction between the factors.

Changing from model (11.1.1) to model (11.1.2) is accomplished by making the substitution

$$\alpha_{gh} \equiv \gamma_g + \xi_h + (\gamma\xi)_{gh}.$$

There is *less* going on here than meets the eye. The only difference between the parameters α_{gh} and $(\gamma\xi)_{gh}$ is the choice of Greek letters and the presence of parentheses. They accomplish exactly the same things for the two models. The parameters γ_g and ξ_h are completely redundant. Anything one could explain with these parameters could be explained equally well with $(\gamma\xi)_{gh}$s. As they stand, models (11.1.1) and (11.1.2) are completely equivalent. The point of using model (11.1.2) is that it lends itself nicely to an interesting reduced model. If we drop the α_{gh}s from model (11.1.1), we drop all of the treatment effects. Testing model (11.1.1) against this reduced model is a test of whether there are any treatment effects. If we drop the $(\gamma\xi)_{gh}$s from model (11.1.2), we get

$$y_{ghj} = \mu + \gamma_g + \xi_h + \beta_j + \varepsilon_{ghj}. \tag{11.1.3}$$

This still has the γ_gs and the ξ_hs in the model. Thus, dropping the $(\gamma\xi)_{gh}$s does not eliminate all of the treatment effects, it only eliminates effects that cannot be explained as the sum of an effect for the first factor plus an effect for the second factor. In other words, it only eliminates the interaction effects. The reduced model (11.1.3) is the model without interaction and consists of *additive* factor effects. The test for interaction is the test of model (11.1.3) against the larger model (11.1.2). By definition, interaction is any effect that can be explained by model (11.1.2) but not by model (11.1.3).

As discussed, testing for interaction is a test of whether the $(\gamma\xi)_{gh}$s can be dropped from model (11.1.2). If there is no interaction, a test for main effects, say the γ_gs, examines whether the γ_gs can be dropped from model (11.1.3), i.e., whether the factor has any effect or whether $\gamma_1 = \gamma_2 = \cdots = \gamma_G$. If interaction is present, the test for main effects is much more complicated. If model (11.1.2) is appropriate, the test for γ main effects examines whether there are differences in the terms $\gamma_g + \overline{(\gamma\xi)}_{g\cdot}$ where we have averaged the interactions over the subscript for the second effect in the model. The results of such a test are difficult to interpret, so it is common practice to begin the analysis by evaluating whether there is evidence that we need to consider interactions. *If interactions are important, they must be dealt with.* Either we give up on model (11.1.2), go back to model (11.1.1), and simply examine the various treatments as best we can or we examine the nature of the interaction directly. Note that we did *not* say that whenever interactions are significant they must be dealt with. Whether an interaction is important or not depends on the particular application. For example, if interactions are statistically significant but are an order of magnitude smaller than the main effects, one *might* be able to draw useful conclusions while ignoring the interactions.

EXAMINING INTERACTIONS

We now present an example that involves examining interaction contrasts.

EXAMPLE 11.1.2. *A* 2×3 *factorial in a* 6×6 *Latin square.*
Fisher (1935, sections 36, 64) presented data on the pounds of potatoes harvested from a piece of ground that was divided into a square consisting of 36 plots. Six treatments were randomly assigned to the plots in such a way that each treatment occurred once in every row and once in every column of the square. The treatments involved two factors, a nitrogen based fertilizer (N) and a phosphorous based fertilizer (P). The nitrogen fertilizer had two levels, none (n_0) and a standard dose (n_1). The phosphorous fertilizer had three levels, none (p_0), a standard dose (p_1), and double the standard dose (p_2). We identify the six treatments for this 2×3 experiment as follows:

A	B	C	D	E	F
n_0p_0	n_0p_1	n_0p_2	n_1p_0	n_1p_1	n_1p_2

The data are presented in Table 11.4 along with row, column, and treatment means. The basic ANOVA is presented in Table 11.5. It is computed from the sample variance of all 36 observations, the row means, the column means, and the treatment means, as in Section 9.3. The ANOVA F test indicates substantial differences between the treatments. Blocking on rows of the square was quite effective with an F ratio of 7.10. Blocking on columns was considerably less effective with an F of only 3.20, but it was still worthwhile.

TABLE 11.4. Potato data

Row	Columns 1	2	3	4	5	6	Row means
1	E(633)	B(527)	F(652)	A(390)	C(504)	D(416)	$520.3\bar{3}$
2	B(489)	C(475)	D(415)	E(488)	F(571)	A(282)	$453.3\bar{3}$
3	A(384)	E(481)	C(483)	B(422)	D(334)	F(646)	$458.3\bar{3}$
4	F(620)	D(448)	E(505)	C(439)	A(323)	B(384)	$453.1\bar{6}$
5	D(452)	A(432)	B(411)	F(617)	E(594)	C(466)	$495.3\bar{3}$
6	C(500)	F(505)	A(259)	D(366)	B(326)	E(420)	396.00
Col. means	513.00	478.00	$454.1\bar{6}$	$453.6\bar{6}$	442.00	$435.6\bar{6}$	462.75
Trts	A	B	C	D	E	F	
Means	345.00	426.50	$477.8\bar{3}$	$405.1\bar{6}$	$520.1\bar{6}$	$601.8\bar{3}$	462.75

TABLE 11.5. Analysis of variance table for potato data

Source	df	SS	MS	F	P
Rows	5	54199	10840	7.10	.001
Columns	5	24467	4893	3.20	.028
Treatments	5	248180	49636	32.51	.000
Error	20	30541	1527		
Total	35	357387			

Figures 11.3 through 11.6 contain residual plots. They show some interesting features but nothing so outstanding that I, personally, would find them disturbing.

Table 11.6 rearranges the treatment means and gives means for nitrogen and phosphorous. Each treatment mean is the average of 6 observations and the nitrogen means are averages over three treatment means, so the nitrogen means are averages over 18 observations. The sample variance of the nitrogen means is 4288.41. Multiplying by 18 gives

$$MS(N) = 18(4288.41) = 77191.38\,.$$

There are 2 nitrogen means, so multiplying by $(2-1)$ gives

$$SS(N) = (2-1)77191.38 = 77191.38\,.$$

The phosphorous means are averages of two treatment means, so they are averages over $2 \times 6 = 12$ observations. The sample variance of the phosphorous means is 6869.65. Multiplying by 12 gives

$$MS(P) = 12(6869.65) = 82435.75\,.$$

There are 3 phosphorous means, so multiplying by $(3-1)$ gives

$$SS(P) = (3-1)82435.75 = 164871.50\,.$$

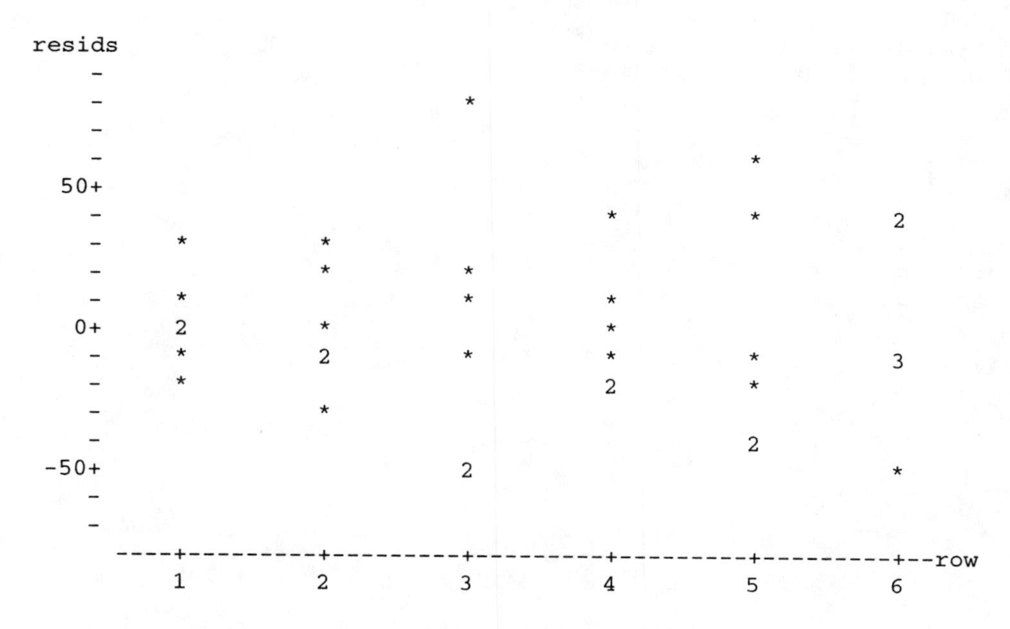

FIGURE 11.3. Plot of residuals versus rows, potato data.

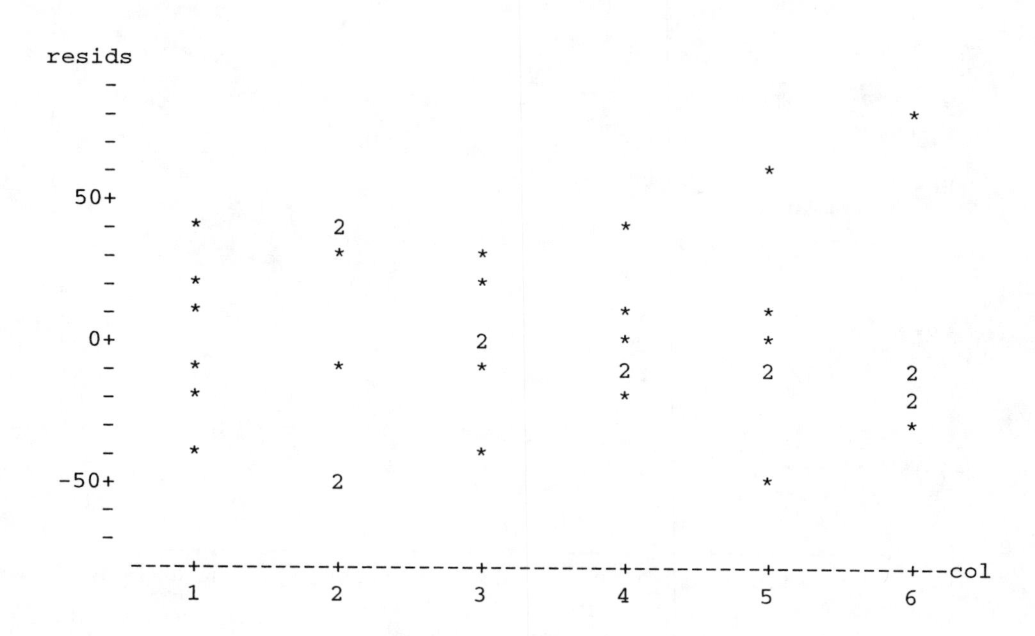

FIGURE 11.4. Plot of residuals versus columns, potato data.

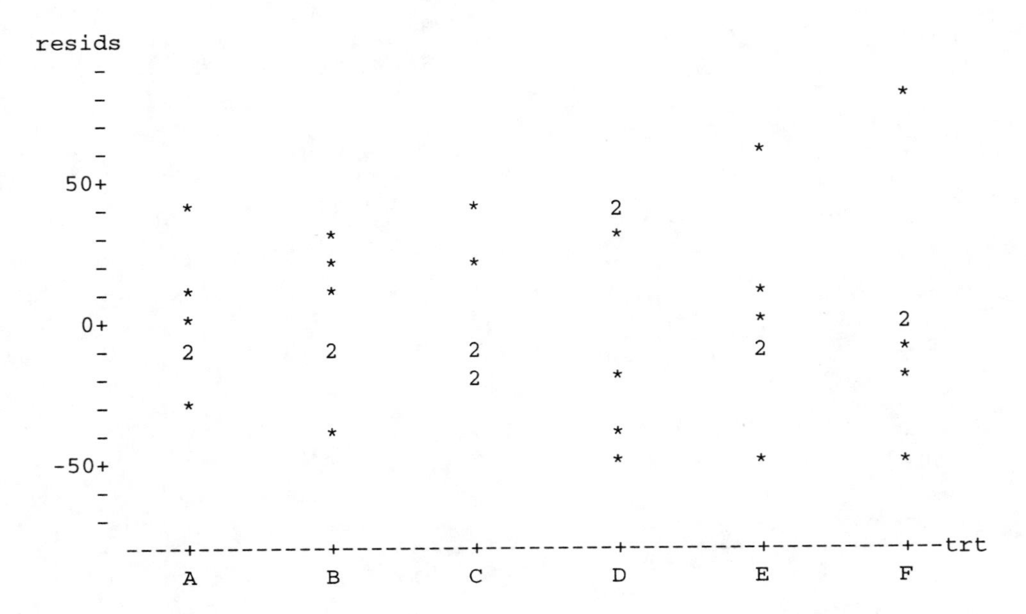

FIGURE 11.5. Plot of residuals versus treatments, potato data.

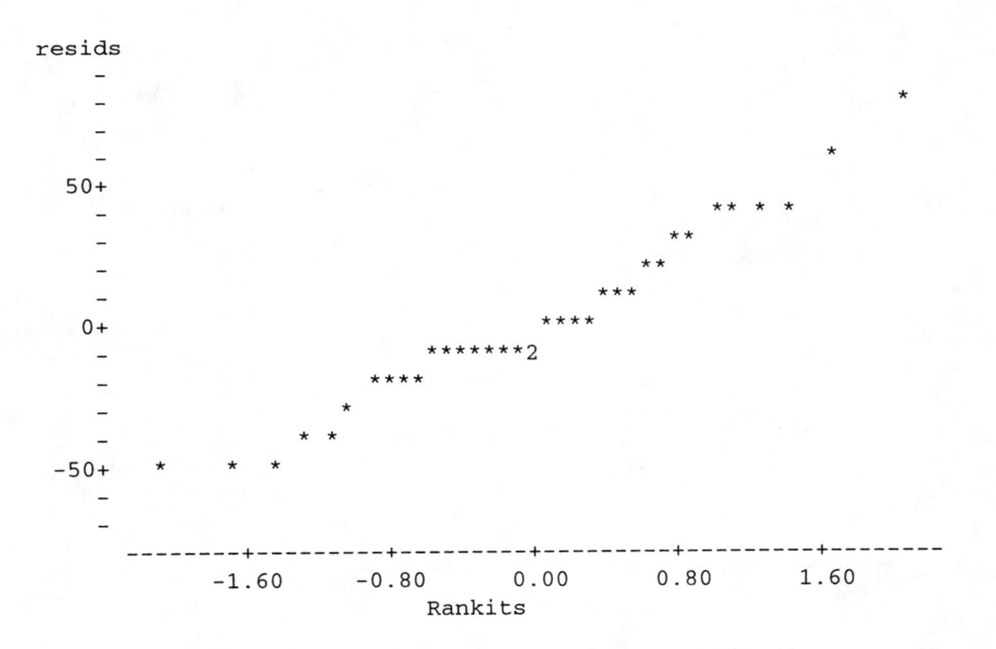

FIGURE 11.6. Normal plot for potato data, $W' = 0.966$.

TABLE 11.6. Treatment means for potato data

Nitrogen	Phosphorous			Nit.
	p_0	p_1	p_2	means
n_0	345.00	426.50	$477.8\bar{3}$	$416.4\bar{4}$
n_1	$405.1\bar{6}$	$520.1\bar{6}$	$601.8\bar{3}$	$509.0\bar{5}$
Phos. means	$375.08\bar{3}$	$473.33\bar{3}$	$539.83\bar{3}$	462.75

The sum of squares for $N * P$ interaction are

$$\begin{aligned} SS(N * P) &= SSTrts - SS(N) - SS(P) \\ &= 248179.92 - 77191.38 - 164871.50 \\ &= 6117.04 . \end{aligned}$$

Similarly,

$$\begin{aligned} df(N * P) &= dfTrts - df(N) - df(P) \\ &= 5 - 1 - 2 \\ &= 2 \end{aligned}$$

or, equivalently,

$$df(N * P) = [df(N)][df(P)] = [1][2] = 2.$$

The mean square for interaction is

$$MS(N * P) = \frac{6117}{2} = 3058.5 .$$

The expanded ANOVA table is given in Table 11.7. The main effects for nitrogen and phosphorous are very significant. The interaction is not significant with a P value of .162 but P is not so large that interaction can be safely ignored. The interaction test has two degrees of freedom so if one contrast with an interesting interpretation accounts for nearly all of the interaction sum of squares, that information might be very useful. We proceed to examine the interaction. However, we also note that the interaction F statistic is an order of magnitude less than the main effect Fs, so a decision to ignore interaction might be based on substantive rather than statistical grounds. A subject matter specialist might argue that even if we found an interpretable interaction, that interaction is so much smaller than the main effects that it is simply not worth bothering about.

TABLE 11.7. Expanded ANOVA table for potato data

Source	df	SS	MS	F	P
Rows	5	54199	10840	7.10	.001
Columns	5	24467	4893	3.20	.028
Nitrogen	1	77191	77191	50.55	.000
Phosphorous	2	164871	82436	53.99	.000
$N * P$	2	6117	3059	2.00	.162
Error	20	30541	1527		
Total	35	357387			

We now begin our discussion of the construction, interpretation, and evaluation of interaction contrasts. We will use notation for contrasts that is analogous to notation used with one-way analysis

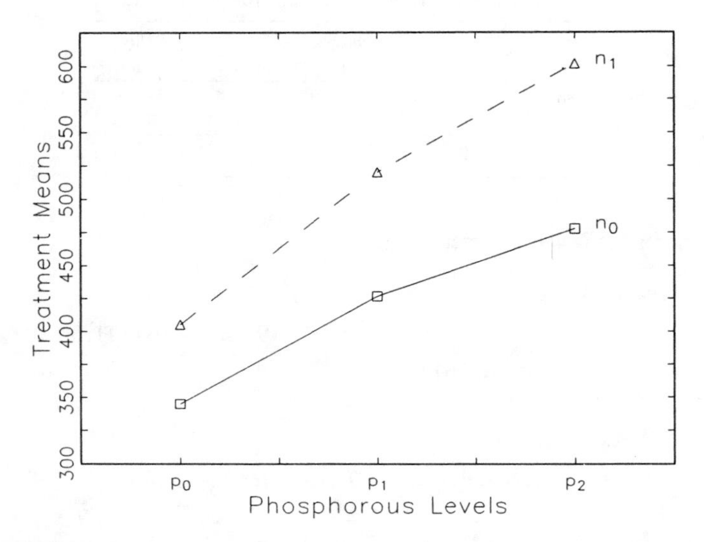

FIGURE 11.7. Interaction plot of nitrogen versus phosphorous, potato data.

of variance. The precise definition of the notation is given at the end of the section. Intuitively, interaction is best evaluated using an interaction plot such as that given in Figure 11.7. The figure has one curve consisting of the three means for treatments that received n_0 and another curve consisting of the three means for treatments that received n_1. If there is no interaction, these curves should be parallel *up to sampling variation*. The deviations of the curves from being parallel suggest the nature of any interaction.

There is only one contrast available for nitrogen,

$$\mu_{n_0} - \mu_{n_1}$$

which we display in tabular form as

N	Contrast
n_0	1
n_1	-1

The phosphorous was applied at 0, 1, and 2 standard doses; these are quantitative factor levels so, as in Section 7.12 and Appendix B.4, we use the linear and quadratic contrasts, i.e.,

$$-\mu_{p_0} + \mu_{p_2}$$

and

$$\mu_{p_0} - 2\mu_{p_1} + \mu_{p_2}.$$

In tabular form we write

	P		
Contrast	p_0	p_1	p_2
Linear	-1	0	1
Quadratic	1	-2	1

These two phosphorous contrasts are orthogonal.

We can construct two orthogonal interaction contrasts from the one nitrogen contrast and the two orthogonal phosphorous contrasts. The interaction contrasts are contrasts in the original treatments so we need six interaction contrast coefficients. The six interaction contrast coefficients are obtained by multiplying the corresponding main effect contrast coefficients. The nitrogen–linear phosphorous interaction contrast coefficients are constructed below from the nitrogen contrast and linear phosphorous contrast.

$N * P$ linear interaction contrast

			P	
		p_0	p_1	p_2
N	Contrasts	-1	0	1
n_0	1	-1	0	1
n_1	-1	1	0	-1

For example, the coefficient of 1 for the combination n_1 and p_0 is obtained by multiplying the marginal coefficient for n_1 (-1) by the marginal coefficient for p_0 (-1). The coefficient of -1 for the combination n_1 and p_2 is obtained by multiplying the marginal coefficient of -1 for n_1 by the marginal coefficient of 1 for p_2. To interpret this contrast, observe that the P linear contrast has been applied to the treatments with n_0 and also to the treatments with n_1. We have then taken the difference between the P linear contrasts at n_0 and at n_1. Thus, this interaction contrast examines whether the P linear contrast is the same at the two levels of nitrogen. In terms of Figure 11.7, we fit two straight lines, one for n_0 and one for n_1. The P linear contrast looks at the slope of such lines, so the $N * P$ linear contrast examines whether the slopes for the two lines are different.

Note that the interaction contrast has coefficients for all six of the original treatments. This is just a contrast in the original treatments but it is constructed in a particular way that insures that it evaluates interaction. The parameter associated with the $N * P$ linear interaction contrast is

$$Par = (-1)\mu_{n_0p_0} + (1)\mu_{n_1p_0} + (0)\mu_{n_0p_1} + (0)\mu_{n_1p_1} + (1)\mu_{n_0p_2} + (-1)\mu_{n_1p_2}.$$

The corresponding estimate uses the 6 treatment means,

$$\begin{aligned} Est &= -345.00 + 405.1\bar{6} + (0)426.50 + (0)520.1\bar{6} + 477.8\bar{3} - 601.8\bar{3} \\ &= -63.8\bar{3}. \end{aligned}$$

The treatment means are averages over 6 observations, so the standard error is

$$SE(Est) = \sqrt{MSE\left[(-1)^2 + (1)^2 + (0)^2 + (0)^2 + (1)^2 + (-1)^2\right]/6} = 31.91.$$

The reference distribution is $t(20)$.

The sum of squares for the contrast is

$$SS(N * P\ Lin) = \frac{(-63.8\bar{3})^2}{\left[(-1)^2 + (1)^2 + (0)^2 + (0)^2 + (1)^2 + (-1)^2\right]/6} = 6112.$$

This yields an F statistic of $6112/1527 = 4.00$ and an unadjusted P value of .059. There is some suggestion of an interaction here. There are fairly clear trends in Figure 11.7 with yields increasing as phosphate increases. This interaction contrast suggests that there may be a difference in the slopes of these trends depending on the level of nitrogen fertilizer used; however the F statistic is not large enough for us to be sure.

We can construct another interaction contrast from the nitrogen contrast and the quadratic contrast in phosphorous. Again, the interaction contrast is obtained by appropriate multiplication of the nitrogen and phosphorous contrast coefficients.

$N * P$ quadratic interaction contrast

			P	
		p_0	p_1	p_2
N	Contrasts	1	−2	1
n_0	1	1	−2	1
n_1	−1	−1	2	−1

The parameter associated with this contrast is

$$Par = (1)\mu_{n_0p_0} + (-1)\mu_{n_1p_0} + (-2)\mu_{n_0p_1} + (2)\mu_{n_1p_1} + (1)\mu_{n_0p_2} + (-1)\mu_{n_1p_2}.$$

This contrast looks at whether the quadratic contrast is the same at nitrogen level n_0 as at n_1. In terms of Figure 11.7, the quadratic contrast measures the curvature in a graph, so the $N * P$ quadratic interaction contrast examines whether the curvature of the n_0 graph differs from the curvature in the n_1 graph. The estimate of the $N * P$ quadratic contrast uses the 6 treatment means,

$$\begin{aligned} Est &= 345.00 - 405.1\bar{6} - (2)426.50 + (2)520.1\bar{6} + 477.8\bar{3} - 601.8\bar{3} \\ &= 3.1\bar{6}. \end{aligned}$$

The treatment means are averages over 6 observations, so the standard error is

$$\begin{aligned} SE(Est) &= \sqrt{MSE\left[(1)^2 + (-1)^2 + (-2)^2 + (2)^2 + (1)^2 + (-1)^2\right]/6} \\ &= 55.26\,. \end{aligned}$$

The reference distribution is $t(20)$.

The sum of squares for the contrast is

$$SS(N * P\ Quad) = \frac{(3.1\bar{6})^2}{\left[(1)^2 + (-1)^2 + (-2)^2 + (2)^2 + (1)^2 + (-1)^2\right]/6} = 5.$$

This yields an F statistic of .003 and a P value of .955. The F statistic is almost disturbingly small but perhaps not as small as first impressions indicate. If we were rejecting for small values of F, the P value would be $1 - .955 = .045$, which is a small P value but not alarmingly small.

There is only one nitrogen contrast and the two phosphorous contrasts were chosen to be orthogonal, so the interaction contrasts are orthogonal and

$$SS(N * P) = 6117 = 6112 + 5 = SS(N * P\ Lin) + SS(N * P\ Quad).$$

The vast bulk of the interaction sum of squares is due to the $N * P$ *Lin* contrast, but $SS(N * P\ Lin)$ is not large enough for us to be sure that the slopes change depending on the level of nitrogen. *In balanced analysis of variance, the method illustrated here for constructing interaction contrasts gives orthogonal interaction contrasts whenever both sets of main effect contrasts are orthogonal.*

In the absence of evidence of interaction, it is necessary to examine contrasts in the main effects. There is one degree of freedom for nitrogen, hence only one contrast: the difference between no nitrogen and the standard dose. From Table 11.7, the difference is highly significant and from the means table we see that yields increase when nitrogen is applied. There are two degrees of freedom for phosphorous, so we must examine phosphorous contrasts explicitly. We use the linear and quadratic contrasts that were used to examine possible interactions.

The quadratic contrast is

$$Par = \mu_{p_0} - 2\mu_{p_1} + \mu_{p_2}.$$

With only three treatments, it is probably better to think of this as a curvature contrast than as a quadratic contrast. If this contrast is substantial, it indicates the lack of a simple linear relationship between increasing phosphorous dose and yield, i.e., there is some curvature in the relationship. The estimate for the quadratic contrast is

$$Est = (1)375.08\bar{3} + (-2)473.33\bar{3} + (1)539.83\bar{3} = -31.75$$

The phosphorous means are averages over 12 observations, so the standard error is

$$SE(Est) = \sqrt{MSE\left[(1)^2 + (-2)^2 + (1)^2\right] / 12} = 27.63\,.$$

The reference distribution is $t(20)$. The sum of squares is

$$SS(P\ Quad) = \frac{(-31.75)^2}{\left[(1)^2 + (-2)^2 + (1)^2\right] / 12} = 2016\,.$$

This yields an F statistic of $2016/ 1527 = 1.32$, which is quite small. There is no evidence for curvature in the relationship between yields and phosphorous dose.

The linear contrast is

$$\begin{aligned} Par &= (-1)\mu_{p_0} + (0)\mu_{p_1} + (1)\mu_{p_2} \\ &= -\mu_{p_0} + \mu_{p_2}. \end{aligned}$$

The corresponding estimate uses the phosphorous means from Table 11.6,

$$Est = (-1)375.08\bar{3} + (0)473.33\bar{3} + (1)539.83\bar{3} = 164.75\,.$$

This indicates that the yields increase by 164.75 as we go from no phosphorous to a double dose. Thus, if a line is appropriate for the relationship between phosphorous dose and yield, each dose increases yields by an estimated $164.75/2 = 82.375$. As before, the phosphorous means are averages over 12 observations, so the standard error of the contrast is

$$SE(Est) = \sqrt{MSE\left[(-1)^2 + (0)^2 + (1)^2\right] / 12} = 15.95\,.$$

The reference distribution is $t(20)$. A 95% confidence interval has endpoints

$$164.75 \pm 2.086(15.95),$$

giving the interval $(131.5, 198.0)$. This is an interval for the mean increase from no dose to a *double* dose, so if a line is appropriate we obtain the 95% confidence interval for the mean increase due to a *single* dose of phosphorous by dividing the entries in this interval by 2, i.e., $(65.7, 99.0)$.

The sum of squares for the linear contrast in phosphorous is

$$SS(P\ Lin) = \frac{(164.75)^2}{\left[(-1)^2 + (0)^2 + (1)^2\right] / 12} = 162855\,.$$

This yields an F statistic of $162855/ 1527 = 106.65$, which is huge. There is clearly a trend as phosphorous dose increases. As seen above, the trend is that yields increase as phosphorous increases. This conclusion is only valid in the range of dosages used in the experiment. It is well known that too much fertilizer can kill plants, so the increasing trend will not go on forever.

Sometimes the analysis of the contrasts is included in the ANOVA table. An example is given in Table 11.8.

TABLE 11.8. ANOVA table including contrasts for potato data

Source	*df*	*SS*	*MS*	*F*	*P*
Rows	5	54199	10840	7.10	.001
Columns	5	24467	4893	3.20	.028
Nitrogen	1	77191	77191	50.55	.000
Phosphorous	2	164871	82436	53.99	.000
(*P lin*)	(1)	(162855)	(162855)	(106.65)	.000
(*P quad*)	(1)	(2016)	(2016)	(1.32)	.264
$N * P$	2	6117	3059	2.00	.162
($N * P$ *Lin*)	(1)	(6112)	(6112)	(4.00)	.059
($N * P$ *Quad*)	(1)	(5)	(5)	(.003)	.955
Error	20	30541	1527		
Total	35	357387			

To summarize, the nitrogen effect is significant and has by an estimate of $509.05-416.44 = 92.61$. The 95% confidence interval for the nitrogen effect has endpoints

$$92.61 \pm 2.086\sqrt{1527\left[1^2+(-1)^2\right]/18},$$

giving the interval $(65.4, 119.8)$. The linear effect in phosphorous acts to increase yields for a single dose an estimated 82.375 pounds with 95% interval $(65.7, 99.0)$. There is no evidence of curvature. Thus the addition of a dose of nitrogen adds, on the average, between 65 and 120 pounds to the yields and each addition of a dose of phosphorous adds, on the average, between 66 and 99 pounds to the yields. There is a suggestion, but no firm evidence, of interaction. Any such interaction is restricted to the linear trends in phosphorous. Such an interaction would indicate that the increase due to a dose of phosphorous differs depending on whether nitrogen was applied.

The entire analysis of Table 11.8 could also be constructed by examining orthogonal contrasts in the original six treatments. Table 11.9 gives the corresponding contrasts, estimates, and sums of squares. Note that the interaction contrasts are precisely the same as those given earlier. For computational convenience we use the main effect contrasts discussed earlier and simply apply them to each level of the other factor rather than devising new main effect contrasts that average over levels of the other factor. While these main effect contrasts give the same sums of squares as in Table 11.8, the parameters and estimates for the phosphorous contrasts in Table 11.9 are twice as large as the parameters and estimates given earlier (the standard errors are also twice as large). Similarly, the parameter and estimate of the nitrogen main effect is three times as large as that computed from the two nitrogen means. We leave the reader to verify that the contrasts are orthogonal. Note also that the interaction contrast coefficients can be obtained by multiplying the coefficients from the corresponding main effect contrasts. □

SOME ADDITIONAL THEORY

We now present additional theory to review the procedure for a Latin square and explicate our notation. The basic Latin square model is

$$y_{ijk} = \mu + \alpha_i + \kappa_j + \rho_k + \varepsilon_{ijk}, \quad \varepsilon_{ijk}\text{s independent } N(0, \sigma^2),$$

where the subscripts i, j, and k indicate treatments, columns, and rows, respectively. With two factors, we can again replace the treatment subscript i with the pair (g, h) and write

$$y_{ghjk} = \mu + \alpha_{gh} + \kappa_j + \rho_k + \varepsilon_{ghjk}, \quad \varepsilon_{ghjk}\text{s independent } N(0, \sigma^2).$$

TABLE 11.9. Orthogonal contrasts in the original six nitrogen–phosphorous treatments

Treatment	Contrasts N	P *Lin*	P *Quad*	$N * P$ *Lin*	$N * P$ *Quad*	Trt. means
n_0p_0	1	−1	1	−1	1	345.00
n_0p_1	1	0	−2	0	−2	426.50
n_0p_2	1	1	1	1	1	$477.8\bar{3}$
n_1p_0	−1	−1	1	1	−1	$405.1\bar{6}$
n_1p_1	−1	0	−2	0	2	$520.1\bar{6}$
n_1p_2	−1	1	1	−1	−1	$601.8\bar{3}$
Est	$-277.8\bar{3}$	329.50	−63.50	$-63.8\bar{3}$	$3.1\bar{6}$	
SS	77191	162855	2016	6112	5	

Again, we can expand the treatment effects α_{gh} to correspond to the factorial treatment structure as

$$y_{ghjk} = \mu + \gamma_g + \xi_h + (\gamma\xi)_{gh} + \kappa_j + \rho_k + \varepsilon_{ghjk}.$$

In constructing interaction contrasts, we discussed nitrogen main effect contrasts in terms of the parameters μ_{n_0} and μ_{n_1}, phosphorous main effect contrasts in terms of the parameters μ_{p_0}, μ_{p_1}, and μ_{p_2}, and interaction contrasts in terms of the parameters $\mu_{n_0p_0}, \mu_{n_1p_0}, \mu_{n_0p_1}, \mu_{n_1p_1}, \mu_{n_0p_2}$, and $\mu_{n_1p_2}$. None of these parameters were explicitly defined. Now that we have specified the model, their definitions can be given.

For Fisher's example, the nitrogen levels n_0 and n_1 are indicated by $g = 1, 2$, respectively. The phosphorous levels p_0, p_1, and p_2 are indicated by $h = 1, 2, 3$, respectively. Columns and rows are $j = 1, \ldots, 6$ and $k = 1, \ldots, 6$, but remember that in a Latin square specifying a treatment and either a row or a column automatically specifies the value of the other index. The estimate of an interaction contrast involves the means $\bar{y}_{gh\cdot\cdot}$. The parameter $\mu_{n_0p_0}$ corresponding to $g = 1, h = 1$ is, by definition,

$$\mu_{n_0p_0} \equiv E(\bar{y}_{11\cdot\cdot}) = \mu + \gamma_1 + \xi_1 + (\gamma\xi)_{11} + \bar{\kappa}_\cdot + \bar{\rho}_\cdot.$$

This is just the mean of *all* the values $E(y_{11jk})$. Similar results hold for the other treatment combinations. It is not difficult to see that in an interaction contrast all of these parameters cancel out except the $(\gamma\xi)_{gh}$ parameters. Interaction contrasts depend only on the interaction parameters $(\gamma\xi)_{gh}$.

Contrasts in the nitrogen main effects are estimated by contrasting the $\bar{y}_{g\cdot\cdot\cdot}$s. By definition,

$$\mu_{n_1} \equiv E(\bar{y}_{2\cdot\cdot\cdot}) = \mu + \gamma_2 + \bar{\xi}_\cdot + \overline{(\gamma\xi)}_{2\cdot} + \bar{\kappa}_\cdot + \bar{\rho}_\cdot.$$

This is just the mean of all the values $E(y_{2hjk})$. It is easily seen that the contrast $\mu_{n_0} - \mu_{n_1}$ reduces to $\gamma_1 + \overline{(\gamma\xi)}_{1\cdot} - \gamma_2 - \overline{(\gamma\xi)}_{2\cdot}$ and if there are no interactions, it reduces further to $\gamma_1 - \gamma_2$, which is a function of only the nitrogen effects. Similarly,

$$\mu_{p_2} \equiv E(\bar{y}_{\cdot 3\cdot\cdot}) = \mu + \bar{\gamma}_\cdot + \xi_3 + \overline{(\gamma\xi)}_{\cdot 3} + \bar{\kappa}_\cdot + \bar{\rho}_\cdot,$$

so a contrast in the phosphorous main effects is a contrast in the $\xi_h + \overline{(\gamma\xi)}_{\cdot h}$ parameters and if no interaction exists, it is a contrast in the ξ_hs.

11.2 Two-way analysis of variance with replication

We have not discussed a two-factor experiment in a balanced completely randomized design (CRD) but the pattern of analysis is the same. In the current section, we give some general formulae relating to the analysis for a CRD, and in the next section, we examine a three-factor experiment in a CRD. We also include some general formulae for RCBs.

The basic analysis of a completely randomized design is always a one-way ANOVA, say,

$$y_{hk} = \mu_h + \varepsilon_{hk}, \quad \varepsilon_{hk}\text{s independent } N(0, \sigma^2)$$

or

$$y_{hk} = \mu + \tau_h + \varepsilon_{hk}, \quad \varepsilon_{hk}\text{s independent } N(0, \sigma^2),$$

where $h = 1, \ldots, T$, $k = 1, \ldots, N$, and the τ_hs indicate treatment effects. With a factorial treatment structure based on two factors, the first having levels $i = 1, \ldots, a$ and the second having levels $j = 1, \ldots, b$, we must have $T = ab$ and we can replace the subscript h with the pair ij to write

$$y_{ijk} = \mu + \tau_{ij} + \varepsilon_{ijk}, \quad \varepsilon_{ijk}\text{s independent } N(0, \sigma^2)$$

$i = 1, \ldots, a, j = 1, \ldots, b, k = 1, \ldots, N$. As shown earlier with the RCB and Latin square designs, the treatment effects can be expanded to illustrate the effects used in the ANOVA table,

$$y_{ijk} = \mu + \alpha_i + \eta_j + (\alpha\eta)_{ij} + \varepsilon_{ijk}, \tag{11.2.1}$$

$i = 1, \ldots, a, j = 1, \ldots, b, k = 1, \ldots, N$. The parameter μ is viewed as a grand mean, α_i is an effect for the ith level of the first factor, η_j is an effect for the jth level of the second factor, and the $(\alpha\eta)_{ij}$s are called the interaction effects.

Model (11.2.1) is grossly over parameterized. There are many more treatment effects than can be clearly defined. The original treatment effects, the τ_{ij} terms, are clearly equivalent to the $(\alpha\eta)_{ij}$ terms, they have the same subscripts. For example, if τ_{ij} has any value, say 5, we can always reconstruct it from the corresponding $\alpha_i + \eta_j + (\alpha\eta)_{ij}$ terms in (11.2.1) by taking $\alpha_i = 0$, $\eta_j = 0$, and $(\alpha\eta)_{ij} = 5$. In fact, we can always set the terms other than the $(\alpha\eta)_{ij}$s to 0 and still reconstruct the τ_{ij}s. Conversely, anything obtained from model (11.2.1) can be obtained from the one-way ANOVA model. If the terms in model (11.2.1) take on any values, say $\mu + \alpha_i + \eta_j + (\alpha\eta)_{ij} = 1 + 2 + 3 + 4$, we can reconstruct this by simply taking $\mu = 0$ and $\tau_{ij} = 10$.

Model (11.2.1) is just an over parameterized version of the one-way ANOVA model. It is valuable because of what happens when we drop the 'interaction' terms, the $(\alpha\eta)_{ij}$s. If we drop these terms, we have a special structure in the factorial effects,

$$y_{ijk} = \mu + \alpha_i + \eta_j + \varepsilon_{ijk}. \tag{11.2.2}$$

In this model, the effect of the first factor and the effect of the second factor are *additive*. The effect of factor 1 is the same, regardless of the level of factor 2, and the effect of factor 2 is the same, regardless of the level of factor 1. This is what is meant when we say that two factors do not interact. Actually, the test for interaction is really just a test of model (11.2.2) against model (11.2.1). If model (11.2.2) gives an adequate fit to the data, then we do not need the interaction terms to explain the data. If the $(\alpha\eta)_{ij}$s are necessary to explain the data, we must have interaction. The idea that interaction exists when the additive effect structure of model (11.2.2) is inadequate is an idea that applies very generally and can be used to extend balanced analysis of variance modeling ideas to situations that are unbalanced. Such models are treated in Chapter 16.

Returning to the analysis of model (11.2.1), compute the sample variance of all observations and get *SSTot*; also compute the two way table of means

i	j 1	2	$\cdots$	b	α means
1	$\bar{y}_{11\bullet}$	$\bar{y}_{12\bullet}$	$\cdots$	$\bar{y}_{1b\bullet}$	$\bar{y}_{1\bullet\bullet}$
2	$\bar{y}_{21\bullet}$	$\bar{y}_{22\bullet}$	$\cdots$	$\bar{y}_{2b\bullet}$	$\bar{y}_{2\bullet\bullet}$
$\vdots$	$\vdots$	$\vdots$	$\ddots$	$\vdots$	$\vdots$
a	$\bar{y}_{a1\bullet}$	$\bar{y}_{a2\bullet}$	$\cdots$	$\bar{y}_{ab\bullet}$	$\bar{y}_{a\bullet\bullet}$
η means	$\bar{y}_{\bullet1\bullet}$	$\bar{y}_{\bullet2\bullet}$	$\cdots$	$\bar{y}_{\bullet b\bullet}$	$\bar{y}_{\bullet\bullet\bullet}$

The *SSTrts* is based on the sample variance of the ab means in the body of the table, these are averages over N observations. The mean square for the first factor, $MS(\alpha)$, is computed from the α means given in the right margin of the table. These means are averages over bN observations. $MS(\alpha)$ is the sample variance of the α means times bN. The sum of squares is obtained from the mean square after multiplying by $a-1$. The mean square for the second factor, $MS(\eta)$, is computed from the η means given along the bottom margin of the table. These means are averages over aN observations. The sum of squares is obtained from the mean square after multiplying by $b-1$. The sum of squares for interaction, $SS(\alpha\eta)$ is what is left of the *SSTrts* after subtracting out $SS(\alpha)$ and $SS(\eta)$. A closed form can be given for this computation. The formulae are given in Table 11.10.

TABLE 11.10. Analysis of variance (CRD)

Source	df	SS
α	$a-1$	$bN\sum_{i=1}^{a}(\bar{y}_{i\bullet\bullet}-\bar{y}_{\bullet\bullet\bullet})^2$
η	$b-1$	$aN\sum_{j=1}^{b}\left(\bar{y}_{\bullet j\bullet}-\bar{y}_{\bullet\bullet\bullet}\right)^2$
$(\alpha\eta)$	$(a-1)(b-1)$	$N\sum_{i,j}\left(\bar{y}_{ij\bullet}-\bar{y}_{i\bullet\bullet}-\bar{y}_{\bullet j\bullet}+\bar{y}_{\bullet\bullet\bullet}\right)^2$
Error	$ab(N-1)$	$\sum_{i,j,k}\left(y_{ijk}-\bar{y}_{ij\bullet}\right)^2$
Total	$abN-1$	$\sum_{i=1}^{a}\sum_{j=1}^{b}\sum_{k=1}^{N}\left(y_{ijk}-\bar{y}_{\bullet\bullet\bullet}\right)^2$

Contrasts in the main effects work just as always but their interpretation is much simpler if there is no interaction. For known λ_is that sum to zero, a contrast in, say the α_is, is

$$Par = \sum_{i=1}^{a} \lambda_i \alpha_i .$$

Such contrasts are only estimable, indeed are only well defined, if there is no interaction as in model (11.2.2). If the interaction model (11.2.1) is necessary, the closest we can come to estimating a main effect contrast in the α_is is to consider

$$Par = \sum_{i=1}^{a} \lambda_i \left(\alpha_i + \overline{(\alpha\eta)}_{i\bullet}\right) .$$

In either case, the estimate of the parameter is

$$Est = \sum_{i=1}^{a} \lambda_i \bar{y}_{i\bullet\bullet} .$$

The sample means in the estimate are averages of bN observations, so

$$SE\left(\sum_{i=1}^{a} \lambda_i \bar{y}_{i\bullet\bullet}\right) = \sqrt{MSE \sum_{i=1}^{a} \frac{\lambda_i^2}{bN}}.$$

The reference distribution is $t(dfE)$. The sum of squares for the contrast is

$$SS\left(\sum_{i=1}^{a} \lambda_i \alpha_i\right) = \left[\sum_{i=1}^{a} \lambda_i \bar{y}_{i\bullet\bullet}\right]^2 \Big/ \left[\sum_{i=1}^{a} \frac{\lambda_i^2}{bN}\right].$$

Main effect contrasts for the other factor are dealt with in a similar fashion.

Interesting contrasts in the interactions are constructed from interesting contrasts in the α_is and η_js. If $\sum_{i=1}^{a} \lambda_i \alpha_i$ and $\sum_{j=1}^{b} \xi_j \eta_j$ are interesting main effect contrasts, an interesting interaction contrast can be constructed as

$$Par = \sum_{i=1}^{a} \sum_{j=1}^{b} \lambda_i \xi_j (\alpha\eta)_{ij}.$$

The estimate of this interaction contrast is

$$Est = \sum_{i=1}^{a} \sum_{j=1}^{b} \lambda_i \xi_j \bar{y}_{ij\bullet}.$$

The sample means in the estimate are averages of N observations, so

$$SE\left(\sum_{i=1}^{a} \sum_{j=1}^{b} \lambda_i \xi_j \bar{y}_{ij\bullet}\right) = \sqrt{MSE \sum_{i=1}^{a} \sum_{j=1}^{b} \frac{(\lambda_i \xi_j)^2}{N}}.$$

The reference distribution is $t(dfE)$. The sum of squares for the contrast is

$$SS\left(\sum_{i=1}^{a} \sum_{j=1}^{b} \lambda_i \xi_j (\alpha\eta)_{ij}\right) = \left[\sum_{i=1}^{a} \sum_{j=1}^{b} \lambda_i \xi_j \bar{y}_{ij\bullet}\right]^2 \Big/ \left[\sum_{i=1}^{a} \sum_{j=1}^{b} \frac{(\lambda_i \xi_j)^2}{N}\right].$$

If we are interested in a set of $a-1$ orthogonal contrasts in the α_is and a set of $b-1$ orthogonal contrasts in the η_js, we can construct a set of $(a-1)(b-1)$ orthogonal interaction contrasts by taking all combinations of an α contrast and an η contrast. As usual with a complete set of orthogonal contrasts, the orthogonal interaction contrasts provide a decomposition of the sums of squares for interaction. *More generally, the interaction contrast coefficients $\lambda_i \xi_j$ can be replaced in all of these formulae by coefficients q_{ij} having the property that the sum over j for any i equals 0 and the sum over i for any j equals 0, i.e., $q_{i\bullet} = 0 = q_{\bullet j}$ for any i and j. The problem with picking q_{ij}s in this manner is that it is difficult to interpret the resulting interaction contrasts.*

The modifications necessary for dealing with RCBs and Latin squares are very natural. The basic model for a RCB is

$$y_{ijk} = \mu + \tau_{ij} + \beta_k + \varepsilon_{ijk}, \quad \varepsilon_{ijk}\text{s independent } N(0, \sigma^2),$$

$i = 1, \ldots, a$, $j = 1, \ldots, b$, $k = 1, \ldots, c$, where the β_ks denote block effects and c is the number of blocks. The model can be expanded to

$$y_{ijk} = \mu + \alpha_i + \eta_j + (\alpha\eta)_{ij} + \beta_k + \varepsilon_{ijk}.$$

TABLE 11.11. Analysis of variance (RCB)

Source	df	SS
α	$a-1$	$bc\sum_{i=1}^{a}(\bar{y}_{i\cdot\cdot}-\bar{y}_{\cdot\cdot\cdot})^2$
η	$b-1$	$ac\sum_{j=1}^{b}\left(\bar{y}_{\cdot j\cdot}-\bar{y}_{\cdot\cdot\cdot}\right)^2$
$(\alpha\eta)$	$(a-1)(b-1)$	$c\sum_{i,j}\left(\bar{y}_{ij\cdot}-\bar{y}_{i\cdot\cdot}-\bar{y}_{\cdot j\cdot}+\bar{y}_{\cdot\cdot\cdot}\right)^2$
Blocks	$c-1$	$ab\sum_{j=1}^{c}(\bar{y}_{\cdot\cdot c}-\bar{y}_{\cdot\cdot\cdot})^2$
Error	$(ab-1)(c-1)$	$\sum_{i,j,k}\left(y_{ijk}-\bar{y}_{ij\cdot}-\bar{y}_{\cdot\cdot k}+\bar{y}_{\cdot\cdot\cdot}\right)^2$
Tot − C	$abc-1$	$\sum_{i=1}^{a}\sum_{j=1}^{b}\left(y_{ijk}-\bar{y}_{\cdot\cdot\cdot}\right)^2$

The RCB ANOVA table is given as Table 11.11. With no interaction, the formulae for contrasts in, say, the η_js are

$$Par = \sum_{j=1}^{b} \lambda_j \eta_j$$

with

$$Est = \sum_{j=1}^{b} \lambda_j \bar{y}_{\cdot j\cdot}\,.$$

The sample means in the estimate are the average of ac observations, so

$$SE\left(\sum_{j=1}^{b} \lambda_j \bar{y}_{\cdot j\cdot}\right) = \sqrt{MSE\sum_{j=1}^{b}\frac{\lambda_j^2}{ac}}.$$

The reference distribution is $t(dfE)$. The sum of squares for the contrast is

$$SS\left(\sum_{j=1}^{b} \lambda_j \eta_j\right) = \left[\sum_{j=1}^{b} \lambda_j \bar{y}_{\cdot j\cdot}\right]^2 \Bigg/ \left[\sum_{j=1}^{b}\frac{\lambda_j^2}{ac}\right].$$

When interaction exists, we can use this same estimate, standard error, reference distribution, and sum of squares, but the parameter differs. In place of η_j, the parameter has η_j plus the average of all the interaction terms with that value of j.

11.3 Multifactor structures

In this section we deal with factorial treatment structures that contain more than two factors. The example involves three factors. In Example 12.2.1 we examine data that involve four factors.

EXAMPLE 11.3.1. Box (1950) considers data on the abrasion resistance of a fabric. The data are weight loss of a fabric that occurs between 1000 and 2000 revolutions of a machine designed to test abrasion resistance. A piece of fabric is put on the machine, it is weighed after 1000 revolutions and again after 2000 revolutions; the measurement is the change in weight between 1000 and 2000 revolutions. Fabrics of several different types are compared. They differ by whether a surface treatment was applied, the type of filler used, and the proportion of filler used. Two pieces of fabric of each type are examined, giving two replications in the analysis of variance. The design is similar to a completely randomized design, but it is not clear which aspects of the manufacturing process could be subjected to randomization. The data are given in Table 11.12.

TABLE 11.12. Abrasion resistance data

		Surface treatment					
			Yes			No	
Proportions		25%	50%	75%	25%	50%	75%
	A	192	217	252	169	187	225
	A	188	222	283	152	196	270
Fill							
	B	127	123	117	82	94	76
	B	105	123	125	82	89	105

The three factors are referred to as '*surf*', '*fill*', and '*prop*', respectively. The factors have 2, 2, and 3 levels, so this is a $2 \times 2 \times 3$ factorial. It can also be viewed as just a one-way ANOVA with 12 treatments. Using the three subscripts *hij* to indicate a treatment by indicating the levels of *surf*, *fill*, and *prop*, respectively, the one-way ANOVA model is

$$y_{hijk} = \mu_{hij} + \varepsilon_{hijk} \tag{11.3.1}$$

$h = 1, 2$, $i = 1, 2$, $j = 1, 2, 3$, $k = 1, 2$. Equivalently, we can break the treatment effects into main effects for each factor, interactions between each pair of factors, and an interaction between all three factors, i.e.,

$$y_{hijk} = \mu + s_h + f_i + p_j + (sf)_{hi} + (sp)_{hj} + (fp)_{ij} + (sfp)_{hij} + \varepsilon_{hijk}. \tag{11.3.2}$$

Here the s, f, and p effects indicate main effects for *surf*, *fill*, and *prop*, respectively. The (sf)s are effects for the two-factor interaction between *surf* and *fill*; (sp) and (fp) are defined similarly. The (sfp)s are three-factor interaction effects. A three-factor interaction can be thought of as a two-factor interaction that changes depending on the level of the third factor. *The main effects, two-factor interactions, and three-factor interaction simply provide the structure that allows us to proceed in a systematic fashion.*

We begin by considering the one-way analysis of variance. The means for the 12 treatments in the one-way ANOVA are given below.

$N = 2$		*surf*			
		Yes		No	
	fill	A	B	A	B
	25%	190.00	116.00	160.50	82.00
prop	50%	219.50	123.00	191.50	91.50
	75%	267.50	121.00	247.50	90.50

In all means tables, the value of N given in the upper left is the number of observations that go into each mean in the table. Finding the sample variance of the 12 means and multiplying by $N = 2$ gives

$$MSTrts = (2)4095.096591 = 8190.193182.$$

The sum of squares is obtained from multiplying the mean square by the degrees of freedom based on all 12 treatments,

$$SSTrts = (12 - 1)MSTrts$$

$$= (11)8190.193182$$
$$= 90092.125 .$$

The sum of squares total is $(n-1)s_y^2 = 92497.6$, i.e., the sample variance of all 24 observations times $(24-1)$. The sum of squares error is most easily obtained as

$$SSE = SSTot - SSTrts = 92497.6 - 90092.1 = 2405.5$$

with

$$dfE = dfTot - dfTrts = (24-1) - (12-1) = 12.$$

The mean squared error is

$$MSE = \frac{2405.5}{12} = 200.5 .$$

The analysis of variance table for the one-way ANOVA model (11.3.1) is given below.

Analysis of variance

Source	df	SS	MS	F
Treatments	11	90092.1	8190.2	40.8
Error	12	2405.5	200.5	
Total	23	92497.6		

The F statistic is very large. If the standard one-way ANOVA assumptions are reasonably valid, there is clear evidence that not all of the treatments have the same effect.

Now consider the standard residual checks for a one-way ANOVA. Figure 11.8 contains the residuals plotted against the predicted values. It has a curious sideways mushroom shape. The variability seems to decrease as the predicted values increase until the last four residual values; these four display more variability than the other residuals. From the means table, the two largest predicted values occur with fill A, prop 75%. The residual pattern is not one that clearly suggests heteroscedastic variances. We simply note the pattern and would bring it to the attention of the experimenter to see if it suggests something to her. In the absence of additional information, we proceed with the analysis. Figure 11.9 contains a normal plot of the residuals. It does not look too bad. Note that with 24 residuals and only 12 dfE, we may want to use dfE as the sample size should we choose to perform a W' test.

TABLE 11.13. Analysis of variance

Source	df	SS	MS	F	P
surf	1	5017.0	5017.0	25.03	0.000
fill	1	70959.4	70959.4	353.99	0.000
prop	2	7969.0	3984.5	19.88	0.000
surf * *fill*	1	57.0	57.0	0.28	0.603
surf * *prop*	2	44.3	22.2	0.11	0.896
fill * *prop*	2	6031.0	3015.5	15.04	0.001
surf * *fill* * *prop*	2	14.3	7.2	0.04	0.965
Error	12	2405.5	200.5		
Total	23	92497.6			

The analysis of variance table for model (11.3.2) simply partitions the 11 degrees of freedom for treatments and the *SSTrts* from the one-way table into various sources. The expanded table is given in Table 11.13. We now consider the analysis of this table; computing the table will be

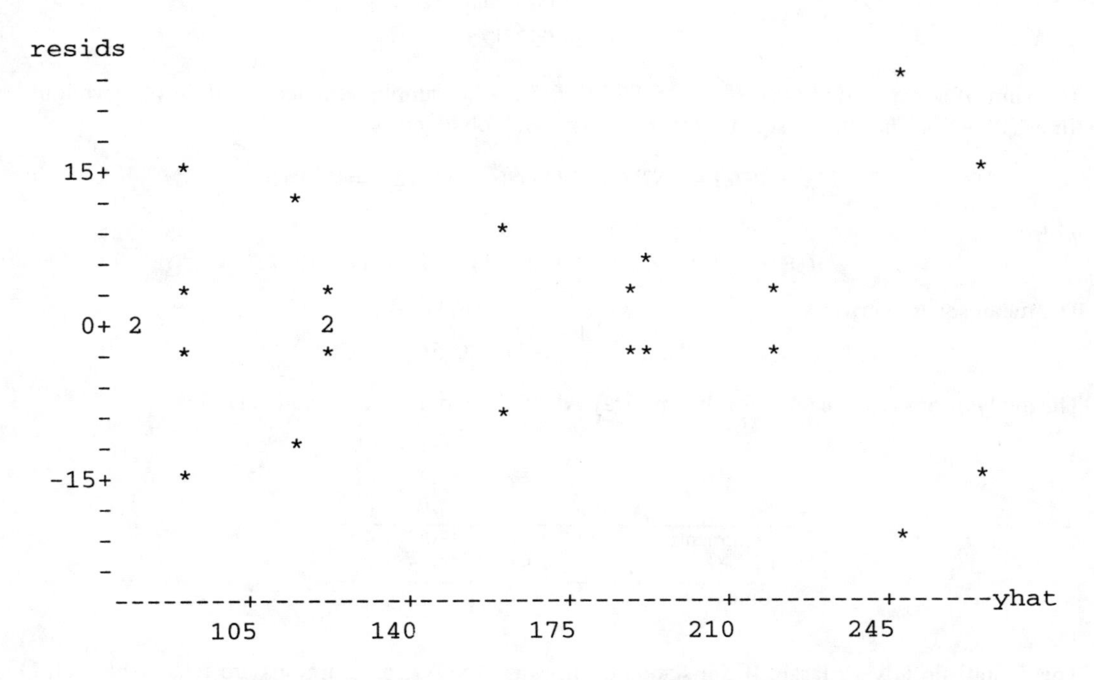

FIGURE 11.8. Plot of residuals versus predicted values, Box data.

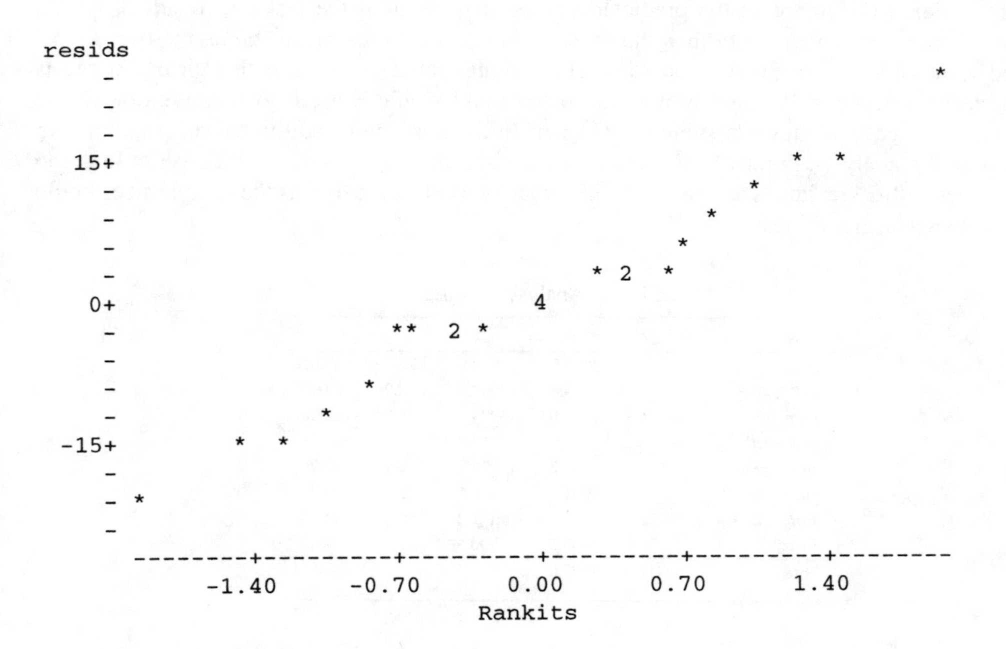

FIGURE 11.9. Normal plot of residuals, $W' = 0.974$, Box data.

dealt with later. We start from the bottom of the table and work our way up. The three-factor interaction has a very large P and only two degrees of freedom, so it is safe to ignore. If the P value were more moderate or the degrees of freedom larger, it would be necessary to investigate the three-way interaction further. It is possible for important effects to be hidden when there are a substantial number of degrees of freedom in the numerator of a test, so typically even nonsignificant interactions need to be investigated. Even when the interaction is not significant, it would be of interest to identify, if possible, one or two interaction contrasts having *natural interpretations* and mean squares substantially larger than the MSE. Since no interaction contrast can have a mean square larger than the sum of squares for the interaction and since in this case 14.31 is much smaller than the MSE, we need pursue the question of three-factor interaction no further.

The most worrisome aspect of the three-factor interaction is that the P value is too large, i.e., the F statistic is too small. If we were to reject for small F statistics, the P value would be $1-.965 = .035$, which is significantly small. This suggests that there may be a systematic pattern in the replications that we may be overlooking, cf. Christensen (1989, 1991). This was precisely our concern after looking at the residual–prediction plot, Figure 11.8. However, a P value of .035 is not overwhelming evidence, so again we note the possible problem with the intention of discussing it with the experimenter.

The two-factor interactions for $surf * fill$ and $surf * prop$ both have very large P values and small degrees of freedom. As with the three-factor interaction, we conclude that they can be ignored. The two-factor interaction for $fill * prop$ is highly significant with 2 degrees of freedom. This suggests that the effects of the three different proportions are different for fill A than for fill B. (Equivalently, it suggests that the effects of the two fills differ depending on which proportion of filler is used.) The interaction requires further investigation to identify the precise effect. We will return to this after considering the main effects.

All of the main effects are highly significant, but in the face of a substantial $fill * prop$ interaction, the main effects for $fill$ and $prop$ are not of direct interest. For example, the main effect for $fill$ indicates only that the effects of fills A and B differ when averaged over the three proportions and two surface treatments. However, we know that the effects of fills A and B differ with the three proportions, so averaging over proportions is generally not a very enlightening thing to do. The main effect for surface treatments is also highly significant. The means table is given below.

$N = 12$	Surface treatment means	
surf	Yes	No
	$172.83\bar{3}$	$143.91\bar{6}$

We know that there is a significant difference between the means, so clearly the fabric with the surface treatment has greater weight loss due to abrasion than the fabric with no surface treatment. A 99% confidence interval for the mean difference in weight loss has endpoints

$$(172.83 - 143.92) \pm t(.995, 12)\sqrt{MSE\left(\frac{1}{12}+\frac{1}{12}\right)}$$

or $28.91 \pm (3.055)5.781$. The confidence interval is $(11.25, 46.57)$, thus we are 99% confident that the mean abrasion weight loss between 1000 and 2000 rotations for surface treated fabric is between eleven and a quarter and forty six and a half units larger than for untreated fabric. (This conclusion is subject to our doubts about the assumptions underlying the ANOVA model, e.g., patterns in the replications.)

We now return to the examination of the $fill * prop$ interaction. The means table for fills and proportions is given below.

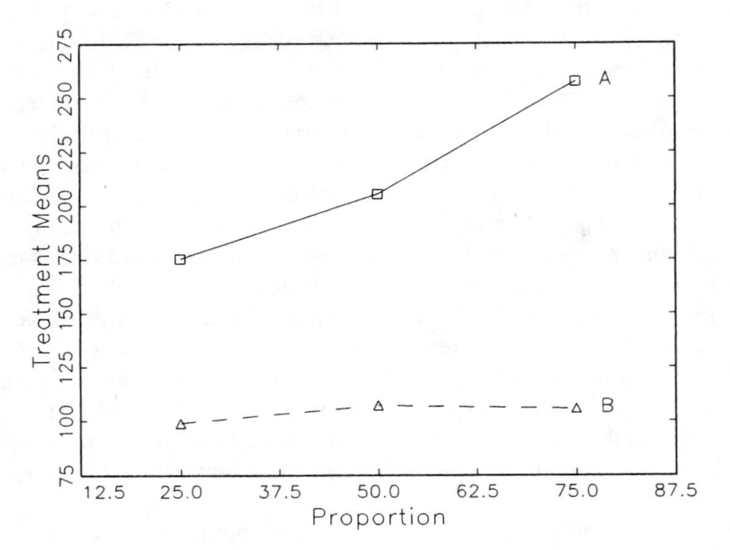

FIGURE 11.10. Interaction plot, proportions on horizontal axis.

$N = 4$	*fill*	
prop	A	B
25%	175.25	99.00
50%	205.50	107.25
75%	257.50	105.75

Interaction is best visualized using an interaction plot such as that given in Figure 11.10.

We now examine interaction contrasts. Interesting interaction contrasts are constructed from interesting contrasts in the main effects. There is only one contrast available for the fills,

$$\mu_A - \mu_B.$$

The proportion levels are quantitative and equally spaced, so we use the linear and quadratic contrasts in the proportions as discussed in Section 7.12. The linear contrast is

$$-\mu_{25} + \mu_{75}$$

and the quadratic contrast is

$$\mu_{25} - 2\mu_{50} + \mu_{75}.$$

With three levels, it is best to think of the linear contrast as a measure of overall trend (increases or decreases in weight loss as proportions increase) and the quadratic contrast as a measure of curvature.

The fill and proportion contrasts are used to construct *fill* $*$ *prop* interaction contrasts. The two proportion contrasts are orthogonal, so the two interaction contrasts will also be orthogonal. The interaction contrast coefficients are obtained by multiplying corresponding elements of the main effect contrasts. The *fill* $*$ *prop linear* interaction contrast coefficients are given below.

Fill * prop linear interaction contrast

		fill	
		A	B
prop	Contrasts	1	−1
25%	−1	−1	1
50%	0	0	0
75%	1	1	−1

Recall that the coefficient of −1 for fill A and proportion 25% is the product of the fill A coefficient 1 and the proportion 25% coefficient −1. The *fill * prop quadratic* coefficients are

Fill * prop quadratic interaction contrast

		fill	
		A	B
prop	Contrasts	1	−1
25%	1	1	−1
50%	−2	−2	2
75%	1	1	−1

The sums of squares for these contrasts are computed from the *fill * prop* means,

$$\begin{aligned} & SS(fill * prop\ lin) \\ &= \frac{[-175.25 + 99.00 + 0 + 0 + 257.50 - 105.75]^2}{\left[(-1)^2 + 1^2 + 0^2 + 0^2 + 1^2 + (-1)^2\right]/4} \\ &= \frac{(75.5)^2}{4/4} \\ &= 5700.25 \end{aligned}$$

and

$$\begin{aligned} & SS(fill * prop\ quad) \\ &= \frac{[175.25 - 99.00 - (2)205.50 + (2)107.25 + 257.50 - 105.75]^2}{\left[1^2 + (-1)^2 + (-2)^2 + 2^2 + 1^2 + (-1)^2\right]/4} \\ &= \frac{(31.5)^2}{12/4} \\ &= 330.75\,. \end{aligned}$$

The *MSE* is 200.5, so the *F* ratios are 28.43 and 1.65, respectively. Based on the *fill * prop quadratic* contrast, there is no evidence that the curvatures are different for fills A and B. However, from the *fill * prop linear* contrast we do have evidence that the slopes are different for fills A and B. From Figure 11.10, the slope for the fill A line is much greater than the slope for fill B. Thus, weight loss increases with proportion at a much faster rate when using fill A. Note also that the overall level of weight loss is much greater for fill A. The *fill * prop linear* contrast accounts for the vast majority of the *fill * prop* interaction sums of squares. Of course with orthogonal interaction contrasts,

$$SS(fill * prop) = 6031 = 5700.25 + 330.75 = SS(fill * prop\ lin) + SS(fill * prop\ quad).$$

We have established that there is no significant *difference* in curvature between fills A and B but we have not established whether curvature exists. To do this we apply the quadratic contrast to the proportion means,

$N = 8$	Proportion means		
prop	25%	50%	75%
Means	137.125	156.375	181.625
Quadratic	1	−2	1

The sum of squares for the contrast is

$$SS(prop\ quadratic) = \frac{[137.125 - (2)156.375 + 181.625]^2}{\left[1^2 + (-2)^2 + 1^2\right]/8} = \frac{6^2}{6/8} = 48.$$

The sum of squares is much smaller than the *MSE*, so there is no evidence of curvature *averaged over fills* in the *fill–proportion* plot. There is no evidence of average curvature and no evidence of a difference in curvatures, so there is no evidence of curvature.

There is no evidence of curvature but there is more to be learned about the individual lines in Figure 11.10. We can apply the linear contrast to the proportion means for fill A and again for fill B. For fill A we get an estimated contrast of

$$-175.25 + 0 + 257.50 = 82.25$$

and a sum of squares

$$\begin{aligned} SS(prop\ lin;\ fill\ A) &= \frac{[-175.25 + 0 + 257.50]^2}{\left[1^2 + 0^2 + (-1)^2\right]/4} \\ &= \frac{(82.25)^2}{2/4} \\ &= 13530.125\,. \end{aligned}$$

Comparing the sum of squares to $MSE = 200.5$ we have a clear linear change in weight loss as proportions change. The estimated contrast indicates that as we increase from 25% to 75% of fill A, the weight loss increases by 82.25 units. If we have only a 25% increase, the weight loss increases only 82.25/2 = 41.125 units. Applying the linear contrast to the means for fill B gives an estimated contrast of

$$-99.00 + 0 + 105.75 = 6.75$$

and a sum of squares

$$SS(prop\ lin;\ fill\ B) = \frac{(6.75)^2}{2/4} = 91.125\,.$$

Comparing the sum of squares to the *MSE* we see no evidence of a linear trend for fill B. Thus the interaction is due to the fact that fill A has a positive slope while fill B is consistent with a 0 slope.

To summarize, we found that the estimated effect of having a surface treatment is to increase the weight loss occurring between 1000 and 2000 revolutions by 173 − 144 = 29 units. As we have just seen, for fill A, increasing the proportion of fill by 25% increases the weight loss by an estimated 41 units. For fill B there is no evidence that changing the fill proportion changes the weight loss. The fill means are 213 for fill A and 104 for fill B, so at the average proportion, i.e., at 50% fill, the estimated weight loss is 213 − 104 = 109 units larger for fill A. Of course, with the difference in proportion slopes, the increased weight loss due to fill A will be less at 25% and greater at 75%. Confidence intervals are readily available for all of these estimates.

Box (1950) reports data not only on fabric weight loss between 1000 and 2000 rotations, but also after the first 1000 rotations and between 2000 and 3000 rotations. The results from the three analysis of variance tables are similar except that there appears to be a *surf* ∗ *fill* interaction after 1000 rotations. In the next chapter, we will examine models that incorporate not only *surf*, *fill*, and *prop* but also *rotations*. From the comment above, the *surf* ∗ *fill* interaction changes with the number of rotations, so we should find a significant three-factor *surf* ∗ *fill* ∗ *rotation* interaction. □

Before going into the details of computing the analysis of variance table for this example, we present some results for arbitrary three-factor models. In general, the model for a three-way ANOVA in a balanced completely randomized design is

$$\begin{aligned} y_{hijk} &= \mu_{hij} + \varepsilon_{hijk} \\ &= \mu + \alpha_h + \beta_i + \gamma_j \\ &\quad + (\alpha\beta)_{hi} + (\alpha\gamma)_{hj} + (\beta\gamma)_{ij} + (\alpha\beta\gamma)_{hij} + \varepsilon_{hijk} \end{aligned}$$

$h = 1, \ldots, a$, $i = 1, \ldots, b$, $j = 1, \ldots, c$, $k = 1, \ldots, N$. The corresponding analysis of variance table is fully determined by the sums of squares that are given below.

Analysis of variance

Source	df	SS
Treatments	$abc - 1$	$N\sum_{hij}\left(\bar{y}_{hij\cdot} - \bar{y}_{\cdot\cdot\cdot\cdot}\right)^2$
Error	$abc(N-1)$	$\sum_{hijk}\left(y_{hijk} - \bar{y}_{hij\cdot}\right)^2$
Total	$abcN - 1$	$\sum_{hijk}\left(y_{hijk} - \bar{y}_{\cdot\cdot\cdot\cdot}\right)^2$

The full factorial analysis of variance table is given in Table 11.14. It consists of a decomposition of the treatment sums of squares and degrees of freedom. If we had a three-factor design in a RCB, the error line above would be broken into blocks and error as in Chapter 9. The breakdown of the treatments line would occur in exactly the same way.

TABLE 11.14. Analysis of variance

Source	df	SS
α	$a-1$	$bcN\sum_h(\bar{y}_{h\cdot\cdot\cdot} - \bar{y}_{\cdot\cdot\cdot\cdot})^2$
β	$b-1$	$acN\sum_i(\bar{y}_{\cdot i\cdot\cdot} - \bar{y}_{\cdot\cdot\cdot\cdot})^2$
γ	$c-1$	$abN\sum_j\left(\bar{y}_{\cdot\cdot j\cdot} - \bar{y}_{\cdot\cdot\cdot\cdot}\right)^2$
$\alpha * \beta$	$(a-1)(b-1)$	$cN\sum_{hi}(\bar{y}_{hi\cdot\cdot} - \bar{y}_{h\cdot\cdot\cdot} - \bar{y}_{\cdot i\cdot\cdot} + \bar{y}_{\cdot\cdot\cdot\cdot})^2$
$\alpha * \gamma$	$(a-1)(c-1)$	$bN\sum_{hj}\left(\bar{y}_{h\cdot j\cdot} - \bar{y}_{h\cdot\cdot\cdot} - \bar{y}_{\cdot\cdot j\cdot} + \bar{y}_{\cdot\cdot\cdot\cdot}\right)^2$
$\beta * \gamma$	$(b-1)(c-1)$	$aN\sum_{ij}\left(\bar{y}_{\cdot ij\cdot} - \bar{y}_{\cdot i\cdot\cdot} - \bar{y}_{\cdot\cdot j\cdot} + \bar{y}_{\cdot\cdot\cdot\cdot}\right)^2$
$\alpha * \beta * \gamma$	$(a-1)(b-1)(c-1)$	see below*
Error	$abc(N-1)$	$\sum_{hijk}\left(y_{hijk} - \bar{y}_{hij\cdot}\right)^2$
Total	$abcN-1$	$\sum_{hijk}\left(y_{hijk} - \bar{y}_{\cdot\cdot\cdot\cdot}\right)^2$

$^*N\sum_{hij}\left(\bar{y}_{hij\cdot} - \bar{y}_{hi\cdot\cdot} - \bar{y}_{h\cdot j\cdot} - \bar{y}_{ij\cdot\cdot} + \bar{y}_{h\cdot\cdot\cdot} + \bar{y}_{\cdot i\cdot\cdot} + \bar{y}_{\cdot\cdot j\cdot} - \bar{y}_{\cdot\cdot\cdot\cdot}\right)^2$

COMPUTING THE ANOVA TABLE

We used tables of means to evaluate the significant effects in the analysis of variance table. (In this example we were able to argue that the insignificant effects could be ignored.) Tables of means are also needed to illustrate the computations for the analysis of variance table. All of the computations can be carried out simply and accurately on hand calculators that have built-in mean and standard deviation functions. The standard deviations are squared to obtain sample variances. After obtaining the means tables, the calculator's mean function is used simply to ensure accuracy. The mean of all observations is

$$\bar{y}_{\cdot\cdot\cdot\cdot} = 158.375 \,.$$

When any sample variance is computed, the mean should also be checked. The mean should be 158.375 for all computations; this is easy to check and helps to verify the computations. It is crucial that numbers not be rounded off; numerical accuracy deteriorates rapidly with round off. However, there should be no occasion to round numbers off and no occasion to write down any numbers other than the final sums of squares and mean squares. Just use the accuracy built into the calculator.

Of course we do not recommend that a hand calculator always be used to do the computations, they quickly become tedious. However, if the process of analysis of variance is well understood, the computations should be recognized as simple and natural. Conversely, learning to do the computations with a hand calculator typically improves the understanding of the process. In constructing tables of means it is best to start with the largest table and work down. In this case, the largest table has twelve means, one for each combination of *surf*, *fill*, and *prop*. These are means averaged over the two replications. Given this three-way table, the two-way table for, say, *surf* – *fill* means can be obtained by averaging the three *prop* means for each level of *surf* and *fill*. Other means tables follow in similar ways. It is vital that the values in the means tables not be rounded off.

In computing sums of squares, we start with the smallest means tables and work up. *Recall that N in the upper left of the means tables denotes the number of observations being averaged to obtain the means.* We begin by computing the mean squares and sums of squares for main effects. Consider the main effect for surface treatments. The means are given below.

$N = 12$	Surface treatment means	
surf	Yes	No
	$172.83\bar{3}$	$143.91\bar{6}$

Finding the sample variance of the two means and multiplying by $N = 12$ gives

$$MS(surf) = (12)418.0868067 = 5017.04168 \,.$$

$N = 12$ is the number of observations that went into each of the surface treatment means. The sum of squares is obtained by multiplying the mean square by the degrees of freedom,

$$SS(surf) = (2 - 1)MS(surf) = 5017.042 \,.$$

Now consider the fills.

$N = 12$	Fill means	
fill	A	B
	212.75	104.00

Finding the sample variance of the two means and multiplying by $N = 12$ gives

$$MS(fill) = (12)5913.28125 = 70959.375 \,.$$

The sum of squares is obtained by multiplying the mean square by the degrees of freedom,

$$SS(fill) = (2 - 1)MS(fill) = 70959.375 \,.$$

Finally, consider the proportions.

$N = 8$	Proportion means		
prop	25%	50%	75%
	137.125	156.375	181.625

Finding the sample variance of the three means and multiplying by $N = 8$ gives

$$MS(prop) = (8)498.0625 = 3984.500\,.$$

The sum of squares is obtained by multiplying the mean square by the degrees of freedom,

$$SS(prop) = (3 - 1)MS(prop) = 7969.000$$

We now examine the two-factor effects. We begin with $surf * fill$. The means table is given below. Note that $N = 6$ is the number of observations in each of the $surf - fill$ means.

$N = 6$	*surf*	
fill	Yes	No
A	$225.6\overline{6}$	$199.8\overline{3}$
B	120.00	88.00

Finding the sample variance of the four means and multiplying by $N = 6$ gives

$$MS(surf - fill\ trts) = (6)4224.081015 = 25344.48609\,.$$

The sum of squares is obtained by multiplying the mean square by the degrees of freedom for $surf - fill$ treatments,

$$SS(surf - fill\ trts) = (4 - 1)MS(surf - fill\ trts) = 76033.458\,.$$

The two-factor interaction sum of squares is

$$\begin{aligned} SS(surf * fill) &= SS(surf - fill\ trts) - SS(surf) - SS(fill) \\ &= 76033.458 - 5017.042 - 70959.375 \\ &= 57.041 \end{aligned}$$

The interaction degrees of freedom are $(2 - 1)(2 - 1) = 1$, which is just the product of the degrees of freedom for *surf* and for *fill*. Finally,

$$MS(surf * fill) = SS(surf * fill)/1 = 57.04\,.$$

Now consider the $surf * prop$ term.

$N = 4$	*surf*	
prop	Yes	No
25%	153.00	121.25
50%	171.25	141.50
75%	194.25	169.00

Finding the sample variance of the six means and multiplying by $N = 4$ gives

$$MS(surf - prop\ trts) = (4)651.51875 = 2606.075\,.$$

The sum of squares is obtained by multiplying the mean square by the $surf - prop$ treatments degrees of freedom,

$$SS(surf - prop\ trts) = (6-1)MS(surf - prop\ trts) = 13030.375\,.$$

The two-factor interaction sum of squares is

$$\begin{aligned} SS(surf * prop) &= SS(surf - prop\ trts) - SS(surf) - SS(prop) \\ &= 13030.375 - 5017.042 - 7969.000 \\ &= 44.333 \end{aligned}$$

The interaction degrees of freedom are $(2-1)(3-1) = 2$, so

$$MS(surf * prop) = SS(surf * prop)/2 = 22.16\,.$$

Finally, consider $fill * prop$.

$N = 4$	*fill*	
prop	A	B
25%	175.25	99.00
50%	205.50	107.25
75%	257.50	105.75

Finding the sample variance of the six means and multiplying by $N = 4$ gives

$$MS(fill - prop\ trts) = (4)4247.96875 = 16991.875\,.$$

The sum of squares is obtained by multiplying the mean square by the $fill - prop$ treatments degrees of freedom,

$$SS(fill - prop\ trts) = (6-1)MS(fill - prop\ trts) = 84959.375\,.$$

The two-factor interaction sum of squares is

$$\begin{aligned} SS(fill * prop) &= SS(fill - prop\ trts) - SS(fill) - SS(prop) \\ &= 84959.375 - 70959.375 - 7969.000 \\ &= 6031.000\,. \end{aligned}$$

The interaction degrees of freedom are $(2-1)(3-1) = 2$, so

$$MS(fill * prop) = SS(fill * prop)/2 = 3015.50\,.$$

The last step in computing the ANOVA table involves the three-factor interaction. The necessary three-way table of means is given below.

$N = 2$		*surf*			
		Yes		No	
	fill	A	B	A	B
	25%	190.00	116.00	160.50	82.00
prop	50%	219.50	123.00	191.50	91.50
	75%	267.50	121.00	247.50	90.50

Finding the sample variance of the 12 means and multiplying by $N = 2$ gives

$$MS(surf - fill - prop\ trts) = (2)4095.096591 = 8190.193182\,.$$

The sum of squares is obtained by multiplying the mean square by the degrees of freedom based on all 12 treatments,

$$\begin{aligned} SS(\mathit{surf} - \mathit{fill} - \mathit{prop\ trts}) &= (12-1)MS(\mathit{surf} - \mathit{fill} - \mathit{prop\ trts}) \\ &= (11)8190.193182 \\ &= 90092.125\,. \end{aligned}$$

Note that these are the mean square and sum of squares computed earlier for the one-way analysis of variance based on 12 treatments. The sum of squares for the three-way interaction is simply what's left of the treatment sum of squares after accounting for all of the other terms, i.e., the main effects and the two-factor interactions,

$$\begin{aligned} & SS(\mathit{surf} * \mathit{fill} * \mathit{prop}) \\ &= SS(\mathit{surf} - \mathit{fill} - \mathit{prop\ trts}) - SS(\mathit{surf} * \mathit{fill}) - SS(\mathit{surf} * \mathit{prop}) - SS(\mathit{fill} * \mathit{prop}) \\ &\quad - SS(\mathit{surf}) - SS(\mathit{fill}) - SS(\mathit{prop}) \\ &= 90092.125 - 57.041 - 44.333 - 6031.000 - 5017.042 - 70959.375 - 7969.000 \\ &= 14.334\,. \end{aligned}$$

The degrees of freedom for three-way interaction are obtained by multiplying the degrees of freedom from the three main effects, i.e., $(2-1)(2-1)(3-1) = 2$. Finally,

$$MS(\mathit{surf} * \mathit{fill} * \mathit{prop}) = SS(\mathit{surf} * \mathit{fill} * \mathit{prop})/2 = 7.17\,.$$

The degrees of freedom for the three-factor interaction can also be computed as the degrees of freedom for treatments minus the degrees of freedom for the other terms.

MINITAB COMMANDS

Minitab commands for generating this analysis are given below. Column c2 contains the data; columns c4, c5, and c6 indicate the levels of the surface, fill and proportion factors that correspond to each observation.

```
MTB > names c4 'surf' c5 'fill' c6 'prop'
MTB > anova c2 = c4|c5|c6;
SUBC> means c4 c5 c6 c4*c5 c4*c6 c5*c6 c4*c5*c6;
SUBC> resid c12;
SUBC> fits c22.
```

11.4 Extensions of Latin squares

Section 9.3 discussed Latin square designs and mentioned that an effective experimental design often requires the use of several small Latin squares. We now present an example of such a design. The example does not actually involve factorial treatment structures but it uses many of the ideas from this chapter.

EXAMPLE 11.4.1. Patterson (1950) and John (1971) considered the milk production of cows that were given three different diets. The three feed regimens were A, good hay; B, poor hay; and C, straw. Eighteen cows were used and milk production was measured during three time periods for each cow. Each cow received a different diet during each time period. The data are given in Table 11.15. The cows were divided into six groups of 3. A 3×3 Latin square design was used for each group of three cows along with the three periods and the three feed treatments. Having eighteen cows, we get 6 Latin squares. The six squares are clearly marked in Table 11.15 by double vertical and horizontal lines. We will not do a complete analysis of these data, rather we point out salient features of the analysis.

TABLE 11.15. Milk production data

	Cow			Cow		
Period	1	2	3	4	5	6
1	A 768	B 662	C 731	A 669	B 459	C 624
2	B 600	C 515	A 680	C 550	A 409	B 462
3	C 411	A 506	B 525	B 416	C 222	A 426

	Cow			Cow		
Period	7	8	9	10	11	12
1	A 1091	B 1234	C 1300	A 1105	B 891	C 859
2	B 798	C 902	A 1297	C 712	A 830	B 617
3	C 534	A 869	B 962	B 453	C 629	A 597

	Cow			Cow		
Period	12	14	15	16	17	18
1	A 941	B 794	C 779	A 933	B 724	C 749
2	B 718	C 603	A 718	C 658	A 649	B 594
3	C 548	A 613	B 515	B 576	C 496	A 612

The basic analysis of variance table for multiple Latin squares is presented in Table 11.16. We go through the table line by line.

TABLE 11.16. Analysis of variance

Source	df	SS
Latin squares	5	1392534
Periods(squares)	12	872013
Cows(squares)	12	318241
Trts	2	121147
Error	22	52770
Total	53	2756704

This example involves 6 Latin squares. Six squares give 5 degrees of freedom for comparisons among squares. From each Latin square obtain the sample mean of the 9 observations in the square. The mean square for Latin squares is 9 times the sample variance of these 6 means. Multiplying by $6 - 1$ gives the sum of squares for Latin squares in the ANOVA table.

Within each individual Latin square, find the sum of squares for periods (rows) as if the other Latin squares did not exist. Each Latin square is 3×3, so there are 2 degrees of freedom for periods within each Latin square. Add together the sums of squares for periods from the six Latin squares to obtain the sum of squares for periods within squares, i.e., periods(squares) in Table 11.16. Similarly, add the degrees of freedom for periods from the six Latin squares to obtain the degrees of freedom for periods within squares. Note that with two degrees of freedom for periods within each square and six squares, we get the total of 12 degrees of freedom given in Table 11.16.

The cows within squares, i.e., cows(squares), line is computed similarly. Within each of the Latin squares, we get 2 degrees of freedom and a sum of squares for cows. Adding these together gives the cows within squares line.

So far we have acted as though the columns are different in every Latin square, as are the rows. This is true for the columns, no cow is ever used in more than one square. It is less clear whether, say, period 1 is the same in the first Latin square as it is in the second and other squares. We will return to this issue later. It is clear, however, that the treatments are the same in every Latin square. To compute the mean for, say, treatment B we average over all 18 observations with treatment B, three from each of the six squares. The means for treatments A and C are computed similarly. The mean square for treatments is 18 times the sample variance of these three numbers. Multiplying by 2 gives the sum of squares for treatments.

The Error line in Table 11.16 is obtained by subtraction. The total sum of squares is 53 times the sample variance of the 54 observations in Table 11.15. From *SSTot*, subtract the other sums of squares to arrive at the *SSE*. The *dfE* is computed similarly using 53 minus the other degrees of freedom.

From Table 11.16, mean squares and F statistics are easily obtained. If this was a classic application of multiple Latin squares, the only F test of real interest would be that for treatments, since the other lines of Table 11.16 denote various forms of blocking. The F statistic for treatments is about 25, so, with 22 degrees of freedom for error, the test is highly significant. One should then compare the three treatments using contrasts and check the validity of the assumptions using residual plots.

The basic analysis of Table 11.16 can be modified in many ways. We now present some of those ways.

As a standard practice, John (1971, section 6.5) computes a square by treatment interaction to examine whether the treatments behave the same in the various Latin squares. In our example with 6 squares and 3 treatments such a term would have $(6-1) \times (3-1) = 10$ degrees of freedom and a sum of squares that would be removed from the error term. Remember that the error term is simply what is left after examining everything else; if we choose to examine an additional term, of necessity it must be removed from the error.

We mentioned earlier that periods might be considered the same from square to square. If so, we should compute three period means across the squares from the 18 observations on each period. A sum of squares for periods can be computed from these three means. The three period means leave us with 2 degrees of freedom for periods as opposed to the 12 degrees of freedom for periods within squares. This 2 degrees of freedom is a subset of the 12. Subtracting the period degrees of freedom and sum of squares from the period(squares) line we obtain a new term, a period by squares interaction, period $*$ square. This new term examines whether the periods behave the same from square to square. The analysis of variance table incorporating this change is presented as Table 11.17.

TABLE 11.17. Analysis of variance

Source	*df*	*SS*
Squares	5	1392534
Cows(squares)	12	318241
Periods	2	814222
Period$*$square	10	57790
Trts	2	121147
Error	22	52770
Total	53	2756704

If the Latin squares were constructed using the complete randomization discussed in Section 9.3, one could argue that the period by squares interaction must really be error and that the 10 degrees of freedom and corresponding sum of squares should be pooled with the current error. Such an analysis is equivalent to simply thinking of the design as one large rectangle with three terms to consider: the 3 periods (rows), the 18 cows (columns), and the 3 treatments. For this design, we simply compute sums of squares for periods, cows, and treatments and subtract them from the total to obtain the error. Such an analysis is illustrated in Table 11.18. The sum of squares for cows in Table 11.18 equals the sum of squares for cows within squares plus the sum of squares for squares from the earlier ANOVA tables. The 17 degrees of freedom for cows are also the 12 degrees of freedom for cows within squares plus the 5 degrees of freedom for squares.

TABLE 11.18. Analysis of variance

Source	df	SS
Cows	17	1710775
Periods	2	814222
Trts	2	121147
Error	32	110560
Total	53	2756704

In this example, choosing between the analyses of Tables 11.17 and 11.18 is easy because of additional structure in the design that we have not yet considered. This particular design was chosen because consuming a particular diet during one period might have an effect that carries over into the next time period. In the three Latin squares on the left of Table 11.15, treatment A is always followed by treatment B, treatment B is always followed by treatment C, and treatment C is always followed by treatment A. In the three Latin squares on the right of Table 11.15, treatment A is always followed by treatment C, treatment B is followed by treatment A, and treatment C is followed by treatment B. This is referred to as a *cross-over* or *change-over* design. Since there are systematic changes in the squares, it is reasonable to investigate whether the period effects differ from square to square and so we should use Table 11.17. In particular, we would like to isolate 2 degrees of freedom from the period by square interaction to look at whether the period effects differ when averaged over the three squares on the left as compared to when they are averaged over the three squares on the right. These issues are addressed in Exercise 11.5.6. □

11.5 Exercises

EXERCISE 11.5.1. The process condition treatments in Exercise 9.5.4 on vinyl thickness had factorial treatment structure. Give the factorial analysis of variance table for the data. The data are repeated below.

Rate	High	Low	Low	High
Temp	Low	High	Low	High
Rep. 1	7.8	11.0	7.4	11.0
Rep. 2	7.6	8.8	7.0	9.2

EXERCISE 11.5.2. Garner (1956) presented data on the tensile strength of fabrics as measured with Scott testing machines. The experimental procedure involved selecting eight 4 × 100 inch strips from available stocks of uniform twill, type I. Each strip was divided into sixteen 4 × 6 inch samples

(with some left over). Each of three operators selected four samples at random and, assigning each sample to one of four machines, tested the samples. The four extra samples from each strip were held in reserve in case difficulties arose in the examination of any of the original samples. It was considered that each 4 × 100 inch strip constituted a relatively homogeneous batch of material. Effects due to operators include differences in the details of preparation of samples for testing and mannerisms of testing. Machine differences include differences in component parts, calibration, and speed. The data are presented in Table 11.19. Entries in Table 11.19 are values of the strengths in excess of 180 pounds.

TABLE 11.19. Tensile strength of uniform twill

	o_1				o_2				o_3			
	m_1	m_2	m_3	m_4	m_1	m_2	m_3	m_4	m_1	m_2	m_3	m_4
s_1	18	7	5	9	12	16	15	9	18	13	10	22
s_2	9	11	12	3	16	4	21	19	25	13	19	12
s_3	7	11	11	1	7	14	12	6	17	20	19	20
s_4	6	4	10	8	15	10	16	12	10	16	12	18
s_5	10	8	6	10	17	12	12	22	18	16	21	22
s_6	7	12	3	15	18	22	14	19	18	23	22	14
s_7	13	5	15	16	14	18	18	9	16	16	10	15
s_8	1	11	8	12	7	13	11	13	15	14	14	11

o = operator, m = machine, s = strip

(a) Identify the design for this experiment and give an appropriate model. List all the assumptions made in the model.

(b) Analyze the data. Give an appropriate analysis of variance table. Examine appropriate contrasts.

(c) Check the assumptions of the model and adjust the analysis appropriately.

EXERCISE 11.5.3. Smith (1988) and Holcomb (1992, p. 105) presented data on knowledge of sexually transmitted diseases. Sixty individuals from each combination of three age groups and gender were given a test measuring their knowledge. The table of mean test scores is given below. In addition, $MSE = 16.52$ on $dfE = 108$. Give an analysis of variance table with sources for main effects, interaction, and error. Analyze the data.

$N = 60$		Age	
Sex	18–24	25–29	30–35
Male	13.40	13.10	14.95
Female	12.90	13.80	13.05

EXERCISE 11.5.4. Baten (1956) presented data on lengths of steel bars. An excessive number of bars had recently failed to meet specifications and the experiment was conducted to identify the causes of this problem. The bars were made with one of two heat treatments (W, L) and cut on one of four screw machines (A, B, C, D) at one of three times of day (8 am, 11 am, 3 pm). The three times were used to investigate the possibility of worker fatigue during the course of the day. The bars were intended to be between 4.380 and 4.390 inches long. The data presented in Table 11.20 are

TABLE 11.20. Steel bar lengths

	Heat treatment W				Heat treatment L			
Machine	A	B	C	D	A	B	C	D
	6	7	1	6	4	6	−1	4
Time 1	9	9	2	6	6	5	0	5
	1	5	0	7	0	3	0	5
	3	5	4	3	1	4	1	4
	6	8	3	7	3	6	2	9
Time 2	3	7	2	9	1	4	0	4
	1	4	1	11	1	1	−1	6
	−1	8	0	6	−2	3	1	3
	5	10	−1	10	6	8	0	4
Time 3	4	11	2	5	0	7	−2	3
	9	6	6	4	3	10	4	7
	6	4	1	8	7	0	−4	0

thousandths of an inch in excess of 4.380. Treating the data as a $2 \times 3 \times 4$ factorial in a completely randomized design, give an analysis of the data.

EXERCISE 11.5.5. Bethea et al. (1985) reported data on an experiment to determine the effectiveness of four adhesive systems for bonding insulation to a chamber. The adhesives were applied both with and without a primer. Tests of peel-strength were conducted on two different thicknesses of rubber. Using two thicknesses of rubber was not part of the original experimental design. *The existence of this factor was only discovered by inquiring about a curious pattern of numbers in the laboratory report.* The data are presented in Table 11.21. Another disturbing aspect of these data is that the values for adhesive system 3 are reported with an extra digit. Presumably, a large number of rubber pieces were available and the treatments were randomly assigned to these pieces, but, given the other disturbing elements in these data, I wouldn't bet the house on it. A subset of these data was examined earlier in Exercise 5.7.7.

TABLE 11.21. Peel-strength of various adhesive systems

	Adhesive				Adhesive			
	1	2	3	4	1	2	3	4
	60	57	19.8	52	73	52	32.0	77
	63	52	19.5	53	79	56	33.0	78
With	57	55	19.7	44	76	57	32.0	70
Primer	53	59	21.6	48	69	58	34.0	74
	56	56	21.1	48	78	52	31.0	74
	57	54	19.3	53	74	53	27.3	81
	59	51	29.4	49	78	52	37.8	77
	48	44	32.2	59	72	42	36.7	76
Without	51	42	37.1	55	72	51	35.4	79
Primer	49	54	31.5	54	75	47	40.2	78
	45	47	31.3	49	71	57	40.7	79
	48	56	33.0	58	72	45	42.6	79
	Thickness A				Thickness B			

(a) Identify the design for this experiment and give an appropriate model. List all the assumptions made in the model.

(b) Check the assumptions of the model and adjust the analysis appropriately.

(c) Analyze the data. Give an appropriate analysis of variance table. Examine appropriate contrasts.

EXERCISE 11.5.6. Consider the milk production data in Table 11.15 and the corresponding analysis of variance in Table 11.17. Relative to the periods, the squares on the left of Table 11.15 always have treatment A followed by B, B followed by C, and C followed by A. The squares on the right always have treatment A followed by C, B followed by A, and C followed by B. Test whether there is an average difference between the squares on the left and those on the right. Test whether there is an interaction between periods and left–right square differences.

EXERCISE 11.5.7. In addition to the data provided in Exercise 11.5.3, Smith (1988) and Holcomb (1992, p. 105) present data on knowledge of sexually transmitted diseases as cross-classified by sex and ethnicity. Forty individuals from each of two ethnic groups and of each sex were given the tests. The table of means is given below and once again use $MSE = 16.52$ with $dfE = 108$. Give an analysis of variance table with sources for main effects, interaction, and error. Analyze the data.

$N = 40$	Ethnic group	
Sex	Black	Hispanic
Male	15.23	12.40
Female	16.77	9.73

EXERCISE 11.5.8. As in Exercise 9.5.8, we consider differences in hydrostatic pressure tests due to operators. Table 11.22 contains two Latin squares. Analyzing these together, give an appropriate analysis of variance table and report on any differences that can be established among the operators.

TABLE 11.22. Hydrostatic pressure tests: operator, yield

Square I				Square II			
C	D	A	B	D	C	B	A
41.0	38.5	39.0	43.0	43.0	40.5	43.5	39.5
D	C	B	A	C	D	A	B
41.0	38.5	41.5	41.0	41.0	39.0	39.5	41.5
A	B	C	D	B	A	D	C
39.5	42.0	41.5	42.0	42.0	41.0	40.5	37.5
B	A	D	C	A	B	C	D
41.5	41.0	40.5	41.5	40.5	42.5	44.0	41.0

Operators are A, B, C, D.

EXERCISE 11.5.9. Exercises 9.5.8, 9.5.9, 9.5.10, and the previous exercise used subsets of data reported in Garner (1956). The experiment was designed to examine differences among operators and machines when using Suter hydrostatic pressure testing machines. No interaction between machines and operators was expected.

A one foot square of cloth was placed in a machine. Water pressure was applied using a lever until the operator observed three droplets of water penetrating the cloth. The pressure was then relieved using the same lever. The observation was the amount of water pressure consumed and it was measured as the number of inches that water rose up a cylindrical tube with radial area of 1 sq. in. Operator differences are due largely to differences in their ability to spot the droplets and their reaction times in relieving the pressure. Machines 1 and 2 were operated with a hand lever. Machines 3 and 4 were operated with at foot lever.

A 52 × 200 inch strip of water repellant cotton Oxford was available for the experiment. From this, four 48 × 48 inch squares were cut successively along the warp (length) of the fabric. It was decided to adjust for heterogeneity in the application of the water repellant along the warp and fill (width) of the fabric, so each 48 × 48 square was divided into four equal parts along the warp and four equal parts along the fill, yielding 16 smaller squares. The design involves four replications of a Graeco-Latin square. In each 48 × 48 square, every operator worked once with every row and column of the larger square and once with every machine. Similarly, every row and column of the 48 × 48 square was used only once on each machine. The data are given in Table 11.23.

TABLE 11.23. Hydrostatic pressure tests: operator, machine, yield

Square 1				Square 2			
A,1	B,3	C,4	D,2	B,2	A,4	D,3	C,1
40.0	43.5	39.0	44.0	39.0	39.0	41.0	41.0
B,2	A,4	D,3	C,1	A,1	B,3	C,4	D,2
40.0	42.0	40.5	38.0	36.5	42.5	40.5	38.5
C,3	D,1	A,2	B,4	D,4	C,2	B,1	A,3
42.0	40.5	38.0	40.0	40.0	39.0	41.5	41.5
D,4	C,2	B,1	A,3	C,3	D,1	A,2	B,4
40.0	36.5	39.0	38.5	41.5	39.5	39.0	44.0
Square 3				**Square 4**			
C,3	D,1	A,2	B,4	D,4	C,2	B,1	A,3
41.0	38.5	39.0	43.0	43.0	40.5	43.5	39.5
D,4	C,2	B,1	A,3	C,3	D,1	A,2	B,4
41.0	38.5	41.5	41.0	41.0	39.0	39.5	41.5
A,1	B,3	C,4	D,2	B,2	A,4	D,3	C,1
39.5	42.0	41.5	42.0	42.0	41.0	40.5	37.5
B,2	A,4	D,3	C,1	A,1	B,3	C,4	D,2
41.5	41.0	40.5	41.5	40.5	42.5	44.0	41.0

Operators are A, B, C, D. Machines are 1, 2, 3, 4.

Analyze the data. Give an appropriate analysis of variance table. Use the Newman–Keuls method to determine differences among operators and use orthogonal contrasts to determine differences among machines. Give a model and check your assumptions.

The cuts along the warp of the fabric were apparently the rows. Should the rows be considered the same from square to square? How would doing this affect the analysis?

Look at the means for each square. Is there any evidence of a trend in the water repellency as we move along the warp of the fabric? How should this be tested?

EXERCISE 11.5.10. For a balanced 3 × 4 with equally spaced quantitative factor levels, give the linear by linear, linear by quadratic, linear by cubic, quadratic by linear, quadratic by quadratic, and quadratic by cubic interaction contrasts. A table or several tables of contrast coefficients is sufficient.

EXERCISE 11.5.11. Consider a balanced design with a 4×4 factorial treatment structure to examine the effects of diet on animal growth. One factor consists of a control diet and three new diets based on beef, pork, and beans. The other factor consists of equally spaced levels of protein content in the diets. Give the following interaction contrasts: the interaction contrasts that examine whether the linear, quadratic, and cubic contrasts in protein content change between the two meat diets, the interaction contrast that examines whether the average linear contrast for the two meat diets differs from the linear contrast for beans, the similar contrasts comparing meat and beans for the quadratic and cubic contrasts, and those that compare how the linear, quadratic, and cubic contrasts change between the control and the average of the other treatments. A table or several tables of contrast coefficients is sufficient.

Chapter 12

Split plots, repeated measures, random effects, and subsampling

In this chapter we examine methods for performing analysis of variance on data that are not completely independent. The first two methods considered are appropriate for similar data but they are based on different assumptions. The data involved have independent groups of observations but the observations within groups are not independent. The first method was developed for analyzing the results of split plot designs. In these models, the lack of independence consists of a constant correlation between observations within the groups. The second method is multivariate analysis of variance. Multivariate ANOVA allows an arbitrary correlation structure among the observations within groups. Section 12.1 introduces split plot models and illustrates the highlights of the analysis. Section 12.2 gives a detailed analysis for a complicated split plot. Section 12.3 introduces multivariate analysis of variance. Section 12.4 considers some special cases of the model examined in Sections 12.1 and 12.2; these are subsampling models and analysis of variance models in which some of the effects are random.

12.1 The analysis of split plot designs

Split plot designs involve simultaneous use of different sized experimental units. They also involve more than one error term.

Suppose we produce an agricultural commodity and are interested in the effects of two factors, an insecticide and a fertilizer. The fertilizer is applied using a tractor and the insecticide is applied via crop dusting. (Crop dusting involves using an airplane to spread the material. If you are of a militaristic bent, you can think of the factors in the experiment as applying mortar fire and napalm from a B-52.) Obviously, you need a fairly large piece of land to use crop dusting, so the number of replications on the crop dusting treatments will be relatively small. On the other hand, different fertilizers can be applied with a tractor to reasonably small pieces of land. If our primary interest is in the main effects of the crop dusted insecticides, we are stuck. Accurate results require a substantial number of large fields to obtain replications on the crop dusting treatments. However, if our primary interest is in the fertilizers or the interaction between fertilizers and insecticides, we can design a good experiment with only a few large fields.

To construct a split plot design, start with the large fields and design an experiment that is appropriate for examining just the insecticides. Depending on the information available about the fields, this can be a CRD, a RCB design, or a Latin Square design. Suppose there are three levels of insecticide to be investigated. If we have three fields in the Gallatin Valley of Montana, three fields near Willmar, Minnesota, and three fields along the Rio Grande River in New Mexico, it is appropriate to set up a block in each state so that we see each insecticide in each location. Alternatively, if we have one field near Bozeman, MT, one near Cedar City, UT, one near Twin Peaks, WA, one near Winters, CA, one near Fields, OR, and one near Grand Marais, MN, a CRD seems more appropriate. We need a valid design for this experiment on insecticides, but often it will not have enough replications to yield a very precise analysis. Each of the large fields used for insecticides is called a *whole plot*. The insecticides are applied to the whole plots, so they are referred to as the *whole plot treatments*.

Regardless of the design for the insecticides, the key to a split plot design is using each whole plot (large field) as a block for examining the *subplot treatments* (fertilizers). If we have four fertilizer treatments, we divide each whole plot into four subplots. The fertilizers are randomly assigned to the subplots. The analysis for the *subplot treatments* is just a modification of the RCB analysis with each whole plot treated as a block.

We have a much more accurate experiment for fertilizers than for insecticides. If the insecticide (whole plot) experiment was set up with 3 blocks each containing 3 whole plots, we have just 3 replications on the insecticides, but each of the 9 whole plots is a block for the fertilizers, so we have 9 replications of the fertilizers. Moreover, fertilizers are compared within whole plots, so they are not subject to the whole plot to whole plot variation.

Perhaps the most important aspect of the design is the interaction. It is easy to set up a mediocre design for insecticides and a good experiment for fertilizers, the difficulty is in getting to look at them together and the primary point in looking at them together is to investigate interaction. The most important single fact in the analysis is that the interaction between insecticides and fertilizers is subject to exactly the same variability as fertilizer comparisons. Thus we have eliminated a major source of variation, the whole plot to whole plot variability. Interaction sums of squares and contrasts are only subject to the subplot variability, i.e., the variability *within* whole plots.

The idea behind split plot designs is very general. The key idea is that an experimental unit (whole plot, large field) is broken up to allow several distinct measurements on the unit. In an example below, the weight loss due to abrasion of *one* piece of fabric is measured after 1000, 2000, and 3000 revolutions of a machine designed to cause abrasion. Another possibility is giving drugs to people and measuring their heart rates after 10, 20, and 30 minutes. When repeated measurements are made on the same experimental unit, these measurements are more likely to be similar than measurements taken on different experimental units. Thus the measurements on the same unit are correlated. This correlation needs to be accounted for in the analysis. This section and the next two discuss appropriate models.

We now consider an example of a very simple split plot design for which only the highlights are discussed. In Section 12.2 a second example considers the detailed analysis of a study with four factors.

EXAMPLE 12.1.1. Garner (1956) presents data on the amount of moisture absorbed by water-repellant cotton Oxford material. Two 24 yard strips of cloth were obtained. Each strip was divided into four 6 yard strips. The 6 yard strips were randomly assigned to one of four laundries. After laundering and drying, the 6 yard strips were further divided into four 1.5 yard strips and randomly assigned to one of four laboratories for determination of dynamic water absorption. The data presented in Table 12.1 are actually the means of two determinations of dynamic absorption made for each 1.5 yard strip. The label 'test' is used to identify different laboratories.

First consider how the experimental design deals with laundries. There are two blocks of material available, the 24 yard strips. These are subdivided into four sections and randomly assigned to

TABLE 12.1. Garner's dynamic absorption data

	Rep. 1				Rep. 2			
	Test				Test			
Laundry	A	B	C	D	A	B	C	D
1	7.20	11.70	15.12	8.10	9.06	11.79	14.38	8.12
2	2.40	7.76	6.13	2.64	2.14	7.76	6.89	3.17
3	2.19	4.92	5.34	2.47	2.69	1.86	4.88	1.86
4	1.22	2.62	5.50	2.74	2.43	3.90	5.27	2.31

laundries. Thus we have a randomized complete block (RCB) design for laundries with two blocks and four treatments. After the 6 yard strips have been laundered, they are further subdivided into 1.5 yard strips and these are randomly assigned to laboratories for testing. Each experimental unit in the RCB design for laundries is split into subunits for further treatment. The 6 yard strips are the whole plot experimental units, laundries are whole plot treatments, and the 24 yard strips are whole plot blocks. The whole plot experimental units (6 yard strips) also serve as blocks for the subplot treatments. The 1.5 yard strips are subplot experimental units and the tests are subplot treatments. Note that every laundry–test treatment combination is observed in every whole plot block (24 yard strip). It is common practice to refer to whole plot blocks as *replications* or just *reps*.

The peculiar structure of the design leads us to analyze the data almost as two separate experiments. There is a whole plot analysis focusing on laundries and a subplot analysis focusing on tests. The subplot analysis also allows us to investigate interaction. As always, the analysis requires the treatment means, which are given below.

$N = 2$	Laundry				Test
Test	1	2	3	4	means
A	8.130	2.270	2.440	1.825	3.6662
B	11.745	7.760	3.390	3.260	6.5388
C	14.750	6.510	5.110	5.385	7.9388
D	8.110	2.905	2.165	2.525	3.9262
Laundry means	10.684	4.861	3.276	3.249	

Consider the effects of the laundries. The analysis for laundries is called the *whole plot analysis*. We have an RCB design in laundries but an RCB analysis requires just one number for each laundry observation (whole plot). The one number used for each whole plot is the mean absorption averaged over the four subplots contained in the wholeplot. With one minor exception, the whole plot analysis is just the RCB analysis of these data. The degrees of freedom are reps., 1; laundries, 3; and *whole plot error*, 3. The minor exception is that the sums of squares and mean squares are all multiplied by the number of subplot treatments, four. Note that this multiplication has no effect on significance tests, e.g., in an F test the numerator mean square and the denominator mean square are both multiplied by the same number, so the multiplications cancel. Multiplying the mean squares and sums of squares by the number of subplot treatments is done simply to maintain consistency in the way we do computations. For example, the mean square for laundries is the sample variance of the laundry means times the number of observations in each mean. The RCB analysis would consider the number of observations in each mean to be the number of replications (2); the split plot analysis considers the number of observations in each mean to be the number of replications times the number of subplot treatments (2×4). Thus the split plot computation is more consistent with our usual methods for analyzing experiments with factorial treatment structure.

Now consider the analysis of the subplot treatments, i.e., the absorption tests. The *subplot analysis* is largely produced by treating each whole plot as a block. Note that we observe every subplot treatment within each whole plot, so the blocks are complete. There will be, however, one notable exception to treating the subplot analysis as an RCB analysis, i.e., the identification of interaction effects.

In an RCB analysis with whole plots taken as subplot blocks there are 8 whole plots, so there are 7 degrees of freedom for subplot blocks. In addition there are 3 degrees of freedom for tests, so the standard RCB analysis would have $3 \times 7 = 21$ degrees of freedom for error. The subplot analysis differs from the standard RCB analysis in the handling of the 21 degrees of freedom for error. A standard RCB analysis takes the block by treatment interaction as error. This is appropriate because the extent to which treatment effects vary from block to block is an appropriate measure of error for treatment effects. However, in a split plot design the subplot blocks are not obtained haphazardly, they have consistencies due to the whole plot treatments. We can identify structure within the subplot block by subplot treatment interaction. Some of the block by treatment interaction can be ascribed to whole plot treatment by subplot treatment interaction. In this experiment the laundry by test interaction has $3 \times 3 = 9$ degrees of freedom. This is extracted from the 21 degrees of freedom for error in the subplot RCB analysis to give a *subplot error* term with only $21 - 9 = 12$ degrees of freedom. Finally, it is of interest to note that the 7 degrees of freedom for subplot blocks correspond to the 7 degrees of freedom in the whole plot analysis: 1 for reps., 3 for laundries, and 3 for whole plot error. In addition, the sum of squares for subplot blocks is also the total of the sums of squares for reps., laundries, and whole plot error.

Table 12.2 combines the whole plot analysis and the subplot analysis into a common analysis of variance table. *Error 1 indicates the whole plot error term and its mean square is used for inferences about laundries and reps.* (if you think it is appropriate to draw inferences about reps.). *Error 2 indicates the subplot error term and its mean square is used for inferences about tests and laundry by test interaction.* A subplot blocks line does not appear in the table; the whole plot analysis replaces it. The means table given earlier suffices for computing the sums of squares for laundries, tests, and laundry * test interaction. The computations are executed in the usual way. A similar two-way means table for reps. and laundries is needed to compute the sums of squares for reps. and error 1. Note that for the given whole plot design, error 1 is computationally equivalent to a rep * laundry interaction.

TABLE 12.2. Analysis of variance for dynamic absorption data

Source	*df*	*SS*	*MS*	*F*	*P*
Reps	1	0.007	0.007	0.01	0.933
Laundry	3	298.330	99.443	124.30	0.001
Error 1	3	2.381	0.794		
Test	3	102.917	34.306	60.08	0.000
Laundry * test	9	26.870	2.986	5.23	0.005
Error 2	12	6.852	0.571		
Total	31	437.356			

From Table 12.2 the laundry * test interaction is clearly significant, so the analysis would typically focus there. On the other hand, while the interaction is statistically important, its *F* statistic is an order of magnitude smaller than the *F* statistic for tests, so the person responsible for the experiment might decide that interaction is not of practical importance. The analysis might then ignore the interaction and focus on the tests. Since I am not responsible for the experiment (only for its inclusion in this book), I will not presume to declare a highly significant interaction unimportant. Figure 12.1 gives the interaction plot. Tests A and D behave very similarly and they behave quite differently from tests B and C. Tests B and C also behave somewhat similarly. This suggests looking at contrasts in the tests that involve A versus D, B versus C, and the sum of A and D versus the sum of B and C. The contrasts' coefficients are given below along with the estimated contrasts when the coefficients are applied to a particular laundry.

$N = 2$	Test contrasts		
Test	A–D	B–C	A,D–B,C
A	1	0	1
B	0	1	−1
C	0	−1	−1
D	−1	0	1
Laundry 1	.020	−3.005	−10.255
Laundry 2	−.635	1.250	−9.095
Laundry 3	.275	−1.720	−3.895
Laundry 4	−.700	−2.125	−4.295

For example, using the table of laundry–test means, the value in this table for the B–C contrast with laundry 4 is

$$(0)1.825 + (1)3.260 + (-1)5.385 + (0)2.525 = -2.125.$$

We can now examine such things as whether the difference between tests A and D is the same for laundry 1 as it is for laundry 2. The estimated interaction contrast is $.020 - (-.635) = .655$. The contrast coefficients are actually

	Laundry			
Test	1	2	3	4
A	1	−1	0	0
B	0	0	0	0
C	0	0	0	0
D	−1	1	0	0

and the laundry–test means are averages over 2 observations, so the sum of squares for the contrast is

$$\frac{.655^2}{4/2} = .2145.$$

As this is an interaction contrast, the sum of squares would be compared to $MSE(2)$, the mean square from the error 2 line.

One option for analyzing an ordinary $a \times b$ two-factor experiment that displays interaction is just to ignore the factorial treatment structure and perform the analysis by constructing contrasts for the ab treatments directly. With a split plot design, that option is no longer available. To handle the factorial as simply ab treatments, one essentially pools the main effects for the two factors and the interaction. In a split plot design, the whole plot treatments are part of the whole plot analysis, so they cannot be pooled with the subplot treatments and interaction which are part of the subplot analysis. The closest we can come to this is pooling the subplot treatments with the interaction. By pooling, we are allowed to look at contrasts *having a fixed level of the whole plot treatments*. Inferences for these contrasts are made in reference to error 2. For example, we earlier applied the test contrasts to each laundry. We can obtain valid F tests for all 12 of these contrasts. In particular, the A,D–B,C contrast applied to laundry 1 has a sum of squares of

$$\frac{-10.255^2}{4/2} = 52.58.$$

Since $MSE(2)$ is .571, the F ratio is huge. Even if we use Scheffé's method to adjust for having looked at the data to determine the contrasts, we reach the clear conclusion that on average tests A and D give very different results from tests B and C for the material sent to laundry 1. *This method works only for contrasts in the subplot treatments applied at a fixed level of the whole-plot*

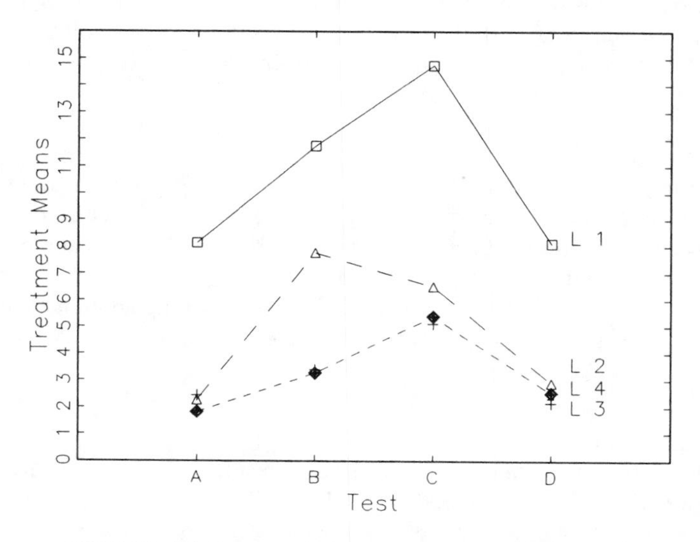

FIGURE 12.1. Interaction plot for dynamic absorption data.

treatments. If we fixed a test (subplot treatment) and looked at a contrast in the laundries for that test, we would not be able to compare the estimated contrast to either error 2 or error 1. Note also that there are 12 of these *orthogonal* contrasts and, when pooled together, the tests and interaction have $3 + 9 = 12$ degrees of freedom.

□

We now examine the assumptions behind this analysis. The basic split plot model is

$$y_{ijk} = \mu + r_i + w_j + \eta_{ij} + s_k + (ws)_{jk} + \varepsilon_{ijk}$$

where $i = 1, \ldots, a$ indicates the replication, $j = 1, \ldots, b$ indicates the whole plot treatment, $k = 1, \ldots, c$ indicates the subplot treatment, μ is a grand mean, r_i, w_j, and s_k indicate rep., whole plot treatment, and subplot treatment effects, and $(ws)_{jk}$ indicates a whole plot–subplot interaction effect. The model has two sets of random error terms: the η_{ij}s, which are errors specific to a given whole plot, and the ε_{ijk}s, which are errors specific to a given subplot. All of the errors are assumed to be independent with the η_{ij}s distributed $N(0, \sigma_w^2)$ and the ε_{ijk}s distributed $N(0, \sigma_s^2)$.

The model assumes normality and equal variances for each set of error terms. These assumptions should be checked using residual plots. We can get error 2 residuals

$$\hat{\varepsilon}_{ijk} = y_{ijk} - \bar{y}_{ij\cdot} - \bar{y}_{\cdot jk} + \bar{y}_{\cdot j\cdot}$$

by treating the η_{ij} error terms as a fixed rep. by whole plot interaction. We can also get error 1 residuals as

$$\hat{\eta}_{ij} = \bar{y}_{ij\cdot} - \bar{y}_{i\cdot\cdot} - \bar{y}_{\cdot j\cdot} + \bar{y}_{\cdot\cdot\cdot}.$$

It is something of an abuse of notation to write the error 1 residuals as $\hat{\eta}_{ij}$s, since they are actually predictors of the $\eta_{ij} + \bar{\varepsilon}_{ij\cdot}$s and not predictors of the η_{ij}s. As usual, the basic analysis can be performed quite easily with a hand calculator, but the residual analysis is a real pain without a computer.

The $MSE(1)$ is an estimate of

$$E[MSE(1)] = \sigma_s^2 + c\sigma_w^2$$

and $MSE(2)$ is an estimate of

$$E[MSE(2)] = \sigma_s^2$$

If there is no whole plot to whole plot variability over and above the variability due to the subplots within the wholeplots, i.e., if $\sigma_w^2 = 0$, then the two error terms are estimating the same thing and their ratio has an F distribution. In other words, we can test $H_0 : \sigma_w^2 = 0$ by rejecting H_0 when

$$MSE(1)/MSE(2) > F(1-\alpha, dfE(1), dfE(2)).$$

In the laundries example we get $.794/.571 = 1.39$ on 3 and 12 degrees of freedom and a P value of $.293$. This is rather like testing for blocks in a randomized complete block design. Both tests merely tell you if you wasted your time. An insignificant test for blocks indicates that blocking was a waste of time. Similarly, an insignificant test for whole plot variability indicates that forming a split plot design was a waste of time. In each case, it is too late to do anything about it. The analysis should follow the design that was actually used. However, the information may be of value in designing future studies.

EXAMPLE 12.1.1 CONTINUED.

Figures 12.2 through 12.5 contain a series of error 1 residual plots. Figure 12.2 has the error 1 residuals versus predicted values. The predicted values are $\bar{y}_{i\bullet\bullet} + \bar{y}_{\bullet j\bullet} - \bar{y}_{\bullet\bullet\bullet}$. Note the wider spread for predicted values near 3. Figure 12.3 plots the error 1 residuals against reps. and shows nothing startling. Figure 12.4 is a plot against laundries. In Figure 12.4 we see that the spread for laundry 3 is much wider than the spread for the other laundries. This seems to be worth discussing with the experimenter. Figure 12.5 contains a normal plot for the error 1 residuals; it looks reasonably straight. Of course there are only eight residuals (and only 3 degrees of freedom), so it is difficult to draw any firm conclusions.

Figures 12.6 through 12.10 contain a series of error 2 residual plots. The predicted values in Figure 12.6 are $\bar{y}_{ij\bullet} + \bar{y}_{\bullet jk} - \bar{y}_{\bullet j\bullet}$. Figures 12.7, 12.8, and 12.9 are plots against reps., laundries, and tests. There is nothing startling. There is more dispersion for laundry 3 than the others but the visual effect is less startling than that in the error 1 plot. Apparently, the small (negative) residuals at laundry 3 belong to one replication and the large residuals to the other. Figure 12.10 contains the normal plot; it looks alright. □

MINITAB COMMANDS

The basic analysis can be obtained from Minitab as follows:

```
MTB > names c1 'y' c2 'reps' c3 'laund' c4 'tests'
MTB > anova c1=c2|c3 c4 c4*c3;
SUBC> test c2 c3 / c2*c3.
SUBC> means c2*c3 c3 c4 c3*c4;
SUBC> resid c10;
SUBC> fits c11.
```

Minitab treats the c2∗c3 term generated by c2|c3 as a fixed rep. by laundry interaction rather than the error 1 term. Minitab only recognizes one error, the error 2 term. Thus it uses the error 2 term in all tests, giving an incorrect whole plot analysis. The 'test' subcommand provides the whole plot analysis by testing reps. and laundries against the c2 ∗ c3 term. The 'resid' and 'fits' commands give the error 2 residuals and the corresponding predicted values. To get the error 1 residuals and their predicted values one can use the rep.–laundry (c2 ∗ c3) means and repeat the whole plot RCB analysis on them.

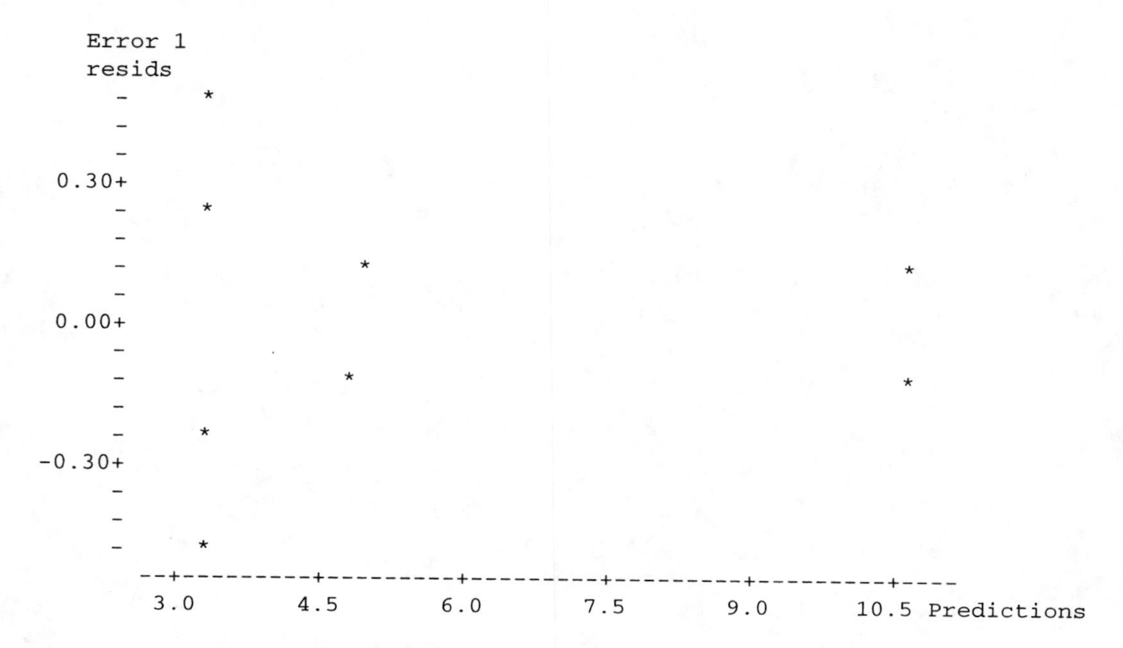

FIGURE 12.2. Whole plot residuals versus predicted values, absorption data.

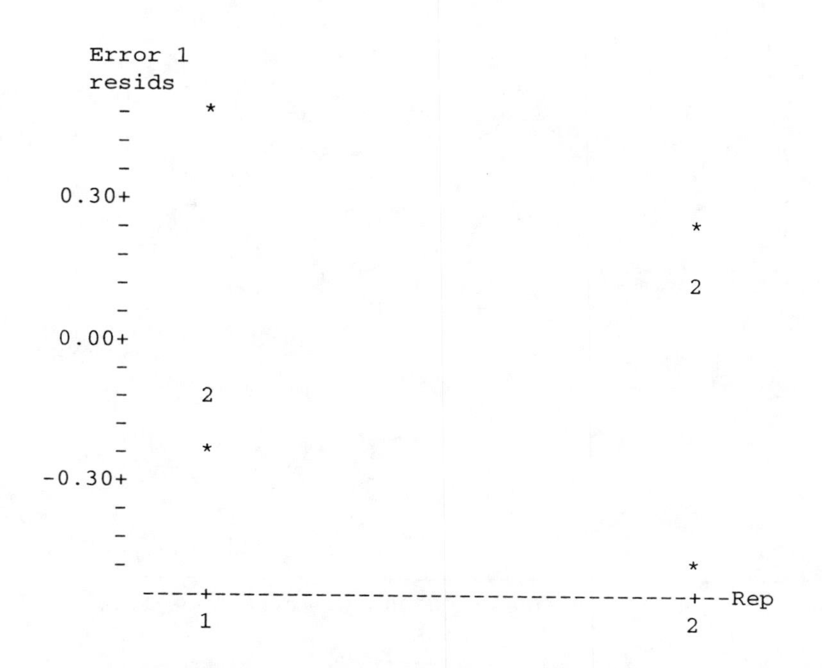

FIGURE 12.3. Whole plot residuals versus blocks, absorption data.

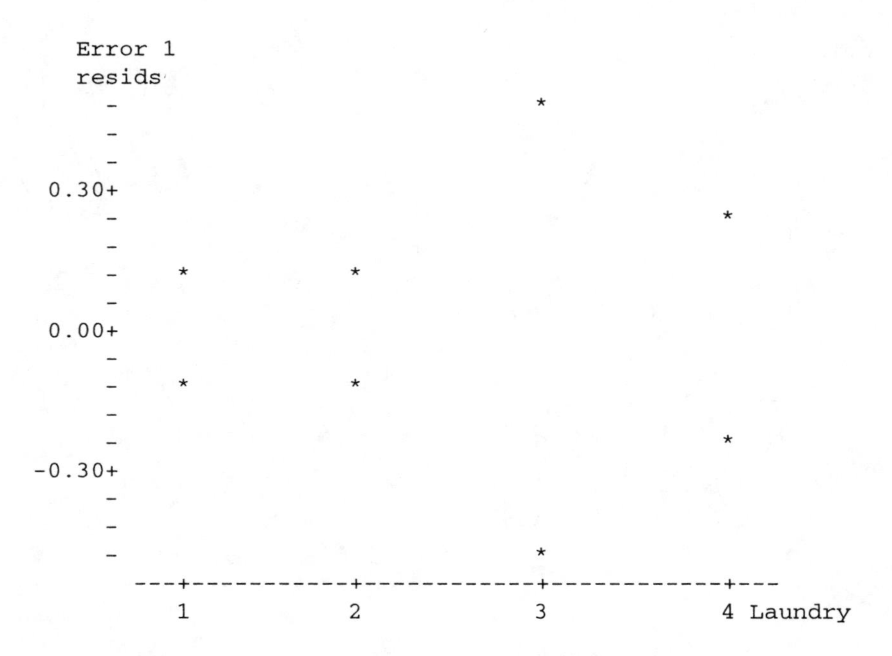

FIGURE 12.4. Whole plot residuals versus laundries, absorption data.

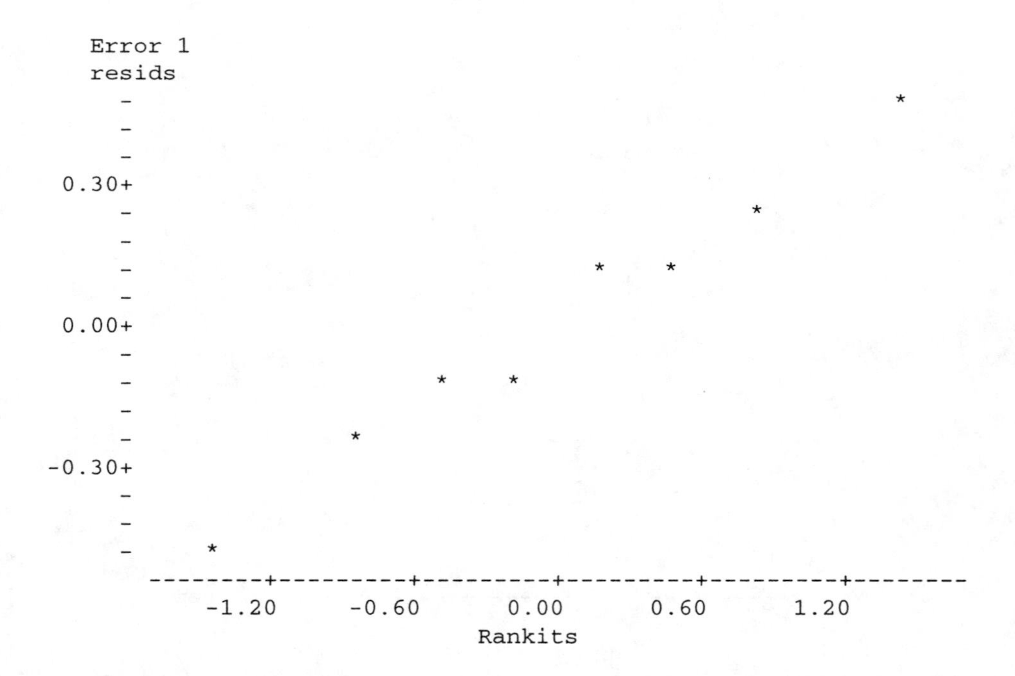

FIGURE 12.5. Normal plot of whole plot residuals, $W' = 0.974$, absorption data.

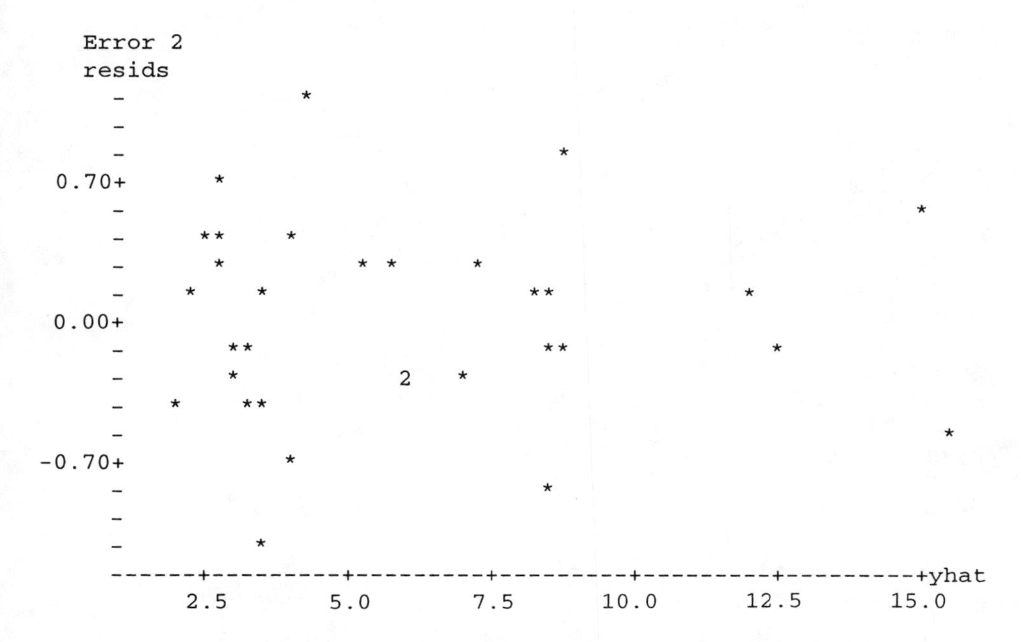

FIGURE 12.6. Subplot residuals versus predicted values, absorption data.

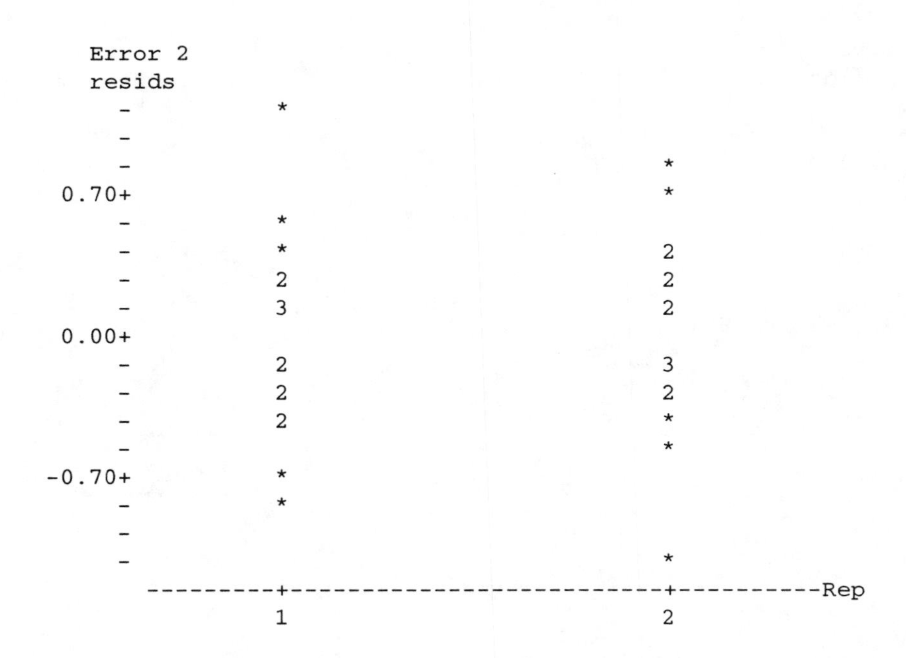

FIGURE 12.7. Subplot residuals versus blocks, absorption data.

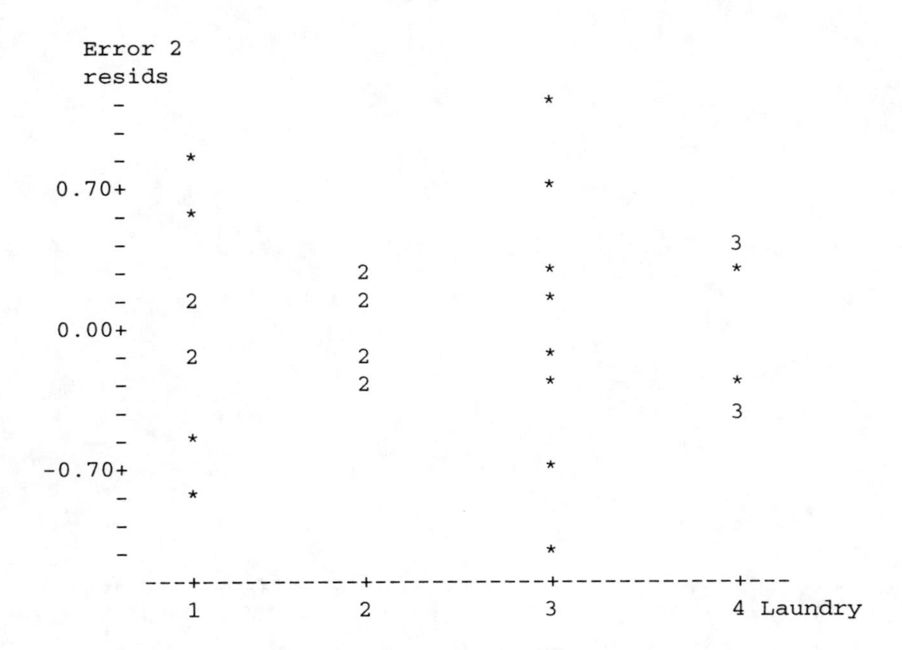

FIGURE 12.8. Subplot residuals versus laundries, absorption data.

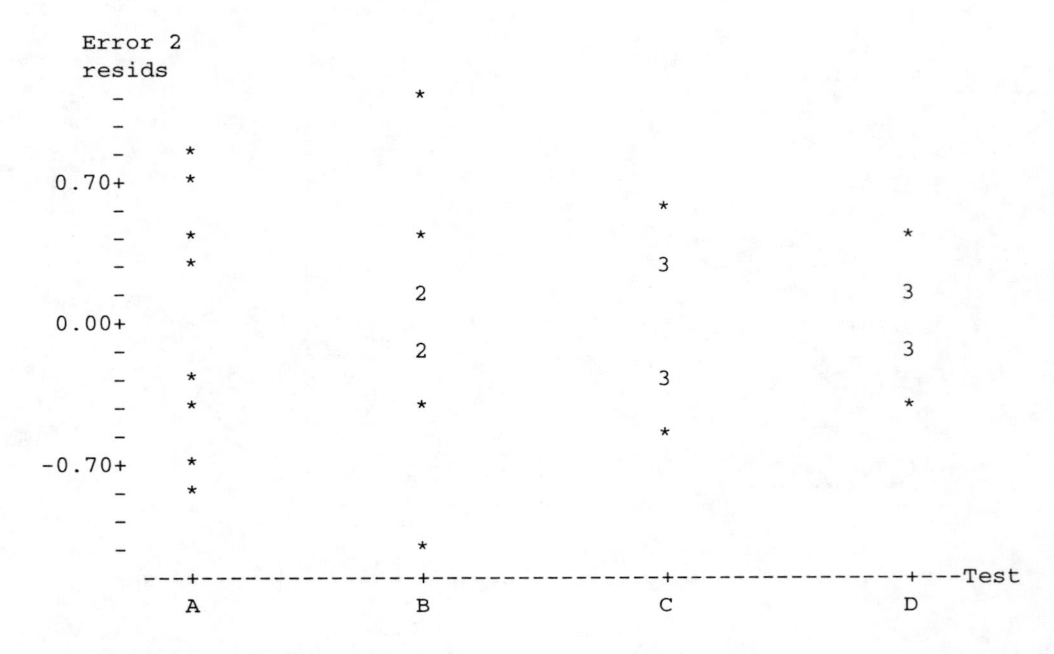

FIGURE 12.9. Subplot residuals versus predicted values, absorption data.

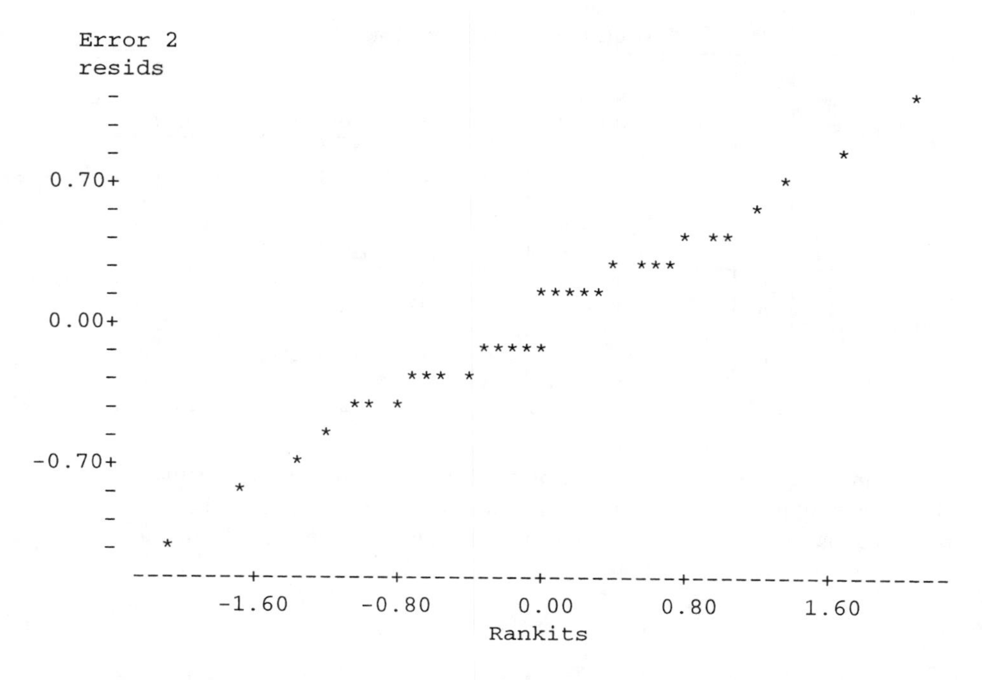

FIGURE 12.10. Normal plot of subplot residuals, $W' = 0.990$, absorption data.

12.2 A four-factor split plot analysis

In this section we consider a split plot analysis involving four factors, detailed examination of three-factor interactions, and a whole plot design that is a CRD.

EXAMPLE 12.2.1. Earlier we considered data from Box (1950) on fabric abrasion. The data consisted of three factors: surface treatment (yes, no), filler (A, B), and proportion of filler (25%, 50%, 75%). These are referred to as *surf*, *fill*, and *prop*. We earlier restricted our attention to the weight loss that occurred between 1000 and 2000 revolutions of a machine designed for evaluating abrasion resistance, but we mentioned that data are also available on each piece of cloth for weight loss after 1000 rotations and weight loss occurring between 2000 and 3000 rotations. The full data are given in Table 12.3. In analyzing the full data, many aspects are just simple extensions of the analysis given earlier in Section 11.3. There are now four factors, *surf*, *fill*, *prop*, and one for rotations, say, *rota*. With four factors, there are many more effects to deal with. There is one more main effect, *rota*, three more two-factor interactions, *surf* ∗ *rota*, *fill* ∗ *rota*, and *prop* ∗ *rota*, three more three-factor interactions, *surf* ∗ *fill* ∗ *rota*, *surf* ∗ *prop* ∗ *rota*, and *fill* ∗ *prop* ∗ *rota*, and a four-factor interaction, *surf* ∗ *fill* ∗ *prop* ∗ *rota*. Calculations of sums of squares for main effects, two-factor interactions and three-factor interactions follow the same pattern as illustrated earlier. The sum of squares for the four-factor interaction is just what is left after accounting for all of the other main effects and interactions. It is found by computing the one-way ANOVA treatment sum of squares from the means of all $2 \times 2 \times 3 \times 3 = 36$ treatment combinations and subtracting out the other sums of squares. The computations will be discussed later.

What makes these data worthy of our further attention is the fact that not all of the observations are independent. Observations on different pieces of fabric may be independent, but the three observations on the same piece of fabric, one after 1000, one after 2000, and one after 3000 revolutions, should behave similarly as compared to observations on different pieces of fabric. In

TABLE 12.3. Abrasion resistance data

Surf. treat.	Fill	25% 1000	25% 2000	25% 3000	50% 1000	50% 2000	50% 3000	75% 1000	75% 2000	75% 3000
Yes	A	194	192	141	233	217	171	265	252	207
	A	208	188	165	241	222	201	269	283	191
	B	239	127	90	224	123	79	243	117	100
	B	187	105	85	243	123	110	226	125	75
No	A	155	169	151	198	187	176	235	225	166
	A	173	152	141	177	196	167	229	270	183
	B	137	82	77	129	94	78	155	76	91
	B	160	82	83	98	89	48	132	105	67

other words, the three observations on one piece of fabric should display positive correlations. The analysis considered in this section assumes that the correlation is the same between any two of the three observations on a piece of fabric. To achieve this, we consider a model that includes two error terms,

$$\begin{aligned} y_{hijkm} &= \mu + s_h + f_i + p_j \\ &\quad + (sf)_{hi} + (sp)_{hj} + (fp)_{ij} + (sfp)_{hij} \\ &\quad + \eta_{hijk} \\ &\quad + r_m + (sr)_{hm} + (fr)_{im} + (pr)_{jm} \\ &\quad + (sfr)_{him} + (spr)_{hjm} + (fpr)_{ijm} + (sfpr)_{hijm} \\ &\quad + \varepsilon_{hijkm}. \end{aligned} \tag{12.2.1}$$

$h = 1, 2$, $i = 1, 2$, $j = 1, 2, 3$, $k = 1, 2$, $m = 1, 2, 3$. The error terms are the η_{hijk}s and the ε_{hijkm}s. These are all assumed to be independent of each other with

$$\eta_{hijk} \sim N(0, \sigma_w^2) \quad \text{and} \quad \varepsilon_{hijkm} \sim N(0, \sigma_s^2). \tag{12.2.2}$$

The η_{hijk}s are error terms due to the use of a particular piece of fabric and the ε_{hijkm}s are error terms due to taking the observations after 1000, 2000, and 3000 rotations. While we have two error terms, and thus two variances, the variances are assumed to be constant for each error term, i.e., all observations have the same variance $\sigma_w^2 + \sigma_s^2$. Observations on the same piece of fabric are correlated because they all involve the same fabric error term η_{hijk}.

Split plot terminology devolves from analyses on plots of ground. In this application, a whole plot is a piece of fabric. The subplots correspond to the three observations on each piece of fabric. The *surf*, *fill*, and *prop* treatments are all applied to an entire piece of fabric, so they are referred to as whole plot treatments. The three levels of rotation are applied 'within' a piece of fabric and are called subplot treatments.

Model (12.2.1) assumes that measurements on different pieces of fabric are independent with the same variance and that for every *given* piece of fabric, regardless of the treatments applied, the three measurements are independent and have the same variance. For example, the model indicates that given the piece of fabric, if the weight loss happens to be above the mean at 1000 rotations, it is just as probable that the weight loss will be below the mean at 2000 rotations. The model is inappropriate if a weight loss above the mean at 1000 rotations predisposes the fabric to have weight loss above the mean at 2000 rotations. Of course, the mean weight losses are allowed to depend on the number of rotations. *If* we were dealing with the *total* weight loss at 1000, 2000, and 3000 rotations, it seems probable that, even after adjusting for the fixed effects of rotations, the observations at 1000 and 2000 rotations would have greater correlation than the observations at 1000 and 3000. It also seems probable that the variance of the observations would increase as the number of rotations increased. Both of these events would be violations of the split plot model. However, we are working with the differences in weight loss. Our data are weight losses due to the first, second, and third one thousand rotations. The split plot model seems at least plausible for the differences. Another possible model, one that we will not address, uses an 'autoregressive' correlation structure, cf. Diggle (1990). In the next section we will briefly consider a more general (multivariate) model that can be applied and includes both the split plot model and the autoregressive structure as special cases. Of course when the split plot model is appropriate, the split plot analysis is more powerful than the general multivariate analysis.

We will concern ourselves with checking the assumptions of equal variances and normality later. We now consider the analysis of variance given in Table 12.4. Just as there are two error terms in model (12.2.1), there are two error terms in the analysis of variance table. Both are used for F tests and it is crucial to understand which error term is used for which F tests. The mean square from error 1 is the whole plot error term and is used for any inferences that exclusively involve whole plot treatments and their interactions. Thus, in Table 12.4, the $MSE(1)$ from the error 1 line is used for all inferences relating exclusively to the whole plot treatment factors *surf*, *fill*, and *prop*. This includes examination of interactions and contrasts. The error 2 line is used for all inferences involving the subplot treatments. This includes all main effects, interactions, and contrasts involving *rota*.

TABLE 12.4. Analysis of variance

Source	*df*	*SS*	*MS*	*F*	*P*
surf	1	24494.2	24494.2	78.58	0.000
fill	1	107802.7	107802.7	345.86	0.000
prop	2	13570.4	6785.2	21.77	0.000
surf * *fill*	1	1682.0	1682.0	5.40	0.039
surf * *prop*	2	795.0	397.5	1.28	0.315
fill * *prop*	2	9884.7	4942.3	15.86	0.000
surf * *fill* * *prop*	2	299.3	149.6	0.48	0.630
Error 1	12	3740.3	311.7		
rota	2	60958.5	30479.3	160.68	0.000
surf * *rota*	2	8248.0	4124.0	21.74	0.000
fill * *rota*	2	18287.7	9143.8	48.20	0.000
prop * *rota*	4	1762.8	440.7	2.32	0.086
surf * *fill* * *rota*	2	2328.1	1164.0	6.14	0.007
surf * *prop* * *rota*	4	686.0	171.5	0.90	0.477
fill * *prop* * *rota*	4	1415.6	353.9	1.87	0.149
surf * *fill* * *prop* * *rota*	4	465.9	116.5	0.61	0.657
Error 2	24	4552.7	189.7		
Total	71	260973.9			

As always, the analysis focuses on the highest order terms involving the factors. The four-factor interaction has a very large P value, .657. Even if all of the sum of squares was ascribed to one contrast, an unadjusted F test would not be significant. There is no evidence of a four-factor interaction.

There are 4 three-factor interaction terms to consider. The sums of squares for $surf * fill * prop$ and $surf * prop * rota$ are so small as to obviate further consideration. The P value for the three-factor interaction $surf * fill * rota$ is sufficiently small to demand examination. In our analysis from Section 11.3 of the 2000 rotation data, we found no $surf * fill$ interaction and we alluded to the facts that a similar analysis for 3000 rotations also showed no $surf * fill$ interaction but the 1000 rotation data did show an interaction. The ANOVA table confirms the existence of a three-factor interaction, thus the *surf* $*$ *fill* interaction depends on the number of rotations. The P value for *fill* $*$ *prop* $*$ *rota* is small enough that it might be of interest if we could find a natural interpretation for it. Recall that in our earlier analysis based on just the 2000 rotation data, we found a *fill* $*$ *prop* interaction; a *fill* $*$ *prop* $*$ *rota* interaction indicates that the *fill* $*$ *prop* interaction changes with the number of rotations. If we do not find an interesting *fill* $*$ *prop* $*$ *rota* interaction, we need to consider two-factor interactions involving *prop*. We need to focus on *prop* because it is the only factor that is not included in the significant *surf* $*$ *fill* $*$ *rota* interaction. The possible interactions are *surf* $*$ *prop*, *fill* $*$ *prop*, and *prop* $*$ *rota*. From Table 12.4, the *surf* $*$ *prop* interaction is clearly insubstantial. We will not consider the *surf* $*$ *prop* interaction further.

We now begin looking at contrasts. To examine contrasts in the *surf* $*$ *fill* $*$ *rota* interaction, we need the appropriate means table.

$N = 6$		*surf*			
		Yes		No	
	fill	A	B	A	B
	1000	235.00	227.00	194.50	$135.1\bar{6}$
rota	2000	$225.6\bar{6}$	120.00	$199.8\bar{3}$	88.00
	3000	$179.3\bar{3}$	$89.8\bar{3}$	164.00	74.00

As in Section 11.3, *the value of N in the upper left of the means table is the number of observations the means are averaged over*. The methods we will use to determine three-factor interaction contrasts are straight-forward generalizations of the methods used for two-factor interaction contrasts, but it is simpler to discuss the process in parts rather than defining the process all at once. Again, we determine interesting contrasts by combining main effect contrasts. Both *surf* and *fill* have only two levels, so the only contrast is the difference between the two levels. These contrasts define the *surf* $*$ *fill* interaction contrast given below.

	surf			
	Yes		No	
fill	A	B	A	B
	1	−1	−1	1

If we apply this contrast to the *rota* 1000 means, we get

$$E_1 = 235.00 - 227.00 - 194.50 + 135.1\bar{6} = -51.3\bar{3}.$$

Applying the contrast to the *rota* 2000 means gives

$$E_2 = 225.6\bar{6} - 120.00 - 199.8\bar{3} + 88.00 = -6.1\bar{6}$$

and applying the contrast to the *rota* 3000 means gives

$$E_3 = 179.3\bar{3} - 89.8\bar{3} - 164.00 + 74.00 = -0.50.$$

Normally, with rotations at equally spaced quantitative levels, we would use the linear and quadratic contrasts in rotations. However, I previously analyzed the data from each number of rotations separately and I discovered no *surf* $*$ *fill* interaction at 2000 and 3000 rotations, so I will use a pair of orthogonal contrasts, one that looks at 2000 versus 3000 rotations, and another that looks at 1000 versus the average of 2000 and 3000.

rota	1000	2000	3000
2000–3000	0	1	−1
1000 vs. others	−2	1	1

To examine the three-factor interaction, we apply these contrasts to the E_1, E_2, and E_3 values that measure the *surf* $*$ *fill* interaction at each level of rotation.

The contrast that examines whether the *surf* $*$ *fill* interaction is the same at 2000 rotations as at 3000 rotations has estimate

$$E_2 - E_3 = (-6.1\bar{6}) - (-0.50) = -5.6\bar{6}.$$

To compute the sum of squares, we need the contrast coefficients as they are applied to each of the means in the *surf* $*$ *fill* $*$ *rota* means table. These are obtained by multiplying the coefficients of the *surf* $*$ *fill* interaction contrast by the coefficients of the *rota* contrast in the manner illustrated in Chapter 11. The contrast coefficients are given below.

		surf			
		Yes		No	
	fill	A	B	A	B
	1000	0	0	0	0
rota	2000	1	−1	−1	1
	3000	−1	1	1	−1

The sum of the squared coefficients is 8 and each mean is the average of 6 observations, so the sum of squares for the contrast is

$$\frac{[-5.6\bar{6}]^2}{8/6} = 24.08.$$

The contrast that examines whether the *surf* $*$ *fill* interaction is the same at 1000 rotations as at the average of 2000 and 3000 rotations has estimate

$$-2E_1 + E_2 + E_3 = (-2)(-51.3\bar{3}) + (-6.1\bar{6}) + (-0.50) = 96.00.$$

To compute the sum of squares, we need the contrast coefficients as they are applied to each of the means in the *surf* $*$ *fill* $*$ *rota* means table.

		surf			
		Yes		No	
	fill	A	B	A	B
	1000	−2	2	2	−2
rota	2000	1	−1	−1	1
	3000	1	−1	−1	1

The sum of the squared coefficients is 24 and each mean is the average of 6 observations, so the sum of squares for the contrast is

$$\frac{[96.00]^2}{24/6} = 2304.00.$$

Note that the two sums of squares add up to $SS(surf * fill * rota)$, i.e.,

$$2328.1 = 24.08 + 2304.00.$$

The overwhelming majority of the sum of squares is due to the difference between the *surf* $*$ *fill* interaction at 1000 rotations and the average *surf* $*$ *fill* interaction at 2000 and 3000 rotations. These interaction contrasts involve *rota*, so to perform tests we compare the sums of squares to $MSE(2)$, which is 189.7. We let the data suggest the contrasts, so using Scheffé's multiple comparison method is appropriate. The test statistic for *rota* 1000 versus the others is

$$F = \frac{2304/2}{189.7} = 6.07,$$

which is significant at the .01 level because $F(.99, 2, 24) = 5.61$.

Re-examining the values $E_1 = -51.3\bar{3}$, $E_2 = -6.1\bar{6}$, $E_3 = -0.50$, there appears to be a substantial *surf* $*$ *fill* interaction at 1000 rotations. This can be ascribed to the fact that the difference between the means for fills A and B is much smaller with the surface treatment $(235.00 - 227.00 = 8.00)$ than without it $(194.50 - 135.1\bar{6} = 59.\bar{3})$. It is also clear from E_2 and E_3 that there is very little *surf* $*$ *fill* interaction at either 2000 or 3000 rotations, i.e., the difference between A and B is about the same for the two surface treatments. Unfortunately, tests for whether *surf* $*$ *fill* interaction exists at the various levels of *rota* are not readily available from the split plot model. *The model does not allow a simple test when we fix a subplot treatment level and examine a contrast in the whole plot treatments for that subplot level.* This is because the comparisons are being made *across* whole plots, so there is no appropriate χ^2 variance estimate. (*The split plot model does let us fix a whole plot treatment level and examine contrasts among the subplot treatments at that whole plot level* because the comparison in made entirely *within* a whole plot.) However, our test in Section 11.3 of the *surf* $*$ *fill* interaction using just the 2000 *rota* data is perfectly appropriate and similar tests using just the 3000 *rota* data and just the 1000 *rota* data would also be appropriate. ANOVA tables for the separate analyses of the 1000, 2000, and 3000 rotation data are given in Section 12.3 as Tables 12.8, 12.9, and 12.10. Note, though, that the separate analyses are not independent, because the observations at 2000 rotations are not independent of the observations at 3000 rotations, etc.

On occasion, when examining contrasts for a fixed subplot treatment, rather than using the *MSE*s from the separate analyses, the degrees of freedom and sums of squares for error 1 and error 2 are pooled and these are used instead. This is precisely the error estimate obtained by pooling the error estimates from the three separate ANOVAs. Such a pooled estimate should be better than the estimates from the separate analyses but it is difficult to quantify the effect of pooling. The three separate ANOVAs are not independent, so pooling the variance estimates does not have the nice properties of the pooled estimate of the variance used in, say, one-way ANOVA. As alluded to above, we cannot get an exact F test for a contrast based on the pooled variance estimate. If the three ANOVA's were independent, the pooled error would have $12+12+12=36$ degrees of freedom, but we do not have independence, so we do not even know an appropriate number of degrees of freedom to use with the pooled estimate, much less the appropriate distribution.

It is not clear that a *fill* $*$ *prop* $*$ *rota* interaction exists but, to be safe, we will examine some reasonable contrasts. If some interpretable contrast has a large sum of squares, it suggests that an important interaction is being hidden within the 4 degrees of freedom test. To examine contrasts in the *fill* $*$ *prop* $*$ *rota* interaction, we need the appropriate means table.

$N=4$		*fill*					
		A			B		
	prop	25%	50%	75%	25%	50%	75%
	1000	182.50	212.25	249.50	180.75	173.50	189.00
rota	2000	175.25	205.50	257.50	99.00	107.25	105.75
	3000	149.50	178.75	186.75	83.75	78.75	83.25

The obvious contrast for the factor *fill* is A versus B. There are equally spaced quantitative levels for *prop*, so we use the linear and quadratic contrasts. Combining the contrasts into *fill* $*$ *prop* interaction contrasts as described in Chapter 11 gives

	fill					
	A			B		
prop	25%	50%	75%	25%	50%	75%
$f * p$ *lin*	1	0	−1	−1	0	1
$f * p$ *quad*	1	−2	1	−1	2	−1

The contrast $f * p$ *lin* examines whether the proportion slope is the same for fill A as for fill B. The contrast $f * p$ *quad* examines whether the proportion curvature is the same for fill A as for fill B. Applying these contrasts to the *fill* $*$ *prop* means at each level of rotations gives

$$\begin{aligned} E_{1l} &= 182.50 - 249.50 - 180.75 + 189.00 = -58.75, \\ E_{2l} &= 175.25 - 257.50 - 99.00 + 105.75 = -75.5, \\ E_{3l} &= 149.50 - 186.75 - 83.75 + 83.25 = -37.75, \end{aligned}$$

$$\begin{aligned} E_{1q} &= 182.50 - (2)212.25 + 249.50 - 180.75 + (2)173.50 - 189.00 \\ &= -15.25, \\ E_{2q} &= 175.25 - (2)205.50 + 257.50 - 99.00 + (2)107.25 - 105.75 \\ &= 31.5, \\ E_{3q} &= 149.50 - (2)178.75 + 186.75 - 83.75 + (2)78.75 - 83.25 \\ &= -30.75\,. \end{aligned}$$

The rotations are also quantitative and equally spaced, so the linear and quadratic contrasts are

rota	1000	2000	3000
Linear	−1	0	1
Quadratic	1	−2	1

We can apply each of these contrasts to the values E_{1l}, E_{2l}, E_{3l}, and to the values E_{1q}, E_{2q}, E_{3q} to obtain 4 orthogonal *fill* $*$ *prop* $*$ *rota* interaction contrasts. Using, say, $f * pl * rq$ to indicate the *fill* by *prop lin* by *rota quad* contrast, we get

$$\begin{aligned} SS(f * pl * rl) &= \frac{[21]^2}{8/4} = 220.50, \\ SS(f * pl * rq) &= \frac{[54.5]^2}{24/4} = 495.04, \\ SS(f * pq * rl) &= \frac{[-15.5]^2}{24/4} = 40.04, \\ SS(f * pq * rq) &= \frac{[-109]^2}{72/4} = 660.06\,. \end{aligned}$$

For example, the numerator of $SS(f * pq * rq)$ comes from

$$E_{1q} - (2)E_{2q} + E_{3q} = -109.$$

Actually, it comes from applying the 18 coefficients

			fill					
			A			B		
	prop		25%	50%	75%	25%	50%	75%
		Contrasts	1	−2	1	−1	2	−1
	1000	1	1	−2	1	−1	2	−1
rota	2000	−2	−2	4	−2	2	−4	2
	3000	1	1	−2	1	−1	2	−1

to the means table. The 72 in the denominator of $SS(f * pq * rq)$ comes from squaring all the coefficients and adding them up. Note that

$$SS(fill * prop * rota) = 1415.6 = 220.50 + 495.04 + 40.04 + 660.06\,.$$

These contrasts were not chosen by looking at the data, so less stringent multiple comparison methods than Scheffé's can be used on them. On the other hand, the contrasts are not particularly informative. None of these contrasts suggests a particularly strong source of interaction. F tests are constructed by dividing each of the four sums of squares by $MSE(2)$. None of the F ratios is significant when compared to $F(.05, 1, 24) = 4.26$. This analysis seems consistent with the hypothesis of no *fill* ∗ *prop* ∗ *rota* interaction.

If we accept the working assumption of no *fill* ∗ *prop* ∗ *rota* interaction, we need to examine the two factor interactions that can be constructed from the three factors. These are *fill* ∗ *prop*, *fill* ∗ *rota*, and *prop* ∗ *rota*. The *fill* ∗ *rota* effects are, however, not worth further consideration because they are subsumed within the *surf* ∗ *fill* ∗ *rota* effects that have already been established as important. Another way of looking at this is that in model (12.2.1), the $(fr)_{im}$ effects are unnecessary in a model that already has $(sfr)_{him}$ effects. Thus we focus our attention on *fill* ∗ *prop* and *prop* ∗ *rota*. From the ANOVA Table 12.4, the *fill* ∗ *prop* interaction is highly significant; the *prop* ∗ *rota* interaction has a P value of .09. We examine the *prop* ∗ *rota* interaction first and then the *fill* ∗ *prop* interaction.

The 4 degrees of freedom for *prop* ∗ *rota* in the interaction F test have the potential of hiding one or two important, *interpretable* interaction contrasts. We explore this possibility by investigating *prop* ∗ *rota* interaction contrasts based on the linear and quadratic contrasts in both *prop* and *rota*. The means table is

$N = 8$		*prop*	
rota	25%	50%	75%
1000	181.625	192.875	219.250
2000	137.125	156.375	181.625
3000	116.625	128.750	135.000

and the sums of squares for the four orthogonal contrasts are

$$\begin{aligned} SS(pl * rl) &= \frac{[-19.25]^2}{4/8} = 741.125, \\ SS(pl * rq) &= \frac{[-33.00]^2}{12/8} = 726.000, \\ SS(pq * rl) &= \frac{[-21.00]^2}{12/8} = 294.000, \\ SS(pq * rq) &= \frac{[-2.75]^2}{36/8} = 1.681\,. \end{aligned}$$

Comparing these to $MSE(2)$, we find that $SS(pl * rl)$ and $SS(pl * rq)$ are not small but neither are they clearly significant. The interaction plot in Figure 12.11 seems to confirm that there is no obvious interaction being overlooked by the four degrees of freedom test. We remain unconvinced that there is any substantial *prop* ∗ *rota* interaction.

Examination of the *fill* ∗ *prop* interaction follows exactly the same pattern as used in Section 11.3. The appropriate means table is given below.

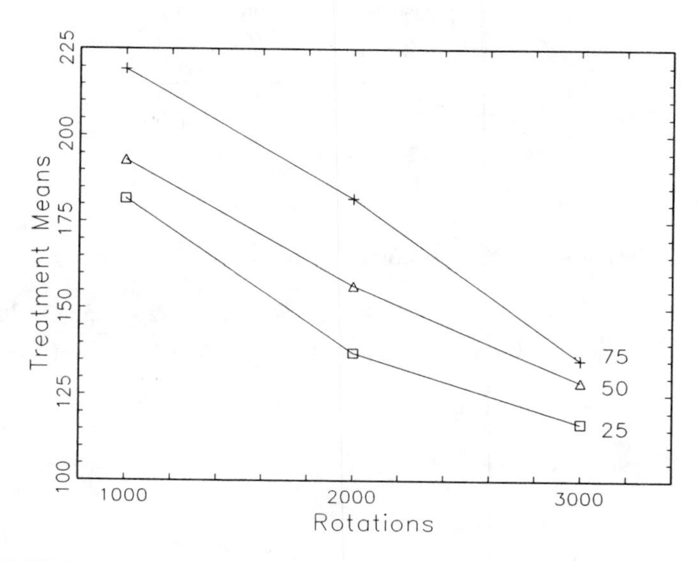

FIGURE 12.11. Interaction plot of proportions versus rotations, Box data.

$N = 12$	*fill*	
prop	A	B
25%	$169.08\bar{3}$	$121.1\bar{6}$
50%	$198.83\bar{3}$	$119.83\bar{3}$
75%	231.250	126.000

Note that the means are now based on 12 observations. There is one main effect contrast for the fills,

$$\mu_A - \mu_B.$$

The proportion levels are quantitative and equally spaced, so we use the linear and quadratic contrasts,

$$-\mu_{25} + \mu_{75}$$

and

$$\mu_{25} - 2\mu_{50} + \mu_{75}.$$

The fill and proportion contrasts are used to construct orthogonal *fill* ∗ *prop* interaction contrasts. The *fill* ∗ *prop linear* interaction contrast coefficients are given below.

Fill ∗ prop linear interaction contrast

		fill	
		A	B
prop	Contrasts	1	−1
25%	−1	−1	1
50%	0	0	0
75%	1	1	−1

The *fill* ∗ *prop quadratic* coefficients are

Fill * prop quadratic interaction contrast

		fill	
		A	B
prop	Contrasts	1	−1
25%	1	1	−1
50%	−2	−2	2
75%	1	1	−1

The sums of squares for these contrasts are computed from the *fill* − *prop* means,

$$\begin{aligned} SS(fill * prop\ lin) &= \frac{\left[-169.08\bar{3} + 121.16\bar{6} + 0 + 0 + 231.250 - 126.000\right]^2}{(1/12)\left[(-1)^2 + 1^2 + 0^2 + 0^2 + 1^2 + (-1)^2\right]} \\ &= \frac{(57.\bar{3})^2}{4/12} \\ &= 9861.33 \end{aligned}$$

and

$$\begin{aligned} &SS(fill * prop\ quad) \\ &= \frac{\left[169.08\bar{3} - 121.16\bar{6} - (2)198.83\bar{3} + (2)119.83\bar{3} + 231.250 - 126.000\right]^2}{(1/12)\left[1^2 + (-1)^2 + (-2)^2 + 2^2 + 1^2 + (-1)^2\right]} \\ &= \frac{(-4.8\bar{3})^2}{12/12} \\ &= 23.36\,. \end{aligned}$$

The *fill* * *prop* interaction is a whole plot effect, so the appropriate error is $MSE(1) = 311.7$ and the F ratios are 31.64 and 0.075, respectively. Based on the *fill* * *prop quadratic* contrast, there is no evidence that the curvatures are different for fills A and B. However, from the *fill* * *prop linear* contrast, there is evidence that the slopes are different for fills A and B. The slope for the fill A line is much greater than the slope for fill B. The fill A means change from 169 to 231, while the fill B means change from 121 to 126. As in Section 11.3, we could use the quadratic contrast on proportion means to establish the lack of evidence for curvature but we take an alternative approach.

The slope for fill B is not significantly different from 0. The linear contrast applied to fill B alone gives

$$SS(prop\ lin;\ fill\ B) = \frac{[-121.16\bar{6} + 126.000]^2}{2/12} = 140.17\,.$$

The quadratic contrast for *fill* B shows no curvature,

$$SS(prop\ quad;\ fill\ B) = \frac{[121.16\bar{6} - (2)119.83\bar{3} + 126.000]^2}{2/12} = 112.70\,.$$

As far as we can tell, the weight loss does not change as a function of proportion of filler when using fill B. Note that both of these are contrasts in the whole plot treatments, so they are compared to $MSE(1)$.

A similar analysis for fill A shows that weight loss increases with proportion and there is again no evidence of curvature. In particular, for fill A, the slope contrast has an estimate of

$$-169.083 + 231.250 = 62.167$$

with a sum of squares of 23188. From the estimated contrast, an increase from 25% fill to 75% fill increases weight loss by 62 units. The relationship appears to be linear, so an increase of 25%

in proportion fill raises weight loss by 31 units. Confidence intervals corresponding to the point estimates are easily obtained.

We have found two important interaction effects, *surf* ∗ *fill* ∗ *rota* from the subplot analysis and *fill* ∗ *prop* from the whole plot analysis. These two interactions are the highest order terms that are significant and include all four of the factors. The only factor contained in both interactions is *fill*, so the simplest overall explanation of the data can be arrived at by giving separate explanations for the two fills. To do this, we need to re-evaluate the *surf* ∗ *fill* ∗ *rota* interaction in terms of how the *surf* ∗ *rota* interaction changes from *fill* A to *fill* B; previously, we focused on how the *surf* ∗ *fill* interaction changed with rotations. One benefit of this change in emphasis is that, as discussed earlier, we can use *MSE*(2) for valid tests of contrasts in the *surf* ∗ *rota* interactions for a fixed level of *fill* because we are fixing a whole plot factor, not a subplot factor.

As before, we compare 2000 rotations with 3000 rotations and compare 1000 rotations with the average of 2000 and 3000. The contrast coefficients for *surf* ∗ (2000 *vs* 3000) *rota* are

Surf ∗ (2000 vs 3000) rota interaction contrast

		surf	
		Yes	No
rota	Contrasts	1	−1
1000	0	0	0
2000	−1	−1	1
3000	1	1	−1

and the *surf* ∗ (1000 *vs others*) *rota* coefficients are

Surf ∗ (1000 vs others) rota interaction contrast

		surf	
		Yes	No
rota	Contrasts	1	−1
1000	−2	−2	2
2000	1	1	−1
3000	1	1	−1

Applying these contrasts to the means for *fill* A in the *surf* ∗ *fill* ∗ *rota* means table gives

$$\begin{aligned} SS(surf * (2000\ vs\ 3000)\ rota;\ fill\ A) &= \frac{[-225.6\bar{6} + 199.8\bar{3} + 179.3\bar{3} - 164.00]^2}{(4/6)} \\ &= 165.69 \end{aligned}$$

and

$$\begin{aligned} &SS(surf * (1000\ vs\ others)\ rota;\ fill\ A) \\ &= \frac{[(-2)235.00 + (2)194.50 + 225.6\bar{6} - 199.8\bar{3} + 179.3\bar{3} - 164.00]^2}{(12/6)} \\ &= 754.01\,. \end{aligned}$$

Applying the contrasts to the means for *fill* B in the *surf* ∗ *fill* ∗ *rota* means table gives

$$\begin{aligned} SS(surf * (2000\ vs\ 3000)\ rota;\ fill\ B) &= \frac{[-120.00 + 88.00 + 89.8\bar{3} - 74.00]^2}{(4/6)} \\ &= 391.72 \end{aligned}$$

and

$$SS(surf * (1000\ vs\ others)\ rota;\ fill\ B)$$
$$= \frac{[(-2)227.00 + (2)135.1\bar{6} + 120.00 - 88.00 + 89.8\bar{3} - 74.00]^2}{(12/6)}$$
$$= 9225.35\,.$$

All of these are compared to $MSE(2) = 189.7$. There is no evidence of interactions involving rotations 2000 and 3000 with surface treatments, regardless of fill type. With fill A, there is marginal evidence of an interaction in which the effect of *surf* is different at 1000 rotations than at 2000 and 3000 rotations. With fill B, there is clear evidence of an interaction where the effect of *surf* is different at 1000 rotations than at 2000 and 3000 rotations.

Using contrasts applied to the $surf * fill * rota$ means table, we can summarize what we know about the three-factor interaction. Recall that any comparisons among rotation levels are evaluated with respect to $MSE(2)$ but any contrast that fixes rotation levels requires an approximate test. The estimated effect of no surface treatment with fill A and either 2000 or 3000 rotations is that the weight loss drops

$$\frac{225.67 - 199.83}{2} + \frac{179.33 - 164.00}{2} = \frac{25.83 + 15.33}{2} = 20.58$$

units. The estimate and interpretation are based on the lack of $rota * surf$ interaction that we found for fill A at 2000 and 3000 rotations. The effect is based on averaging over 2000 and 3000 rotations, so the standard error is computed in the usual way except that the appropriate variance estimate is obtained from the whole plot error of a split plot analysis that eliminates the 1000 rotation data. $MSE(1)$ overestimates the variance because it involves averaging over the 1000 rotation data as well as the 2000 and 3000 rotation data. An underestimate of the variance is $MSE(2)$. Either of the two variance estimates shows a significant effect, so the appropriate variance estimate would also show an effect. The estimated effect of no surface treatment with fill A at 1000 rotations is that the weight loss drops by

$$235.00 - 194.5 = 40.5\,.$$

This effect is for fixed rotations but an approximate test, based on pooling $SSE(1)$ and $SSE(2)$ to obtain a variance estimate for use in the usual standard error formula, indicates that the effect is clearly significant. The separate analysis on 1000 rotations would yield the same result.

We have no evidence of $surf * rota$ interaction for fill A at 2000 or 3000 rotations, so the estimated effect of going from 2000 to 3000 rotations is a drop of

$$\frac{225.67 - 179.33}{2} + \frac{199.83 - 164.00}{2} = \frac{46.33 + 35.83}{2} = 41.09$$

in weight loss, regardless of surface treatment. This is a comparison among rotation levels, so it can be compared to $MSE(2)$. The result is highly significant. The estimated effect of going from 1000 to 2000 rotations with the surface treatment is an insignificant drop of

$$235.00 - 225.67 = 9.33$$

in weight loss, while the estimated effect of going from 1000 to 2000 rotations without the surface treatment is an *increase* of

$$199.83 - 194.5 = 5.33,$$

but again the change is not significant.

For fill B the estimated effect of no surface treatment at either 2000 or 3000 rotations is a decreased weight loss of

$$\frac{120.00 - 88.00}{2} + \frac{89.83 - 74.00}{2} = \frac{32.00 + 15.83}{2} = 23.92\,.$$

This is comparable to the effect seen with fill A. The estimated effect of no surface treatment with fill B at 1000 rotations is that the weight loss drops by

$$227.00 - 135.17 = 91.83\,.$$

This is a much larger effect than we saw with fill A. Not only is the drop significant but the difference from fill A is significant. Of course, these are only approximate tests.

With fill B, the estimated effect of going from 2000 to 3000 rotations is a drop of

$$\frac{120.00 - 89.83}{2} + \frac{88.00 - 74.00}{2} = \frac{30.17 + 14.00}{2} = 22.09\,.$$

This is about half of the estimated effect seen with fill A. The effect is of marginal significance. The estimated effect of going from 1000 to 2000 rotations with the surface treatment is a drop of

$$227.00 - 120.00 = 107.00$$

in weight loss. This is about 10 times larger than the effect that was seen with fill A and is highly significant. The estimated effect of going from 1000 to 2000 rotations without the surface treatment is a decrease of

$$135.17 - 88.00 = 47.17\,.$$

The decrease is much smaller than with a surface treatment but it is highly significant.

In addition, from the *fill * prop* interaction, we have learned that for fill A an increase of 25% fill raises weight loss by an estimated 31 units, while there is no evidence of a relationship between proportion and fill B.

Figures 12.12 and 12.13 contain residual plots for the error 1 residuals. Figures 12.14 and 12.15 contain residual plots for the error 2 residuals. I see no serious problems in any of the plots.

With some computer programs, it may not be possible to obtain a split plot analysis but it may be possible to obtain a full factorial analysis. In such cases, it is a simple matter to reconstruct the split plot ANOVA table. A full factorial analysis for these data involves treating the replications as another factor, say, *rep*. The full factorial ANOVA table is given in Table 12.5. The split plot error 1 term involves pooling all terms that involve *rep* but do not involve both *rep* and *rota*. The error 2 involves pooling all terms that involve both *rep* and *rota*. The computations are illustrated below. If you add up the sums of squares in the tabulations given below, you will notice some round off error after the decimal place.

Components of error 1

Source	*df*	*SS*
rep	1	0.2
*surf * rep*	1	53.4
*fill * rep*	1	800.0
*prop * rep*	2	49.4
*surf * fill * rep*	1	84.5
*surf * prop * rep*	2	1589.7
*fill * prop * rep*	2	166.6
*surf * fill * prop * rep*	2	996.6
Error 1	12	3740.3

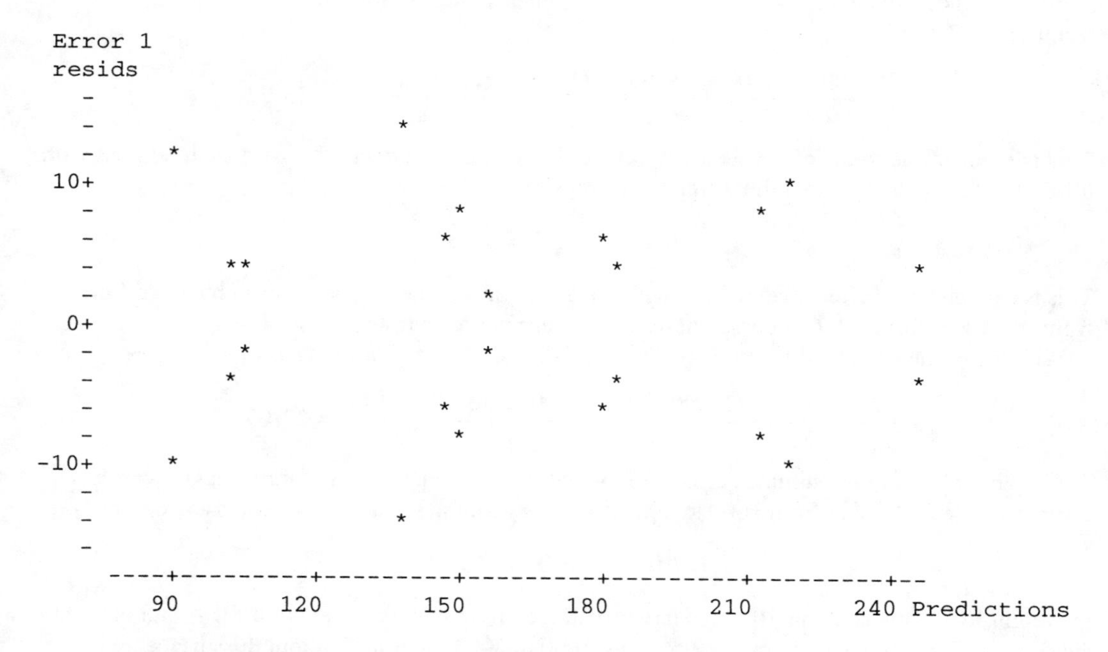

FIGURE 12.12. Wholeplot residuals versus predicted values, Box data.

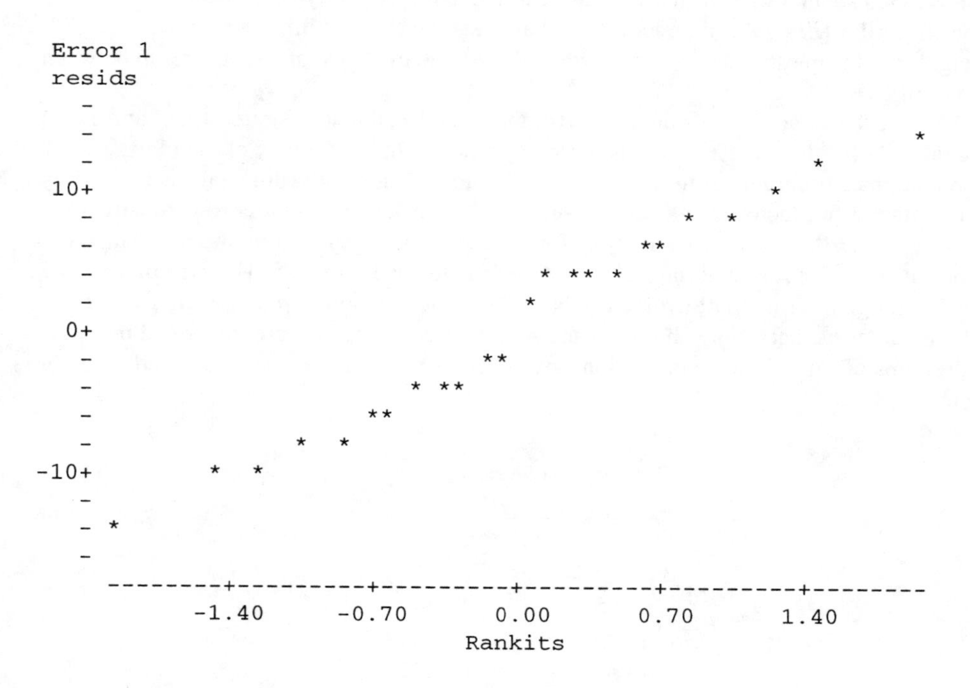

FIGURE 12.13. Normal plot of wholeplot residuals, $W' = 0.980$, Box data.

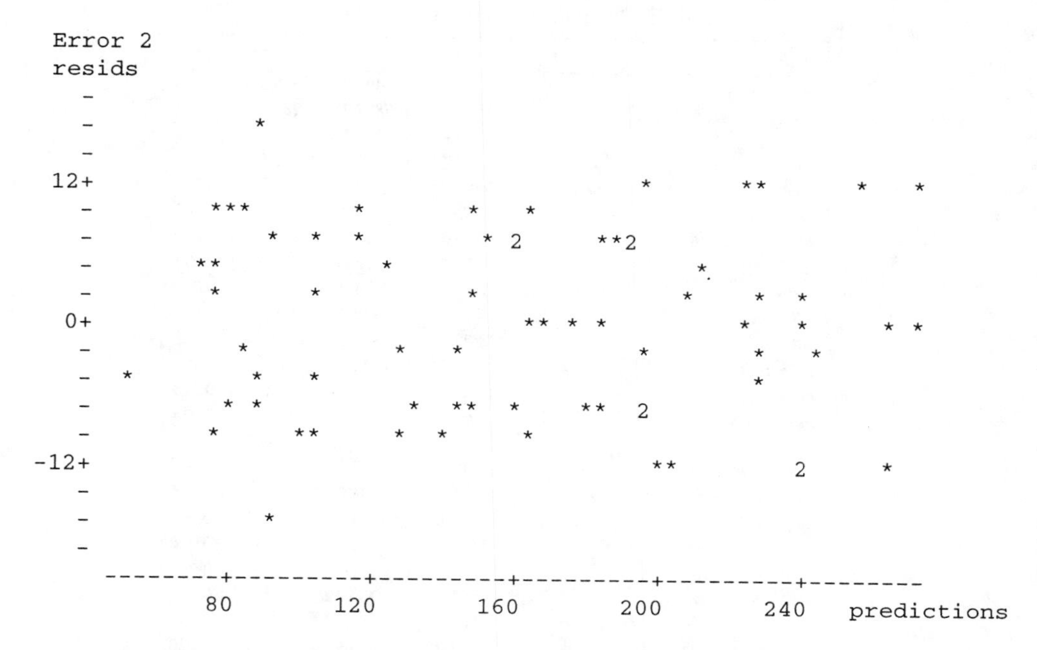

FIGURE 12.14. Subplot residuals versus predicted values, Box data.

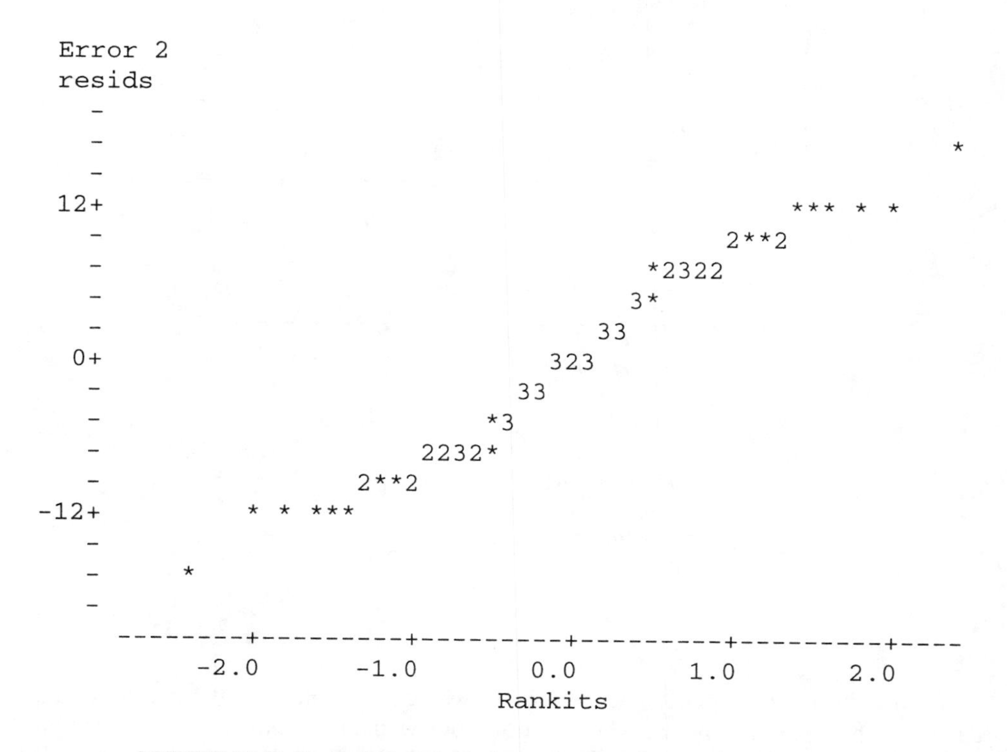

FIGURE 12.15. Normal plot of subplot residuals, $W' = 0.980$, Box data.

TABLE 12.5. Full factorial analysis of variance

Source	*df*	*SS*	*MS*	*F*	*P*
surf	1	24494.2	24494.2	745.54	0.000
fill	1	107802.7	107802.7	3281.25	0.000
prop	2	13570.4	6785.2	206.52	0.000
rep	1	0.2	0.2	0.01	0.938
rota	2	60958.5	30479.3	927.71	0.000
surf * *fill*	1	1682.0	1682.0	51.20	0.002
surf * *prop*	2	795.0	397.5	12.10	0.020
surf * *rep*	1	53.4	53.4	1.63	0.271
surf * *rota*	2	8248.0	4124.0	125.52	0.000
fill * *prop*	2	9884.7	4942.3	150.43	0.000
fill * *rep*	1	800.0	800.0	24.35	0.008
fill * *rota*	2	18287.7	9143.8	278.32	0.000
prop * *rep*	2	49.4	24.7	0.75	0.528
prop * *rota*	4	1762.8	440.7	13.41	0.014
rep * *rota*	2	435.5	217.8	6.63	0.054
surf * *fill* * *prop*	2	299.3	149.6	4.55	0.093
surf * *fill* * *rep*	1	84.5	84.5	2.57	0.184
surf * *fill* * *rota*	2	2328.1	1164.0	35.43	0.003
surf * *prop* * *rep*	2	1589.7	794.8	24.19	0.006
surf * *prop* * *rota*	4	686.0	171.5	5.22	0.069
surf * *rep* * *rota*	2	364.4	182.2	5.55	0.070
fill * *prop* * *rep*	2	166.6	83.3	2.54	0.194
fill * *prop* * *rota*	4	1415.6	353.9	10.77	0.020
fill * *rep* * *rota*	2	32.2	16.1	0.49	0.645
prop * *rep* * *rota*	4	2028.6	507.2	15.44	0.011
surf * *fill* * *prop* * *rep*	2	996.6	498.3	15.17	0.014
surf * *fill* * *prop* * *rota*	4	465.9	116.5	3.55	0.124
surf * *fill* * *rep* * *rota*	2	85.8	42.9	1.31	0.366
surf * *prop* * *rep* * *rota*	4	1208.8	302.2	9.20	0.027
fill * *prop* * *rep* * *rota*	4	265.9	66.5	2.02	0.256
surf * *fill* * *prop* * *rep* * *rota*	4	131.4	32.9		
Total	71	260973.9			

Components of error 2

Source	*df*	*SS*
rep * *rota*	2	435.5
surf * *rep* * *rota*	2	364.4
fill * *rep* * *rota*	2	32.2
prop * *rep* * *rota*	4	2028.6
surf * *fill* * *rep* * *rota*	2	85.8
surf * *prop* * *rep* * *rota*	4	1208.8
fill * *prop* * *rep* * *rota*	4	265.9
surf * *fill* * *prop* * *rep* * *rota*	4	131.4
Error 2	24	4552.7

Any of the standard designs can be used for the whole plot treatments. If a RCB design had been used in the whole plot treatments, the replication index would have indicated the whole plot blocks, the *rep* line would be listed separately in the ANOVA, and the *rep* line would not be included in the error 1 term. Otherwise, the analysis is the same. □

Computing techniques

Minitab can be used to generate the split plot analysis. Let c1 be a column containing the weight losses. The columns c2, c3, c4, c5, and c6 contain integers indicating the surface treatment, fill type, fill proportion, replication, and number of rotations corresponding to the weight loss in c1. In the ANOVA model given below c5(c2 c3 c4) is an effect 'nested' within all combinations of the whole plot treatments. This term corresponds to the whole plot error term. The original ANOVA table given by Minitab will use error 2 for all tests but a subcommand provides the appropriate whole plot F tests. Columns c10 and c11 contain the subplot residuals and subplot predicted values, respectively. Another analysis performed on the whole plot means is necessary to obtain the whole plot residuals and predicted values. I was unable to get all of the means tables from one split plot analysis; the example provides most of the means tables for the subplot analysis. The last command given below provides the full factorial ANOVA table with the five-factor interaction listed as error.

```
MTB > names c1 'y' c2 'surf' c3 'fill' c4 'prop' c5 'rep' c6 'rota'
MTB > anova c1= c2|c3|c4|c6 c5(c2 c3 c4);
SUBC> test c2 c3 c4 c2*c3 c2*c4 c3*c4 c2*c3*c4 / c5(c2 c3 c4);
SUBC> resid c10;
SUBC> fits c11;
SUBC> means c6*c2 c6*c3 c6*c4 c6*c2*c4 c6*c3*c4 c6*c2*c3*c4.
MTB > anova c1 = c2|c3|c4|c5|c6 - c2*c3*c4*c5*c6
```

Computation of the ANOVA table

We now consider the computation of the whole plot analysis. This corresponds to the error 1 line and everything above it in Table 12.4. The whole plot analysis is based on averaging over the three levels of rotations. Table 12.6 gives the means $\bar{y}_{hijk\cdot}$ averaged over rotations. Once again, in all means tables the value of N in the upper left indicates the number of observations over which the table entries are averaged. The sum of squares total for whole plots is the sample variance of the 24 means in Table 12.6 multiplied by $N = 3$ and the degrees of freedom. The degrees of freedom are the number of means minus one, so

$$SS(Whole\ Plot\ Total) = (24-1)(3)2351.71900 = 162268.611\,.$$

TABLE 12.6. Whole plot means

surf: Yes

$N = 3$		*fill*					
			A			B	
	prop	25	50	75	25	50	75
rep	1	175.67	207.00	241.33	152.00	142.00	153.33
	2	187.00	221.33	247.67	125.67	158.67	142.00

surf: No

		fill					
			A			B	
	prop	25	50	75	25	50	75
rep	1	158.33	187.00	208.67	98.67	100.33	107.33
	2	155.33	180.00	227.33	108.33	78.33	101.33

Table 12.7 contains the whole plot treatment means. These are averages over the 2 replications in Table 12.6. The mean square for whole plot treatments is just the sample variance of the 12 whole

TABLE 12.7. Means table for whole plot treatments

N = 6		*surf*			
		Yes		No	
	fill	A	B	A	B
	25%	181.33	138.83	156.83	103.50
prop	50%	214.17	150.33	183.50	89.33
	75%	244.50	147.67	218.00	104.33

plot treatment means times $N = 6$,

$$MS(Whole\ Plot\ Treatments) = (6)2401.943606 = 14411.66163$$

and the sum of squares is the mean square times the degrees of freedom, i.e., the number of means minus 1,

$$SS(Whole\ Plot\ Treatments) = (12 - 1)14411.66163 = 158528.278\,.$$

The sum of squares for whole plot error is obtained by subtraction,

$$\begin{aligned} SSE(1) &= SS(Whole\ Plot\ Total) - SS(Whole\ Plot\ Treatments) \\ &= 162268.611 - 158528.278 \\ &= 3740.3\,. \end{aligned}$$

The degrees of freedom for error are also obtained by subtraction, 23 − 11 = 12. An abbreviated analysis of variance table is given below.

Whole plot analysis of variance

Source	*df*	*SS*
Whole plot treatments	11	158528.3
Error 1	12	3740.3
Total	23	162268.6

The $MSE(1)$ is an estimate of $\sigma_w^2 + 3\sigma_s^2$, where σ_w^2 and σ_s^2 are identified in (12.2.2) and the multiplier 3 comes from the fact that there are three levels of the subplot treatment, rotations. The whole plot residuals

$$\bar{y}_{hijk\cdot} - \bar{y}_{hij\cdot\cdot}$$

can be used in plots to evaluate the assumptions of the whole plot analysis. The sum of squares of these residuals multiplied by 3 yields $SSE(1)$.

The sum of squares for whole plot treatments is decomposed into terms for *surf*, *fill*, *prop*, *surf* ∗ *fill*, *surf* ∗ *prop*, *fill* ∗ *prop*, and *surf* ∗ *fill* ∗ *prop*. As mentioned earlier, the methods for computing the ANOVA table entries for the whole plot effects are the same as usual. We simply present the appropriate means tables and allow the reader to reconstruct the whole plot ANOVA table. The means tables are given below.

N = 36	Surface treatment means	
surf	Yes	No
	179.47	142.58

$N = 36$	Fill means	
fill	A	B
	199.72	122.33

$N = 24$	Proportion means		
prop	25%	50%	75%
	145.12	159.33	178.62

$N = 18$	*surf*	
fill	Yes	No
A	213.33	186.11
B	145.61	99.06

$N = 12$	*surf*	
prop	Yes	No
25%	160.08	130.17
50%	182.25	136.42
75%	196.08	161.17

$N = 12$	*fill*	
prop	A	B
25%	169.08	121.17
50%	198.83	119.83
75%	231.25	126.00

We now consider computation of the subplot analysis. The subplot analysis corresponds to everything below the error 1 line in Table 12.4 The sum of squares total is computed in the usual way; it is the sample variance of all 72 observations times 71. The sum of squares and degrees of freedom for error 2 are obtained by subtraction. The error 2 sum of squares is the sum of squares total minus the sums of squares for everything above the error 2 line in Table 12.4. The degrees of freedom for error 2 are obtained similarly by subtraction. The $MSE(2)$ is an estimate of σ_s^2, which is defined in (12.2.2).

The sums of squares for *rota*, *surf* ∗ *rota*, *fill* ∗ *rota*, *prop* ∗ *rota*, *surf* ∗ *fill* ∗ *rota*, *surf* ∗ *prop* ∗ *rota*, *fill* ∗ *prop* ∗ *rota*, and *surf* ∗ *fill* ∗ *prop* ∗ *rota* are found in a manner similar to that used for the whole plot effects but using the means tables given below.

$N = 24$	Rotation means		
rota	1000	2000	3000
	197.92	158.375	26.792

As usual, $SS(rota)$ is found from the means table above. From the means table below $SS(surf - rota\ trts)$ can be found.

N = 12	*surf*	
rota	Yes	No
1000	231.00	164.83
2000	172.83	143.92
3000	134.58	119.00

Then

$$SS(surf * rota) = SS(surf - rota\ trts) - SS(surf) - SS(rota)$$

where $SS(surf)$ is from the whole plot analysis.

The analysis of variance table is completed by performing similar computations on the following means tables.

N = 12	*fill*	
rota	A	B
1000	214.75	181.08
2000	212.75	104.00
3000	171.67	81.92

N = 8		*prop*	
rota	25%	50%	75%
1000	181.625	192.875	219.250
2000	137.125	156.375	181.625
3000	116.625	128.750	135.000

N = 6		*surf*			
		Yes		No	
	fill	A	B	A	B
	1000	235.00	227.00	194.50	135.17
rota	2000	225.67	120.00	199.83	88.00
	3000	179.33	89.83	164.00	74.00

N = 4		*surf*					
		Yes			No		
	prop	25%	50%	75%	25%	50%	75%
	1000	207.00	235.25	250.75	156.25	150.50	187.75
rota	2000	153.00	171.25	194.25	121.25	141.50	169.00
	3000	120.25	140.25	143.25	113.00	117.25	126.75

N = 4		*fill*					
		A			B		
	prop	25%	50%	75%	25%	50%	75%
	1000	182.50	212.25	249.50	180.75	173.50	189.00
rota	2000	175.25	205.50	257.50	99.00	107.25	105.75
	3000	149.50	178.75	186.75	83.75	78.75	83.25

From the table below, the sum of squares for the 24 different treatment combinations can be computed. The four-factor interaction is obtained from this sum of squares by subtracting out the sums of squares for the 4 main effects, the 6 two-factor interactions, and the 4 three-factor interactions.

surf: Yes

$N = 2$		*fill*					
		A			B		
	prop	25%	50%	75%	25%	50%	75%
	1000	201.00	237.00	267.00	213.00	233.50	234.50
rota	2000	190.00	219.50	267.50	116.00	123.00	121.00
	3000	153.00	186.00	199.00	87.50	94.50	87.50

surf: No

		fill					
		A			B		
	prop	25%	50%	75%	25%	50%	75%
	1000	164.00	187.50	232.00	148.50	113.50	143.50
rota	2000	160.50	191.50	247.50	82.00	91.50	90.50
	3000	146.00	171.50	174.50	80.00	63.00	79.00

12.3 Multivariate analysis of variance

The multivariate approach to analyzing data that contain repeated measurements on each subject involves using the repeated measures as separate dependent variables in a collection of standard analyses of variance. The method of analysis, known as multivariate analysis of variance (MANOVA), then combines results from the several ANOVAs. A detailed discussion of MANOVA is beyond the scope of this book, but we present a short introduction to some of the underlying ideas. In particular, we will not illustrate the important but somewhat sophisticated methods for examining contrasts in MANOVA. The discussion in Christensen (1990a, section I.5) is quite relevant in that it makes extensive comparisons to split plot analyses and makes detailed use of contrasts. Unfortunately, the mathematical level of Christensen (1990a) is much higher than the level of this book. Almost all statistics books on multivariate analysis deal with MANOVA. Johnson and Wichern (1988) is one reasonable place to look for more information on the subject.

The discussion in this section makes some use of matrices. Matrices are reviewed in Appendix A.

EXAMPLE 12.3.1. Consider again the Box (1950) data on the abrasion resistance of a fabric. We began in Section 11.3 by analyzing the weight losses obtained between 1000 and 2000 revolutions of the testing machine. In the split plot analysis we combined these data for 2000 rotations with the data for 1000 and 3000 rotations. In the multivariate approach, we revert to the earlier analysis and fit separate ANOVA models for the data from 1000 rotations, 2000 rotations, and 3000 rotations. Again, the three factors are referred to as '*surf*', '*fill*', and '*prop*', respectively. The variables $y_{hijk,1}$, $y_{hijk,2}$, and $y_{hijk,3}$ denote the data from 1000, 2000, and 3000 rotations respectively. We fit the models

$$\begin{aligned} y_{hijk,1} &= \mu_{hijk,1} + \varepsilon_{hijk,1} \\ &= \mu_1 + s_{h,1} + f_{i,1} + p_{j,1} \\ &\quad + (sf)_{hi,1} + (sp)_{hj,1} + (fp)_{ij,1} + (sfp)_{hij,1} + \varepsilon_{hijk,1}, \end{aligned}$$

$$\begin{aligned} y_{hijk,2} &= \mu_{hijk,2} + \varepsilon_{hijk,2} \\ &= \mu_2 + s_{h,2} + f_{i,2} + p_{j,2} \\ &\quad + (sf)_{hi,2} + (sp)_{hj,2} + (fp)_{ij,2} + (sfp)_{hij,2} + \varepsilon_{hijk,2}, \end{aligned}$$

and

$$\begin{aligned} y_{hijk,3} &= \mu_{hijk,3} + \varepsilon_{hijk,3} \\ &= \mu_3 + s_{h,3} + f_{i,3} + p_{j,3} \\ &\quad + (sf)_{hi,3} + (sp)_{hj,3} + (fp)_{ij,3} + (sfp)_{hij,3} + \varepsilon_{hijk,3} \end{aligned}$$

$h = 1, 2$, $i = 1, 2$, $j = 1, 2, 3$, $k = 1, 2$. As in standard ANOVA models, for fixed $m = 1, 2, 3$, the $\varepsilon_{hijk,m}$s are independent $N(0, \sigma_{mm})$ random variables. We are now using a double subscript in σ_{mm} to denote a variance rather than writing σ_m^2. As usual, the errors on a common dependent variable, say $\varepsilon_{hijk,m}$ and $\varepsilon_{h'i'j'k',m}$, are independent when $(h, i, j, k) \neq (h', i', j', k')$, but we also assume that the errors on different dependent variables, say $\varepsilon_{hijk,m}$ and $\varepsilon_{h'i'j'k',m'}$ are independent when $(h, i, j, k) \neq (h', i', j', k')$. However, not all of the errors for all the variables are assumed independent. Two observations (or errors) on the same experimental unit are *not* assumed to be independent. For fixed h, i, j, k the errors for any two variables are possibly correlated with, say, $Cov(\varepsilon_{hijk,m}, \varepsilon_{hijk,m'}) = \sigma_{mm'}$.

The models for each variable are of the same form but the parameters differ for the different dependent variables $y_{hijk,m}$. All the parameters have an additional subscript to indicate which dependent variable they belong to. The essence of the procedure is simply to fit each of the models individually and then to combine results. Fitting individually gives three separate sets of residuals, $\hat{\varepsilon}_{hijk,m} = y_{hijk,m} - \bar{y}_{hij\cdot,m}$ $m = 1, 2, 3$, three separate sets of residual plots, and three separate ANOVA tables. The three ANOVA tables are given as Tables 12.8, 12.9, and 12.10. All of the means tables necessary for these analyses were given in Section 12.2 in the subsection on Computation of the ANOVA Table. Each variable can be analyzed in detail using the ordinary methods for multifactor designs illustrated in Section 11.3. Residual plots for the analyses on y_1 and y_3 are given in Figures 12.16 through 12.19. Residual plots for y_2 were given in Section 11.3 as Figures 11.8 and 11.9.

TABLE 12.8. Analysis of variance for y_1

Source	*df*	*SS*	*MS*	*F*	*P*
surf	1	26268.2	26268.2	97.74	0.000
fill	1	6800.7	6800.7	25.30	0.000
prop	2	5967.6	2983.8	11.10	0.002
*surf * fill*	1	3952.7	3952.7	14.71	0.002
*surf * prop*	2	1186.1	593.0	2.21	0.153
*fill * prop*	2	3529.1	1764.5	6.57	0.012
*surf * fill * prop*	2	478.6	239.3	0.89	0.436
Error	12	3225.0	268.8		
Total	23	51407.8			

The key to multivariate analysis of variance is to combine results *across* the three variables y_1, y_2, and y_3. Recall that the mean squared errors are just the sums of the squared residuals divided by the error degrees of freedom, e.g.,

$$MSE_{mm} \equiv s_{mm} = \frac{1}{dfE} \sum_{hijk} \hat{\varepsilon}^2_{hijk,m}.$$

TABLE 12.9. Analysis of variance for y_2

Source	*df*	*SS*	*MS*	*F*	*P*
surf	1	5017.0	5017.0	25.03	0.000
fill	1	70959.4	70959.4	353.99	0.000
prop	2	7969.0	3984.5	19.88	0.000
surf ∗ *fill*	1	57.0	57.0	0.28	0.603
surf ∗ *prop*	2	44.3	22.2	0.11	0.896
fill ∗ *prop*	2	6031.0	3015.5	15.04	0.001
surf ∗ *fill* ∗ *prop*	2	14.3	7.2	0.04	0.965
Error	12	2405.5	200.5		
Total	23	92497.6			

TABLE 12.10. Analysis of variance for y_3

Source	*df*	*SS*	*MS*	*F*	*P*
surf	1	1457.0	1457.0	6.57	0.025
fill	1	48330.4	48330.4	217.83	0.000
prop	2	1396.6	698.3	3.15	0.080
surf ∗ *fill*	1	0.4	0.4	0.00	0.968
surf ∗ *prop*	2	250.6	125.3	0.56	0.583
fill ∗ *prop*	2	1740.3	870.1	3.92	0.049
surf ∗ *fill* ∗ *prop*	2	272.2	136.1	0.61	0.558
Error	12	2662.5	221.9		
Total	23	56110.0			

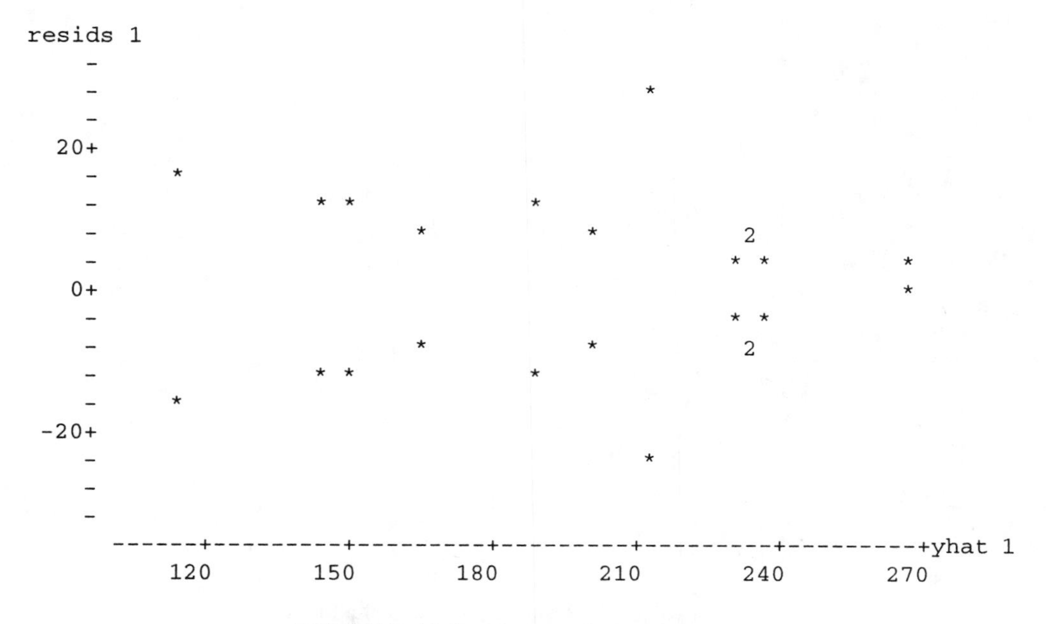

FIGURE 12.16. Residual–prediction plot for y_1.

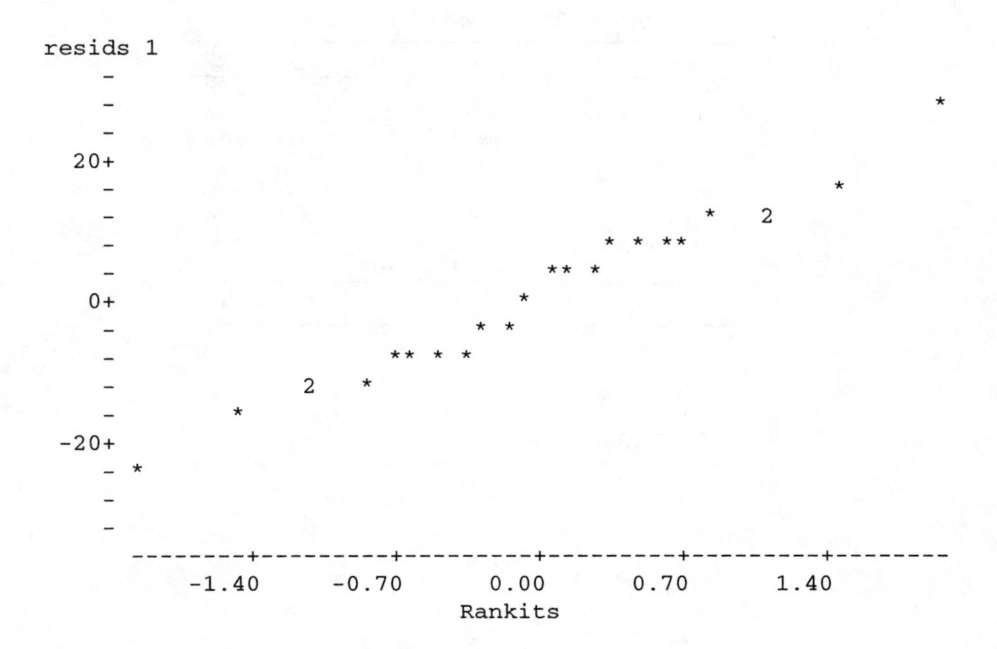

FIGURE 12.17. Normal plot for y_1, $W' = 0.974$.

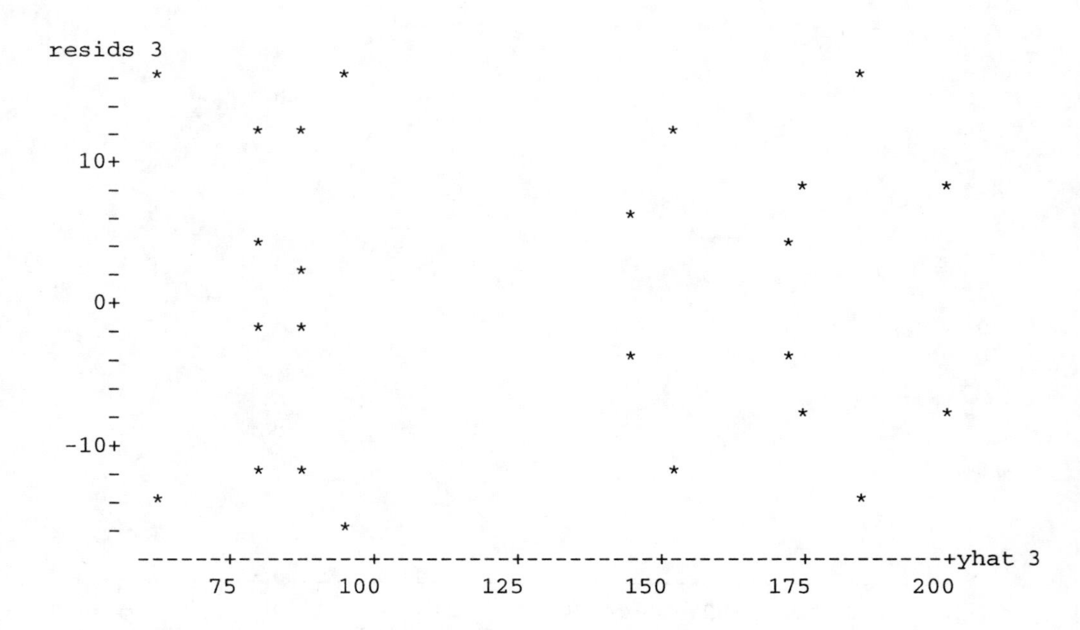

FIGURE 12.18. Residual–prediction plot for y_3.

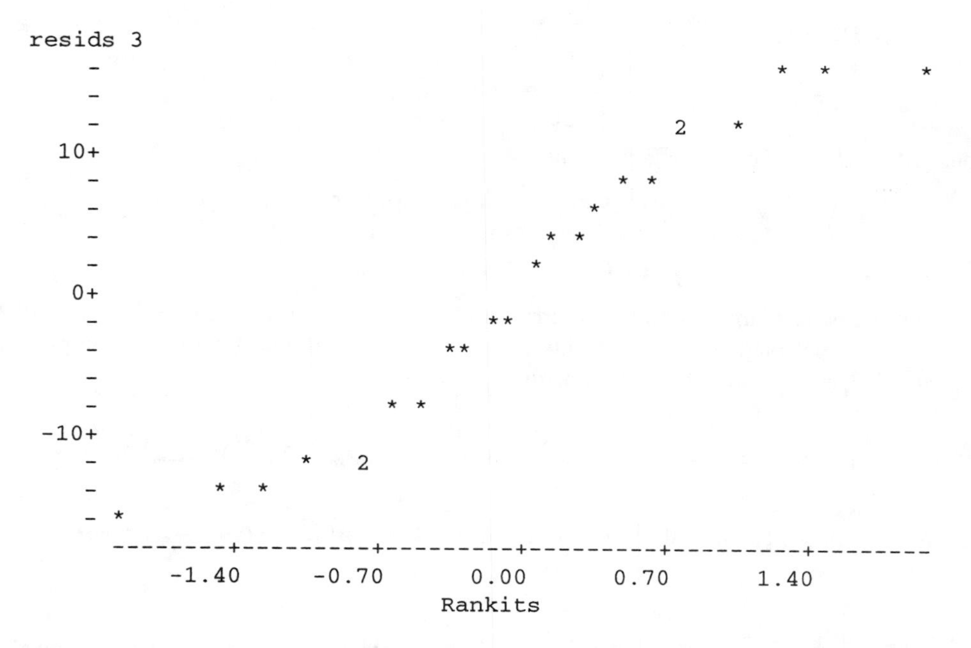

FIGURE 12.19. Normal plot for y_3, $W' = 0.939$.

This provides an estimate of σ_{mm}. We can also use the residuals to estimate covariances between the three variables. The estimate of $\sigma_{mm'}$ is

$$MSE_{mm'} \equiv s_{mm'} = \frac{1}{dfE} \sum_{hijk} \hat{\varepsilon}_{hijk,m} \hat{\varepsilon}_{hijk,m'} .$$

We now form the estimates into a matrix of estimated covariances

$$S = \begin{bmatrix} s_{11} & s_{12} & s_{13} \\ s_{21} & s_{22} & s_{23} \\ s_{31} & s_{32} & s_{33} \end{bmatrix} .$$

Note that $s_{mm'} = s_{m'm}$, e.g., $s_{12} = s_{21}$. The matrix S provides an estimate of the covariance matrix

$$\Sigma \equiv \begin{bmatrix} \sigma_{11} & \sigma_{12} & \sigma_{13} \\ \sigma_{21} & \sigma_{22} & \sigma_{23} \\ \sigma_{31} & \sigma_{32} & \sigma_{33} \end{bmatrix} .$$

The key difference between this analysis and the split plot analysis is that this analysis makes no assumptions about the variances and covariances in Σ. The split plot analysis assumes that

$$\sigma_{11} = \sigma_{22} = \sigma_{33} = \sigma_w^2 + \sigma_s^2$$

and that for $m \neq m'$,

$$\sigma_{mm'} = \sigma_w^2 .$$

Similarly, we can construct a matrix that contains sums of squares error and sums of cross products error. Write

$$e_{mm'} \equiv \sum_{hijk} \hat{\varepsilon}_{hijk,m} \hat{\varepsilon}_{hijk,m'}$$

where $e_{mm} = SSE_{mm}$ and

$$E \equiv \begin{bmatrix} e_{11} & e_{12} & e_{13} \\ e_{21} & e_{22} & e_{23} \\ e_{31} & e_{32} & e_{33} \end{bmatrix}.$$

Obviously, $E = (dfE)S$. For Box's fabric data

$$E = \begin{bmatrix} 3225.00 & -80.50 & 1656.50 \\ -80.50 & 2405.50 & -112.00 \\ 1656.50 & -112.00 & 2662.50 \end{bmatrix}.$$

The diagonal elements of this matrix are the error sums of squares from Tables 12.8, 12.9, and 12.10.

We can use similar methods for every line in the analysis of variance table. For example, each variable $m = 1, 2, 3$ has a sum of squares for *surf* $*$ *prop*, say,

$$SS(s * p)_{mm} \equiv h(s * p)_{mm} = 4 \sum_{h=1}^{2} \sum_{j=1}^{3} \left(\bar{y}_{h \cdot j \cdot, m} - \bar{y}_{h \cdot\cdot\cdot, m} - \bar{y}_{\cdot\cdot j \cdot, m} + \bar{y}_{\cdot\cdot\cdot\cdot, m} \right)^2.$$

Similar to ACOVA, we can include cross products using $SS(s * p)_{mm'} \equiv h(s * p)_{mm'}$, where

$$h(s * p)_{mm'} = 4 \sum_{h=1}^{2} \sum_{j=1}^{3} \left(\bar{y}_{h \cdot j \cdot, m} - \bar{y}_{h \cdot\cdot\cdot, m} - \bar{y}_{\cdot\cdot j \cdot, m} + \bar{y}_{\cdot\cdot\cdot\cdot, m} \right) \left(\bar{y}_{h \cdot j \cdot, m'} - \bar{y}_{h \cdot\cdot\cdot, m'} - \bar{y}_{\cdot\cdot j \cdot, m'} + \bar{y}_{\cdot\cdot\cdot\cdot, m'} \right)$$

and create a matrix

$$H(s * p) \equiv \begin{bmatrix} h(s * p)_{11} & h(s * p)_{12} & h(s * p)_{13} \\ h(s * p)_{21} & h(s * p)_{22} & h(s * p)_{23} \\ h(s * p)_{31} & h(s * p)_{32} & h(s * p)_{33} \end{bmatrix}.$$

For the fabric data

$$H(s * p) = \begin{bmatrix} 1186.0833 & -33.166667 & 526.79167 \\ -33.166667 & 44.333333 & -41.583333 \\ 526.79167 & -41.583333 & 250.58333 \end{bmatrix}.$$

Note that the diagonal elements of $H(s * p)$ are the *surf* $*$ *prop* interaction sums of squares from Tables 12.8, 12.9, and 12.10. Table 12.11 contains the H matrices for all of the sources in the analysis of variance.

In the standard (univariate) analysis of y_2 that was performed in Section 11.3, the test for *surf* $*$ *prop* interactions was based on

$$F = \frac{MS(s * p)_{22}}{MSE_{22}} = \frac{SS(s * p)_{22}}{SSE_{22}} \frac{1/df(s * p)}{1/dfE} = \frac{h(s * p)_{22}}{e_{22}} \frac{dfE}{df(s * p)}.$$

The last two equalities are given to emphasize that the test depends on the $y_{hijk,2}$s only through $h(s * p)_{22} (e_{22})^{-1}$. Similarly, a multivariate test of *surf* $*$ *prop* is a function of the matrices

$$H(s * p)E^{-1},$$

where E^{-1} is the matrix inverse of E. A major difference between the univariate and multivariate procedures is that there is no uniform agreement on how to use $H(s * p)E^{-1}$ to construct a test. The 'generalized likelihood ratio' test, also known as Wilks' lambda is

$$\Lambda(s * p) \equiv \frac{1}{|I + H(s * p)E^{-1}|}$$

TABLE 12.11. MANOVA statistics

$H(GRANDMEAN)$	=	$\begin{bmatrix} 940104.17 & 752281.25 & 602260.42 \\ 752281.25 & 601983.37 & 481935.13 \\ 602260.42 & 481935.13 & 385827.04 \end{bmatrix}$
$H(s)$	=	$\begin{bmatrix} 26268.167 & 11479.917 & 6186.5833 \\ 11479.917 & 5017.0417 & 2703.7083 \\ 6186.5833 & 2703.7083 & 1457.0417 \end{bmatrix}$
$H(f)$	=	$\begin{bmatrix} 6800.6667 & 21967.500 & 18129.500 \\ 21967.500 & 70959.375 & 58561.875 \\ 18129.500 & 58561.875 & 48330.375 \end{bmatrix}$
$H(p)$	=	$\begin{bmatrix} 5967.5833 & 6818.2500 & 2646.9583 \\ 6818.2500 & 7969.0000 & 3223.7500 \\ 2646.9583 & 3223.7500 & 1396.5833 \end{bmatrix}$
$H(s*f)$	=	$\begin{bmatrix} 3952.6667 & 474.83333 & 38.500000 \\ 474.83333 & 57.041667 & 4.6250000 \\ 38.500000 & 4.6250000 & 0.37500000 \end{bmatrix}$
$H(s*p)$	=	$\begin{bmatrix} 1186.0833 & -33.166667 & 526.79167 \\ -33.166667 & 44.333333 & -41.583333 \\ 526.79167 & -41.583333 & 250.58333 \end{bmatrix}$
$H(f*p)$	=	$\begin{bmatrix} 3529.0833 & 4275.5000 & 2374.1250 \\ 4275.5000 & 6031.0000 & 2527.2500 \\ 2374.1250 & 2527.2500 & 1740.2500 \end{bmatrix}$
$H(s*f*p)$	=	$\begin{bmatrix} 478.58333 & 4.4166667 & 119.62500 \\ 4.4166667 & 14.333333 & -57.750000 \\ 119.62500 & -57.750000 & 272.25000 \end{bmatrix}$
E	=	$\begin{bmatrix} 3225.00 & -80.50 & 1656.50 \\ -80.50 & 2405.50 & -112.00 \\ 1656.50 & -112.00 & 2662.50 \end{bmatrix}$

where I indicates a 3×3 identity matrix and $|A|$ denotes the determinant of a matrix A. Roy's maximum root statistic is the maximum eigenvalue of $H(s*p)E^{-1}$, say, $\phi_{max}(s*p)$. On occasion, Roy's statistic is taken as

$$\theta_{max}(s*p) \equiv \frac{\phi_{max}(s*p)}{1+\phi_{max}(s*p)}.$$

A third statistic is the Lawley–Hotelling trace,

$$T^2(s*p) \equiv dfE \operatorname{tr}\left[H(s*p)E^{-1}\right]$$

and a final statistic is Pillai's trace,

$$V(s*p) \equiv \operatorname{tr}\left[H(s*p)\,(E+H(s*p))^{-1}\right].$$

Similar test statistics Λ, ϕ, θ, T^2 and V can be constructed for all of the other main effects and interactions. It can be shown that for H terms with only one degree of freedom, these test statistics are equivalent to each other and to an F statistic. In such cases, we only present T^2 and the F value.

Table 12.12 presents the test statistics for each term. When the F statistic is exactly correct, it is given in the table. In other cases, the table presents F statistic approximations. The approximations

TABLE 12.12. Multivariate statistics

Effect	Statistics	F	df	P
GRAND MEAN	$T^2 = 6836.64$	1899.07	3, 10	0.000
surf	$T^2 = 137.92488$	38.31	3, 10	0.000
fill	$T^2 = 612.96228$	170.27	3, 10	0.000
prop	$\Lambda = 0.13732$	5.66	6, 20	0.001
	$T^2 = 65.31504$	8.16	6, 18	0.000
	$V = 0.97796$	3.51	6, 22	0.014
	$\phi_{max} = 5.28405$			
surf $*$ *fill*	$T^2 = 21.66648$	6.02	3, 10	0.013
surf $*$ *prop*	$\Lambda = 0.71068$	0.62	6, 20	0.712
	$T^2 = 4.76808$	0.60	6, 18	0.730
	$V = 0.29626$	0.64	6, 22	0.699
	$\phi_{max} = 0.37102$			
fill $*$ *prop*	$\Lambda = 0.17843$	4.56	6, 20	0.005
	$T^2 = 46.03092$	5.75	6, 18	0.002
	$V = 0.95870$	3.38	6, 22	0.016
	$\phi_{max} = 3.62383$			
surf $*$ *fill* $*$ *prop*	$\Lambda = 0.75452$	0.50	6, 20	0.798
	$T^2 = 3.65820$	0.46	6, 18	0.831
	$V = 0.26095$	0.55	6, 22	0.765
	$\phi_{max} = 0.20472$			

are commonly used and discussed; see, for example, Rao (1973, chapter 8) or Christensen (1990a, section I.2). Degrees of freedom for the F approximations and P values are also given.

Each effect in Table 12.12 corresponds to a combination of a whole plot effect and a whole plot by subplot interaction from the split plot analysis. For example, the multivariate effect *surf* corresponds to combining the effects *surf* and *surf* $*$ *rota* from the univariate analysis. The highest order terms in the table that are significant are the *fill* $*$ *prop* and the *surf* $*$ *fill* terms. Relative to the split plot analysis, these suggest the presence of *fill* $*$ *prop* interaction or *fill* $*$ *prop* $*$ *rota* interaction and *surf* $*$ *fill* interaction or *surf* $*$ *fill* $*$ *rota* interaction. In Section 12.2, we found the merest suggestion of a *fill* $*$ *prop* $*$ *rota* interaction but clear evidence of a *fill* $*$ *prop* interaction; we also found clear evidence of a *surf* $*$ *fill* $*$ *rota* interaction. However, the split plot results were obtained under different, and perhaps less appropriate, assumptions. To complete the multivariate analysis, MANOVA contrasts are needed. The MANOVA assumptions also suggest some alternative residual analysis. We will not discuss either of these subjects. □

MINITAB COMMANDS

Minitab commands for obtaining Tables 12.8 through 12.12 and the corresponding residuals and predicted values are given below. The three columns c1, c2, and c3 contain y_1, y_2, and y_3, respectively. Columns c4, c5, and c6 contain indices for *surf*, *fill*, and *prop*. Some older versions of Minitab do not allow the last three subcommands that provide the multivariate analysis of variance. Incidentally, Minitab uses T^2 / dfE as its definition for the Lawley–Hotelling trace.

```
MTB > anova c1-c3 = c4|c5|c6;
SUBC> means c4 c5 c6 c4*c5 c4*c6 c5*c6 c4*c5*c6;
SUBC> resid c11 c12 c13;
SUBC> fits c21 c22 c23;
SUBC> manova c4 c5 c6 c4*c5 c4*c6 c5*c6 c4*c5*c6;
SUBC> sscp;
SUBC> eigen.
```

12.4 Random effects models

In this section we consider two special cases of split plot models. First we consider a model in which several observations are taken on the same experimental unit. Repeat observations on a unit involve some random errors but they do not involve the error associated with unit to unit variation. Such models are called subsampling models. The second class of models are those in which some treatment effects in an ANOVA can actually be considered as random. For simplicity, both discussions are restricted to balanced models. Unbalanced models are much more difficult to deal with and typically require a knowledge of linear model theory, cf. Christensen (1987, especially chapter XII).

12.4.1 SUBSAMPLING

It is my impression that many of the disasters that occur in planning and analyzing studies occur because people misunderstand subsampling. The following is both a true story and part of the folklore of the statistics program at the University of Minnesota. A graduate student wanted to study the effects of two drugs on mice. The student collected 200 observations in the following way. Two mice were randomly assigned to each drug. From each mouse, tissue samples were collected at 50 sites. The experimental units were the mice because the drugs were applied to the mice, not to the tissue sites. There are two sources of variation: mouse to mouse variation and within mouse variation. The 50 observations (*subsamples*) on each mouse are very useful in reducing the within mouse variation but do nothing to reduce mouse to mouse variation. Relative to the mouse to mouse variation, which is likely to be larger than the within mouse variation, there are only two observations that have the same treatment. As a result, each of the two treatment groups provides only one degree of freedom for estimating the variance that applies to treatment comparisons. In other words, the experiment provides two degrees of freedom for (the appropriate) error. Obviously a lot of work went into collecting the 200 observations. The work was wasted! Moreover, the problem in the design of this experiment could easily have been compounded by an analysis that ignored the subsampling problem. If subsampling is ignored in the analysis of such data, the *MSE* is inappropriately small and effects look more significant than they really are. (Fortunately, none of the many statistics students that were approached to analyze these data were willing to do it incorrectly.)

Another example comes from Montana State University. A Range science graduate student wanted to compare two types of mountain meadows. He had located two such meadows and was planning to take extensive measurements on each. It had not occurred to him that this procedure would look only at within meadow variation and that there was variation between meadows that he was ignoring.

Consider the subsampling model

$$y_{ijk} = \mu_i + \eta_{ij} + \varepsilon_{ijk} \tag{12.4.1}$$

where $i = 1, \ldots, a$ is the number of treatments, $j = 1, \ldots, n_i$ is the number of replications on different experimental units, and $k = 1, \ldots, N$ is the number of subsamples on each experimental unit. We assume that the ε_{ijk}s are independent $N(0, \sigma_s^2)$ random variables, that the η_{ij}s are independent $N(0, \sigma_w^2)$, and that the η_{ij}s and ε_{ijk}s are independent. The ηs indicate errors (variability) that occur from experimental unit to experimental unit, whereas the εs indicate errors (variability) that occur in measurements taken on a given experimental unit. Model (12.4.1) can be viewed as a special case of a split plot model in which there are no subplot treatments. If there are no subplot treatments, interest lies exclusively in the whole plot analysis. The whole plot analysis can be conducted in the usual way by taking the data to be the averages over the subsamples (subplots).

We can be more formal by using model (12.4.1) to obtain

$$\bar{y}_{ij\bullet} = \mu_i + e_{ij} \tag{12.4.2}$$

where we define

$$e_{ij} \equiv \eta_{ij} + \bar{\varepsilon}_{ij\bullet}$$

and have $i = 1, \ldots, a$, $j = 1, \ldots, n_i$. Using Proposition 1.2.11, it is not difficult to see that the e_{ij}s are independent $N(0, \sigma_w^2 + \sigma_s^2/N)$, so that model (12.4.2) is just an unbalanced one-way ANOVA model and can be analyzed as such. If desired, the methods of the next subsection can be used to estimate the between unit (whole plot) variance σ_w^2 and the within unit (subplot) variance σ_s^2. Note that our analysis in Example 12.1.1 was actually on a model similar to (12.4.2). The data analyzed were averages of two repeat measurements of dynamic absorption.

Model (12.4.2) also helps to formalize the benefits of subsampling. We have N subsamples which lead to $\text{Var}(e_{ij}) = \sigma_w^2 + \sigma_s^2 / N$. If we did not take subsamples, the variance would be $\sigma_w^2 + \sigma_s^2$, so we have reduced one of the terms in the variance by subsampling. If the within unit variance σ_s^2 is large relative to the between unit variance σ_w^2, subsampling can be very beneficial. If the between unit variance σ_w^2 is substantial when compared to the within unit variance σ_s^2, subsampling has very limited benefits. In this latter case, it is important to obtain a substantial number of true replications involving the between unit variability with subsampling based on convenience (rather than importance).

Model (12.4.1) was chosen to have unequal numbers of units on each treatment but a balanced number of subsamples. This was done to suggest the generality of the procedure. Subsamples can be incorporated into any balanced multifactor design and, as long as the number of subsamples is constant for each unit, a simple analysis can be obtained by averaging the subsamples for each unit and using the averages as data. Christensen (1987, section XI.4) provides a closely related discussion that is not too mathematical.

12.4.2 RANDOM EFFECTS

In Chapter 5 we considered data on an electrical characteristic of ceramic assemblies used in manufacturing phonographs. The analysis looked for differences between four specific ceramic strips. An alternative approach to those data is to think of the four ceramic strips as being a random sample from the population of ceramic strips that are involved in making the assemblies. If we do that, we have two sources of variability, variability among the observations on a given strip and variability between different ceramic strips. Our goal in this section is to estimate the variances and test whether there is any variability between strips.

Consider a balanced one-way ANOVA model

$$y_{ij} = \mu + \alpha_i + \varepsilon_{ij}$$

where $i = 1, \ldots, a$ and $j = 1, \ldots, N$. As usual, we assume that the ε_{ij}s are independent $N(0, \sigma^2)$ random variables, but now, rather than assuming that the α_is are fixed treatment effects, we assume that they are *random treatment effects*. In particular, assume that the α_is are independent $N(0, \sigma_A^2)$ random variables that are also independent of the ε_{ij}s. This model can be viewed as a split plot model in which there are no whole plot factors or subplot factors.

The analysis revolves around the analysis of variance table and the use of Proposition 1.2.11. As usual, begin with the summary statistics $\bar{y}_{i\cdot}$ and s_i^2, $i = 1, \ldots, a$. In comparing the observations within a single strip, there is no strip to strip variability. The sample variances s_i^2 each involve comparisons only within a given strip, so each provides an estimate of the within strip variance, σ^2. In particular, $E\left(s_i^2\right) = \sigma^2$. Clearly, if we pool these estimates we continue to get an estimate of σ^2. In particular,

$$E(MSE) = \sigma^2.$$

We now examine *MSTrts*. Before proceeding note that by independence of the α_is and the ε_{ij}s,

$$\begin{aligned}
\text{Var}(y_{ij}) &= \text{Var}(\mu + \alpha_i + \varepsilon_{ij}) \\
&= \text{Var}(\alpha_i) + \text{Var}(\varepsilon_{ij}) \\
&= \sigma_A^2 + \sigma^2.
\end{aligned}$$

Thus $\text{Var}(y_{ij})$ is the sum of two *variance components* σ_A^2 and σ^2. Moreover,

$$\begin{aligned}\text{Var}(\bar{y}_{i\cdot}) &= \text{Var}(\mu + \alpha_i + \bar{\varepsilon}_{i\cdot}) \\ &= \text{Var}(\alpha_i) + \text{Var}(\bar{\varepsilon}_{i\cdot}) \\ &= \sigma_A^2 + \frac{\sigma^2}{N}\end{aligned}$$

because $\bar{\varepsilon}_{i\cdot}$ is the sample mean of N independent random variables that have variance σ^2. It is easily seen that $E(\bar{y}_{i\cdot}) = \mu$. As in the discussion of Chapter 5, the $\bar{y}_{i\cdot}$s form a random sample of size a, but now the population that they are sampled from is $N\left(\mu, \sigma_A^2 + \sigma^2/N\right)$. Note that, unlike Chapter 5, this result holds all the time and does not require the validity of some null hypothesis. Clearly, the sample variance of the $\bar{y}_{i\cdot}$s provides an estimate of $\sigma_A^2 + \sigma^2/N$. The *MSTrts* is N times the sample variance of the $\bar{y}_{i\cdot}$s, so *MSTrts* provides an unbiased estimate of $N\sigma_A^2 + \sigma^2$.

We already have an estimate of σ^2. To obtain an estimate of σ_A^2 use the results of the previous paragraph and take

$$\hat{\sigma}_A^2 = \frac{MSTrts - MSE}{N}.$$

It is a simple exercise to show that

$$E\left(\hat{\sigma}_A^2\right) = \sigma_A^2.$$

The usual F statistic is *MSTrts/ MSE*. Clearly, it is a (biased) estimate of

$$\frac{N\sigma_A^2 + \sigma^2}{\sigma^2} = 1 + \frac{N\sigma_A^2}{\sigma^2}.$$

If $H_0 : \sigma_A^2 = 0$ holds, the F statistic should be about 1. In general, if H_0 holds,

$$\frac{MSTrts}{MSE} \sim F(a-1, dfE)$$

and the usual F test can be interpreted as a test of $H_0 : \sigma_A^2 = 0$.

EXAMPLE 12.4.1. For the electrical characteristic data of Chapter 5, the analysis of variance table is given below.

Analysis of variance table: electrical characteristic data

Source	df	SS	MS	F	P
Treatments	3	10.873	3.624	6.45	0.002
Error	24	13.477	0.562		
Total	27	24.350			

The F statistic shows strong evidence that variability exists between ceramic strips. The estimate of within strip variability is $MSE = .562$. With 7 observations on each ceramic strip, the estimate of between strip variability is

$$\hat{\sigma}_A^2 = \frac{MSTrts - MSE}{N} = \frac{3.624 - 0.562}{7} = .437.$$

While in many ways this random effects analysis seems more appropriate for the relatively undifferentiated strips being considered, this analysis also seems less informative for these data than the analysis in Chapter 5. It was clear in Chapter 5 that most of the between strip 'variation' was due to a single strip, number 4. Are we to consider this strip an outlier in the population of ceramic strips? Having three sample means that are quite close and one that is substantially different certainly calls in question the assumption that the random treatment effects are normally distributed. Most importantly, some kind of analysis that looks at individual sample means is necessary to have any chance of identifying an odd strip. □

The ideas behind the analysis of the balanced one-way ANOVA model generalize nicely to other balanced models. Consider the balanced two-way with replication,

$$y_{ijk} = \mu + \alpha_i + \beta_j + \gamma_{ij} + \varepsilon_{ijk}$$

where $i = 1, \ldots, a, j = 1, \ldots, b$, and $k = 1, \ldots, N$. Assume that the ε_{ijk}s are independent $N(0, \sigma^2)$ random variables, that the γ_{ij}s are independent $N(0, \sigma_\gamma^2)$, and that the ε_{ijk}s and γ_{ij}s are independent. This model involves two variance components σ_γ^2 and σ^2.

While we will not provide proofs, the following results hold. *MSE* still estimates σ^2. $MS(\gamma)$ estimates

$$E[MS(\gamma)] = \sigma^2 + N\sigma_\gamma^2.$$

The usual interaction test is a test of $H_0 : \sigma_\gamma^2 = 0$. For main effects, $MS(\beta)$ estimates

$$E[MS(\beta)] = \sigma^2 + N\sigma_\gamma^2 + \frac{aN}{b-1}\sum_{j=1}^{b}(\beta_j - \bar{\beta}_\bullet)^2.$$

When the β_js are all equal, $MS(\beta)$ estimates $\sigma^2 + N\sigma_\gamma^2$. It follows that to obtain an F test for equality of the β_js, the test must reject when $MS(\beta)$ is much larger than $MS(\gamma)$. In particular, an α level test rejects if

$$\frac{MS(\beta)}{MS(\gamma)} > F(1 - \alpha, a - 1, [a - 1][b - 1]).$$

This is just the usual result *except that the MSE has been replaced by the* $MS(\gamma)$. The analysis of contrasts in the β_js also follows the standard pattern with $MS(\gamma)$ used in place of *MSE*. Similar results hold for investigating the α_is.

The moral of this analysis is that one needs to think very carefully about whether to model interactions as fixed effects or random effects. It would seem that if you do not care about interactions, if they are just an annoyance in evaluating the main effects, you probably should treat them as random and use the interaction mean square as the appropriate estimate of variability. A related way of thinking is to stipulate that you do not care about any main effects unless they are large enough to show up above any interaction. In particular, this is essentially what is done in a randomized complete block design. An RCB takes the block by treatment interaction as the error and only treatment effects that are strong enough to show up above any block by treatment interaction are deemed significant. On the other hand, if interactions are something of direct interest, they should typically be treated as fixed effects.

12.5 Exercises

EXERCISE 12.5.1. In Exercises 9.5.3, 9.5.4, and 11.5.1 we considered data from Cornell (1988) on scaled vinyl thicknesses. Exercise 9.5.3 involved five blends of vinyl and we discussed the fact that the production process was set up eight times with a group of five blends run on each setting. The eight production settings where those in Exercise 9.5.4. The complete data are displayed in Table 12.13.

(a) Identify the design for this experiment and give an appropriate model. List all the assumptions made in the model.

TABLE 12.13. Cornell's scaled vinyl thickness values

		Replication 1				Replication 2			
	Rate	High	Low	Low	High	High	Low	Low	High
	Temp	Low	High	Low	High	Low	High	Low	High
	1	8	12	7	12	7	10	8	11
	2	6	9	7	10	5	8	6	9
Blend	3	10	13	9	14	11	12	10	12
	4	4	6	5	6	5	3	4	5
	5	11	15	9	13	10	11	7	9

(b) Analyze the data. Give an appropriate analysis of variance table. Examine appropriate contrasts using the LSD method with an α of .05.

(c) Check the assumptions of the model and adjust the analysis appropriately.

(d) Discuss the relationship between the current analysis and those conducted earlier.

EXERCISE 12.5.2. Wilm (1945) presented data involving the effect of logging on soil moisture deficits under a forest. Treatments consist of five intensities of logging. Treatments were identified as the volume of the trees left standing after logging that were larger than 9.6 inches in diameter. The logging treatments were uncut, 6000 board-feet, 4000 board-feet, 2000 board-feet, 0 board-feet. The experiment was conducted by selecting four blocks (A,B,C,D) of forest. These were subdivided into five plots. Within each block each of the treatments were randomly assigned to a plot. Soil moisture deficits were measured in each of three consecutive years, 1941, 1942, and 1943. The data are presented in Table 12.14.

TABLE 12.14. Soil moisture deficits as affected by logging

		Block			
Treatment	Year	A	B	C	D
	41	2.40	0.98	1.38	1.37
Uncut	42	3.32	1.91	2.36	1.62
	43	2.59	1.44	1.66	1.75
	41	1.76	1.65	1.69	1.11
6000	42	2.78	2.07	2.98	2.50
	43	2.27	2.28	2.16	2.06
	41	1.43	1.30	0.18	1.66
4000	42	2.51	1.48	1.83	2.36
	43	1.54	1.46	0.16	1.84
	41	1.24	0.70	0.69	0.82
2000	42	3.29	2.00	1.38	1.98
	43	2.67	1.44	1.75	1.56
	41	0.79	0.21	0.01	0.16
None	42	1.70	1.44	2.65	2.15
	43	1.62	1.26	1.36	1.87

Treatments are volumes of timber left standing in trees with diameters greater than 9.6 inches. Volumes are measured in board-feet.

(a) Identify the design for this experiment and give an appropriate model. List all the assumptions made in the model.

(b) Analyze the data. Give an appropriate analysis of variance table. Examine appropriate contrasts. In particular, use the treatment contrast that compares the uncut plots to the average of the other plots and use orthogonal polynomial contrasts to examine differences among the other four treatments. Discuss the reasonableness of this procedure in which the 'uncut' treatment is excluded when fitting the polynomial contrasts.

(c) Check the assumptions of the model and adjust the analysis appropriately. What assumptions are difficult to check? Identify any such assumptions that are particularly suspicious.

EXERCISE 12.5.3. Day and del Priore (1953) report data from an experiment on the noise generated by various reduction gear designs. The data were collected because of the Navy's interest in building quiet submarines. Primary interest focused on the direction of lubricant application. Lubricants were applied either inmesh (I) or tangent (T) and either at the top (T) or the bottom (B). Thus the direction TB indicates tangent, bottom while IT is inmesh, top.

Four additional factors were considered. Load was 25%, 100%, or 125%. The temperature of the input lubricant was 90, 120, or 160 degrees F. The volume of lubricant flow was .5 gpm, 1 gpm, or 2 gpm. The speed was either 300 rpm or 1200 rpm. Temperature and volume were of less interest than direction; speed and load were of even less interest. It was considered that load, temperature, and volume would not interact but that speed might interact with the other factors. There was little idea whether direction would interact with other factors. As a result, a split plot design with whole plots in a 3×3 Latin Square was used. The factors used in defining the whole plot Latin Square were load, temperature, and volume. The subplot factors were speed and the direction factors.

The data are presented in Table 12.15. The four observations with 100% load, 90 degree temperature, .5 gpm volume, and lubricant applied tangentially were not made. Substitutes for these values were used. As an approximate analysis, treat the substitute values as real values but subtract four degrees of freedom from the subplot error. Analyze the data.

TABLE 12.15. Gear test data

		Volume					
Load	Direction	.5 gpm		1 gpm		2 gpm	
		Temp 120		Temp 90		Temp 160	
25%	TB	92.7	81.4	91.3	68.0	86.9	78.2
	IT	95.9	79.2	87.7	77.7	90.7	97.9
	IB	92.7	85.5	93.6	76.2	92.1	80.2
	TT	92.2	81.4	92.9	72.2	90.6	85.8
		Temp 90		Temp 160		Temp 120	
100%	TB	94.2*	80.2*	89.7	86.0	91.9	84.8
	IT	88.6	83.7	87.8	86.6	85.4	79.5
	IB	89.8	83.9	90.4	79.3	85.7	86.9
	TT	89.8*	75.4*	90.4	85.0	82.6	79.0
		Temp 160		Temp 120		Temp 90	
125%	TB	88.7	94.2	90.3	86.7	88.4	75.8
	IT	92.1	91.1	90.3	83.5	86.3	71.2
	IB	91.7	89.2	90.4	86.6	88.3	87.9
	TT	93.4	86.2	89.7	83.0	88.6	84.5
		300	1200	300	1200	300	1200
		Speed		Speed		Speed	

* indicates a replacement for missing data

EXERCISE 12.5.4. In Exercise 11.5.4 and Table 11.20 we presented Baten's (1956) data on lengths of steel bars. The bars were made with one of two heat treatments (W, L) and cut on one of four screw machines (A, B, C, D) at one of three times of day (8 am, 11 am, 3 pm). There are distressing aspects to Baten's article. First, he never mentions what the heat treatments are. Second, he does not discuss how the four screw machines differ or whether the same person operates the same machine all the time. If the machines were largely the same and one specific person always operates the same machine all the time, then machine differences would be due to operators rather than machines. If the machines were different and one person operates the same machine all the time, it becomes impossible to tell whether machine differences are due to machines or operators. Most importantly, Baten does not discuss how the replications were obtained. In particular, consider the role of day to day variation in the analysis.

If the 12 observations on a heat treatment–machine combination are all taken on the same day, there is no replication in the experiment that accounts for day to day variation. In that case the average of the four numbers for each heat treatment–machine–time combination gives essentially one observation and for each heat treatment–machine combination the three time means are correlated. To obtain an analysis, the heat–machine interaction and the heat–machine–time interaction would have to be used as the two error terms.

Suppose the 12 observations on a heat treatment–machine combination are taken on four different days with one observation obtained on each day for each time period. Then the three observations on a given day are correlated but the observations on different days are independent. This leads to a traditional split plot analysis.

Finally, suppose that the 12 observations on a heat treatment–machine combination are all taken on 12 different days. Yet another analysis is appropriate.

Compare the results of these three different methods of analyzing the experiment. If the day to day variability is no larger than the within day variability, there should be little difference. When considering the analysis that assumes 12 observations taken on four different days, treat the order of the four heat treatment–machine–time observations as indicating the day. For example, with heat treatment W and machine A, take 9, 3, and 4 as the three time observations on the second day.

EXERCISE 12.5.5. People who really want to test their skill may wish to examine the data presented in Snedecor and Haber (1946) and repeated in Table 12.16. The experiment was to examine the effects of three cutting dates on asparagus. Six blocks were used. One plot was assigned a cutting date of June 1 (a), one a cutting date of June 15 (b), and the last a cutting date of July 1 (c). Data were collected on these plots for 10 years.

Try to come up with an intelligible summary of the data that would be of use to someone growing asparagus. In particular, the experiment was planned to run for the effective lifetime of the planting, normally 20 years or longer. The experiment was cut short due to lack of labor but interest remained in predicting behavior ten years after the termination of data collection. The linear and quadratic contrast coefficients for ten groups are

Contrast	Coefficients									
Linear	−9	−7	−5	−3	−1	1	3	5	7	9
Quadratic	6	2	−1	−3	−4	−4	−3	−1	2	6

As most effects seem to be significant, I would be inclined to focus on effects that seem relatively large rather than on statistically significant effects.

EXERCISE 12.5.6. For a balanced $3 \times 4 \times 3$ with equally spaced quantitative factor levels find the contrast coefficients for the linear by linear by linear, the linear by quadratic by linear, and the quadratic by cubic by linear three-way interaction contrasts.

EXERCISE 12.5.7. Reanalyze the data of Exercise 5.7.1 assuming that the five laboratories are a random sample from a population of laboratories. Include estimates of both variance components.

TABLE 12.16.

Year	Treatments a	b	c	a	b	c	a	b	c
29	201	301	362	185	236	341	209	226	357
30	230	296	353	216	256	328	219	212	354
31	324	543	594	317	397	487	357	358	560
32	512	778	755	448	639	622	496	545	685
33	399	644	580	361	483	445	344	415	520
34	891	1147	961	783	998	802	841	833	871
35	449	585	535	409	525	478	418	451	538
36	595	807	548	566	843	510	622	719	578
37	632	804	565	629	841	576	636	735	634
38	527	749	353	527	823	299	530	731	413
29	219	330	427	225	307	382	219	342	464
30	222	301	391	239	297	321	216	287	364
31	348	521	599	347	463	502	356	557	584
32	487	742	802	512	711	684	508	768	819
33	372	534	573	405	577	467	377	529	612
34	773	1051	880	786	1066	763	780	969	1028
35	382	570	540	415	610	468	407	526	651
36	505	737	577	549	779	548	595	772	660
37	534	791	524	559	741	621	626	826	673
38	434	614	343	433	706	352	518	722	424

Blocks are indicated by vertical and horizontal lines.

EXERCISE 12.5.8. Reanalyze the data of Example 11.1.1 assuming that the disk by window interaction is a random effect. Include estimates of both variance components.

Chapter 13

Multiple regression: introduction

Multiple regression involves predicting values of a dependent variable from the values on a collection of other (predictor) variables. In particular, linear combinations of the predictor variables are used in modeling the dependent variable.

13.1 Example of inferential procedures

In our discussion of simple linear regression, we considered data from *The Coleman Report*. The data given were only two of six variables reported in Mosteller and Tukey (1977). We now consider the entire collection of variables. Recall that the data are from schools in the New England and Mid-Atlantic states. The variables are y, the mean verbal test score for sixth graders; x_1, staff salaries per pupil; x_2, percentage of sixth graders whose fathers have white collar jobs; x_3, a composite measure of socioeconomic status; x_4, the mean of verbal test scores given to the teachers; and x_5,- the mean educational level of the sixth grader's mothers (one unit equals two school years). The dependent variable y is the same as in the simple linear regression example and the variable x_3 was used as the sole predictor variable in the earlier example. The data are given in Table 13.1.

It is of interest to examine the correlations between y and the predictor variables.

	x_1	x_2	x_3	x_4	x_5
Correlation with y	0.192	0.753	0.927	0.334	0.733

Of the five variables, x_3, the one used in the simple linear regression, has the highest correlation. Thus it explains more of the y variability than any other single variable. Variables x_2 and x_5 also have reasonably high correlations with y. Low correlations exist between y and both x_1 and x_4. Interestingly, x_1 and x_4 turn out to be more important in explaining y than either x_2 or x_5. However, the explanatory power of x_1 and x_4 only manifests itself after x_3 has been fitted to the data.

We assume the data satisfy the model

$$y_i = \beta_0 + \beta_1 x_{i1} + \beta_2 x_{i2} + \beta_3 x_{i3} + \beta_4 x_{i4} + \beta_5 x_{i5} + \varepsilon_i, \tag{13.1.1}$$

$i = 1, \ldots, 20$, where the ε_is are unobservable independent $N(0, \sigma^2)$ random variables and the βs are fixed unknown parameters. Fitting model (13.1.1) with a computer program typically yields

TABLE 13.1. *Coleman Report* data

School	y	x_1	x_2	x_3	x_4	x_5
1	37.01	3.83	28.87	7.20	26.60	6.19
2	26.51	2.89	20.10	−11.71	24.40	5.17
3	36.51	2.86	69.05	12.32	25.70	7.04
4	40.70	2.92	65.40	14.28	25.70	7.10
5	37.10	3.06	29.59	6.31	25.40	6.15
6	33.90	2.07	44.82	6.16	21.60	6.41
7	41.80	2.52	77.37	12.70	24.90	6.86
8	33.40	2.45	24.67	−0.17	25.01	5.78
9	41.01	3.13	65.01	9.85	26.60	6.51
10	37.20	2.44	9.99	−0.05	28.01	5.57
11	23.30	2.09	12.20	−12.86	23.51	5.62
12	35.20	2.52	22.55	0.92	23.60	5.34
13	34.90	2.22	14.30	4.77	24.51	5.80
14	33.10	2.67	31.79	−0.96	25.80	6.19
15	22.70	2.71	11.60	−16.04	25.20	5.62
16	39.70	3.14	68.47	10.62	25.01	6.94
17	31.80	3.54	42.64	2.66	25.01	6.33
18	31.70	2.52	16.70	−10.99	24.80	6.01
19	43.10	2.68	86.27	15.03	25.51	7.51
20	41.01	2.37	76.73	12.77	24.51	6.96

parameter estimates, standard errors for the estimates, t ratios for testing whether the parameters are zero, and an analysis of variance table.

Predictor	$\hat{\beta}_k$	$SE(\hat{\beta}_k)$	t	P
Constant	19.95	13.63	1.46	0.165
x_1	−1.793	1.233	−1.45	0.168
x_2	0.04360	0.05326	0.82	0.427
x_3	0.55576	0.09296	5.98	0.000
x_4	1.1102	0.4338	2.56	0.023
x_5	−1.811	2.027	−0.89	0.387

Analysis of Variance

Source	df	SS	MS	F	P
Regression	5	582.69	116.54	27.08	0.000
Error	14	60.24	4.30		
Total	19	642.92			

From just these two tables of statistics much can be learned. In particular, the estimated regression equation is

$$\hat{y} = 19.9 - 1.79x_1 + 0.0436x_2 + 0.556x_3 + 1.11x_4 - 1.81x_5.$$

As discussed in simple linear regression, this equation *describes* the relationship between y and the predictor variables for the *current* data; *it does not imply a causal relationship.* If we go out and increase the percentage of sixth graders whose fathers have white collar jobs by 1%, i.e., increase x_2 by one unit, we cannot infer that mean verbal test scores will tend to increase by .0436 units. In fact, we cannot think about any of the variables in a vacuum. No variable has an effect in the equation apart from the *observed values* of all the other variables. If we conclude that some variable can be eliminated from the model, we cannot conclude that the variable has no effect on y, we can only conclude that the variable is not necessary to explain *these* data. The same variable may be very important in explaining other, rather different, data collected on the same variables. All too often, people choose to interpret the estimated regression coefficients as if the predictor variables cause the value of y; the estimated regression coefficients simply describe an observed relationship. Frankly, since the coefficients do not describe a causal relationship, many people, including the author, find regression coefficients to be remarkably uninteresting quantities. What this model is good at is predicting values of y for new cases that are similar to those in the current data. In particular, such new cases should have predictor variables with values similar to those in the current data.

The t statistics for testing $H_0 : \beta_k = 0$ versus $H_A : \beta_k \neq 0$ were reported previously. For example, the test of $H_0 : \beta_4 = 0$ versus $H_A : \beta_4 \neq 0$ has

$$t = \frac{1.1102}{.4338} = 2.56.$$

The significance level of the test is the P value,

$$P = \Pr[|t(dfE)| \geq 2.56] = .023.$$

The value .023 indicates a reasonable amount of evidence that variable x_4 is needed in the model. We can be reasonably sure that dropping x_4 from the model harms the explanatory (predictive) power of the model. In particular, with a P value of .023, the test of $H_0 : \beta_4 = 0$ versus $H_A : \beta_4 \neq 0$ is rejected at the $\alpha = .05$ level (because $.05 > .023$), but the test is not rejected at the $\alpha = .01$ level (because $.023 > .01$).

A 95% confidence interval for β_3 has endpoints $\hat{\beta}_3 \pm t(.975, dfE)\, SE(\hat{\beta}_3)$. From a t table, $t(.975, 14) = 2.145$ and from the computer output the endpoints are

$$.55576 \pm 2.145(.09296).$$

The confidence interval is $(.356, .755)$, so we are 95% 'confident' that the hypothetical parameter β_3 is between .356 and .755. (In using the word 'hypothetical,' I am letting my bias against parameters show.) As will be discussed later, simultaneous 95% confidence intervals for the regression parameters can be obtained by changing the multiplier $t(.975, dfE)$.

The primary value of the analysis of variance table is that it gives the degrees of freedom, the sum of squares, and the mean square for error. The mean squared error is the estimate of σ^2, and the sum of squares error and degrees of freedom for error are vital for comparing various regression models. The degrees of freedom for error are $n - 1 -$ (the number of predictor variables). The minus 1 is an adjustment for fitting the intercept β_0.

The analysis of variance table also gives the test for whether any of the x variables help to explain y. This test is rarely of interest because it is almost always highly significant. It is a poor scholar who cannot find any predictor variables that are related to the measurement of primary interest. (Ok, I admit to being a little judgmental here.) The test of

$$H_0 : \beta_1 = \cdots = \beta_5 = 0$$

versus

$$H_A : \text{not all } \beta_k\text{s equal to } 0$$

is based on

$$F = \frac{MSReg}{MSE} = \frac{116.5}{4.303} = 27.08$$

and is rejected for large values of F. The numerator and denominator degrees of freedom come from the ANOVA table. As implied, the corresponding P value in the ANOVA table is infinitesimal, zero to three decimal places. Thus these x variables, as a group, help to explain the variation in the y variable. In other words, it is possible to predict the mean verbal test scores for a school's sixth grade class from the five x variables measured. Of course, the fact that some predictive ability exists does not mean that the predictive ability is sufficient to be useful.

The coefficient of determination, R^2, measures the percentage of the total variability in y that is explained by the x variables. If this number is large, it suggests a substantial predictive ability. In this example

$$R^2 \equiv \frac{SSReg}{SSTot} = \frac{582.69}{642.92} = .906,$$

so 90.6% of the total variability is explained by the regression model. This is a large percentage, suggesting that the five x variables have substantial predictive power. However, we saw in Section 7.1 that a large R^2 does not imply that the model is good in absolute terms. It may be possible to show that this model does not fit the data adequately. In other words, this model is explaining much of the variability but we may be able to establish that it is not explaining as much of the variability as it ought. Conversely, a model with a low R^2 value may be the perfect model but the data may simply have a great deal of variability. Moreover, even an R^2 of .906 may be inadequate for the predictive purposes of the researcher, while in some situations an R^2 of .3 may be perfectly adequate. It depends on the purpose of the research. Finally, it must be recognized that a large R^2 may be just an unrepeatable artifact of a particular data set. The coefficient of determination is a useful tool but it must be used with care. Recall from Section 7.1 that the R^2 was .86 when using just x_3 to predict y.

MINITAB COMMANDS

To obtain the statistics in this section, use the Minitab command 'regress.' The form of the command has regress, the y variable, the number of predictors (excluding the intercept), and a list of the predictor variables.

```
MTB > names c1 'x1' c2 'x2' c3 'x3' c4 'x4' c5 'x5' c6 'y'
MTB > regress c6 on 5 c1-c5
```

GENERAL STATEMENT OF THE MULTIPLE REGRESSION MODEL

In general we consider a dependent variable y which is a random variable that we are interested in examining. We also consider $p-1$ nonrandom predictor variables $x_1, \ldots, x_{p-1}$. The general multiple (linear) regression model relates n observations on y to a linear combination of the corresponding observations on the x_js plus a random error ε. In particular, we assume

$$y_i = \beta_0 + \beta_1 x_{i1} + \cdots + \beta_{p-1} x_{i,p-1} + \varepsilon_i,$$

where the subscript $i = 1, \ldots, n$ indicates different observations and the ε_is are independent $N(0, \sigma^2)$ random variables. The β_js and σ^2 are unknown constants and the fundamental parameters of the regression model.

Estimates of the β_js are obtained by the method of least squares. The least squares estimates are those that minimize

$$\sum_{i=1}^{n} \left(y_i - \beta_0 - \beta_1 x_{i1} - \beta_2 x_{i2} - \cdots - \beta_{p-1} x_{i,p-1}\right)^2 .$$

In this function the y_is and the x_{ij}s are all known quantities. Least squares estimates have a number of interesting statistical properties. If the errors are independent with mean zero, constant variance, and are normally distributed, the least squares estimates are maximum likelihood estimates and minimum variance unbiased estimates. If the errors are merely uncorrelated with mean zero and constant variance, the least squares estimates are best (minimum variance) linear unbiased estimates.

In checking assumptions we often use the predictions $\hat{y}$ corresponding to the observed values of the predictor variables, i.e.,

$$\hat{y}_i = \hat{\beta}_0 + \hat{\beta}_1 x_{i1} + \cdots + \hat{\beta}_{p-1} x_{i,p-1},$$

$i = 1, \ldots, n$. An interesting fact about the coefficient of determination R^2 is that it is the square of the sample correlation between the data y_i and the corresponding predicted values $\hat{y}_i$. As usual, residuals are the values

$$\hat{\varepsilon}_i = y_i - \hat{y}_i.$$

The other fundamental parameter to be estimated, besides the β_js, is the variance σ^2. The sum of squares error is

$$SSE = \sum_{i=1}^{n} \hat{\varepsilon}_i^2$$

and the estimate of σ^2 is

$$MSE = SSE/(n-p).$$

Details of the estimation procedures are given in Section 15.3.

13.2 Regression surfaces and prediction

One of the most valuable aspects of regression analysis is its ability to provide good predictions of future observations. Of course, to obtain a prediction for a new value y we need to know the corresponding values of the predictor variables, the x_js. Moreover, to obtain good predictions, the values of the x_js need to be similar to those on which the regression model was fitted. Typically, a fitted regression model is only an approximation to the true relationship between y and the predictor variables. These approximations can be very good, but, because they are only approximations, they are not valid for predictor variables that are dissimilar to those on which the approximation was based. Trying to predict for x_j values that are far from the original data is always difficult. Even if the regression model is true and not an approximation, the variance of such predictions is large. When the model is only an approximation, the approximation is typically invalid for such predictor variables and the predictions can be utter nonsense.

The regression surface is the set of all values z that satisfy

$$z = \beta_0 + \beta_1 x_1 + \beta_2 x_2 + \beta_3 x_3 + \beta_4 x_4 + \beta_5 x_5$$

for some values of the predictor variables. The estimated regression surface is

$$z = \hat{\beta}_0 + \hat{\beta}_1 x_1 + \hat{\beta}_2 x_2 + \hat{\beta}_3 x_3 + \hat{\beta}_4 x_4 + \hat{\beta}_5 x_5.$$

There are two problems of interest. The first is estimating the value z on the regression surface for a fixed set of predictor variables. The second is predicting the value of a new observation to be obtained with a fixed set of predictor variables. For any set of predictor variables, the estimate of the regression surface and the prediction are identical. What differs are the standard errors associated with the different problems.

Consider estimation and prediction at

$$(x_1, x_2, x_3, x_4, x_5) = (2.07, 9.99, -16.04, 21.6, 5.17).$$

These are the minimum values for each of the variables, so there will be substantial variability in estimating the regression surface at this point. The estimator (predictor) is

$$\hat{y} = \hat{\beta}_0 + \sum_{j=1}^{5} \hat{\beta}_j x_j = 19.9 - 1.79(2.07) + 0.0436(9.99)$$
$$+ 0.556(-16.04) + 1.11(21.6) - 1.81(5.17) = 22.375.$$

For constructing 95% t intervals, the percentile needed is $t(.975, 14) = 2.145$.

The 95% confidence interval for the point $\beta_0 + \sum_{j=1}^{5} \beta_j x_j$ on the regression surface uses the standard error for the regression surface which is

$$SE(Surface) = 1.577.$$

The standard error is obtained from the regression program and depends on the specific value of $(x_1, x_2, x_3, x_4, x_5)$. The formula for the standard error is given in Section 15.4. This interval has endpoints

$$22.375 \pm 2.145(1.577)$$

which gives the interval

$$(18.992, 25.757).$$

The 95% prediction interval is

$$(16.785, 27.964).$$

This is about 4 units wider than the confidence interval for the regression surface. The standard error for the prediction interval can be computed from the standard error for the regression surface.

$$SE(Prediction) = \sqrt{MSE + SE(Surface)^2}.$$

In this example,

$$SE(Prediction) = \sqrt{4.303 + (1.577)^2} = 2.606.$$

and the prediction interval endpoints are

$$22.375 \pm 2.145(2.606).$$

We mentioned earlier that even if the regression model is true, the variance of predictions is large when the x_j values for the prediction are far from the original data. We can use this fact to identify situations in which the predictions are unreliable because the locations are too far away. Let $p - 1$ be the number of predictor variables so that, including the intercept, there are p regression parameters. Let n be the number of observations. A sensible rule of thumb is that we should start worrying about the validity of the prediction whenever

$$\frac{SE(Surface)}{\sqrt{MSE}} \geq \sqrt{\frac{2p}{n}}$$

and we should be very concerned about the validity of the prediction whenever

$$\frac{SE(Surface)}{\sqrt{MSE}} \geq \sqrt{\frac{3p}{n}}.$$

Recall that for simple linear regression we suggested that leverages greater than $4/n$ cause concern and those greater than $6/n$ cause considerable concern. In general, leverages greater than $2p/n$ and

$3p/n$ cause these levels of concern. The simple linear regression guidelines are based on having $p = 2$. We are comparing $SE(Surface)/\sqrt{MSE}$ to the square roots of these guidelines. In our example, $p = 6$ and $n = 20$, so

$$\frac{SE(Surface)}{\sqrt{MSE}} = \frac{1.577}{\sqrt{4.303}} = .760 < .775 = \sqrt{\frac{2p}{n}}.$$

The location of this prediction is near the boundary of those locations for which we feel comfortable making predictions.

MINITAB COMMANDS

To obtain the predictions in this section, use the Minitab subcommand 'predict.'

```
MTB > names c1 'x1' c2 'x2' c3 'x3' c4 'x4' c5 'x5' c6 'y'
MTB > regress c6 on 5 c1-c5;
SUBC> predict 2.07 9.99 -16.04 21.6 5.17 .
```

13.3 Comparing regression models

A frequent goal in regression analysis is to find the simplest model that provides an adequate explanation of the data. In examining the full model with all five x variables, there is little evidence that any of x_1, x_2, or x_5 are needed in the regression model. The t tests reported in Section 13.1 for the corresponding regression parameters gave P values of .168, .427 and .387. We could drop *any one* of the three variables without significantly harming the model. While this does not imply that all three variables can be dropped without harming the model, dropping the three variables makes an interesting point of departure.

Fitting the reduced model

$$y_i = \beta_0 + \beta_3 x_{i3} + \beta_4 x_{i4} + \varepsilon_i$$

gives

Predictor	$\hat{\beta}_k$	$SE(\hat{\beta}_k)$	t	P
Constant	14.583	9.175	1.59	0.130
x_3	0.54156	0.05004	10.82	0.000
x_4	0.7499	0.3666	2.05	0.057

Analysis of Variance

Source	df	SS	MS	F	P
Regression	2	570.50	285.25	66.95	0.000
Error	17	72.43	4.26		
Total	19	642.92			

We can test whether this reduced model is an adequate explanation of the data as compared to the full model. The sum of squares for error from the full model was reported in Section 13.1 as $SSE(F) = 60.24$ with degrees of freedom $dfE(F) = 14$ and mean squared error $MSE(F) = 4.30$. For the reduced model we have $SSE(R) = 72.43$ and $dfE(R) = 17$. The test statistic for the adequacy of the reduced model is

$$F = \frac{[SSE(R) - SSE(F)]/[dfE(R) - dfE(F)]}{MSE(F)} = \frac{[72.43 - 60.24]/[17 - 14]}{4.30} = 0.94.$$

F has $[dfE(R) - dfE(F)]$ and $dfE(F)$ degrees of freedom in the numerator and denominator, respectively. Here F is less than 1, so it is not significant. In particular, 0.94 is less than $F(.95, 3, 14)$, so a formal $\alpha = .05$ level F test does not reject the adequacy of the reduced model. In other words, the $.05$ level test of $H_0 : \beta_1 = \beta_2 = \beta_5 = 0$ is not rejected.

This test lumps the three variables x_1, x_2, and x_5 together into one big test. It is possible that the uselessness of two of these variables could hide the fact that one of them is (marginally) significant when added to the model with x_3 and x_4. To fully examine this possibility, we need to fit three additional models. Each variable should be added, in turn, to the model with x_3 and x_4. We consider in detail only one of these three models, the model with x_1, x_3, and x_4. From fitting this model, the t statistic for testing whether x_1 is needed in the model turns out to be -1.47. This has a P value of .162, so there is little indication that x_1 is useful. We could also construct an F statistic as illustrated previously. The sum of squares for error in the model with x_1, x_3, and x_4 is 63.84 on 16 degrees of freedom, so

$$F = \frac{[72.43 - 63.84]/[17 - 16]}{63.84/16} = 2.16 .$$

Note that, up to round off error, $F = t^2$. The tests are equivalent and the P value for the F statistic is also .162. F tests are only equivalent to a corresponding t test when the numerator of the F statistic has one degree of freedom. Methods similar to these establish that neither x_2 nor x_5 are important when added to the model that contains x_3 and x_4.

In testing the reduced model with only x_3 and x_4 against the full five-variable model, we observed that one might miss recognizing a variable that was (marginally) significant. In this case we did not miss anything important. However, if we had taken the reduced model as containing only x_3 and tested it against the full five-variable model, we would have missed the importance of x_4. The F statistic for this test turns out to be only 1.74.

In the model with x_1, x_3, and x_4, the t test for x_4 turns out to have a P value of .021. As seen in the table given previously, if we drop x_1 and use the model with only x_3, and x_4, the P value for x_4 goes to .057. Thus dropping a weak variable, x_1, can make a reasonably strong variable, x_4, look weaker. There is a certain logical inconsistency here. If x_4 is important in the x_1, x_3, x_4 model or the full five-variable model (P value .023), it is illogical that dropping some of the other variables could make it unimportant. Even though x_1 is not particularly important by itself, it augments the evidence that x_4 is useful. The problem in these apparent inconsistencies is that the x variables are all related to each other, this is known as the problem of *collinearity*.

Although a reduced model may be an adequate substitute for a full model on a particular set of data, it does *not* follow that the reduced model will be an adequate substitute for the full model with any data collected on the variables in the full model.

GENERAL DISCUSSION

Suppose that we want to compare two regression models, say,

$$y_i = \beta_0 + \beta_1 x_{i1} + \cdots + \beta_{q-1} x_{i,q-1} + \cdots + \beta_{p-1} x_{i,p-1} + \varepsilon_i \qquad (13.3.1)$$

and

$$y_i = \beta_0 + \beta_1 x_{i1} + \cdots + \beta_{q-1} x_{i,q-1} + \varepsilon_i . \qquad (13.3.2)$$

For convenience, in this subsection we refer to equations such as (13.3.1) and (13.3.2) simply as (1) and (2). The key fact here is that all of the variables in model (2) are also in model (1). In this comparison, we dropped the last variables $x_{i,q}, \ldots, x_{i,p-1}$ for notational convenience only; the discussion applies to dropping any group of variables from model (1). *Throughout, we assume that model (1) gives an adequate fit to the data and then compare how well model (2) fits the data with how well model (1) fits*. Before applying the results of this subsection, the validity of the model (1) assumptions should be evaluated.

We want to know if the variables $x_{i,q}, \ldots, x_{i,p-1}$ are needed in the model, i.e., whether they are useful predictors. In other words, we want to know if model (2) is an adequate model, whether it gives an adequate explanation of the data. The variables $x_q, \ldots, x_{p-1}$ are extraneous if and only if $\beta_q = \cdots = \beta_{p-1} = 0$. The test we develop can be considered as a test of

$$H_0 : \beta_q = \cdots = \beta_{p-1} = 0.$$

versus

$$H_A : \text{not all of } \beta_q, \ldots, \beta_{p-1} \text{ are } 0.$$

Parameters are very tricky things; you never get to see the value of a parameter. I strongly prefer the interpretation of testing one model against another model rather than the interpretation of testing whether $\beta_q = \cdots = \beta_{p-1} = 0$. If the assumption that model (1) is true fails, interpretations based on parameters have little meaning. In practice, useful regression models are rarely correct models, although they can be *very* good approximations. Typically, we do not really care whether model (1) is true, only whether it is useful, but dealing with parameters in an incorrect model becomes tricky.

In practice, we are looking for a (relatively) succinct way of summarizing the data. The smaller the model the more succinct the summarization. However, we do not want to eliminate useful explanatory variables, so we test the smaller (more succinct) model against the larger model to see if the smaller model gives up significant explanatory power. Note that the larger model always has at least as much explanatory power as the smaller model because the larger model includes all the variables in the smaller model plus some more.

We want to compare how well models (1) and (2) fit the data. A natural measure of how well any linear model fits a set of data is the sum of squared errors (*SSE*). Small values of the *SSE* indicate a good fit. Referring to model (1) as the full (*F*) model and model (2) as the reduced (*R*) model, we can compare how well they fit by the measure $SSE(R) - SSE(F)$. As mentioned, the full model contains the reduced model as a special case, so it has at least as much explanatory power as the reduced model. It follows that $SSE(R)$ is never less than $SSE(F)$.

We can base a formal test on the comparison of *SSE*s. The full model is assumed to fit, so $MSE(F)$ estimates σ^2. If the variables $x_q, \ldots, x_{p-1}$ do not add much explanatory power to the model, the errors for the full and reduced models should be about the same. In other words, if the reduced model holds, the $MSE(R)$ is also an estimate of σ^2. Break $SSE(R)$ into two parts

$$SSE(R) = SSE(F) + [SSE(R) - SSE(F)]. \tag{13.3.3}$$

If the reduced model fits, then $SSE(R) - SSE(F)$ divided by its degrees of freedom is also an estimate of σ^2. The degrees of freedom for $SSE(R) - SSE(F)$ are $dfE(R) - dfE(F)$, i.e., $(n-q) - (n-p) = p - q$. Conversely, if $\beta_q, \ldots, \beta_{p-1}$ are not all zero, $SSE(R) - SSE(F)$ measures how much the variables $x_q, \ldots, x_{p-1}$ add to the model. In this case, the difference $SSE(R) - SSE(F)$ will be biased upward and $[SSE(R) - SSE(F)] / (p - q)$ estimates something bigger than σ^2.

To repeat, we assume throughout that model (1) holds so that $MSE(F)$ is always an estimate of σ^2. If model (2) holds, $MSE(R)$ is also an estimate of σ^2. In any case, using equation (3) we can write

$$MSE(R) = \xi\, MSE(F) + (1 - \xi) \left(\frac{SSE(R) - SSE(F)}{p - q} \right)$$

where $\xi = (n - p) / (n - q)$ is between 0 and 1. The point is that when model (2) is correct, an estimate of σ^2, namely $MSE(R)$, has been written as a weighted average of another estimate of σ^2, $MSE(F)$, and the term $[SSE(R) - SSE(F)] / (p - q)$. If $MSE(R)$ and $MSE(F)$ are estimates of σ^2, then

$[SSE(R) - SSE(F)]/(p-q)$ must also be a legitimate estimate of σ^2. If model (2) does not hold, $MSE(R)$ estimates something larger than σ^2 and so must $[SSE(R) - SSE(F)]/(p-q)$. The F ratio

$$F \equiv \frac{[SSE(R) - SSE(F)]/(p-q)}{MSE(F)}$$

estimates the number 1 if model (2) holds because it is the ratio of two estimates of σ^2. The F ratio estimates something larger than 1 if model (2) does not hold. If the observed value of F is much larger than 1, it suggests that model (2) is not valid. An F ratio close to 1 suggests that model (2) is adequate. (It does not suggest that model (2) is correct, only that it is adequate for explaining these data.)

The problem remains to quantify what we mean by an F ratio much larger than 1. Even when model (2) is absolutely correct, the variability in the data causes variability in the F ratio. By quantifying the variability in the F ratio when model (2) is correct, we get an idea of what F ratios are consistent with model (2) and what F values are so large as to be inconsistent with model (2). In particular, if model (2) is correct with independent homoscedastic normally distributed errors, the F statistic has an F distribution. The F distribution depends on two parameters, the degrees of freedom for the estimate of σ^2 in the numerator of the F ratio, $p-q$, and the degrees of freedom for the estimate of σ^2 in the denominator, $n-p$.

Applying this discussion to the model comparison problem yields the following test: Reject the hypothesis

$$H_0 : \beta_q = \cdots = \beta_{p-1} = 0$$

in favor of

$$H_A : \text{not all of } \beta_q, \ldots, \beta_{p-1} \text{ are } 0$$

at the α level if

$$F \equiv \frac{[SSE(R) - SSE(F)]/(p-q)}{MSE(F)} > F(1-\alpha, p-q, n-p).$$

Note that we reject the adequacy of model (2) when the observed F ratio is larger than $(1-\alpha)100$ percent of the F ratios that occur when model (2) holds.

The notation $SSE(R) - SSE(F)$ focuses on the ideas of full and reduced models. Other notations that focus on variables and parameters are also commonly used. One can view the model comparison procedure as fitting model (2) first and then seeing how much better model (1) fits. The notation based on this refers to the (extra) *sum of squares for regressing* on $x_q, \ldots, x_{p-1}$ *after* regressing on $x_1, \ldots, x_{q-1}$ and is written

$$SSR(x_q, \ldots, x_{p-1}|x_1, \ldots, x_{q-1}) \equiv SSE(R) - SSE(F).$$

This notation assumes that the model contains an intercept. Alternatively, one can think of fitting the parameters $\beta_q, \ldots, \beta_{p-1}$ after fitting the parameters $\beta_0, \ldots, \beta_{q-1}$. The relevant notation refers to the *reduction in sum of squares* (for error) due to fitting $\beta_q, \ldots, \beta_{p-1}$ after $\beta_0, \ldots, \beta_{q-1}$ and is written

$$R(\beta_q, \ldots, \beta_{p-1}|\beta_0, \ldots, \beta_{q-1}) \equiv SSE(R) - SSE(F).$$

Note that it makes perfect sense to refer to $SSR(x_q, \ldots, x_{p-1}|x_1, \ldots, x_{q-1})$ as the reduction in sum of squares for fitting $x_q, \ldots, x_{p-1}$ after $x_1, \ldots, x_{q-1}$.

It was mentioned earlier that the degrees of freedom for $SSE(R) - SSE(F)$ is $p-q$. Note that $p-q$ is the number of variables to the left of the vertical bar in $SSR(x_q, \ldots, x_{p-1}|x_1, \ldots, x_{q-1})$ and the number of parameters to the left of the vertical bar in $R(\beta_q, \ldots, \beta_{p-1}|\beta_0, \ldots, \beta_{q-1})$.

A point that is quite clear when thinking of model comparisons is that if you change either model (1) or (2), the test statistic and thus the test changes. This point continues to be clear when dealing with the notations $SSR(x_q, \ldots, x_{p-1}|x_1, \ldots, x_{q-1})$ and $R(\beta_q, \ldots, \beta_{p-1}|\beta_0, \ldots, \beta_{q-1})$. If you change any variable on either side of the vertical bar, you change $SSR(x_q, \ldots, x_{p-1}|x_1, \ldots, x_{q-1})$. Similarly, the parametric notation $R(\beta_q, \ldots, \beta_{p-1}|\beta_0, \ldots, \beta_{q-1})$ is also perfectly precise, but confusion can easily arise when dealing with parameters if one is not careful. For example, when testing, say, $H_0: \beta_1 = \beta_3 = 0$ versus the alternative that they are not both zero, the tests are completely different in the three models

$$y_i = \beta_0 + \beta_1 x_{i1} + \beta_3 x_{i3} + \varepsilon_i, \tag{13.3.4}$$

$$y_i = \beta_0 + \beta_1 x_{i1} + \beta_2 x_{i2} + \beta_3 x_{i3} + \varepsilon_i, \tag{13.3.5}$$

and

$$y_i = \beta_0 + \beta_1 x_{i1} + \beta_2 x_{i2} + \beta_3 x_{i3} + \beta_4 x_{i4} + \varepsilon_i . \tag{13.3.6}$$

In model (4) the test is based on $SSR(x_1, x_3) \equiv R(\beta_1, \beta_3|\beta_0)$, i.e., the sum of squares for regression ($SSReg$) in the model with only x_1 and x_3 as predictor variables. In model (5) the test uses

$$SSR(x_1, x_3|x_2) \equiv R(\beta_1, \beta_3|\beta_0, \beta_2).$$

Model (6) uses $SSR(x_1, x_3|x_2, x_4) \equiv R(\beta_1, \beta_3|\beta_0, \beta_2, \beta_4)$. In all cases we are testing $\beta_1 = \beta_3 = 0$ *after* fitting all the other parameters in the model. In general, we think of testing $H_0: \beta_q = \cdots = \beta_{p-1} = 0$ after fitting $\beta_0, \ldots, \beta_{q-1}$.

If the reduced model is obtained by dropping out only one variable, e.g., if $q - 1 = p - 2$, the parametric hypothesis is $H_0: \beta_{p-1} = 0$ versus $H_A: \beta_{p-1} \neq 0$. We have just developed an F test for this and we have earlier used a t test for the hypothesis. In multiple regression, just as in simple linear regression, the F test is equivalent to the t test. It follows that the t test must be considered as a test for the parameter *after fitting all* of the other parameters in the model. In particular, the t tests reported when fitting a regression tell you only whether a variable can be dropped relative to the model that contains all the other variables. These t tests cannot tell you whether more than one variable can be dropped from the fitted model. If you drop any variable from a regression model, all of the t tests change. It is only for notational convenience that we are discussing testing $\beta_{p-1} = 0$; the results hold for all β_k.

The SSR notation can also be used to find SSEs. Consider models (4), (5), and (6) and suppose we know $SSR(x_2|x_1, x_3)$, $SSR(x_4|x_1, x_2, x_3)$, and the SSE from model (6). We can easily find the SSEs for models (4) and (5). By definition,

$$\begin{aligned} SSE(5) &= [SSE(5) - SSE(6)] + SSE(6) \\ &= SSR(x_4|x_1, x_2, x_3) + SSE(6). \end{aligned}$$

Also

$$\begin{aligned} SSE(4) &= [SSE(4) - SSE(5)] + SSE(5) \\ &= SSR(x_2|x_1, x_3) + \{SSR(x_4|x_1, x_2, x_3) + SSE(6)\} . \end{aligned}$$

Moreover, we see that

$$\begin{aligned} SSR(x_2, x_4|x_1, x_3) &= SSE(4) - SSE(6) \\ &= SSR(x_2|x_1, x_3) + SSR(x_4|x_1, x_2, x_3). \end{aligned}$$

Note also that we can change the order of the variables.

$$SSR(x_2, x_4|x_1, x_3) = SSR(x_4|x_1, x_3) + SSR(x_2|x_1, x_3, x_4).$$

13.4 Sequential fitting

Multiple regression analysis is largely impractical without the aid of a computer. One specifies a regression model and the computer returns the vital statistics for that model. Many computer programs actually fit a sequence of models rather than fitting the model all at once.

EXAMPLE 13.4.1. Suppose you want to fit the model

$$y_i = \beta_0 + \beta_1 x_{i1} + \beta_2 x_{i2} + \beta_3 x_{i3} + \beta_4 x_{i4} + \varepsilon_i.$$

Many regression programs actually fit the sequence of models

$$\begin{aligned}
y_i &= \beta_0 + \beta_1 x_{i1} + \varepsilon_i, \\
y_i &= \beta_0 + \beta_1 x_{i1} + \beta_2 x_{i2} + \varepsilon_i, \\
y_i &= \beta_0 + \beta_1 x_{i1} + \beta_2 x_{i2} + \beta_3 x_{i3} + \varepsilon_i, \\
y_i &= \beta_0 + \beta_1 x_{i1} + \beta_2 x_{i2} + \beta_3 x_{i3} + \beta_4 x_{i4} + \varepsilon_i.
\end{aligned}$$

The sequence is determined by the order in which the variables are specified. If the identical model is specified in the form

$$y_i = \beta_0 + \beta_3 x_{i3} + \beta_1 x_{i1} + \beta_4 x_{i4} + \beta_2 x_{i2} + \varepsilon_i,$$

the end result is exactly the same but the sequence of models is

$$\begin{aligned}
y_i &= \beta_0 + \beta_3 x_{i3} + \varepsilon_i, \\
y_i &= \beta_0 + \beta_3 x_{i3} + \beta_1 x_{i1} + \varepsilon_i, \\
y_i &= \beta_0 + \beta_3 x_{i3} + \beta_1 x_{i1} + \beta_4 x_{i4} + \varepsilon_i, \\
y_i &= \beta_0 + \beta_3 x_{i3} + \beta_1 x_{i1} + \beta_4 x_{i4} + \beta_2 x_{i2} + \varepsilon_i.
\end{aligned}$$

Frequently, programs that fit sequences of models also provide sequences of sums of squares. Thus the first sequence of models yields

$$SSR(x_1),\ SSR(x_2|x_1),\ SSR(x_3|x_1,x_2),\ \text{and } SSR(x_4|x_1,x_2,x_3)$$

while the second sequence yields

$$SSR(x_3),\ SSR(x_1|x_3),\ SSR(x_4|x_3,x_1),\ \text{and } SSR(x_2|x_3,x_1,x_4).$$

These can be used in a variety of ways. For example, as shown at the end of the previous section, to test

$$y_i = \beta_0 + \beta_1 x_{i1} + \beta_3 x_{i3} + \varepsilon_i$$

against

$$y_i = \beta_0 + \beta_1 x_{i1} + \beta_2 x_{i2} + \beta_3 x_{i3} + \beta_4 x_{i4} + \varepsilon_i$$

we need $SSR(x_2,x_4|x_3,x_1)$. This is easily obtained from the second sequence as

$$SSR(x_2,x_4|x_3,x_1) = SSR(x_4|x_3,x_1) + SSR(x_2|x_3,x_1,x_4).$$

□

EXAMPLE 13.4.2. If we fit the model

$$y_i = \beta_0 + \beta_1 x_{i1} + \beta_2 x_{i2} + \beta_3 x_{i3} + \beta_4 x_{i4} + \beta_5 x_{i5} + \varepsilon_i$$

to the school data, we get the sequential sums of squares listed below.

Source	df	$Seq\ SS$	Notation
x_1	1	23.77	$SSR(x_1)$
x_2	1	343.23	$SSR(x_2 \vert x_1)$
x_3	1	186.34	$SSR(x_3 \vert x_1, x_2)$
x_4	1	25.91	$SSR(x_4 \vert x_1, x_2, x_3)$
x_5	1	3.43	$SSR(x_5 \vert x_1, x_2, x_3, x_4)$

Recall that the *MSE* for the five-variable model is 4.30 on 14 degrees of freedom.

From the sequential sums of squares we can test a variety of hypotheses related to the full model. For example, we can test whether variable x_5 can be dropped from the five-variable model. The F statistic is $3.43/4.30$, which is less than 1, so the effect of x_5 is insignificant. This test is equivalent to the t test for x_5 given in Section 13.1 when fitting the five-variable model. We can also test whether we can drop both x_4 and x_5 from the full model. The F statistic is

$$F = \frac{(25.91 + 3.43)/2}{4.30} = 3.41.$$

$F(.95, 2, 14) = 3.74$, so this F statistic provides little evidence that the pair of variables is needed. Similar tests can be constructed for dropping x_3, x_4, and x_5, for dropping x_2, x_3, x_4, and x_5, and for dropping x_1, x_2, x_3, x_4, and x_5 from the full model. The last of these is just the ANOVA table F test.

We can also make a variety of tests related to 'full' models that do not include all five variables. In the previous paragraph, we found little evidence that the pair x_4 and x_5 help explain the data in the five-variable model. We now test whether x_4 can be dropped when we have already dropped x_5. In other words, we test whether x_4 adds explanatory power to the model that contains x_1, x_2, and x_3. The numerator has one degree of freedom and is $SSR(x_4|x_1, x_2, x_3) = 25.91$. The usual denominator mean square for this test is the *MSE* from the model with x_1, x_2, x_3, and x_4, i.e., $\{14(4.303) + 3.43\}/15$. (For numerical accuracy we have added another significant digit to the *MSE* from the five-variable model. The *SSE* from the model without x_5 is just the *SSE* from the five-variable model plus the sequential sum of squares $SSR(x_5|x_1, x_2, x_3, x_4)$.) Alternatively, we could construct the test using the same numerator mean square but the *MSE* from the five-variable model in the denominator of the test. Using this second denominator, the F statistic is $25.91/4.30 = 6.03$. Corresponding F percentiles are $F(.95, 1, 14) = 4.60$ and $F(.99, 1, 14) = 8.86$, so x_4 may be contributing to the model. If we had used the *MSE* from the model with x_1, x_2, x_3, and x_4, the F statistic would be equivalent to the t statistic for dropping x_4 that is obtained when fitting this four-variable model.

If we wanted to test whether x_2 and x_3 can be dropped from the model that contains x_1, x_2, and x_3, the usual denominator is $[14(4.303) + 25.91 + 3.43]/16 = 5.60$. (The *SSE* for the model without x_4 or x_5 is just the *SSE* from the five-variable model plus the sequential sum of squares for x_4 and x_5.) Again, we could alternatively use the *MSE* from the five-variable model in the denominator. Using the first denominator, the test uses

$$F = \frac{(343.23 + 186.34)/2}{5.60} = 47.28.$$

This is much larger than $F(.999, 2, 16) = 10.97$, so there is overwhelming evidence that variables x_2 and x_3 cannot be dropped from the x_1, x_2, x_3 model.

The argument for basing tests on the *MSE* from the five-variable model is that it is less subject to bias than the other *MSE*s. In the test given in the previous paragraph, the *MSE* from the usual 'full' model incorporates the sequential sums of squares for x_4 and x_5. A reason for doing this is that we have tested x_4 and x_5 and are not convinced that they are important. As a result, their sums of squares are incorporated into the error. Even though we may not have established an overwhelming case for the importance of either variable, there is some evidence that x_4 is a useful predictor when added to the first three variables. The sum of squares for x_4 may or may not be large enough to convince us of its importance but it is large enough to change the *MSE* from 4.30 in the five-variable model to 5.60 in the x_1, x_2, x_3 model. In general, if you test terms and pool them with the error whenever the test is insignificant, you are biasing the *MSE* that results from this pooling. □

In general, *when given the ANOVA table and the sequential sums of squares, we can test any model in the sequence against any reduced model that is part of the sequence. We* cannot *use these statistics to obtain a test involving a model that is not part of the sequence.*

13.5 Reduced models and prediction

Fitted regression models are, not surprisingly, very dependent on the observed values of the predictor variables. We have already discussed the fact that fitted regression models are particularly good for making predictions but only for making predictions on new cases with predictor variables that are similar to those used in fitting the model. Fitted models are not good at predicting observations with predictor variable values that are far from those in the observed data. We have also discussed the fact that in evaluating a reduced model we are evaluating whether the reduced model is an adequate explanation of the data. An adequate reduced model should serve well as a prediction equation but only for new cases with predictor variables similar to those in the original data. It should not be overlooked that *when using a reduced model for prediction, new cases need to be similar to the observed data on* all *predictor variables and not just on the predictor variables in the reduced model.*

Good prediction from reduced models requires that new cases be similar to observed cases on all predictor variables because of the process of selecting reduced models. Predictor variables are eliminated from a model if they are not necessary to explain the data. This can happen in two ways. If a predictor variable is truly unrelated to the dependent variable, it is both proper and beneficial to eliminate that variable. The other possibility is that a predictor variable may be related to the dependent variable but that the relationship is hidden by the nature of the observed predictor variables. In the Mosteller and Tukey data, suppose the true response depends on both x_3 and x_5. We know that x_3 is clearly the best single predictor but the observed values of x_5 and x_3 are closely related; the sample correlation between them is .819. Because of their high correlation *in these data*, much of the actual dependence of y on x_5 could be accounted for by the regression on x_3 alone. Variable x_3 acts as a surrogate for x_5. As long as we try to predict new cases that have values of x_5 and x_3 similar to those in the original data, a reduced model based on x_3 should work well. Variable x_3 should continue to act as a surrogate. On the other hand, if we tried to predict a new case that had an x_3 value similar to that in the observed data but where the pair x_3, x_5 was not similar to x_3, x_5 pairs in the observed data, the reduced model that uses x_3 as a surrogate for x_5 would be inappropriate. Predictions could be very bad and, if we thought only about the fact that the x_3 value is similar to those in the original data, we might expect the predictions to be good. Unfortunately, when we eliminate a variable from a regression model, we typically have no idea if it is eliminated because the variable really has no effect on y or because its effect is being masked by some other set of predictor variables. For further discussion of these issues see Mandel (1989a, b).

Of course there is reason to hope that predictions will typically work well for reduced models. If the data come from an observational study in which the cases are some kind of sample from a population, there is reason to expect that future cases that are *sampled in the same way* will behave similarly to those in the original study. In addition, if the data come from an experiment in which the predictor variables are under the control of the investigator, it is reasonable to expect the investigator to select values of the predictor variables that cover the full range over which predictions will be made. Nonetheless, regression models give good approximations and good predictions only within the range of the observed data and, when a reduced model is used, the definition of the range of the observed data includes the values of all predictor variables that were in the full model. In fact, *even this statement is too weak.* When using a reduced model or even when using the full model for prediction, *new cases need to be similar to the observed cases in all relevant ways.* If there is some unmeasured predictor that is related to y and if the observed predictors are highly correlated with this unmeasured variable, then for good prediction a new case needs to have a value of the unmeasured variable that is similar to those for the observed cases. In other words, *the variables in any model may be acting as surrogates for some unmeasured variables and to obtain good predictions the new cases must be similar on both the observed predictor variables and on these unmeasured variables.*

13.6 Partial correlation coefficients and added variable plots

Partial correlation coefficients measure the linear relationship between two variables after adjusting for a group of other variables. The square of a partial correlation coefficient is also known as a *coefficient of partial determination.* The squared sample partial correlation coefficient between y and x_1 after adjusting for x_2, x_3, and x_4 is

$$r^2_{y1\bullet 234} = \frac{SSR(x_1|x_2,x_3,x_4)}{SSE(x_2,x_3,x_4)},$$

where $SSE(x_2,x_3,x_4)$ is the sum of squares error from a model with an intercept and the three predictors x_2,x_3,x_4. The squared sample partial correlation coefficient between y and x_2 given x_1, x_3, and x_4 is

$$r^2_{y2\bullet 134} = \frac{SSR(x_2|x_1,x_3,x_4)}{SSE(x_1,x_3,x_4)}.$$

Alternatively, the sample partial correlation $r_{y2\bullet 134}$ is precisely the ordinary sample correlation computed between the residuals from fitting

$$y_i = \beta_0 + \beta_1 x_{i1} + \beta_3 x_{i3} + \beta_4 x_{i4} + \varepsilon_i \tag{13.6.1}$$

and the residuals from fitting

$$x_{i2} = \gamma_0 + \gamma_1 x_{i1} + \gamma_3 x_{i3} + \gamma_4 x_{i4} + \varepsilon_i. \tag{13.6.2}$$

The residuals are $y_i - \hat{y}_i$ and $x_{i2} - \hat{x}_{i2}$, where $\hat{y}_i$ was defined near the end of Section 13.2 and $\hat{x}_{i2}$ is defined similarly.

The information in $r^2_{y2\bullet 134}$ is equivalent to the information in the F statistic for testing $H_0: \beta_2 = 0$ versus $H_A: \beta_2 \neq 0$ in the model

$$y_i = \beta_0 + \beta_1 x_{i1} + \beta_2 x_{i2} + \beta_3 x_{i3} + \beta_4 x_{i4} + \varepsilon_i. \tag{13.6.3}$$

To see this, observe that

$$\begin{aligned} F &= \frac{SSR(x_2|x_1,x_3,x_4)/1}{[SSE(x_1,x_3,x_4) - SSR(x_2|x_1,x_3,x_4)]/(n-5)} \\ &= (n-5)\frac{SSR(x_2|x_1,x_3,x_4)/SSE(x_1,x_3,x_4)}{1 - SSR(x_2|x_1,x_3,x_4)/SSE(x_1,x_3,x_4)} \\ &= (n-5)\frac{r^2_{y2\bullet 134}}{1 - r^2_{y2\bullet 134}}. \end{aligned}$$

EXAMPLE 13.6.1. In the school data,

$$r_{y3\bullet 1245} = .8477.$$

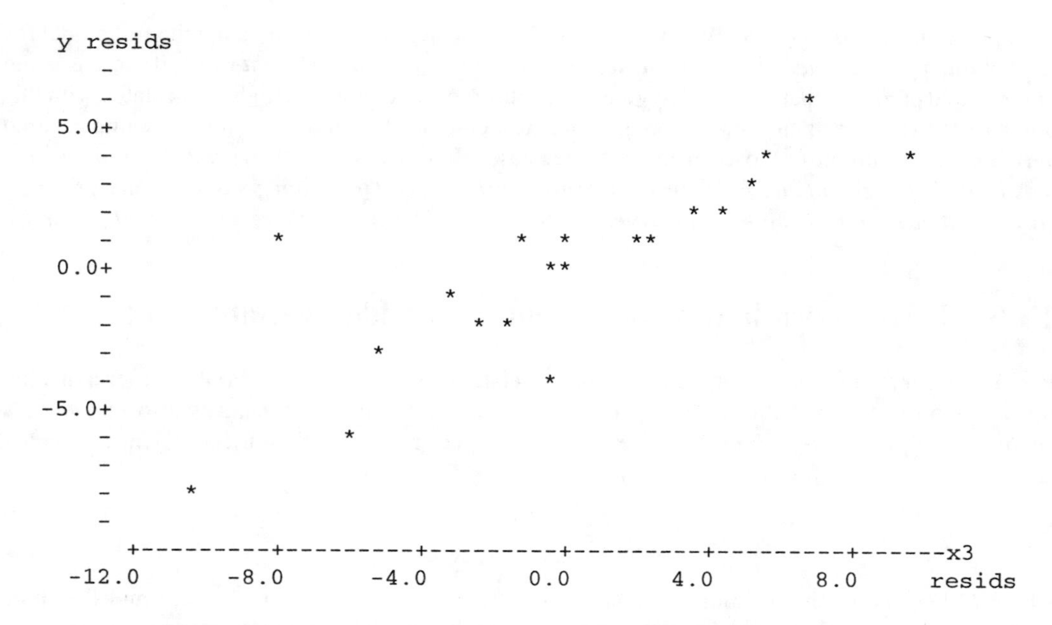

FIGURE 13.1. Added variable plot: y residuals versus x_3 residuals; school data.

Thus even after adjusting for all of the other predictor variables, socioeconomic status has major predictive abilities for mean verbal test scores. □

Actually, the residuals from models (13.6.1) and (13.6.2) give the basis for the perfect plot to evaluate whether adding variable x_2 will improve model (13.6.1). Simply plot the $y_i - \hat{y}_i$s from model (13.6.1) against the $x_{i2} - \hat{x}_{i2}$s from model (13.6.2). If there seems to be no relationship between the $y_i - \hat{y}_i$s and the $x_{i2} - \hat{x}_{i2}$s, x_2 will not be important in model (13.6.3). If the plot looks clearly linear, x_2 will be important in model (13.6.3). When a linear relationship exists in the plot but is due to the existence of a few points, those points are the dominant cause for x_2 being important in model (13.6.3). The reason these *added variable* plots work is because the least squares estimate of β_2 from model (13.6.3) is identical to the least squares estimate of β_2 from the regression through the origin

$$(y_i - \hat{y}_i) = \beta_2(x_{i2} - \hat{x}_{i2}) + \varepsilon_i,$$

see Christensen (1987, exercise 9.2).

EXAMPLE 13.6.2. For the school data, Figure 13.1 gives the added variable plot to determine whether the variable x_3 adds to the model that already contains x_1, x_2, x_4, and x_5. A clear linear relationship exists, so x_3 will improve the model. Here the entire data support the linear relationship, but there are a couple of unusual cases. The second smallest x_3 residual has an awfully large y residual and the largest x_3 residual has a somewhat surprisingly small y residual. □

13.7 Collinearity

Collinearity exists when the predictor variables $x_1, \ldots, x_{p-1}$ are correlated. We have n observations on each of these variables, so we can compute the sample correlations between them. Since the x variables are assumed to be fixed and not random, there is some question as to what a correlation between two x variables means. Actually, we are concerned with whether the observed variables are *orthogonal*, but this turns out to be equivalent to having sample correlations of zero between the x variables. Nonzero sample correlations indicate nonorthogonality. Thus we need not concern ourselves with the interpretation of sample correlations between nonrandom samples.

In regression, it is almost unheard of to have x variables that display no collinearity (correlation). In other words, observed x variables are almost never orthogonal. The key ideas in dealing with collinearity were previously incorporated into the discussion of comparing regression models. In fact, the methods discussed earlier were built around dealing with the collinearity of the x variables. This section merely reviews a few of the main ideas.

1. The estimate of any parameter, say $\hat{\beta}_2$, depends on *all* the variables that are included in the model.

2. The sum of squares for any variable, say x_2, depends on *all* the other variables that are included in the model. For example, none of $SSR(x_2)$, $SSR(x_2|x_1)$, and $SSR(x_2|x_3, x_4)$ would typically be equal.

3. Suppose the model
$$y_i = \beta_0 + \beta_1 x_{i1} + \beta_2 x_{i2} + \beta_3 x_{i3} + \varepsilon_i$$
is fitted and we obtain t statistics for each parameter. If the t statistic for testing $H_0 : \beta_1 = 0$ versus $H_A : \beta_1 \neq 0$ is small, we are led to the model
$$y_i = \beta_0 + \beta_2 x_{i2} + \beta_3 x_{i3} + \varepsilon_i.$$
If the t statistic for testing $H_0 : \beta_2 = 0$ versus $H_A : \beta_2 \neq 0$ is small, we are led to the model
$$y_i = \beta_0 + \beta_1 x_{i1} + \beta_3 x_{i3} + \varepsilon_i.$$
However, if the t statistics for both tests are small, we are *not* led to the model
$$y_i = \beta_0 + \beta_3 x_{i3} + \varepsilon_i.$$
To arrive at the model containing only the intercept and x_3, one must at some point use the model containing only the intercept and x_3 as a reduced model.

4. A moderate amount of collinearity has little effect on predictions and therefore little effect on SSE, R^2, and the explanatory power of the model. Collinearity increases the variance of the $\hat{\beta}_k$s, making the estimates of the parameters less reliable. (I told you not to rely on parameters anyway.) Depending on circumstances, sometimes a large amount of collinearity can have an effect on predictions. Just by chance, one may get a better fit to the data than can be justified scientifically.

The complications associated with points 1 through 4 all vanish if the sample correlations between the x variables are *all zero*.

Many computer programs will print out a matrix of correlations between the variables. One would like to think that if all the correlations between the x variables are reasonably small, say less than .3 or .4, then the problems of collinearity would not be serious. Unfortunately, that is simply not true. To avoid difficulties with collinearity, not only do all the correlations need to be small but *all of the partial correlations among the x variables must be small.* Thus, small correlations alone do not ensure small collinearity.

EXAMPLE 13.7.1. The correlations among predictors for the Coleman data are given below.

	x_1	x_2	x_3	x_4	x_5
x_1	1.000	0.181	0.230	0.503	0.197
x_2	0.181	1.000	0.827	0.051	0.927
x_3	0.230	0.827	1.000	0.183	0.819
x_4	0.503	0.051	0.183	1.000	0.124
x_5	0.197	0.927	0.819	0.124	1.000

Note that x_3 is highly correlated with x_2 and x_5. Since x_3 is highly correlated with y, the fact that x_2 and x_5 are also quite highly correlated with y is not surprising. Recall that the correlations with y were given at the beginning of Section 13.1. Moreover, since x_3 is highly correlated with x_2 and x_5, it is also not surprising that x_2 and x_5 have little to add to a model that already contains x_3. We have seen that it is the two variables x_1 and x_4, i.e., the variables that do not have high correlations with either x_3 or y, that have the greater impact on the regression equation.

Having regressed y on x_3, the sample correlations between y and any of the other variables are no longer important. Having done this regression, it is more germane to examine the partial correlations between y and the other variables that adjust for x_3. However, as we will see in our discussion of model selection in Chapter 14, even this has its drawbacks. □

As long as points 1 through 4 are kept in mind, a moderate amount of collinearity is not a big problem. For severe collinearity, there are four common approaches: a) classical ridge regression, b) generalized inverse regression, c) principal components regression, and 4) canonical regression. Classical ridge regression is probably the best known of these methods (and in my opinion, the worst). The other three methods are closely related and seem quite reasonable. Principal components regression is discussed in Chapter 15.

13.8 Exercises

EXERCISE 13.8.1. Younger (1979, p. 533) presents data from a sample of 12 discount department stores that advertize on television, radio, and in the newspapers. The variables x_1, x_2, and x_3 represent the respective amounts of money spent on these advertising activities during a certain month while y gives the store's revenues during that month. The data are given in Table 13.2. Complete the following tasks using multiple regression.

(a) Give the theoretical model along with the relevant assumptions.

(b) Give the fitted model, i.e., repeat (a) substituting the estimates for the unknown parameters.

(c) Test $H_0 : \beta_2 = 0$ versus $H_A : \beta_2 \neq 0$ at $\alpha = 0.05$.

(d) Test the hypothesis $H_0 : \beta_1 = \beta_2 = \beta_3 = 0$.

(e) Give a 99% confidence interval for β_2.

(f) Test whether the reduced model $y_i = \beta_0 + \beta_1 x_{i1} + \varepsilon_i$ is an adequate explanation of the data as compared to the full model.

(g) Test whether the reduced model $y_i = \beta_0 + \beta_1 x_{i1} + \varepsilon_i$ is an adequate explanation of the data as compared to the model $y_i = \beta_0 + \beta_1 x_{i1} + \beta_2 x_{i2} + \varepsilon_i$.

(h) Write down the ANOVA table for the 'full' model used in (g).

(i) Construct an added variable plot for adding variable x_3 to a model that already contains variables x_1 and x_2. Interpret the plot.

(j) Compute the sample partial correlation $r_{y3\bullet 12}$. What does this value tell you?

TABLE 13.2. Younger's advertising data

Obs.	y	x_1	x_2	x_3	Obs.	y	x_1	x_2	x_3
1	84	13	5	2	7	34	12	7	2
2	84	13	7	1	8	30	10	3	2
3	80	8	6	3	9	54	8	5	2
4	50	9	5	3	10	40	10	5	3
5	20	9	3	1	11	57	5	6	2
6	68	13	5	1	12	46	5	7	2

EXERCISE 13.8.2. The information below relates y, a second measurement on wood volume, to x_1, a first measurement on wood volume, x_2, the number of trees, x_3, the average age of trees, and x_4, the average volume per tree. Note that $x_4 = x_1/x_2$. Some of the information has not been reported, so that you can figure it out on your own.

Predictor	$\hat{\beta}_k$	$SE(\hat{\beta}_k)$	t	P
Constant	23.45	14.90		0.122
x_1	0.93209	0.08602		0.000
x_2		0.4721	1.5554	0.126
x_3	−0.4982	0.1520		0.002
x_4	3.486	2.274		0.132

Analysis of Variance

Source	df	SS	MS	F	P
Regression	4	887994			0.000
Error					
Total	54	902773			

Source	df	Sequential SS
x_1	1	883880
x_2	1	183
x_3	1	3237
x_4	1	694

(a) How many observations are in the data?

(b) What is R^2 for this model?

(c) What is the mean squared error?

(d) Give a 95% confidence interval for β_2.

(e) Test the null hypothesis $\beta_3 = 0$ with $\alpha = .05$.

(f) Test the null hypothesis $\beta_1 = 1$ with $\alpha = .05$.

(g) Give the F statistic for testing the null hypothesis $\beta_3 = 0$.

(h) Give $SSR(x_3|x_1, x_2)$ and find $SSR(x_3|x_1, x_2, x_4)$.

(i) Test the model with only variables x_1 and x_2 against the model with all of variables x_1, x_2, x_3, and x_4.

(j) Test the model with only variables x_1 and x_2 against the model with variables x_1, x_2, and x_3.

(k) Should the test in part (g) be the same as the test in part (j)? Why or why not?

(l) For estimating the point on the regression surface at $(x_1, x_2, x_3, x_4) = (100, 25, 50, 4)$, the standard error of the estimate for the point on the surface is 2.62. Give the estimated point on the surface, a 95% confidence interval for the point on the surface, and a 95% prediction interval for a new point with these x values.

(m) Test the null hypothesis $\beta_1 = \beta_2 = \beta_3 = \beta_4 = 0$ with $\alpha = .05$.

EXERCISE 13.8.3. Atkinson (1985) and Hader and Grandage (1958) have presented Prater's data on gasoline. The variables are y, the percentage of gasoline obtained from crude oil; x_1, the crude oil gravity ^{o}API; x_2, crude oil vapor pressure measured in lbs/in^2; x_3, the temperature, in ^{o}F, at which 10% of the crude oil is vaporized; and x_4, the temperature, in ^{o}F, at which all of the crude oil is vaporized. The data are given in Table 13.3. Find a good model for predicting gasoline yield from the other four variables.

TABLE 13.3. Prater's gasoline–crude oil data

y	x_1	x_2	x_3	x_4	y	x_1	x_2	x_3	x_4
6.9	38.4	6.1	220	235	24.8	32.2	5.2	236	360
14.4	40.3	4.8	231	307	26.0	38.4	6.1	220	365
7.4	40.0	6.1	217	212	34.9	40.3	4.8	231	395
8.5	31.8	0.2	316	365	18.2	40.0	6.1	217	272
8.0	40.8	3.5	210	218	23.2	32.2	2.4	284	424
2.8	41.3	1.8	267	235	18.0	31.8	0.2	316	428
5.0	38.1	1.2	274	285	13.1	40.8	3.5	210	273
12.2	50.8	8.6	190	205	16.1	41.3	1.8	267	358
10.0	32.2	5.2	236	267	32.1	38.1	1.2	274	444
15.2	38.4	6.1	220	300	34.7	50.8	8.6	190	345
26.8	40.3	4.8	231	367	31.7	32.2	5.2	236	402
14.0	32.2	2.4	284	351	33.6	38.4	6.1	220	410
14.7	31.8	0.2	316	379	30.4	40.0	6.1	217	340
6.4	41.3	1.8	267	275	26.6	40.8	3.5	210	347
17.6	38.1	1.2	274	365	27.8	41.3	1.8	267	416
22.3	50.8	8.6	190	275	45.7	50.8	8.6	190	407

EXERCISE 13.8.4. Dixon and Massey (1983) report data from the Los Angeles Heart Study supervised by J. M. Chapman. The variables are y, weight in pounds; x_1, age in years; x_2, systolic blood pressure in millimeters of mercury; x_3, diastolic blood pressure in millimeters of mercury; x_4, cholesterol in milligrams per dl; x_5, height in inches. The data from 60 men are given in Table 13.4. Find a good model for predicting weight from the other variables.

EXERCISE 13.8.5. Table 13.5 contains a subset of the pollution data analyzed by McDonald and Schwing (1973). The data are from various years in the early 1960s. They relate air pollution to

TABLE 13.4. L. A. heart study data

i	x_1	x_2	x_3	x_4	x_5	y	i	x_1	x_2	x_3	x_4	x_5	y
1	44	124	80	254	70	190	31	42	136	82	383	69	187
2	35	110	70	240	73	216	32	28	124	82	360	67	148
3	41	114	80	279	68	178	33	40	120	85	369	71	180
4	31	100	80	284	68	149	34	40	150	100	333	70	172
5	61	190	110	315	68	182	35	35	100	70	253	68	141
6	61	130	88	250	70	185	36	32	120	80	268	68	176
7	44	130	94	298	68	161	37	31	110	80	257	71	154
8	58	110	74	384	67	175	38	52	130	90	474	69	145
9	52	120	80	310	66	144	39	45	110	80	391	69	159
10	52	120	80	337	67	130	40	39	106	80	248	67	181
11	52	130	80	367	69	162	41	40	130	90	520	68	169
12	40	120	90	273	68	175	42	48	110	70	285	66	160
13	49	130	75	273	66	155	43	29	110	70	352	66	149
14	34	120	80	314	74	156	44	56	141	100	428	65	171
15	37	115	70	243	65	151	45	53	90	55	334	68	166
16	63	140	90	341	74	168	46	47	90	60	278	69	121
17	28	138	80	245	70	185	47	30	114	76	264	73	178
18	40	115	82	302	69	225	48	64	140	90	243	71	171
19	51	148	110	302	69	247	49	31	130	88	348	72	181
20	33	120	70	386	66	146	50	35	120	88	290	70	162
21	37	110	70	312	71	170	51	65	130	90	370	65	153
22	33	132	90	302	69	161	52	43	122	82	363	69	164
23	41	112	80	394	69	167	53	53	120	80	343	71	159
24	38	114	70	358	69	198	54	58	138	82	305	67	152
25	52	100	78	336	70	162	55	67	168	105	365	68	190
26	31	114	80	251	71	150	56	53	120	80	307	70	200
27	44	110	80	322	68	196	57	42	134	90	243	67	147
28	31	108	70	281	67	130	58	43	115	75	266	68	125
29	40	110	74	336	68	166	59	52	110	75	341	69	163
30	36	110	80	314	73	178	60	68	110	80	268	62	138

mortality rates for various standard metropolitan statistical areas in the United States. The dependent variable y is the total age-adjusted mortality rate per 100,000 as computed for different metropolitan areas. The predictor variables are, in order, mean annual precipitation in inches, mean January temperature in degrees F, mean July temperature in degrees F, population per household, median school years completed by those over 25, percent of housing units that are sound and with all facilities, population per sq. mile in urbanized areas, percent non-white population in urbanized areas, relative pollution potential of sulphur dioxide, annual average of percent relative humidity at 1 pm. Find a good predictive model for mortality.

Alternatively, you can obtain the complete data from the internet statistical service STATLIB by e-mailing to 'statlib@lib.stat.cmu.edu' the one line message

```
send pollution from datasets
```

The message you receive in return includes the data consisting of 16 variables on 60 cases.

EXERCISE 13.8.6. Obtain the data set 'bodyfat' from STATLIB by e-mailing the one line message

```
send bodyfat from datasets
```

to the address 'statlib@lib.stat.cmu.edu'. The message you receive in return includes data for 15 variables along with a description of the data.

TABLE 13.5. Pollution data

x_1	x_2	x_3	x_4	x_5	x_6	x_7	x_8	x_9	x_{10}	y
36	27	71	3.34	11.4	81.5	3243	8.8	42.6	59	921.870
35	23	72	3.14	11.0	78.8	4281	3.5	50.7	57	997.875
44	29	74	3.21	9.8	81.6	4260	.8	39.4	54	962.354
47	45	79	3.41	11.1	77.5	3125	27.1	50.2	56	982.291
43	35	77	3.44	9.6	84.6	6441	24.4	43.7	55	1071.289
53	45	80	3.45	10.2	66.8	3325	38.5	43.1	54	1030.380
43	30	74	3.23	12.1	83.9	4679	3.5	49.2	56	934.700
45	30	73	3.29	10.6	86.0	2140	5.3	40.4	56	899.529
36	24	70	3.31	10.5	83.2	6582	8.1	42.5	61	1001.902
36	27	72	3.36	10.7	79.3	4213	6.7	41.0	59	912.347
52	42	79	3.39	9.6	69.2	2302	22.2	41.3	56	1017.613
33	26	76	3.20	10.9	83.4	6122	16.3	44.9	58	1024.885
40	34	77	3.21	10.2	77.0	4101	13.0	45.7	57	970.467
35	28	71	3.29	11.1	86.3	3042	14.7	44.6	60	985.950
37	31	75	3.26	11.9	78.4	4259	13.1	49.6	58	958.839
35	46	85	3.22	11.8	79.9	1441	14.8	51.2	54	860.101
36	30	75	3.35	11.4	81.9	4029	12.4	44.0	58	936.234
15	30	73	3.15	12.2	84.2	4824	4.7	53.1	38	871.766
31	27	74	3.44	10.8	87.0	4834	15.8	43.5	59	959.221
30	24	72	3.53	10.8	79.5	3694	13.1	33.8	61	941.181
31	45	85	3.22	11.4	80.7	1844	11.5	48.1	53	891.708
31	24	72	3.37	10.9	82.8	3226	5.1	45.2	61	871.338
42	40	77	3.45	10.4	71.8	2269	22.7	41.4	53	971.122
43	27	72	3.25	11.5	87.1	2909	7.2	51.6	56	887.466
46	55	84	3.35	11.4	79.7	2647	21.0	46.9	59	952.529
39	29	75	3.23	11.4	78.6	4412	15.6	46.6	60	968.665
35	31	81	3.10	12.0	78.3	3262	12.6	48.6	55	919.729
43	32	74	3.38	9.5	79.2	3214	2.9	43.7	54	844.053
11	53	68	2.99	12.1	90.6	4700	7.8	48.9	47	861.833
30	35	71	3.37	9.9	77.4	4474	13.1	42.6	57	989.265
50	42	82	3.49	10.4	72.5	3497	36.7	43.3	59	1006.490
60	67	82	2.98	11.5	88.6	4657	13.5	47.3	60	861.439
30	20	69	3.26	11.1	85.4	2934	5.8	44.0	64	929.150
25	12	73	3.28	12.1	83.1	2095	2.0	51.9	58	857.622
45	40	80	3.32	10.1	70.3	2682	21.0	46.1	56	961.009
46	30	72	3.16	11.3	83.2	3327	8.8	45.3	58	923.234
54	54	81	3.36	9.7	72.8	3172	31.4	45.5	62	1113.156
42	33	77	3.03	10.7	83.5	7462	11.3	48.7	58	994.648
42	32	76	3.32	10.5	87.5	6092	17.5	45.3	54	1015.023
36	29	72	3.32	10.6	77.6	3437	8.1	45.5	56	991.290
37	38	67	2.99	12.0	81.5	3387	3.6	50.3	73	893.991
42	29	72	3.19	10.1	79.5	3508	2.2	38.8	56	938.500
41	33	77	3.08	9.6	79.9	4843	2.7	38.6	54	946.185
44	39	78	3.32	11.0	79.9	3768	28.6	49.5	53	1025.502
32	25	72	3.21	11.1	82.5	4355	5.0	46.4	60	874.281

(a) Using the body density measurements as a dependent variable, perform a multiple regression using all of the other variables except body fat as predictor variables. What variables can be safely eliminated from the analysis? Discuss any surprising or expected results in terms of the variables that seem to be most important.

(b) Using the body fat measurements as a dependent variable, perform a multiple regression using all of the other variables except density as predictor variables. What variables can be safely eliminated from the analysis? Discuss any surprising or expected results in terms of the variables that seem to be most important.

Chapter 14

Regression diagnostics and variable selection

In this chapter we continue our discussion of multiple regression. In particular, we focus on checking the assumptions of regression models by looking at diagnostic statistics. If problems with assumptions become apparent, one way to deal with them is to try transformations. The discussion of transformations in Section 7.10 continues to apply. Among the methods discussed there, only the circle of transformations depends on having a simple linear regression model. The other methods apply with multiple regression as well as analysis of variance models. In particular, the discussion of transforming x at the end of Section 7.10 takes on new importance in multiple regression because multiple regression involves several predictor variables, each of which is a candidate for transformation. Incidentally, the modified Box–Tidwell procedure evaluates each predictor variable separately, so it involves adding only one predictor variable $x_{ij}\log(x_{ij})$ to the multiple regression model at a time.

This chapter also examines methods for choosing good reduced models. Variable selection methods fall into two categories: best subset selection methods and stepwise regression methods. Finally, we examine the interplay between influential cases and model selection techniques. We continue to illustrate techniques on the data from *The Coleman Report* given in the previous chapter.

14.1 Diagnostics

Table 14.1 contains a variety of measures for checking the assumptions of the multiple regression model with five predictor variables that was fitted in Chapter 13 to the *Coleman Report* data. The table includes case indicators, the data y, the predicted values $\hat{y}$, the leverages, the standardized residuals r, the standardized deleted residuals t, and Cook's distances C. All of these, except for Cook's distance, were introduced in Section 7.9. Recall that leverages measure the distance between the predictor variables for a particular case and the rest of the predictor variables in the data. Cases with leverages near 1 dominate any fitted regression. As a rule of thumb, leverages greater than $2p/n$ cause concern and leverages greater than $3p/n$ cause (at least mild) consternation. Here n is the number of observations in the data and p is the number of regression parameters, including the intercept. The standardized deleted residuals t contain essentially the same information as the standardized residuals r but t values can be compared to a $t(dfE-1)$ distribution to obtain a formal test of whether a case is consistent with the other data. (A formal test based on the r values requires a more exotic distribution than the $t(dfE-1)$.) Cook's distance for case i is defined as

$$C_i = \frac{\sum_{j=1}^{n}\left(\hat{y}_j - \hat{y}_{j[i]}\right)^2}{pMSE}, \tag{14.1.1}$$

where $\hat{y}_j$ is the predictor of the jth case and $\hat{y}_{j[i]}$ is the predictor of the jth case when case i has been removed from the data. Cook's distance measures the effect of deleting case i on the prediction of all of the original observations.

TABLE 14.1. Diagnostics, full data

Case	y	$\hat{y}$	Leverage	r	t	C
1	37.01	36.66	0.482	0.23	0.23	0.008
2	26.51	26.86	0.486	−0.24	−0.23	0.009
3	36.51	40.46	0.133	−2.05	−2.35	0.107
4	40.70	41.17	0.171	−0.25	−0.24	0.002
5	37.10	36.32	0.178	0.42	0.40	0.006
6	33.90	33.99	0.500	−0.06	−0.06	0.001
7	41.80	41.08	0.239	0.40	0.38	0.008
8	33.40	33.83	0.107	−0.22	−0.21	0.001
9	41.01	40.39	0.285	0.36	0.34	0.008
10	37.20	36.99	0.618	0.16	0.16	0.007
11	23.30	25.51	0.291	−1.26	−1.29	0.110
12	35.20	33.45	0.403	1.09	1.10	0.133
13	34.90	35.95	0.369	−0.64	−0.62	0.040
14	33.10	33.45	0.109	−0.18	−0.17	0.001
15	22.70	24.48	0.346	−1.06	−1.07	0.099
16	39.70	38.40	0.157	0.68	0.67	0.014
17	31.80	33.24	0.291	−0.82	−0.81	0.046
18	31.70	26.70	0.326	2.94	4.56	0.694
19	43.10	41.98	0.285	0.64	0.63	0.027
20	41.01	40.75	0.223	0.14	0.14	0.001

Figures 14.1 and 14.2 are plots of the standardized residuals versus normal scores and against the predicted values. The largest standardized residual, that for case 18, appears to be somewhat unusually large. To test whether the data from case 18 are consistent with the other data, we can compare the standardized deleted residual to a $t(dfE-1)$ distribution. From Table 14.1, the t residual is 4.56. The corresponding P value for a two-sided test is .0006. Actually, we chose to perform the test on the t residual for case 18 because it was the largest of the 20 t residuals. Because the test is based on the largest of the t values, it is appropriate to multiply the P value by the number of t statistics considered. This gives $20 \times .0006 = .012$, which is still a very small P value. There is considerable evidence that the data of case 18 are inconsistent, for whatever reason, with the other data. This fact cannot be discovered from a casual inspection of the raw data.

The only point of any concern with respect to the leverages is case 10. Its leverage is .618, while $2p/n = .6$. This is only a mildly high leverage and case 10 seems well behaved in all other respects; in particular, C_{10} is small, so deleting case 10 has very little effect on predictions.

We now reconsider the analysis with case 18 deleted. The regression equation is

$$y = 34.3 - 1.62x_1 + 0.0854x_2 + 0.674x_3 + 1.11x_4 - 4.57x_5$$

and $R^2 = .963$. Table 14.2 contains the estimated regression coefficients, standard errors, t statistics, and P values. Table 14.3 contains the analysis of variance. Table 14.4 contains diagnostics. Note that

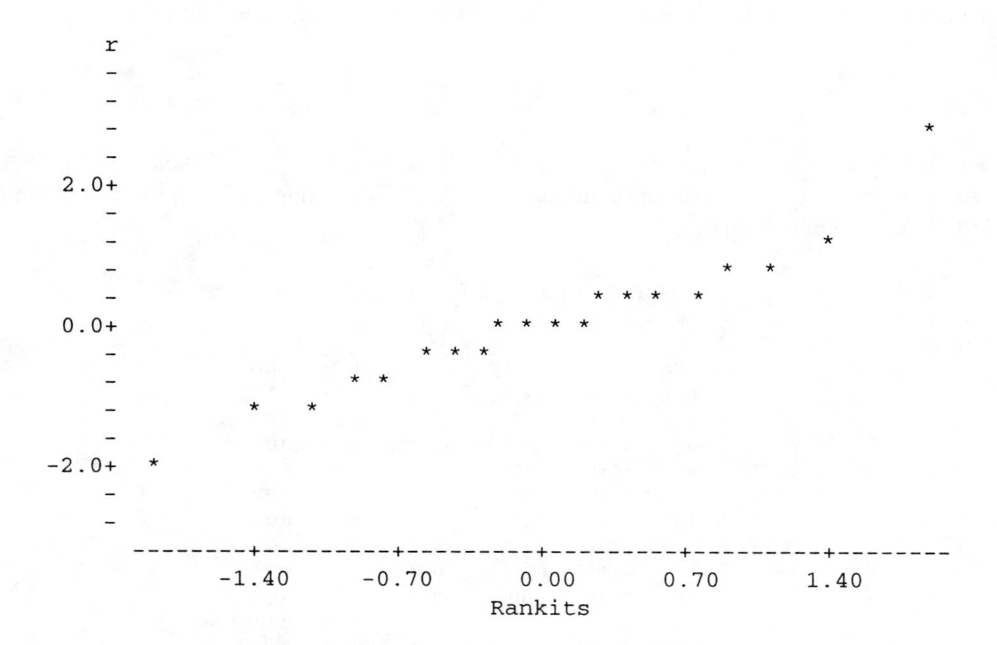

FIGURE 14.1. Normal plot, full data, $W' = 0.903$.

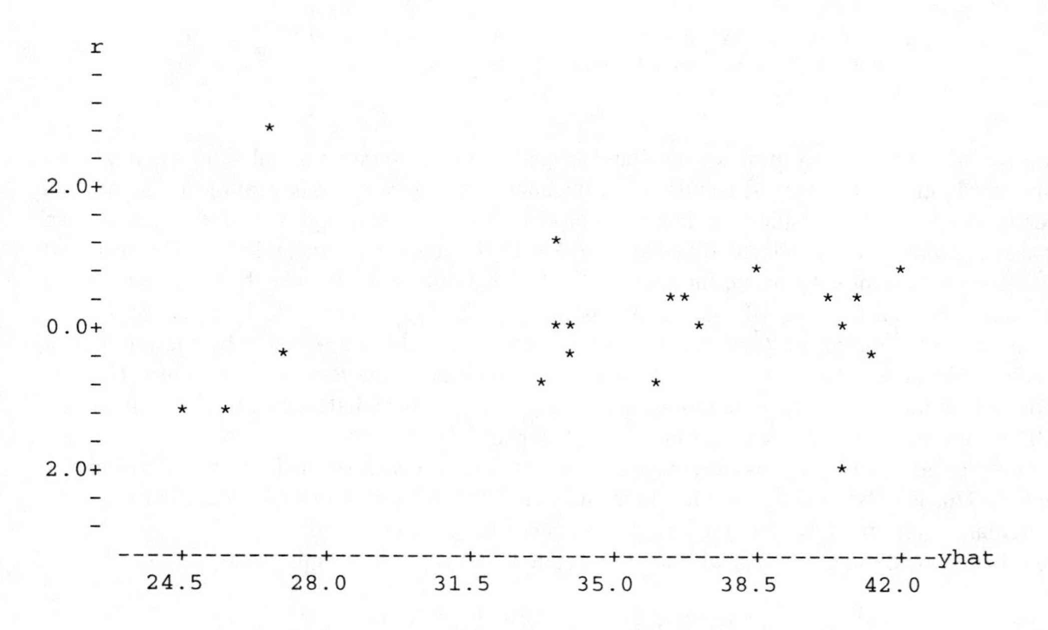

FIGURE 14.2. Standardized residuals versus predicted values.

the *MSE* is less than half of its previous value when case 18 was included in the analysis. Also, the regression parameter t statistics in Table 14.2 are all much more significant. The actual regression coefficients have changed a bit but not greatly. Predictions have not changed radically either, as can be seen by comparing the predictions given in Tables 14.1 and 14.4. Although the predictions have not changed radically, they have changed more than they would have if we deleted any observation other than case 18. From the definition of Cook's distance given in equation (14.1.1), C_{18} is precisely the sum of the squared differences between the predictions in Tables 14.1 and 14.4 divided by 6 times the *MSE* from the full data. From Table 14.1, Cook's distance when dropping case 18 is much larger than Cook's distance from dropping any other case.

TABLE 14.2. Regression, case 18 deleted

Predictor	$\hat{\beta}$	$SE(\hat{\beta})$	t	P
Constant	34.287	9.312	3.68	0.003
x_1	-1.6173	0.7943	-2.04	0.063
x_2	0.08544	0.03546	2.41	0.032
x_3	0.67393	0.06516	10.34	0.000
x_4	1.1098	0.2790	3.98	0.002
x_5	-4.571	1.437	-3.18	0.007

TABLE 14.3. Analysis of variance, case 18 deleted

Source	df	SS	MS	F	P
Regression	5	607.74	121.55	68.27	0.000
Error	13	23.14	1.78		
Total	18	630.88			

Consider again Table 14.4 containing the diagnostic statistics when case 18 has been deleted. Case 10 has moderately high leverage but seems to be no real problem. Figures 14.3 and 14.4 give the normal plot and the standardized residual versus predicted value plot, respectively, with case 18 deleted. Figure 14.4 is particularly interesting. At first glance, it appears to have a horn shape opening to the right. But there are only three observations on the left of the plot and many on the right, so one would *expect* a horn shape because of the data distribution. Looking at the right of the plot, we see that in spite of the data distribution, much of the horn shape is due to a single very small residual. If we mentally delete that residual, the remaining residuals contain a hint of an upward opening parabola. The potential outlier is case 3. From Table 14.4, the standardized deleted residual for case 3 is -5.08 which yields a two-sided raw P value of .0001 and if we adjust for having 19 t statistics, the P value is .0019, still an extremely small value. Note also that in Table 14.1, when case 18 was included in the data, the standardized deleted residual for case 3 was somewhat large but not nearly so extreme.

With cases 3 and 18 deleted, the regression equation becomes

$$y = 29.8 - 1.70x_1 + 0.0851x_2 + 0.666x_3 + 1.18x_4 - 4.07x_5.$$

The R^2 for these data is .988. The regression parameters are in Table 14.5, the analysis of variance is in Table 14.6, and the diagnostics are in Table 14.7.

Deleting the outlier, case 3, again causes a drop in the *MSE*, from 1.78 with only case 18 deleted to 0.61 with both cases 3 and 18 deleted. This creates a corresponding drop in the standard errors for all regression coefficients and makes them all appear to be more significant. The actual estimates of

TABLE 14.4. Diagnostics, case 18 deleted

Case	y	$\hat{y}$	Leverage	r	t	C
1	37.01	36.64	0.483	0.39	0.37	0.023
2	26.51	26.89	0.486	−0.39	−0.38	0.024
3	36.51	40.21	0.135	−2.98	−5.08	0.230
4	40.70	40.84	0.174	−0.12	−0.11	0.001
5	37.10	36.20	0.179	0.75	0.73	0.020
6	33.90	33.59	0.504	0.33	0.32	0.018
7	41.80	41.66	0.248	0.12	0.12	0.001
8	33.40	33.65	0.108	−0.20	−0.19	0.001
9	41.01	41.18	0.302	−0.15	−0.15	0.002
10	37.20	36.79	0.619	0.50	0.49	0.068
11	23.30	23.69	0.381	−0.37	−0.35	0.014
12	35.20	34.54	0.435	0.66	0.64	0.055
13	34.90	35.82	0.370	−0.87	−0.86	0.074
14	33.10	32.38	0.140	0.58	0.57	0.009
15	22.70	22.36	0.467	0.35	0.33	0.017
16	39.70	38.25	0.158	1.18	1.20	0.044
17	31.80	32.82	0.295	−0.91	−0.90	0.058
18		24.28	0.483			
19	43.10	41.44	0.292	1.48	1.56	0.151
20	41.01	41.00	0.224	0.00	0.00	0.000

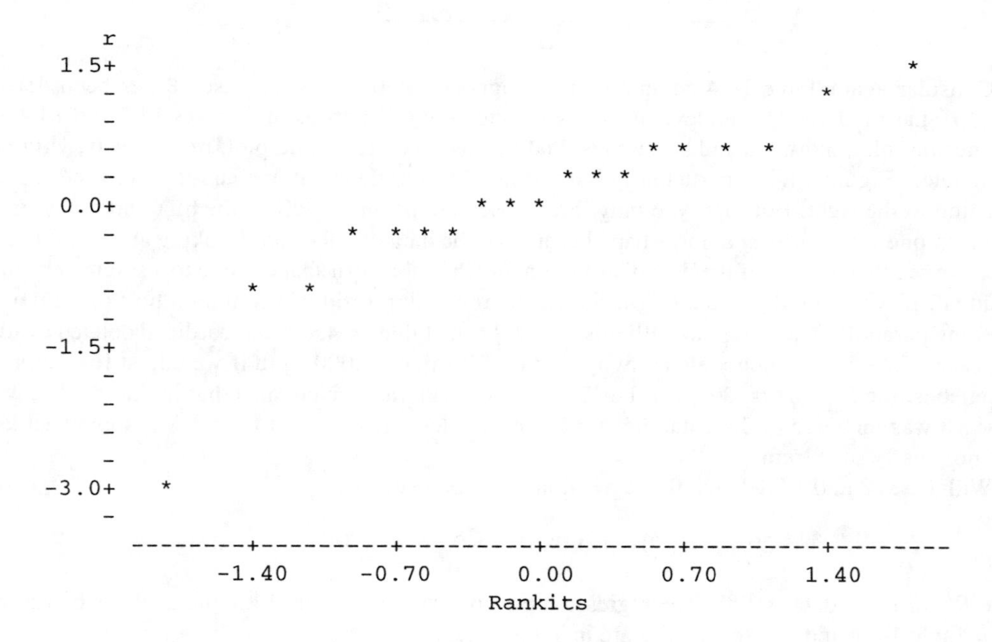

FIGURE 14.3. Normal plot, case 18 deleted, $W' = 0.854$.

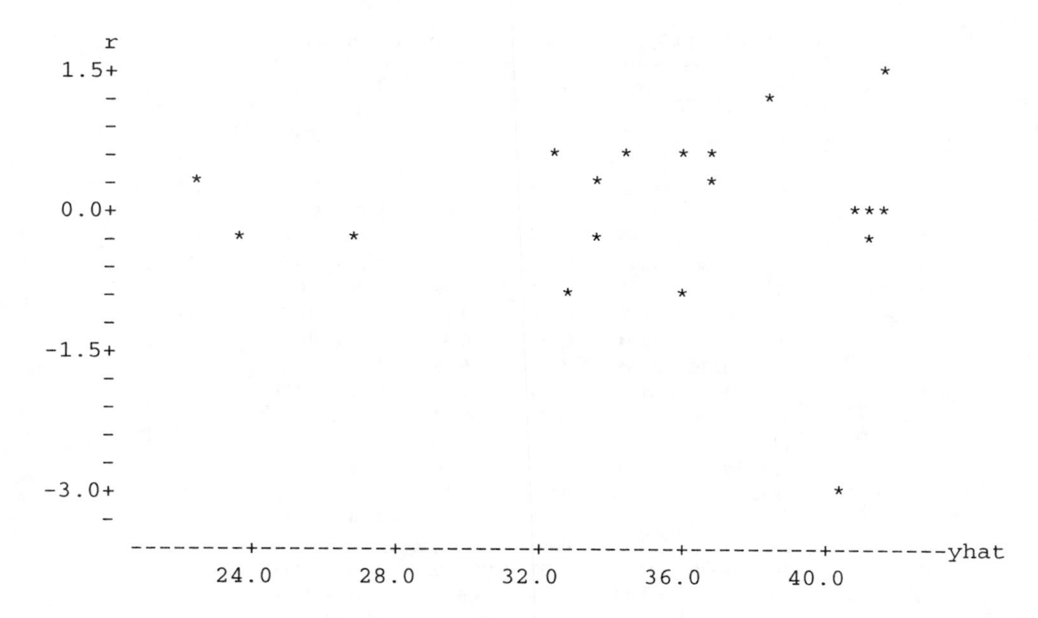

FIGURE 14.4. Standardized residuals versus predicted values, case 18 deleted.

TABLE 14.5. Regression, cases 3 and 18 deleted

Predictor	$\hat{\beta}$	$SE(\hat{\beta})$	t	P
Constant	29.758	5.532	5.38	0.000
x_1	−1.6985	0.4660	−3.64	0.003
x_2	0.08512	0.02079	4.09	0.001
x_3	0.66617	0.03824	17.42	0.000
x_4	1.1840	0.1643	7.21	0.000
x_5	−4.0668	0.8487	−4.79	0.000

the regression coefficients do not change much from Table 14.2 to Table 14.5. The largest changes seem to be in the constant and in the coefficient for x_5.

From Table 14.7, the leverages, t statistics, and Cook's distances seem reasonable. Figures 14.5 and 14.6 contain a normal plot and a plot of standardized residuals versus predicted values. Both plots look good. In particular, the suggestion of lack of fit in Figure 14.4 appears to be unfounded. Once again, Figure 14.6 could be misinterpreted as a horn shape but the 'horn' is due to the distribution of the predicted values.

Ultimately, someone must decide whether or not to delete unusual cases based on subject matter considerations. There is only moderate statistical evidence that case 18 is unusual and case 3 does not look severely unusual unless one previously deletes case 18. Are there subject matter reasons

TABLE 14.6. Analysis of variance, cases 3 and 18 deleted

Source	df	SS	MS	F	P
Regression	5	621.89	124.38	203.20	0.000
Error	12	7.34	0.61		
Total	17	629.23			

TABLE 14.7. Diagnostics, cases 3 and 18 deleted

Case	y	$\hat{y}$	Leverage	r	t	C
1	37.01	36.83	0.485	0.33	0.31	0.017
2	26.51	26.62	0.491	−0.20	−0.19	0.007
3		40.78	0.156			
4	40.70	41.43	0.196	−1.04	−1.05	0.044
5	37.10	36.35	0.180	1.07	1.07	0.041
6	33.90	33.67	0.504	0.42	0.41	0.030
7	41.80	42.11	0.261	−0.46	−0.44	0.012
8	33.40	33.69	0.108	−0.39	−0.38	0.003
9	41.01	41.56	0.311	−0.84	−0.83	0.053
10	37.20	36.94	0.621	0.54	0.52	0.078
11	23.30	23.66	0.381	−0.58	−0.57	0.035
12	35.20	34.24	0.440	1.65	1.79	0.356
13	34.90	35.81	0.370	−1.47	−1.56	0.212
14	33.10	32.66	0.145	0.60	0.59	0.010
15	22.70	22.44	0.467	0.46	0.44	0.031
16	39.70	38.72	0.171	1.38	1.44	0.066
17	31.80	33.02	0.298	−1.85	−2.10	0.243
18		24.50	0.486			
19	43.10	42.22	0.332	1.37	1.43	0.155
20	41.01	41.49	0.239	−0.70	−0.68	0.025

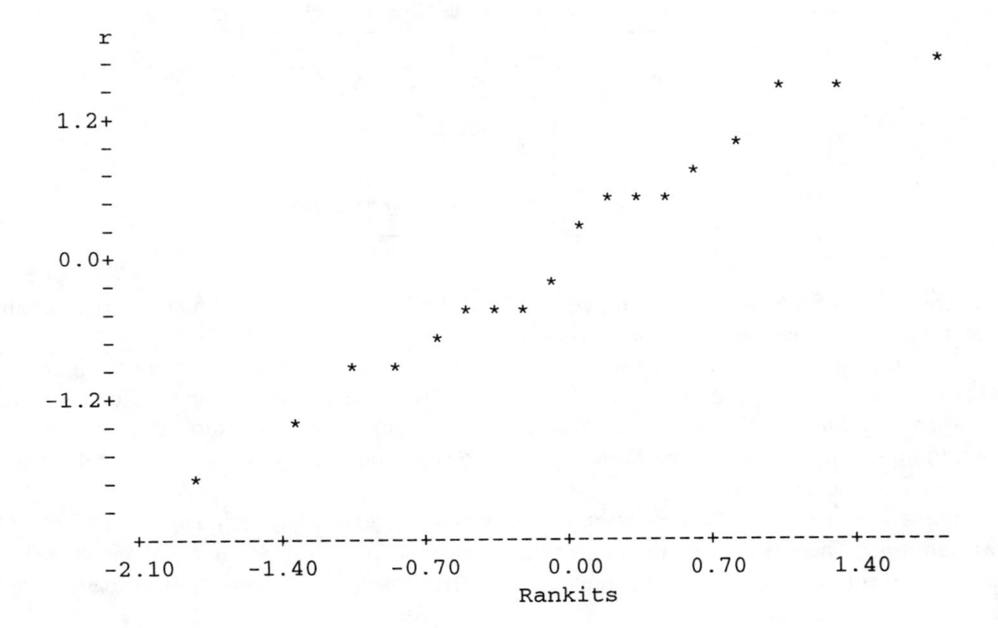

FIGURE 14.5. Normal plot, cases 3 and 18 deleted, $W' = 0.980$.

for these schools to be unusual? Will the data be more or less representative of the appropriate population if these data are deleted?

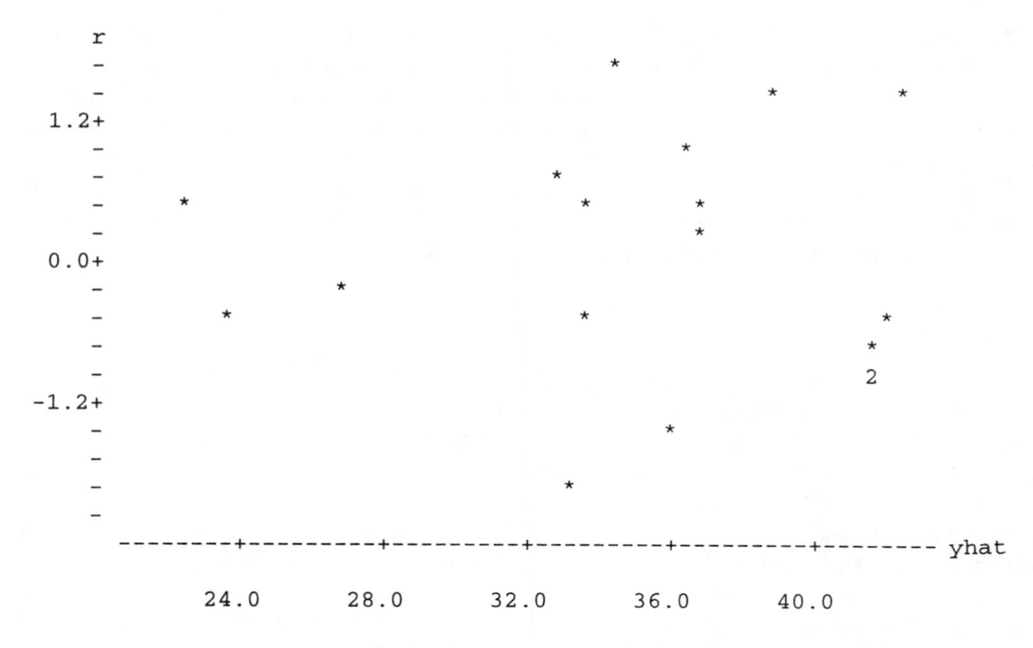

FIGURE 14.6. Standardized residuals versus predicted values, cases 3 and 18 deleted.

MINITAB COMMANDS

Below are given the Minitab commands for obtaining the diagnostics. On the 'regress' line, 5 predictors are specified, so the next 5 columns are taken to contain the predictor variables. The standardized residuals are placed in the next column listed and the predicted values are placed in the column listed after that. Thus the standardized residuals are in c21 and the predicted values are in c22. The subcommands 'tresid', 'hi', and 'cookd' indicate the standardized deleted residuals, leverages, and Cook distances, respectively. The t statistics are in c23. The leverages are in c24. The Cook distances are in c25.

```
MTB > regress c8 on 5 c2-c6 c21 c22;
SUBC> tresid c23;
SUBC> hi c24;
SUBC> cookd c25.
```

14.2 Best subset model selection methods

In this section and the next, we examine methods for identifying good reduced models *relative to a given (full) model.* Reduced models are of interest because a good reduced model provides an adequate explanation of the current data and, typically, the reduced model is more understandable because it is more succinct. Even more importantly, *for data collected in a similar fashion, a good reduced model often provides* better *predictions and parameter estimates than the full model*, cf. the subsection below on Mallows's C_p statistic and Christensen (1987, section XIV.7). Of course difficulties with predictions arise when a good reduced model is used with new cases that are not similar to those on which the reduced model was fitted and evaluated. In particular, a good fitted reduced model should not be used for prediction of a new case unless *all* of the predictor variables in the new case are similar to those in the original data. *It is not enough that new cases be similar on just the variables in the reduced model*. It is not sufficient that they be similar on all of the variables in the full model because some important variable may not have been measured for the full model, yet a new case with a very different value of this unmeasured variable can act very differently.

This section presents three methods for examining all possible reduced models. These methods are based on defining a criterion for a best model and then finding the models that are best by this criterion. Section 14.3 considers three methods of making sequential selections of variables. Obviously, it is better to consider all reduced models whenever feasible rather than making sequential selections. Sequential methods are flawed but they are cheap and easy.

14.2.1 R^2 STATISTIC

The fundamental statistic in comparing all possible reduced models is the R^2 statistic. This is appropriate but we should recall some of the weaknesses of R^2. The numerical size of R^2 is more related to predictive ability than to model adequacy. The perfect model can have small predictive ability and thus a small R^2, while demonstrably inadequate models can still have substantial predictive ability and thus a high R^2. Fortunately, we are typically more interested in prediction than in finding the perfect model, especially since our models are typically empirical approximations for which no perfect model exists. In addition, when considering transformations of the dependent variable, the R^2 values for different models are not comparable.

In the present context, the most serious drawback of R^2 is that it typically goes up when more predictor variables are added to a model. (It cannot go down.) Thus it is not really appropriate to compare the R^2 values of two models with different numbers of predictors. However, we can use R^2 to compare models with *the same* number of predictor variables. In fact, for models with the same number of predictors, we can use R^2 to order them from best to worse; the largest R^2 value then corresponds to the best model. R^2 is the fundamental model comparison statistic for best subset methods in that, *for comparing models with the same number of predictors*, the other methods considered give the same relative orderings for models as R^2. The essence of the other methods is to develop a criterion for comparing models that have *different* numbers of predictors.

Table 14.8 contains the two best models for the school data based on the R^2 statistic for each number of predictor variables. The best single variable is x_3; the second best is x_2. This information could be obtained from the correlations between y and the predictor variables given in Section 13.1. Note the drastic difference between the R^2 for using x_3 and that for x_2. The best pair of variables for predicting y are x_3 and x_4, while the second best pair is x_3 and x_5. The best three-variable model contains x_1, x_3, and x_4. Note that the largest R^2 values go up very little when a forth or fifth variable is added. Moreover, all the models in Table 14.8 that contain three or more variables include x_3 and x_4. We could conduct F tests to compare models with different numbers of predictor variables, as long as the smaller models are contained in the large ones.

Any models that we think are good candidates should be examined for influential and outlying observations, consistency with assumptions, and subject matter implications. Any model that makes particularly good sense to a subject matter specialist warrants special consideration. Models that make particularly poor sense to subject matter specialists may be dumb luck but they may also be the springboard for new insights into the process generating the data. We also need to concern ourselves with the role of observations that are influential or outlying in the original (full) model. We will examine this in more detail later. Finally, recall that when making predictions based on reduced models, the point at which we are making the prediction generally needs to be consistent with the original data on all variables, not just the reduced model variables.

When we drop a variable, we do not conclude that the variable is not important, we conclude that it is not important *for this set of data*. For different data, a dropped variable may become important. We cannot presume to make predictions from a reduced model for new cases that are substantially different from the original data.

TABLE 14.8. Best subset regression, R^2 statistic

			Included variables				
Vars.	R^2	$\sqrt{MSE}$	x_1	x_2	x_3	x_4	x_5
1	86.0	2.2392			X		
1	56.8	3.9299		X			
2	88.7	2.0641			X	X	
2	86.2	2.2866			X		X
3	90.1	1.9974	X		X	X	
3	88.9	2.1137			X	X	X
4	90.2	2.0514	X		X	X	X
4	90.1	2.0603	X	X	X	X	
5	90.6	2.0743	X	X	X	X	X

14.2.2 ADJUSTED R^2 STATISTIC

The adjusted R^2 statistic is simply an adjustment of R^2 that allows comparisons to be made between models with different numbers of predictor variables. Let p be the number of predictor variables in a regression equation (including the intercept), then the adjusted R^2 is defined to be

$$\text{Adj } R^2 \equiv 1 - \frac{n-1}{n-p}\left(1 - R^2\right).$$

For the school example with all predictor variables, this becomes

$$.873 = 1 - \frac{20-1}{20-6}(1 - .9063),$$

or, as it is commonly written, 87.3%.

It is not too difficult to see that

$$\text{Adj } R^2 = 1 - \frac{MSE}{s_y^2}$$

where s_y^2 is the sample variance of the y_is, i.e., $s_y^2 = SSTot/(n-1)$. This is a much simpler statement than the defining relationship. For the school example with all predictor variables, this is

$$.873 = 1 - \frac{4.30}{(642.92)/19}.$$

Note that *when comparing two models, the model with the smaller MSE has the larger adjusted* R^2.

R^2 is always between 0 and 1, but while the adjusted R^2 cannot get above 1, it can get below 0. It is possible to find models that have $MSE > s_y^2$. In these cases, the adjusted R^2 is actually less than 0.

Models with large adjusted R^2s are precisely models with small mean squared errors. At first glance, this seems like a reasonable way to choose models, but upon closer inspection the idea seems flawed. The problem is that when comparing some model with a reduced model, the adjusted R^2 is greater for the larger model whenever the mean squared error of the larger model is less than the numerator mean square for testing the adequacy of the smaller model. In other words, the adjusted R^2 is greater for the larger model whenever the F statistic for comparing the models is greater than 1. Typically, we want the F statistic to be substantially larger than 1 before concluding that the extra variables in the larger model are important.

To see that the adjusted R^2 is larger for the larger model whenever $F > 1$, consider the simplest example, that of comparing the full model to the model that contains just an intercept. For the school data, the mean squared error for the intercept model is

$$SSTot/19 = 642.92/19 = (SSReg + SSE)/19$$

$$= (5MSReg + 14MSE)/19 = \frac{5}{19}116.54 + \frac{14}{19}4.30.$$

Thus $SSTot/19$ is a weighted average of $MSReg$ and MSE. The $MSReg$ is greater than the MSE ($F > 1$), so the weighted average of the terms must be greater than the smaller term, MSE. The weighted average is $SSTot/19$, which is the mean squared error for the intercept model, while MSE is the mean squared error for the full model. Thus $F > 1$ implies that the mean squared error for the smaller model is greater than the mean squared error for the larger model and recall that the model with the smaller mean squared error has the higher adjusted R^2.

In general, the mean squared error for the smaller model is a weighted average of the mean square for the variables being added and the mean squared error of the larger model. If the mean square for the variables being added is greater than the mean squared error of the larger model, i.e., if $F > 1$, the mean squared error for the smaller model must be greater than that for the larger model. If we add variables to a model whenever the F statistic is greater than 1, we will include a lot on unnecessary variables.

Table 14.9 contains the 6 best fitting models as judged by the adjusted R^2 criterion. As advertised, the ordering of the models from best to worst is consistent whether one maximizes the adjusted R^2 or minimizes the MSE (or equivalently, $\sqrt{MSE}$). The best model based on the adjusted R^2 is the model with variables x_1, x_3, and x_4, but a number of the best models are given. Presenting a number of the best models reinforces the idea that selection of one or more final models should be based on many more considerations than just the value of one model selection statistic. Moreover, the *best* model as determined by the adjusted R^2 often contains too many variables.

TABLE 14.9. Best subset regression, adjusted R^2 statistic

Vars.	Adj. R^2	$\sqrt{MSE}$	x_1	x_2	x_3	x_4	x_5
			Included variables				
3	88.2	1.9974	X		X	X	
4	87.6	2.0514	X		X	X	X
4	87.5	2.0603	X	X	X	X	
2	87.4	2.0641			X	X	
5	87.3	2.0743	X	X	X	X	X
3	86.8	2.1137			X	X	X

Note also that the two models in Table 14.9 with three variables are precisely the two three-variable models with the highest R^2 values from Table 14.8. The same is true about the two four-variable models that made this list. As indicated earlier, when the number of variables is fixed, ordering models by their R^2s is equivalent to ordering models by their adjusted R^2s. The comments about model checking and prediction made in the previous subsection continue to apply.

14.2.3 MALLOWS'S C_p STATISTIC

Mallows's C_p statistic estimates a measure of the difference between the fitted regression surface from a reduced model and the actual regression surface. The idea is to compare the points

$$z_i = \beta_0 + \beta_1 x_{i1} + \beta_2 x_{i2} + \beta_3 x_{i3} + \ldots + \beta_{p-1} x_{i,p-1}$$

on the actual regression surface of the full model (F) to the corresponding predictions $\hat{y}_{iR}$ from some reduced model (R) with, say, r predictor variables (including the constant). The comparisons are made at the locations of the original data. The model comparison is based on the sum of standardized squared differences,

$$\kappa \equiv \sum_{i=1}^{n} (\hat{y}_{iR} - z_i)^2 / \sigma^2.$$

The term σ^2 serves only to provide some standardization. Small values of κ indicate good reduced models. Note that κ is not directly useful because it is unknown. It depends on the z_i values and they depend on the unknown full model regression parameters. However, if we think of the $\hat{y}_{iR}$s as functions of the random variables y_i, the comparison value κ is a function of the y_is and thus is a random variable with an expected value. Mallows's C_p statistic is an estimate of the expected value of κ. In particular, Mallows's C_p statistic is

$$C_p = \frac{SSE(R)}{MSE(F)} - (n - 2r).$$

For a derivation of this statistic, see Christensen (1987, section XIV.1). The smaller the C_p value, the better the model (up to the variability of the estimation). If the C_p statistic is computed for the full model, the result is always p, the number of predictor variables including the intercept.

In multiple regression, estimated regression surfaces are identical to prediction surfaces, so models with Mallows's C_p statistics that are substantially less than p can be viewed as reduced models that are estimated to be better at prediction than the full model. Of course this comparison between predictions from the full and reduced models is restricted to the actual combinations of predictor variables in the observed data.

Table 14.10 contains the best six models based on the C_p statistic. The best model is the one with variables x_3 and x_4, but the model including x_1, x_3, and x_4 has essentially the same value of C_p. There is a substantial increase in C_p for any of the other four models. Clearly, we would focus attention on the two best models to see if they are adequate in terms of outliers, influential observations, agreement with assumptions, and subject matter implications. As always, predictions can only be made with safety from the reduced models when the new cases are to be obtained in a similar fashion to the original data. In particular, new cases must have similar values to those in the original data for all of the predictor variables, not just those in the reduced model. Note that the ranking of the best models is different here than for the adjusted R^2. The full model is not included here, while it was in the adjusted R^2 table. Conversely, the model with x_2, x_3, and x_4 is included here but was not included in the adjusted R^2 table. Note also that among models with three variables, the C_p rankings agree with the R^2 rankings and the same holds for four-variable models.

TABLE 14.10. Best subset regression, C_p statistic

			Included variables				
Vars	C_p	$\sqrt{MSE}$	x_1	x_2	x_3	x_4	x_5
2	2.8	2.0641			X	X	
3	2.8	1.9974	X		X	X	
3	4.6	2.1137			X	X	X
4	4.7	2.0514	X		X	X	X
3	4.8	2.1272		X	X	X	
4	4.8	2.0603	X	X	X	X	

It is my impression that Mallows's C_p statistic is the most popular method for selecting a best subset of the predictor variables. It is certainly my favorite. Mallows's C_p statistic is closely related to Akaike's information criterion (AIC), which is a general criterion for model selection. AIC and the relationship between C_p and AIC are examined in Christensen (1990b, section IV.8).

14.2.4 A COMBINED SUBSET SELECTION TABLE

Table 14.11 lists the three best models based on R^2 for each number of predictor variables. In addition, the adjusted R^2 and C_p values for each model are listed in the table. It is easy to identify the best models based on any of the model selection criteria. The output is extensive enough to include a few notably bad models. Rather than asking for the best 3, one might ask for the best 4, or 5, or 6 models for each number of predictor variables but it is difficult to imagine a need for any more extensive summary of the models when beginning a search for good reduced models.

TABLE 14.11. Best subset regression

		Adj.			Included variables				
Vars.	R^2	R^2	C_p	$\sqrt{MSE}$	x_1	x_2	x_3	x_4	x_5
1	86.0	85.2	5.0	2.2392			X		
1	56.8	54.4	48.6	3.9299		X			
1	53.7	51.2	53.1	4.0654					X
2	88.7	87.4	2.8	2.0641			X	X	
2	86.2	84.5	6.7	2.2866			X		X
2	86.0	84.4	6.9	2.2993		X	X		
3	90.1	88.2	2.8	1.9974	X		X	X	
3	88.9	86.8	4.6	2.1137			X	X	X
3	88.7	86.6	4.8	2.1272		X	X	X	
4	90.2	87.6	4.7	2.0514	X		X	X	X
4	90.1	87.5	4.8	2.0603	X	X	X	X	
4	89.2	86.3	6.1	2.1499		X	X	X	X
5	90.6	87.3	6.0	2.0743	X	X	X	X	X

Note that the model with x_1, x_3, and x_4 is the best model as judged by adjusted R^2 and is nearly the best model as judged by the C_p statistic. (The model with x_3 and x_4 has a slightly smaller C_p value.) The model with x_2, x_3, x_4 has essentially the same C_p statistic as the model with x_1, x_2, x_3, x_4 but the later model has a larger adjusted R^2.

MINITAB COMMANDS

Below are given Minitab commands for obtaining Table 14.11.

```
MTB > breg c8 on c2-c6;
SUBC> best 3.
```

14.3 Stepwise model selection methods

Best subset selection methods evaluate all the possible subsets of variables from a full model and identify the best reduced regression models based on some criterion. Evaluating all possible models is the most reasonable way to proceed in variable selection but the computational demands of evaluating every model can be staggering. Every additional variable in a model doubles the number of reduced models that can be constructed. In our example with five variables, there are $2^5 = 32$ reduced models to be considered; in an example with 8 variables there are $2^8 = 256$ reduced models to be fitted. Years ago, when computation was slow and expensive, fitting large numbers of models was not practical, and even now, when one has a very large number of predictor variables, fitting all models can easily overwhelm a computer. (Actually, improved computer algorithms allow us to avoid fitting all models, but even with the improved algorithms, computational limits can be exceeded.)

An alternative to fitting all models is to evaluate the variables one at a time and look at a sequence of models. Stepwise variable selection methods do this. The best of these methods begin with a full model and sequentially identify variables that can be eliminated. In some procedures, variables that have been eliminated may be put back into the model if they meet certain criteria. The virtue of starting with the full model is that if you start with an adequate model and only do reasonable things, you should end up with an adequate model. A less satisfactory procedure is to begin with no variables and see which ones can be added into the model. This begins with an inadequate model and there is no guarantee that an adequate model will ever be achieved. We consider three methods: backwards elimination in which variables are deleted from the full model, forward selection in which variables are added to a model, typically the model that includes only the intercept, and stepwise methods in which variables can be both added and deleted. Because these methods only consider the deletion or addition of one variable at a time, they may never find the best models as determined by best subset selection methods.

14.3.1 BACKWARDS ELIMINATION

Backwards elimination begins with the full model and sequentially eliminates from the model the least important variable. The importance of a variable is judged by the size of the t (or equivalent F) statistic for dropping the variable from the model, i.e., the t statistic for testing whether the corresponding regression coefficient is 0. After the variable with the smallest absolute t statistic is dropped, the model is refitted and the t statistics recalculated. Again, the variable with the smallest absolute t statistic is dropped. The process ends when all of the absolute values of the t statistics are greater than some predetermined level. The predetermined level can be a fixed number for all steps or it can change depending on the step. When allowing it to change depending on the step, we could set up the process so that it stops when all of the P values are below a fixed level.

Table 14.12 illustrates backwards elimination for the school data. In this example, the predetermined level for stopping the procedure is 2. If all $|t|$ statistics are greater than 2, elimination of variables halts. Step 1 includes all 5 predictor variables. The table gives estimated regression coefficients, t statistics, the R^2 value, and the square root of the MSE. In step 1, the smallest absolute t statistic is 0.82, so variable x_2 is eliminated from the model. The statistics in step 2 are similar to those in step 1 but now the model includes only variables x_1, x_3, x_4, and x_5. In step 2, the smallest absolute t statistic is $|-0.41|$, so variable x_5 is eliminated from the model. Step 3 is based on the model with x_1, x_3, and x_4. The smallest absolute t statistic is the -1.47 for variable x_1, so x_1 is dropped. Step 4 uses the model with only x_3 and x_4. At this step, the t statistics are both greater than 2, so the process halts. Note that the intercept is not considered for elimination.

TABLE 14.12. Backwards elimination of y on 5 predictors, with $N = 20$

Step		Const.	x_1	x_2	x_3	x_4	x_5	R^2	$\sqrt{MSE}$
1	$\hat{\beta}$	19.95	-1.8	0.044	0.556	1.11	-1.8	90.63	2.07
	t_{obs}		-1.45	0.82	5.98	2.56	-0.89		
2	$\hat{\beta}$	15.47	-1.7		0.582	1.03	-0.5	90.18	2.05
	t_{obs}		-1.41		6.75	2.46	-0.41		
3	$\hat{\beta}$	12.12	-1.7		0.553	1.04		90.07	2.00
	t_{obs}		-1.47		11.27	2.56			
4	$\hat{\beta}$	14.58			0.542	0.75		88.73	2.06
					10.82	2.05			

The final model given in Table 14.12 happens to be the best model as determined by the C_p statistic and the model at stage 3 is the second best model as determined by the C_p statistic. This is a fortuitous event; there is no reason that this should happen other than these data being particularly clear about the most important variables.

14.3.2 FORWARD SELECTION

Forward selection begins with an initial model and adds variables to this model one at a time. Most often, the initial model contains only the intercept, but many computer programs have options for including other variables in the initial model. To determine which variable to add at any step in the process, a candidate variable is added to the current model and the t statistic is computed for the candidate variable. This is done for each candidate variable and the candidate variable with the largest $|t|$ statistic is added to the model. The procedure stops when none of the absolute t statistics is greater than a predetermined level. The predetermined level can be a fixed number for all steps or it can change with the step. When allowing it to change depending on the step, we could set the process so that it stops when none of the P values for the candidate variables is below a fixed level.

Table 14.13 gives an abbreviated summary of the procedure for the school data using 2 as the predetermined $|t|$ level for stopping the process. At the first step, the five models $y_i = \gamma_{0j} + \gamma_j x_{ij} + \varepsilon_i$, $j = 1, \ldots, 5$ are fitted to the data. The variable x_j with the largest absolute t statistic for testing $\gamma_j = 0$ is added to the model. Table 14.13 indicates that this was variable x_3. At step 2, the four models $y_i = \beta_{0j} + \beta_{3j}x_{i3} + \beta_j x_{ij} + \varepsilon_i$, $j = 1, 2, 4, 5$ are fitted to the data and the variable x_j with the largest absolute t statistic for testing $\beta_j = 0$ is added to the model. In the example, the largest absolute t statistic belongs to x_4. At this point, the table stops, indicating that when the three models $y_i = \eta_{0j} + \eta_{3j}x_{i3} + \eta_{4j}x_{i4} + \eta_j x_{ij} + \varepsilon_i$, $j = 1, 2, 5$ were fitted to the model, none of the absolute t statistics for testing $\eta_j = 0$ were greater than 2.

TABLE 14.13. Forward selection of y on 5 predictors, with $N = 20$

Step		Const.	x_1	x_2	x_3	x_4	x_5	R^2	$\sqrt{MSE}$
1	$\hat{\beta}$	33.32			0.560			85.96	2.24
	t_{obs}				10.50				
2	$\hat{\beta}$	14.58			0.542	0.75		88.73	2.06
	t_{obs}				10.82	2.05			

The final model selected is the model with predictor variables x_3 and x_4. This is the same model obtained from backwards elimination and the model that has the smallest C_p statistic. Again, this is a fortuitous circumstance. There is no assurance that such agreement between methods will occur.

Rather than using t statistics, the decisions could be made using the equivalent F statistics. The stopping value of 2 for t statistics corresponds to a stopping value of 4 for F statistics. In addition, this same procedure can be based on sample correlations and partial correlations. The decision in step 1 is equivalent to adding the variable that has the largest absolute sample correlation with y. The decision in step 2 is equivalent to adding the variable that has the largest absolute sample partial correlation with y after adjusting for x_3. Step 3 is not shown in the table, but the computations for step 3 must be made in order to know that the procedure stops after step 2. The decision in step 3 is equivalent to adding the variable that has the largest absolute sample partial correlation with y after adjusting for x_3 and x_4, provided this value is large enough.

The author has a hard time imagining any situation where forward selection is a reasonable thing to do, except possibly as a screening device when there are more predictor variables than there are observations. In such a case, the full model cannot be fitted meaningfully, so best subset methods and backwards elimination do not work.

14.3.3 STEPWISE METHODS

Stepwise methods alternate between forward selection and backwards elimination. Suppose you have just arrived at a model by dropping a variable. A stepwise method will then check to see if any variable can be added to the model. If you have just arrived at a model by adding a variable, a stepwise method then checks to see if any variable can be dropped. The value of the absolute t statistic required for dropping a variable is allowed to be different from the value required for adding a variable. Stepwise methods often start with an initial model that contains only an intercept, but many computer programs allow starting the process with the full model. In the school example, the stepwise method beginning with the intercept model gives the same results as forward selection and the stepwise method beginning with the full model gives the same results as backwards elimination. (The absolute t statistics for both entering and removing were set at 2.) Other initial models can also be used. Christensen (1987, section XIV.2) discusses some alternative rules for conducting stepwise regression.

MINITAB COMMANDS

Minitab's 'stepwise' command provides stepwise variable selection with a default in which variables are only added if the greatest F statistic is greater than 4 and only removed if the smallest F statistic is less than 4. There are options for forcing variables out of the model, forcing variables into the model, specifying an initial model, and setting the comparison values for the F statistics. Backwards elimination is obtained by entering all the variables into the initial model and resetting the F value for entering a variable to a very large number, say, 100000. Forward selection is obtained by setting the F value for removing a variable to 0. The commands are given below.

```
MTB > names c2 'x1' c3 'x2' c4 'x3' c5 'x4' c6 'x5' c8 'y'
MTB > note        STEPWISE:  starting with full model
MTB > stepwise c8 on c2-c6;
SUBC> enter c2-c6.
MTB > note        STEPWISE:  starting with intercept model
MTB > stepwise c8 on c2-c6
MTB > note        BACKWARDS ELIMINATION
MTB > stepwise c8 on c2-c6;
SUBC> enter c2-c6;
SUBC> fenter = 100000.
MTB > note        FORWARD SELECTION
MTB > stepwise c8 on c2-c6;
SUBC> fremove = 0.
```

14.4 Model selection and case deletion

In this section we examine how the results of the previous two sections change when influential cases are deleted. Before beginning, we make a crucial point. *Both variable selection and the elimination of outliers cause the resulting model to appear better than it probably should. Both tend to give MSEs that are unrealistically small. It follows that confidence and prediction intervals are unrealistically narrow and test statistics are unrealistically large.* Outliers tend to be cases with large residuals; any policy of eliminating the largest residuals obviously makes the *SSE*, which is the sum of the squared residuals, and the *MSE* smaller. Some large residuals occur by chance even when the model is correct. Systematically eliminating these large residuals makes the estimate of the variance too small. Variable selection methods tend to identify as good reduced models those with small *MSE*s. The most extreme case is that of using the adjusted R^2 criterion, which identifies as the best model the one with the smallest *MSE*. Confidence and prediction intervals based on models that are arrived at after variable selection or outlier deletion should be viewed as the smallest reasonable intervals available, with the understanding that more appropriate intervals would probably be wider. Tests performed after variable selection or outlier deletion should be viewed as giving the greatest reasonable evidence against the null hypothesis, with the understanding that more appropriate tests would probably display a lower level of significance.

Recall that in Section 14.1, case 18 was identified as an influential point in the school data and then case 3 was identified as highly influential. Table 14.14 gives the results of a best subset selection when case 18 has been eliminated. The full model is the best model as measured by either the C_p statistic or the adjusted R^2 value. This is a far cry from the full data analysis in which the models with x_3, x_4 and x_1, x_3, x_4 had the smallest C_p statistics. These two models are only the seventh and fifth best models in Table 14.14. The two closest competitors to the full model in Table 14.14 involve dropping one of variables x_1 and x_2. The fourth and fifth best models involve dropping x_2 and one of variables x_1 and x_5. In this case, the adjusted R^2 ordering of the five best models agrees with the C_p ordering.

TABLE 14.14. Best subset regression with case 18 deleted

Vars	R^2	Adj. R^2	C_p	$\sqrt{MSE}$	x_1	x_2	x_3	x_4	x_5
					Included variables				
1	89.6	89.0	21.9	1.9653			X		
1	56.0	53.4	140.8	4.0397		X			
1	53.4	50.6	150.2	4.1595					X
2	92.3	91.3	14.3	1.7414			X	X	
2	91.2	90.1	18.2	1.8635			X		X
2	89.8	88.6	23.0	2.0020		X	X		
3	93.7	92.4	11.4	1.6293			X	X	X
3	93.5	92.2	12.1	1.6573	X		X	X	
3	92.3	90.8	16.1	1.7942		X	X	X	
4	95.2	93.8	8.1	1.4766		X	X	X	X
4	94.7	93.2	9.8	1.5464	X		X	X	X
4	93.5	91.6	14.1	1.7143	X	X	X	X	
5	96.3	94.9	6.0	1.3343	X	X	X	X	X

Table 14.15 gives the best subset summary when cases 3 and 18 have both been eliminated. Once again, the best model as judged by either C_p or adjusted R^2 is the full model. The second best model drops x_1 and the third best model drops x_2. However, the subsequent ordering changes substantially.

Now consider backwards elimination and forward selection with influential observations deleted. In both cases, we continue to use the $|t|$ value 2 as the cutoff to stop addition and removal of variables.

Table 14.16 gives the results of a backwards elimination when case 18 is eliminated and when cases 3 and 18 are eliminated. In both situations, all five of the variables remain in the model. The regression coefficients are similar in the two models with the largest difference being in the coefficients for x_5. Recall that when all of the cases were included, the backwards elimination model included only variables x_3 and x_4, so we see a substantial difference due to the deletion of one or two cases.

The results of forward selection are given in Table 14.17. With case 18 deleted, the process stops with a model that includes x_3 and x_4. With case 3 also deleted, the model includes x_1, x_3, and x_4. While these happen to agree quite well with the results from the complete data, they agree poorly with the results from best subset selection and from backwards elimination, both of which indicate

TABLE 14.15. Best subset regression with cases 3 and 18 deleted

Vars	R^2	Adj. R^2	C_p	$\sqrt{MSE}$	x_1	x_2	x_3	x_4	x_5
1	92.2	91.7	66.5	1.7548			X		
1	57.9	55.3	418.8	4.0688		X			
1	55.8	53.0	440.4	4.1693					X
2	95.3	94.7	36.1	1.4004			X	X	
2	93.2	92.2	58.3	1.6939			X		X
2	92.3	91.2	67.6	1.8023		X	X		
3	96.6	95.8	25.2	1.2412	X		X	X	
3	96.1	95.2	30.3	1.3269			X	X	X
3	95.3	94.3	38.0	1.4490		X	X	X	
4	97.5	96.8	17.3	1.0911		X	X	X	X
4	97.2	96.3	20.8	1.1636	X		X	X	X
4	96.6	95.6	27.0	1.2830	X	X	X	X	
5	98.8	98.3	6.0	0.78236	X	X	X	X	X

TABLE 14.16. Backwards elimination

Step		Const.	x_1	x_2	x_3	x_4	x_5	R^2	$\sqrt{MSE}$
Case 18 deleted									
1	$\hat{\beta}$	34.29	−1.62	0.085	0.674	1.11	−4.6	96.33	1.33
	t_{obs}		−2.04	2.41	10.34	3.98	−3.18		
Cases 18 and 3 deleted									
1	$\hat{\beta}$	29.76	−1.70	0.085	0.666	1.18	−4.07	98.83	0.782
	t_{obs}		−3.64	4.09	17.42	7.21	−4.79		

that all variables are important. Forward selection gets hung up after a few variables and cannot deal with the fact that adding several variables (rather than one at a time) improves the fit of the model substantially.

TABLE 14.17. Forward selection

Step		Const.	x_1	x_2	x_3	x_4	x_5	R^2	$\sqrt{MSE}$
Case 18 deleted									
1	$\hat{\beta}$	32.92			0.604			89.59	1.97
	t_{obs}				12.10				
2	$\hat{\beta}$	14.54			0.585	0.74		92.31	1.74
	t_{obs}				13.01	2.38			
Cases 18 and 3 deleted									
1	$\hat{\beta}$	33.05			0.627			92.17	1.75
	t_{obs}				13.72				
2	$\hat{\beta}$	13.23			0.608	0.79		95.32	1.40
	t_{obs}				16.48	3.18			
3	$\hat{\beta}$	10.86	−1.66		0.619	1.07		96.57	1.24
	t_{obs}		−2.26		18.72	4.23			

14.5 Exercises

EXERCISE 14.5.1. Reconsider the advertising data of Exercise 13.8.1.

(a) Are there any high leverage points? Why or why not?

(b) Test whether each case is an outlier using an overall significance level no greater than $\alpha = .05$. Completely state the appropriate reference distribution.

(c) Discuss the importance of Cook's distances in regard to these data.

(d) Using only analysis of variance tables, compute R^2, the adjusted R^2, and the C_p statistic for $y_i = \beta_0 + \beta_1 x_{i1} + \beta_2 x_{i2} + \varepsilon_i$. Show your work.

(e) In the three-variable model, which if any variable would be deleted by a backwards elimination method? Why?

EXERCISE 14.5.2. Consider the information given below on diagnostic statistics for the wood data of Exercise 13.8.2.

(a) Are there any outliers in the predictor variables? Why are these considered outliers?

(b) Are there any outliers in the dependent variable? If so, why are these considered outliers?

(c) What are the most influential observations in terms of the predictive ability of the model?

Obs.	Leverage	r	t	C	Obs.	Leverage	r	t	C
1	.085	−0.25	−0.25	.001	29	.069	0.27	0.26	.001
2	.055	1.34	1.35	.021	30	.029	0.89	0.89	.005
3	.021	0.57	0.57	.001	31	.204	0.30	0.30	.005
4	.031	0.35	0.35	.001	32	.057	0.38	0.37	.002
5	.032	2.19	2.28	.032	33	.057	0.05	0.05	.000
6	.131	0.20	0.19	.001	34	.085	−2.43	−2.56	.109
7	.027	1.75	1.79	.017	35	.186	−2.17	−2.26	.215
8	.026	1.23	1.24	.008	36	.184	1.01	1.01	.046
9	.191	0.52	0.52	.013	37	.114	0.85	0.85	.019
10	.082	0.47	0.46	.004	38	.022	0.19	0.19	.000
11	.098	−3.39	−3.82	.250	39	.022	−0.45	−0.45	.001
12	.066	0.32	0.32	.001	40	.053	−1.15	−1.15	.015
13	.070	−0.09	−0.09	.000	41	.053	0.78	0.78	.007
14	.059	0.08	0.08	.000	42	.136	−0.77	−0.76	.018
15	.058	−0.91	−0.91	.010	43	.072	−0.78	−0.77	.009
16	.085	−0.09	−0.09	.000	44	.072	−0.27	−0.26	.001
17	.113	1.28	1.29	.042	45	.072	−0.40	−0.40	.002
18	.077	−1.05	−1.05	.018	46	.063	−0.62	−0.62	.005
19	.167	0.38	0.38	.006	47	.025	0.46	0.46	.001
20	.042	0.24	0.23	.000	48	.021	0.18	0.18	.000
21	.314	−0.19	−0.19	.003	49	.050	−0.44	−0.44	.002
22	.099	0.56	0.55	.007	50	.161	−0.66	−0.66	.017
23	.093	0.47	0.46	.004	51	.042	−0.44	−0.43	.002
24	.039	−0.60	−0.60	.003	52	.123	−0.26	−0.26	.002
25	.098	−1.07	−1.07	.025	53	.460	1.81	1.86	.558
26	.033	0.14	0.13	.000	54	.055	0.50	0.50	.003
27	.042	1.19	1.19	.012	55	.093	−1.03	−1.03	.022
28	.185	−1.41	−1.42	.090					

EXERCISE 14.5.3. Consider the following information on best subset regression for the wood data of Exercise 13.8.2.

Best subset regression of wood data

Vars	R^2	Adj. R^2	C_p	$\sqrt{MSE}$	x_1	x_2	x_3	x_4
					Included variables			
1	97.9	97.9	12.9	18.881	X			
1	63.5	62.8	1064.9	78.889				X
1	32.7	31.5	2003.3	107.04			X	
2	98.3	98.2	3.5	17.278	X		X	
2	97.9	97.8	14.3	18.969	X	X		
2	97.9	97.8	14.9	19.061	X			X
3	98.3	98.2	5.3	17.419	X	X	X	
3	98.3	98.2	5.4	17.430	X		X	X
3	98.0	97.9	13.7	18.763	X	X		X
4	98.4	98.2	5.0	17.193	X	X	X	X

(a) In order, what are the three best models as measured by the C_p criterion?

(b) What is the mean squared error for the model with variables x_1, x_3, and x_4?

(c) In order, what are the three best models as measured by the adjusted R^2 criterion? (Yes, it is possible to distinguish between the best four!)

(d) What do you think are the best models and what would you do next?

EXERCISE 14.5.4. Consider the following information on stepwise regression for the wood data of Exercise 13.8.2.

Stepwise regression

STEP	1	2	3
Constant	23.45	41.87	43.85
x_1	0.932	1.057	1.063
t	10.84	38.15	44.52
x_2	0.73	0.09	
t	1.56	0.40	
x_3	−0.50	−0.50	−0.51
t	−3.28	−3.27	−3.36
x_4	3.5		
t	1.53		
$\sqrt{MSE}$	17.2	17.4	17.3
R^2	98.36	98.29	98.28

(a) What is being given in the rows labeled x_1, x_2, x_3, and x_4? What is being given in the rows labeled t?

(b) Is this table for forward selection, backwards elimination, stepwise regression, or some other procedure?

(c) Describe the results of the procedure.

EXERCISE 14.5.5. Reanalyze the Prater data of Atkinson (1985) and Hader and Grandage (1958) from Exercise 13.8.3. Examine residuals and influential observations. Explore the use of the various model selection methods.

EXERCISE 14.5.6. Reanalyze the Chapman data of Exercise 13.8.4. Examine residuals and influential observations. Explore the use of the various model selection methods.

EXERCISE 14.5.7. Reanalyze the pollution data of Exercise 13.8.5. Examine residuals and influential observations. Explore the use of various model selection methods.

EXERCISE 14.5.8. Repeat Exercise 13.8.6 on the body fat data with special emphasis on diagnostics and model selection.

Chapter 15

Multiple regression: matrix formulation

In this chapter we use matrices to write regression models. Properties of matrices are reviewed in Appendix A. The economy of notation achieved through using matrices allows us to arrive at some interesting new insights and to derive several of the important properties of regression analysis.

15.1 Random vectors

In this section we discuss vectors and matrices that are made up of random variables rather than just numbers. For simplicity, we focus our discussion on vectors that contain 3 rows, but the results are completely general.

Let y_1, y_2, and y_3 be random variables. From these, we can construct a 3×1 random vector, say

$$Y = \begin{bmatrix} y_1 \\ y_2 \\ y_3 \end{bmatrix}.$$

The expected value of the random vector is just the vector of expected values of the random variables. For the random variables write $E(y_i) = \mu_i$, then

$$E(Y) \equiv \begin{bmatrix} E(y_1) \\ E(y_2) \\ E(y_3) \end{bmatrix} = \begin{bmatrix} \mu_1 \\ \mu_2 \\ \mu_3 \end{bmatrix} \equiv \mu.$$

In other words, expectation of a random vector is performed elementwise. In fact, the expected value of any random matrix (a matrix consisting of random variables) is the matrix made up of the expected values of the elements in the random matrix. Thus if w_{ij}, $i = 1, 2, 3$, $j = 1, 2$ is a collection of random variables and we write

$$W = \begin{bmatrix} w_{11} & w_{12} \\ w_{21} & w_{22} \\ w_{31} & w_{33} \end{bmatrix},$$

then

$$E(W) \equiv \begin{bmatrix} E(w_{11}) & E(w_{12}) \\ E(w_{21}) & E(w_{22}) \\ E(w_{31}) & E(w_{33}) \end{bmatrix}$$

We also need a concept for random vectors that is analogous to the variance of a random variable. This is the *covariance matrix*, sometimes called the dispersion matrix, the variance matrix, or the variance-covariance matrix. The covariance matrix is simply a matrix consisting of all the variances and covariances associated with the vector Y. Write

$$\mathrm{Var}(y_i) = E(y_i - \mu_i)^2 \equiv \sigma_{ii}$$

and

$$Cov(y_i, y_j) = E[(y_i - \mu_i)(y_j - \mu_j)] \equiv \sigma_{ij}.$$

Two subscripts are used on σ_{ii} to indicate that it is the variance of y_i *rather than* writing $\mathrm{Var}(y_i) = \sigma_i^2$.

The covariance matrix of our 3×1 vector Y is

$$Cov(Y) = \begin{bmatrix} \sigma_{11} & \sigma_{12} & \sigma_{13} \\ \sigma_{21} & \sigma_{22} & \sigma_{23} \\ \sigma_{31} & \sigma_{32} & \sigma_{33} \end{bmatrix}.$$

When Y is 3×1, the covariance matrix is 3×3. If Y were 20×1, $Cov(Y)$ would be 20×20. The covariance matrix is always symmetric because $\sigma_{ij} = \sigma_{ji}$ for any i, j. The variances of the individual random variables lie on the diagonal that runs from the top left to the bottom right. The covariances lie off the diagonal.

In general, if Y is an $r \times 1$ random vector and $E(Y) = \mu$, then $Cov(Y) = E[(Y - \mu)(Y - \mu)']$. In other words, $Cov(Y)$ is the expected value of the random matrix $(Y - \mu)(Y - \mu)'$.

15.2 Matrix formulation of regression models

SIMPLE LINEAR REGRESSION IN MATRIX FORM

The usual model for simple linear regression is

$$y_i = \beta_0 + \beta_1 x_i + \varepsilon_i \quad i = 1, \ldots, n, \tag{15.2.1}$$

$E(\varepsilon_i) = 0$, $\mathrm{Var}(\varepsilon_i) = \sigma^2$, and $Cov(\varepsilon_i, \varepsilon_j) = 0$ for $i \neq j$. In matrix terms this can be written as

$$\begin{bmatrix} y_1 \\ y_2 \\ \vdots \\ y_n \end{bmatrix} = \begin{bmatrix} 1 & x_1 \\ 1 & x_2 \\ \vdots & \vdots \\ 1 & x_n \end{bmatrix} \begin{bmatrix} \beta_0 \\ \beta_1 \end{bmatrix} + \begin{bmatrix} \varepsilon_1 \\ \varepsilon_2 \\ \vdots \\ \varepsilon_n \end{bmatrix}$$

$$Y_{n\times 1} = X_{n\times 2} \quad \beta_{2\times 1} + e_{n\times 1}$$

Multiplying and adding the matrices on the right-hand side gives

$$\begin{bmatrix} y_1 \\ y_2 \\ \vdots \\ y_n \end{bmatrix} = \begin{bmatrix} \beta_0 + \beta_1 x_1 + \varepsilon_1 \\ \beta_0 + \beta_1 x_2 + \varepsilon_2 \\ \vdots \\ \beta_0 + \beta_1 x_n + \varepsilon_n \end{bmatrix}.$$

These two vectors are equal if and only if the corresponding elements are equal, which occurs if and only if model (15.2.1) holds. The conditions on the ε_is translate into matrix terms as

$$E(e) = 0$$

where 0 is the $n \times 1$ matrix containing all zeros and

$$Cov(e) = \sigma^2 I$$

where I is the $n \times n$ identity matrix. By definition, the covariance matrix $Cov(e)$ has the variances of the ε_is down the diagonal. The variance of each individual ε_i is σ^2, so all the diagonal elements of $Cov(e)$ are σ^2, just as in $\sigma^2 I$. The covariance matrix $Cov(e)$ has the covariances of distinct ε_is as its off-diagonal elements. The covariances of distinct ε_is are all 0, so all the off-diagonal elements of $Cov(e)$ are zero, just as in $\sigma^2 I$.

EXAMPLE 15.2.1. In matrix terms, the model for regressing weights on heights using the data of Exercise 7.13.10 is

$$\begin{bmatrix} y_1 \\ y_2 \\ y_3 \\ y_4 \\ y_5 \\ y_6 \\ y_7 \\ y_8 \\ y_9 \\ y_{10} \\ y_{11} \\ y_{12} \end{bmatrix} = \begin{bmatrix} 1 & 65 \\ 1 & 65 \\ 1 & 65 \\ 1 & 65 \\ 1 & 66 \\ 1 & 66 \\ 1 & 63 \\ 1 & 63 \\ 1 & 63 \\ 1 & 72 \\ 1 & 72 \\ 1 & 72 \end{bmatrix} \begin{bmatrix} \beta_0 \\ \beta_1 \end{bmatrix} + \begin{bmatrix} \varepsilon_1 \\ \varepsilon_2 \\ \varepsilon_3 \\ \varepsilon_4 \\ \varepsilon_5 \\ \varepsilon_6 \\ \varepsilon_7 \\ \varepsilon_8 \\ \varepsilon_9 \\ \varepsilon_{10} \\ \varepsilon_{11} \\ \varepsilon_{12} \end{bmatrix}.$$

The observed data for this example are

$$\begin{bmatrix} y_1 \\ y_2 \\ y_3 \\ y_4 \\ y_5 \\ y_6 \\ y_7 \\ y_8 \\ y_9 \\ y_{10} \end{bmatrix} = \begin{bmatrix} 120 \\ 140 \\ 130 \\ 135 \\ 150 \\ 135 \\ 110 \\ 135 \\ 120 \\ 170 \\ 185 \\ 160 \end{bmatrix}.$$

□

THE GENERAL LINEAR MODEL

The general linear model is a generalization of the matrix form for the simple linear regression model. The general linear model is

$$Y = X\beta + e, \quad E(e) = 0, \quad Cov(e) = \sigma^2 I.$$

Y is an $n \times 1$ vector of observable random variables. X is an $n \times p$ matrix of known constants. β is a $p \times 1$ vector of unknown (regression) parameters. e is an $n \times 1$ vector of unobservable random errors. It will be assumed that $n \geq p$. Regression is any general linear model where the rank of X is p.

EXAMPLE 15.2.2. *Multiple regression*
In non-matrix form, the multiple regression model is

$$y_i = \beta_0 + \beta_1 x_{i1} + \beta_2 x_{i2} + \cdots + \beta_{p-1} x_{i,p-1} + \varepsilon_i, \quad i = 1, \ldots, n, \tag{15.2.2}$$

where

$$E(\varepsilon_i) = 0, \quad \mathrm{Var}(\varepsilon_i) = \sigma^2, \quad Cov(\varepsilon_i, \varepsilon_j) = 0, \ i \neq j.$$

In matrix terms this can be written as

$$\begin{bmatrix} y_1 \\ y_2 \\ \vdots \\ y_n \end{bmatrix} = \begin{bmatrix} 1 & x_{11} & x_{12} & \cdots & x_{1,p-1} \\ 1 & x_{21} & x_{22} & \cdots & x_{2,p-1} \\ \vdots & \vdots & \vdots & \ddots & \vdots \\ 1 & x_{n1} & x_{n2} & \cdots & x_{n,p-1} \end{bmatrix} \begin{bmatrix} \beta_0 \\ \beta_1 \\ \beta_2 \\ \vdots \\ \beta_{p-1} \end{bmatrix} + \begin{bmatrix} \varepsilon_1 \\ \varepsilon_2 \\ \vdots \\ \varepsilon_n \end{bmatrix}$$

$$Y_{n \times 1} = X_{n \times p} \quad \beta_{p \times 1} + e_{n \times 1}$$

Multiplying and adding the right-hand side gives

$$\begin{bmatrix} y_1 \\ y_2 \\ \vdots \\ y_n \end{bmatrix} = \begin{bmatrix} \beta_0 + \beta_1 x_{11} + \beta_2 x_{12} + \cdots + \beta_{p-1} x_{1,p-1} + \varepsilon_1 \\ \beta_0 + \beta_1 x_{21} + \beta_2 x_{22} + \cdots + \beta_{p-1} x_{2,p-1} + \varepsilon_2 \\ \vdots \\ \beta_0 + \beta_1 x_{n1} + \beta_2 x_{n2} + \cdots + \beta_{p-1} x_{n,p-1} + \varepsilon_n \end{bmatrix},$$

which is just (15.2.2). The conditions on the ε_is translate into

$$E(e) = 0,$$

where 0 is the $n \times 1$ matrix consisting of all zeros, and

$$Cov(e) = \sigma^2 I,$$

where I is the $n \times n$ identity matrix. □

Analysis of variance and analysis of covariance models can also be written as general linear models. This will be discussed further at the end of Chapter 16.

15.3 Least squares estimation of regression parameters

The regression estimates given by standard computer programs are least squares estimates. For simple linear regression, the least squares estimates are the values of β_0 and β_1 that minimize

$$\sum_{i=1}^{n} (y_i - \beta_0 - \beta_1 x_i)^2 . \tag{15.3.1}$$

For multiple regression, the least squares estimates of the β_js minimize

$$\sum_{i=1}^{n} \left(y_i - \beta_0 - \beta_1 x_{i1} - \beta_2 x_{i2} - \cdots - \beta_{p-1} x_{i,p-1}\right)^2 .$$

In matrix terms these can both be written as minimizing

$$(Y - X\beta)'(Y - X\beta). \tag{15.3.2}$$

The form in (15.3.2) is just the sum of the squares of the elements in the vector $(Y - X\beta)$. See also Exercise 15.8.1.

We now give the general form for the least squares estimate of β.

Proposition 15.3.1. $\hat{\beta} = \left(X'X\right)^{-1} X'Y$ is the least squares estimate of β.

PROOF: *The proof is optional material.*

Note that $\left(X'X\right)^{-1}$ exists only because in a regression problem the rank of X is p. The proof stems from rewriting the function to be minimized.

$$\begin{aligned}(Y - X\beta)'(Y - X\beta) &= \left(Y - X\hat{\beta} + X\hat{\beta} - X\beta\right)'\left(Y - X\hat{\beta} + X\hat{\beta} - X\beta\right) \\ &= \left(Y - X\hat{\beta}\right)'\left(Y - X\hat{\beta}\right) + \left(Y - X\hat{\beta}\right)'\left(X\hat{\beta} - X\beta\right) \\ &\quad + \left(X\hat{\beta} - X\beta\right)'\left(Y - X\hat{\beta}\right) + \left(X\hat{\beta} - X\beta\right)'\left(X\hat{\beta} - X\beta\right).\end{aligned} \tag{15.3.3}$$

Now consider one of the two middle terms, say $\left(X\hat{\beta} - X\beta\right)'\left(Y - X\hat{\beta}\right)$. Using the definition of $\hat{\beta}$ given in the proposition,

$$\begin{aligned}\left(X\hat{\beta} - X\beta\right)'\left(Y - X\hat{\beta}\right) &= \left[X\left(\hat{\beta} - \beta\right)\right]'\left(Y - X\hat{\beta}\right) \\ &= \left(\hat{\beta} - \beta\right)'X'\left(Y - X\left(X'X\right)^{-1}X'Y\right) \\ &= \left(\hat{\beta} - \beta\right)'X'\left(I - X\left(X'X\right)^{-1}X'\right)Y\end{aligned}$$

but

$$X'\left(I - X\left(X'X\right)^{-1}X'\right) = X' - \left(X'X\right)\left(X'X\right)^{-1}X' = X' - X' = 0.$$

Thus

$$\left(X\hat{\beta} - X\beta\right)'\left(Y - X\hat{\beta}\right) = 0$$

and similarly

$$\left(Y - X\hat{\beta}\right)'\left(X\hat{\beta} - X\beta\right) = 0.$$

Eliminating the two middle terms in (15.3.3) gives

$$(Y - X\beta)'(Y - X\beta) = \left(Y - X\hat{\beta}\right)'\left(Y - X\hat{\beta}\right) + \left(X\hat{\beta} - X\beta\right)'\left(X\hat{\beta} - X\beta\right).$$

This form is easily minimized. The first of the terms on the right-hand side does not depend on β, so the β that minimizes $(Y - X\beta)'(Y - X\beta)$ is the β that minimizes the second term $\left(X\hat{\beta} - X\beta\right)'\left(X\hat{\beta} - X\beta\right)$. The second term is non-negative because it is the sum of squares of the elements in the vector $X\hat{\beta} - X\beta$ and it is minimized by making it zero. This is accomplished by choosing $\beta = \hat{\beta}$. □

EXAMPLE 15.3.2. *Simple linear regression*

We now show that Proposition 15.3.1 gives the usual estimates for simple linear regression. Readers should refamiliarize themselves with the results in Section 7.3. They should also be warned that the algebra in the first half of the example is a bit more sophisticated than that used elsewhere in this book.

Assume the model

$$y_i = \beta_0 + \beta_1 x_i + \varepsilon_i \quad i = 1, \ldots, n.$$

and write

$$X = \begin{bmatrix} 1 & x_1 \\ 1 & x_2 \\ \vdots & \vdots \\ 1 & x_n \end{bmatrix}$$

so

$$X'X = \begin{bmatrix} n & \sum_{i=1}^n x_i \\ \sum_{i=1}^n x_i & \sum_{i=1}^n x_i^2 \end{bmatrix}.$$

Inverting this matrix gives

$$(X'X)^{-1} = \frac{1}{n\sum_{i=1}^n x_i^2 - \left(\sum_{i=1}^n x_i\right)^2} \begin{bmatrix} \sum_{i=1}^n x_i^2 & -\sum_{i=1}^n x_i \\ -\sum_{i=1}^n x_i & n \end{bmatrix}.$$

The denominator in this term can be simplified by observing that

$$n\sum_{i=1}^n x_i^2 - \left(\sum_{i=1}^n x_i\right)^2 = n\left(\sum_{i=1}^n x_i^2 - n\bar{x}_\bullet^2\right) = n\sum_{i=1}^n (x_i - \bar{x})^2.$$

Note also that

$$X'Y = \begin{bmatrix} \sum_{i=1}^n y_i \\ \sum_{i=1}^n x_i y_i \end{bmatrix}.$$

Finally, we get

$$\begin{aligned}
\hat{\beta} &= (X'X)^{-1} X'Y \\
&= \frac{1}{n\sum_{i=1}^n (x_i - \bar{x})^2} \begin{bmatrix} \sum_{i=1}^n x_i^2 \sum_{i=1}^n y_i - \sum_{i=1}^n x_i \sum_{i=1}^n x_i y_i \\ -\sum_{i=1}^n x_i \sum_{i=1}^n y_i + n\sum_{i=1}^n x_i y_i \end{bmatrix} \\
&= \frac{1}{\sum_{i=1}^n (x_i - \bar{x})^2} \begin{bmatrix} \bar{y}\sum_{i=1}^n x_i^2 - \bar{x}\sum_{i=1}^n x_i y_i \\ \left(\sum_{i=1}^n x_i y_i\right) - n\bar{x}\bar{y} \end{bmatrix} \\
&= \frac{1}{\sum_{i=1}^n (x_i - \bar{x})^2} \begin{bmatrix} \bar{y}\sum_{i=1}^n x_i^2 - n\bar{x}^2\bar{y} - \left\{\bar{x}\left(\sum_{i=1}^n x_i y_i\right) - \left(n\bar{x}^2\bar{y}\right)\right\} \\ \hat{\beta}_1 \sum_{i=1}^n (x_i - \bar{x})^2 \end{bmatrix} \\
&= \frac{1}{\sum_{i=1}^n (x_i - \bar{x})^2} \begin{bmatrix} \bar{y}\left(\sum_{i=1}^n x_i^2 - n\bar{x}^2\right) - \bar{x}\left(\sum_{i=1}^n x_i y_i - n\bar{x}\bar{y}\right) \\ \hat{\beta}_1 \sum_{i=1}^n (x_i - \bar{x})^2 \end{bmatrix} \\
&= \begin{bmatrix} \bar{y} - \hat{\beta}_1 \bar{x} \\ \hat{\beta}_1 \end{bmatrix} = \begin{bmatrix} \hat{\beta}_0 \\ \hat{\beta}_1 \end{bmatrix}.
\end{aligned}$$

As usual, the alternative regression model

$$y_i = \beta_{*0} + \beta_1 (x_i - \bar{x}) + \varepsilon_i \quad i = 1, \ldots, n$$

is easier to work with. Write the model in matrix form as

$$Y = Z\beta_* + e$$

where

$$Z = \begin{bmatrix} 1 & (x_1 - \bar{x}) \\ 1 & (x_2 - \bar{x}) \\ \vdots & \vdots \\ 1 & (x_n - \bar{x}) \end{bmatrix}$$

and

$$\beta_* = \begin{bmatrix} \beta_{*0} \\ \beta_1 \end{bmatrix}.$$

We need to compute $\hat{\beta}_* = (Z'Z)^{-1} Z'Y$. Observe that

$$Z'Z = \begin{bmatrix} n & 0 \\ 0 & \sum_{i=1}^n (x_i - \bar{x})^2 \end{bmatrix},$$

$$(Z'Z)^{-1} = \begin{bmatrix} \frac{1}{n} & 0 \\ 0 & 1 \Big/ \sum_{i=1}^n (x_i - \bar{x})^2 \end{bmatrix},$$

$$Z'Y = \begin{bmatrix} \sum_{i=1}^n y_i \\ \sum_{i=1}^n (x_i - \bar{x}) y_i \end{bmatrix},$$

and

$$\hat{\beta}_* = (Z'Z)^{-1} Z'Y = \begin{bmatrix} \bar{y} \\ \sum_{i=1}^n (x_i - \bar{x}) y_i \Big/ \sum_{i=1}^n (x_i - \bar{x})^2 \end{bmatrix} = \begin{bmatrix} \hat{\beta}_{*0} \\ \hat{\beta}_1 \end{bmatrix}.$$

These are the usual estimates. □

Recall that least squares estimates have a number of other properties. If the errors are independent with mean zero, constant variance, and are normally distributed, the least squares estimates are maximum likelihood estimates (cf. subsection 19.2.2) and minimum variance unbiased estimates. If the errors are merely uncorrelated with mean zero and constant variance, the least squares estimates are best (minimum variance) linear unbiased estimates.

In multiple regression, simple algebraic expressions for the parameter estimates are not possible. The only nice equations for the estimates are the matrix equations.

We now find expected values and covariance matrices for the data Y and the least squares estimate $\hat{\beta}$. Two simple rules about expectations and covariance matrices can take one a long way in the theory of regression. These are matrix analogues of Proposition 1.2.11. In fact, to prove these matrix results, one really only needs Proposition 1.2.11, cf. Exercise 15.8.3.

Proposition 15.3.3. Let A be a fixed $r \times n$ matrix, let c be a fixed $r \times 1$ vector, and let Y be an $n \times 1$ random vector, then

1. $E(AY + c) = A\,E(Y) + c$

2. $Cov(AY + c) = ACov(Y)A'$.

Applying these results allows us to find the expected value and covariance matrix for Y in a linear model. The linear model has $Y = X\beta + e$ where $X\beta$ is a fixed vector (even though β is unknown), $E(e) = 0$, and $Cov(e) = \sigma^2 I$. Applying the proposition gives

$$E(Y) = E(X\beta + e) = X\beta + E(e) = X\beta$$

and

$$Cov(Y) = Cov(e) = \sigma^2 I.$$

We can also find the expected value and covariance matrix of the least squares estimate $\hat{\beta}$. In particular, we show that $\hat{\beta}$ is an *unbiased* estimate of β by showing

$$E(\hat{\beta}) = E\left((X'X)^{-1} X'Y\right) = (X'X)^{-1} X'E(Y) = (X'X)^{-1} X'X\beta = \beta.$$

To find variances and standard errors we need $Cov(\hat{\beta})$. To obtain this matrix, we use the rules in Proposition A.7.1. In particular, recall that the inverse of a symmetric matrix is symmetric and that $X'X$ is symmetric.

$$\begin{aligned}
Cov(\hat{\beta}) &= Cov\left[(X'X)^{-1} X'Y\right] \\
&= \left[(X'X)^{-1} X'\right] Cov(Y) \left[(X'X)^{-1} X'\right]' \\
&= \left[(X'X)^{-1} X'\right] Cov(Y) X \left[(X'X)^{-1}\right]' \\
&= (X'X)^{-1} X' Cov(Y) X (X'X)^{-1} \\
&= \sigma^2 (X'X)^{-1} X'X (X'X)^{-1} \\
&= \sigma^2 (X'X)^{-1}.
\end{aligned}$$

15.4 Inferential procedures

We begin by examining the analysis of variance table for the regression model (15.2.2). We then discuss tests, confidence intervals, and prediction intervals.

There are two frequently used forms of the ANOVA table:

Source	df	SS	MS
β_0	1	$n\bar{y}^2 \equiv C$	$n\bar{y}^2$
Regression	$p-1$	$\hat{\beta}'X'X\hat{\beta} - C$	$SSReg/(p-1)$
Error	$n-p$	$Y'Y - C - SSReg$	$SSE/(n-p)$
Total	n	$Y'Y$	

and the more often used form

Source	df	SS	MS
Regression	$p-1$	$\hat{\beta}'X'X\hat{\beta} - C$	$SSReg/(p-1)$
Error	$n-p$	$Y'Y - C - SSReg$	$SSE/(n-p)$
Total	$n-1$	$Y'Y - C$	

Note that $Y'Y = \sum_{i=1}^n y_i^2$, $C = n\bar{y}^2 = \left(\sum_{i=1}^n y_i\right)^2/n$, and $\hat{\beta}'X'X\hat{\beta} = \hat{\beta}'X'Y$. The difference between the two tables is that the first includes a line for the intercept or grand mean while in the second the total has been corrected for the grand mean.

The coefficient of determination is

$$R^2 = \frac{SSReg}{Y'Y - C}.$$

This is the ratio of the variability explained by the predictor variables to the total variability of the data. Note that $(Y'Y - C)/(n-1) = s_y^2$, the sample variance of the ys without adjusting for any structure except the existence of a possibly nonzero mean.

EXAMPLE 15.4.1. *Simple linear regression*

For simple linear regression, we know that

$$SSReg = \hat{\beta}_1^2 \sum_{i=1}^{n} (x_i - \bar{x})^2 = \hat{\beta}_1 \sum_{i=1}^{n} (x_i - \bar{x})^2 \hat{\beta}_1$$

We will examine the alternative model

$$y_i = \beta_{*0} + \beta_1 (x_i - \bar{x}) + \varepsilon_i.$$

Note that $C = n\hat{\beta}_{*0}^2$, so the general form for $SSReg$ reduces to the simple linear regression form because

$$\begin{aligned} SSReg &= \hat{\beta}_*' Z' Z \hat{\beta}_* - C \\ &= \begin{bmatrix} \hat{\beta}_{*0} \\ \hat{\beta}_1 \end{bmatrix}' \begin{bmatrix} n & 0 \\ 0 & \sum_{i=1}^{n} (x_i - \bar{x})^2 \end{bmatrix} \begin{bmatrix} \hat{\beta}_{*0} \\ \hat{\beta}_1 \end{bmatrix} - C \\ &= \hat{\beta}_1^2 \sum_{i=1}^{n} (x_i - \bar{x})^2 . \end{aligned}$$

The same result can be obtained from $\hat{\beta}' X' X \hat{\beta} - C$ but the algebra is more tedious. □

To obtain tests and confidence regions we need to make additional distributional assumptions. In particular, we assume that the y_is have independent normal distributions. Equivalently, we take

$$\varepsilon_1, \ldots, \varepsilon_n \text{ indep. } N(0, \sigma^2).$$

To test the hypothesis

$$H_0 : \beta_1 = \beta_2 = \cdots = \beta_{p-1} = 0$$

versus

$$H_A : \text{ not all of } \beta_1, \beta_2, \ldots, \beta_{p-1} \text{ are zero,}$$

use the analysis of variance table test statistic

$$F = \frac{MSReg}{MSE}.$$

Under H_0,

$$F \sim F(p-1, n-p).$$

We can also perform a variety of t tests for individual regression parameters β_k. The procedures fit into the general techniques of Chapter 3 based on identifying 1) the parameter, 2) the estimate, 3) the standard error of the estimate, and 4) the distribution of $(Est - Par)/SE(Est)$. The parameter of interest is β_k. Having previously established that

$$E \begin{bmatrix} \hat{\beta}_0 \\ \hat{\beta}_1 \\ \vdots \\ \hat{\beta}_{p-1} \end{bmatrix} = \begin{bmatrix} \beta_0 \\ \beta_1 \\ \vdots \\ \beta_{p-1} \end{bmatrix},$$

it follows that for any $k = 0, \ldots, p-1$,

$$E(\hat{\beta}_k) = \beta_k.$$

This shows that $\hat{\beta}_k$ is an unbiased estimate of β_k. Before obtaining the standard error of $\hat{\beta}_k$, it is necessary to identify its variance. The covariance matrix of $\hat{\beta}$ is $\sigma^2 \left(X'X\right)^{-1}$, so the variance of $\hat{\beta}_k$ is the $(k+1)$st diagonal element of $\sigma^2 \left(X'X\right)^{-1}$. The $(k+1)$st diagonal element is appropriate because the first diagonal element is the variance of $\hat{\beta}_0$ not $\hat{\beta}_1$. If we let a_k be the $(k+1)$st diagonal element of $\left(X'X\right)^{-1}$ and estimate σ^2 with MSE, we get a standard error for $\hat{\beta}_k$ of

$$SE\left(\hat{\beta}_k\right) = \sqrt{MSE}\sqrt{a_k}.$$

Under normal errors, the appropriate reference distribution is

$$\frac{\hat{\beta}_k - \beta_k}{SE(\hat{\beta}_k)} \sim t(n-p).$$

Standard techniques now provide tests and confidence intervals. For example, a 95% confidence interval for β_k is

$$\hat{\beta}_k \pm t(.975, n-p)\, SE(\hat{\beta}_k)$$

where $t(.975, n-p)$ is the 97.5th percentile of a t distribution with $n-p$ degrees of freedom.

A $(1-\alpha)100\%$ simultaneous confidence region for $\beta_0, \beta_1, \ldots, \beta_{p-1}$ consists of all the β vectors that satisfy

$$\frac{\left(\hat{\beta}-\beta\right)' X'X \left(\hat{\beta}-\beta\right) \Big/ p}{MSE} \leq F(1-\alpha, p, n-p).$$

This region also determines joint $(1-\alpha)100\%$ confidence intervals for the individual β_ks with limits

$$\hat{\beta}_k \pm \sqrt{pF(1-\alpha, p, n-p)}\, SE(\hat{\beta}_k).$$

These intervals are an application of Scheffé's method of multiple comparisons.

We can also use the Bonferroni method to obtain joint $(1-\alpha)100\%$ confidence intervals with limits

$$\hat{\beta}_k \pm t\left(1 - \frac{\alpha}{2p}, n-p\right)\, SE(\hat{\beta}_k).$$

Finally, we consider estimation of the point on the surface that corresponds to a given set of predictor variables and the prediction of a new observation with a given set of predictor variables. Let the predictor variables be $x_1, x_2, \ldots, x_{p-1}$. Combine these into the row vector

$$x' = \left(1, x_1, x_2, \ldots, x_{p-1}\right).$$

The point on the surface that we are trying to estimate is the parameter $x'\beta = \beta_0 + \sum_{j=1}^{p-1} \beta_j x_j$. The least squares estimate is $x'\hat{\beta}$. The variance of the estimate is

$$\mathrm{Var}\left(x'\hat{\beta}\right) = Cov\left(x'\hat{\beta}\right) = x' Cov\left(\hat{\beta}\right) x = \sigma^2 x' \left(X'X\right)^{-1} x,$$

so the standard error is

$$SE\left(x'\hat{\beta}\right) = \sqrt{MSE}\sqrt{x' \left(X'X\right)^{-1} x}.$$

This is the standard error of the estimated regression surface. The appropriate reference distribution is

$$\frac{x'\hat{\beta} - x'\beta}{SE\left(x'\hat{\beta}\right)} \sim t(n-p)$$

and a $(1 - \alpha)100\%$ confidence interval is

$$x'\hat{\beta} \pm t\left(1 - \frac{\alpha}{2}, n - p\right) SE(x'\hat{\beta}).$$

When predicting a new observation, the point prediction is just the estimate of the point on the surface but the standard error must incorporate the additional variability associated with a new observation. The original observations were assumed to be independent with variance σ^2. It is reasonable to assume that a new observation is independent of the previous observations and has the same variance. Thus, in the prediction we have to account for the variance of the new observation, which is σ^2, plus the variance of the estimate $x'\hat{\beta}$, which is $\sigma^2 x' \left(X'X\right)^{-1} x$. This leads to a variance for the prediction of $\sigma^2 + \sigma^2 x' \left(X'X\right)^{-1} x$ and a standard error of

$$\sqrt{MSE + MSE\, x' (X'X)^{-1} x} = \sqrt{MSE \left[1 + x' (X'X)^{-1} x\right]}.$$

The $(1 - \alpha)100\%$ prediction interval is

$$x'\hat{\beta} \pm t\left(1 - \frac{\alpha}{2}, n - p\right) \sqrt{MSE \left[1 + x' (X'X)^{-1} x\right]}.$$

Results of this section constitute the theory behind most of the applications in Sections 13.1 and 13.2.

15.5 Residuals, standardized residuals, and leverage

Let $x_i' = (1, x_{i1}, \ldots, x_{i,p-1})$ be the ith row of X, then

$$\hat{y}_i = \hat{\beta}_0 + \hat{\beta}_1 x_{i1} + \cdots + \hat{\beta}_{p-1} x_{i,p-1} = x_i'\hat{\beta}$$

and the corresponding residual is

$$\hat{\varepsilon}_i = y_i - \hat{y}_i = y_i - x_i'\hat{\beta}.$$

The vector of predicted values is

$$\hat{Y} = \begin{bmatrix} \hat{y}_1 \\ \vdots \\ \hat{y}_n \end{bmatrix} = \begin{bmatrix} x_1'\hat{\beta} \\ \vdots \\ x_n'\hat{\beta} \end{bmatrix} = X\hat{\beta}.$$

The vector of residuals is

$$\begin{aligned} \hat{e} &= Y - \hat{Y} \\ &= Y - X\hat{\beta} \\ &= Y - X(X'X)^{-1}X'Y \\ &= \left(I - X(X'X)^{-1}X'\right) Y \\ &= (I - M)\, Y \end{aligned}$$

where

$$M \equiv X(X'X)^{-1}X'.$$

M is called the perpendicular projection operator (matrix) onto $C(X)$, the column space of X. M is the key item in the analysis of the general linear model, cf. Christensen (1987). Note that M is symmetric, i.e., $M = M'$, and that $MM = M$. Using these facts, observe that

$$\begin{aligned} SSE &= \sum_{i=1}^{n} \hat{\varepsilon}_i^2 \\ &= \hat{e}'\hat{e} \\ &= [(I - M)Y]'\,[(I - M)Y] \\ &= Y'(I - M' - M + M'M)Y \\ &= Y'(I - M)Y. \end{aligned}$$

Another common way of writing SSE is

$$SSE = \left[Y - X\hat{\beta}\right]'\left[Y - X\hat{\beta}\right].$$

Having identified M, we can define the standardized residuals. First we find the covariance matrix of the residual vector $\hat{e}$:

$$\begin{aligned} Cov(\hat{e}) &= Cov([I - M]Y) \\ &= [I - M]Cov(Y)[I - M]' \\ &= [I - M]\sigma^2 I[I - M]' \\ &= \sigma^2\left(I - M - M' + MM'\right) \\ &= \sigma^2\,(I - M). \end{aligned}$$

The last equality follows from $M = M'$ and $MM = M$. Typically, the covariance matrix is not diagonal, so the residuals are not uncorrelated.

The variance of a particular residual $\hat{\varepsilon}_i$ is σ^2 times the ith diagonal element of $(I - M)$. The ith diagonal element of $(I - M)$ is the ith diagonal element of I, 1, minus the ith diagonal element of M, say, m_{ii}. Thus

$$\text{Var}(\hat{\varepsilon}_i) = \sigma^2(1 - m_{ii})$$

and the standard error of $\hat{\varepsilon}_i$ is

$$SE(\hat{\varepsilon}_i) = \sqrt{MSE(1 - m_{ii})}.$$

The ith standardized residual is defined as

$$r_i \equiv \frac{\hat{\varepsilon}_i}{\sqrt{MSE(1 - m_{ii})}}.$$

The leverage of the ith case is defined to be m_{ii}, the ith diagonal element of M. Some people like to think of M as the 'hat' matrix because it transforms Y into $\hat{Y}$, i.e., $\hat{Y} = X\hat{\beta} = MY$. More common than the name 'hat matrix' is the consequent use of the notation h_i for the ith leverage. This notation was used in Chapter 7 but the reader should realize that $h_i \equiv m_{ii}$. In any case, the leverage can be interpreted as a measure of how unusual x_i' is relative to the other rows of the X matrix, cf. Christensen (1987, section XIII.1).

15.6 Principal components regression

In Section 13.7 we dealt with the issue of collinearity. Four points were emphasized as the effects of collinearity.

1. The estimate of any parameter, say $\hat{\beta}_2$, depends on *all* the variables that are included in the model.

2. The sum of squares for any variable, say x_2, depends on *all* the other variables that are included in the model. For example, none of $SSR(x_2)$, $SSR(x_2|x_1)$, and $SSR(x_2|x_3, x_4)$ would typically be equal.

3. In a model such as $y_i = \beta_0 + \beta_1 x_{i1} + \beta_2 x_{i2} + \beta_3 x_{i3} + \varepsilon_i$, small t statistics for both $H_0 : \beta_1 = 0$ and $H_0 : \beta_2 = 0$ are not sufficient to conclude that an appropriate model is $y_i = \beta_0 + \beta_3 x_{i3} + \varepsilon_i$. To arrive at a reduced model, one must compare the reduced model to the full model.

4. A moderate amount of collinearity has little effect on predictions and therefore little effect on SSE, R^2, and the explanatory power of the model. Collinearity increases the variance of the $\hat{\beta}_j$s, making the estimates of the parameters less reliable. Depending on circumstances, sometimes a large amount of collinearity can have an effect on predictions. Just by chance one may get a better fit to the data than can be justified scientifically.

At its worst, collinearity involves near redundancies among the predictor variables. An exact redundancy among the predictor variables occurs when we can find a p vector $d \neq 0$ so that $Xd = 0$. When this happens the rank of X is not p, so we cannot find $(X'X)^{-1}$ and we cannot find the estimates of β in Proposition 15.3.1. *Near* redundancies occur when we can find a vector d that is not too small, say with $d'd = 1$, having $Xd \doteq 0$. Principal components (PC) regression is a method designed to identify near redundancies among the predictor variables. Having identified near redundancies, they can be eliminated if we so choose. In Section 13.7 we mentioned that having small collinearity requires more than having small correlations among all the predictor variables, it requires all partial correlations among the predictor variables to be small as well. For this reason, eliminating near redundancies cannot always be accomplished by simply dropping well chosen predictor variables from the model.

The basic idea of principal components is to find new variables that are linear combinations of the x_js and that are *best able to (linearly) predict the entire set of x_js*, see Christensen (1990a, chapter III). Thus the first principal component variable is the one linear combination of the x_js that is best able to predict all of the x_js. The second principal component variable is the linear combination of the x_js that is best able to predict all the x_js among those linear combinations having a sample correlation of 0 with the first principal component variable. The third principal component variable is the best predictor that has sample correlations of 0 with the first two principal component variables. The remaining principal components are defined similarly. With $p - 1$ predictor variables, there are $p - 1$ principal component variables. The full collection of principal component variables always predicts the full collection of x_js perfectly. The last few principal component variables are least able to predict the original x_j variables, so they are the least useful. They are also the aspects of the predictor variables that are most redundant, see Christensen (1987, section XIV.5). The best (linear) predictors used in defining principal components can be based on either the covariances between the x_js or the correlations between the x_js. Unless the x_js are measured on the same scale (with similarly sized measurements), it is generally best to use principal components defined using the correlations.

For *The Coleman Report* data, a matrix of sample correlations between the x_js was given in Example 13.7.1. Principal components are derived from the eigenvalues and eigenvectors of this matrix. (Alternatively, one could use eigenvalues and eigenvectors of the matrix of sample covariances.) An eigenvector corresponding to the largest eigenvalue determines the first principal component variable. The eigenvalues are given in Table 15.1 along with proportions and cumulative proportions.

The proportions in Table 15.1 are simply the eigenvalues divided by the sum of the eigenvalues. The cumulative proportions are the sum of the first group of eigenvalues divided by the sum of all

TABLE 15.1. Eigen analysis of the correlation matrix

Eigenvalue	2.8368	1.3951	0.4966	0.2025	0.0689
Proportion	0.567	0.279	0.099	0.041	0.014
Cumulative	0.567	0.846	0.946	0.986	1.000

the eigenvalues. In this example, the sum of the eigenvalues is

$$5 = 2.8368 + 1.3951 + 0.4966 + 0.2025 + 0.0689.$$

The sum of the eigenvalues must equal the sum of the diagonal elements of the original matrix. The sum of the diagonal elements of a correlation matrix is the number of variables in the matrix. The third eigenvalue in Table 15.1 is $.4966$. The proportion is $.4966/5 = .099$. The cumulative proportion is $(2.8368 + 1.3951 + .4966)/5 = .946$. With an eigenvalue proportion of 9.9%, the third principal component variable accounts for 9.9% of the variance associated with predicting the x_js. Taken together, the first three principal components account for 94.6% of the variance associated with predicting the x_js because the third cumulative eigenvalue proportion is .946.

TABLE 15.2. Principal component variable coefficients

Variable	PC1	PC2	PC3	PC4	PC5
x_1	−0.229	−0.651	0.723	0.018	−0.024
x_2	−0.555	0.216	0.051	−0.334	0.729
x_3	−0.545	0.099	−0.106	0.823	−0.060
x_4	−0.170	−0.701	−0.680	−0.110	0.075
x_5	−0.559	0.169	−0.037	−0.445	−0.678

For the school data, the principal component (PC) variables are determined by the coefficients in Table 15.2. The first principal component variable is

$$\begin{aligned} PC1_i = &-0.229(x_{i1} - \bar{x}_{\bullet 1})/s_1 - 0.555(x_{i2} - \bar{x}_{\bullet 2})/s_2 \\ &- 0.545(x_{i3} - \bar{x}_{\bullet 3})/s_3 - 0.170(x_{i4} - \bar{x}_{\bullet 5})/s_4 - 0.559(x_{i5} - \bar{x}_{\bullet 5})/s_5 \end{aligned} \quad (15.6.1)$$

for $i = 1, \ldots, 20$ where s_1 is the sample standard deviation of the x_{i1}s, etc. The columns of coefficients given in Table 15.2 are actually eigenvectors for the correlation matrix of the x_js. The PC1 coefficients are an eigenvector corresponding to the largest eigenvalue, the PC2 coefficients are an eigenvector corresponding to the second largest eigenvalue, etc.

We can now perform a regression on the new principal component variables. The estimates, standard errors, and t tests are given in Table 15.3. The analysis of variance is given in Table 15.4. The value of R^2 is .906. The analysis of variance table and R^2 are identical to those for the original predictor variables given in Section 13.1. The plot of standardized residuals versus predicted values from the principal component regression is given in Figure 15.1. This is identical to the plot given in Figure 14.2 for the original variables. All of the predicted values and all of the standardized residuals are identical.

Since Table 15.4 and Figure 15.1 are unchanged, any usefulness associated with principal component regression must come from Table 15.3. The principal component variables display no collinearity. Thus, contrary to the warnings given earlier about the effects of collinearity, we can make final conclusions about the importance of variables directly from Table 15.3. We do not have to worry about fitting one model after another or about which variables are included in which models. From examining Table 15.3, it is clear that the important variables are PC1, PC3, and PC4. We can

TABLE 15.3. Regression analysis on principal component scores

Predictor	$\hat{\gamma}$	$SE(\hat{\gamma})$	t	P
Constant	35.0825	0.4638	75.64	0.000
PC1	−2.9419	0.2825	−10.41	0.000
PC2	0.0827	0.4029	0.21	0.840
PC3	−2.0457	0.6753	−3.03	0.009
PC4	4.380	1.057	4.14	0.001
PC5	1.433	1.812	0.79	0.442

TABLE 15.4. Analysis of variance for regression on principal component scores

Source	df	SS	MS	F	P
Regression	5	582.69	116.54	27.08	0.000
Error	14	60.24	4.30		
Total	19	642.92			

construct a reduced model with these three; the estimated regression surface is simply

$$\hat{y} = 35.0825 - 2.9419(\text{PC1}) - 2.0457(\text{PC3}) + 4.380(\text{PC4}). \qquad (15.6.2)$$

In equation (15.6.2), we merely used the estimated regression coefficients from Table 15.3. Refitting the reduced model is unnecessary because there is no collinearity.

To get predictions for a new set of x_js, just compute the corresponding PC1, PC3, and PC4 variables using formulae similar to those in equation (15.6.1) and make the predictions using the fitted model in equation (15.6.2). When using equations like (15.6.1) to obtain new values of the principal component variables, continue to use the $\bar{x}_{\bullet j}$s and s_js computed from only the original observations.

As an alternative to this prediction procedure, we could use the definitions of the principal component variables, e.g., equation (15.6.1), and substitute for PC1, PC3, and PC4 in equation (15.6.2) to obtain estimated coefficients on the original x_j variables.

$$\begin{aligned}
\hat{y} &= 35.0825 + [-2.9419, -2.0457, 4.380]\begin{bmatrix}\text{PC1}\\ \text{PC3}\\ \text{PC4}\end{bmatrix}\\
&= 35.0825 + [-2.9419, -2.0457, 4.380] \times\\
&\quad \begin{bmatrix} -0.229 & -0.555 & -0.545 & -0.170 & -0.559\\ 0.723 & 0.051 & -0.106 & -0.680 & -0.037\\ 0.018 & -0.334 & 0.823 & -0.110 & -0.445 \end{bmatrix}\begin{bmatrix}(x_1-\bar{x}_{\bullet 1})/s_1\\ (x_2-\bar{x}_{\bullet 2})/s_2\\ (x_3-\bar{x}_{\bullet 3})/s_3\\ (x_4-\bar{x}_{\bullet 4})/s_4\\ (x_5-\bar{x}_{\bullet 5})/s_5\end{bmatrix}\\
&= 35.0825 + [-0.72651, 0.06550, 5.42492, 1.40940, -0.22889] \times\\
&\quad \begin{bmatrix}(x_1-2.731)/0.454\\ (x_2-40.91)/25.90\\ (x_3-3.14)/9.63\\ (x_4-25.069)/1.314\\ (x_5-6.255)/0.654\end{bmatrix}.
\end{aligned}$$

Obviously this can be simplified into a form $\hat{y} = 35.0825 + \tilde{\beta}_1 x_1 + \tilde{\beta}_2 x_2 + \tilde{\beta}_3 x_3 + \tilde{\beta}_4 x_4 + \tilde{\beta}_5 x_5$, which in turn simplifies the process of making predictions and provides new estimated regression coefficients

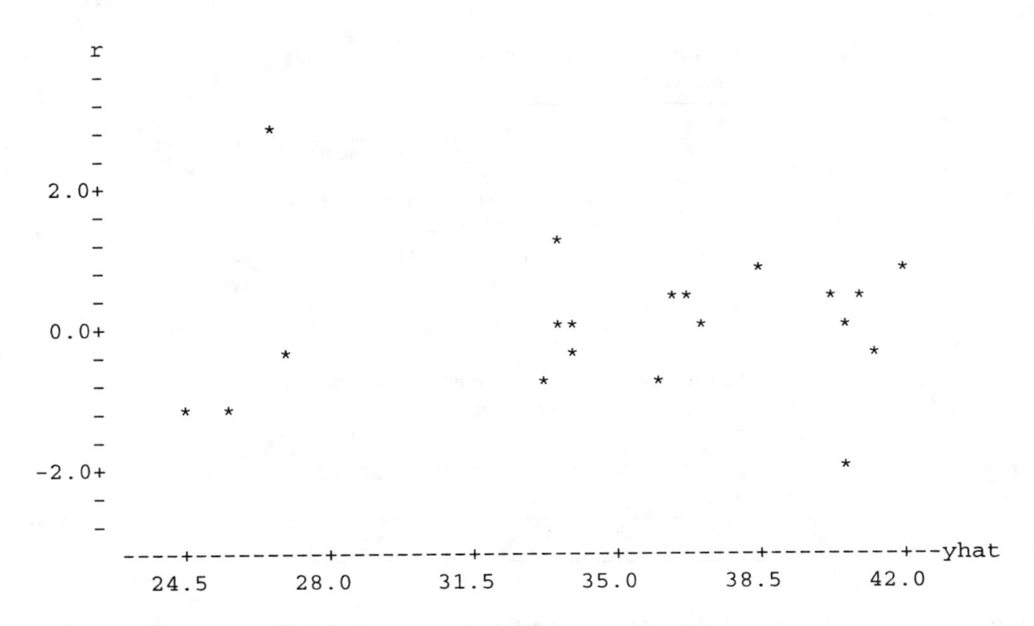

FIGURE 15.1. Standardized residuals versus predicted values for principal component regression.

for the x_js that correspond to the fitted principal component model. These PC regression estimates of the original β_js can be compared to the least squares estimates. Many computer programs for performing PC regression report these estimates of the β_js and their corresponding standard errors.

It was mentioned earlier that collinearity tends to increase the variance of regression coefficients. The fact that the later principal component variables are more nearly redundant is reflected in Table 15.3 by the fact that the standard errors for their estimated regression coefficients increase.

One rationale for using PC regression is that you just don't believe in using nearly redundant variables. The exact nature of such variables can be changed radically by small errors in the x_js. For this reason, one might choose to ignore PC5 because of its small eigenvalue proportion, regardless of any importance it may display in Table 15.3. If the t statistic for PC5 appeared to be significant, it could be written off as a chance occurrence or, perhaps more to the point, as something that is unlikely to be reproducible. If you don't believe redundant variables, i.e., if you don't believe that they are themselves reproducible, any predictive ability due to such variables will not be reproducible either.

When considering PC5, the case is pretty clear. PC5 accounts for only about 1.5% of the variability involved in predicting the x_js. It is a very poorly defined aspect of the predictor variables x_j and, anyway, it is not a significant predictor of y. The case is less clear when considering PC4. This variable has a significant effect for explaining y, but it accounts for only 4% of the variability in predicting the x_js, so PC4 is reasonably redundant within the x_js. If this variable is measuring some reproducible aspect of the original x_j data, it should be included in the regression. If it is not reproducible, it should not be included. From examining the PC4 coefficients in Table 15.2, we see that PC4 is roughly the average of the percent white collar fathers x_2 and the mothers' education x_5 contrasted with the socio- economic variable x_3. (Actually, this comparison is between the variables after they have been adjusted for their means and standard deviation as in equation (15.6.1).) If PC4 strikes the investigator as a meaningful, reproducible variable, it should be included in the regression.

In our discussion, we have used PC regression both to eliminate questionable aspects of the predictor variables and as a method for selecting a reduced model. We dropped PC5 primarily because it was poorly defined. We dropped PC2 solely because it was not a significant predictor.

Some people might argue against this second use of PC regression and choose to take a model based on PC1, PC2, PC3, and possibly PC4.

On occasion, PC regression is based on the sample covariance matrix of the x_js rather than the sample correlation matrix. Again, eigenvalues and eigenvectors are used, but in using relationships like equation (15.6.1), the s_js are deleted. The eigenvalues and eigenvectors for the covariance matrix typically differ from those for the correlation matrix. The relationship between estimated principal component regression coefficients and original least squares regression coefficient estimates is somewhat simpler when using the covariance matrix.

It should be noted that PC regression is just as sensitive to violations of the assumptions as regular multiple regression. Outliers and high leverage points can be very influential in determining the results of the procedure. Tests and confidence intervals rely on the independence, homoscedasticity, and normality assumptions. Recall that in the full principal components regression model, the residuals and predicted values are identical to those from the regression on the original predictor variables. Moreover, highly influential points in the original predictor variables typically have a large influence on the coefficients in the principal component variables.

MINITAB COMMANDS

Minitab commands for the principal components regression analysis are given below. The basic command is 'pca.' The 'scores' subcommand places the principal component variables into columns c12 through c16. The 'coef' subcommand places the eigenvectors into columns c22 through c26. If one wishes to define principal components using the covariances rather than the correlations, simply include a pca subcommand with the word 'covariance.'

```
MTB > pca c2-c6;
SUBC> scores c12-c16;
SUBC> coef c22-c26.
MTB > regress c8 on 5 c12-c16 c17 c18
MTB > plot c17 c18
```

15.7 Weighted least squares

In general, weighted regression is a method for dealing with observations that have nonconstant variances and nonzero correlations. In this section, we deal with the simplest form of weighted regression in which we continue to assume zero correlations between observations. This is the form used for logistic regression in Section 8.7.

Our standard regression model has

$$Y = X\beta + e, \quad E(e) = 0, \quad Cov(e) = \sigma^2 I.$$

We now consider a model for data that do not all have the same variance. In this model, we assume that the *relative* sizes of the variances are known but that the variances themselves are unknown. In this simplest form of weighted regression, we have a covariance structure that changes from $Cov(e) = \sigma^2 I$ to $Cov(e) = \sigma^2 D(w)^{-1}$. Here $D(w)$ is a diagonal matrix with *known* weights $w = (w_1, \ldots, w_n)'$ along the diagonal. The covariance matrix involves $D(w)^{-1}$, which is just a diagonal matrix having diagonal entries that are $1/w_1, \ldots, 1/w_n$. The variance of an observation y_i is σ^2/w_i. If w_i is large relative to the other weights, the relative variance of y_i is small, so it contains more information than other observations and we should place more weight on it. Conversely, if w_i is relatively small, the variance of y_i is large, so it contains little information and we should place little weight on it. For all cases, w_i is a measure of how much relative weight should be placed on case i. Note that the weights are relative, so we could multiply or divide them all by a constant and obtain essentially the same analysis. Obviously, in standard regression the weights are all taken to be 1.

In matrix form, our new model is

$$Y = X\beta + e, \quad E(e) = 0, \quad Cov(e) = \sigma^2 D(w)^{-1}. \tag{15.7.1}$$

In this model all the observations are uncorrelated because the covariance matrix is diagonal. We do not know the variance of any observation because σ^2 is unknown. However, we do know the relative sizes of the variances because we know the weights w_i. It should be noted that when model (15.7.1) is used to make predictions, it is necessary to specify weights for any future observations.

Before giving a general discussion of weighted regression models, we examine some examples of their application. A natural application of weighted regression is to data for a one-way analysis of variance with treatments that are quantitative levels of some factor. With a quantitative factor, we can perform either a one-way ANOVA or a regression on the data. However, if for some reason the full data are not available, we can still obtain an appropriate simple linear regression by performing a weighted regression analysis on the treatment means. The next examples explore the relationships between regression on the full data and weighted regression on the treatment means.

In the weighted regression, the weights turn out to be the treatment group sample sizes from the ANOVA. In a standard unbalanced ANOVA $y_{ij} = \mu_i + \varepsilon_{ij}$, $i = 1, \ldots, a$, $j = 1, \ldots, N_i$, the sample means have $\text{Var}(\bar{y}_{i\cdot}) = \sigma^2/N_i$. Thus, if we perform a regression on the means, the observations have different variances. In particular, from our earlier discussion of variances and weights, it is appropriate to take the sample sizes as the weights, i.e., $w_i = N_i$.

EXAMPLE 15.7.1. In Section 7.12 we considered data on the ASI indices given in Table 7.10. A simple linear regression on the full data gives the following results for the parameters.

Predictor	$\hat{\beta}_k$	$SE(\hat{\beta}_k)$	t	P
Constant	285.30	15.96	17.88	0.000
x (plates)	12.361	1.881	6.57	0.000

The analysis of variance table for the simple linear regression is given below. The usual error line would have 33 degrees of freedom but, as in Section 7.12, we have broken this into two components, one for lack of fit and one for pure error.

Source	df	SS	MS
Regression	1	42780	42780
Lack of fit	3	1212	404
Pure error	30	31475	1049
Total	34	75468	

In Section 7.12 we also considered these data as a one-way ANOVA. As such, we examined the sample sizes and sample means given below.

Plate	4	6	8	10	12
N	7	7	7	7	7
$\bar{y}_{i\cdot}$	333.2143	368.0571	375.1286	407.3571	437.1714

Of course in Section 7.12 we also used the sample variances from each group in obtaining an error term.

As mentioned in Section 7.12, one can get the same estimated line by just fitting a simple linear regression to the means. For an unbalanced ANOVA, getting the correct regression line from the means requires a weighted regression. In this balanced case, if we use a weighted regression we get not only the same fitted line but also some interesting relationships in the ANOVA tables. Below are given the parameter estimate table and the ANOVA table for the weighted regression on the means. The weights are the sample sizes for each mean.

Predictor	$\hat{\beta}_k$	$SE(\hat{\beta}_k)$	t	P
Constant	285.30	10.19	27.99	0.000
x (plates)	12.361	1.201	10.29	0.002

Analysis of variance: weighted simple linear regression

Source	df	SS	MS	F	P
Regression	1	42780	42780	105.88	0.002
Error	3	1212	404		
Total	4	43993			

Note that the estimated regression coefficients are identical to those given earlier. The standard errors and thus the other entries in the parameter estimate table differ. In the ANOVA tables, the regression lines agree while the error line from the weighted regression is identical to the lack of fit line in the ANOVA table for the full data. In the weighted regression, all standard errors use the lack of fit as an estimate of the variance. In the regression on the full data, the standard errors use a variance estimate obtained from pooling the lack of fit and the pure error. The ultimate point is that by using weighted regression on the summarized data, we can still get most relevant summary statistics for simple linear regression. Of course, this assumes that the simple linear regression model is correct, and unfortunately the weighted regression does not allow us to test for lack of fit.

If we had taken all the weights to be one, i.e., if we had performed a standard regression on the means, the parameter estimate table would be the same but the ANOVA table would not display the identities discussed above. The sums of squares would all have been off by a factor of 7. □

MINITAB COMMANDS

To get the weighted regression from Minitab, suppose that c1 contains the plate lengths, c2 contains the sample sizes, and c3 contains the means. The commands are as follows:

```
MTB > regress c3 on 1 c1;
SUBC> weights c2.
```

A complete set of commands for generating an analysis such as this are given for the next example.

UNBALANCED WEIGHTS

We now examine an unbalanced one-way ANOVA and again compare a simple linear regression including identification of pure error and lack of fit to a weighted regression on sample means.

EXAMPLE 15.7.2. Consider the data of Exercise 7.13.1 and Table 7.14. These involve ages of truck tractors and the costs of maintaining the tractors. The analysis on the full data yields the tables given below.

Predictor	$\hat{\beta}_k$	$SE(\hat{\beta}_k)$	t	P
Constant	323.6	146.9	2.20	0.044
Age	131.72	35.61	3.70	0.002

Source	df	SS	MS
Regression	1	1099635	1099635
Lack of fit	5	520655	104131
Pure error	10	684752	68475
Total	16	2305042	

The weighted regression analysis is based on the sample means and sample sizes given below. The means serve as the y variable, the ages are the x variable, and the sample sizes are the weights.

Age	0.5	1.0	4.0	4.5	5.0	5.5	6.0
N_i	2	3	3	3	3	1	2
$\bar{y}_{i\cdot}$	172.5	664.3	633.0	900.3	1202.0	987.0	1068.5

The estimates and ANOVA table for the weighted regression are given below.

Predictor	$\hat{\beta}_k$	$SE(\hat{\beta}_k)$	t	P
Constant	323.6	167.3	1.93	0.111
Age	131.72	40.53	3.25	0.023

Analysis of variance: weighted simple linear regression

Source	df	SS	MS	F	P
Regression	1	1099635	1099635	10.56	0.023
Error	5	520655	104131		
Total	6	1620290			

Note that, as in the previous example, the regression estimates agree with those from the full data, that the regression sum of squares from the ANOVA table agrees with the full data, and that the lack of fit line from the full data ANOVA agrees with the error line from the weighted regression. *For an unbalanced ANOVA, you cannot obtain a correct simple linear regression analysis from the treatment means without using weighted regression.* □

MINITAB COMMANDS

The Minitab commands for this analysis are given below. To obtain pure error and lack of fit from the full data one fits both the simple linear regression and the one-way ANOVA. The ages from Table 7.14 are in c1 and the costs are in c2.

```
MTB > regress c2 on 1 c1
MTB > note      THE AGES IN c1 ARE NOT INTEGERS, SO WE
MTB > note      MULTIPLY THEM BY TWO TO MAKE INTEGER
MTB > note      GROUP LABELS FOR THE ONE-WAY
MTB > let c3=2*c1
MTB > oneway c2 c3
MTB > note      PUT THE MEANS FROM THE ONE-WAY INTO c6, THE
MTB > note      AGES INTO c7, AND THE SAMPLE SIZES INTO c8.
MTB > set c6
DATA> 172.5 664.333333 633 900.3333333 1202 987 1068.5
DATA> set c7
DATA> .5 1 4 4.5 5 5.5 6
DATA> end
MTB > set c8
DATA> 2 3 3 3 3 1 2
```

```
DATA> end
MTB > note      DO THE WEIGHTED REGRESSION
MTB > regress c6 on 1 c7;
SUBC> weight c8.
```

THEORY

The analysis of the weighted regression model (15.7.1) is based on changing it into a standard regression model. The trick is to create a new diagonal matrix that has entries $\sqrt{w_i}$. In a minor abuse of notation, we write this matrix as $D(\sqrt{w})$. We now multiply model (15.7.1) by this matrix to obtain

$$D(\sqrt{w})Y = D(\sqrt{w})X\beta + D(\sqrt{w})e. \qquad (15.7.2)$$

It is not difficult to see that

$$E\big(D(\sqrt{w})e\big) = D(\sqrt{w})E(e) = D(\sqrt{w})0 = 0$$

and

$$Cov\big(D(\sqrt{w})e\big) = D(\sqrt{w})Cov(e)D(\sqrt{w})' = D(\sqrt{w})\left[\sigma^2 D(w)^{-1}\right] D(\sqrt{w}) = \sigma^2 I.$$

Thus equation (15.7.2) defines a standard regression model. For example, by Proposition 15.3.1 the least squares regression estimates from model (15.7.2) are

$$\begin{aligned}\hat{\beta} &= \left([D(\sqrt{w})X]'[D(\sqrt{w})X]\right)^{-1}[D(\sqrt{w})X]'[D(\sqrt{w})Y] \\ &= (X'D(w)X)^{-1}X'D(w)Y.\end{aligned}$$

The estimate of β given above is referred to as a weighted least squares estimate because rather than minimizing $[Y - X\beta]'[Y - X\beta]$, the estimates are obtained by minimizing

$$\left[D(\sqrt{w})Y - D(\sqrt{w})X\beta\right]'\left[D(\sqrt{w})Y - D(\sqrt{w})X\beta\right] = [Y - X\beta]'D(w)[Y - X\beta].$$

Thus the original minimization problem has been changed into a similar minimization problem that incorporates the weights. The sum of squares for error from model (15.7.2) is

$$SSE = \left[D(\sqrt{w})Y - D(\sqrt{w})X\hat{\beta}\right]'\left[D(\sqrt{w})Y - D(\sqrt{w})X\hat{\beta}\right] = \left[Y - X\hat{\beta}\right]'D(w)\left[Y - X\hat{\beta}\right].$$

The dfE are unchanged from a standard model and MSE is simply SSE divided by dfE. Standard errors are found in much the same manner as usual except now

$$Cov\big(\hat{\beta}\big) = \sigma^2(X'D(w)X)^{-1}.$$

Because the $D(w)$ matrix is diagonal, it is very simple to modify a computer program for standard regression to allow the analysis of models like (15.7.1). Of course to make a prediction, a weight must now be specified for the new observation. Essentially the same idea of rewriting model (15.7.1) as the standard regression model (15.7.2) works even when $D(w)$ is not a diagonal matrix, cf. Christensen (1987, sections II.7 and III.8).

In Section 8.7 we used weighted regression to analyze binomial data. We used observed proportions $\hat{p}_i$ to define a dependent variable $\log[\hat{p}_i/(1-\hat{p}_i)]$ and we used weights $w_i = N_i\hat{p}_i(1-\hat{p}_i)$. Here N_i is the number of trials involved in the ith binomial. In this application, the weights are functions of the data, whereas in the previous discussion the weights were fixed. Ideally, the weights for binomial regression would be the fixed quantities $N_i p_i(1-p_i)$, but there is no hope of knowing these weights, so we estimate them and rely on the law of large numbers. The example in Section 8.7 involved only one predictor variable, but exactly the same methods apply when more than one predictor variable is available. However, as suggested in Section 8.7, it is probably better to use maximum likelihood methods to analyze binomial data than these weighted regression methods.

15.8 Exercises

EXERCISE 15.8.1. Show that the form (15.3.2) simplifies to the form (15.3.1) for simple linear regression.

EXERCISE 15.8.2. Show that $Cov(Y) = E[(Y - \mu)(Y - \mu)']$.

EXERCISE 15.8.3. Use Proposition 1.2.11 to show that $E(AY + c) = A\,E(Y) + c$ and $Cov(AY + c) = ACov(Y)A'$.

EXERCISE 15.8.4. Using eigenvalues, discuss the level of collinearity in:

(a) the Younger data from Exercise 13.8.1,

(b) the Prater data from Exercise 13.8.3,

(c) the Chapman data of Exercise 13.8.4,

(d) the pollution data from Exercise 13.8.5,

(e) the body fat data of Exercise 13.8.6.

EXERCISE 15.8.5. Do a principal components regression for the Younger data from Exercise 13.8.1.

EXERCISE 15.8.6. Do a principal components regression for the Prater data from Exercise 13.8.3.

EXERCISE 15.8.7. Do a principal components regression for the Chapman data of Exercise 13.8.4.

EXERCISE 15.8.8. Do a principal components regression on for the pollution data of Exercise 13.8.5.

EXERCISE 15.8.9. Do a principal components regression on for the body fat data of Exercise 13.8.6.

Chapter 16
Unbalanced multifactor analysis of variance

Multifactor unbalanced analysis of variance models are typically not amenable to analysis by hand calculations unless the models are equivalent to a one-way ANOVA model. In general, these models need to be treated with methods similar to multiple regression. In particular, tests are based on model comparisons and the order in which effects are fitted can be important. In this chapter we discuss general cases of analysis of variance with unequal numbers of observations on the treatment combinations and some special cases in which relatively simple calculations can be made.

16.1 Unbalanced two-way analysis of variance

Unbalanced two-way analysis of variance situations can be divided into two categories: proportional numbers in which a simple analysis can still be obtained and the general case in which methods similar to regression analysis must be used. This section involves a lot of model comparisons, so for simplicity, equations such as (16.1.1) are referred to simply as (1).

16.1.1 PROPORTIONAL NUMBERS

Consider a randomized complete block design where one of the treatments is a standard treatment, e.g., a placebo or control. In such a situation, interest often focuses on comparing each of the other treatments to the standard. If the analysis is to focus on the standard treatment, it may be wise to include extra observations on the standard treatment. For example, with four treatments including the standard, it might be wise to have blocks of five units where two experimental units are randomly chosen to receive the standard treatment, while the other treatments are randomly assigned to the other units. This procedure destroys the balance in the usual randomized complete block design but if all the blocks are handled similarly, a simple analysis can still be salvaged. In general situations with unequal numbers of observations on the treatments, a simple analysis is just not possible.

The experiment described in the previous paragraph is a special case of a two-way ANOVA with proportional numbers. Suppose there are a levels of the first factor and b levels of the second and suppose there are N_{ij} observations on the i,j treatment combination. The N_{ij} numbers are said to be proportional if for any pair $i = 1, \ldots, a$ and $j = 1, \ldots, b$,

$$N_{ij} = \frac{N_{i\bullet}N_{\bullet j}}{N_{\bullet\bullet}}. \tag{16.1.1}$$

Here

$$N_{i\bullet} = \sum_{j=1}^{b} N_{ij}, \quad N_{\bullet j} = \sum_{i=1}^{a} N_{ij}, \quad N_{\bullet\bullet} = \sum_{i=1}^{a}\sum_{j=1}^{b} N_{ij}.$$

In any two-way ANOVA with proportional numbers, the analysis is analogous to a balanced two-way ANOVA. If there is interaction, treat the problem as a large one-way ANOVA. If there is no interaction, examine the main effects separately. Contrasts and the sums of squares for each factor are simply computed as in a one-way analysis of variance, ignoring the other factor. The only difference is that there are, for example, $N_{i\bullet}$ observations on the ith level of the first factor. These values differ from level to level, so the one-way ANOVA on the first factor is a one-way ANOVA with unequal numbers. Similarly, the one-way ANOVA on the second factor is also a one-way with unequal numbers, the $N_{\bullet j}$s being the numbers of observations on group j.

Return now to the idea of taking extra observations on a standard treatment in a blocking design. Suppose that there are a treatments and b blocks with the *first* treatment applied to N experimental units in each block and the other treatments applied to one unit in each block. To show that the numbers are proportional, observe that the number of observations on treatment i in block j is

$$N_{ij} = \begin{cases} N & \text{if } i = 1 \\ 1 & \text{otherwise} \end{cases}.$$

The number of observations on treatment i is

$$N_{i\bullet} = \begin{cases} bN & \text{if } i = 1 \\ b & \text{otherwise} \end{cases}.$$

The number of observations in block j is

$$N_{\bullet j} = N + (a - 1).$$

The total number of observations is

$$N_{\bullet\bullet} = b[N + (a - 1)].$$

For $i = 1$, equation (1) becomes $N = (bN)[N + (a - 1)]/ b[N + (a - 1)]$, and for $i \neq 1$, equation (1) becomes $1 = b[N + (a - 1)]/ b[N + (a - 1)]$. Thus the sum of squares for blocks can be computed simply from the block means as in a balanced one-way ANOVA and the sum of squares for treatments can be computed from the treatment means as in an unbalanced one-way ANOVA. The standard treatment has more observations on it than the other treatments, so an unbalanced one-way analysis of the treatment means is required.

Minitab's 'ancova' and 'glm' commands can be used to analyze proportional numbers ANOVAs.

16.1.2 GENERAL CASE

If the N_{ij}s are not proportional, the analysis proceeds by fitting various models. We illustrate with an example.

Bailey (1953), Scheffé (1959), and Christensen (1987) examine data on infant female rats that were given to foster mothers for nursing. The variable of interest is the weight of the rat at 28 days. Weights were measured in grams. A subset of the data is given in Table 16.1. There are four groups: rats with genotype F were given to foster mothers of genotype A, rats with genotype F were given to

TABLE 16.1. Infant rats weight gains with foster mothers

Genotype of foster mother	Genotype of litter F	I
A	48.0	68.0
	49.3	36.3
	51.7	37.0
	60.3	
J	40.5	54.5
	51.3	42.8
		50.2

foster mothers of genotype J, rats with genotype I were given to foster mothers of genotype A, and rats with genotype I were given to foster mothers of genotype J. The astute reader will recognize the groups as having factorial treatment structure and this will be exploited as the example progresses.

As a first step, we wish to establish whether the data display interaction. We begin by fitting the model with interaction.

$$\begin{aligned} y_{ijk} &= \mu_{ij} + \varepsilon_{ijk} \\ &= \mu + \alpha_i + \eta_j + \gamma_{ij} + \varepsilon_{ijk}, \end{aligned} \tag{16.1.2}$$

$$\varepsilon_{ij}\text{s independent } N(0, \sigma^2)$$

$i = 1, ..., a, j = 1, ..., b, k = 1, ..., N_{ij}$. Using model (2) we have the sample sizes $N_{11} = 4$, $N_{12} = 3$, $N_{21} = 2$, $N_{22} = 3$. We also fit the model with no interaction:

$$y_{ijk} = \mu + \alpha_i + \eta_j + \varepsilon_{ijk}, \quad \varepsilon_{ij}\text{s independent } N(0, \sigma^2) \tag{16.1.3}$$

$i = 1, ..., a, j = 1, ..., b, k = 1, ..., N_{ij}$. In the context of our example, the α_is are mother effects and the η_js are litter effects.

We wish to evaluate how well model (3) fits as compared to model (2). A measure of how well any model fits is the sum of squared errors. Since the no interaction model (3) is a special case of the interaction model (2), the error from model (3) must be at least as great as the error from model (2), i.e., $SSE(3) \geq SSE(2)$. However, if $SSE(3)$ is much greater than $SSE(2)$, it suggests that the special case, model (3), is an inadequate substitute for the *full* model (2). In particular, large values of $SSE(3) - SSE(2)$ suggest that the *reduced* model (3) is inadequate to explain the data that were explained by the full model (2). It can be established that if the reduced model is true,

$$\frac{SSE(3) - SSE(2)}{dfE(3) - dfE(2)}$$

is an estimate of σ^2 and this is independent of the estimate of σ^2 from the full model

$$\frac{SSE(2)}{dfE(2)}.$$

A test of whether model (3) is an adequate substitute for model (2) rejects model (3) if

$$F = \frac{[SSE(3) - SSE(2)] \big/ [dfE(3) - dfE(2)]}{SSE(2)/dfE(2)}$$

is too large. The F statistic is compared to an $F(dfE(3) - dfE(2), dfE(2))$ distribution. When the data are balanced, this is exactly the analysis of variance F test for no interaction that was discussed in earlier chapters.

Table 16.2 gives two analysis of variance tables for the two-way ANOVA model with interaction. In the first ANOVA table, mothers are fitted to the data before litters. In the second table, litters are fitted before mothers. The rows for mother∗litter interaction are identical in both tables. The sum of squares for mother∗litter interaction in the table is obtained by differencing the error sums of squares for models (3) and (2). The F statistic is very small, less than 1, so there is no evidence of interaction and we proceed with an analysis of model (3). In particular, we now examine the main effects.

TABLE 16.2. Analyses of variance for rat weight gains

Source	df	Seq SS	MS	F
Mothers	1	14.4	14.4	0.13
Litters	1	8.7	8.7	0.08
Mothers∗litters	1	50.9	50.9	0.47
Error	8	875.7	109.5	
Total	11	949.7		

Source	df	Seq SS	MS	F
Litters	1	12.6	12.6	0.12
Mothers	1	10.5	10.5	0.10
Litters∗mothers	1	50.9	50.9	0.47
Error	8	875.7	109.5	
Total	11	949.7		

The effect of litters can measured in two ways. First, by comparing the no interaction model (3) with a model that eliminates the effect for litters

$$y_{ijk} = \mu + \alpha_i + \varepsilon_{ijk}. \tag{16.1.4}$$

The difference in the error sums of squares for these models is the sum of squares reported for litters in the first of the two ANOVA tables in Table 16.2. Note that this model comparison *assumes* that there is an effect for mothers because the α_is are included in both models. The corresponding F statistic is very small, so *there is no evidence for differences in litters* **after accounting** *for any differences due to mothers.*

Alternatively, we could assume that there are no mother effects and base our evaluation of litter effects on comparing the model with litter effects but no mother effects,

$$y_{ijk} = \mu + \eta_j + \varepsilon_{ijk}, \tag{16.1.5}$$

to the model that contains no treatment effects,

$$y_{ijk} = \mu + \varepsilon_{ijk}. \tag{16.1.6}$$

The difference in the error sums of squares for these models is the sum of squares reported for litters in the second of the two ANOVA tables in Table 16.2. Once again, the appropriate F statistic is very small, so *there is no evidence for differences in litters* **when ignoring** *any differences due to mothers.*

Similarly, the effect of mothers can be measured by comparing the no interaction model (3) with model (5) that eliminates the effect for mothers. The difference in the error sums of squares for these models is the sum of squares reported for mothers in the second of the two ANOVA tables in Table 16.2. This model comparison *assumes* that there is an effect for litters because the η_js are included in both models. Additionally, we could assume that there are no litter effects and base our evaluation of mother effects on comparing model (4) with model (6). The difference in the error sums of squares for these models is the sum of squares reported for mothers in the first of the two ANOVA tables in Table 16.2. Both of the corresponding F statistics are very small, so there is no evidence of a mother effect whether accounting for or ignoring effects due to litters.

Table 16.2 results from doing model comparisons in two sequential fitting schemes. The first ANOVA table results from fitting the sequence of models (6), (4), (3), (2). The second ANOVA results from fitting (6), (5), (3), (2).

Generally, if there were an effect for mothers after accounting for litters and an effect for litters after accounting for mothers, both mothers and litters would have to appear in the final model, i.e.,

$$y_{ijk} = \mu + \alpha_i + \eta_j + \varepsilon_{ijk},$$

because neither effect could be dropped.

If there were an effect for mothers after accounting for litters but no effect for litters after accounting for mothers we could drop the effect of litters from the model. Then *if the effect for mothers was still apparent* when litters were ignored, a final model

$$y_{ijk} = \mu + \alpha_i + \varepsilon_{ijk}$$

that includes mother effects but not litter effects would be appropriate. Similar reasoning with the roles of mothers and litters reversed would lead one to the model

$$y_{ijk} = \mu + \eta_j + \varepsilon_{ijk}.$$

Unfortunately, with unequal numbers ANOVA, it is possible to get contradictory results. If there were an effect for mothers after accounting for litters but no effect for litters after accounting for mothers we could drop the effect of litters from the model and consider the model

$$y_{ijk} = \mu + \alpha_i + \varepsilon_{ijk}.$$

However, it is possible that in this model *there may be no apparent effect for mothers* (when litters are ignored), so dropping mothers is suggested and we get the model

$$y_{ijk} = \mu + \varepsilon_{ijk}.$$

This model contradicts our first conclusion which was that there is an effect for mothers, albeit one that only shows up after adjusting for litters. These issues are discussed more extensively in Christensen (1987, section VII.5).

We return now to the analysis of the actual data in Table 16.1. It is necessary to consider the validity of our assumptions. Table 16.3 contains many of the standard diagnostic statistics used in regression analysis. The diagnostics are computed from the two-way with interaction model (2). Model (2) is equivalent to a one-way ANOVA model, so the leverage of y_{ijk} in Table 16.3 is just $1/N_{ij}$.

Figures 16.1, 16.2, and 16.3 contain diagnostic plots. Figure 16.1 is a normal plot of the standardized residuals, Figure 16.2 is a plot of the standardized residuals versus the predicted values, and Figure 16.3 is a plot of the Cook's distances against case numbers. The plots identify one potential outlier. From Table 16.3 this is easily identified as the observed value of 68.0 for foster mother A and litter I. This case has by far the largest standardized residual r, standardized deleted residual t, and Cook's distance C. We can test whether this case is consistent with the other data. The t residual of 4.56 has an unadjusted P value of $.001$. If we use a Bonferroni adjustment for having made $n = 12$ tests, the P value is $12 \times .001 = .012$. There is substantial evidence that this case does not belong with the other data.

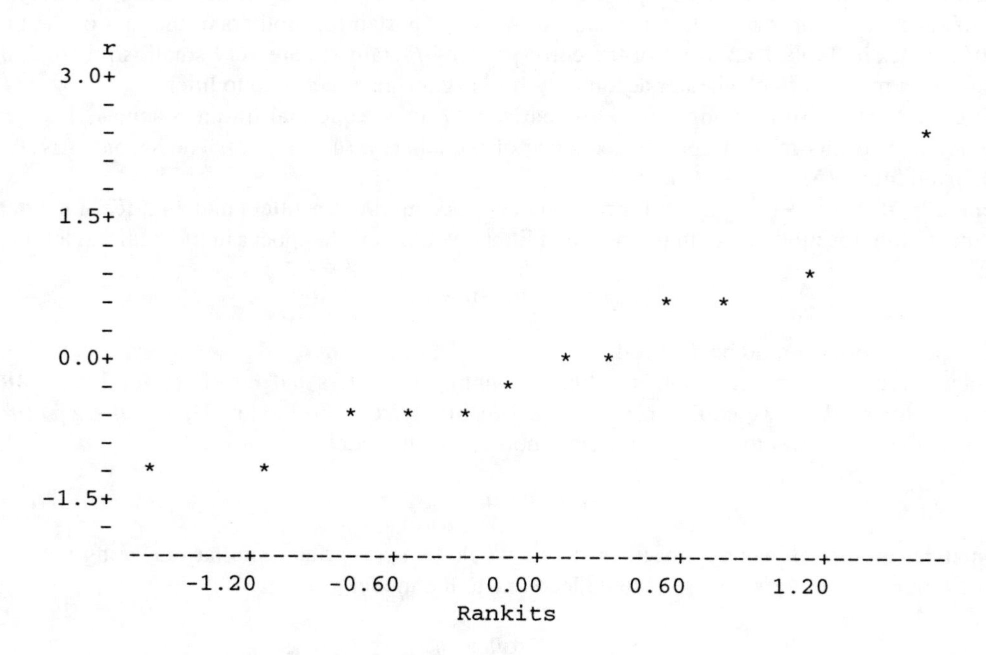

FIGURE 16.1. Normal plot, $W' = 0.916$.

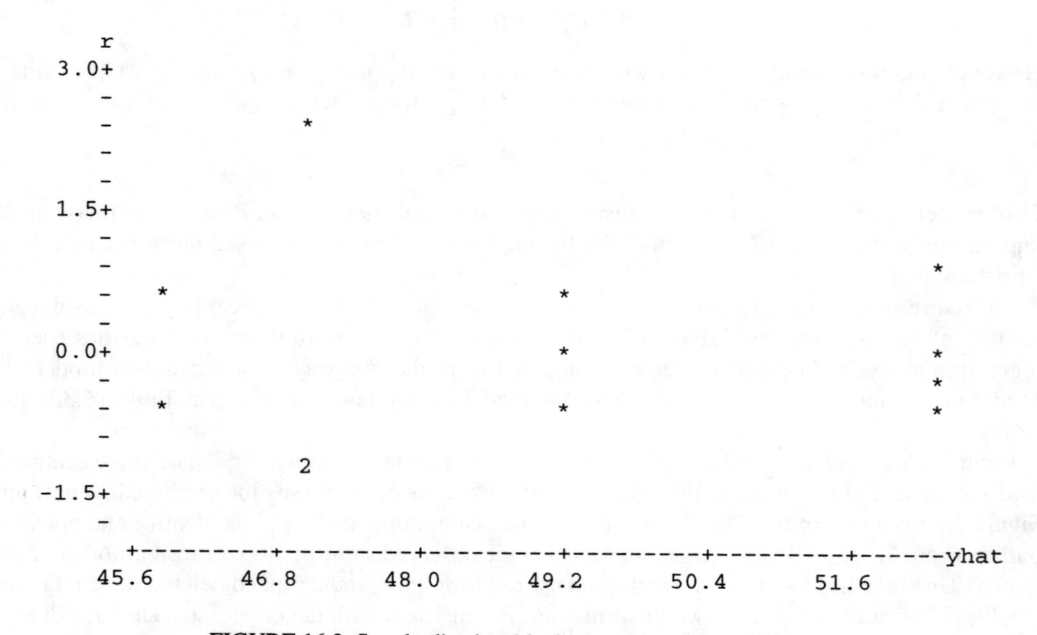

FIGURE 16.2. Standardized residuals versus predicted values.

TABLE 16.3. Diagnostics for rat weight gains

Mother	Litter	y	$\hat{y}$	Leverage	r	t	C
A	F	48.0	52.325	0.25	−0.48	−0.45	0.019
A	F	49.3	52.325	0.25	−0.33	−0.31	0.009
A	F	51.7	52.325	0.25	−0.07	−0.06	0.000
A	F	60.3	52.325	0.25	0.88	0.87	0.065
A	I	68.0	47.100	0.33	2.45	4.56	0.748
A	I	36.3	47.100	0.33	−1.26	−1.32	0.200
A	I	37.0	47.100	0.33	−1.18	−1.22	0.175
J	F	40.5	45.900	0.50	−0.73	−0.71	0.133
J	F	51.3	45.900	0.50	0.73	0.71	0.133
J	I	54.5	49.167	0.33	0.62	0.60	0.049
J	I	42.8	49.167	0.33	−0.75	−0.72	0.069
J	I	50.2	49.167	0.33	0.12	0.11	0.002

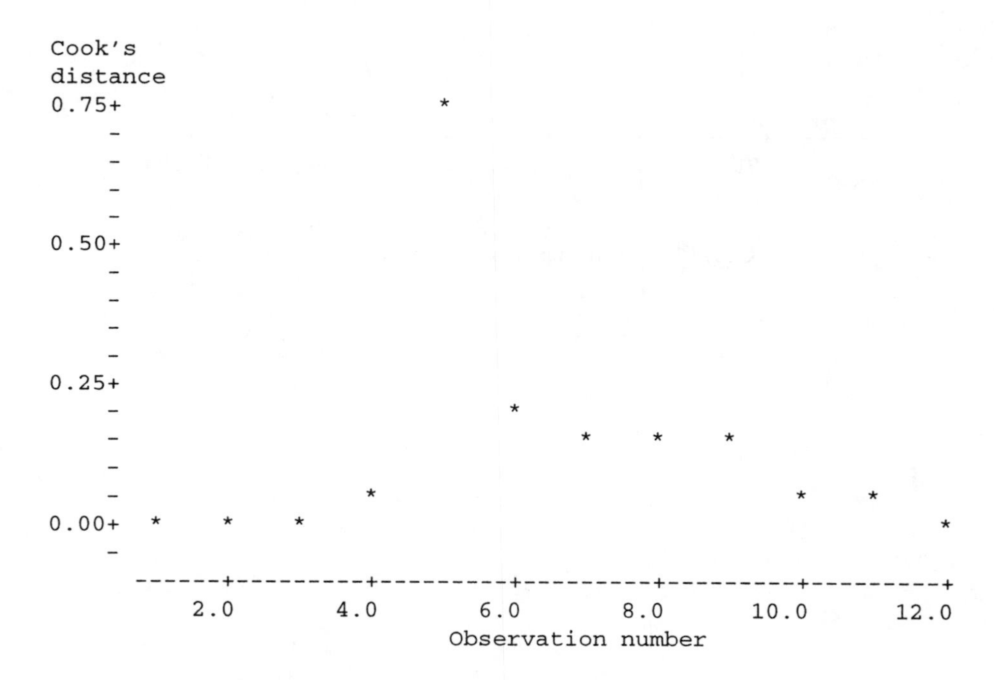

FIGURE 16.3. Cook's distances versus observation number.

We now consider the results of an analysis with the outlier deleted. Table 16.4 contains two ANOVA tables similar to those in Table 16.2. With the outlier deleted, the *MSE* has been reduced to less than a third of its previous value. The *P* value for the interaction test is .031, which is substantial evidence of interaction. In the face of such interaction, there is little point in pursuing the main effects, rather we revert to thinking of this as a one-way ANOVA.

Diagnostics for the interaction model with the outlier deleted are given in Table 16.5. The leverages are again $1/N_{ij}$, so they are not reported, but note that the leverages for mother A, litter I change because of the deletion of a case in that group. The standardized residuals, standardized deleted residuals, and Cook's distances look reasonable. The normal plot in Figure 16.4 looks tolerable and the plot of r versus $\hat{y}$ given in Figure 16.5 also looks reasonable except perhaps that there is too little variability with mother A, litter I (the two smallest $\hat{y}$ values).

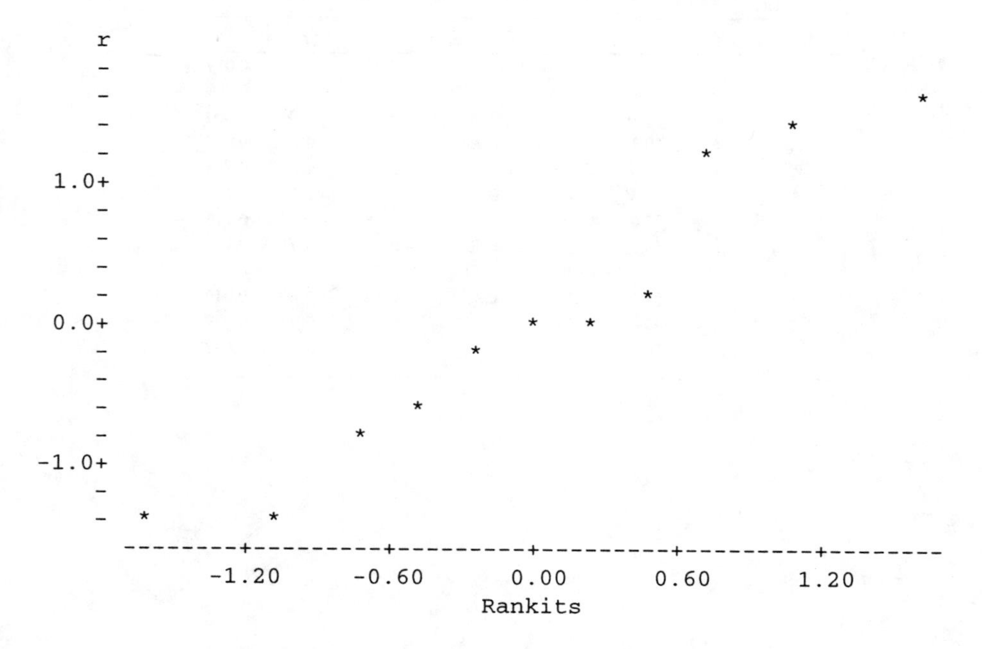

FIGURE 16.4. Normal plot, outlier deleted, $W' = 0.955$.

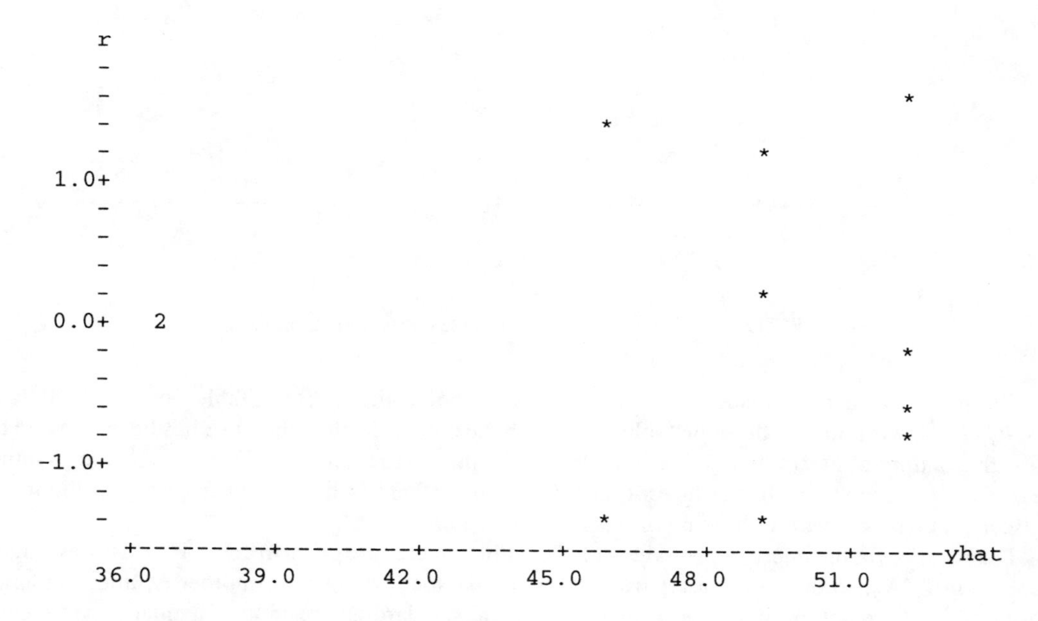

FIGURE 16.5. Standardized residuals versus predicted values, outlier deleted.

TABLE 16.4. Analysis of variance for rat weight gains with outlier deleted

Source	df	Seq SS	MS	F	P
Mothers	1	1.58	1.58	0.06	
Litters	1	113.81	113.81	3.61	
Mothers*litters	1	226.60	226.60	7.20	0.031
Error	7	220.46	31.49		
Total	10	562.45			

Source	df	Seq SS	MS	F	P
Litters	1	98.95	98.95	3.14	
Mothers	1	16.44	16.44	0.52	
Litters*mothers	1	226.60	226.60	7.20	0.031
Error	7	220.46	31.49		
Total	10	562.45			

TABLE 16.5. Diagnostics for rat weight gains, outlier deleted

Mother	Litter	y	$\hat{y}$	r	t	C
A	F	48.0	52.325	−0.89	−0.87	0.07
A	F	49.3	52.325	−0.62	−0.59	0.03
A	F	51.7	52.325	−0.13	−0.12	0.00
A	F	60.3	52.325	1.64	1.94	0.22
A	I					
A	I	36.3	36.650	−0.09	−0.08	0.00
A	I	37.0	36.650	0.09	0.08	0.00
J	F	40.5	45.900	−1.36	−1.47	0.46
J	F	51.3	45.900	1.36	1.47	0.46
J	I	54.5	49.167	1.16	1.20	0.17
J	I	42.8	49.167	−1.39	−1.51	0.24
J	I	50.2	49.167	0.23	0.21	0.01

Glancing at the predicted values in Table 16.5, it seems that the predicted values are all quite comparable except those from mother A and litter I. We proceed to test whether the other three treatments have the same mean. Thinking of the two-way with interaction as a one-way ANOVA with four treatments, we can construct a new one-way model with two treatments in which all combinations except mother A and litter I are combined into a single common treatment. This forms a reduced model relative to the one-way ANOVA with 4 treatments. The ANOVA table for the reduced model is given in Table 16.6. The dfE, SSE, and MSE for the one-way with 4 treatments come from Table 16.4. The F test for comparing the two models is

$$F = \frac{(277.57 - 220.46)/(9-7)}{31.49} = .91.$$

The F is small, so there is no evidence for any differences between the 3 groups other than mother A, litter I. However, from the F test in Table 16.6, we see that there is substantial evidence of a difference between mother A, litter I and the 3 other groups (the data suggest this test).

With the outlier included, there is no evidence for any differences due to mothers or litters. With the outlier deleted, the mother, litter combination that contained the outlier becomes the only distinct treatment group. Since the deletion of the outlier for mother A, litter I was the direct cause of establishing that mother A, litter I is different from the other treatments, I am leery of this conclusion.

□

TABLE 16.6. Analysis of variance for rat weight gains with outlier deleted and three groups treated as one

Source	*df*	*SS*	*MS*	*F*	*P*
Mom-A, lit-I vs others	1	284.88	284.88	9.24	0.014
Error	9	277.57	30.84		
Total	10	562.45			

An important application of general methods for analyzing unbalanced two-ways is the analysis of randomized complete block designs with missing observations. Exercise 16.6.4 involves an RCB design with a potential outlier. Analyzing the data after deleting the outlier involves analyzing an unbalanced two-way ANOVA. Note, however, that in an RCB the treatments should be adjusted for blocks but blocks *should not* be considered after adjusting for treatments. Block effects exist by design and thus should always be adjusted for and never considered for elimination.

MINITAB COMMANDS

Below are given Minitab commands for the initial analysis of variance on the rat data as contained in Tables 16.2 and 16.3.

```
MTB > glm c1 = c2|c3;
SUBC> fits c5;
SUBC> sresid c6;
SUBC> cookd c7;
SUBC> hi c8;
SUBC> tresid c9.
MTB > glm c1 = c3|c2;
```

Minitab's glm command reports both sequential and adjusted sums of squares for each effect. We have reported and analyzed the sequential sums of squares. The adjusted sum of squares for a main effect is the (sequential) sum of squares for fitting that main effect after the other main effect. To delete the outlier in Minitab, just replace the 68.0 with an asterisk (*) and repeat the commands given above. Computing Table 16.6 requires constructing a new column of indices to identify the two new treatments in the one-way ANOVA.

16.2 Balanced incomplete block designs

Other than proportional numbers, the best behaved unbalanced two-way anova models are probably those for balanced incomplete block (BIB) designs. In a balanced incomplete block design, the blocks are incomplete; they do not contain every treatment in every block. Such designs are useful when blocks need to be smaller than the number of treatments in order to maintain homogeneity of the experimental material within the blocks. Balanced incomplete block designs are balanced in the sense that *every pair of treatments occurs together in the same block some fixed number of times*, say, λ.

Balanced incomplete block designs are not balanced in the same way that balanced ANOVAs are balanced. In particular, BIBs are not sufficiently balanced to allow the analysis of blocks and treatments to be performed as separate one-way ANOVAs. Typically, in a BIB the analysis of blocks is conducted ignoring treatments and the analysis of treatments is conducted after adjusting for blocks. This is the only order of fitting models that we need to consider. Blocks are designed to have effects and these effects are of no intrinsic interest, so there is no reason to worry about fitting treatments first and then examining blocks after adjusting for treatments. Blocks are nothing more than an adjustment factor.

The virtue of the form of balance in a BIB is that the analysis of treatments after adjusting for blocks can be based on methods *analogous* to one-way ANOVA by computing adjusted treatment means and by using an adjusted (effective) number of observations on each treatment. The analysis being discussed here is known as the *intrablock* analysis of a BIB; it is appropriate when the block effects are viewed as fixed effects. If the block effects are viewed as random effects with mean 0, there is an alternative analysis that is known as the recovery of *interblock* information. Cochran and Cox (1957) discuss this analysis; we will not.

EXAMPLE 16.2.1. A simple balanced incomplete block design is given below for four treatments A, B, C, D in four blocks of three units each.

Block	Treatments		
1	A	B	C
2	B	C	D
3	C	D	A
4	D	A	B

Note that every pair of treatments occurs together in the same block exactly $\lambda = 2$ times. Thus, for example, the pair A, B occurs in blocks 1 and 4. There are $b = 4$ blocks each containing $k = 3$ experimental units. There are $t = 4$ treatments and each treatment is observed $r = 3$ times. □

There are two relationships that must be satisfied by the numbers of blocks, b, units per block, k, treatments, t, replications per treatment, r, and λ. Recall that λ is the number of times two treatments occur together in a block. First, the total number of observations is the number of blocks times the number of units per block, but the total number of observations is also the number of treatments times the number of replications per treatment, thus

$$bk = rt.$$

The other key relationship in balanced incomplete block designs involves the number of comparisons that can be made between a given treatment and the other treatments *within the same block*. Again, there are two ways to count this. The number of comparisons is the number of other treatments, $t-1$, multiplied by the number of times each other treatment is in the same block as the given treatment, λ. Alternatively, the number of comparisons within blocks is the number of other treatments within each block, $k - 1$, times the number of blocks in which the given treatment occurs, r. Thus we have

$$(t - 1)\lambda = r(k - 1).$$

In Example 16.2.1, these relationships reduce to

$$(4)3 = 3(4)$$

and

$$(4 - 1)2 = 3(3 - 1).$$

The nice thing about balanced incomplete block designs is that the theory behind them works out so simply that the computations can all be done on a hand calculator. I know, I did it once, see Christensen (1987, section IX.4). But once was enough for this lifetime! We will rely on a computer program to provide the more difficult computations and content ourselves with using the simple structure of BIB designs to allow us to examine orthogonal contrasts. Examination of contrasts, orthogonal or otherwise, is something that generally cannot be done easily in unbalanced two-way ANOVAs. We illustrate the techniques with an example.

EXAMPLE 16.2.2. John (1961) reported data on the number of dishes washed prior to losing the suds in the wash basin. Dishes were soiled in a standard way and washed one at a time. Three operators and three basins were available for the experiment, so at any one time only three treatments could be applied. Operators worked at the same speed, so no effect for operators was necessary nor should there be any effect due to basins. Nine detergent treatments were evaluated in a balanced incomplete block design. The treatments and numbers of dishes washed are given in Table 16.7. There were $b = 12$ blocks with $k = 3$ units in each block. Each of the $t = 9$ treatments was replicated $r = 4$ times. Each pair of treatments occurred together $\lambda = 1$ time. The three treatments assigned to a block were randomly assigned to basins as were the operators. The blocks were run in random order.

TABLE 16.7. Balanced incomplete block design investigating detergents; data are numbers of dishes washed

Block	Treatment, Observation			Mean
1	A, 19	B, 17	C, 11	$15.\bar{6}$
2	D, 6	E, 26	F, 23	$18.\bar{3}$
3	G, 21	H, 19	J, 28	$22.\bar{6}$
4	A, 20	D, 7	G, 20	$15.\bar{6}$
5	B, 17	E, 26	H, 19	$20.\bar{6}$
6	C, 15	F, 23	J, 31	$23.\bar{0}$
7	A, 20	E, 26	J, 31	$25.\bar{6}$
8	B, 16	F, 23	G, 21	$20.\bar{0}$
9	C, 13	D, 7	H, 20	$13.\bar{3}$
10	A, 20	F, 24	H, 19	$21.\bar{0}$
11	B, 17	D, 6	J, 29	$17.\bar{3}$
12	C, 14	E, 24	G, 21	$19.\bar{6}$
				$19.41\bar{6}$

The analysis of variance is given in Table 16.8. Computation of the ANOVA table and the adjusted means will be discussed later. The F test for treatment effects is clearly significant. We now need to examine contrasts in the treatments.

TABLE 16.8. Analysis of variance

Source	df	Seq SS	MS	F	P
Blocks	11	412.750	37.523	45.54	0.000
Trts	8	1086.815	135.852	164.85	0.000
Error	16	13.185	0.824		
Total	35	1512.750			

The treatments were constructed with a structure that leads to interesting orthogonal contrasts. Treatments A, B, C, and D all consisted of detergent I using, respectively, 3, 2, 1, and 0 doses of an additive. Similarly, treatments E, F, G, and H used detergent II with 3, 2, 1, and 0 doses of the additive. Treatment J was a control. Except for the control, the treatment structure is factorial in detergents and levels of additive.

Table 16.9 gives the treatments, the adjusted treatment means, and a series of orthogonal contrasts in the treatments, along with the estimates and sums of squares for the contrasts. The adjusted treatment means have been rounded off to three decimal places, so some slight numerical inaccuracies will result from their use. The contrasts chosen are essentially those used in the analysis of a factorial experiment, except the first contrast compares the control to the other treatments. The sum of squares for the first contrast is much larger than the *MSE* from Table 16.8. The second contrast looks at the main effect for detergents; it compares detergent I with detergent II. Again, there is a clear effect. In a factorial experiment, we would have three degrees of freedom for main effects comparing the four levels of additive. These 3 degrees of freedom can be broken into linear, quadratic, and cubic contrasts. The linear and quadratic contrasts are given in the table; these are just the standard balanced ANOVA linear and quadratic contrasts for equally spaced levels given in Appendix B.4. Each contrast is applied to both detergents. The cubic contrast will be considered separately. With one degree of freedom for detergents and 3 degrees of freedom for additive levels, there are $1 \times 3 = 3$ degrees of freedom for detergent–additive level interaction. Again, these can be broken into contrasts: in particular, the detergent by linear contrast, the detergent by quadratic contrast, and the detergent by cubic contrast. The first two of these are given in the table, the detergent by cubic contrast will be considered separately.

TABLE 16.9. Adjusted treatment means and orthogonal contrasts

Treat.	Adj. mean	Ctrl vs others	Det. I vs II	Linear	Quad.	Det. vs linear	Det. vs quad.
A	19.750	−1	1	−3	−1	−3	−1
B	17.194	−1	1	−1	1	−1	1
C	13.194	−1	1	1	1	1	1
D	6.528	−1	1	3	−1	3	−1
E	25.306	−1	−1	−3	−1	3	1
F	22.972	−1	−1	−1	1	1	−1
G	21.083	−1	−1	1	1	−1	−1
H	19.194	−1	−1	3	−1	−3	1
J	29.528	8	0	0	0	0	0
	Est	91.003	−31.889	−63.891	3.665	−23.441	4.555
	SS	345.06	381.34	306.15	5.04	41.21	7.78

In balanced incomplete block designs, contrasts are orthogonal if and only if the contrasts would be orthogonal in a standard balanced ANOVA. Estimates are computed for contrasts in the usual way using the adjusted means. In other words, they are computed just as in a balanced one-way analysis of variance treating the adjusted treatment means as the treatment means. Sums of squares and standard errors for contrasts are computed in almost the same way, but in a balanced one-way ANOVA the sample size for each treatment is N, while in a BIB this is replaced by an effective sample size for the treatment, $\lambda t/k$. For example, treatment A is actually observed $r = 4$ times, but because of the balanced incomplete block design, the effective sample size of the adjusted treatment mean for A is

$$\frac{\lambda t}{k} = \frac{1(9)}{3} = 3.$$

To illustrate, compare treatment A with the control, treatment J. The estimate is $19.750 - 29.528 = -9.778$, the standard error is

$$\sqrt{MSE[1^2 + (-1)^2]/3} = .741,$$

and the sum of squares for the comparison is

$$\frac{[-9.778]^2}{[1^2 + (-1)^2]/3} = 143.41.$$

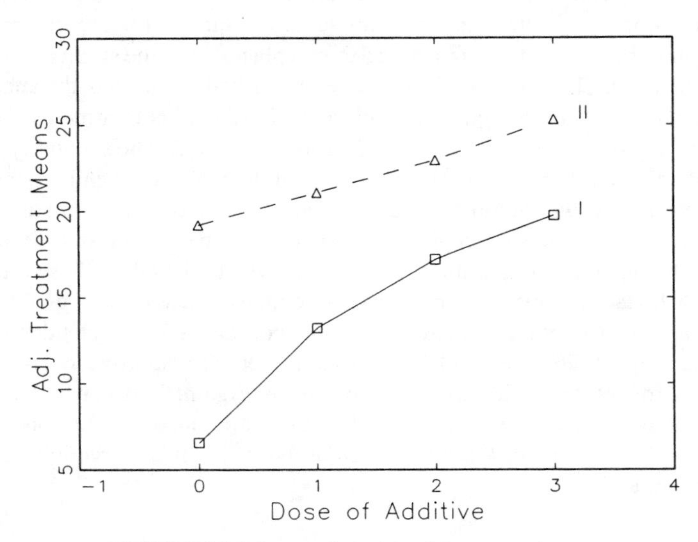

FIGURE 16.6. Interaction plot of detergent data.

Similar computations give the sums of squares in Table 16.9.

We begin our examination of Table 16.9 by considering the missing cubic contrasts. $SSTrt$ = 1086.815 with 8 degrees of freedom and adding the six orthogonal contrast sums of squares in Table 16.9 gives 1086.58. That leaves 2 degrees of freedom, those for the cubic main effect and the detergent by cubic interaction. The sum of squares for these 2 degrees of freedom are

$$1086.815 - 1086.58 = .24$$

where we have a small amount of rounding error perpetuated in the calculations. With a MSE of .824, neither of these contrasts can be important because even if the sum of squares for one of them is .24 and the other is 0, the larger sum of squares is still much smaller than the MSE.

We should now examine the sums of squares for the interaction contrasts. As we are fitting polynomials, the initial interest is in the detergent by quadratic contrast. The F statistic for this contrast is

$$F = \frac{7.78}{.824} = 9.44$$

so there is a clear interaction effect. Another way to think about this is that the t statistic for this contrast has an absolute value of $\sqrt{9.44} = 3.07$, which is very large. An interaction plot is given in Figure 16.6. Our conclusion from these tests is that there is no evidence of any cubic effects but that the quadratic curvature is different for detergent I than for detergent II. From the figure, there seems to be very little quadratic curvature in detergent II.

An alternative to the contrasts in Table 16.9 is to fit a separate parabola for each detergent. Appropriate contrasts are given in Table 16.10. Our earlier argument assures that we need not consider cubic effects. The last contrast has a very small sum of squares, so there is no evidence of quadratic curvature in detergent II, i.e., as far as we can tell there is at most a linear relationship between numbers of dishes washed and the amount of additive. The linear contrast has a large sum of squares for detergent II, so there is indeed a linear relationship. The quadratic contrast in detergent I has a large sum of squares, so there is clear evidence of a quadratic curvature in the relationship between numbers of dishes washed and the amount of additive for detergent I. From inspection of the adjusted treatment means or equivalently of the plots in Figure 16.6, suds last longer when there

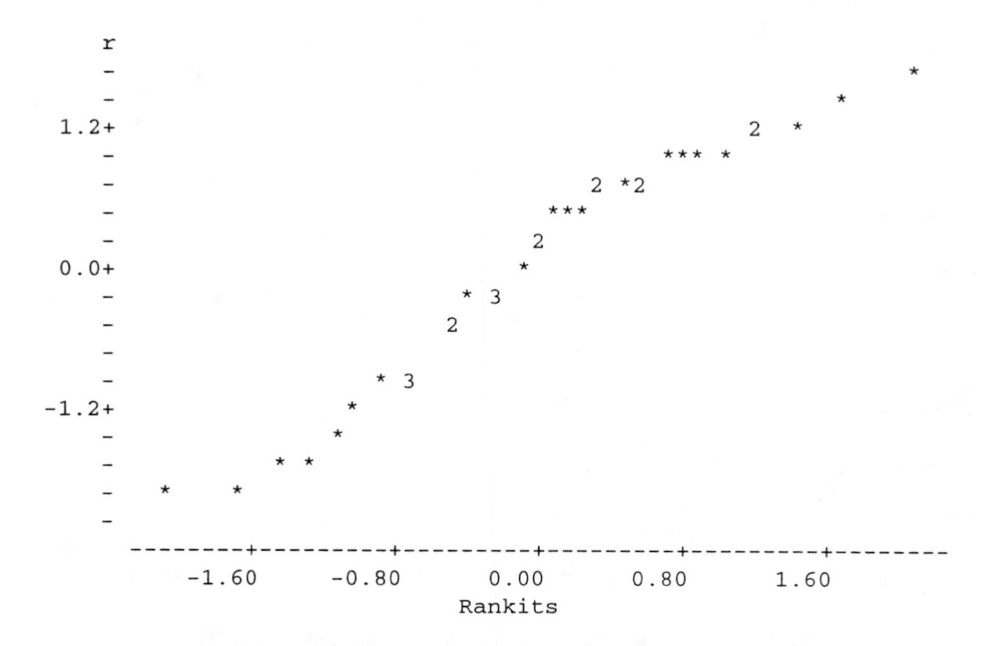

FIGURE 16.7. Normal plot, $W' = 0.953$, dish data.

is more additive (up to a triple dose). Detergent II works uniformly better than detergent I. The effect of a dose of the additive is greater at low levels for detergent I than at high levels but the effect of a dose is apparently steady for detergent II. The control is easily better than any of the new treatments.

TABLE 16.10. Adjusted treatment means and contrasts

Treat.	Adj. mean	Ctrl vs others	Det. I vs II	Det. I linear	Det. I quad.	Det. II linear	Det. II quad.
A	19.750	−1	1	−3	−1	0	0
B	17.194	−1	1	−1	1	0	0
C	13.194	−1	1	1	1	0	0
D	6.528	−1	1	3	−1	0	0
E	25.306	−1	−1	0	0	−3	−1
F	22.972	−1	−1	0	0	−1	1
G	21.083	−1	−1	0	0	1	1
H	19.194	−1	−1	0	0	3	−1
J	29.528	8	0	0	0	0	0
	Est	91.003	−31.889	−43.666	4.110	−20.225	−.445
	SS	345.06	381.34	286.01	12.67	61.36	0.16

As always, we need to evaluate our assumptions. Figures 16.7 and 16.8 contain a normal plot and a plot of the residuals versus the predicted values. The normal plot looks less than thrilling but is not too bad. The fifth percentile of W' for 36 observations is .940, whereas the observed value is .953. Alternatively, the residuals have only 16 degrees of freedom and $W'(.95, 16) = .886$. The data are counts, so a square root or log transformation might be appropriate, but we continue with the current analysis. The plot of standardized residuals versus predicted values looks good.

Table 16.11 contains diagnostic statistics for the example. Note that the leverages are all identical. Some of the standardized deleted residuals (ts) are near 2 but none are so large as to indicate an

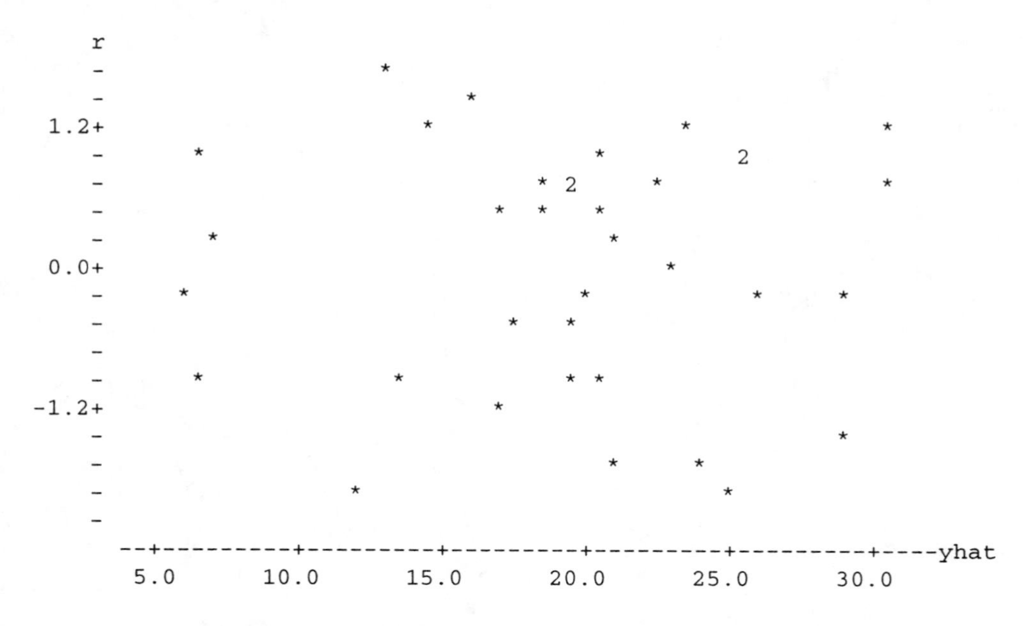

FIGURE 16.8. Standardized residuals versus predicted values, dishes.

outlier. The Cook's distances bring to one's attention exactly the same points as the standardized residuals and the *t*s. □

MINITAB COMMANDS

The analysis can be obtained from Minitab's 'glm' command. Each variable has 36 entries. The dependent variable is in c3. Variable c1 contains integers from 1 to 12 indicating blocks, c2 contains integers from 1 to 9 indicating treatments. The pairing of these values is determined by Table 16.7. The commands used are given below.

```
MTB > names c1 'Blocks' c2 'Trts' c3 'y'
MTB > glm c3=c1 c2;
SUBC> means c1 c2;
SUBC> sresids c11;
SUBC> fits c12;
SUBC> tresid c13;
SUBC> hi c14;
SUBC> cookd c15.
```

None of the numbers obtained from the subcommand 'means' are simple averages. They are 'adjusted means'; the adjusted block means will do us little good in this section, but the adjusted treatment means can be used in almost the same way as treatment means in a balanced ANOVA.

COMPUTING ADJUSTED TREATMENT MEANS AND THE ANALYSIS OF VARIANCE TABLE

The lines for total and blocks in the analysis of variance table are computed just as in a one-way ANOVA. In particular, the mean square for blocks is the sample variance of the block means times k, the number of observations in each block. In the process of computing the mean square for blocks,

TABLE 16.11. Diagnostics for the detergent data

Block	Trt.	y	$\hat{y}$	Leverage	r	t	C
1	A	19	18.7	0.56	0.49	0.48	0.01
1	B	17	16.1	0.56	1.41	1.46	0.12
1	C	11	12.1	0.56	−1.90	−2.09	0.22
2	D	6	6.6	0.56	−0.98	−0.98	0.06
2	E	26	25.4	0.56	1.04	1.04	0.07
2	F	23	23.0	0.56	−0.06	−0.06	0.00
3	G	21	20.5	0.56	0.86	0.85	0.05
3	H	19	18.6	0.56	0.67	0.66	0.03
3	J	28	28.9	0.56	−1.53	−1.60	0.15
4	A	20	19.6	0.56	0.61	0.60	0.02
4	D	7	6.4	0.56	0.98	0.98	0.06
4	G	20	21.0	0.56	−1.59	−1.68	0.16
5	B	17	17.3	0.56	−0.49	−0.48	0.01
5	E	26	25.4	0.56	0.98	0.98	0.06
5	F	19	19.3	0.56	−0.49	−0.48	0.01
6	C	15	14.3	0.56	1.16	1.18	0.08
6	F	23	24.1	0.56	−1.77	−1.92	0.20
6	J	31	30.6	0.56	0.61	0.60	0.02
7	A	20	20.6	0.56	−0.92	−0.91	0.05
7	E	26	26.1	0.56	−0.18	−0.18	0.00
7	J	31	30.3	0.56	1.10	1.11	0.08
8	B	16	16.8	0.56	−1.29	−1.31	0.10
8	F	23	22.6	0.56	0.73	0.72	0.03
8	G	21	20.7	0.56	0.55	0.54	0.02
9	C	13	13.6	0.56	−0.92	−0.91	0.05
9	D	7	6.9	0.56	0.18	0.18	0.00
9	H	20	19.6	0.56	0.73	0.72	0.03
10	A	20	20.1	0.56	−0.18	−0.18	0.00
10	F	24	23.3	0.56	1.10	1.11	0.08
10	H	19	19.6	0.56	−0.92	−0.91	0.05
11	B	17	16.8	0.56	0.37	0.36	0.01
11	D	6	6.1	0.56	−0.18	−0.18	0.00
11	J	29	29.1	0.56	−0.18	−0.18	0.00
12	C	14	13.0	0.56	1.65	1.76	0.17
12	E	24	25.1	0.56	−1.84	−2.00	0.21
12	G	21	20.9	0.56	0.18	0.18	0.00

one must compute the grand mean of all of the observations. The grand mean is also used in finding adjusted treatment means. For the example, the necessary means are given in Table 16.7.

To find an adjusted treatment mean, locate every block in which the treatment is observed. From each block, compute the difference between the observation on the treatment and the block mean. For treatment C in Table 16.7, these values are: block 1, $11 - 15.\bar{6}$; block 6, $15 - 23.\bar{0}$; block 9, $13 - 13.\bar{3}$; block 12, $14 - 19.\bar{6}$. Add these numbers and divide by the effective number of observations on a treatment, $\lambda t/k$. For treatment C in the example, this gives

$$\frac{1}{1(9)/3}\left[-4.\bar{6} + (-8) + -0.\bar{3} + (-5.\bar{6})\right] = -6.\bar{2}.$$

The adjusted treatment mean is obtained by adding the grand mean to this value. In the example the grand mean is $19.41\bar{6}$, so the adjusted mean for treatment C is

$$-6.\bar{2} + 19.41\bar{6} = 13.19\bar{4}.$$

In Table 16.9, this was rounded off to 13.194. The mean square for treatments (adjusted for blocks) is the sample variance of the adjusted treatment means times the effective number of observations on a treatment. The degrees of freedom for treatments is, as always, the number of treatments minus one. □

COMPUTING TECHNIQUES

One difficulty with using computer programs for unbalanced analysis of variance is figuring out what the program is giving you. In the previous section we discussed the fact that Minitab's glm command gives both sequential and adjusted sums of squares. Some programs provide as many as four different sums of squares for each effect. I never use anything except the sequential sums of squares but I frequently need to read the program's manual to find out what it gives and how to get what I want. Minitab's glm command also has a subcommand to report means. These are adjusted means, so for BIBs the mean reported for a treatment is *not* the mean of all of the observations that have that treatment. Similarly for a block. I am not aware of any universally accepted definition for the term 'adjusted mean', so I had to figure out what Minitab was reporting and how to use their adjusted means. It turns out that Minitab's adjusted means for BIB's are very easy to use. The point, however, is that if I was using a different program I would have to find out their definition of an adjusted mean and how to use it. To do this well, you need to examine the method of calculation and compare that to the methods given for this example. As an alternative, when using a program other than Minitab, you might try reproducing the results of the data analysis given here. If you get the same adjusted means as reported here, you are probably safe.

SPECIAL CASES

Balanced lattice designs are BIBs with $t = k^2$, $r = k + 1$, and $b = k(k + 1)$. Table 16.12 gives an example for $k = 3$. These designs can be viewed as $k + 1$ squares in which each treatment occurs once. Each row of a square is a block, each block contains k units, there are k rows in a square, so all of the $t = k^2$ treatments can appear in each square. To achieve a BIB, $k + 1$ squares are required, so there are $r = k + 1$ replications of each treatment. With $k + 1$ squares and k blocks (rows) per square, there are $b = k(k + 1)$ blocks. The analysis follows the standard form for a BIB. In fact, the design in Example 16.2.2 is a balanced lattice with $k = 3$.

TABLE 16.12. Balanced lattice design for 9 treatments

Block				Block			
1	A	B	C	7	A	H	F
2	D	E	F	8	D	B	I
3	G	H	I	9	G	E	C
4	A	D	G	10	A	E	I
5	B	E	H	11	G	B	F
6	C	F	I	12	D	H	C

Youden squares are a generalization of BIBs that allows a second form of blocking and a very similar analysis. These designs are discussed in Section 16.4.

16.3 Unbalanced multifactor analysis of variance

The material of this section is essentially example 7.6.1 from Christensen (1987). Minor editorial changes have been made. It is reprinted with the kind permission of Springer-Verlag.

Table 16.13 below is derived from Scheffé (1959) and gives the moisture content (in grams) for samples of a food product made with three kinds of salt (A), three amounts of salt (B), and two additives (C). The amounts of salt, as measured in moles, are equally spaced. The two numbers listed for some treatment combinations are replications. We wish to analyze these data.

TABLE 16.13. Moisture content of a food product

A (salt)		1			2			3		
B (amount salt)		1	2	3	1	2	3	1	2	3
	1	8	17	22	7	26	34	10	24	39
			13	20	10	24		9		36
C (additive)										
	2	5	11	16	3	17	32	5	16	33
		4	10	15	5	19	29	4		34

We will consider these data as a three-factor ANOVA. From the structure of the replications the ANOVA has unequal numbers. The general model for a three-factor ANOVA with replications is

$$y_{ijkm} = G + A_i + B_j + C_k + [AB]_{ij} + [AC]_{ik} + [BC]_{jk} + [ABC]_{ijk} + e_{ijkm}.$$

Our first priority is to find out which interactions are important. Table 16.14 contains the sum of squares for error and the degrees of freedom for error for the ANOVA models that include all of the main effects. Each model is identified in the table by the highest order terms in the model. (For example, $[AB][AC]$ indicates the model with only the $[AB]$ and $[AC]$ interactions. In $[AB][AC]$, the grand mean and all of the main effects are redundant; it does not matter whether these terms are included in the model. Similarly, $[AB][C]$ indicates the model with the $[AB]$ interaction and the C main effect. In $[AB][C]$, the grand mean and the A and B main effects are redundant.) Readers familiar with methods for fitting log-linear models (cf. Christensen, 1990b or Fienberg, 1980) will notice a correspondence between Table 16.14 and similar displays used in fitting three-dimensional contingency tables. The analogies between selecting log-linear models and selecting models for unbalanced ANOVA are pervasive.

All of the models have been compared to the full model using F statistics in Table 16.14. It takes neither a genius nor an F table to see that the only models that fit the data are the models that include the $[AB]$ interaction. There are a number of other model comparisons that can be made among models that include $[AB]$. These are $[AB][AC][BC]$ versus $[AB][AC]$, $[AB][AC][BC]$ versus $[AB][BC]$, $[AB][AC][BC]$ versus $[AB][C]$, $[AB][AC]$ versus $[AB][C]$, and $[AB][BC]$ versus $[AB][C]$. None of the comparisons show any lack of fit. The last two comparisons are illustrated below.

$$[AB][AC] \text{ versus } [AB][C]$$

$$R(AC|AB, C) = 45.75 - 45.18 = 0.57$$

$$F = (0.57/2)/2.3214 = .123$$

$$[AB][BC] \text{ versus } [AB][C]$$

TABLE 16.14. Statistics for fitting models to the data of Table 16.13

Model	SSE	dfE	F^*
$[ABC]$	32.50	14	
$[AB][AC][BC]$	39.40	18	.743
$[AB][AC]$	45.18	20	.910
$[AB][BC]$	40.46	20	.572
$[AC][BC]$	333.2	22	16.19
$[AB][C]$	45.75	22	.713
$[AC][B]$	346.8	24	13.54
$[BC][A]$	339.8	24	13.24
$[A][B][C]$	351.1	26	11.44

* The F statistics are for testing each model against the model with a three-factor interaction, i.e., $[ABC]$. The denominator of each F statistic is $MSE([ABC]) = 32.50/14 = 2.3214$.

$$R(BC|AB, C) = 45.75 - 40.46 = 5.29$$

$$F = (5.29/2)/2.3214 = 1.139$$

Note that, by analogy to the commonly accepted practice for balanced ANOVAs, the denominator in each test is $MSE([ABC])$, i.e., the estimate of pure error from the full model.

The smallest model that seems to fit the data adequately is $[AB][C]$. The F statistics for comparing $[AB][C]$ to the larger models are all extremely small. Writing out the model $[AB][C]$, it is

$$y_{ijkm} = G + A_i + B_j + C_k + [AB]_{ij} + e_{ijkm}.$$

We need to examine the $[AB]$ interaction. Since the levels of B are quantitative, a model that is equivalent to $[AB][C]$ is a model that includes the main effects for C, but, instead of fitting an interaction in A and B, fits a separate regression equation in the levels of B for each level of A. Let $x_j, j = 1, 2, 3$ denote the levels of B. There are three levels of B, so the most general polynomial we can fit is a second-degree polynomial in x_j. Since the levels of salt were equally spaced, it does not matter much what we use for the x_js. The computations were performed using $x_1 = 1, x_2 = 2, x_3 = 3$. In particular, the model $[AB][C]$ was reparameterized as

$$y_{ijkm} = A_{i0} + A_{i1}x_j + A_{i2}x_j^2 + C_k + e_{ijkm}. \qquad (16.3.1)$$

With a notation similar to that used in Table 16.14, the SSE and the dfE are reported in Table 16.15 for model (16.3.1) and three reduced models. Note that the SSE and dfE reported in Table 16.15 for $[A_0][A_1][A_2][C]$ are identical to the values reported in Table 16.14 for $[AB][C]$. This, of course, must be true if the models are merely reparameterizations of one another. First we want to establish whether the quadratic effects are necessary in the regressions. To do this we test

$$[A_0][A_1][A_2][C] \text{ versus } [A_0][A_1][C]$$

$$R(A_2|A_1, A_0, C) = 59.88 - 45.75 = 14.23$$

$$F = (14.23/3)/2.3214 = 2.04.$$

Since $F(.95, 3, 14) = 3.34$, there is no evidence of any nonlinear effects.

At this point it might be of interest to test whether there are any linear effects. This is done by testing $[A_0][A_1][C]$ against $[A_0][C]$. The statistics needed for this test are in Table 16.15. Instead of actually doing the test, recall that no models in Table 16.14 fit the data unless they included the $[AB]$

TABLE 16.15. Additional statistics for fitting models to the data of Table 16.13

Model	SSE	dfE
$[A_0][A_1][A_2][C]$	45.75	22
$[A_0][A_1][C]$	59.98	25
$[A_0][A_1]$	262.0	26
$[A_0][C]$	3130.	28

interaction. If we eliminated the linear effects we would have a model that involved none of the $[AB]$ interaction. (The model $[A_0][C]$ is identical to the ANOVA model $[A][C]$.) We already know that such models do not fit.

Finally, we have never explored the possibility that there is no main effect for C. This can be done by testing

$$[A_0][A_1][C] \text{ versus } [A_0][A_1]$$

$$R(C|A_1, A_0) = 262.0 - 59.88 = 202$$

$$F = (202/1)/2.3214 = 87.$$

Obviously, there is a substantial main effect for C, the type of food additive.

Our conclusion is that the model $[A_0][A_1][C]$ is the smallest model that has been considered that adequately fits the data. This model indicates that there is an effect for the type of additive and a linear relationship between amount of salt and moisture content. The slope and intercept of the line may depend on the type of salt. (The intercept of the line also depends on the type of additive.) Table 16.16 contains parameter estimates and standard errors for the model. All estimates in the example use the side condition $C_1 = 0$.

TABLE 16.16. Parameter estimates and standard errors for the model $y_{ijkm} = A_{i0} + A_{i1}x_j + C_k + e_{ijkm}$

Parameter	Estimate	SE
A_{10}	3.35	1.375
A_{11}	5.85	.5909
A_{20}	−3.789	1.237
A_{21}	13.24	.5909
A_{30}	−4.967	1.231
A_{31}	14.25	.5476
C_1	0.	none
C_2	−5.067	.5522

Note that, in lieu of the F test, the test for the main effect C could be performed by looking at $t = -5.067/.5522 = -9.176$. Moreover, we should have $t^2 = F$. The t statistic squared is 84, while the F statistic reported earlier is 87. The difference is due to the fact that the SE reported uses the MSE for the model being fitted, while in performing the F test we used the $MSE([ABC])$.

Are we done yet? No. The parameter estimates suggest some additional questions. Are the slopes for salts 2 and 3 the same, i.e., is $A_{21} = A_{31}$? In fact, are the entire lines for salts 2 and 3 the same, i.e., are $A_{21} = A_{31}, A_{20} = A_{30}$? We can fit models that incorporate these assumptions.

Model	SSE	dfE
$[A_0][A_1][C]$	59.88	25
$[A_0][A_1][C], A_{21} = A_{31}$	63.73	26
$[A_0][A_1][C], A_{21} = A_{31}, A_{20} = A_{30}$	66.97	27

It is a small matter to check that there is no lack of fit displayed by any of these models. The smallest model that fits the data is now $[A_0][A_1][C], A_{21} = A_{31}, A_{20} = A_{30}$. Thus there seems to be no difference between salts 2 and 3, but salt 1 has a different regression than the other two salts. (We did not actually test whether salt 1 is different, but if salt 1 had the same slope as the other two then there would be no interaction and we know that interaction exists.) There is also an effect for the food additives. The parameter estimates and standard errors for the final model are given in Table 16.17.

TABLE 16.17. Parameter estimates and standard errors for the model $y_{ijkm} = A_{i0} + A_{i1}x_j + C_k + e_{ijkm}$, $A_{21} = A_{31}, A_{20} = A_{30}$

Parameter	Estimate	SE
A_{10}	3.395	1.398
A_{11}	5.845	.6008
A_{20}	−4.466	.9030
A_{21}	13.81	.4078
C_1	0.	none
C_2	−5.130	.5602

Are we done yet? Probably not. We have not even considered the validity of the assumptions. Are the errors normally distributed? Are the variances the same for every treatment combination? Some methods for addressing these questions are discussed in Christensen (1987, chapter XIII) (and elsewhere in this book). Technically, we need to ask whether $C_1 = C_2$ in this new model. A quick look at the estimate and standard error for C_2 answers the question in the negative. We also have not asked whether $A_{10} = A_{20}$. Personally, given that the slopes are different, I find this last question so uninteresting that I would be loath to examine it. However, a look at the estimates and standard errors suggest that the answer is no.

Exercise 16.6.7 examines the process of fitting the more unusual models found in this section.

16.4 Youden squares

Consider the data on mangold roots in Table 16.18. There are five rows, four columns, and five treatments. If we ignore the columns, the rows and the treatments form a balanced incomplete block design, every pair of treatments occurs together three times. The key feature of Youden squares is that additionally the treatments are set up in such a way that every treatment occurs once in each column. Since every row also occurs once in each column, the analysis for columns can be conducted independently of the analysis for rows and treatments. Columns are balanced relative to both treatments and rows.

Table 16.19 contains the analysis of variance for these data. The line for columns is computed as in a one-way ANOVA, ignoring both the rows and the treatments. The line for rows is computed as in a one-way ANOVA, ignoring both the columns and the treatments. The means necessary for these computations are given in Table 16.18. The adjusted treatment means and the ANOVA table line for

TABLE 16.18. Mangold root data

Row	Columns 1	2	3	4	Row means
1	D(376)	E(371)	C(355)	B(356)	364.50
2	B(316)	D(338)	E(336)	A(356)	336.50
3	C(326)	A(326)	B(335)	D(343)	332.50
4	E(317)	B(343)	A(330)	C(327)	329.25
5	A(321)	C(332)	D(317)	E(318)	322.00
Col. means	331.2	342.0	334.6	340.0	336.95

treatments are computed as in a balanced incomplete block, adjusting for the rows and ignoring the columns. The adjusted treatment means are given below.

Adjusted treatment means

Treatment	A	B	C	D	E
Mean	340.4	333.5	334.8	341.9	334.2

From the ANOVA table, there is no evidence for a difference between treatments.

TABLE 16.19. Analysis of variance

Source	df	Seq SS	MS	F	P
Rows	4	4247.2	1061.8	6.87	
Column	3	367.0	122.3	0.79	
Trts	4	224.1	56.0	0.36	0.829
Error	8	1236.7	154.6		
Total	19	6075.0			

Evaluation of assumptions is carried out as in all unbalanced ANOVAs. Diagnostic statistics are given in Table 16.20. The diagnostic statistics look reasonably good.

Figures 16.9 and 16.10 contain a normal plot for the standardized residuals and a plot of standardized residuals against predicted values, respectively. The normal plot looks very reasonable. The predicted value plot may indicate increasing variability as predicted values increase. One could attempt to find a transformation that would improve the plot but there is so little evidence of any difference between treatments that it hardly seems worth the bother.

The reader may note that the data in this section consist of the first four columns of the Latin square examined in Example 9.3.1. Dropping one column (or row) from a Latin square is a simple way to produce a Youden square. As Youden square designs do not give a square array of numbers (recall our example had 4 columns and 5 rows), one presumes that the name Youden *square* derives from this relationship to Latin squares. Table 16.21 presents an alternative method of presenting the data in Table 16.18 that is often used. □

MINITAB COMMANDS

The Minitab commands for the mangold root analysis are given below.

```
MTB > names c1 'y' c2 'Rows' c3 'Cols' c4 'Trts'
MTB > glm c1 = c2 c3 c4;
SUBC> means c4;
SUBC> fits c11;
```

TABLE 16.20. Diagnostics

Row	Col	Trt	y	$\hat{y}$	Leverage	r	t	C
1	1	D	376	364.5	0.6	1.46	1.59	0.27
2	1	B	316	326.8	0.6	−1.37	−1.47	0.24
3	1	C	326	323.9	0.6	0.27	0.25	0.01
4	1	E	317	322.0	0.6	−0.64	−0.61	0.05
5	1	A	321	318.8	0.6	0.28	0.26	0.01
1	2	E	371	367.7	0.6	0.42	0.40	0.02
2	2	D	338	345.9	0.6	−1.01	−1.01	0.13
3	2	A	326	340.3	0.6	−1.81	−2.21	0.41
4	2	B	343	332.1	0.6	1.38	1.48	0.24
5	2	C	332	324.0	0.6	1.02	1.02	0.13
1	3	C	355	360.8	0.6	−0.74	−0.71	0.07
2	3	E	336	330.9	0.6	0.65	0.63	0.05
3	3	B	335	326.1	0.6	1.14	1.16	0.16
4	3	A	330	331.5	0.6	−0.19	−0.18	0.00
5	3	D	317	323.7	0.6	−0.86	−0.84	0.09
1	4	B	356	365.0	0.6	−1.14	−1.17	0.16
2	4	A	356	342.4	0.6	1.73	2.04	0.37
3	4	D	343	339.8	0.6	0.41	0.38	0.02
4	4	C	327	331.3	0.6	−0.55	−0.53	0.04
5	4	E	318	321.5	0.6	−0.44	−0.42	0.02

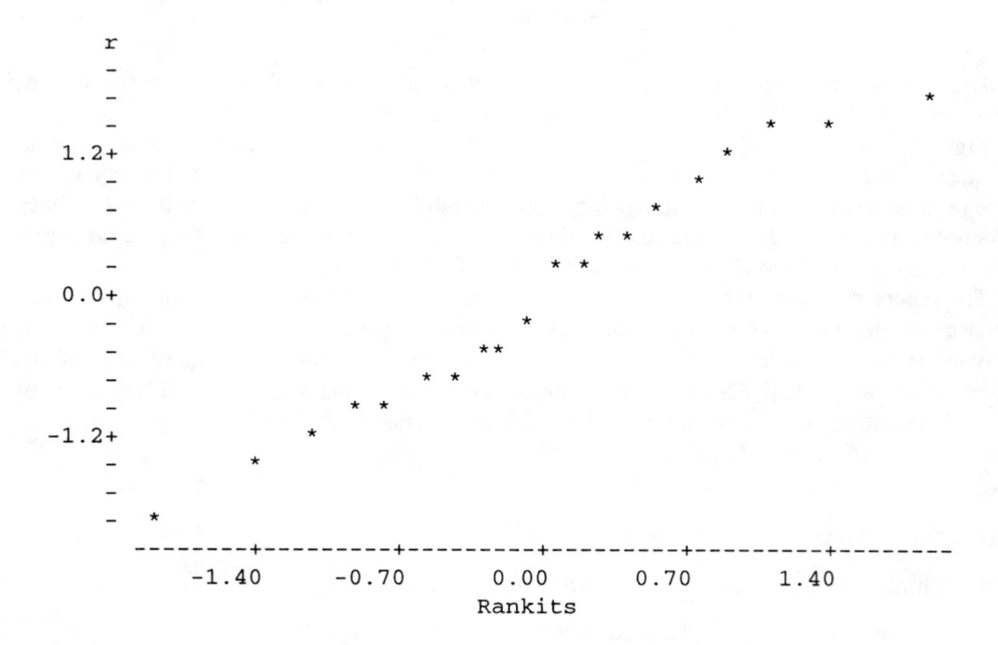

FIGURE 16.9. Normal plot, $W' = .978$.

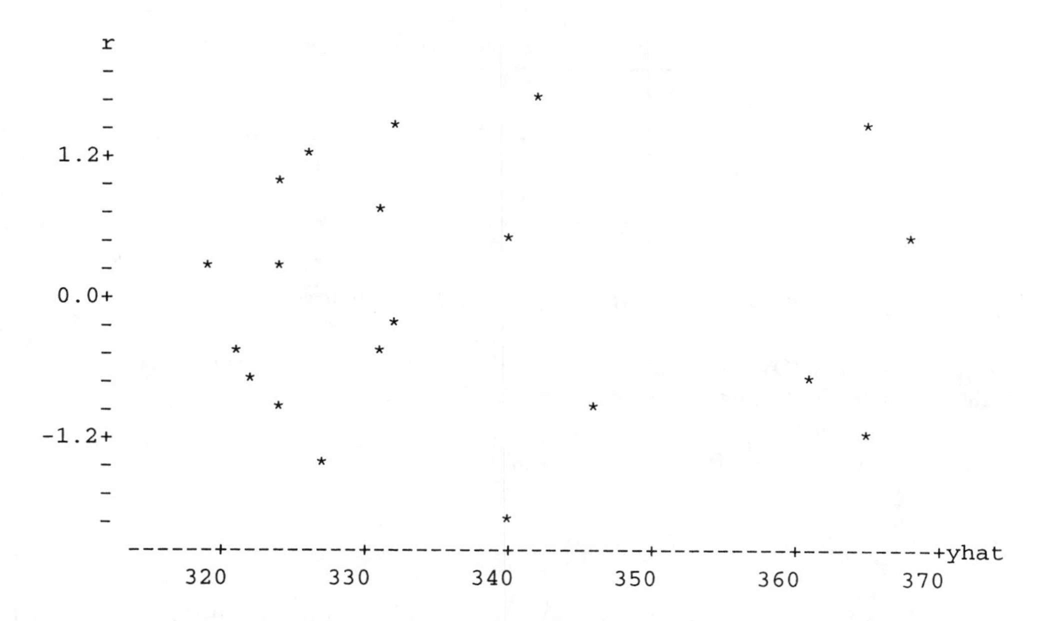

FIGURE 16.10. Standardized residuals versus predicted values.

TABLE 16.21. Mangold root data: column(observation)

Row	Treatments A	B	C	D	E
1		4(356)	3(355)	1(376)	2(371)
2	4(356)	1(316)		2(338)	3(336)
3	2(326)	3(335)	1(326)	4(343)	
4	3(330)	2(343)	4(327)		1(317)
5	1(321)		2(332)	3(317)	4(318)

```
SUBC> sresids c12;
SUBC> tresids c13;
SUBC> hi c14;
SUBC> cookd c15.
```

BALANCED LATTICE SQUARES

The key idea in *balanced lattice square designs* is that if you look at every row as a block, the treatments form a balanced incomplete block design and simultaneously if every column is viewed as a block, the treatments again form a balanced incomplete block design. In other words, each pair of treatments occurs together in the same row *or* column the same number of times. Balanced lattice square designs are similar to balanced lattices in that the number of treatments is $t = k^2$ and that the treatments are arranged in $k \times k$ squares. Table 16.22 gives an example for $k = 3$. If k is odd, one can typically get by with $(k+1)/2$ squares. If k is even, $k+1$ squares are generally needed. See Cochran and Cox (1957) for the analysis of such designs.

TABLE 16.22. Balanced lattice square design for 9 treatments

Row	Column 1	Column 2	Column 3	Row	Column 4	Column 5	Column 6
1	A	B	C	4	A	F	H
2	D	E	F	5	I	B	D
3	G	H	I	6	E	G	C

16.5 Matrix formulation of analysis of variance

Consider a one-way ANOVA with three treatments, 3 observations on the first two treatments and 2 observations on the third treatment group. The model can be written as

$$y_{ij} = \mu_i + \varepsilon_{ij}$$

or

$$y_{ij} = \mu + \alpha_i + \varepsilon_{ij},$$

both with $i = 1, 2, 3, j = 1, \dots, N_i$, and $N_1 = 3, N_2 = 3, N_3 = 2$. In matrix form, the first model can be written as

$$\begin{bmatrix} y_{11} \\ y_{12} \\ y_{13} \\ y_{21} \\ y_{22} \\ y_{23} \\ y_{31} \\ y_{32} \end{bmatrix} = \begin{bmatrix} \mu_1 + \varepsilon_{11} \\ \mu_1 + \varepsilon_{12} \\ \mu_1 + \varepsilon_{13} \\ \mu_2 + \varepsilon_{21} \\ \mu_2 + \varepsilon_{22} \\ \mu_2 + \varepsilon_{23} \\ \mu_3 + \varepsilon_{31} \\ \mu_3 + \varepsilon_{32} \end{bmatrix}.$$

It can also be written in the form $Y = X\beta + e$ that we used in Chapter 15 for regression models. In this form it is written

$$\begin{bmatrix} y_{11} \\ y_{12} \\ y_{13} \\ y_{21} \\ y_{22} \\ y_{23} \\ y_{31} \\ y_{32} \end{bmatrix} = \begin{bmatrix} 1 & 0 & 0 \\ 1 & 0 & 0 \\ 1 & 0 & 0 \\ 0 & 1 & 0 \\ 0 & 1 & 0 \\ 0 & 1 & 0 \\ 0 & 0 & 1 \\ 0 & 0 & 1 \end{bmatrix} \begin{bmatrix} \mu_1 \\ \mu_2 \\ \mu_3 \end{bmatrix} + \begin{bmatrix} \varepsilon_{11} \\ \varepsilon_{12} \\ \varepsilon_{13} \\ \varepsilon_{21} \\ \varepsilon_{22} \\ \varepsilon_{23} \\ \varepsilon_{31} \\ \varepsilon_{32} \end{bmatrix}.$$

The one-way ANOVA model $y_{ij} = \mu + \alpha_i + \varepsilon_{ij}$ can be written as

$$\begin{bmatrix} y_{11} \\ y_{12} \\ y_{13} \\ y_{21} \\ y_{22} \\ y_{23} \\ y_{31} \\ y_{32} \end{bmatrix} = \begin{bmatrix} 1 & 1 & 0 & 0 \\ 1 & 1 & 0 & 0 \\ 1 & 1 & 0 & 0 \\ 1 & 0 & 1 & 0 \\ 1 & 0 & 1 & 0 \\ 1 & 0 & 1 & 0 \\ 1 & 0 & 0 & 1 \\ 1 & 0 & 0 & 1 \end{bmatrix} \begin{bmatrix} \mu \\ \alpha_1 \\ \alpha_2 \\ \alpha_3 \end{bmatrix} + \begin{bmatrix} \varepsilon_{11} \\ \varepsilon_{12} \\ \varepsilon_{13} \\ \varepsilon_{21} \\ \varepsilon_{22} \\ \varepsilon_{23} \\ \varepsilon_{31} \\ \varepsilon_{32} \end{bmatrix}.$$

Suppose we want to examine a model in which $\alpha_2 = \alpha_3$. We can rewrite the model as

$$\begin{bmatrix} y_{11} \\ y_{12} \\ y_{13} \\ y_{21} \\ y_{22} \\ y_{23} \\ y_{31} \\ y_{32} \end{bmatrix} = \mu \begin{bmatrix} 1 \\ 1 \\ 1 \\ 1 \\ 1 \\ 1 \\ 1 \\ 1 \end{bmatrix} + \alpha_1 \begin{bmatrix} 1 \\ 1 \\ 1 \\ 0 \\ 0 \\ 0 \\ 0 \\ 0 \end{bmatrix} + \alpha_2 \begin{bmatrix} 0 \\ 0 \\ 0 \\ 1 \\ 1 \\ 1 \\ 0 \\ 0 \end{bmatrix} + \alpha_3 \begin{bmatrix} 0 \\ 0 \\ 0 \\ 0 \\ 0 \\ 0 \\ 1 \\ 1 \end{bmatrix} + \begin{bmatrix} \varepsilon_{11} \\ \varepsilon_{12} \\ \varepsilon_{13} \\ \varepsilon_{21} \\ \varepsilon_{22} \\ \varepsilon_{23} \\ \varepsilon_{31} \\ \varepsilon_{32} \end{bmatrix}.$$

If $\alpha_2 = \alpha_3$ this becomes

$$\begin{bmatrix} y_{11} \\ y_{12} \\ y_{13} \\ y_{21} \\ y_{22} \\ y_{23} \\ y_{31} \\ y_{32} \end{bmatrix} = \mu \begin{bmatrix} 1 \\ 1 \\ 1 \\ 1 \\ 1 \\ 1 \\ 1 \\ 1 \end{bmatrix} + \alpha_1 \begin{bmatrix} 1 \\ 1 \\ 1 \\ 0 \\ 0 \\ 0 \\ 0 \\ 0 \end{bmatrix} + \alpha_2 \begin{bmatrix} 0 \\ 0 \\ 0 \\ 1 \\ 1 \\ 1 \\ 0 \\ 0 \end{bmatrix} + \alpha_2 \begin{bmatrix} 0 \\ 0 \\ 0 \\ 0 \\ 0 \\ 0 \\ 1 \\ 1 \end{bmatrix} + \begin{bmatrix} \varepsilon_{11} \\ \varepsilon_{12} \\ \varepsilon_{13} \\ \varepsilon_{21} \\ \varepsilon_{22} \\ \varepsilon_{23} \\ \varepsilon_{31} \\ \varepsilon_{32} \end{bmatrix}$$

$$\begin{bmatrix} y_{11} \\ y_{12} \\ y_{13} \\ y_{21} \\ y_{22} \\ y_{23} \\ y_{31} \\ y_{32} \end{bmatrix} = \mu \begin{bmatrix} 1 \\ 1 \\ 1 \\ 1 \\ 1 \\ 1 \\ 1 \\ 1 \end{bmatrix} + \alpha_1 \begin{bmatrix} 1 \\ 1 \\ 1 \\ 0 \\ 0 \\ 0 \\ 0 \\ 0 \end{bmatrix} + \alpha_2 \left(\begin{bmatrix} 0 \\ 0 \\ 0 \\ 1 \\ 1 \\ 1 \\ 0 \\ 0 \end{bmatrix} + \begin{bmatrix} 0 \\ 0 \\ 0 \\ 0 \\ 0 \\ 0 \\ 1 \\ 1 \end{bmatrix} \right) + \begin{bmatrix} \varepsilon_{11} \\ \varepsilon_{12} \\ \varepsilon_{13} \\ \varepsilon_{21} \\ \varepsilon_{22} \\ \varepsilon_{23} \\ \varepsilon_{31} \\ \varepsilon_{32} \end{bmatrix}$$

$$\begin{bmatrix} y_{11} \\ y_{12} \\ y_{13} \\ y_{21} \\ y_{22} \\ y_{23} \\ y_{31} \\ y_{32} \end{bmatrix} = \mu \begin{bmatrix} 1 \\ 1 \\ 1 \\ 1 \\ 1 \\ 1 \\ 1 \\ 1 \end{bmatrix} + \alpha_1 \begin{bmatrix} 1 \\ 1 \\ 1 \\ 0 \\ 0 \\ 0 \\ 0 \\ 0 \end{bmatrix} + \alpha_2 \begin{bmatrix} 0 \\ 0 \\ 0 \\ 1 \\ 1 \\ 1 \\ 1 \\ 1 \end{bmatrix} + \begin{bmatrix} \varepsilon_{11} \\ \varepsilon_{12} \\ \varepsilon_{13} \\ \varepsilon_{21} \\ \varepsilon_{22} \\ \varepsilon_{23} \\ \varepsilon_{31} \\ \varepsilon_{32} \end{bmatrix}.$$

Bringing the model back into the standard form $Y = X\beta + e$ gives

$$\begin{bmatrix} y_{11} \\ y_{12} \\ y_{13} \\ y_{21} \\ y_{22} \\ y_{23} \\ y_{31} \\ y_{32} \end{bmatrix} = \begin{bmatrix} 1 & 1 & 0 \\ 1 & 1 & 0 \\ 1 & 1 & 0 \\ 1 & 0 & 1 \\ 1 & 0 & 1 \\ 1 & 0 & 1 \\ 1 & 0 & 1 \\ 1 & 0 & 1 \end{bmatrix} \begin{bmatrix} \mu \\ \alpha_1 \\ \alpha_2 \end{bmatrix} + \begin{bmatrix} \varepsilon_{11} \\ \varepsilon_{12} \\ \varepsilon_{13} \\ \varepsilon_{21} \\ \varepsilon_{22} \\ \varepsilon_{23} \\ \varepsilon_{31} \\ \varepsilon_{32} \end{bmatrix}.$$

This technique is used in Exercise 16.6.7.

EXAMPLE 16.5.1. Consider again the data of Table 16.1. In specifying models, we will need to play games with subscripts. We are considering the data as an unbalanced one-way ANOVA,

$$y_{hk} = \mu_h + \varepsilon_{hk} \tag{16.5.1}$$

where $h = 1, 2, 3, 4$ and $k = 1, \ldots, N_h$. We denote the treatments as follows.

	Treatments			
	A		J	
Index	F	I	F	I
h	1	2	3	4
(i,j)	(1, 1)	(1, 2)	(2, 1)	(2, 2)

The index h identifies a treatment in model (16.5.1). The pair (i,j) can also be used to identify the same treatment.

Using model (16.5.1) we have the sample sizes $N_1 = 4$, $N_2 = 3$, $N_3 = 2$, $N_4 = 3$ and the model can be written in the matrix form $Y = X\beta + e$ as

$$\begin{bmatrix} y_{11} \\ y_{12} \\ y_{13} \\ y_{14} \\ y_{21} \\ y_{22} \\ y_{23} \\ y_{31} \\ y_{32} \\ y_{21} \\ y_{22} \\ y_{23} \end{bmatrix} = \begin{bmatrix} 48.0 \\ 49.3 \\ 51.7 \\ 60.3 \\ 68.0 \\ 36.3 \\ 37.0 \\ 40.5 \\ 51.3 \\ 54.5 \\ 42.8 \\ 50.2 \end{bmatrix} = \begin{bmatrix} 1 & 0 & 0 & 0 \\ 1 & 0 & 0 & 0 \\ 1 & 0 & 0 & 0 \\ 1 & 0 & 0 & 0 \\ 0 & 1 & 0 & 0 \\ 0 & 1 & 0 & 0 \\ 0 & 1 & 0 & 0 \\ 0 & 0 & 1 & 0 \\ 0 & 0 & 1 & 0 \\ 0 & 0 & 0 & 1 \\ 0 & 0 & 0 & 1 \\ 0 & 0 & 0 & 1 \end{bmatrix} \begin{bmatrix} \mu_1 \\ \mu_2 \\ \mu_3 \\ \mu_4 \end{bmatrix} + \begin{bmatrix} \varepsilon_{11} \\ \varepsilon_{12} \\ \varepsilon_{13} \\ \varepsilon_{14} \\ \varepsilon_{21} \\ \varepsilon_{22} \\ \varepsilon_{23} \\ \varepsilon_{31} \\ \varepsilon_{32} \\ \varepsilon_{21} \\ \varepsilon_{22} \\ \varepsilon_{23} \end{bmatrix}.$$

Alternatively, we can replace the single subscript h with the equivalent pair of subscripts ij and rewrite the one-way ANOVA model as

$$y_{ijk} = \mu_{ij} + \varepsilon_{ijk}.$$

where $i = 1, 2$, $j = 1, 2$, and $k = 1, \ldots, N_{ij}$ with $N_{11} = 4$, $N_{12} = 3$, $N_{21} = 2$, $N_{22} = 3$. The matrix form of this model is very similar to that of the previous model; the only change is that we now use two subscripts to identify treatments instead of one subscript.

$$\begin{bmatrix} y_{111} \\ y_{112} \\ y_{113} \\ y_{114} \\ y_{121} \\ y_{122} \\ y_{123} \\ y_{211} \\ y_{212} \\ y_{221} \\ y_{222} \\ y_{223} \end{bmatrix} = \begin{bmatrix} 48.0 \\ 49.3 \\ 51.7 \\ 60.3 \\ 68.0 \\ 36.3 \\ 37.0 \\ 40.5 \\ 51.3 \\ 54.5 \\ 42.8 \\ 50.2 \end{bmatrix} = \begin{bmatrix} 1 & 0 & 0 & 0 \\ 1 & 0 & 0 & 0 \\ 1 & 0 & 0 & 0 \\ 1 & 0 & 0 & 0 \\ 0 & 1 & 0 & 0 \\ 0 & 1 & 0 & 0 \\ 0 & 1 & 0 & 0 \\ 0 & 0 & 1 & 0 \\ 0 & 0 & 1 & 0 \\ 0 & 0 & 0 & 1 \\ 0 & 0 & 0 & 1 \\ 0 & 0 & 0 & 1 \end{bmatrix} \begin{bmatrix} \mu_{11} \\ \mu_{12} \\ \mu_{21} \\ \mu_{22} \end{bmatrix} + \begin{bmatrix} \varepsilon_{111} \\ \varepsilon_{112} \\ \varepsilon_{113} \\ \varepsilon_{114} \\ \varepsilon_{121} \\ \varepsilon_{122} \\ \varepsilon_{123} \\ \varepsilon_{211} \\ \varepsilon_{212} \\ \varepsilon_{221} \\ \varepsilon_{222} \\ \varepsilon_{223} \end{bmatrix}.$$

The key portions of the model, the Y data and the X matrix, are identical in the two models. That is why the models are equivalent.

As discussed earlier, the one-way ANOVA model using the two subscripts ij can be written as an equivalent model that involves a grand mean, main effects, and an interaction, i.e.,

$$\begin{aligned} y_{ijk} &= \mu_{ij} + \varepsilon_{ijk} \\ &= \mu + \alpha_i + \eta_j + \gamma_{ij} + \varepsilon_{ijk}, \end{aligned}$$

The Y vector and the e vector remain unchanged when using the second parameterization, but the $X\beta$ portion of the model becomes

$$X\beta = \begin{bmatrix} 1 & 1 & 0 & 1 & 0 & 1 & 0 & 0 & 0 \\ 1 & 1 & 0 & 1 & 0 & 1 & 0 & 0 & 0 \\ 1 & 1 & 0 & 1 & 0 & 1 & 0 & 0 & 0 \\ 1 & 1 & 0 & 1 & 0 & 1 & 0 & 0 & 0 \\ 1 & 1 & 0 & 0 & 1 & 0 & 1 & 0 & 0 \\ 1 & 1 & 0 & 0 & 1 & 0 & 1 & 0 & 0 \\ 1 & 1 & 0 & 0 & 1 & 0 & 1 & 0 & 0 \\ 1 & 0 & 1 & 1 & 0 & 0 & 0 & 1 & 0 \\ 1 & 0 & 1 & 1 & 0 & 0 & 0 & 1 & 0 \\ 1 & 0 & 1 & 0 & 1 & 0 & 0 & 0 & 1 \\ 1 & 0 & 1 & 0 & 1 & 0 & 0 & 0 & 1 \\ 1 & 0 & 1 & 0 & 1 & 0 & 0 & 0 & 1 \end{bmatrix} \begin{bmatrix} \mu \\ \alpha_1 \\ \alpha_2 \\ \eta_1 \\ \eta_2 \\ \gamma_{11} \\ \gamma_{12} \\ \gamma_{21} \\ \gamma_{22} \end{bmatrix}. \qquad (16.5.2)$$

Note that the last four columns of this X matrix are identical to the columns in the one-way ANOVA X matrices given earlier for these data. In fact, these four are the only columns in the X matrix that matter. All of the other columns (the first five) can be obtained by adding together various of the last four columns. (The column space of this new X matrix is the same as in the one-way ANOVA model.)

The only reason for considering the interaction model is that it leads naturally to the model without interaction,

$$y_{ijk} = \mu + \alpha_i + \eta_j + \varepsilon_{ijk}.$$

In matrix form, the no interaction model becomes

$$\begin{bmatrix} y_{111} \\ y_{112} \\ y_{113} \\ y_{114} \\ y_{121} \\ y_{122} \\ y_{123} \\ y_{211} \\ y_{212} \\ y_{221} \\ y_{222} \\ y_{223} \end{bmatrix} = \begin{bmatrix} 48.0 \\ 49.3 \\ 51.7 \\ 60.3 \\ 68.0 \\ 36.3 \\ 37.0 \\ 40.5 \\ 51.3 \\ 54.5 \\ 42.8 \\ 50.2 \end{bmatrix} = \begin{bmatrix} 1 & 1 & 0 & 1 & 0 \\ 1 & 1 & 0 & 1 & 0 \\ 1 & 1 & 0 & 1 & 0 \\ 1 & 1 & 0 & 1 & 0 \\ 1 & 1 & 0 & 0 & 1 \\ 1 & 1 & 0 & 0 & 1 \\ 1 & 1 & 0 & 0 & 1 \\ 1 & 0 & 1 & 1 & 0 \\ 1 & 0 & 1 & 1 & 0 \\ 1 & 0 & 1 & 0 & 1 \\ 1 & 0 & 1 & 0 & 1 \\ 1 & 0 & 1 & 0 & 1 \end{bmatrix} \begin{bmatrix} \mu \\ \alpha_1 \\ \alpha_2 \\ \eta_1 \\ \eta_2 \end{bmatrix} + \begin{bmatrix} \varepsilon_{111} \\ \varepsilon_{112} \\ \varepsilon_{113} \\ \varepsilon_{114} \\ \varepsilon_{121} \\ \varepsilon_{122} \\ \varepsilon_{123} \\ \varepsilon_{211} \\ \varepsilon_{212} \\ \varepsilon_{221} \\ \varepsilon_{222} \\ \varepsilon_{223} \end{bmatrix}.$$

Note that the X matrix for this model consists of the first five columns from the X matrix for the interaction model given in (16.5.2). We just dropped the columns that corresponded to the interaction terms, the γ_{ij}s. This model is not equivalent to the interaction model or any version of the one-way ANOVA model. The columns of the previous X matrices can be added together to obtain every column in this X matrix but we cannot go the other way. From these five columns, we cannot reconstruct the columns of the X matrices for either the interaction model or the one-way ANOVA. □

16.6 Exercises

EXERCISE 16.6.1. In Example 16.2.2, find the sums of squares for the cubic main effect contrast and the detergent by cubic interaction contrast. Also find the sums of squares for the separate cubic effect contrasts in detergents I and II.

EXERCISE 16.6.2. Cochran and Cox (1957) presented data from Pauline Paul on the effect of cold storage on roast beef tenderness. Treatments are labeled A through F and consist of 0, 1, 2, 4, 9, and 18 days of storage respectively. The data are tenderness scores and are presented in Table 16.23. Determine the values of t, r, b, k, and λ. Analyze the data.

TABLE 16.23. Beef tenderness scores

Block	Trt, Score		Block	Trt, Score	
1	A, 7	B, 17	9	A, 17	C, 27
2	C, 26	D, 25	10	B, 23	E, 27
3	E, 33	F, 29	11	D, 29	F, 30
4	A, 25	E, 40	12	A, 11	F, 27
5	B, 25	D, 34	13	B, 24	C, 21
6	C, 34	F, 32	14	D, 26	E, 32
7	A, 10	D, 25	15	B, 26	F, 37
8	C, 24	E, 26	16		

EXERCISE 16.6.3. The balanced incomplete block data of Table 16.24 were presented in Finney (1964) and Bliss (1947). The observations are serum calcium values of dogs after they have been injected with a dose of parathyroid extract. The doses are the treatments and they have factorial structure. One factor involves using either the standard preparation (S) or a test preparation (T). The other factor is the amount of a dose; it is either low (L) or high (H). Low doses are .125 cc and high doses are .205 cc. Each dog is subjected to three injections at about 10 day intervals. Serum calcium is measured on the day after an injection. Analyze the data. Look at contrasts and residuals. Should day effects be isolated? Can this be done conveniently? If so, do so.

EXERCISE 16.6.4. Inman et al. (1992) report data on the percentages of Manganese (Mn) in various samples as determined by a spectrometer. Ten samples were used and the percentage of Mn in each sample was determined by each of 4 operators. The data are given in Table 16.25. The operators actually made two readings; the data presented are the averages of the two readings for each sample–operator combination.

Treating the samples as blocks, analyze the data. Include in your analysis an evaluation of whether any operators are significantly different. Identify a potential outlier, delete that outlier, reanalyze the data, and compare the results of the two analyses.

EXERCISE 16.6.5. Nelson (1993) presents data on the average access times for various disk drives. The disk drives are five brands of half-height fixed drives. The performance of disk drives depends on the computer where they are installed, so computers were used as blocks. The computers could only hold four disk drives, so a balanced incomplete block design was used. The data are given in Table 16.26. Analyze them.

EXERCISE 16.6.6. Write models (10.1.1), (10.1.2), and (10.1.3) in matrix form. Use a regression program on the heart weight data of Table 10.1 to find 95% and 99% prediction intervals for a male

TABLE 16.24. Serum calcium for dogs after parathyroid extract injections

Dog	Day I	Day II	Day III
1	TL, 14.7	TH, 15.4	SH, 14.8
2	TL, 15.1	TH, 15.0	SH, 15.8
3	TH, 14.4	SH, 13.8	TL, 14.4
4	TH, 16.2	TL, 14.0	SH, 13.0
5	TH, 15.8	SH, 16.0	TL, 15.0
6	TH, 15.8	TL, 14.3	SL, 14.8
7	TH, 17.0	TL, 16.5	SL, 15.0
8	TL, 13.6	SL, 15.3	TH, 17.2
9	TL, 14.0	TH, 13.8	SL, 14.0
10	TL, 13.0	SL, 13.4	TH, 13.8
11	SL, 13.8	SH, 17.0	TH, 16.0
12	SL, 12.0	SH, 13.8	TH, 14.0
13	SH, 14.6	TH, 15.4	SL, 14.0
14	SH, 13.0	SL, 14.0	TH, 14.0
15	SH, 15.2	TH, 16.2	SL, 15.0
16	SH, 15.0	SL, 14.5	TL, 14.0
17	SH, 15.0	SL, 14.0	TL, 14.6
18	SL, 15.8	TL, 15.0	SH, 15.2
19	SL, 13.2	SH, 16.0	TL, 14.9
20	SL, 14.2	TL, 14.1	SH, 15.0

TABLE 16.25. Percentage of manganese concentrations

Sample	Operator 1	2	3	4
1	.615	.620	.600	.600
2	.635	.635	.660	.630
3	.590	.605	.600	.590
4	.745	.740	.735	.745
5	.695	.695	.680	.695
6	.640	.635	.635	.630
7	.655	.665	.650	.650
8	.640	.645	.620	.610
9	.670	.675	.670	.665
10	.655	.660	.645	.650

and a female each with body weight of 3.0. (Hint: In Minitab use the subcommand 'noconstant' to

TABLE 16.26. Access times (ms) for disk drives

Computer	Brand 1	2	3	4	5
A	35	42	31	30	—
B	41	45	—	32	40
C	—	40	42	33	39
D	32	—	33	35	36
E	40	38	35	—	37

eliminate the intercept.)

EXERCISE 16.6.7. Using the notation of Section 16.3, write the models $[A_0][A_1][C]$, $[A_0][A_1][C]\ A_{21} = A_{31}$, and $[A_0][A_1][C]\ A_{21} = A_{31}, A_{20} = A_{30}$ in matrix form. (Hint: To obtain $[A_0][A_1][C]\ A_{21} = A_{31}$ from $[A_0][A_1][C]$, replace the two columns of X corresponding to A_{21} and A_{31} with one column consisting of their sum.) Use a regression program to fit these three models. (Hint: In Minitab use the subcommand 'noconstant' to eliminate the intercept, and to impose the side condition $C_1 = 0$, drop the column corresponding to C_1.)

Chapter 17

Confounding and fractional replication in 2^n factorial systems

Confounding is a method of designing a factorial experiment that allows incomplete blocks, i.e., blocks of smaller size than the full number of factorial treatments. In *fractional replication* an experiment has fewer observations than the full factorial number of treatments. A basic idea in experimental design is ensuring adequate replication to provide a good estimate of error. Fractional replication not only fails to get replication – it fails to provide even an observation on every factor combination. Not surprisingly, fractional replications present new challenges in analyzing data.

In this chapter, we will informally use concepts of modular arithmetic, e.g., $7 \ mod \ 5 = 2$ where 2 is the remainder when 7 is divided by 5. Modular arithmetic is crucial to more advanced discussions of confounding and fractional replication, but its use in this chapter will be minimal. To help minimize modular arithmetic, we will refer to 0 as an even number.

Previously, we have used the notation $2 \times 2 \times 2$ to indicate the presence of three factors each at two levels. It is a natural extension of this notation to write

$$2 \times 2 \times 2 = 2^3.$$

A 2^n factorial system involves n factors each at 2 levels, so there are 2^n treatment combinations. 2^n factorials and fractional replications of 2^n factorials are often used in experiments designed to screen large numbers of factors to see which factors have substantial effects.

In a 2^n experiment, the treatments have $2^n - 1$ degrees of freedom and they are broken down into $2^n - 1$ effects each with one degree of freedom.

EXAMPLE 17.0.1. *A 2^4 factorial structure*
Consider a 2^4 experiment with factors A, B, C, D at levels $a_0, a_1, b_0, b_1, c_0, c_1$, and d_0, d_1, respectively. There are $2^4 = 16$ treatment combinations, so there are 15 degrees of freedom for treatments. The treatments line in the ANOVA table can be broken down as follows

Source	df	Source	df
A	1	ABC	1
B	1	ABD	1
C	1	ACD	1
D	1	BCD	1
AB	1	$ABCD$	1
AC	1		
AD	1		
BC	1		
BD	1		
CD	1		

Since each effect has a single degree of freedom, it can be identified with a contrast among the 16 treatments. □

EFFECT CONTRASTS IN 2^n FACTORIALS

The simplest way to understand confounding and fractional replication in 2^n systems is in terms of contrasts corresponding to the different effects. As was just seen, each effect in a 2^n has one degree of freedom and thus each effect corresponds to a single contrast. We now review the correspondence between contrasts and factorial effects.

EXAMPLE 17.0.2. *A 2^2 experiment*
Consider a 2^2 experiment with factors A and B at levels a_0, a_1 and b_0, b_1, respectively. The coefficients of the contrasts that correspond to the main effects and interaction are given below

Treatment	A	B	AB
a_0b_0	1	1	1
a_0b_1	1	-1	-1
a_1b_0	-1	1	-1
a_1b_1	-1	-1	1

In Example 11.1.1, we examined a 2^2 experiment and showed that these contrasts give the same sums of squares as the analysis of variance table methods for obtaining the sums of squares for A, B, and AB.

The contrast coefficients are determined by the subscripts in the treatment combinations. The A contrast coefficient is 1 for any treatment that has an a subscript of 0 and -1 for any treatment that has an a subscript of 1. In other words, the A contrast is 1 for a_0b_0 and a_0b_1 and -1 for a_1b_0 and a_1b_1. Similarly the B contrast is 1 for any treatment that has a b subscript of 0 and -1 for any treatment that has an b subscript of 1, i.e., a_0b_0 and a_1b_0 have coefficients of 1 and a_0b_1 and a_1b_1 have coefficients of -1. The AB contrast involves both factors, so the subscripts are added. Treatments with an even total, 0 or 2, get contrast coefficients of 1, while treatments with an odd total for the subscripts get -1. Thus a_0b_0 and a_1b_1 get 1s and a_0b_1 and a_1b_0 get -1s. Actually, the key is modular arithmetic. For 2^n factorials, the contrast coefficients are determined by an appropriate sum of the subscripts modulo 2. Thus *any sum that is an even number is 0 mod 2 and any odd sum is 1 mod 2.* □

EXAMPLE 17.0.3. *A 2^3 experiment*
Consider a 2^3 experiment with factors A, B, and C. The contrast coefficients for main effects and interactions are given below.

Treatment	A	B	C	AB	AC	BC	ABC
$a_0b_0c_0$	1	1	1	1	1	1	1
$a_0b_0c_1$	1	1	−1	1	−1	−1	−1
$a_0b_1c_0$	1	−1	1	−1	1	−1	−1
$a_0b_1c_1$	1	−1	−1	−1	−1	1	1
$a_1b_0c_0$	−1	1	1	−1	−1	1	−1
$a_1b_0c_1$	−1	1	−1	−1	1	−1	1
$a_1b_1c_0$	−1	−1	1	1	−1	−1	1
$a_1b_1c_1$	−1	−1	−1	1	1	1	−1

Once again the contrast coefficients are determined by the subscripts of the treatment combinations. The A contrast has 1s for a_0s and -1s for a_1s; similarly for B and C. The AB contrast is determined by the sum of the a and b subscripts. The sum of the a and b subscripts is even, either 0 or 2, for the treatments $a_0b_0c_0$, $a_0b_0c_1$, $a_1b_1c_0$, $a_1b_1c_1$, so all have AB contrast coefficients of 1. The sum of the a and b subscripts is 1 for the treatments $a_0b_1c_0$, $a_0b_1c_1$, $a_1b_0c_0$, $a_1b_0c_1$, so all have coefficients of -1. The AC contrast is determined by the sum of the a and c subscripts and the BC contrast is determined by the sum of the b and c subscripts. The ABC contrast is determined by the sum of the a, b, and c subscripts. The sum of the a, b, and c subscripts is even, either 0 or 2, for the treatments $a_0b_0c_0$, $a_0b_1c_1$, $a_1b_0c_1$, $a_1b_1c_0$, so all have ABC coefficients of 1. The sum of the a, b, and c subscripts is odd, 1 or 3, for the treatments $a_0b_0c_1$, $a_0b_1c_0$, $a_1b_0c_0$, $a_1b_1c_1$, so all have coefficients of -1. □

EXAMPLE 17.0.4. *A 2^4 experiment*

Consider a 2^4 experiment with factors A, B, C, and D. The contrast coefficients are given in Tables 17.1 and 17.2. Again the contrast coefficients are determined by the subscripts of the treatments. The A, B, C, and D contrasts are determined by the subscripts of a, b, c, and d, respectively. The AB, AC, AD, BC, BD, and CD contrasts are determined by the sums of the appropriate pair of subscripts. The ABC, ABD, ACD, and BCD contrasts are determined by the sum of the three appropriate subscripts. The coefficients of the $ABCD$ contrast are determined by the sum of all four subscripts. As before, the contrast coefficient is 1 if the appropriate value or sum equals 0 mod 2 (even) and is -1 if it equals 1 mod 2 (odd). □

TABLE 17.1. Main effect and second-order interaction contrast coefficients for a 2^4 factorial

Treatment	A	B	C	D	AB	AC	AD	BC	BD	CD
$a_0b_0c_0d_0$	1	1	1	1	1	1	1	1	1	1
$a_0b_0c_0d_1$	1	1	1	−1	1	1	−1	1	−1	−1
$a_0b_0c_1d_0$	1	1	−1	1	1	−1	1	−1	1	−1
$a_0b_0c_1d_1$	1	1	−1	−1	1	−1	−1	−1	−1	1
$a_0b_1c_0d_0$	1	−1	1	1	−1	1	1	−1	−1	1
$a_0b_1c_0d_1$	1	−1	1	−1	−1	1	−1	−1	1	−1
$a_0b_1c_1d_0$	1	−1	−1	1	−1	−1	1	1	−1	−1
$a_0b_1c_1d_1$	1	−1	−1	−1	−1	−1	−1	1	1	1
$a_1b_0c_0d_0$	−1	1	1	1	−1	−1	−1	1	1	1
$a_1b_0c_0d_1$	−1	1	1	−1	−1	−1	1	1	−1	−1
$a_1b_0c_1d_0$	−1	1	−1	1	−1	1	−1	−1	1	−1
$a_1b_0c_1d_1$	−1	1	−1	−1	−1	1	1	−1	−1	1
$a_1b_1c_0d_0$	−1	−1	1	1	1	−1	−1	−1	−1	1
$a_1b_1c_0d_1$	−1	−1	1	−1	1	−1	1	−1	1	−1
$a_1b_1c_1d_0$	−1	−1	−1	1	1	1	−1	1	−1	−1
$a_1b_1c_1d_1$	−1	−1	−1	−1	1	1	1	1	1	1

TABLE 17.2. Higher order interaction contrast coefficients for a 2^4 factorial

Treatment	ABC	ABD	ACD	BCD	$ABCD$
$a_0b_0c_0d_0$	1	1	1	1	1
$a_0b_0c_0d_1$	1	−1	−1	−1	−1
$a_0b_0c_1d_0$	−1	1	−1	−1	−1
$a_0b_0c_1d_1$	−1	−1	1	1	1
$a_0b_1c_0d_0$	−1	−1	1	−1	−1
$a_0b_1c_0d_1$	−1	1	−1	1	1
$a_0b_1c_1d_0$	1	−1	−1	1	1
$a_0b_1c_1d_1$	1	1	1	−1	−1
$a_1b_0c_0d_0$	−1	−1	−1	1	−1
$a_1b_0c_0d_1$	−1	1	1	−1	1
$a_1b_0c_1d_0$	1	−1	1	−1	1
$a_1b_0c_1d_1$	1	1	−1	1	−1
$a_1b_1c_0d_0$	1	1	−1	−1	1
$a_1b_1c_0d_1$	1	−1	1	1	−1
$a_1b_1c_1d_0$	−1	1	1	1	−1
$a_1b_1c_1d_1$	−1	−1	−1	−1	1

Most books on experimental design contain a discussion of confounding and fractional replication for 2^n treatment structures. Daniel (1976), Box, Hunter, and Hunter (1978), and Box and Draper (1987) are excellent books that focus on industrial applications.

17.1 Confounding

Confounding involves creating blocks that are smaller than the total number of treatments. Thus, confounding is a method for arriving at an incomplete block design. However, we will see that the analysis of confounding designs remains simple. For example, the analysis is considerably simpler than the balanced incomplete block analysis of the previous chapter.

EXAMPLE 17.1.1. *Confounding in a 2^3 experiment*
Suppose we have three drugs that we wish to investigate simultaneously. Each drug is a factor; the levels are either no dose of the drug or a standard dose. There are $2^3 = 8$ treatment combinations. The drugs will be applied to a certain type of animal. To reduce variation, we may want to use different litters of animals as blocks. However, it may be difficult to find litters containing 8 animals. On the other hand, litters of size 4 may be readily available. In such a case, we want to use four treatments on one litter and the other four treatments on a different litter. There are 70 ways to do this. We need a systematic method of choosing the treatments for each litter that allows us to perform as complete an analysis as possible.

To examine the application of the treatments from a 2^3 factorial in blocks of size 4, recall that the 2^3 has 8 treatments, so $1/2$ the treatments will go in each block. The table of contrast coefficients for a 2^3 factorial is repeated below.

Treatment	A	B	C	AB	AC	BC	ABC
$a_0b_0c_0$	1	1	1	1	1	1	1
$a_0b_0c_1$	1	1	−1	1	−1	−1	−1
$a_0b_1c_0$	1	−1	1	−1	1	−1	−1
$a_0b_1c_1$	1	−1	−1	−1	−1	1	1
$a_1b_0c_0$	−1	1	1	−1	−1	1	−1
$a_1b_0c_1$	−1	1	−1	−1	1	−1	1
$a_1b_1c_0$	−1	−1	1	1	−1	−1	1
$a_1b_1c_1$	−1	−1	−1	1	1	1	−1

We need to divide the treatments into two groups of size 4 but every contrast does this. The two groups of four are those treatments that have contrast coefficients of 1 and those that have -1. Thus we can use any contrast to define the blocks. Unfortunately, the contrast we choose will be lost to us because it will be *confounded* with blocks. In other words, we will not be able to tell what effects are due to blocks (litters) and what effects are due to the defining contrast. We choose to define blocks using the ABC contrast because it is the highest order interaction. Typically, it is the least painful to lose. The ABC contrast defines two groups of treatments

ABC coefficients	
$ABC(1)$	$ABC(-1)$
$a_0b_0c_0$	$a_0b_0c_1$
$a_0b_1c_1$	$a_0b_1c_0$
$a_1b_0c_1$	$a_1b_0c_0$
$a_1b_1c_0$	$a_1b_1c_1$

Each group of treatments is used in a separate block. The four treatments labeled $ABC(1)$ will be randomly assigned to the animals in one randomly chosen litter and the four treatments labeled $ABC(-1)$ will be randomly assigned to the animals in another litter. Recall that all information about ABC has been lost because it is confounded with blocks.

As indicated earlier, we could choose any of the contrasts to define the blocks. Typically, we use high order interactions because they are the effects that are most difficult to interpret and thus the most comfortable to live without. For illustrative purposes, we also give the blocks defined by the BC contrast.

BC coefficients	
$BC(1)$	$BC(-1)$
$a_0b_0c_0$	$a_0b_0c_1$
$a_0b_1c_1$	$a_0b_1c_0$
$a_1b_0c_0$	$a_1b_0c_1$
$a_1b_1c_1$	$a_1b_1c_0$

If the subjects of the drug study are humans, it will be difficult to obtain 'litters' of four, but it may be practical to use identical twins. We now have 8 treatments that need to be divided into blocks of 2 units. Each block will consist of 1/ 4 of the treatments. Since each contrast divides the treatments into two groups of four, if we use two contrasts we can divide each group of four into 2 groups of two. We take as our first contrast ABC and as our second contrast AB. The four treatments with ABC coefficients of 1 are $a_0b_0c_0$, $a_0b_1c_1$, $a_1b_0c_1$, and $a_1b_1c_0$. These can be divided into 2 groups of two, depending on whether their AB coefficient is 1 or -1. The two groups are $a_0b_0c_0$, $a_1b_1c_0$ and $a_0b_1c_1$, $a_1b_0c_1$. Similarly, the $ABC(-1)$ group, $a_0b_0c_1$, $a_0b_1c_0$, $a_1b_0c_0$, and $a_1b_1c_1$ can be divided into $a_0b_0c_1$, $a_1b_1c_1$ and $a_0b_1c_0$, $a_1b_0c_0$ based on the AB coefficients. In tabular form we get

$ABC(1)$		$ABC(-1)$	
$AB(1)$	$AB(-1)$	$AB(1)$	$AB(-1)$
$a_0b_0c_0$	$a_1b_0c_1$	$a_0b_0c_1$	$a_1b_0c_0$
$a_1b_1c_0$	$a_0b_1c_1$	$a_1b_1c_1$	$a_0b_1c_0$

To get blocks of size 2, we confounded two contrasts, ABC and AB. Thus we have lost all information on both of these contrasts. It turns out that we have also lost all information on another contrast, C. Exactly the same four blocks would be obtained if we confounded ABC and C.

$ABC(1)$		$ABC(-1)$	
$C(1)$	$C(-1)$	$C(-1)$	$C(1)$
$a_0b_0c_0$	$a_1b_0c_1$	$a_0b_0c_1$	$a_1b_0c_0$
$a_1b_1c_0$	$a_0b_1c_1$	$a_1b_1c_1$	$a_0b_1c_0$

Similarly, if we had confounded AB and C, we would obtain the same four blocks. Note that with four blocks, there are three degrees of freedom for blocks. Each contrast has one degree of freedom, so there must be three contrasts confounded with blocks.

Given the two defining contrasts ABC and AB, there is a simple way to identify the other contrast that is confounded with blocks. The confounding is determined by a form of modular multiplication where any even power is treated as 0; thus $A^2 = A^0 = 1$ and $B^2 = 1$. Multiplying the defining contrasts gives

$$ABC \times AB = A^2B^2C = C,$$

so C is also confounded with blocks.

Typically, we want to retain information on all main effects. The choice of ABC and AB for defining contrasts is poor because it leads to complete loss of information on the main effect C. We would do better to choose AB and BC. In that case, the other confounded contrast is

$$AB \times BC = AB^2C = AC,$$

which is another two-factor interaction. Using this confounding scheme, we get information on all main effects. The blocking scheme is given below.

$AB(1)$		$AB(-1)$	
$BC(1)$	$BC(-1)$	$BC(1)$	$BC(-1)$
$a_0b_0c_0$	$a_0b_0c_1$	$a_1b_0c_0$	$a_1b_0c_1$
$a_1b_1c_1$	$a_1b_1c_0$	$a_0b_1c_1$	$a_0b_1c_0$

□

EXAMPLE 17.1.2. *Confounding in a 2^4 experiment*

The $ABCD$ contrast was given in Table 17.2. Dividing the treatments into two groups based on their $ABCD$ coefficients defines two blocks of size 8,

$ABCD$ coefficients	
$ABCD(1)$	$ABCD(-1)$
$a_0b_0c_0d_0$	$a_0b_0c_0d_1$
$a_0b_0c_1d_1$	$a_0b_0c_1d_0$
$a_0b_1c_0d_1$	$a_0b_1c_0d_0$
$a_0b_1c_1d_0$	$a_0b_1c_1d_1$
$a_1b_0c_0d_1$	$a_1b_0c_0d_0$
$a_1b_0c_1d_0$	$a_1b_0c_1d_1$
$a_1b_1c_0d_0$	$a_1b_1c_0d_1$
$a_1b_1c_1d_1$	$a_1b_1c_1d_0$

To define four blocks of size 4 requires choosing two defining contrasts. To obtain four blocks, $ABCD$ is not a good choice for a defining contrast because if we choose the second contrast as a three-factor effect, we also confound a main effect, e.g.,

$$ABCD \times ABC = A^2B^2C^2D = D.$$

Similarly, if we choose the second contrast as a two-factor effect, we lose another two-factor effect, e.g.,

$$ABCD \times AB = A^2B^2CD = CD$$

However, if we choose two three-factor effects as defining contrasts, we lose only one two-factor effect, e.g.,

$$ABC \times BCD = AB^2C^2D = AD.$$

The four blocks for the confounding scheme based on ABC and BCD are given below.

$ABC(1)$		$ABC(-1)$	
$BCD(1)$	$BCD(-1)$	$BCD(1)$	$BCD(-1)$
$a_0b_0c_0d_0$	$a_0b_0c_0d_1$	$a_0b_0c_1d_1$	$a_0b_0c_1d_0$
$a_0b_1c_1d_0$	$a_0b_1c_1d_1$	$a_0b_1c_0d_1$	$a_0b_1c_0d_0$
$a_1b_0c_1d_1$	$a_1b_0c_1d_0$	$a_1b_0c_0d_0$	$a_1b_0c_0d_1$
$a_1b_1c_0d_1$	$a_1b_1c_0d_0$	$a_1b_1c_1d_0$	$a_1b_1c_1d_1$

The treatment groups can be checked against the contrasts given in Table 17.2.

If we wanted blocks of size 2 we would need three defining contrasts, say ABC, BCD, and ACD. Blocks of size 2 imply the existence of 8 blocks, so 7 degrees of freedom must be confounded with blocks. To obtain the other confounded contrasts, multiply each pair of defining contrasts and multiply all three defining contrasts together. Multiplying the pairs gives $ABC \times BCD = AD$, $ABC \times ACD = BD$, and $BCD \times ACD = AB$. Multiplying all three together gives $ABC \times BCD \times ACD = AD \times ACD = C$. □

Consider the problem of creating 16 blocks for a 2^n experiment. Since $16 = 2^4$, we need 4 defining contrasts. With 16 blocks there are 15 degrees of freedom for blocks, hence 15 contrasts confounded with blocks. Four of these 15 are the defining contrasts. Multiplying distinct pairs of defining contrasts gives 6 implicitly confounded contrasts. There are 4 distinct triples that can be made from the defining contrasts; multiplying the triples gives 4 more confounded contrasts. Multiplying all four of the defining contrasts gives the fifteenth and last confounded contrast.

We now consider the analysis of data obtained from a confounded 2^n design.

EXAMPLE 17.1.3. *Analysis of a 2^3 in blocks of four with replication*
Yates (1935) presented data on a 2^3 agricultural experiment involving yields of peas when various fertilizers were applied. The three factors were a nitrogen fertilizer (N), a phosphorous fertilizer (P), and a potash fertilizer (K). Each factor consisted of two levels, none of the fertilizer and a standard dose. It was determined that homogenous blocks of land were best obtained by creating six squares each containing four plots. Thus we have $2^3 = 8$ treatments, blocks of size 4, and six available blocks. By confounding one treatment contrast, we can obtain blocks of size 4. With six blocks, we can have 3 replications of the treatments. The confounded contrast was chosen to be the NPK interaction, so the treatments in one block are $n_0p_0k_0$, $n_1p_1k_0$, $n_1p_0k_1$, and $n_0p_1k_1$ and the treatments in the other block are $n_1p_0k_0$, $n_0p_1k_0$, $n_0p_0k_1$, and $n_1p_1k_1$. The data are given in Table 17.3. The table displays the original geographical layout of the experiment with lines identifying blocks and replications. Each pair of rows in the table are a replication with the left and right halves identifying blocks. In each replication, the set of four treatments to be applied to a block is randomly decided and then within each block the four treatments are randomly assigned to the four plots.

The analysis of these data is straightforward; it follows the usual pattern. The mean square and sum of squares for blocks is obtained from the six block means. Each block mean is the average of 4 observations. The sum of squares for a main effect, say, N, can be obtained from the two nitrogen means, each based on 12 observations, or equivalently, it can be obtained from the contrast

TABLE 17.3. Yates's confounded pea data

$n_0p_0k_0$(56.0)	$n_1p_1k_0$(59.0)	$n_0p_0k_1$(55.0)	$n_1p_1k_1$(55.8)
$n_0p_1k_1$(53.2)	$n_1p_0k_1$(57.2)	$n_1p_0k_0$(69.5)	$n_0p_1k_0$(62.8)
$n_0p_1k_1$(48.8)	$n_0p_0k_0$(51.5)	$n_0p_1k_0$(56.0)	$n_1p_1k_1$(58.5)
$n_1p_0k_1$(49.8)	$n_1p_1k_0$(52.0)	$n_0p_0k_1$(55.5)	$n_1p_0k_0$(59.8)
$n_0p_1k_0$(44.2)	$n_1p_1k_1$(48.8)	$n_1p_0k_1$(57.0)	$n_1p_1k_0$(62.8)
$n_0p_0k_1$(45.5)	$n_1p_0k_0$(62.0)	$n_0p_0k_0$(46.8)	$n_0p_1k_1$(49.5)

Three replications with *NPK* confounded in each.

Treatment	N
$n_0p_0k_0$	1
$n_0p_0k_1$	1
$n_0p_1k_0$	1
$n_0p_1k_1$	1
$n_1p_0k_0$	−1
$n_1p_0k_1$	−1
$n_1p_1k_0$	−1
$n_1p_1k_1$	−1

applied to the 8 treatment means which are obtained by averaging over the 3 replications. The contrast for *NPK* was confounded with blocks, so it should not appear in the analysis; the one degree of freedom for *NPK* is part of the five degrees of freedom for blocks.

TABLE 17.4. Analysis of variance

Source	*df*	*SS*	*MS*	*F*	*P*
Blocks	5	343.30	68.66	4.45	0.016
N	1	189.28	189.28	12.26	0.004
P	1	8.40	8.40	0.54	0.475
K	1	95.20	95.20	6.17	0.029
NP	1	21.28	21.28	1.38	0.263
NK	1	33.14	33.14	2.15	0.169
PK	1	0.48	0.48	0.03	0.863
Error	12	185.29	15.44		
Total	23	876.36			

The complete analysis of variance is given in Table 17.4. It is the result of fitting the model

$$y_{hijk} = \mu + \beta_h + \nu_i + \rho_j + \kappa_k + (\nu\rho)_{ij} + (\nu\kappa)_{ik} + (\rho\kappa)_{jk} + \varepsilon_{hijk} \quad (17.1.1)$$

where β_h, $h = 1, \ldots, 6$ indicates a block effect and ν, ρ, and κ indicate effects relating to N, P, and K respectively. Every effect in the analysis of variance has one degree of freedom, so there is no need to investigate contrasts beyond what is given in the ANOVA table. The only effects that appear significant are those for N and K. The evidence for an effect due to the nitrogen-based fertilizer is quite clear. The means for the nitrogen treatments are

N	n_0	n_1
	52.067	57.683

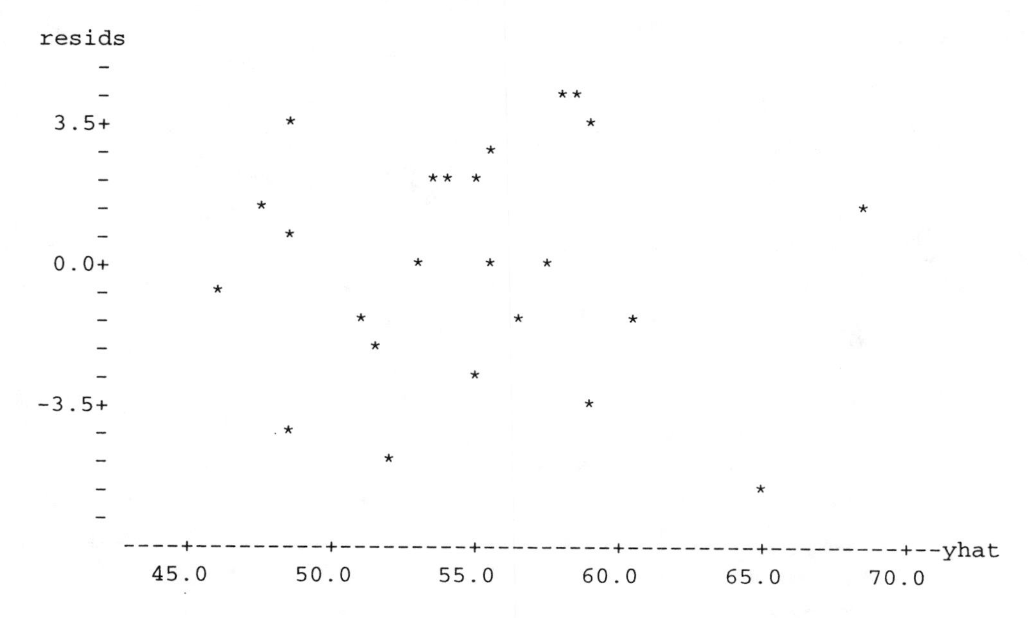

FIGURE 17.1. Plot of residuals versus predicted values, pea data.

so the addition of nitrogen increases yields. The evidence for an affect due to potash is somewhat less clear. The means are

K	k_0	k_1
	56.867	52.883

Surprisingly (to a city boy like me), application of potash actually decreases pea yields.

Fitting the analysis of variance model (17.1.1) provides residuals that can be evaluated in the usual way. Figures 17.1 and 17.2 contain residual plots. Except for some slight curvature at the very ends of the normal plot, the residuals look good. Remember that these are plots of the residuals, not the standardized residuals, so residual values greater than 3 do not necessarily contain a suggestion of outlying points. □

MINITAB COMMANDS

Minitab's 'glm' command gives a simple way of obtaining the analysis of these data. The glm command does not recognize the orthogonality in the design, so it reports two types of sums of squares for each term. However, the orthogonality ensures that the values are identical for the two types. This particular analysis could also be run using the 'ancova' command, but the other analyses in this chapter require glm.

```
MTB > names c1 'y' c2 'Blocks' c3 'N' c4 'P' c5 'K'
MTB > glm c1=c2 c3|c4|c5 - c3*c4*c5;
SUBC> resid c10;
SUBC> fits c11;
SUBC> means c3 c5.
```

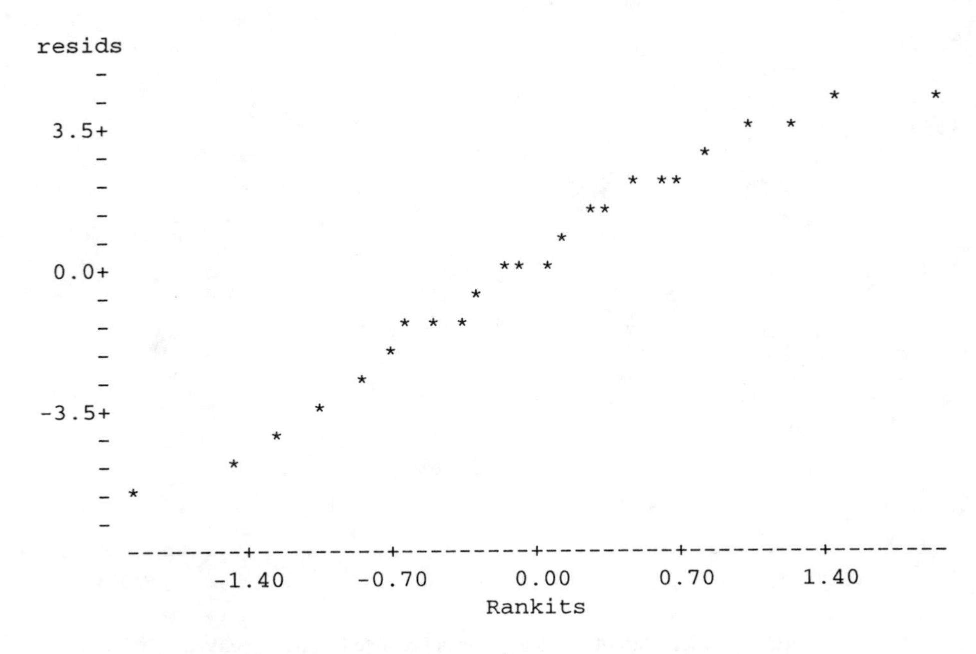

FIGURE 17.2. Normal plot of residuals, pea data, $W' = .980$.

SPLIT PLOT ANALYSIS

EXAMPLE 17.1.4. *Split plot analysis of a 2^3 in blocks of four with replication*
Example 17.1.3 gives the standard analysis of confounded data with replications. However, more information can be extracted. Yates' design is very similar to a split plot design. We have three replications and we can think of each block as a whole plot. The whole plots are randomly assigned one of two treatments but the treatments have a peculiar structure. One 'treatment' applied to a whole plot consists of having the $NPK(1)$ treatments assigned to the whole plot and the other 'treatment' is having the $NPK(-1)$ treatments assigned to a whole plot. The one degree of freedom for whole plot treatments is due to the difference between these sets of treatments. This difference is just the NPK interaction. The analysis in Example 17.1.3 is the subplot analysis and remains unchanged when we consider applying a split plot analysis to the data. Recall that in a subplot analysis each whole plot is treated like a block; that is precisely what we did in Example 17.1.3. We need only perform the whole plot analysis to complete the split plot analysis.

The 6 blocks in Example 17.1.3 are now considered as whole plots. The 5 degrees of freedom for blocks are the 5 degrees of freedom total for the whole plot analysis. These 5 degrees of freedom can be decomposed into 2 degrees of freedom for comparing the three replications, 1 degree of freedom for NPK interaction, i.e., whole plot treatments, and 2 degrees of freedom for replication by whole plot treatment interaction. *The replication by whole plot treatment interaction is just the error term for the whole plot analysis.* Note that in this model, rather than thinking about having 6 fixed block effects, we have 3 fixed replication effects, a fixed NPK effect, and a random error term that distinguishes the blocks within the replications.

The necessary means for the whole plot analysis are given below.

$N = 4$	*NPK* level		
Rep.	*NPK*(1)	*NPK*(−1)	Rep. means
1	56.350	60.775	58.5625
2	50.525	57.450	53.9875
3	54.025	50.125	52.0750
NPK means	$53.63\bar{3}$	$56.11\bar{6}$	54.8750

The three rep. means are averages over the 8 observations in each replication. The mean square for replications is obtained in the usual way as 8 times the sample variance of the rep. means. The mean square for *NPK* interaction is obtained from the *NPK*(1) mean and the *NPK*(−1) mean. Each of these means is averaged over 12 observations. The mean square for *NPK* is 12 times the sample variance of the *NPK* means. The interaction (whole plot error) sum of squares is found by subtracting the *NPK* and replication sums of squares from the blocks sum of squares found in Example 17.1.3. The whole plot analysis of variance is given in Table 17.5. There is no evidence for an *NPK* interaction.

TABLE 17.5. Whole plot analysis of variance

Source	*df*	*SS*	*MS*	*F*
Reps	2	177.80	88.90	1.38
NPK	1	37.00	37.00	0.58
Error	2	128.50	64.25	
Total	5	343.30	68.66	

In this experiment, there are not enough whole plots to provide a very powerful test for the *NPK* interaction. There are only 2 degrees of freedom in the denominator of the F test. If we had 5 replications of a 2^3 in blocks of size 2 rather than blocks of size 4, i.e., if we confounded three contrasts with blocks, the whole plot analysis would have 4 degrees of freedom for reps., three 1 degree of freedom effects for the confounded contrasts, and 12 degrees of freedom for whole plot error. The 12 error degrees of freedom come from pooling the rep. by effect interactions. □

PARTIAL CONFOUNDING

In Example 17.1.3, we considered a 2^3 in blocks of 4 with three replications. The same contrast *NPK* was confounded in all replications, so, within the subplot analysis, all information was lost on the *NPK* contrast. Something must be lost when a treatment effect is confounded with blocks, but with multiple replications it is not necessary to give up all the information on *NPK*. Consider a 2^3 experiment with factors A, B, and C that is conducted in blocks of size 4 with two replications. It would be natural to confound *ABC* with blocks. However, instead of confounding *ABC* with blocks in both replications, we could pick another contrast, say, *BC*, to confound in the second replication. The design is given below.

Replication 1		Replication 2	
ABC(1)	*ABC*(−1)	*BC*(1)	*BC*(−1)
$a_0b_0c_0$	$a_0b_0c_1$	$a_0b_0c_0$	$a_0b_0c_1$
$a_0b_1c_1$	$a_0b_1c_0$	$a_0b_1c_1$	$a_0b_1c_0$
$a_1b_0c_1$	$a_1b_0c_0$	$a_1b_0c_0$	$a_1b_0c_1$
$a_1b_1c_0$	$a_1b_1c_1$	$a_1b_1c_1$	$a_1b_1c_0$

In the first replication we give up all the information on *ABC* but retain information on *BC*. In the second replication we give up all the information on *BC* but retain information on *ABC*. Thus we have partial information available on both ABC and BC.

The process of confounding different contrasts in different replications is known as *partial confounding* because contrasts are only partially confounded with blocks. In some replications they are confounded but in others they are not. Thus partial information is available on contrasts that are partially confounded.

When more than one confounded contrast is needed to define blocks of an appropriate size, some contrasts can be totally confounded, while others are only partially confounded. We now consider an example using these ideas.

EXAMPLE 17.1.5. We again analyze pea yield data with the same 2^3 fertilizer treatments considered in Examples 17.1.3 and 17.1.4. This analysis involves blocks of size 2 with two replications. Both replications have NPK confounded, but the first rep. has PK (and thus N) confounded, while the second has NK (and thus P) confounded. The data are given in Table 17.6 with lines used to identify blocks. Again, pairs of rows denote replications.

TABLE 17.6. Pea data with partial confounding

$n_1p_0k_0$(59.8)	$n_0p_0k_1$(55.5)	$n_1p_0k_1$(57.2)	$n_0p_1k_1$(53.2)
$n_1p_1k_1$(58.5)	$n_0p_1k_0$(56.0)	$n_1p_1k_0$(59.0)	$n_0p_0k_0$(56.0)
$n_0p_1k_1$(49.5)	$n_0p_0k_0$(46.8)	$n_0p_1k_0$(62.8)	$n_1p_0k_0$(69.5)
$n_1p_1k_0$(62.8)	$n_1p_0k_1$(57.0)	$n_1p_1k_1$(55.8)	$n_0p_0k_1$(55.0)

The first replication has NPK, PK, and N confounded. The second replication has NPK, NK, and P confounded.

The mean square and sum of squares for blocks is obtained from the eight block means. Each of these means is the average of 2 observations. The sums of squares for error is obtained by subtraction. The sums of squares for treatment effects are more complicated. We use contrasts to compute them. First, NPK is confounded in both replications, so no information is available on it. The effects of K and NP are not confounded in either replication, so they can be estimated from both. The contrasts, means, and sums of squares are given below.

Treatment	K	NP	Means
$n_0p_0k_0$	1	1	51.40
$n_0p_0k_1$	−1	1	55.25
$n_0p_1k_0$	1	−1	59.40
$n_0p_1k_1$	−1	−1	51.35
$n_1p_0k_0$	1	−1	64.65
$n_1p_0k_1$	−1	−1	57.10
$n_1p_1k_0$	1	1	60.90
$n_1p_1k_1$	−1	1	57.15
SS	60.0625	15.210	

The sum of squares for K is

$$SS(K) = \frac{[-15.50]^2}{8/2} = 60.0625$$

where -15.50 is the estimated contrast, 8 is the sum of the squared contrast coefficients, and the means are averages of 2 observations, one from each replication.

The effects of N and PK are confounded in the first replication but P and NK are not. Thus P and NK can be evaluated from the first replication.

Treatment	P	NK	Rep. 1
$n_0p_0k_0$	1	1	56.0
$n_0p_0k_1$	1	-1	55.5
$n_0p_1k_0$	-1	1	56.0
$n_0p_1k_1$	-1	-1	53.2
$n_1p_0k_0$	1	-1	59.8
$n_1p_0k_1$	1	1	57.2
$n_1p_1k_0$	-1	-1	59.0
$n_1p_1k_1$	-1	1	58.5
SS	0.405	0.005	

The sum of squares for P is

$$SS(P) = \frac{[-1.8]^2}{8/1} = 0.405$$

where -1.8 is the estimated contrast and the treatment 'means' used in the contrasts are just the observations from the first replication.

Note that the sums of squares for the partially confounded contrasts involve a multiplier of 1, because they are averages of one observation, rather than the multiplier of 2 that is used for the contrasts that have full information. This is the price paid for partial confounding. An estimated contrast with full information would have a sum of squares that is twice as large as the same estimated contrast that is confounded in half of the replications. This sort of thing happens generally when using partial confounding but the size of the multiplicative effect depends on the exact experimental design. In this example the factor is 2 because the sample sizes of the means are twice as large for full information contrasts as for partial information contrasts.

Similarly, the effects of N and PK can be obtained from the second replication but not the first.

Treatment	N	PK	Rep. 2
$n_0p_0k_0$	-1	-1	46.8
$n_0p_0k_1$	-1	1	55.0
$n_0p_1k_0$	-1	1	62.8
$n_0p_1k_1$	-1	-1	49.5
$n_1p_0k_0$	1	-1	69.5
$n_1p_0k_1$	1	1	57.0
$n_1p_1k_0$	1	1	62.8
$n_1p_1k_1$	1	-1	55.8
SS	120.125	32.000	

The analysis of variance table is given in Table 17.7. Too much blocking has been built into this experiment. The F statistic for blocks is only 0.85, so the differences between blocks are not as substantial as the error. The whole point of blocking is that the differences between blocks should be greater than error so that by isolating the block effects we reduce the experimental error. The excessive blocking has also reduced the degrees of freedom for error to 2, thus ensuring a poor estimate of the variance. A poor error estimate reduces power; it is difficult to establish that effects are significant. For example, the F statistic for N is 4.92; this would be significant at the .05 level if there were 11 degrees of freedom for error, but with only 2 degrees of freedom the P value is just .157. □

In general, *effects are evaluated from all replications in which they are not confounded.* If we had 3 replications with NPK confounded in each and PK and N confounded only in the first, the sums of squares for N and PK would be based on treatment means averaged over both rep. 2 and rep. 3.

TABLE 17.7. Analysis of variance

Source	*df*	*SS*	*MS*	*F*	*P*
Blocks	7	145.12	20.73	0.85	—
N	1	120.12	120.12	4.92	0.157
P	1	0.40	0.40	0.02	0.909
K	1	60.06	60.06	2.46	0.257
NP	1	15.21	15.21	0.62	0.513
NK	1	0.01	0.01	0.00	0.990
PK	1	32.00	32.00	1.31	0.371
Error	2	48.79	24.40		
Total	15	421.72			

MINITAB COMMANDS

Minitab's glm command can again provide the analysis, however, the column with the block indices must appear immediately after the equals sign in the statement of the glm model. Minitab provides both sequential and adjusted sums of squares. For all effects other than blocks, these are the same. *When blocks are listed first in the model, the* sequential *sum of squares is appropriate for blocks.*

```
MTB > names c1 'Blocks' c2 'N' c3 'P' c4 'K' c5 'y'
MTB > glm c5=c1 c2|c3|c4 - c2*c3*c4
```

17.2 Fractional replication

Consider a chemical process in which there are seven factors that affect the process. These could be percentages of four chemical components, temperature of the reaction, choice of catalyst, and choice of different machines for conducting the process. If all of these factors are at just two levels, there are $2^7 = 128$ treatment combinations. If obtaining data on a treatment combination is expensive, running 128 treatment combinations may be prohibitive. If there are just three levels for each factor, there are $3^7 = 2187$ treatment combinations. In fractional replication, one uses only a fraction of the treatment combinations. Of course, if we give up the full factorial, we must lose information in the analysis of the experiment. In fractional replication the treatments are chosen in a systematic fashion so that we lose only the higher order interactions. (I, for one, do not look forward to trying to interpret 6- and 7-factor interactions anyway.) To be more precise, we do not actually give up the high order interactions, we give up our ability to distinguish them from the main effects and lower order interactions. Effects are *aliased* with other effects. The contrasts we examine involve both a main effect or lower order interaction and some higher order interactions. Of course we assume the high order interactions are of no interest and treat the contrasts as if they involve only the main effects and low order interactions. That is the whole point of fractional replication. While fractional replication is of most interest when there are large numbers of factors, we will illustrate the techniques with smaller numbers of factors.

EXAMPLE 17.2.1. *A 1/2 rep. of a 2^3 experiment*

We now examine the construction of a 1/2 replicate of a 2^3 factorial. The 2^3 has 8 treatments, so a 1/2 replicate involves 4 treatments. Recall the table of contrast coefficients for a 2^3 factorial.

Treatment	A	B	C	AB	AC	BC	ABC
$a_0b_0c_0$	1	1	1	1	1	1	1
$a_0b_0c_1$	1	1	−1	1	−1	−1	−1
$a_0b_1c_0$	1	−1	1	−1	1	−1	−1
$a_0b_1c_1$	1	−1	−1	−1	−1	1	1
$a_1b_0c_0$	−1	1	1	−1	−1	1	−1
$a_1b_0c_1$	−1	1	−1	−1	1	−1	1
$a_1b_1c_0$	−1	−1	1	1	−1	−1	1
$a_1b_1c_1$	−1	−1	−1	1	1	1	−1

Every contrast divides the treatments into two groups of four. We can use any contrast to define the 1/2 rep. We choose to use ABC because it is the highest order interaction. The ABC contrast defines two groups of treatments

ABC coefficients	
$ABC(1)$	$ABC(-1)$
$a_0b_0c_0$	$a_0b_0c_1$
$a_0b_1c_1$	$a_0b_1c_0$
$a_1b_0c_1$	$a_1b_0c_0$
$a_1b_1c_0$	$a_1b_1c_1$

These are just the two blocks obtained when confounding a 2^3 into two blocks of 4. Each of these groups of four treatments comprises a 1/2 replicate. It is irrelevant which of the two groups is actually used. These 1/2 replicates are referred to as resolution III designs because the defining contrast involves three factors.

The 1/2 rep. involves only four treatments, so there can be at most three orthogonal treatment contrasts. All seven effects cannot be estimated. The aliasing of effects is determined by the modular multiplication illustrated earlier. To determine the aliases, multiply each effect by the defining contrast ABC. For example, to find the alias of A, multiply

$$A \times ABC = A^2BC = BC$$

where any even power is treated as 0, so $A^2 = A^0 = 1$. Thus A and BC are aliased; we cannot tell them apart; they are two names for the same contrast. Similarly,

$$BC \times ABC = AB^2C^2 = A.$$

The aliasing structure for the entire 1/2 rep. based on ABC is given below.

Effect	$\times ABC$	Alias
A	=	BC
B	=	AC
C	=	AB
AB	=	C
AC	=	B
BC	=	A
ABC		—

In this experiment, we completely lose any information about the defining contrast, ABC. In addition, we cannot tell the main effects from the two-factor interactions. If we had no interest in two-factor interactions, this design would be fine. Generally, if there are only three factors each at two levels, there is little reason not to perform the entire experiment. As mentioned earlier, fractional replication is primarily of interest when there are many factors so that even a fractional replication involves many observations.

Another way to examine aliasing is by looking at the table of contrast coefficients when we use only 1/2 of the treatments. We consider the 1/2 rep. in which the treatments have ABC coefficients of 1. The contrasts, when restricted to the treatments actually used, have the following coefficients:

Treatment	A	B	C	AB	AC	BC	ABC
$a_0b_0c_0$	1	1	1	1	1	1	1
$a_0b_1c_1$	1	−1	−1	−1	−1	1	1
$a_1b_0c_1$	−1	1	−1	−1	1	−1	1
$a_1b_1c_0$	−1	−1	1	1	−1	−1	1

All of the columns other than ABC still define contrasts in the four treatment combinations; each column has two 1s and two −1s. However, the contrast defined by A is identical to the contrast defined by its alias, BC. In fact, this is true for each contrast and its alias.

Consider now the choice of a different contrast to define the 1/2 rep. Instead of ABC, we might choose AB. Again, we lose all information about the defining contrast AB and we have aliases involving the other effects. The AB contrast defines two groups of treatments

AB coefficients	
$AB(1)$	$AB(-1)$
$a_0b_0c_0$	$a_0b_1c_0$
$a_0b_0c_1$	$a_0b_1c_1$
$a_1b_1c_0$	$a_1b_0c_0$
$a_1b_1c_1$	$a_1b_0c_1$

Each of these groups of four treatments comprises a 1/2 replicate. Again, it is irrelevant which of the two groups is actually used. Both groups determine resolution II designs because the defining contrast involves two factors.

The aliasing is determined by modular multiplication with the defining contrast AB. To find the alias of A multiply

$$A \times AB = A^2B = B.$$

The alias of BC is

$$BC \times AB = AB^2C = AC.$$

The aliasing structure for the entire 1/2 rep. based on AB is given below.

Effect	$\times AB$	Alias
A	=	B
B	=	A
C	=	ABC
AB		—
AC	=	BC
BC	=	AC
ABC	=	C

With this 1/2 rep., we do not even get to estimate all of the main effects because A is aliased with B. □

EXAMPLE 17.2.2. *A 1/4 replicate of a 2^4 experiment*
In Example 17.1.2 we considered confounding ABC and BCD in a 2^4 experiment. The four blocks are given below.

$ABC(1)$		$ABC(-1)$	
$BCD(1)$	$BCD(-1)$	$BCD(1)$	$BCD(-1)$
$a_0b_0c_0d_0$	$a_0b_0c_0d_1$	$a_0b_0c_1d_1$	$a_0b_0c_1d_0$
$a_0b_1c_1d_0$	$a_0b_1c_1d_1$	$a_0b_1c_0d_1$	$a_0b_1c_0d_0$
$a_1b_0c_1d_1$	$a_1b_0c_1d_0$	$a_1b_0c_0d_0$	$a_1b_0c_0d_1$
$a_1b_1c_0d_1$	$a_1b_1c_0d_0$	$a_1b_1c_1d_0$	$a_1b_1c_1d_1$

With ABC and BCD defining the blocks,

$$ABC \times BCD = AB^2C^2D = AD$$

is also confounded with blocks. Any one of these blocks can be used as a $1/4$ replicate of the 2^4 experiment. The smallest defining contrast is the two-factor effect AD, so this $1/4$ replicate is a resolution II design.

The aliasing structure of the $1/4$ rep. must account for all three of the defining contrasts. An effect, say A, is aliased with

$$A \times ABC = A^2BC = BC,$$

$$A \times BCD = ABCD,$$

and

$$A \times AD = A^2D = D.$$

Thus we cannot tell main effects A from main effects D, from BC interaction, or from $ABCD$ interaction. After all, what do you expect when you take four observations to learn about 16 treatments? Similar computations show that

$$B = AC = CD = ABD$$

and

$$C = AB = BD = ACD.$$

This is the complete aliasing structure for the $1/4$ rep. There are 4 observations, so there are 3 degrees of freedom for treatment effects. We can label these effects as A, B, and C with the understanding that we cannot tell aliases apart, so we have no idea if an effect referred to as A is really due, entirely or in part, to D, BC, or $ABCD$. □

Fractional replication is primarily of value when you have large numbers of treatments, require information only on low order effects, can assume that high order effects are negligible, and can find a design that aliases low order effects with high order effects.

EXAMPLE 17.2.3. *Fractional replication of a 2^8 experiment*
A 2^8 experiment involves eight factors, A through H, and 256 treatments. It may be impractical to take that many observations. Consider first a $1/8 = 2^{-3}$ replication. This involves only $2^{8-3} = 32$ treatment combinations, a much more manageable number than 256. A $1/8 = 2^{-3}$ rep. requires 3 defining contrasts, say $ABCD$, $EFGH$, and $CDEF$. Multiplying pairs of the defining contrasts and multiplying all three of the contrasts give the other contrasts that implicitly define the $1/8$ rep. The other implicit defining contrasts are $ABED$, $CDGH$, $ABCDEFGH$, and $ABGH$. Note that the smallest defining contrast has four terms, so this is a resolution IV design.

The aliases of an effect are obtained from multiplying the effect by all 7 of the defining contrasts; e.g., for A the aliases are

$$A = A(ABCD) = A(EFGH) = A(CDEF) = A(ABED)$$
$$= A(CDGH) = A(ABCDEFGH) = A(ABGH)$$

or simplifying

$$A = BCD = AEFGH = ACDEF = BED = ACDGH = BCDEFGH = BGH.$$

With a resolution IV design, it is easily seen that main effects are only aliased with three-factor and higher order effects. A two-factor effect, say AB, has aliases

$$AB = AB(ABCD) = AB(EFGH) = AB(CDEF) = AB(ABED)$$
$$= AB(CDGH) = AB(ABCDEFGH) = AB(ABGH)$$

or simplifying

$$AB = CD = ABEFGH = ABCDEF = ED = ABCDGH = CDEFGH = GH.$$

Unfortunately, at least some two-factor effects are aliased with other two-factor effects in a resolution IV design.

If we had constructed a 1/4 replicate, we could have chosen the defining contrasts in such a way that two-factor effects were only aliased with three-factor and higher order effects. For example, the defining contrasts $ABCDE$ and $DEFGH$ determine such a design. The additional defining contrast is $ABCDE(DEFGH) = ABCFGH$. The smallest defining effect involves 5 factors, so this has resolution V. In computing aliases, a two-factor term is multiplied by a five-factor or greater term. The result is at least a three-factor term. Thus two-factor effects are aliased with 3 or higher order effects. Similarly, main effects are aliased with 4 or higher order effects. A 1/4 replicate of a 2^8 experiment can provide information on all main effects and all two-factor interactions under the assumption of no 3 or higher order interaction effects. □

As mentioned, the 1/4 replicate given above is known as a resolution V design because the smallest defining contrast involved a five-factor interaction. As should now be clear, in general, *the resolution of a 2^n fractional replication is the order of the smallest defining contrast.* To keep main effects from being aliased with one another, one needs a resolution III or higher design. To keep both main effects and two-factor effects from being aliased with one another, one needs a resolution V or higher design.

FRACTIONAL REPLICATION WITH CONFOUNDING

The two concepts of fractional replication and confounding can be combined in designing an experiment. To illustrate fractional replication with confounding we consider a subset of the 2^3 data in Table 17.6. The subset is given in Table 17.8. This is the first half of the first replication in Table 17.6. The fractional replication is based on NPK. All of the observations have NPK contrast coefficients of -1. The confounding is based on PK. The first block has PK contrast coefficients of 1 and the second block has PK contrast coefficients of -1.

TABLE 17.8. Pea data

$n_1p_0k_0(59.8)$	$n_0p_0k_1(55.5)$
$n_1p_1k_1(58.5)$	$n_0p_1k_0(56.0)$

The fractional replication is based on NPK. Confounding is based on PK.

The aliasing structure for the 1/2 rep. based on NPK is given below.

Effect	$\times NPK$	Alias
N	=	PK
P	=	NK
K	=	NP
NP	=	K
NK	=	P
PK	=	N
NPK		—

Blocks are confounded with PK and PK is aliased with N, so N is also confounded with blocks. With only 4 observations, we can compute sums of squares for only 3 effects. Ignoring the two-factor interactions, those effects are blocks, P, and K.

Perhaps the simplest way to perform the analysis of such designs is to begin by ignoring the blocking. If the blocking is ignored, the analysis is just that of a fractional factorial and can be conducted as discussed in the next section. After computing all the sums of squares ignoring blocks, go back and isolate the effects that are confounded with blocks. In this example, the fractional factorial ignoring blocks gives sums of squares for $N = PK$, $P = NK$, and $NP = K$. Then observe that the sum of squares for N is really the sum of squares for blocks.

17.3 Analysis of unreplicated experiments

One new problem we have in a fractional replication is that there is no natural estimate of error because there is no replication. We don't even have observations on every factor combination, much less multiple observations on treatments. We present two ways to proceed, one is to assume that higher order interactions do not exist, the other is based on a graphical display of the effects that is similar in spirit to a normal plot.

EXAMPLE 17.3.1. We consider a 1/2 rep. of a 2^5 that was reported by Hare (1988). The issue is excessive variability in the taste of a dry soup mix. The source of variability was identified as a particular component of the mix called the 'intermix' containing flavorful ingredients such as salt and vegetable oil.

From each batch of intermix, the original data are groups of 5 samples taken every 15 minutes throughout a day of processing. Thus each batch yields data for a balanced one-way analysis of variance with $N = 5$. The data actually analyzed are derived from the ANOVAs on different batches. There are two sources of variability in the original observations, the variability within a group of 5 samples and variability that occurs between 15 minute intervals. From the analysis of variance data, the within group variability is estimated with the MSE and summarized as the estimated 'capability' standard deviation

$$s_c = \sqrt{MSE}.$$

The 'process' standard deviation was defined as the standard deviation of an individual observation. The standard deviation of an observation incorporates both the between group and the within group sources of variability. The estimated process standard deviation is taken as

$$s_p = \sqrt{MSE + \frac{MSTrts - MSE}{5}},$$

where the 5 is the number of samples taken at each time, cf. Subsection 12.4.2. These two statistics, s_c and s_p, are available from every batch of soup mix prepared and provide the data for analyzing batches. The 1/2 rep. of a 2^5 involves different ways of making batches of soup mix. The factors in the design are discussed later. For now, we analyze only the data on s_p.

The two blocks obtained by confounding *ABCDE* in a 2^5 are reported in Table 17.9. Table 17.9 also presents an alternative form of identifying the treatments. In the alternative form, only the letters with a subscript of 1 are reported. Hare's experiment used the block consisting of treatments with *ABCDE* contrast coefficients of 1.

TABLE 17.9. 1/2 reps. from a 2^5 based on *ABCDE*

ABCDE(1)	Treatment	*ABCDE*(−1)	Treatment
$a_0b_0c_0d_0e_0$	(1)	$a_0b_0c_0d_0e_1$	e
$a_0b_0c_0d_1e_1$	de	$a_0b_0c_0d_1e_0$	d
$a_0b_0c_1d_0e_1$	ce	$a_0b_0c_1d_0e_0$	c
$a_0b_0c_1d_1e_0$	cd	$a_0b_0c_1d_1e_1$	cde
$a_0b_1c_0d_0e_1$	be	$a_0b_1c_0d_0e_0$	b
$a_0b_1c_0d_1e_0$	bd	$a_0b_1c_0d_1e_1$	bde
$a_0b_1c_1d_0e_0$	bc	$a_0b_1c_1d_0e_1$	bce
$a_0b_1c_1d_1e_1$	bcde	$a_0b_1c_1d_1e_0$	bcd
$a_1b_0c_0d_0e_1$	ae	$a_1b_0c_0d_0e_0$	a
$a_1b_0c_0d_1e_0$	ad	$a_1b_0c_0d_1e_1$	ade
$a_1b_0c_1d_0e_0$	ac	$a_1b_0c_1d_0e_1$	ace
$a_1b_0c_1d_1e_1$	acde	$a_1b_0c_1d_1e_0$	acd
$a_1b_1c_0d_0e_0$	ab	$a_1b_1c_0d_0e_1$	abe
$a_1b_1c_0d_1e_1$	abde	$a_1b_1c_0d_1e_0$	abd
$a_1b_1c_1d_0e_1$	abce	$a_1b_1c_1d_0e_0$	abc
$a_1b_1c_1d_1e_0$	abcd	$a_1b_1c_1d_1e_1$	abcde

There are five factors involved in the experiment. Intermix is made in a large mixer. Factor *A* is the number of ports for adding vegetable oil to the mixer. This was set at either 1 (a_0) or 3 (a_1). Factor *B* is the temperature of the mixer. The mixer can be cooled by circulating water through the mixer jacket (b_0) or the mixer can be used at room temperature (b_1). Factor *C* is the mixing time, 60 seconds (c_0) or 80 seconds (c_1). Factor *D* is the size of the intermix batch, either 1500 pounds (d_0) or 2000 pounds (d_1). Factor *E* is the delay between making the intermix and using it in the final soup mix. The delay is either 1 day (e_0) or 7 days (e_1). Table 17.10 contains the data along with the aliases for a 1/2 rep. of a 2^5 based on *ABCDE*. The order in which the treatments were run was randomized and they are listed in that order. Batch number 7 contains the standard operating conditions. Note that this is a resolution V design: all main effects are confounded with four-factor interactions and all two-factor interactions are confounded with three-factor interactions. If we are prepared to assume that there are no three- or four-factor interactions, we have estimates of all the main effects and two-factor interactions.

One way to perform the analysis is to compute the sums of squares for each contrast, however the simplest way to obtain an analysis is to trick a computer program into doing most of the work. If we could drop one of the factors, we would have observed a full factorial (without replication) on the remaining factors. For example, *if we dropped factor E, and thus dropped the e terms from all the treatment combinations in Table 17.10, we would have observations on all 16 of the treatment combinations in the* 2^4 *defined by A, B, C, and D.* It is easy to find computer programs that will analyze a full factorial. Table 17.11 gives the results of an analysis in which we have ignored the presence of factor *E*. Table 17.11 contains two columns labeled 'Source'. The one on the left gives the sources from the full factorial on *A*, *B*, *C*, and *D*; the one on the right replaces the higher order interactions from the full factorial with their lower order aliases. Table 17.11 also contains a ranking of the sizes of the sums of squares from smallest to largest.

One method of analysis is to assume that no higher order interactions exist and form an error term by pooling the estimable terms that involve only higher order interactions. A particular term involves only higher order interactions if the term and all of its aliases are high order interactions. What we

TABLE 17.10. Hare's 1/2 rep. from a 2^5 based on $ABCDE$

Batch	Treatment		s_c	s_p	Aliases		
1	$a_0b_0c_0d_1e_1$	de	.43	.78	A	=	$BCDE$
2	$a_1b_0c_1d_1e_1$	acde	.52	1.10	B	=	$ACDE$
3	$a_1b_1c_0d_0e_0$	ab	.58	1.70	C	=	$ABDE$
4	$a_1b_0c_1d_0e_0$	ac	.55	1.28	D	=	$ABCE$
5	$a_0b_1c_0d_0e_1$	be	.58	.97	E	=	$ABCD$
6	$a_0b_0c_1d_0e_1$	ce	.60	1.47	AB	=	CDE
7	$a_0b_1c_0d_1e_0$	bd	1.04	1.85	AC	=	BDE
8	$a_1b_1c_1d_1e_0$	abcd	.53	2.10	AD	=	BCE
9	$a_0b_1c_1d_1e_1$	bcde	.38	.76	AE	=	BCD
10	$a_1b_1c_0d_1e_1$	abde	.41	.62	BC	=	ADE
11	$a_0b_0c_1d_1e_0$	cd	.66	1.09	BD	=	ACE
12	$a_0b_0c_0d_0e_0$	(1)	.55	1.13	BE	=	ACD
13	$a_1b_0c_0d_0e_1$	ae	.65	1.25	CD	=	ABE
14	$a_1b_1c_1d_0e_1$	abce	.72	.98	CE	=	ABD
15	$a_1b_0c_0d_1e_0$	ad	.48	1.36	DE	=	ABC
16	$a_0b_1c_1d_0e_0$	bc	.68	1.18			

TABLE 17.11. ANOVA for s_p

Source	Source	df	SS	Rank
A	A	1	0.0841	10
B	B	1	0.0306	7
C	C	1	0.0056	4
D	D	1	0.0056	3
AB	AB	1	0.0009	1
AC	AC	1	0.0361	8
AD	AD	1	0.0036	2
BC	BC	1	0.0182	5
BD	BD	1	0.1056	12
CD	CD	1	0.0210	6
ABC	DE	1	0.3969	13
ABD	CE	1	0.0729	9
ACD	BE	1	0.6561	14
BCD	AE	1	0.0930	11
$ABCD$	E	1	0.8836	15
Total	Total	15	2.4140	

mean by high order interactions is intentionally left ill defined to maintain flexibility. In this design, unless you consider second-order interactions as higher order, there are no terms involving only higher order interactions. Most often, higher order interactions are taken to be interactions that only involve three or more factors, but in designs like this, one *might* be willing to consider two-factor interactions as higher order to obtain an error term for testing main effects. (I personally would not be willing to do it with these data.) Often terms that involve only three and higher order interactions are pooled into an error, but in designs with more factors and many high order interactions, one might wish to estimate three-factor interactions and use only terms involving four or more factors in a pooled error.

If we assume away all two-factor and higher order interactions for the present data, the ANOVA table becomes that displayed in Table 17.12. With this error term, only factor E appears to be important. As we will see later, most of the important effects in these data seem to be interactions, so the error term based on no interactions is probably inappropriate.

TABLE 17.12. Analysis of variance on s_p for Hare's data

Source	df	SS	MS	F
A	1	0.0841	0.0841	0.60
B	1	0.0306	0.0306	0.22
C	1	0.0056	0.0056	0.04
D	1	0.0056	0.0056	0.04
E	1	0.8836	0.8836	6.29
Error	10	1.4044	0.1404	
Total	15	2.4140		

Rather than assuming away higher order interactions, Daniel (1959) proposed an alternative method of analysis based on an idea similar to normal plotting. Recall that in a normal plot, the data from a single sample are ordered from smallest to largest and plotted against the *expected order statistics* from a standard normal distribution. In other words, the smallest observation in a sample of size, say, 13 is plotted against the expected value for the smallest observation in a sample of size 13 from a $N(0, 1)$ distribution. The second smallest observation is plotted against the expected value for the second smallest observation in a sample of size 13 from a $N(0, 1)$ distribution, and so on. This plot should approximate a straight line if the data are truly normal, the slope of the plot estimates the standard deviation of the population, and the intercept estimates the population mean. One approach to a graphical analysis of 2^n experiments is to perform a normal plot on the estimated contrasts. Daniel (1959) used a plot of the absolute values of the estimated contrasts. Here we discuss a graphical method of analysis for unreplicated and fractional factorials that applies a similar idea to the sums of squares in Table 17.11.

Assume a standard ANOVA model with independent $N(0, \sigma^2)$ errors. The analysis looks for departures from the assumption that none of the factors have an effect on the observations. Under the assumption of no effects, every mean square gives an estimate of σ^2 and every sum of squares has the distribution

$$\frac{SS}{\sigma^2} \sim \chi^2(1),$$

where the degrees of freedom in the χ^2 are 1 because each effect has 1 degree of freedom. Moreover, the sums of squares are independent, so in the absence of treatment effects, the sums of squares form a random sample from a $\sigma^2\chi^2(1)$ distribution. If we order the sums of squares from smallest to largest, the ordered sums of squares should estimate the expected order statistics from a $\sigma^2\chi^2(1)$ distribution. Plotting the ordered sums of squares against the expected order statistics, we should get an approximate straight line through the origin with a slope of 1. In practice, we cannot obtain expected values from a $\sigma^2\chi^2(1)$ distribution because we do not know σ^2. Instead, we plot the ordered sums of squares against the expected order statistics of a $\chi^2(1)$ distribution. This plot should be an approximate straight line through the origin with a slope of σ^2.

Table 17.13 contains the statistics necessary for the χ^2 plot of the 15 effects from Hare's data. Figure 17.3 contains the plot. The $\chi^2(1)$ scores in Table 17.13 are approximate expected order statistics. They are computed by applying the inverse of the $\chi^2(1)$ cumulative distribution function to the values $i/(n + 1)$, where i goes from 1 to 15 and $n = 15$. This is easily done in Minitab. Table 17.13 also contains partial sums of the ordered sums of squares; these values will be used in the next section.

The key to the graphical analysis is that nonnegligible treatment effects cause the mean square to estimate something larger than σ^2. The sums of squares for nonnegligible effects should show up in the plot as inappropriately large values. The lower 12 observations in Figure 17.3 seem to fit roughly on a line, but the three largest observations seem to be inconsistent with the others. These three

TABLE 17.13. $\chi^2(1)$ scores, ordered sums of squares, and partial sums of the sums of squares for Hare's (1988) data

$\chi^2(1)$ scores	Ordered SS	Partial sums
0.00615	0.0009	0.0009
0.02475	0.0036	0.0045
0.07711	0.0056	0.0101
0.07711	0.0056	0.0157
0.16181	0.0182	0.0339
0.23890	0.0210	0.0549
0.33539	0.0306	0.0855
0.45494	0.0361	0.1216
0.60283	0.0729	0.1945
0.78703	0.0841	0.2786
1.02008	0.0930	0.3716
1.32330	0.1056	0.4772
1.73715	0.3969	0.8741
2.35353	0.6561	1.5302
3.46977	0.8836	2.4138

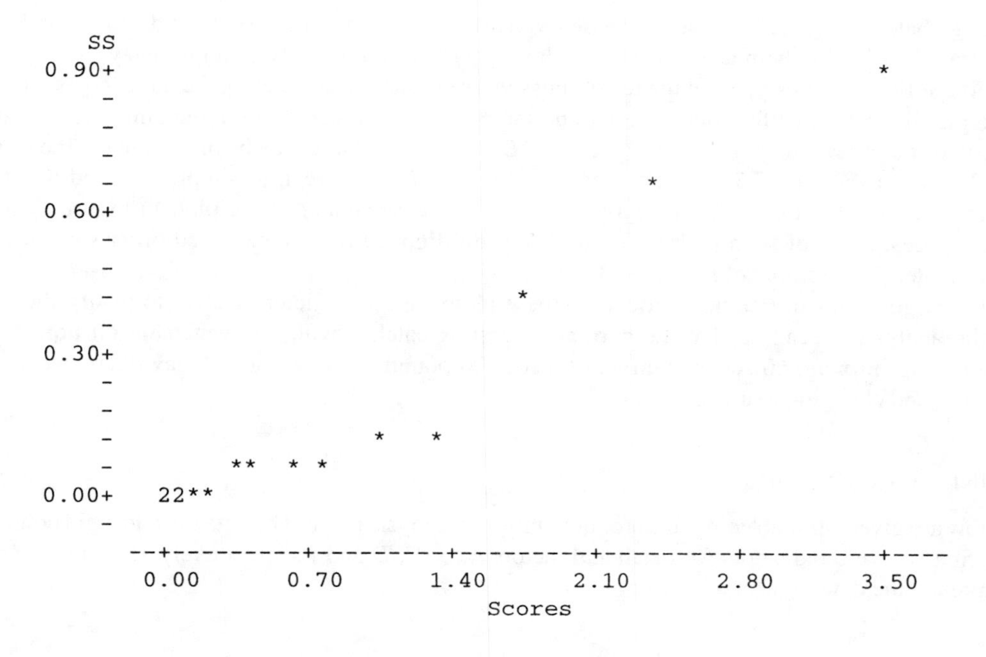

FIGURE 17.3. $\chi^2(1)$ plot of sums of squares.

observations correspond to the most important effects in the data. From the rankings in Table 17.11, we see that the important effects are E, BE, and DE.

We need to evaluate the meaning of the important effects. There is no question of breaking things down into contrasts because all of the effects already have only one degree of freedom. We need only interpret the meanings of the specific effects. The largest effect is due to E, the delay in using the intermix. However, this effect is complicated by interactions involving the delay.

The means for the four combinations of B and E are given below.

$N = 4$	B	
E	b_0	b_1
e_0	1.215	1.7075
e_1	1.150	0.8325

The BE interaction is due to the fact that running the mixer at room temperature, b_1, increases variability if the intermix is used after one day, e_0, but decreases variability if the intermix is used a week later, e_1. However, the variability under delay is smaller for both B levels than the variability for immediate use with either B level. This suggests delaying use of the intermix.

The means for the four combinations of D and E are given below.

$N = 4$	D	
E	d_0	d_1
e_0	1.3225	1.6000
e_1	1.1675	0.8150

A large batch weight, d_1, causes increased variability when the intermix is used immediately but decreased variability with use delayed to 7 days. Again, it is uniformly better to delay.

Figure 17.4 contains a plot of the remaining sums of squares after deleting the three largest effects. The plot indicates that the four largest values are somewhat larger than the remaining effects. The fourth through seventh largest effects are BD, AE, A, and CE. These may be important but the results are less clear. Figure 17.5 is an alternative to Figure 17.4. Figure 17.4 simply dropped the three largest cases in Figure 17.3 to give a better view of the remainder of the plot. In Figure 17.5 the three largest sums of squares from Figure 17.3 are dropped but the expected order statistics are recomputed for a sample of size $15 - 3 = 12$.

The suggestions of treatment effects in these plots are not sufficiently clear to justify their use in the analysis. Recalling that the current process is batch 7 with one vegetable oil port, room temperature mixing, 60 seconds mixing time, 2000 pound batches, and a 1 day delay, we would recommend changing to a 7 day delay. □

MINITAB COMMANDS

Below are given Minitab commands for obtaining the analysis given. The data file had eight columns, the first six were indicators for batch and factors A, B, C, D and E, respectively. Columns 7 and 8 contained the data on s_c and s_p.

```
MTB > names c8 'y' c2 'a' c3 'b' c4 'c' c5 'd' c6 'e'
MTB > anova c8=c2|c3|c4|c5 - c2*c3*c4*c5
MTB > note     AFTER SEEING THE ANOVA, ENTER THE SUMS
```

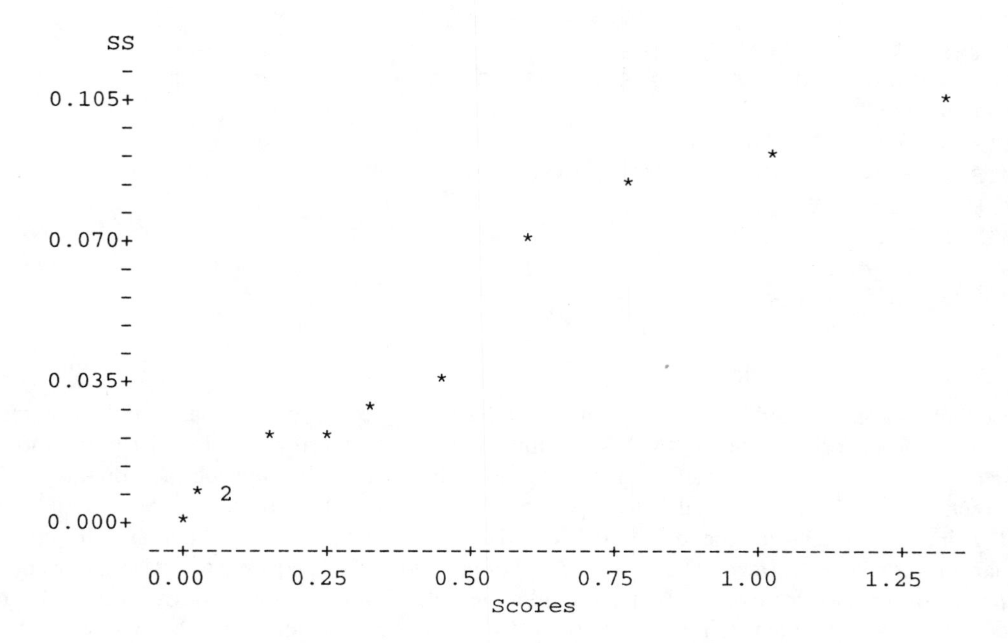

FIGURE 17.4. $\chi^2(1)$ plot of sums of squares, largest 3 cases deleted.

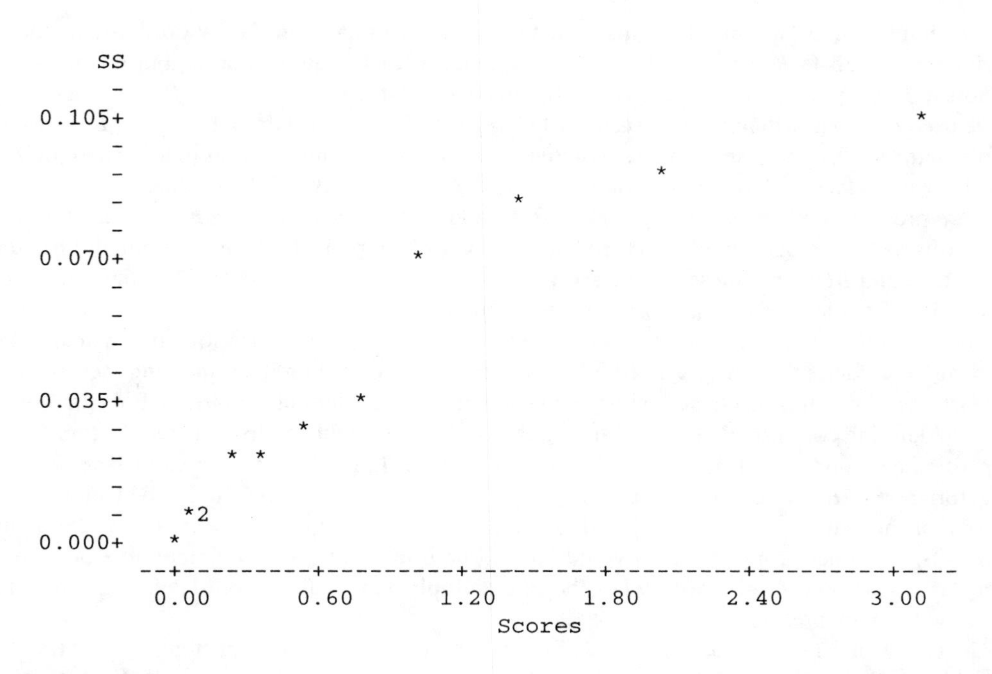

FIGURE 17.5. $\chi^2(1)$ plot of sums of squares, largest 3 cases deleted, expected order statistics recomputed.

```
MTB > note      OF SQUARES INTO c10.
MTB > set c10
DATA> 841 306 56 56 9 361 36 182
DATA> 1056 210 3969 729 6561 930 8836
DATA> end
MTB > let c10=c10/10000
MTB > note      CONSTRUCT CHI-SQUARED SCORES AND PLOT.
MTB > rank c10 c11
MTB > let c11=c11/16
MTB > invcdf c11 c12;
SUBC> chisquare 1.
MTB > plot c10 c12
```

Note that $c6$ was not used in the anova command. Factor E was dropped to deal with the fractional nature of the factorial. Minitab's ANOVA command requires an error term to exist in the model. The command given above specifies a full factorial model ($c2|c3|c4|c5$) but subtracts out the $ABCD$ interaction ($c2 * c3 * c4 * c5$) and sets it equal to the error. Thus, Minitab's error term is actually the $ABCD$ interaction. The command 'set $c10$' is used to create a data column that contains the sums of squares for the various effects. The commands involving $c11$ and $c12$ are used to get the approximate expected order statistics from a $\chi^2(1)$ and to plot the ordered sums of squares against the expected order statistics. After identifying the important effects, the ANOVA command can be repeated with various factors deleted to obtain the necessary means tables.

COMPUTING TECHNIQUES

The technique of computing the sums of squares in a fractional factorial by dropping factors and performing a full factorial analysis on the remaining factors is quite general, but when choosing factors to be dropped *every defining contrast must involve at least one dropped factor*. For example, if we used $ABCD$ as a defining contrast in a $1/2$ rep. of a 2^5, we must drop A, B, C, or D to compute a full factorial. Dropping any of these will give all 16 treatment combinations in a 2^4 based on E and the other three factors. On the other hand, dropping E does not give all 16 treatments combinations that are present in a 2^4 based on factors A, B, C, and D, nor does dropping E give the appropriate sums of squares. In particular, a factorial analysis with factors A, B, C, and D normally has terms for both A and BCD, but these are aliased in the $1/2$ rep. based on $ABCD$. Thus the full factorial cannot be computed. (If you are confused, do Exercise 17.6.7.)

In a $1/4$ rep, two defining contrasts are used with another contrast also lost. In the analysis of a $1/4$ rep, two factors are dropped and a full factorial is computed on the remaining factors. Again, at least one of the dropped factors must be in each of the three defining contrasts. For example, in a 2^5 with defining contrasts BCD, CDE and implicitly BE, we could not drop the two factors C and D because the defining contrast BE contains neither of these. In particular, dropping C and D leads to a factorial on A, B, and E, but the main effects for B and E are aliased. Similarly, the factor A cannot be one of those dropped to obtain a full factorial. Since A is not contained in any of the defining contrasts, the other dropped factor would have to be in all three. This is impossible because if a factor is in two defining contrasts, when they are multiplied to obtain the third defining contrast that factor will not be present.

Generally, in a $1/2^s$ replication of a 2^n factorial there are $2^{n-s} - 1$ distinct groups of effects that are aliased. We need to find the sum of squares for each group. To do this we drop s appropriately chosen factors and compute a full factorial analysis on the remaining factors. The effects in this analysis of a 2^{n-s} factorial represent all of the alias groups in the $1/2^s$ replication. We merely have to identify the lowest order, and thus most interesting, effects in each group of aliases. Having at least one dropped factor in every defining contrast ensures that the effects arising in the 2^{n-s} factorial are all aliased only with effects that involve a dropped factor and thus are not aliased with any other effect in the 2^{n-s} factorial. Therefore all the effects in the 2^{n-s} factorial are in distinct alias groups and we have sums of squares for every alias group.

17.4 More on graphical analysis

Normal and χ^2 graphs of effects give valuable information on the relative sizes of effects but it is difficult to judge which effects are truly important and which could be the result of random variation. Many such plots give the false impression that there are a number of important effects. The problem is that we tend to see what we look for. In a normal plot of, say, regression residuals, we look for a straight line and are concerned if the residuals obviously contradict the assumption of normality. When analyzing a saturated linear model, we expect to see important effects, so instead of looking for an overall line, we focus on the extreme order statistics. Doing so can easily lead us astray.

In this section we consider two methods for evaluating significant effects in a χ^2 plot. The first is based on using Cochran's (1941) test for homogeneity (equality) of variances. This method was originally suggested by Holms and Berrettoni (1969). The second is similar in spirit to the simulation envelopes suggested by Atkinson (1981) for evaluating normal plots of regression residuals. The methods are introduced in relation to Hare's data but they apply quite generally. Both methods provide envelopes for the χ^2 plots. A χ^2 plot that goes outside the envelope suggests the existence of significant effects. Box and Meyer (1986), Lenth (1989), and Berk and Picard (1991) recently proposed alternative methods for the analysis of contrasts in unreplicated factorials.

MULTIPLE MAXIMUM TESTS

Cochran's test for homogeneity of variances applies when there are, say, $n \geq 2$ independent $\chi^2(r)$ estimates of a variance. In a balanced one-way ANOVA with a treatments, N observations per group, and independent $N(0, \sigma^2)$ errors, Cochran's test applies to the individual group variances s_i^2. Cochran's n equals a and his r is $N - 1$. In this case, Cochran's test statistic is the maximum of the variance estimates divided by the sum of the variance estimates. The test is rejected for large values of the statistic. In analyzing Hare's unreplicated factorial, if there are no significant effects, the 15 sums of squares in Hare's data are independent $\chi^2(1)$ estimates of σ^2. Thus Cochran's procedure can be applied with $n = 15$ and $r = 1$ to test the hypothesis that all the sums of squares are estimating the same variance.

Cochran's test is best suited for detecting a single variance that is larger than the others. (Under the alternative, large terms *other than* the maximum get included in the total, making it more difficult to detect unusual behavior in the maximum.) In analyzing unreplicated linear models we often expect more than one significant effect. In the spirit of the multiple range tests used for comparing pairs of means in analysis of variance, we use Cochran's test repeatedly to evaluate the ordered sums of squares. Thus we define C_j as the jth smallest of the sums of squares divided by the sum of the j smallest sums of squares. From Table 17.13, the values of C_j are obtained by taking the ordered sums of squares and dividing by the partial sums of the ordered sums of squares. Each value of C_j is then compared to an appropriate percentile of Cochran's distribution based on having j estimates of the variance. Note that such a procedure does not provide any control of the experimentwise error rate for the multiple comparisons. Weak control can be achieved by first performing an overall test for equality of variances and then evaluating individual C_js only if this overall test is significant. One such choice could be Cochran's test for the entire collection of sums of squares, but that seems like a poor selection. As mentioned, Cochran's test is best at detecting a single unusual variance; having more than one large variance (as we often expect to have) reduces the power of Cochran's test. To control the experimentwise error rate, it is probably better to use alternative tests for equality of variances such as Bartlett's (1937) or Hartley's (1938) tests, see Snedecor and Cochran (1980).

Table 17.14 gives the values of the C_j statistics and various percentiles of Cochran's distribution. Note that the 13th largest effect exceeds the .15 percentage point and is almost significant at the .10 level. While a significance level of .15 is not commonly thought to be very impressive, in an unreplicated experiment one would not expect to have a great deal of power, so the use of larger α levels may be appropriate. Having concluded that the 13th effect is significant, it is logical to conclude that all larger effects are also significant. Once we have a single significant effect, testing the larger effects makes little sense. For larger effects, the sum of squares in the denominator of Cochran's statistic is biased by the inclusion of the sum of squares for the 13th effect. Moreover, if the 13th effect is so large as to be identified as significant, effects that are even larger should also be significant. Note that for Hare's data the test of the 14th effect is also significant at the .15 level. This simply compounds the evidence for the significance of the 14th effect.

TABLE 17.14. Percentiles for Cochran's statistic with $r = 1$ and Cochran's statistics for Hare's data

j	.01	.05	.10	.15	C_j
2	0.9999	0.9985	0.9938	0.9862	0.80000
3	0.9933	0.9670	0.9344	0.9025	0.55446
4	0.9677	0.9065	0.8533	0.8096	0.35669
5	0.9279	0.8413	0.7783	0.7311	0.53687
6	0.8826	0.7808	0.7141	0.6668	0.38251
7	0.8377	0.7270	0.6598	0.6139	0.35789
8	0.7945	0.6798	0.6138	0.5696	0.29688
9	0.7549	0.6385	0.5742	0.5320	0.37481
10	0.7176	0.6020	0.5399	0.4997	0.30187
11	0.6837	0.5697	0.5100	0.4716	0.25027
12	0.6528	0.5411	0.4834	0.4469	0.22129
13	0.6248	0.5152	0.4598	0.4249	0.45407
14	0.5987	0.4921	0.4386	0.4053	0.42877
15	0.5749	0.4711	0.4196	0.3876	0.36606

The commonly available tables for Cochran's distribution are inadequate for the analysis just given. The α level percentage points in Table 17.14 were obtained by evaluating the inverse of the cumulative distribution function of a $Beta(r/2, r(j-1))$ distribution at the point $1 - \alpha/j$. These are easily obtained in Minitab. Cochran (1941) notes that these values are generally a good approximation to the true percentage points and that they are exact whenever the true percentage point is greater than .5. Moreover, the true significance level corresponding to a nominal significance level of α in Table 17.14 is at most α and at least $\alpha - \alpha^2/2$, so the true significance level associated with the .15 values listed in Table 17.14 is between .13875 and .15 for $j = 10, \ldots, 15$ and is exactly .15 for $j = 2, \ldots, 9$.

Most tests for equality of variances, including Cochran's test, are notoriously sensitive to nonnormality – so much so that they are rarely used in practice. However the analysis of variance F test is not noted for extreme sensitivity to nonnormality, even though it is a test for the equality of two variances. This is probably because the numerator mean square is computed from sample means and sample means tend to be reasonably normal. The current application of Cochran's test should benefit in the same way. The sums of squares in this example are essentially computed from the difference between two sample means each based on 8 observations. Thus the sensitivity to nonnormality should be mitigated. Of course for nonnormal data the sums of squares are unlikely

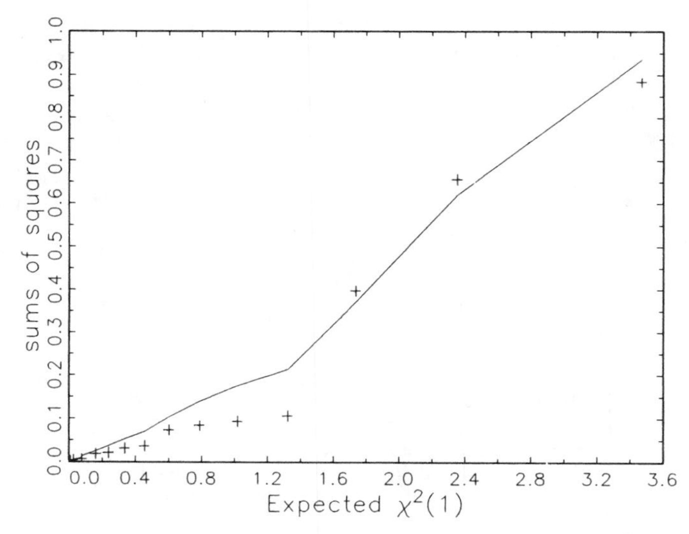

FIGURE 17.6. $\chi^2(1)$ plot for Hare's data (+) with 15% Cochran upper envelope.

to be independent, but the estimated effects are still uncorrelated.

The multiple maximum procedure is easily incorporated into χ^2 plots. Figure 17.6 contains the $\chi^2(1)$ plot for Hare's data along with an upper envelope. The upper envelope is the product of the Cochran 15% points and the partial sums from Table 17.13. The 13th and 14th largest sums of squares exceed the upper envelope, indicating that the corresponding maximum tests are rejected.

SIMULATION ENVELOPES

Figure 17.7 contains the $\chi^2(1)$ plot for Hare's data with a simulation envelope. Actually, the plot uses the standardized sums of squares, i.e., the sums of squares divided by the total sum of squares. Obviously, dividing each sum of squares by the same number has no effect on the visual interpretation of the plot. The simulation envelope is based on 99 analyses of randomly generated standard normal data. The upper envelope is the maximum for each order statistic from the 99 replications and the lower envelope is the minimum of the replications. Performing 99 analyses of a 2^4 experiment is time consuming; it is computationally more efficient just to take 99 random samples from a $\chi^2(1)$ distribution, standardize them, and order them. Unlike Atkinson's (1981) residual envelopes for regression, having divided all the sums of squares by the total, the same envelopes can be used for any subsequent analysis of 15 sums of squares each having one degree of freedom.

There are two prominent features in Figure 17.7. The 12th effect is barely below the lower envelope and the 14th effect is barely above the upper envelope. All of the sums of squares have been divided by the total, so the low value for the 12th effect indicates that the sum of squares for the 12th effect has been divided by a number that is too large. In other words, the sum of squares total is too large to be consistent with the 12 smallest sums of squares. This indicates that there must be significant effects among the 3 largest terms. The simulation envelope does not indicate which of the 3 larger terms are real effects; violation of the envelope only suggests that the envelope is inappropriate, i.e., that there are real effects. Visual interpretation of the graph must be used to identify the important effects.

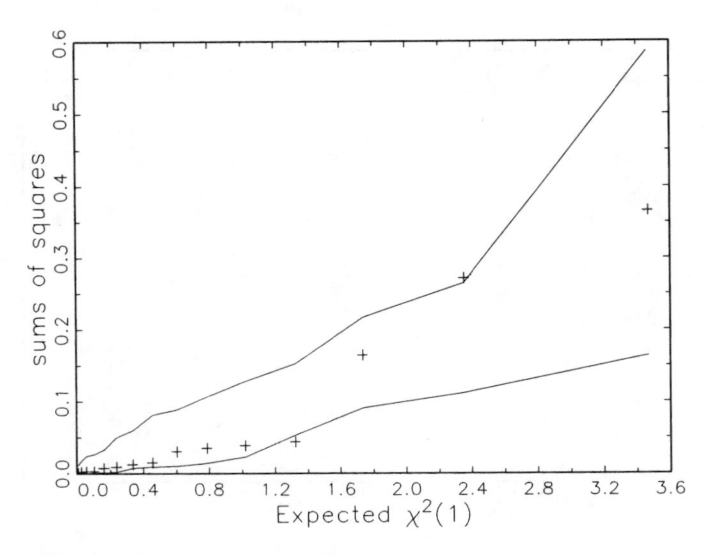

FIGURE 17.7. $\chi^2(1)$ plot for Hare's data (+) with simulated envelope.

Outliers

It is frequently suggested that outliers can be spotted from patterns in the plots. Consider a single outlier. Every effect contrast is the sum of 8 observations minus the sum of the other 8, thus if the outlier is not on the treatment $a_0b_0c_0d_0e_0$, the outlier is added into either 7 or 8 effect contrasts and is subtracted in the others. In a normal plot, such an outlier should cause a jump in the level of the line involving about half of the effects. In a χ^2 plot, a single outlier should cause all of the effects to look large, i.e., the intercept of the plot should not be zero. If two outliers exist in the data, they should cancel each other in about half of the effect contrasts and compound each other in the other effects. Thus, half of the effects should appear to be important in the plot. In my experience the natural variation in the plots is large enough that it is difficult to identify even very large outliers based on these facts.

17.5 Augmenting designs for factors at two levels

When running a sequence of experiments, one may find that a particular 2^q experiment or fractional replication is inadequate to answer the relevant questions. In such cases it is often possible to add more points to the design to gain the necessary information. After running an initial fractional replication, a particular factor, say A, may be identified as being of primary importance. It is then of interest to estimate the main effect A and all two-factor interactions involving A without their being aliased with other two-factor interactions.

Box and Draper (1987, p. 156) suggest a method for adding points to the original design to achieve this end. Christensen and Huzurbazar (1996) also treat this method and provide a proof. Consider the resolution III, 1/16th replication of a 2^7 experiment defined by $ABD(1), ACE(1), BCF(1), ABCG(1)$. Box and Draper suggest augmenting the design with the 1/16th replication $ABD(-1)$, $ACE(-1)$, $BCF(1)$, $ABCG(-1)$. The idea is that 1s have been changed to -1s in all defining contrasts that include A.

Together, the two 1/16th replications define a 1/8th replication. The nature of this 1/8th replication can be explored by adding an imaginary factor H to the experiment. As H is imaginary, obviously it can have no effect on the responses. Any effects that involve H are simply error. For example, the DH interaction examines whether the effect for D changes from the low to high level of H. As there are no low and high levels of H, any observed change must be due to random error. The treatments included in the augmented 1/8th rep. of the 2^7 given previously are identical to those obtained from the 1/16th rep. of the 2^8 defined by $ABDH(1), ACEH(1), BCF(1), ABCGH(1)$. Here H has been added to any defining effect that includes A. In fact, if we want to consider the augmented 1/8th rep. as occurring in two blocks, the imaginary factor H is the effect confounded with blocks.

Now consider the aliasing structure of this 2^8 design. In constructing any 2^q design in 16 blocks, there are $16 - 1$ effects confounded with the blocks, i.e., with the possible fractional replications. All of these effects are involved in determining the aliases of the 1/16th rep. To get the complete set of 15 defining effects we must multiply the nominal defining effects together and multiply products of the nominal defining effects with other nominal defining effects. Using an asterisk $(*)$ to identify the nominal defining effects and performing the multiplications systematically, the complete set of defining effects is

$$ABDH^*, ACEH^*, BCDE, BCF^*, ACDFH, ABEFH, DEF, ABCGH^*,$$

$$CDG, BEG, AFGH, ADEGH, BDFG, CEFG, ABCDEFGH.$$

This is still a resolution III design but now every defining effect that includes A has at least three other factors, one of which is the imaginary H. Thus multiplying A times the defining effects we see that A is not aliased with any two-factor or main effects. Moreover, a two-factor effect involving A, say AB, is only aliased with other two-factor effects that involve H. In the case of AB, the only two-factor effect that it is aliased with is DH. But H is imaginary, so two-factor effects involving A are not aliased with any real two-factor effects.

Box and Draper (1987) and Christensen and Huzurbazar (1996) provide a similar solution to the problem of augmenting a design to allow estimation of all main effects unaliased with two-factor effects. Again consider the resolution III, 1/16th replication of a 2^7 experiment defined by $ABD(1), ACE(1), BCF(1), ABCG(1)$. For this problem they suggest adding the 1/16th fraction defined by $ABD(-1), ACE(-1), BCF(-1), ABCG(1)$. Here 1s have been changed to -1s in the defining effects that involve an odd number of factors. This augmented 1/8th rep. design is equivalent to adding an imaginary factor H and using the 1/16th rep. of the 2^8 experiment defined by $ABDH(1), ACEH(1), BCFH(1), ABCG(1)$. In this approach, any defining effect with an odd number of terms has H added to it. As before, any effects that involve H are error. Adding H in this manner has changed the resolution III design into a resolution IV design. Thus all main effects are aliased with three-factor or higher terms. As before, if we view the augmented 1/8th rep. as occurring in two blocks, H is the effect confounded with blocks.

17.6 Exercises

EXERCISE 17.6.1. Analyze Hare's s_c data that was given in Table 17.10.

EXERCISE 17.6.2. To consider the effect of a possible outlier, reanalyze Hare's s_p data, changing the largest value, the 2.10 in Table 17.10, into the second largest value, 1.85.

EXERCISE 17.6.3. Reanalyze Hare's s_c data after identifying and deleting the possible outlier. Does having an outlier in that particular batch suggest anything?

EXERCISE 17.6.4. Consider a 2^6 factorial. Give a good design for performing this in blocks of sixteen. Try to avoid confounding main effects and two-factor interactions with blocks.

EXERCISE 17.6.5. Consider a 2^6 factorial. Give a good design for performing a $1/4$ replication. Try to avoid aliasing main effects with each other and with two-factor interactions. Also try to avoid aliasing two-factor interactions with other two-factor interactions.

EXERCISE 17.6.6. Consider a 2^6 factorial. Give a good design for performing a $1/2$ replication in blocks of 16. Do not confound main effects or two-factor interactions with blocks. Try to avoid aliasing main effects with each other and with two-factor interactions. Also try to avoid aliasing two-factor interactions with other two-factor interactions.

EXERCISE 17.6.7. Consider a $1/2$ rep. of a 2^5 with factors A, B, C, D, E and $ABCD$ defining the $1/2$ rep. Write down the treatment combinations in the $1/2$ rep. Now drop factor D from the analysis. In particular, write down all the treatment combinations in the $1/2$ rep. but delete all the d terms from the treatments. Does this list contain all the treatments in a 2^4 on factors A, B, C, E? Now return to the original $1/2$ rep., drop factor E, and list the treatment combinations. Does this list contain all the treatments in a 2^4 on factors A, B, C, D?

Chapter 18
Nonlinear regression

Most relationships between predictor variables and the mean values of observations are nonlinear. Fortunately, the linear regression models of Chapters 13, 14, and 15 can be used in a wide variety of situations. Taylor's theorem from calculus indicates that linear models can make good approximate models to nonlinear relationships. However, when you have good knowledge about the relationship between mean values and predictor variables, nonlinear regression provides a way to use that knowledge and thus can provide much better models. The biggest difficulty with nonlinear regression is that to use it you need detailed knowledge about the process generating the data, i.e., you need a good idea about the appropriate nonlinear relationship between the predictor variables and the mean values of the observations. Nonlinear regression is a technique with wide applicability in the biological and physical sciences.

From a statistical point of view, nonlinear regression models are much more difficult to work with than linear regression models. It is harder to obtain estimates of the parameters. It is harder to do good statistical inference once those parameter estimates are obtained. Section 1 introduces nonlinear regression models. In section 2 we discuss parameter estimation. Section 3 examines methods for statistical inference. Section 4 considers the choice that is sometimes available between doing nonlinear regression and doing linear regression on transformed data. For a much more extensive treatment of nonlinear regression see Seber and Wild (1989).

18.1 Introduction and examples

In Chapters 13, 14, and 15 we considered linear regression models

$$y_i = \beta_0 + \beta_1 x_{i1} + \cdots + \beta_{p-1} x_{ip-1} + \varepsilon_i \tag{18.1.1}$$

$i = 1, \ldots, n$. Written in vector form these are

$$y_i = x_i' \beta + \varepsilon_i \tag{18.1.2}$$

where $x_i' = (1, x_{i1}, \ldots, x_{ip-1})$ and $\beta = (\beta_0, \ldots, \beta_{p-1})'$ These models are linear in the sense that $E(y_i) = x_i'\beta$ where the unknown parameters, the β_is, are multiplied by constants, the x_{ij}s, and added together. In this chapter we consider an important generalization of this model, *nonlinear regression*. A nonlinear regression model is simply a model for $E(y_i)$ that does not combine the parameters of the model in a linear fashion.

EXAMPLE 18.1.1. *Some nonlinear regression models*
Almost any nonlinear function can be made into a nonlinear regression model. Consider the following four nonlinear functions of parameters β_i and a single predictor variable x:

$$\begin{aligned} f_1(x; \beta_0, \beta_1, \beta_2) &= \beta_0 + \beta_1 \sin(\beta_2 x) \\ f_2(x; \beta_0, \beta_1, \beta_2) &= \beta_0 + \beta_1 e^{\beta_2 x} \\ f_3(x; \beta_0, \beta_1, \beta_2) &= \beta_0 / [1 + \beta_1 e^{\beta_2 x}] \\ f_4(x; \beta_0, \beta_1, \beta_2, \beta_3) &= \beta_0 + \beta_1 [e^{\beta_2 x} - e^{\beta_3 x}]. \end{aligned}$$

Each of these can be made into a nonlinear regression model. Using f_4, we can write a model for data pairs (y_i, x_i), $i = 1, \ldots, n$:

$$\begin{aligned} y_i &= \beta_0 + \beta_1 [e^{\beta_2 x_i} - e^{\beta_3 x_i}] + \varepsilon_i \\ &\equiv f_4(x_i; \beta_0, \beta_1, \beta_2, \beta_3) + \varepsilon_i . \end{aligned}$$

Similarly, for $k = 1, 2, 3$ we can write models

$$y_i = f_k(x_i; \beta_0, \beta_1, \beta_2) + \varepsilon_i.$$

As usual, we assume that the ε_is are independent $N(0, \sigma^2)$ random variables. As alluded to earlier, the problem is to find an appropriate function $f(\bullet)$ for the data at hand. □

In general, for s predictor variables and p regression parameters we can write a nonlinear regression model that generalizes model (18.1.1) as

$$y_i = f(x_{i1}, \ldots, x_{is}; \beta_0, \beta_1, \ldots, \beta_{p-1}) + \varepsilon_i, \quad \varepsilon_i\text{s indep. } N(0, \sigma^2)$$

$i = 1, \ldots, n$. This is quite an awkward way to write $f(\bullet)$, so we write the model in vector form as

$$y_i = f(x_i; \beta) + \varepsilon_i, \quad \varepsilon_i\text{s indep. } N(0, \sigma^2) \tag{18.1.3}$$

where $x_i = (x_{i1}, \ldots, x_{is})'$ and $\beta = (\beta_0, \beta_1, \ldots, \beta_{p-1})'$ are vectors defined similarly to model (18.1.2). Note that

$$E(y_i) = f(x_i; \beta).$$

EXAMPLE 18.1.2. Pritchard, Downie, and Bacon (1977) reported data from Jaswal et al. (1969) on the initial rate r of benzene oxidation over a vanadium pentoxide catalyst. The predictor variables involve three temperatures, T, for the reactions, different oxygen and benzene concentrations, x_1 and x_2, and the observed number of moles of oxygen consumed per mole of benzene, x_4. Based on chemical theory, a steady state adsorption model was proposed. One algebraically simple form of this model is

$$y_i = \exp[\beta_0 + \beta_1 x_{i3}] \frac{1}{x_{i2}} + \exp[\beta_2 + \beta_3 x_{i3}] \frac{x_{i4}}{x_{i1}} + \varepsilon_i. \tag{18.1.4}$$

where $y = 100/r$ and the temperature is involved through $x_3 = 1/T - 1/648$. The data are given in Table 18.1.

The function giving the mean structure for model (18.1.4) is

$$f(x; \beta) \equiv f(x_1, x_2, x_3, x_4; \beta_0, \beta_2, \beta_3, \beta_4) = \exp[\beta_0 + \beta_1 x_3] \frac{1}{x_2} + \exp[\beta_2 + \beta_3 x_3] \frac{x_4}{x_1}. \tag{18.1.5}$$

□

TABLE 18.1. Benzene oxidation data

Obs.	x_1	x_2	T	x_4	$r = 100/y$	Obs.	x_1	x_2	T	x_4	$r = 100/y$
1	134.5	19.1	623	5.74	218	28	30.0	20.0	648	5.64	294
2	108.0	20.0	623	5.50	189	29	16.3	20.0	648	5.61	233
3	68.6	19.9	623	5.44	192	30	16.5	20.0	648	5.63	222
4	49.5	20.0	623	5.55	174	31	20.4	12.5	648	5.70	188
5	41.7	20.0	623	5.45	152	32	20.5	16.6	648	5.67	231
6	29.4	19.9	623	6.31	139	33	20.8	20.0	648	5.63	239
7	22.5	20.0	623	5.39	118	34	21.3	30.0	648	5.63	301
8	17.2	19.9	623	5.60	120	35	19.6	43.3	648	5.62	252
9	17.0	19.7	623	5.61	122	36	20.6	20.0	648	5.72	217
10	22.8	20.0	623	5.54	132	37	20.5	30.0	648	5.43	276
11	41.3	20.0	623	5.52	167	38	20.3	42.7	648	5.60	467
12	59.6	20.0	623	5.53	208	39	16.0	19.1	673	5.88	429
13	119.7	20.0	623	5.50	216	40	23.5	20.0	673	6.01	475
14	158.2	20.0	623	5.48	294	41	132.8	20.0	673	6.48	1129
15	23.3	20.0	648	5.65	229	42	107.7	20.0	673	6.26	957
16	40.8	20.0	648	5.95	296	43	68.5	20.0	673	6.40	745
17	140.3	20.0	648	5.98	547	44	47.2	19.7	673	5.82	649
18	140.8	19.9	648	5.96	582	45	42.5	20.3	673	5.86	742
19	141.2	20.0	648	5.64	480	46	30.1	20.0	673	5.87	662
20	140.0	19.7	648	5.56	493	47	11.2	20.0	673	5.87	373
21	121.2	19.96	648	6.06	513	48	17.1	20.0	673	5.84	440
22	104.7	19.7	648	5.63	411	49	65.8	20.0	673	5.85	662
23	40.8	20.0	648	6.09	349	50	108.2	20.0	673	5.86	724
24	22.6	20.0	648	5.88	226	51	123.5	20.0	673	5.85	915
25	55.2	20.0	648	5.64	338	52	160.0	20.0	673	5.81	944
26	55.4	20.0	648	5.64	351	53	66.4	20.0	673	5.87	713
27	29.5	20.0	648	5.63	295	54	66.5	20.0	673	5.88	736

18.2 Estimation

We used least squares estimation to obtain $\hat{\beta}_i$s in linear regression; we will continue to use least squares estimation in nonlinear regression. For the linear regression model (18.1.2), least squares estimates minimize

$$SSE(\beta) \equiv \sum_{i=1}^{n} [y_i - E(y_i)]^2 = \sum_{i=1}^{n} \left[y_i - x_i'\beta\right]^2 .$$

For the nonlinear regression model (18.1.3), least squares estimates minimize

$$SSE(\beta) \equiv \sum_{i=1}^{n} [y_i - E(y_i)]^2 = \sum_{i=1}^{n} [y_i - f(x_i; \beta)]^2 . \qquad (18.2.1)$$

As shown below, in nonlinear regression with independent $N(0, \sigma^2)$ errors, the least squares estimates are also maximum likelihood estimates. Not surprisingly, finding the minimum of a function like (18.2.1) involves extensive use of calculus. We present in detail the Gauss–Newton algorithm for finding the least squares estimates and briefly mention an alternative method for finding the estimates.

18.2.1 The Gauss–Newton algorithm

The Gauss–Newton algorithm produces a series of vectors β^r that we hope will converge to $\hat{\beta}$. The algorithm requires some initial value for the vector β, say β^0. This can be thought of as a guess for the least squares estimate $\hat{\beta}$. We use matrix methods similar to those in Chapter 15 to present the algorithm.

In matrix notation write $Y = (y_1, \ldots, y_n)'$, $e = (\varepsilon_1, \ldots, \varepsilon_n)'$, and

$$F(X; \beta) \equiv \begin{bmatrix} f(x_1; \beta) \\ \vdots \\ f(x_n; \beta) \end{bmatrix}.$$

We can now write model (18.1.3) as

$$Y = F(X; \beta) + e, \quad \varepsilon_i\text{s indep. } N(0, \sigma^2). \tag{18.2.2}$$

Given β^r, the algorithm defines β^{r+1}. Define the matrix Z_r as the $n \times p$ matrix of partial derivatives $\partial f(x_i; \beta) / \partial \beta_j$ evaluated at β^r. Note that to find the ith row of Z_r, we need only differentiate to find the p partial derivatives $\partial f(x; \beta) / \partial \beta_j$ and evaluate these p functions at $x = x_i$ and $\beta = \beta^r$. For β values that are sufficiently close to β^r, a vector version of Taylor's theorem from calculus gives the approximation

$$F(X; \beta) \doteq F(X; \beta^r) + Z_r(\beta - \beta^r). \tag{18.2.3}$$

Here, because β^r is known, $F(X; \beta^r)$ and Z_r are known. Substituting the approximation (18.2.3) into equation (18.2.2), we get the *approximate* model

$$\begin{aligned} Y &= F(X; \beta^r) + Z_r(\beta - \beta^r) + e \\ &= F(X; \beta^r) + Z_r\beta - Z_r\beta^r + e. \end{aligned}$$

Rearranging terms gives

$$[Y - F(X; \beta^r) + Z_r\beta^r] = Z_r\beta + e. \tag{18.2.4}$$

If Z_r has full column rank, this is simply a linear regression model. The dependent variable vector is $Y - F(X; \beta^r) + Z_r\beta^r$, the matrix of predictor variables (design matrix) is Z_r, the parameter vector is β, and the error vector is e. Using least squares to estimate β gives us

$$\begin{aligned} \beta^{r+1} &= (Z_r'Z_r)^{-1}Z_r'[Y - F(X; \beta^r) + Z_r\beta^r] \\ &= (Z_r'Z_r)^{-1}Z_r'[Y - F(X; \beta^r)] + (Z_r'Z_r)^{-1}Z_r'Z_r\beta^r \\ &= (Z_r'Z_r)^{-1}Z_r'[Y - F(X; \beta^r)] + \beta^r \end{aligned} \tag{18.2.5}$$

From linear regression theory, the value β^{r+1} minimizes the function

$$SSE_r(\beta) \equiv \left[\{Y - F(X; \beta^r) + Z_r\beta^r\} - Z_r\beta\right]' \left[\{Y - F(X; \beta^r) + Z_r\beta^r\} - Z_r\beta\right].$$

Actually, we wish to minimize the function defined in (18.2.1). In matrix form, (18.2.1) is

$$SSE(\beta) = [Y - F(X; \beta)]' [Y - F(X; \beta)].$$

From (18.2.3), we have $SSE_r(\beta) \doteq SSE(\beta)$ for βs near β^r. If β^r is near the least squares estimate $\hat{\beta}$, the minimum of $SSE_r(\beta)$ should be close to the minimum of $SSE(\beta)$. While β^{r+1} minimizes $SSE_r(\beta)$ exactly, β^{r+1} is merely an approximation to the estimate $\hat{\beta}$ that minimizes $SSE(\beta)$. However, when β^r is close to $\hat{\beta}$, the approximation (18.2.3) is good. At the end of this subsection, we give a geometric argument that β^r converges to the least squares estimate.

EXAMPLE 18.2.1. *Multiple linear regression*
Suppose we treat model (18.1.2) as a nonlinear regression model. Then $f(x_i; \beta) = x_i'\beta$, $F(X; \beta) = X\beta$,

$\partial f(x_i; \beta)/\partial \beta_j = x_{ij}$, where $x_{i0} = 1$, and $Z_r = X$. From standard linear regression theory we know that $\hat{\beta} = (X'X)^{-1}X'Y$. Using the Gauss–Newton algorithm (18.2.5) with any β^0,

$$\begin{aligned} \beta^1 &= (Z_r'Z_r)^{-1}Z_r'[Y - F(X; \beta^0) + Z_0\beta^0] \\ &= (X'X)^{-1}X'[Y - X\beta^0 + X\beta^0] \\ &= (X'X)^{-1}X'Y \\ &= \hat{\beta}. \end{aligned}$$

Thus, for a linear regression problem, the Gauss–Newton algorithm arrives at $\hat{\beta}$ in only one iteration. □

EXAMPLE 18.2.2. To perform the analysis on the benzene oxidation data, we need the partial derivatives of the function (18.1.5):

$$\begin{aligned} \frac{\partial f(x; \beta)}{\partial \beta_0} &= \exp[\beta_0 + \beta_1 x_3]\frac{1}{x_2} \\ \frac{\partial f(x; \beta)}{\partial \beta_1} &= \exp[\beta_0 + \beta_1 x_3]\frac{x_3}{x_2} \\ \frac{\partial f(x; \beta)}{\partial \beta_2} &= \exp[\beta_2 + \beta_3 x_3]\frac{x_4}{x_1} \\ \frac{\partial f(x; \beta)}{\partial \beta_3} &= \exp[\beta_2 + \beta_3 x_3]\frac{x_3 x_4}{x_1}. \end{aligned}$$

With $\beta^1 = (.843092, 11427.598, .039828, 2018.7689)'$, we illustrate one step of the algorithm. The dependent variable in model (18.2.4) is

$$Y - F(X; \beta^1) + Z_1\beta^1 = \begin{bmatrix} 0.458716 \\ 0.529101 \\ 0.520833 \\ 0.574713 \\ 0.657895 \\ \vdots \\ 0.140252 \\ 0.135870 \end{bmatrix} - \begin{bmatrix} 0.297187 \\ 0.295806 \\ 0.330450 \\ 0.367968 \\ 0.389871 \\ \vdots \\ 0.142284 \\ 0.142300 \end{bmatrix} + \begin{bmatrix} 0.391121 \\ 0.375497 \\ 0.382850 \\ 0.387393 \\ 0.391003 \\ \vdots \\ 0.005125 \\ 0.005123 \end{bmatrix} = \begin{bmatrix} 0.552649 \\ 0.608792 \\ 0.573233 \\ 0.594137 \\ 0.659027 \\ \vdots \\ 0.003093 \\ -0.001307 \end{bmatrix}.$$

The design matrix in model (18.2.4) is

$$Z_1 = \begin{bmatrix} 0.246862 & 0.0000153 & 0.050325 & 0.0000031 \\ 0.235753 & 0.0000146 & 0.060052 & 0.0000037 \\ 0.236938 & 0.0000147 & 0.093512 & 0.0000058 \\ 0.235753 & 0.0000146 & 0.132214 & 0.0000082 \\ 0.235753 & 0.0000146 & 0.154117 & 0.0000095 \\ \vdots & \vdots & \vdots & \vdots \\ 0.060341 & -0.0000035 & 0.081942 & -0.0000047 \\ 0.060341 & -0.0000035 & 0.081958 & -0.0000047 \end{bmatrix}.$$

Fitting model (18.2.4) gives the estimate $\beta^2 = (1.42986, 12717, -0.15060, 9087.3)'$. Eventually, the sequence converges to $\hat{\beta}' = (1.3130, 11908, -.23463, 10559.5)$. □

In practice, methods related to Marquardt (1963) are often used to find the least squares estimates. These involve use of a statistical procedure known as ridge regression, cf. Seber and Wild (1989,

p. 624). Marquardt's method involves modifying model (18.2.4) to estimate $\beta - \beta^r$ by subtracting $Z_r\beta^r$ from both sides of the equality. Now, rather than using the least squares estimate $\beta^{r+1} - \beta^r = (Z_r'Z_r)^{-1}Z_r'[Y - F(X;\beta^r)]$, the simplest form of ridge regression (cf. Christensen, 1987, section XIV.6) uses the estimate

$$\beta^{r+1} - \beta^r = (Z_r'Z_r + kI_p)^{-1}Z_r'[Y - F(X;\beta^r)]$$

where I_p is a $p \times p$ identity matrix and k is a number that needs to be determined. More complicated forms of ridge regression involve replacing I_p with a diagonal matrix.

When the sequence of values β^r stops changing (converges), β^r is the least squares estimate. We will use a geometric argument to justify this statement. The argument applies to both the Gauss–Newton algorithm and the Marquardt method. By definition, $SSE(\beta)$ is the squared length of the vector $Y - F(X;\beta)$, i.e., it is the square of the distance between Y and $F(X;\beta)$. Geometrically, $\hat{\beta}$ is the value of β that makes $Y - F(X;\beta)$ as short a vector as possible. Y can be viewed as either a point in R^n or as a vector in R^n. For now, think of it as a point. $Y - F(X;\beta)$ is as short as possible when the line connecting Y and $F(X;\beta)$ is perpendicular to the surface $F(X;\beta)$. By definition, a line is perpendicular to a surface if it is perpendicular to the tangent plane of the surface at the point of intersection between the line and the surface. Thus in Figure 18.1, β^r has $Y - F(X;\beta^r)$ as short as possible but β^1 does not have $Y - F(X;\beta^1)$ as short as possible. We will show that when β^r converges, the line connecting Y and $F(X;\beta^r)$ is perpendicular to the tangent plane at β^r and thus $Y - F(X;\beta^r)$ is as short as possible. To do this technically, i.e., using vectors, we need to subtract $F(X;\beta^r)$ from everything. Thus we want to show that $Y - F(X;\beta^r)$ is a vector that is perpendicular to the surface $F(X;\beta) - F(X;\beta^r)$. From (18.2.3), the tangent plane to the surface $F(X;\beta)$ at β^r is $F(X;\beta^r) + Z_r(\beta - \beta^r)$, so the tangent plane to the surface $F(X;\beta) - F(X;\beta^r)$ is just $Z_r(\beta - \beta^r)$. Thus we need to show that when β^r converges, $Y - F(X;\beta^r)$ is perpendicular to the plane defined by Z_r. Algebraically, this means showing that

$$0 = Z_r'[Y - F(X;\beta^r)].$$

From the Gauss–Newton algorithm, at convergence we have $\beta^{r+1} = \beta^r$ and by (18.2.5) $\beta^{r+1} = (Z_r'Z_r)^{-1}Z_r'[Y - F(X;\beta^r)] + \beta^r$, so we must have

$$0 = (Z_r'Z_r)^{-1}Z_r'[Y - F(X;\beta^r)]. \tag{18.2.6}$$

This occurs precisely when $0 = Z_r'[Y - F(X;\beta^r)]$ because you can go back and forth between the two equations by multiplying with $(Z_r'Z_r)$ and $(Z_r'Z_r)^{-1}$, respectively. Thus β^r is the value that makes $Y - F(X;\beta)$ as short a vector as possible and $\beta^r = \hat{\beta}$. Essentially the same argument applies to the Marquardt method except equation (18.2.6) is replaced by $0 = (Z_r'Z_r + kI_p)^{-1}Z_r'[Y - F(X;\beta^r)]$.

The problem with this geometric argument – and indeed with the algorithms themselves – is that sometimes there is more than one β for which $Y - F(X;\beta)$ is perpendicular to the surface $F(X;\beta)$. If you start with an unfortunate choice of β^0, the sequence might converge to a value that does not minimize $SSE(\beta)$ over *all* β but only in a region around β^0. In fact, sometimes the sequence β^r might not even converge.

18.2.2 MAXIMUM LIKELIHOOD ESTIMATION

Nonlinear regression is a problem in which least squares estimates are maximum likelihood estimates. We now show this. The density of a random variable y with distribution $N(\mu, \sigma^2)$ is

$$\phi(y) = \frac{1}{\sqrt{2\pi}\sqrt{\sigma^2}} \exp[-(y-\mu)^2/2\sigma^2].$$

The joint density of independent random variables is obtained by multiplying the densities of the individual random variables. From model (18.1.3), the y_is are independent $N(f(x_i;\beta), \sigma^2)$ random

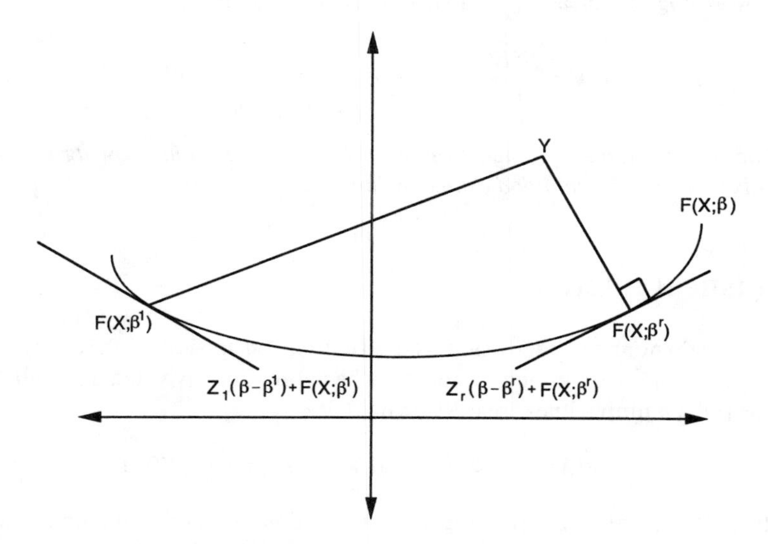

FIGURE 18.1. The geometry of nonlinear least squares estimation.

variables, so

$$\begin{aligned}\phi(Y) &\equiv \phi(y_1, \ldots, y_n) = \prod_{i=1}^{n} \phi(y_i) \\ &= \prod_{i=1}^{n} \frac{1}{\sqrt{2\pi}\sqrt{\sigma^2}} \exp[-\{y_i - f(x_i; \beta)\}^2 / 2\sigma^2] \\ &= \left[\frac{1}{\sqrt{2\pi}}\right]^n \left[\sqrt{\sigma^2}\right]^{-n} \exp\left[-\frac{1}{2\sigma^2} \sum_{i=1}^{n} \{y_i - f(x_i; \beta)\}^2\right] \\ &= \left[\frac{1}{\sqrt{2\pi}}\right]^n \left[\sqrt{\sigma^2}\right]^{-n} \exp\left[-\frac{1}{2\sigma^2} SSE(\beta)\right].\end{aligned}$$

The density is a function of Y for fixed values of β and σ^2. The likelihood is exactly the same function except that the likelihood is a function of β and σ^2 for fixed values of the observations y_i. Thus, the likelihood function is

$$L(\beta, \sigma^2) = \left[\frac{1}{\sqrt{2\pi}}\right]^n \left[\sqrt{\sigma^2}\right]^{-n} \exp\left[-\frac{1}{2\sigma^2} SSE(\beta)\right].$$

The maximum likelihood estimates of β and σ^2 are those values that maximize $L(\beta, \sigma^2)$. For *any* given value of σ^2, the likelihood is a simple function of $SSE(\beta)$. In fact, the likelihood is maximized by whatever value of β that minimizes $SSE(\beta)$, i.e., the least squares estimate $\hat{\beta}$. Moreover, the function $SSE(\beta)$ does not involve σ^2, so $\hat{\beta}$ does not involve σ^2 and the maximum of $L(\beta, \sigma^2)$ occurs wherever the maximum of $L(\hat{\beta}, \sigma^2)$ occurs. This is now a function of σ^2 alone. Differentiating with respect to σ^2, it is not difficult to see that the maximum likelihood estimate of σ^2 is

$$\hat{\sigma}^2 = \frac{SSE(\hat{\beta})}{n} = \frac{1}{n} \sum_{i=1}^{n} \left[y_i - f(x_i; \hat{\beta})\right]^2.$$

Alternatively, by analogy to linear regression, an estimate of σ^2 is

$$MSE = \frac{SSE(\hat{\beta})}{n-p} = \frac{1}{n-p} \sum_{i=1}^{n} \left[y_i - f(x_i; \hat{\beta})\right]^2.$$

Incidentally, *these exact same arguments apply to linear regression, showing that least squares estimates are also maximum likelihood estimates in linear regression.*

18.3 Statistical inference

Statistical inference for nonlinear regression is based entirely on versions of the central limit theorem. It requires a large sample size for the procedures to be approximately valid. The entire analysis can be conducted as if the multiple linear regression model

$$\left[Y - F(X; \hat{\beta}) + Z_*\hat{\beta}\right] = Z_*\beta + e, \quad \varepsilon_i\text{s indep. } N(0, \sigma^2) \tag{18.3.1}$$

were valid. Here Z_* is just like Z_r from the previous section except that the partial derivatives are evaluated at $\hat{\beta}$ rather than β^r. In other words, Z_* is the $n \times p$ matrix of partial derivatives $\partial f(x_i; \beta) / \partial \beta_j$ evaluated at $\hat{\beta}$. Actually, model (18.3.1) is simply the linear model (18.2.4) from the Gauss–Newton algorithm evaluated when β^r has converged to $\hat{\beta}$.

EXAMPLE 18.3.1. *Inference on regression parameters*
For the benzene oxidation data, $\hat{\beta}' = (1.3130, 11908, -.23463, 10559.5)$. It follows that the dependent variable for model (18.3.1) is

$$Y - F(X; \hat{\beta}) + Z_*\hat{\beta} = \begin{bmatrix} 0.458716 \\ 0.529101 \\ 0.520833 \\ 0.574713 \\ 0.657895 \\ \vdots \\ 0.140252 \\ 0.135870 \end{bmatrix} - \begin{bmatrix} 0.471786 \\ 0.466022 \\ 0.511129 \\ 0.559092 \\ 0.587341 \\ \vdots \\ 0.132084 \\ 0.132091 \end{bmatrix} + \begin{bmatrix} 0.86150 \\ 0.82922 \\ 0.85132 \\ 0.86824 \\ 0.88009 \\ \vdots \\ 0.02715 \\ 0.02714 \end{bmatrix} = \begin{bmatrix} 0.84843 \\ 0.89230 \\ 0.86102 \\ 0.88386 \\ 0.95064 \\ \vdots \\ 0.03532 \\ 0.03092 \end{bmatrix}$$

and the design matrix for model (18.3.1) is

$$Z_* = \begin{bmatrix} 0.406880 & 0.0000252 & 0.064906 & 0.0000040 \\ 0.388570 & 0.0000241 & 0.077452 & 0.0000048 \\ 0.390523 & 0.0000242 & 0.120606 & 0.0000075 \\ 0.388570 & 0.0000241 & 0.170522 & 0.0000106 \\ 0.388570 & 0.0000241 & 0.198771 & 0.0000123 \\ \vdots & \vdots & \vdots & \vdots \\ 0.093918 & -0.0000054 & 0.038166 & -0.0000022 \\ 0.093918 & -0.0000054 & 0.038174 & -0.0000022 \end{bmatrix}.$$

The size of the values in the second and fourth columns could easily cause numerical instability, but there were no signs of such problems in this analysis. Note also that the two small columns of Z_* correspond to the large values of $\hat{\beta}$. Fitting this model gives $SSE = 0.0810169059$ with $dfE = 54 - 4 = 50$, so $MSE = 0.0016203381$. The parameters, estimates, large sample standard errors, t statistics, P values, and 95% confidence intervals for the parameters are given below.

Par	*Est*	Asymptotic *SE(Est)*	*t*	*P*	95% Confidence interval
β_0	1.3130	.0600724	21.86	0.000	(1.1923696, 1.433687)
β_1	11908	1118.1335	10.65	0.000	(9662.1654177, 14153.831076)
β_2	−.23463	.0645778	−3.63	0.001	(−.3643371, −.104921)
β_3	10559.5	1311.4420	8.05	0.000	(7925.4156062, 13193.622791)

Generally, $Cov(\hat{\beta})$ is estimated with

$$MSE(Z_*'Z_*)^{-1} = 0.0016203381 \begin{bmatrix} 2 & -30407 & -2 & 24296 \\ -30407 & 771578688 & 24805 & -720666240 \\ -2 & 24805 & 3 & -30620 \\ 24296 & -720666240 & -30620 & 1061435136 \end{bmatrix}$$

however, here we begin to see some numerical instability, at least in the reporting of this matrix. For example, using this matrix, $SE(\hat{\beta}_0) = .0600724 = \sqrt{0.0016203381(2)}$. The 2 in the matrix has been rounded off because of the large numbers in other entries of the matrix. In reality, $SE(\hat{\beta}_0) = .0600724 = \sqrt{0.0016203381(2.22712375)}$. □

The primary complication from using model (18.3.1) involves forming confidence intervals for points on the regression surface and prediction intervals. Suppose we want to predict a new value y_0 for a given vector of predictor variable values, say x_0. Unfortunately, model (18.3.1) is not set up to predict y_0 but rather to provide a prediction of $y_0 - f(x_0;\hat{\beta}) + z_{*0}'\hat{\beta}$, where z_{*0}' is $(\partial f(x_0;\beta)/\partial\beta_0, \ldots, \partial f(x_0;\beta)/\partial\beta_{p-1})$ evaluated at $\hat{\beta}$. Happily, a simple modification of the prediction interval for $y_0 - f(x_0;\hat{\beta}) + z_{*0}'\hat{\beta}$ produces a prediction interval for y_0. As in Section 15.4, the $(1-\alpha)100\%$ prediction interval has endpoints $z_{*0}'\hat{\beta} \pm W_p$, where

$$W_p \equiv t\left(1-\frac{\alpha}{2}, n-p\right)\sqrt{MSE\left[1 + z_{*0}'(Z_*'Z_*)^{-1}z_{*0}\right]}.$$

In other words, the prediction interval is

$$z_{*0}'\hat{\beta} - W_p < y_0 - f(x_0;\hat{\beta}) + z_{*0}'\hat{\beta} < z_{*0}'\hat{\beta} + W_p.$$

To make this into an interval for y_0, simply add $f(x_0;\hat{\beta}) - z_{*0}'\hat{\beta}$ to each term, giving the interval

$$f(x_0;\hat{\beta}) - W_p < y_0 < f(x_0;\hat{\beta}) + W_p.$$

Similarly, the $(1-\alpha)100\%$ confidence interval from model (18.3.1) for a point on the surface gives a confidence interval for $z_{*0}'\beta$ rather than for $f(x_0;\beta)$. Defining

$$W_s \equiv t\left(1-\frac{\alpha}{2}, n-p\right)\sqrt{MSE\, z_{*0}'(Z_*'Z_*)^{-1}z_{*0}'},$$

the confidence interval for $z_{*0}'\beta$ is

$$z_{*0}'\hat{\beta} - W_s < z_{*0}'\beta < z_{*0}'\hat{\beta} + W_s.$$

As in (18.2.3), $f(x_0;\beta) \doteq f(x_0;\hat{\beta}) + z_{*0}'(\beta - \hat{\beta})$ or equivalently

$$f(x_0;\beta) - f(x_0;\hat{\beta}) + z_{*0}'\hat{\beta} \doteq z_{*0}'\beta.$$

We can substitute into the confidence interval to get

$$z_{*0}'\hat{\beta} - W_s < f(x_0;\beta) - f(x_0;\hat{\beta}) + z_{*0}'\hat{\beta} < z_{*0}'\hat{\beta} + W_s$$

and again adding $f(x_0; \hat{\beta}) - z'_{*0}\hat{\beta}$ to each term, gives

$$f(x_0; \hat{\beta}) - W_s < f(x_0; \beta) < f(x_0; \hat{\beta}) + W_s.$$

EXAMPLE 18.3.2. *Prediction*
For the benzene oxidation data, we choose to make a prediction at $x'_0 = (x_{01}, x_{02}, x_{03}, x_{04}) = (100, 20, 0, 5.7)$. Using x_0 and $\hat{\beta}$ to evaluate the partial derivatives, the vector used for making predictions in model (18.3.1) is $z'_{*0} = (0.185871, 0, 0.0450792, 0)$ and the prediction (estimate of the value on the surface at z_{*0}) for model (18.3.1) is $z'_{*0}\hat{\beta} = 0.233477$. The standard error of the surface is 0.00897 and the standard error for prediction is $\sqrt{.0016203381 + .00897^2}$. Model (18.3.1) gives the 95% confidence interval for the surface as (0.21545,0.25150) and the 95% prediction interval as (0.15062,0.31633). The actual prediction (estimate of the value on the surface at x_0) is $f(x_0; \hat{\beta}) = 0.230950$. The confidence interval and prediction interval need to be adjusted by $f(x_0; \hat{\beta}) - z'_{*0}\hat{\beta} = 0.230950 - 0.233477 = -0.002527$. This term needs to be added to the endpoints of the intervals, giving a 95% confidence interval for the surface of (0.21292,0.24897) and a 95% prediction interval of (0.14809,0.31380). Actually, our interest is in $r = 100/y$ rather than y, so a 95% prediction interval for r is $(100/.31380, 100/.14809)$, which is (318.7,675.3). □

We can also test full models against reduced models. Again write the full model as

$$y_i = f(x_i; \beta) + \varepsilon_i, \quad \varepsilon_i\text{s indep. } N(0, \sigma^2) \tag{18.3.2}$$

which, when fitted, gives $SSE(\hat{\beta})$ and write the reduced model as

$$y_i = f_0(x_i; \gamma) + \varepsilon_i \tag{18.3.3}$$

with $\gamma' = (\gamma_0, \ldots, \gamma_{q-1})$. When fitted, model (18.3.3) gives $SSE(\hat{\gamma})$. The simplest way of ensuring that model (18.3.3) is a reduced model relative to model (18.3.2) is by specifying constraints on the parameters.

EXAMPLE 18.3.3. In Section 1 we considered the model $y_i = \beta_0 + \beta_1[e^{\beta_2 x_i} - e^{\beta_3 x_i}] + \varepsilon_i$ with $p = 4$. If we specify $H_0 : \beta_1 = 4; 2\beta_2 = \beta_3$, the reduced model is $y_i = \beta_0 + 4[e^{\beta_2 x_i} - e^{2\beta_2 x_i}] + \varepsilon_i$. The parameters do not mean the same things in the reduced model as in the original model, so we can rewrite the reduced model as $y_i = \gamma_0 + 4[e^{\gamma_1 x_i} - e^{2\gamma_1 x_i}] + \varepsilon_i$ with $q = 2$. This particular reduced model can also be rewritten as $y_i = \gamma_0 + 4[e^{\gamma_1 x_i}(1 - e^{\gamma_1 x_i})] + \varepsilon_i$, which is beginning to look quite different from the full model. □

Corresponding to model (18.3.3), there is a linear model similar to model (18.3.1),

$$[Y - F_0(X; \hat{\gamma}) + Z_{*0}\hat{\gamma}] = Z_{*0}\gamma + e.$$

Alas, this model will typically *not* be a reduced model relative to model (18.3.1). In fact, the dependent variables (left-hand sides of the equations) are not even the same. Nonetheless, because model (18.3.3) is a reduced version of model (18.3.2), we can test the models in the usual way by using sums of squares error, cf. Section 13.3. Reject the reduced model with an α level test if

$$\frac{[SSE(\hat{\gamma}) - SSE(\hat{\beta})]/(p - q)}{SSE(\hat{\beta})/(n - p)} > F(1 - \alpha, p - q, n - p). \tag{18.3.4}$$

Of course, as in all of inference for nonlinear regression, the test is only a large sample approximation. The test statistic does not have exactly an F distribution when the reduced model is true.

EXAMPLE 18.3.4. *Testing a reduced model*
Consider the reduced model obtained from (18.1.4) by setting $\beta_0 = \beta_2$ and $\beta_1 = \beta_3$. We can rewrite the model as

$$y_i = \exp[\gamma_0 + \gamma_1 x_{i3}]\frac{1}{x_{i2}} + \exp[\gamma_0 + \gamma_1 x_{i3}]\frac{x_{i4}}{x_{i1}} + \varepsilon_i.$$

This model has $q = 2$ parameters. The partial derivatives of the function

$$f_0(x;\gamma) \equiv f(x_1, x_2, x_3, x_4; \gamma_0, \gamma_1) = \exp[\gamma_0 + \gamma_1 x_3]\frac{1}{x_2} + \exp[\gamma_0 + \gamma_1 x_3]\frac{x_4}{x_1}$$

are

$$\frac{\partial f_0(x;\gamma)}{\partial \gamma_0} = \exp[\gamma_0 + \gamma_1 x_3]\frac{1}{x_2} + \exp[\gamma_0 + \gamma_1 x_3]\frac{x_4}{x_1}$$

$$\frac{\partial f_0(x;\gamma)}{\partial \gamma_1} = \exp[\gamma_0 + \gamma_1 x_3]\frac{x_3}{x_2} + \exp[\gamma_0 + \gamma_1 x_3]\frac{x_3 x_4}{x_1}.$$

Fitting the model gives estimated parameters $\hat{\gamma}' = (.267172, 12155.54478)$ with $SSE = .4919048545$ on $dfE = 54 - 2 = 52$ for $MSE = .0094597087$. From inequality (18.3.4) and Example 18.3.1, the test statistic is

$$\frac{[.4919048545 - .0810169059]/[4-2]}{0.0016203381} = 126.79\,.$$

With an F statistic this large, the test will be rejected for any reasonable α level. □

EXAMPLE 18.3.5. *Diagnostics*
Table 18.2 contains the standard diagnostic quantities from model (18.3.1). We use these quantities in the usual way but possible problems are discussed at the end of the section.

The predicted values from model (18.3.1) are denoted $\hat{d}$. Given that there are 54 cases, none of the standardized residuals r or standardized deleted residuals t look exceptionally large.

Figures 18.2 through 18.9 contain index plots and standardized residual plots. Figures 18.2 and 18.3 are index plots of the leverages and Cook's distances, respectively. They simply plot the value against the observation number for each case. The symbol being plotted is the last digit of the observation number. Neither plot looks too bad to me (at least at 12:15 a.m. while I am doing this). However, there are some leverages that exceed the $3p/n = 3(4)/54 = .222$ rule.

Figure 18.4 contains a normal plot of the standardized residuals. This does not look too bad to me either and the Wilk–Francia statistic of $W' = .964$ is not significantly low.

Now we get to the real fun. Figures 18.5 through 18.9 contain plots of the standardized residuals versus the $\hat{d}$s, i.e., the predicted values from model (18.3.1), and versus the predictor variables x_1, x_2, x_3, x_4. The plot versus $\hat{d}$ is most notable for the large empty space in the middle. To some extent in the plot versus x_1 and very clearly in the plots versus x_3 and x_4, we see signs of heteroscedastic variances. This calls in question all of the inferential procedures that we have illustrated because the analysis *assumes* that the variance is the same for each observation. For more on how to analyze these data, see Pritchard et al. (1977) and Carroll and Ruppert (1984). □

Unlike linear regression, where the procedure is dominated by the predictor variables, nonlinear regression is very parameter oriented. This is perhaps excusable because in nonlinear regression there is usually some specific theory suggesting the regression model and that theory may give meaning to the parameters. Nonetheless, one can create big statistical problems or remove statistical problems simply by the choice of the parameterization. For example, model (18.1.4) can be rewritten as

$$y_i = \gamma_0 \exp[\gamma_1 x_{i3}]\frac{1}{x_2} + \gamma_2 \exp[\gamma_3 x_{i3}]\frac{x_4}{x_1} + \varepsilon_i \qquad (18.3.5)$$

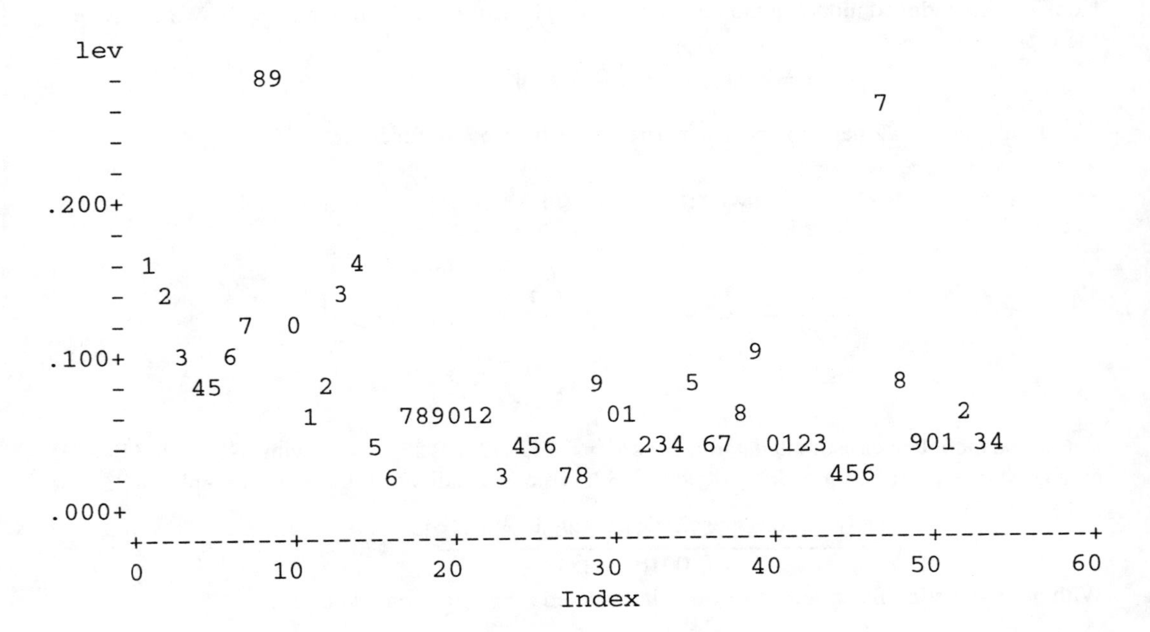

FIGURE 18.2. Index plot of leverage versus observation number.

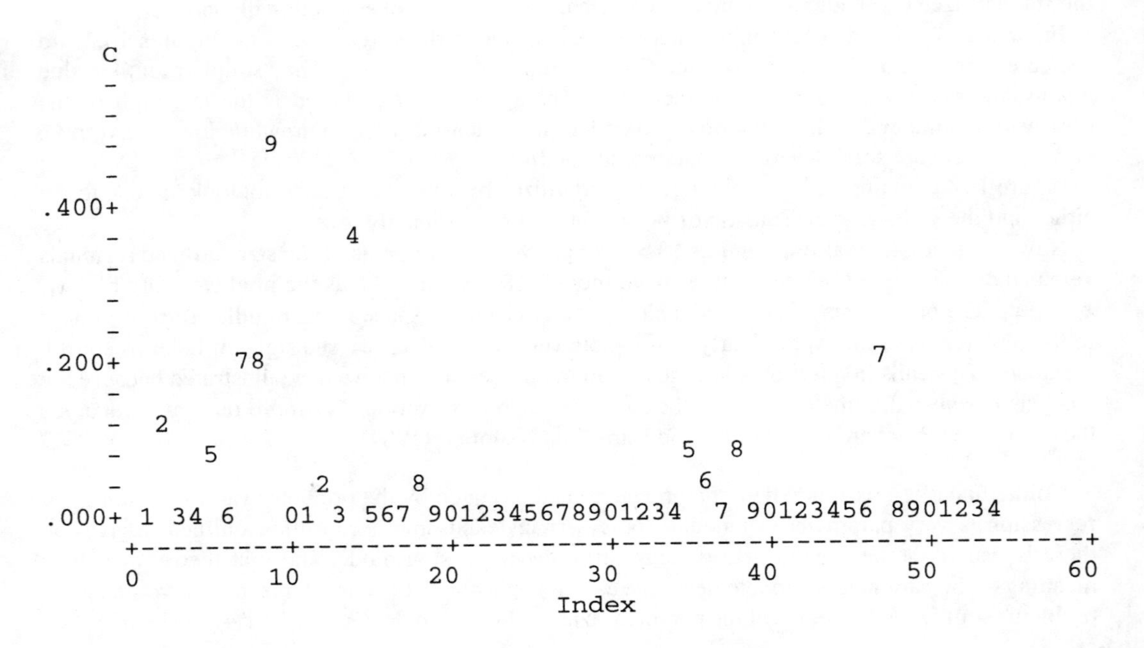

FIGURE 18.3. Index plot of Cook's distance versus observation number.

TABLE 18.2. Diagnostic statistics

Obs.	$\hat{e}$	$\hat{d}$	Leverage	r	t	C
1	−.01307	.47179	.15936	−0.35412	−0.35103	.005944
2	.06308	.46602	.13152	1.68152	1.71375	.107045
3	.00971	.51113	.10156	0.25436	0.25195	.001828
4	.01562	.55909	.07718	0.40396	0.40054	.003412
5	.07055	.58734	.07046	1.81795	1.86227	.062624
6	.00248	.71694	.09688	0.06491	0.06427	.000113
7	.09455	.75290	.12257	2.50767	2.65506	.219614
8	−.05236	.88569	.27329	−1.52579	−1.54686	.218863
9	−.07670	.89638	.28149	−2.24800	−2.34713	.494925
10	−.00054	.75812	.12675	−0.01435	−0.01419	.000007
11	.00696	.59184	.06980	0.17923	0.17748	.000603
12	−.04892	.52968	.08901	−1.27316	−1.28133	.039595
13	.00451	.45845	.13811	0.12073	0.11951	.000584
14	−.10112	.44125	.15431	−2.73157	−2.93174	.340371
15	.05903	.37765	.03341	1.49170	1.51071	.019231
16	.03663	.30121	.02807	0.92310	0.92172	.006153
17	−.03676	.21958	.05558	−0.93981	−0.93868	.012994
18	−.04846	.22028	.05636	−1.23931	−1.24613	.022932
19	−.00913	.21746	.05675	−0.23346	−0.23122	.000820
20	−.01727	.22011	.05887	−0.44225	−0.43864	.003058
21	−.03085	.22579	.05273	−0.78755	−0.78450	.008630
22	.01208	.23123	.05285	0.30838	0.30559	.001327
23	−.01739	.30392	.02775	−0.43803	−0.43446	.001369
24	.05084	.39164	.03765	1.28753	1.29625	.016213
25	.02918	.26668	.03552	0.73818	0.73478	.005018
26	.01852	.26638	.03561	0.46839	0.46472	.002025
27	.00218	.33681	.02682	0.05485	0.05430	.000021
28	.00558	.33455	.02671	0.14058	0.13920	.000136
29	−.02888	.45806	.07150	−0.74455	−0.74120	.010672
30	−.00527	.45572	.06992	−0.13582	−0.13450	.000347
31	.01355	.51837	.06789	0.34853	0.34546	.002212
32	−.00978	.44268	.04494	−0.24865	−0.24631	.000727
33	.01847	.39994	.04064	0.46855	0.46486	.002325
34	−.00073	.33295	.04625	−0.01853	−0.01836	.000004
35	.08420	.31262	.07329	2.17301	2.26054	.093364
36	.05536	.40547	.04282	1.40571	1.41991	.022102
37	.02892	.33340	.04651	0.73583	0.73239	.006603
38	−.09110	.30523	.06575	−2.34133	−2.45639	.096455
39	−.02390	.25700	.09852	−0.62540	−0.62158	.010687
40	.00620	.20433	.04176	0.15727	0.15571	.000269
41	−.02641	.11498	.04883	−0.67271	−0.66896	.005807
42	−.01452	.11901	.04522	−0.36911	−0.36588	.001613
43	−.00003	.13425	.03438	−0.00065	−0.00063	.000000
44	.00550	.14858	.02983	0.13875	0.13740	.000148
45	−.01729	.15206	.02610	−0.43515	−0.43159	.001268
46	−.02705	.17811	.02799	−0.68170	−0.67801	.003346
47	−.05209	.32019	.26019	−1.50456	−1.52442	.199053
48	−.01409	.24136	.08364	−0.36561	−0.36245	.003051
49	.01876	.13230	.03552	0.47447	0.47078	.002073
50	.02082	.11730	.04672	0.52980	0.52598	.003439
51	−.00508	.11437	.04941	−0.12939	−0.12809	.000217
52	−.00366	.10959	.05414	−0.09355	−0.09259	.000125
53	.00817	.13208	.03565	0.20665	0.20468	.000395
54	.00378	.13209	.03565	0.09558	0.09465	.000084

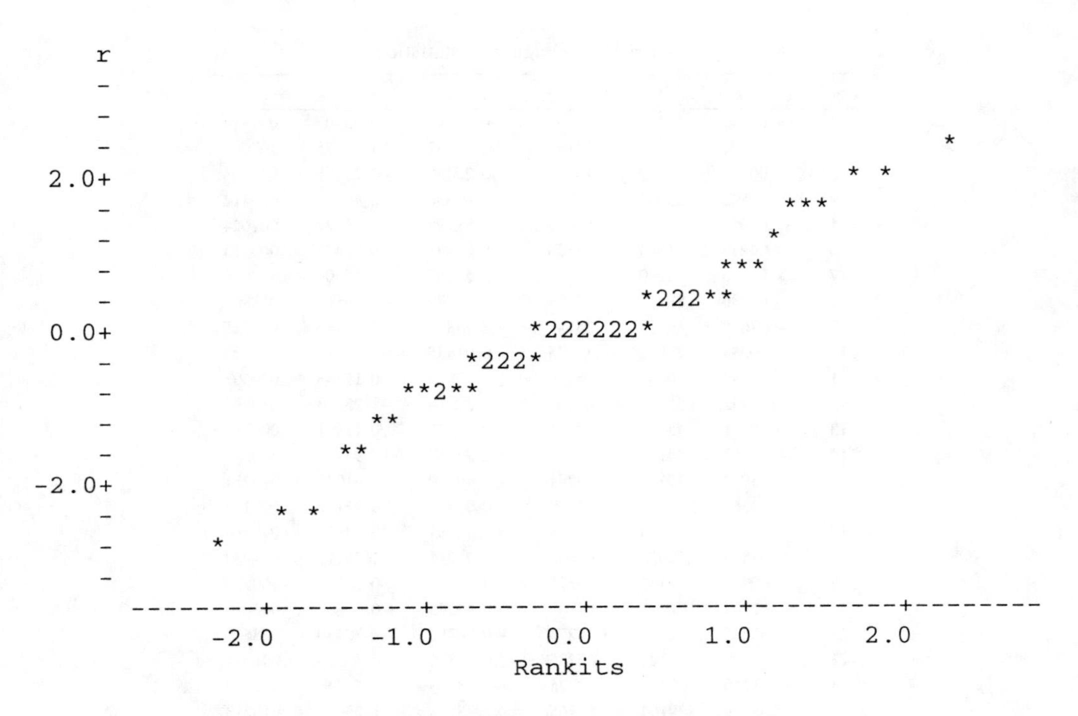

FIGURE 18.4. Rankit plot of standardized residuals, $W' = .964$.

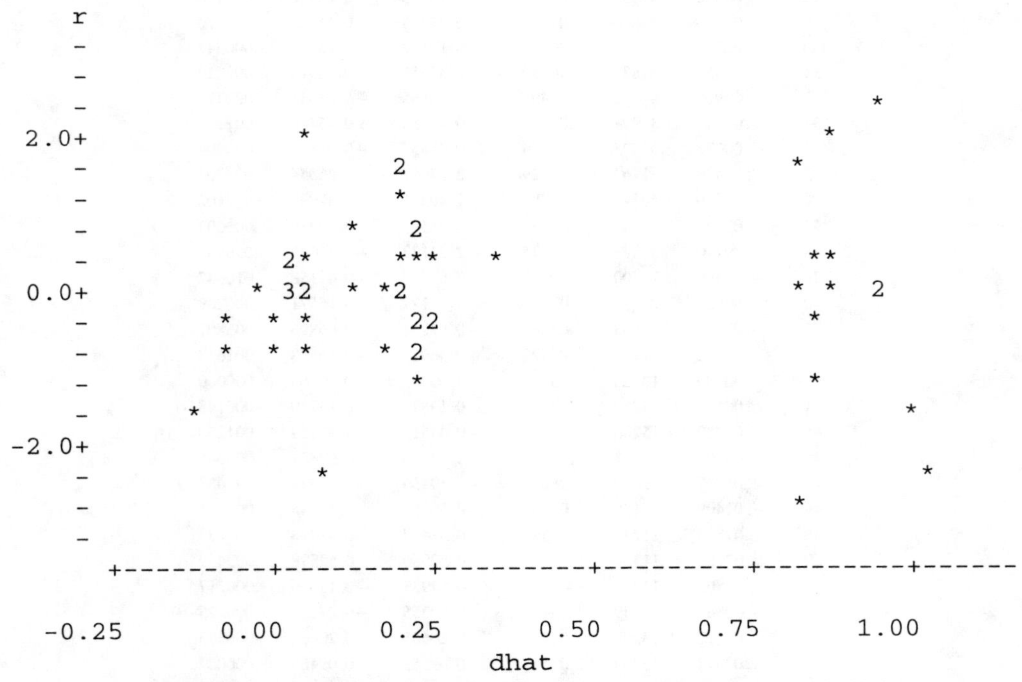

FIGURE 18.5. Standardized residuals versus predicted values from model (18.3.1).

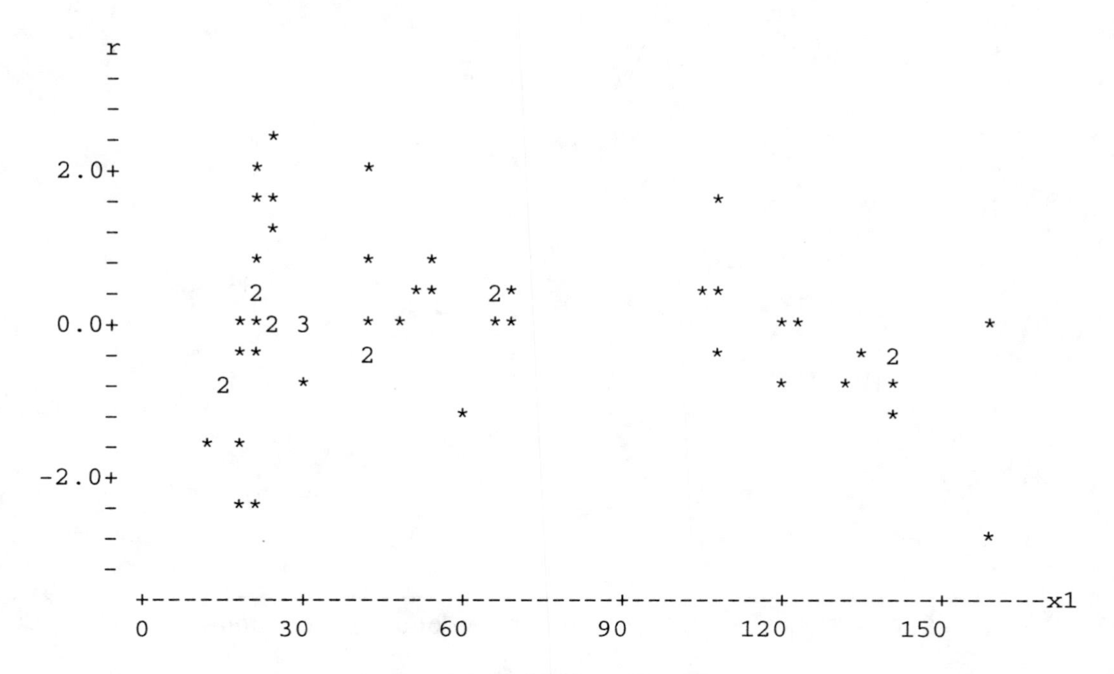

FIGURE 18.6. Standardized residuals versus x_1.

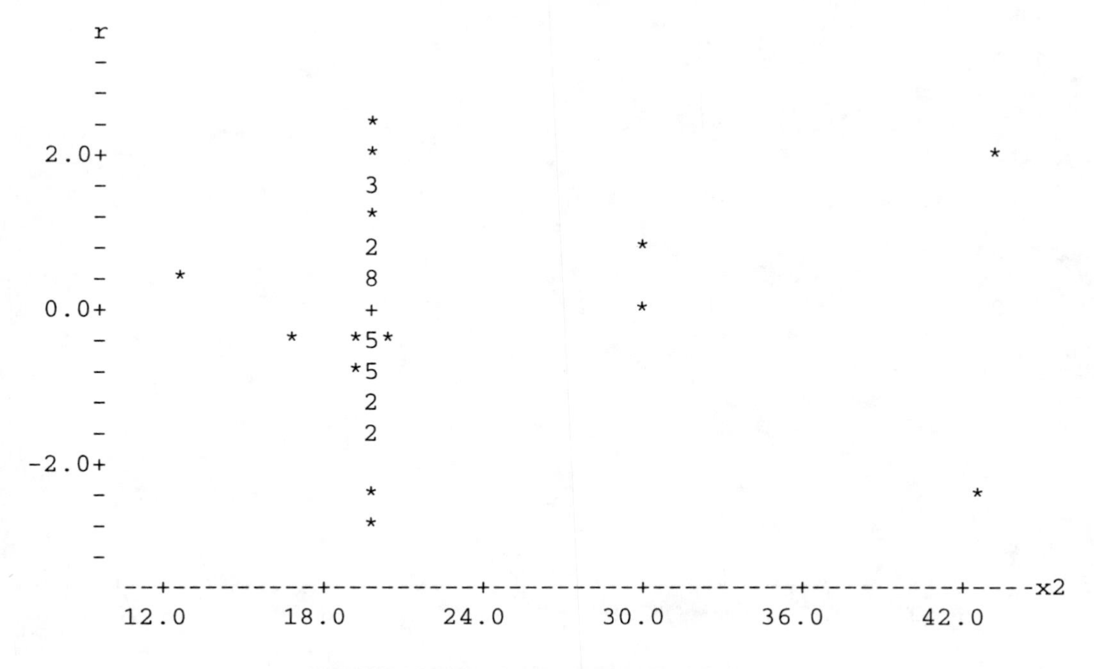

FIGURE 18.7. Standardized residuals versus x_2.

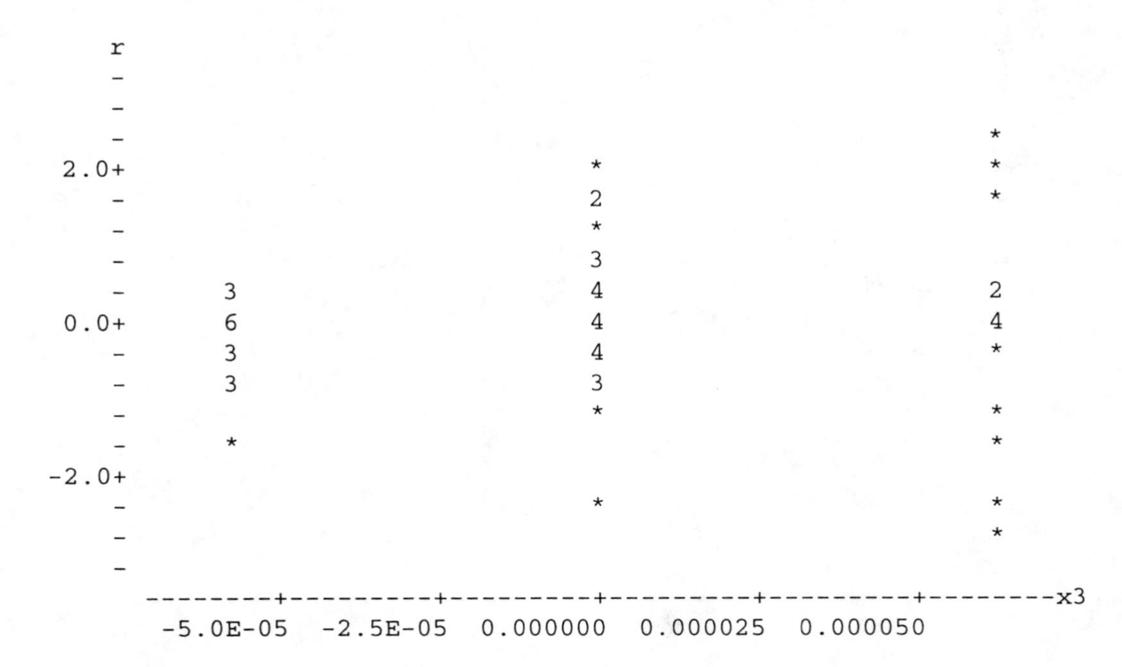

FIGURE 18.8. Standardized residuals versus x_3.

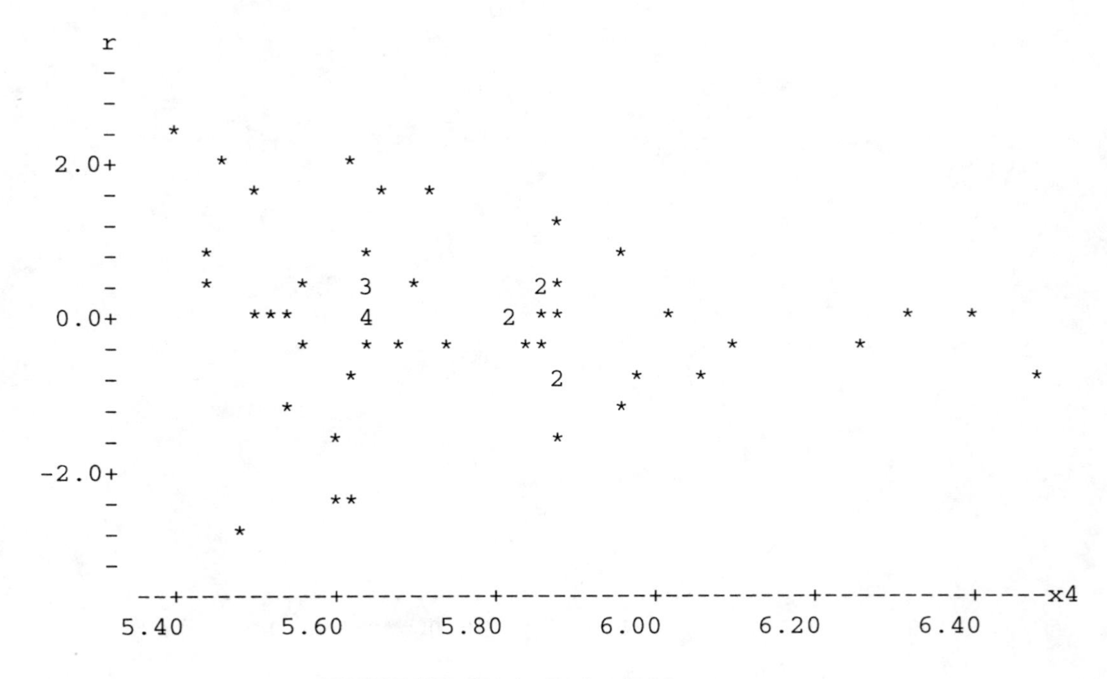

FIGURE 18.9. Standardized residuals versus x_4.

If, say, $\gamma_0 = 0$, the entire term $\gamma_0 \exp[\gamma_1 x_{i3}]/x_2$ vanishes. This term is the only place in which the parameter γ_1 appears. So if $\gamma_0 = 0$, it will be impossible to learn about γ_1. More to the point, if γ_0 is near zero, it will be very difficult to learn about γ_1. (Of course, one could argue that from the viewpoint of prediction, one may not care much what γ_1 is if γ_0 is very near zero and x_3 is of moderate size.) In any case, unlike linear regression, the value of one parameter can affect what we learn about other parameters. (In linear regression, the values of some predictor variables effect what we can learn about the parameters for other predictor variables, but it is not the parameters themselves that create the problem. In fact, in nonlinear regression, as the benzene example indicates, the predictor variables are not necessarily associated with any particular parameter.) In model (18.1.4) we have ameliorated the problem of γ_0 near 0 by using the parameter β_0. When γ_0 approaches zero, β_0 approaches negative infinity, so this problem with the coefficient of x_{i3}, i.e. γ_1 or β_1, will not arise for finite β_0. However, unlike (18.3.5), model (18.1.4) cannot deal with the possibility of $\gamma_0 < 0$. Similar problems can occur with γ_2.

All of the methods in this section depend crucially on the quality of the approximation in (18.2.3) when $\beta^r = \hat{\beta}$. If this approximation is poor, these methods can be very misleading. In particular, Cook and Tsai (1985, 1990) discuss problems with residual analysis when the approximation is poor and diagnostics for the quality of the normal approximation. St. Laurent and Cook (1992) discuss concepts of leverage for nonlinear regression. For large samples, the true value of β should be close to $\hat{\beta}$ and the approximation should be good. (This conclusion also depends on having the standard errors for functions of $\hat{\beta}$ small in large samples.) But it is very difficult to tell what constitutes a 'large sample.' As a practical matter, the quality of the approximation depends a great deal on the amount of curvature found in $f(x; \beta)$ near $\beta = \hat{\beta}$. This curvature is conveniently measured by the second partial derivatives $\partial^2 f(x; \beta)/\partial\beta_j \partial\beta_k$ evaluated at $\hat{\beta}$. A good analysis of nonlinear regression data should include an examination of curvature, but such an examination is beyond the scope of this book, cf. Seber and Wild (1989, chapter 4).

18.4 Linearizable models

Some nonlinear relationships can be changed into linear relationships. The nonlinear regression model (18.1.3) indicates that

$$E(y_i) = f(x_i; \beta).$$

Sometimes $f(x_i; \beta)$ can be written as

$$f(x_i; \beta) = f(x_i'\beta).$$

If f is invertible, we get

$$f^{-1}[E(y_i)] = x_i'\beta.$$

Often it is not too clear whether we should be modeling $f^{-1}[E(y_i)] = x_i'\beta$ or $E[f^{-1}(y_i)] = x_i'\beta$. As we saw, the first of these comes from nonlinear regression. The second equality suggests the *linear* regression model,

$$f^{-1}(y_i) = x_i'\beta + \varepsilon_i. \qquad (18.4.1)$$

It can be very difficult to choose between analyzing the nonlinear model (18.1.3) and the linear model (18.4.1). The decision is often based on which model gives better approximations to the assumption of independent identically distributed mean zero normal errors.

EXAMPLE 18.4.1. In Section 7.10 we analyzed the Hooker data using a linear model $\log(y_i) = \beta_0 + \beta_1 x_i + \varepsilon_i$. Exponentiating both sides gives $y_i = \exp[\beta_0 + \beta_1 x_i + \varepsilon_i]$, which we can rewrite as

$y_i = \exp[\beta_0 + \beta_1 x_i]\xi_i$, where ξ_i is a multiplicative error term with $\xi_i = \exp(\varepsilon_i)$. Alternatively, we could fit a nonlinear regression model

$$y_i = \exp[\beta_0 + \beta_1 x_i] + \varepsilon_i. \tag{18.4.2}$$

The difference between these two models is that in the first model (the linearized model) the errors on the original scale are multiplied by the regression structure $\exp[\beta_0 + \beta_1 x_i]$, whereas in the nonlinear model the errors are additive, i.e., are added to the regression structure. To fit the nonlinear model (18.4.2), we need the partial derivatives of $f(x; \beta_0, \beta_1) \equiv \exp[\beta_0 + \beta_1 x]$, namely $\partial f(x; \beta_0, \beta_1)/\partial\beta_0 = \exp[\beta_0 + \beta_1 x]$ and $\partial f(x; \beta_0, \beta_1)/\partial\beta_1 = \exp[\beta_0 + \beta_1 x]x$. As mentioned earlier, the choice between using the linearized model from Section 7.10 or the nonlinear regression model (18.4.2) is often based on which model seems to have better residual plots, etc. Exercise 18.5.1 asks for this comparison. □

18.5 Exercises

EXERCISE 18.5.1. Fit the nonlinear regression (18.4.2) to the Hooker data and compare the fit of this model to the fit of the linearized model described in Section 7.10.

EXERCISE 18.5.2. For pregnant women, Day (1966) modeled the relationship between weight z and week of gestation x with

$$E(y) = \beta_0 + \exp[\beta_1 + \beta_2 x]$$

where $y = 1/\sqrt{z - z_0}$ and z_0 is the initial weight of the woman. For a woman with initial weight of 138 pounds, the following data were recorded:

Week	Weight	Week	Weight
8	140.50	34	163.75
12	139.25	35	168.75
14	138.75	36	170.00
15	140.00	37	171.25
19	147.25	38	173.00
23	150.50	39	174.00
27	156.75	40	174.00
31	162.75	42	174.50

Fit the model $y_i = \beta_0 + \exp[\beta_1 + \beta_2 x_i] + \varepsilon_i$. Test whether each parameter is equal to zero, give 95% confidence intervals for each parameter, give 95% prediction intervals and surface confidence intervals for $x = 21$ weeks, and check the diagnostic quantities. Test the reduced model defined by $H_0 : \beta_0 = 0; \beta_1 = 0$.

EXERCISE 18.5.3. Following Bliss and James (1966) fit the model $y_i = (x_i\beta_0)/(x_i + \beta_1) + \varepsilon_i$ to the following data on the relationship between reaction velocity y and concentration of substrate x.

x	.138	.220	.291	.560	.766	1.460
y	.148	.171	.234	.324	.390	.493

Test whether each parameter is equal to zero, give 99% confidence intervals for each parameter, give 99% prediction intervals and surface confidence intervals for $x = .5$, and check the diagnostic quantities.

EXERCISE 18.5.4. Bliss and James (1966) give data on the median survival time z of house flies following application of the pesticide DDT at a level of molar concentration x. Letting $y = 100/z$, fit the model $y_i = \beta_0 + \beta_1 x_i/(x_i + \beta_2) + \varepsilon_i$ to the data given below.

x	z	x	z
.200	99	.0150	172
.100	115	.0100	188
.075	119	.0075	284
.050	112	.0070	227
.0375	126	.0060	275
.0250	149	.0050	525
.0200	152	.0025	948

Test whether each parameter is equal to zero, give 99% confidence intervals for each parameter, give 95% prediction intervals and surface confidence intervals for a concentration of $x = .03$, and check the diagnostic quantities. Find the SSE and test the reduced model defined by $H_0 : \beta_0 = 0, \beta_2 = .0125$. Test $H_0 : \beta_2 = .0125$.

Appendix A: Matrices

A matrix is a rectangular array of numbers. Such arrays have *rows* and *columns*. The numbers of rows and columns are referred to as the *dimensions* of a matrix. A matrix with, say, 5 rows and 3 columns is referred to as a 5×3 matrix.

EXAMPLE A.0.1. Three matrices are given below along with their dimensions.

$$\underset{3 \times 2}{\begin{bmatrix} 1 & 4 \\ 2 & 5 \\ 3 & 6 \end{bmatrix}}, \quad \underset{2 \times 2}{\begin{bmatrix} 20 & 80 \\ 90 & 140 \end{bmatrix}}, \quad \underset{4 \times 1}{\begin{bmatrix} 6 \\ 180 \\ -3 \\ 0 \end{bmatrix}}.$$

□

Let r be an arbitrary positive integer. A matrix with r rows and r columns, i.e., an $r \times r$ matrix, is called a *square matrix*. The second matrix in Example A.0.1 is square. A matrix with only one column, i.e., an $r \times 1$ matrix, is a *vector*, sometimes called a *column vector*. The third matrix in Example A.0.1 is a vector. A $1 \times r$ matrix is sometimes called a *row vector*.

An arbitrary matrix A is often written

$$A = [a_{ij}]$$

where a_{ij} denotes the element of A in the ith row and jth column. Two matrices are equal if they have the same dimensions and all of their elements (entries) are equal. Thus for $r \times c$ matrices $A = [a_{ij}]$ and $B = [b_{ij}]$, $A = B$ if and only if $a_{ij} = b_{ij}$ for every $i = 1, \ldots, r$ and $j = 1, \ldots, c$.

EXAMPLE A.0.2. Let

$$A = \begin{bmatrix} 20 & 80 \\ 90 & 140 \end{bmatrix} \text{ and } B = \begin{bmatrix} b_{11} & b_{12} \\ b_{21} & b_{22} \end{bmatrix}.$$

If $B = A$, then $b_{11} = 20, b_{12} = 80, b_{21} = 90$, and $b_{22} = 140$. □

The *transpose* of a matrix A, denoted A', changes the rows of A into columns of a new matrix A'. If A is an $r \times c$ matrix, the transpose A' is a $c \times r$ matrix. In particular, if we write $A' = [\tilde{a}_{ij}]$, then the element in row i and column j of A' is defined to be $\tilde{a}_{ij} = a_{ji}$.

EXAMPLE A.0.3.

$$\begin{bmatrix} 1 & 4 \\ 2 & 5 \\ 3 & 6 \end{bmatrix}' = \begin{bmatrix} 1 & 2 & 3 \\ 4 & 5 & 6 \end{bmatrix}$$

and

$$\begin{bmatrix} 20 & 80 \\ 90 & 140 \end{bmatrix}' = \begin{bmatrix} 20 & 90 \\ 80 & 140 \end{bmatrix}.$$

The transpose of a column vector is a row vector,

$$\begin{bmatrix} 6 \\ 180 \\ -3 \\ 0 \end{bmatrix}' = [6 \quad 180 \quad -3 \quad 0].$$

□

A.1 Matrix addition and subtraction

Two matrices can be added (or subtracted) if they have the same dimensions, that is, if they have the same number of rows and columns. Addition and subtraction is performed elementwise.

EXAMPLE A.1.1.

$$\begin{bmatrix} 1 & 4 \\ 2 & 5 \\ 3 & 6 \end{bmatrix} + \begin{bmatrix} 2 & 8 \\ 4 & 10 \\ 6 & 12 \end{bmatrix} = \begin{bmatrix} 1+2 & 4+8 \\ 2+4 & 5+10 \\ 3+6 & 6+12 \end{bmatrix} = \begin{bmatrix} 3 & 12 \\ 6 & 15 \\ 9 & 18 \end{bmatrix}.$$

$$\begin{bmatrix} 20 & 80 \\ 90 & 140 \end{bmatrix} - \begin{bmatrix} -15 & -75 \\ 80 & 130 \end{bmatrix} = \begin{bmatrix} 35 & 155 \\ 10 & 10 \end{bmatrix}.$$

□

In general, if A and B are $r \times c$ matrices with $A = [a_{ij}]$ and $B = [b_{ij}]$, then

$$A + B = [a_{ij} + b_{ij}] \text{ and } A - B = [a_{ij} - b_{ij}].$$

A.2 Scalar multiplication

Any matrix can be multiplied by a scalar. Multiplication by a scalar (a *real number*) is elementwise.

EXAMPLE A.2.1. Scalar multiplication gives

$$\frac{1}{10}\begin{bmatrix} 20 & 80 \\ 90 & 140 \end{bmatrix} = \begin{bmatrix} 20/10 & 80/10 \\ 90/10 & 140/10 \end{bmatrix} = \begin{bmatrix} 2 & 8 \\ 9 & 14 \end{bmatrix}.$$

$$2[6 \quad 180 \quad -3 \quad 0] = [12 \quad 360 \quad -6 \quad 0].$$

□

In general, if λ is any number and $A = [a_{ij}]$, then

$$\lambda A = [\lambda a_{ij}].$$

A.3 Matrix multiplication

Two matrices can be multiplied together if the number of columns in the first matrix is the same as the number of rows in the second matrix. In the process of multiplication, the rows of the first matrix are matched up with the columns of the second matrix.

EXAMPLE A.3.1.

$$\begin{bmatrix} 1 & 4 \\ 2 & 5 \\ 3 & 6 \end{bmatrix} \begin{bmatrix} 20 & 80 \\ 90 & 140 \end{bmatrix} = \begin{bmatrix} (1)(20)+(4)(90) & (1)(80)+(4)(140) \\ (2)(20)+(5)(90) & (2)(80)+(5)(140) \\ (3)(20)+(6)(90) & (3)(80)+(6)(140) \end{bmatrix}$$
$$= \begin{bmatrix} 380 & 640 \\ 490 & 860 \\ 600 & 1080 \end{bmatrix}.$$

The entry in the first row and column of the product matrix, $(1)(20)+(4)(90)$, matches the elements in the first row of the first matrix, $(1\ 4)$, with the elements in the first column of the second matrix, $\begin{pmatrix} 20 \\ 90 \end{pmatrix}$. The 1 in $(1\ 4)$ is matched up with the 20 in $\begin{pmatrix} 20 \\ 90 \end{pmatrix}$ and these numbers are multiplied. Similarly, the 4 in $(1\ 4)$ is matched up with the 90 in $\begin{pmatrix} 20 \\ 90 \end{pmatrix}$ and the numbers are multiplied. Finally, the two products are added to obtain the entry $(1)(20) + (4)(90)$. Similarly, the entry in the third row, second column of the product, $(3)(80) + (6)(140)$, matches the elements in the third row of the first matrix, $(3\ 6)$, with the elements in the second column of the second matrix, $\begin{pmatrix} 80 \\ 140 \end{pmatrix}$. After multiplying and adding we get the entry $(3)(80)+(6)(140)$. To carry out this matching, the number of columns in the first matrix must equal the number of rows in the second matrix. The matrix product has the same number of rows as the first matrix and the same number of columns as the second because each row of the first matrix can be matched with each column of the second. □

EXAMPLE A.3.2. We illustrate another matrix multiplication commonly performed in statistics, multiplying a matrix on its left by the transpose of that matrix, i.e., computing $A'A$.

$$\begin{bmatrix} 1 & 4 \\ 2 & 5 \\ 3 & 6 \end{bmatrix}' \begin{bmatrix} 1 & 4 \\ 2 & 5 \\ 3 & 6 \end{bmatrix} = \begin{bmatrix} 1 & 2 & 3 \\ 4 & 5 & 6 \end{bmatrix} \begin{bmatrix} 1 & 4 \\ 2 & 5 \\ 3 & 6 \end{bmatrix}$$
$$= \begin{bmatrix} 1+4+9 & 4+10+18 \\ 4+10+18 & 16+25+36 \end{bmatrix}$$
$$= \begin{bmatrix} 14 & 32 \\ 32 & 77 \end{bmatrix}.$$
□

Notice that in matrix multiplication the roles of the first matrix and the second matrix are *not* interchangeable. In particular, if we reverse the order of the matrices in Example A.3.1, the matrix product

$$\begin{bmatrix} 20 & 80 \\ 90 & 140 \end{bmatrix} \begin{bmatrix} 1 & 4 \\ 2 & 5 \\ 3 & 6 \end{bmatrix}$$

is undefined because the first matrix has two columns while the second matrix has three rows. Even when the matrix products are defined for both AB and BA, the results of the multiplication typically differ. If A is $r \times s$ and B is $s \times r$, then AB is an $r \times r$ matrix and BA is and $s \times s$ matrix. When $r \neq s$, clearly $AB \neq BA$, but even when $r = s$ we still can not expect AB to equal BA.

EXAMPLE A.3.3. Consider two square matrices, say,

$$A = \begin{bmatrix} 1 & 2 \\ 3 & 4 \end{bmatrix} \quad B = \begin{bmatrix} 0 & 2 \\ 1 & 2 \end{bmatrix}.$$

Multiplication gives

$$AB = \begin{bmatrix} 2 & 6 \\ 4 & 14 \end{bmatrix}$$

and

$$BA = \begin{bmatrix} 6 & 8 \\ 7 & 10 \end{bmatrix},$$

so $AB \neq BA$. □

In general if $A = [a_{ij}]$ is an $r \times s$ matrix and $B = [b_{ij}]$ is a $s \times c$ matrix, then

$$AB = [d_{ij}]$$

is the $r \times c$ matrix with

$$d_{ij} = \sum_{\ell=1}^{s} a_{i\ell} b_{\ell j}.$$

A useful result is that the transpose of the product AB is the product, in reverse order, of the transposed matrices, i.e. $(AB)' = B'A'$.

EXAMPLE A.3.4. As seen in Example A.3.1,

$$AB \equiv \begin{bmatrix} 1 & 4 \\ 2 & 5 \\ 3 & 6 \end{bmatrix} \begin{bmatrix} 20 & 80 \\ 90 & 140 \end{bmatrix} = \begin{bmatrix} 380 & 640 \\ 490 & 860 \\ 600 & 1080 \end{bmatrix} \equiv C.$$

The transpose of this matrix is

$$C' = \begin{bmatrix} 380 & 490 & 600 \\ 640 & 860 & 1080 \end{bmatrix} = \begin{bmatrix} 20 & 90 \\ 80 & 140 \end{bmatrix} \begin{bmatrix} 1 & 2 & 3 \\ 4 & 5 & 6 \end{bmatrix} = B'A'.$$

□

Let $a = (a_1, \ldots, a_n)'$ be a vector. A very useful property of vectors is that

$$a'a = \sum_{i=1}^{n} a_i^2 \geq 0.$$

A.4 Special matrices

If $A = A'$, then A is said to be *symmetric*. If $A = [a_{ij}]$ and $A = A'$, then $a_{ij} = a_{ji}$. The entry in row i and column j is the same as the entry in row j and column i. Only square matrices can be symmetric.

EXAMPLE A.4.1. The matrix

$$A = \begin{bmatrix} 4 & 3 & 1 \\ 3 & 2 & 6 \\ 1 & 6 & 5 \end{bmatrix}$$

has $A = A'$. A is symmetric about the diagonal that runs from the upper left to the lower right. □

For any $r \times c$ matrix A, the product $A'A$ is always symmetric. This was illustrated in Example .3.2. More generally, write $A = [a_{ij}]$, $A' = [\tilde{a}_{ij}]$ with $\tilde{a}_{ij} = a_{ji}$, and

$$A'A = \left[d_{ij}\right] = \left[\sum_{\ell=1}^{c} \tilde{a}_{i\ell} a_{\ell j}\right].$$

Note that

$$d_{ij} = \sum_{\ell=1}^{c} \tilde{a}_{i\ell} a_{\ell j} = \sum_{\ell=1}^{c} a_{\ell i} a_{\ell j} = \sum_{\ell=1}^{c} \tilde{a}_{j\ell} a_{\ell i} = d_{ji}$$

so the matrix is symmetric.

Diagonal matrices are square matrices with all off diagonal elements equal to zero.

EXAMPLE A.4.2. The matrices

$$\begin{bmatrix} 1 & 0 & 0 \\ 0 & 2 & 0 \\ 0 & 0 & 3 \end{bmatrix}, \quad \begin{bmatrix} 20 & 0 \\ 0 & -3 \end{bmatrix}, \text{ and } \begin{bmatrix} 1 & 0 & 0 \\ 0 & 1 & 0 \\ 0 & 0 & 1 \end{bmatrix}$$

are diagonal. □

In general, a diagonal matrix is a square matrix $A = \left[a_{ij}\right]$ with $a_{ij} = 0$ for $i \neq j$. Obviously, diagonally matrices are symmetric.

An *identity matrix* is a diagonal matrix with all 1s along the diagonal, i.e., $a_{ii} = 1$ for all i. The third matrix in Example A.4.2 above is a 3×3 identity matrix. The identity matrix gets it name because any matrix multiplied by an identity matrix remains unchanged.

EXAMPLE A.4.3.

$$\begin{bmatrix} 1 & 4 \\ 2 & 5 \\ 3 & 6 \end{bmatrix} \begin{bmatrix} 1 & 0 \\ 0 & 1 \end{bmatrix} = \begin{bmatrix} 1 & 4 \\ 2 & 5 \\ 3 & 6 \end{bmatrix}.$$

$$\begin{bmatrix} 1 & 0 & 0 \\ 0 & 1 & 0 \\ 0 & 0 & 1 \end{bmatrix} \begin{bmatrix} 1 & 4 \\ 2 & 5 \\ 3 & 6 \end{bmatrix} = \begin{bmatrix} 1 & 4 \\ 2 & 5 \\ 3 & 6 \end{bmatrix}.$$

□

An $r \times r$ identity matrix is denoted I_r with the subscript deleted if the dimension is clear.

A *zero matrix* is a matrix that consists entirely of zeros. Obviously, the product of any matrix multiplied by a zero matrix is zero.

EXAMPLE A.4.4.

$$\begin{bmatrix} 0 & 0 \\ 0 & 0 \\ 0 & 0 \end{bmatrix}, \quad \begin{bmatrix} 0 \\ 0 \\ 0 \\ 0 \end{bmatrix}.$$

□

Often a zero matrix is denoted by 0 where the dimension of the matrix, and the fact that it is a matrix rather than a scalar, must be inferred from the context.

A.5 Linear dependence and rank

Consider the matrix

$$A = \begin{bmatrix} 1 & 2 & 5 & 1 \\ 2 & 2 & 10 & 6 \\ 3 & 4 & 15 & 1 \end{bmatrix}.$$

Note that each column of A can be viewed as a vector. The *column space* of A, denoted $C(A)$, is the collection of all vectors that can be written as *a linear combination of the columns of* A. In other words, $C(A)$ is the set of all vectors that can be written as

$$\lambda_1 \begin{bmatrix} 1 \\ 2 \\ 3 \end{bmatrix} + \lambda_2 \begin{bmatrix} 2 \\ 2 \\ 4 \end{bmatrix} + \lambda_3 \begin{bmatrix} 5 \\ 10 \\ 15 \end{bmatrix} + \lambda_4 \begin{bmatrix} 1 \\ 6 \\ 1 \end{bmatrix} = A \begin{bmatrix} \lambda_1 \\ \lambda_2 \\ \lambda_3 \\ \lambda_4 \end{bmatrix} = A\lambda$$

for some vector $\lambda = (\lambda_1, \lambda_2, \lambda_3, \lambda_4)'$.

The columns of any matrix A are *linearly dependent* if they contain redundant information. Specifically, let x be some vector in $C(A)$. The columns of A are linearly dependent if we can find two distinct vectors λ and γ such that $x = A\lambda$ and $x = A\gamma$. Thus two distinct linear combinations of the columns of A give rise to the same vector x. Note that $\lambda \neq \gamma$ because λ and γ are distinct. Note also that, using a distributive property of matrix multiplication, $A(\lambda - \gamma) = A\lambda - A\gamma = 0$, where $\lambda - \gamma \neq 0$. This condition is frequently used as an alternative definition for linear dependence, i.e., the columns of A are linearly dependent if there exists a vector $\delta \neq 0$ such that $A\delta = 0$. If the columns of A are not linearly dependent, they are *linearly independent*.

EXAMPLE A.5.1. Observe that the example matrix A given at the beginning of the section has

$$\begin{bmatrix} 1 & 2 & 5 & 1 \\ 2 & 2 & 10 & 6 \\ 3 & 4 & 15 & 1 \end{bmatrix} \begin{bmatrix} 5 \\ 0 \\ -1 \\ 0 \end{bmatrix} = \begin{bmatrix} 0 \\ 0 \\ 0 \end{bmatrix},$$

so the columns of A are linearly dependent. □

The *rank* of A is the smallest number of columns of A that can generate $C(A)$. It is also the maximum number of linearly independent columns in A.

EXAMPLE A.5.2. The matrix

$$A = \begin{bmatrix} 1 & 2 & 5 & 1 \\ 2 & 2 & 10 & 6 \\ 3 & 4 & 15 & 1 \end{bmatrix}$$

has rank 3 because the columns

$$\begin{bmatrix} 1 \\ 2 \\ 3 \end{bmatrix}, \begin{bmatrix} 2 \\ 2 \\ 4 \end{bmatrix}, \begin{bmatrix} 1 \\ 6 \\ 1 \end{bmatrix}$$

generate $C(A)$. We saw in Example A.5.1 that the column $(5, 10, 15)'$ was redundant. None of the other three columns are redundant; they are linearly independent. In other words, the only way to get

$$\begin{bmatrix} 1 & 2 & 1 \\ 2 & 2 & 6 \\ 3 & 4 & 1 \end{bmatrix} \delta = \begin{bmatrix} 0 \\ 0 \\ 0 \end{bmatrix}$$

is to take $\delta = (0, 0, 0)'$. □

A.6 Inverse matrices

The *inverse* of a square matrix A is the matrix A^{-1} such that

$$AA^{-1} = A^{-1}A = I.$$

The inverse of A exists only if the columns of A are linearly independent. Typically, it is difficult to find inverses without the aid of a computer. For a 2×2 matrix

$$A = \begin{bmatrix} a_{11} & a_{12} \\ a_{21} & a_{22} \end{bmatrix},$$

the inverse is given by

$$A^{-1} = \frac{1}{a_{11}a_{22} - a_{12}a_{21}} \begin{bmatrix} a_{22} & -a_{12} \\ -a_{21} & a_{11} \end{bmatrix}. \qquad \text{(A.6.1)}$$

To confirm that this is correct, multiply AA^{-1} to see that it gives the identity matrix. Moderately complicated formulae exist for computing the inverse of 3×3 matrices. Inverses of larger matrices become very difficult to compute by hand. Of course computers are ideally suited for finding such things.

One use for inverse matrices is in solving systems of equations.

EXAMPLE A.6.1. Consider the system of equations

$$\begin{aligned} 2x + 4y &= 20 \\ 3x + 4y &= 10. \end{aligned}$$

We can write this in matrix form as

$$\begin{bmatrix} 2 & 4 \\ 3 & 4 \end{bmatrix} \begin{bmatrix} x \\ y \end{bmatrix} = \begin{bmatrix} 20 \\ 10 \end{bmatrix}.$$

Multiplying on the left by the inverse of the coefficient matrix gives

$$\begin{bmatrix} 2 & 4 \\ 3 & 4 \end{bmatrix}^{-1} \begin{bmatrix} 2 & 4 \\ 3 & 4 \end{bmatrix} \begin{bmatrix} x \\ y \end{bmatrix} = \begin{bmatrix} 2 & 4 \\ 3 & 4 \end{bmatrix}^{-1} \begin{bmatrix} 20 \\ 10 \end{bmatrix}.$$

Using the definition of the inverse on the left-hand side of the equality and the formula in (A.6.1) on the right-hand side gives

$$\begin{bmatrix} 1 & 0 \\ 0 & 1 \end{bmatrix} \begin{bmatrix} x \\ y \end{bmatrix} = \begin{bmatrix} -1 & 1 \\ 3/4 & -1/2 \end{bmatrix} \begin{bmatrix} 20 \\ 10 \end{bmatrix}$$

or

$$\begin{bmatrix} x \\ y \end{bmatrix} = \begin{bmatrix} -10 \\ 10 \end{bmatrix}.$$

Thus $(x, y) = (-10, 10)$ is the solution for the two equations, i.e., $2(-10) + 4(10) = 20$ and $3(-10) + 4(10) = 10$. □

More generally a system of equations, say,

$$\begin{aligned} a_{11}\,y_1 + a_{12}\,y_2 + a_{13}\,y_3 &= c_1 \\ a_{21}\,y_1 + a_{22}\,y_2 + a_{23}\,y_3 &= c_2 \\ a_{31}\,y_1 + a_{32}\,y_2 + a_{33}\,y_3 &= c_3 \end{aligned}$$

in which the a_{ij}s and c_is are known and the y_is are variables, can be written in matrix form as

$$\begin{bmatrix} a_{11} & a_{12} & a_{13} \\ a_{21} & a_{22} & a_{23} \\ a_{31} & a_{32} & a_{33} \end{bmatrix} \begin{bmatrix} y_1 \\ y_2 \\ y_3 \end{bmatrix} = \begin{bmatrix} c_1 \\ c_2 \\ c_3 \end{bmatrix}$$

or

$$AY = C.$$

To find Y simply observe that $AY = C$ implies $A^{-1}AY = A^{-1}C$ and $Y = A^{-1}C$. Of course this argument assumes that A^{-1} exists, which is not always the case. Moreover, the procedure obviously extends to larger sets of equations.

On a computer, there are better ways of finding solutions to systems of equations than finding the inverse of a matrix. In fact, inverses are often found by solving systems of equations. For example, in a 3×3 case the first column of A^{-1} can be found as the solution to

$$\begin{bmatrix} a_{11} & a_{12} & a_{13} \\ a_{21} & a_{22} & a_{23} \\ a_{31} & a_{32} & a_{33} \end{bmatrix} \begin{bmatrix} y_1 \\ y_2 \\ y_3 \end{bmatrix} = \begin{bmatrix} 1 \\ 0 \\ 0 \end{bmatrix}.$$

For a special type of square matrix, called an *orthogonal matrix*, the transpose is also the inverse. In other words, a square matrix P is an orthogonal matrix if

$$P'P = I = PP'.$$

To establish that P is orthogonal, it is enough to show either that $P'P = I$ or that $PP' = I$. Orthogonal matrices are particularly useful in discussions of eigenvalues and principal component regression.

A.7 A list of useful properties

The following proposition summarizes many of the key properties of matrices and the operations performed on them.

Proposition A.7.1. Let A, B, and C be matrices of appropriate dimensions and let λ be a scalar.

$$\begin{aligned} A + B &= B + A \\ (A + B) + C &= A + (B + C) \\ (AB)C &= A(BC) \\ C(A + B) &= CA + CB \\ \lambda(A + B) &= \lambda A + \lambda B \\ (A')' &= A \\ (A + B)' &= A' + B' \\ (AB)' &= B'A' \\ (A^{-1})^{-1} &= A \\ (A')^{-1} &= (A^{-1})' \\ (AB)^{-1} &= B^{-1}A^{-1}. \end{aligned}$$

The last equality only holds when A and B both have inverses. The second to the last property implies that the inverse of a symmetric matrix is symmetric because then $A^{-1} = (A')^{-1} = (A^{-1})'$. This is a very important property.

A.8 Eigenvalues and eigenvectors

Let A be a square matrix. A scalar ϕ is an eigenvalue of A and $x \neq 0$ is an eigenvector for A corresponding to ϕ if

$$Ax = \phi x.$$

EXAMPLE A.8.1. Consider the matrix

$$A = \begin{bmatrix} 3 & 1 & -1 \\ 1 & 3 & -1 \\ -1 & -1 & 5 \end{bmatrix}.$$

The value 3 is an eigenvalue and any nonzero multiple of the vector $(1, 1, 1)'$ is a corresponding eigenvector. For example,

$$\begin{bmatrix} 3 & 1 & -1 \\ 1 & 3 & -1 \\ -1 & -1 & 5 \end{bmatrix} \begin{bmatrix} 1 \\ 1 \\ 1 \end{bmatrix} = \begin{bmatrix} 3 \\ 3 \\ 3 \end{bmatrix} = 3 \begin{bmatrix} 1 \\ 1 \\ 1 \end{bmatrix}.$$

Similarly, if we consider a multiple, say, $4(1, 1, 1)'$,

$$\begin{bmatrix} 3 & 1 & -1 \\ 1 & 3 & -1 \\ -1 & -1 & 5 \end{bmatrix} \begin{bmatrix} 4 \\ 4 \\ 4 \end{bmatrix} = \begin{bmatrix} 12 \\ 12 \\ 12 \end{bmatrix} = 3 \begin{bmatrix} 4 \\ 4 \\ 4 \end{bmatrix}.$$

The value 2 is also an eigenvalue with eigenvectors that are nonzero multiples of $(1, -1, 0)'$.

$$\begin{bmatrix} 3 & 1 & -1 \\ 1 & 3 & -1 \\ -1 & -1 & 5 \end{bmatrix} \begin{bmatrix} 1 \\ -1 \\ 0 \end{bmatrix} = \begin{bmatrix} 2 \\ -2 \\ 0 \end{bmatrix} = 2 \begin{bmatrix} 1 \\ -1 \\ 0 \end{bmatrix}.$$

Finally, 6 is an eigenvalue with eigenvectors that are nonzero multiples of $(1, 1, -2)'$. □

Proposition A.8.2. Let A be a symmetric matrix, then for a diagonal matrix $D(\phi_i)$ consisting of eigenvalues there exists an orthogonal matrix P whose columns are corresponding eigenvectors such that

$$A = PD(\phi_i)P'.$$

EXAMPLE A.8.3. Consider again the matrix

$$A = \begin{bmatrix} 3 & 1 & -1 \\ 1 & 3 & -1 \\ -1 & -1 & 5 \end{bmatrix}.$$

In writing $A = PD(\phi_i)P'$, the diagonal matrix is

$$D(\phi_i) = \begin{bmatrix} 3 & 0 & 0 \\ 0 & 2 & 0 \\ 0 & 0 & 6 \end{bmatrix}.$$

The orthogonal matrix is

$$P = \begin{bmatrix} \frac{1}{\sqrt{3}} & \frac{1}{\sqrt{2}} & \frac{1}{\sqrt{6}} \\ \frac{1}{\sqrt{3}} & \frac{-1}{\sqrt{2}} & \frac{1}{\sqrt{6}} \\ \frac{1}{\sqrt{3}} & 0 & \frac{-2}{\sqrt{6}} \end{bmatrix}.$$

We leave it to the reader to verify that $PD(\phi_i)P' = A$ and that $P'P = I$.

Note that the columns of P are multiples of the vectors identified as eigenvectors in Example A.8.1; hence the columns of P are also eigenvectors. The multiples of the eigenvectors were chosen so that $PP' = I$ and $P'P = I$. Moreover, the first column of P is an eigenvector corresponding to 3, which is the first eigenvalue listed in $D(\phi_i)$. Similarly, the second column of P is an eigenvector corresponding to 2 and the third column corresponds to the third listed eigenvalue, 6.

With a 3×3 matrix A having three *distinct* eigenvalues, any matrix P with eigenvectors for columns would have $P'P$ a diagonal matrix, but the multiples of the eigenvectors must be chosen so that the diagonal entries of $P'P$ are all 1. □

EXAMPLE A.8.4. Consider the matrix

$$B = \begin{bmatrix} 5 & -1 & -1 \\ -1 & 5 & -1 \\ -1 & -1 & 5 \end{bmatrix}.$$

This matrix is closely related to the matrix in Example A.8.1. The matrix B has 3 as an eigenvalue with corresponding eigenvectors that are multiples of $(1, 1, 1)'$, just like the matrix A. Once again 6 is an eigenvalue with corresponding eigenvector $(1, 1, -2)'$ and once again $(1, -1, 0)'$ is an eigenvector, but now, unlike A, $(1, -1, 0)$ also corresponds to the eigenvalue 6. We leave it to the reader to verify these facts. The point is that in this matrix, 6 is an eigenvalue that has two linearly independent eigenvectors. In such cases, any nonzero linear combination of the two eigenvectors is also an eigenvector. For example, it is easy to see that

$$3\begin{bmatrix} 1 \\ -1 \\ 0 \end{bmatrix} + 2\begin{bmatrix} 1 \\ 1 \\ -2 \end{bmatrix} = \begin{bmatrix} 5 \\ -1 \\ -4 \end{bmatrix}$$

is an eigenvector corresponding to the eigenvalue 6.

To write $B = PD(\phi)P'$ as in Proposition A.8.2, $D(\phi)$ has 3, 6, and 6 down the diagonal and one choice of P is that given in Example A.8.3. However, because one of the eigenvalues occurs more than once in the diagonal matrix, there are many choices for P. □

Generally, if we need eigenvalues or eigenvectors we get a computer to find them for us.

Two frequently used functions of a square matrix are the determinant and the trace.

Definition A.8.5.

a) The determinant of a square matrix is the product of the eigenvalues of the matrix.

b) The trace of a square matrix is the sum of the eigenvalues of the matrix.

In fact, one can show that the trace of a square matrix also equals the sum of the diagonal elements of that matrix.

Appendix B: Tables

B.1 Tables of the t distribution

TABLE B.1. Percentage points of the t distribution

	α levels							
Two-sided	.20	.10	.05	.04	.02	.01	.002	.001
One-sided	.10	.05	.025	.02	.01	.005	.001	.0005
	Percentiles							
df	0.90	0.95	0.975	0.98	0.99	0.995	0.999	0.9995
1	3.078	6.314	12.7062	15.8946	31.8206	63.6570	318.317	636.607
2	1.886	2.920	4.3027	4.8487	6.9646	9.9248	22.327	31.598
3	1.638	2.353	3.1824	3.4819	4.5407	5.8409	10.215	12.924
4	1.533	2.132	2.7764	2.9985	3.7470	4.6041	7.173	8.610
5	1.476	2.015	2.5706	2.7565	3.3649	4.0322	5.893	6.869
6	1.440	1.943	2.4469	2.6122	3.1427	3.7075	5.208	5.959
7	1.415	1.895	2.3646	2.5168	2.9980	3.4995	4.785	5.408
8	1.397	1.860	2.3060	2.4490	2.8965	3.3554	4.501	5.041
9	1.383	1.833	2.2622	2.3984	2.8214	3.2499	4.297	4.781
10	1.372	1.812	2.2281	2.3593	2.7638	3.1693	4.144	4.587
11	1.363	1.796	2.2010	2.3281	2.7181	3.1058	4.025	4.437
12	1.356	1.782	2.1788	2.3027	2.6810	3.0546	3.930	4.318
13	1.350	1.771	2.1604	2.2816	2.6503	3.0123	3.852	4.221
14	1.345	1.761	2.1448	2.2638	2.6245	2.9769	3.787	4.140
15	1.341	1.753	2.1315	2.2485	2.6025	2.9467	3.733	4.073
16	1.337	1.746	2.1199	2.2354	2.5835	2.9208	3.686	4.015
17	1.333	1.740	2.1098	2.2239	2.5669	2.8982	3.646	3.965
18	1.330	1.734	2.1009	2.2137	2.5524	2.8784	3.611	3.922
19	1.328	1.729	2.0930	2.2047	2.5395	2.8610	3.579	3.883
20	1.325	1.725	2.0860	2.1967	2.5280	2.8453	3.552	3.850
21	1.323	1.721	2.0796	2.1894	2.5176	2.8314	3.527	3.819
22	1.321	1.717	2.0739	2.1829	2.5083	2.8188	3.505	3.792
23	1.319	1.714	2.0687	2.1769	2.4999	2.8073	3.485	3.768
24	1.318	1.711	2.0639	2.1716	2.4922	2.7969	3.467	3.745
25	1.316	1.708	2.0595	2.1666	2.4851	2.7874	3.450	3.725
26	1.315	1.706	2.0555	2.1620	2.4786	2.7787	3.435	3.707
27	1.314	1.703	2.0518	2.1578	2.4727	2.7707	3.421	3.690
28	1.313	1.701	2.0484	2.1539	2.4671	2.7633	3.408	3.674
29	1.311	1.699	2.0452	2.1503	2.4620	2.7564	3.396	3.659
30	1.310	1.697	2.0423	2.1470	2.4573	2.7500	3.385	3.646
31	1.309	1.696	2.0395	2.1438	2.4528	2.7441	3.375	3.633
32	1.309	1.694	2.0369	2.1409	2.4487	2.7385	3.365	3.622
33	1.308	1.692	2.0345	2.1382	2.4448	2.7333	3.356	3.611
34	1.307	1.691	2.0323	2.1356	2.4412	2.7284	3.348	3.601
35	1.306	1.690	2.0301	2.1332	2.4377	2.7238	3.340	3.591

TABLE B.2. Percentage points of the t distribution

	α levels							
Two-sided	.20	.10	.05	.04	.02	.01	.002	.001
One-sided	.10	.05	.025	.02	.01	.005	.001	.0005
	Percentiles							
df	0.90	0.95	0.975	0.98	0.99	0.995	0.999	0.9995
36	1.306	1.688	2.0281	2.1309	2.4345	2.7195	3.333	3.582
37	1.305	1.687	2.0262	2.1287	2.4314	2.7154	3.326	3.574
38	1.304	1.686	2.0244	2.1267	2.4286	2.7116	3.319	3.566
39	1.304	1.685	2.0227	2.1247	2.4258	2.7079	3.313	3.558
40	1.303	1.684	2.0211	2.1229	2.4233	2.7045	3.307	3.551
41	1.303	1.683	2.0196	2.1212	2.4208	2.7012	3.301	3.544
42	1.302	1.682	2.0181	2.1195	2.4185	2.6981	3.296	3.538
43	1.302	1.681	2.0167	2.1179	2.4163	2.6951	3.291	3.532
44	1.301	1.680	2.0154	2.1164	2.4142	2.6923	3.286	3.526
45	1.301	1.679	2.0141	2.1150	2.4121	2.6896	3.281	3.520
46	1.300	1.679	2.0129	2.1136	2.4102	2.6870	3.277	3.515
47	1.300	1.678	2.0117	2.1123	2.4083	2.6846	3.273	3.510
48	1.299	1.677	2.0106	2.1111	2.4066	2.6822	3.269	3.505
49	1.299	1.677	2.0096	2.1099	2.4049	2.6800	3.265	3.500
50	1.299	1.676	2.0086	2.1087	2.4033	2.6778	3.261	3.496
51	1.298	1.675	2.0076	2.1076	2.4017	2.6757	3.258	3.492
52	1.298	1.675	2.0067	2.1066	2.4002	2.6737	3.255	3.488
53	1.298	1.674	2.0058	2.1055	2.3988	2.6718	3.251	3.484
54	1.297	1.674	2.0049	2.1046	2.3974	2.6700	3.248	3.480
55	1.297	1.673	2.0041	2.1036	2.3961	2.6682	3.245	3.476
56	1.297	1.673	2.0033	2.1027	2.3948	2.6665	3.242	3.473
57	1.297	1.672	2.0025	2.1018	2.3936	2.6649	3.239	3.470
58	1.296	1.672	2.0017	2.1010	2.3924	2.6633	3.237	3.466
59	1.296	1.671	2.0010	2.1002	2.3912	2.6618	3.234	3.463
60	1.296	1.671	2.0003	2.0994	2.3902	2.6604	3.232	3.460
70	1.294	1.667	1.9944	2.0927	2.3808	2.6480	3.211	3.435
80	1.292	1.664	1.9901	2.0878	2.3739	2.6387	3.195	3.416
90	1.291	1.662	1.9867	2.0840	2.3685	2.6316	3.183	3.402
100	1.290	1.660	1.9840	2.0809	2.3642	2.6259	3.174	3.391
110	1.289	1.659	1.9818	2.0784	2.3607	2.6213	3.166	3.381
120	1.289	1.658	1.9799	2.0763	2.3578	2.6174	3.160	3.373
150	1.287	1.655	1.9759	2.0718	2.3515	2.6090	3.145	3.357
200	1.286	1.653	1.9719	2.0672	2.3451	2.6006	3.131	3.340
250	1.285	1.651	1.9695	2.0645	2.3414	2.5956	3.123	3.330
300	1.284	1.650	1.9679	2.0627	2.3388	2.5923	3.118	3.323
350	1.284	1.649	1.9668	2.0614	2.3371	2.5900	3.114	3.319
400	1.284	1.649	1.9659	2.0605	2.3357	2.5882	3.111	3.315
∞	1.282	1.645	1.9600	2.0537	2.3263	2.5758	3.090	3.291

B.2 Tables of the χ^2 distribution

TABLE B.3. Percentage points of the χ^2 distribution

	α levels							
Two-sided	0.002	0.01	0.02	0.04	0.05	0.10	0.20	0.40
One-sided	0.001	0.005	0.01	0.02	0.025	0.05	0.10	0.20
	Percentiles							
df	0.001	0.005	0.01	0.02	0.025	0.05	0.10	0.20
1	0.000	0.000	0.000	0.001	0.001	0.004	0.016	0.064
2	0.002	0.010	0.020	0.040	0.051	0.103	0.211	0.446
3	0.024	0.072	0.115	0.185	0.216	0.352	0.584	1.005
4	0.091	0.207	0.297	0.429	0.484	0.711	1.064	1.649
5	0.210	0.412	0.554	0.752	0.831	1.145	1.610	2.343
6	0.381	0.676	0.872	1.134	1.237	1.635	2.204	3.070
7	0.598	0.989	1.239	1.564	1.690	2.167	2.833	3.822
8	0.857	1.344	1.646	2.032	2.180	2.733	3.490	4.594
9	1.152	1.735	2.088	2.532	2.700	3.325	4.168	5.380
10	1.479	2.156	2.558	3.059	3.247	3.940	4.865	6.179
11	1.834	2.603	3.053	3.609	3.816	4.575	5.578	6.989
12	2.214	3.074	3.571	4.178	4.404	5.226	6.304	7.807
13	2.617	3.565	4.107	4.765	5.009	5.892	7.042	8.634
14	3.041	4.075	4.660	5.368	5.629	6.571	7.790	9.467
15	3.483	4.601	5.229	5.985	6.262	7.261	8.547	10.307
16	3.942	5.142	5.812	6.614	6.908	7.962	9.312	11.152
17	4.416	5.697	6.408	7.255	7.564	8.672	10.085	12.002
18	4.905	6.265	7.015	7.906	8.231	9.390	10.865	12.857
19	5.407	6.844	7.633	8.567	8.907	10.117	11.651	13.716
20	5.921	7.434	8.260	9.237	9.591	10.851	12.443	14.578
21	6.447	8.034	8.897	9.915	10.283	11.591	13.240	15.445
22	6.983	8.643	9.542	10.600	10.982	12.338	14.041	16.314
23	7.529	9.260	10.196	11.293	11.689	13.091	14.848	17.187
24	8.085	9.886	10.856	11.992	12.401	13.848	15.659	18.062
25	8.649	10.520	11.524	12.697	13.120	14.611	16.473	18.940
26	9.222	11.160	12.198	13.409	13.844	15.379	17.292	19.820
27	9.803	11.808	12.879	14.125	14.573	16.151	18.114	20.703
28	10.391	12.461	13.565	14.847	15.308	16.928	18.939	21.588
29	10.986	13.121	14.256	15.574	16.047	17.708	19.768	22.475
30	11.588	13.787	14.953	16.306	16.791	18.493	20.599	23.364
31	12.196	14.458	15.655	17.042	17.539	19.281	21.434	24.255
32	12.811	15.134	16.362	17.783	18.291	20.072	22.271	25.148
33	13.431	15.815	17.074	18.527	19.047	20.867	23.110	26.042
34	14.057	16.501	17.789	19.275	19.806	21.664	23.952	26.938
35	14.688	17.192	18.509	20.027	20.569	22.465	24.797	27.836

TABLE B.4. Percentage points of the χ^2 distribution

	α levels							
Two-sided	.40	.20	.10	.05	.04	.02	.01	.002
One-sided	.20	.10	.05	.025	.02	.01	.005	.001
	Percentiles							
df	0.80	0.90	0.95	0.975	0.98	0.99	0.995	0.999
1	1.642	2.706	3.841	5.024	5.412	6.635	7.879	10.828
2	3.219	4.605	5.991	7.378	7.824	9.210	10.597	13.816
3	4.642	6.251	7.815	9.348	9.837	11.345	12.838	16.266
4	5.989	7.779	9.488	11.143	11.668	13.277	14.860	18.467
5	7.289	9.236	11.070	12.833	13.388	15.086	16.750	20.515
6	8.558	10.645	12.592	14.449	15.033	16.812	18.548	22.458
7	9.803	12.017	14.067	16.013	16.622	18.475	20.278	24.322
8	11.030	13.362	15.507	17.535	18.168	20.090	21.955	26.125
9	12.242	14.684	16.919	19.023	19.679	21.666	23.589	27.877
10	13.442	15.987	18.307	20.483	21.161	23.209	25.188	29.588
11	14.631	17.275	19.675	21.920	22.618	24.725	26.757	31.264
12	15.812	18.549	21.026	23.337	24.054	26.217	28.300	32.910
13	16.985	19.812	22.362	24.736	25.471	27.688	29.819	34.528
14	18.151	21.064	23.685	26.119	26.873	29.141	31.319	36.124
15	19.311	22.307	24.996	27.488	28.259	30.578	32.801	37.697
16	20.465	23.542	26.296	28.845	29.633	32.000	34.267	39.254
17	21.615	24.769	27.587	30.191	30.995	33.409	35.718	40.789
18	22.760	25.989	28.869	31.526	32.346	34.805	37.156	42.312
19	23.900	27.204	30.143	32.852	33.687	36.191	38.582	43.819
20	25.038	28.412	31.410	34.170	35.020	37.566	39.997	45.315
21	26.171	29.615	32.671	35.479	36.343	38.932	41.401	46.797
22	27.301	30.813	33.924	36.781	37.660	40.290	42.796	48.270
23	28.429	32.007	35.172	38.076	38.968	41.638	44.181	49.726
24	29.553	33.196	36.415	39.364	40.270	42.980	45.559	51.179
25	30.675	34.382	37.653	40.647	41.566	44.314	46.928	52.622
26	31.795	35.563	38.885	41.923	42.856	45.642	48.290	54.054
27	32.912	36.741	40.113	43.195	44.140	46.963	49.645	55.477
28	34.027	37.916	41.337	44.461	45.419	48.278	50.994	56.893
29	35.139	39.087	42.557	45.722	46.693	49.588	52.336	58.303
30	36.250	40.256	43.773	46.979	47.962	50.892	53.672	59.703
31	37.359	41.422	44.985	48.232	49.226	52.192	55.003	61.100
32	38.466	42.585	46.194	49.480	50.487	53.486	56.328	62.486
33	39.572	43.745	47.400	50.725	51.743	54.775	57.648	63.868
34	40.676	44.903	48.602	51.966	52.995	56.061	58.964	65.246
35	41.778	46.059	49.802	53.204	54.244	57.342	60.275	66.622

TABLE B.5. Percentage points of the χ^2 distribution

	α levels							
Two-sided	0.002	0.01	0.02	0.04	0.05	0.10	0.20	0.40
One-sided	0.001	0.005	0.01	0.02	0.025	0.05	0.10	0.20
	Percentiles							
df	0.001	0.005	0.01	0.02	0.025	0.05	0.10	0.20
36	15.32	17.887	19.233	20.783	21.336	23.269	25.64	28.74
37	15.96	18.586	19.960	21.542	22.106	24.075	26.49	29.64
38	16.61	19.289	20.691	22.304	22.878	24.884	27.34	30.54
39	17.26	19.996	21.426	23.069	23.654	25.695	28.20	31.44
40	17.92	20.707	22.164	23.838	24.433	26.509	29.05	32.34
41	18.58	21.421	22.906	24.609	25.215	27.326	29.91	33.25
42	19.24	22.138	23.650	25.383	25.999	28.144	30.76	34.16
43	19.91	22.859	24.398	26.159	26.785	28.965	31.62	35.06
44	20.58	23.584	25.148	26.939	27.575	29.787	32.49	35.97
45	21.25	24.311	25.901	27.720	28.366	30.612	33.35	36.88
46	21.93	25.041	26.657	28.504	29.160	31.439	34.22	37.80
47	22.61	25.775	27.416	29.291	29.956	32.268	35.08	38.71
48	23.30	26.511	28.177	30.080	30.755	33.098	35.95	39.62
49	23.98	27.249	28.941	30.871	31.555	33.930	36.82	40.53
50	24.67	27.991	29.707	31.664	32.357	34.764	37.69	41.45
51	25.37	28.735	30.475	32.459	33.162	35.600	38.56	42.36
52	26.06	29.481	31.246	33.256	33.968	36.437	39.43	43.28
53	26.76	30.230	32.018	34.055	34.776	37.276	40.31	44.20
54	27.47	30.981	32.793	34.856	35.586	38.116	41.18	45.12
55	28.17	31.735	33.570	35.659	36.398	38.958	42.06	46.04
56	28.88	32.490	34.350	36.464	37.212	39.801	42.94	46.96
57	29.59	33.248	35.131	37.270	38.027	40.646	43.82	47.88
58	30.30	34.008	35.913	38.078	38.843	41.492	44.70	48.80
59	31.02	34.770	36.698	38.888	39.662	42.339	45.58	49.72
60	31.74	35.535	37.485	39.699	40.482	43.188	46.46	50.64
70	39.04	43.275	45.442	47.893	48.758	51.739	55.33	59.90
80	46.52	51.172	53.540	56.213	57.153	60.391	64.28	69.21
90	54.16	59.196	61.754	64.635	65.647	69.126	73.29	78.56
100	61.92	67.328	70.065	73.142	74.222	77.930	82.36	87.94
110	69.79	75.550	78.458	81.723	82.867	86.792	91.47	97.36
120	77.76	83.852	86.923	90.367	91.573	95.705	100.62	106.81
150	102.11	109.142	112.668	116.608	117.984	122.692	128.28	135.26
200	143.84	152.241	156.432	161.100	162.728	168.279	174.84	183.00
250	186.55	196.161	200.939	206.249	208.098	214.392	221.81	231.01
300	229.96	240.663	245.972	251.864	253.912	260.878	269.07	279.21
350	273.90	285.608	291.406	297.831	300.064	307.648	316.55	327.56
400	318.26	330.903	337.155	344.078	346.482	354.641	364.21	376.02

TABLE B.6. Percentage points of the χ^2 distribution

	α levels							
Two-sided	.40	.20	.10	.05	.04	.02	.01	.002
One-sided	.20	.10	.05	.025	.02	.01	.005	.001
	Percentiles							
df	0.80	0.90	0.95	0.975	0.98	0.99	0.995	0.999
36	42.88	47.212	50.998	54.437	55.489	58.619	61.58	67.99
37	43.98	48.363	52.192	55.668	56.731	59.893	62.89	69.35
38	45.08	49.513	53.384	56.896	57.969	61.163	64.18	70.71
39	46.17	50.660	54.572	58.120	59.204	62.429	65.48	72.06
40	47.27	51.805	55.759	59.342	60.437	63.691	66.77	73.41
41	48.36	52.949	56.942	60.561	61.665	64.950	68.05	74.75
42	49.46	54.090	58.124	61.777	62.892	66.207	69.34	76.09
43	50.55	55.230	59.303	62.990	64.115	67.459	70.62	77.42
44	51.64	56.369	60.481	64.201	65.337	68.709	71.89	78.75
45	52.73	57.505	61.656	65.410	66.555	69.957	73.17	80.08
46	53.82	58.640	62.829	66.616	67.771	71.201	74.44	81.40
47	54.91	59.774	64.001	67.820	68.985	72.443	75.70	82.72
48	55.99	60.907	65.171	69.022	70.196	73.682	76.97	84.03
49	57.08	62.038	66.339	70.222	71.406	74.919	78.23	85.35
50	58.16	63.167	67.505	71.420	72.613	76.154	79.49	86.66
51	59.25	64.295	68.669	72.616	73.818	77.386	80.75	87.97
52	60.33	65.422	69.832	73.810	75.021	78.616	82.00	89.27
53	61.41	66.548	70.993	75.002	76.222	79.843	83.25	90.57
54	62.50	67.673	72.153	76.192	77.422	81.070	84.50	91.88
55	63.58	68.796	73.312	77.381	78.619	82.292	85.75	93.17
56	64.66	69.919	74.469	78.568	79.815	83.515	87.00	94.47
57	65.74	71.040	75.624	79.752	81.009	84.733	88.24	95.75
58	66.82	72.160	76.777	80.935	82.200	85.949	89.47	97.03
59	67.90	73.279	77.931	82.118	83.392	87.167	90.72	98.34
60	68.97	74.397	79.082	83.298	84.581	88.381	91.96	99.62
70	79.72	85.527	90.531	95.023	96.387	100.424	104.21	112.31
80	90.41	96.578	101.879	106.628	108.069	112.328	116.32	124.84
90	101.05	107.565	113.145	118.135	119.648	124.115	128.30	137.19
100	111.67	118.499	124.343	129.563	131.144	135.811	140.18	149.48
110	122.25	129.385	135.480	140.917	142.562	147.416	151.95	161.59
120	132.81	140.233	146.567	152.211	153.918	158.950	163.65	173.62
150	164.35	172.580	179.579	185.798	187.675	193.202	198.35	209.22
200	216.61	226.022	233.997	241.062	243.192	249.455	255.28	267.62
250	268.60	279.052	287.884	295.694	298.045	304.951	311.37	324.93
300	320.40	331.787	341.393	349.870	352.419	359.896	366.83	381.34
350	372.05	384.305	394.624	403.720	406.454	414.466	421.89	437.43
400	423.59	436.647	447.628	457.298	460.201	468.707	476.57	492.99

B.3 Tables of the W' statistic

TABLE B.7. Percentiles of the W' statistic

n	.01	.05	n	.01	.05
5	0.69	0.77	36	0.91	0.940
6	0.70	0.79	38	0.915	0.942
7	0.72	0.81	40	0.918	0.946
8	0.75	0.82	45	0.928	0.951
9	0.75	0.83	50	0.931	0.952
10	0.78	0.83	55	0.938	0.958
11	0.79	0.85	60	0.943	0.961
12	0.79	0.86	65	0.945	0.961
13	0.81	0.870	70	0.953	0.966
14	0.82	0.877	75	0.954	0.968
15	0.82	0.883	80	0.957	0.970
16	0.83	0.886	85	0.958	0.970
17	0.84	0.896	90	0.960	0.972
18	0.85	0.896	95	0.961	0.972
19	0.86	0.902	100	0.962	0.974
20	0.86	0.902	120	0.970	0.978
22	0.87	0.910	140	0.973	0.981
24	0.88	0.915	160	0.976	0.983
26	0.89	0.923	180	0.978	0.985
28	0.89	0.924	200	0.981	0.986
30	0.89	0.928	250	0.984	0.988
32	0.90	0.933	300	0.987	0.991
34	0.91	0.936			

This table was obtained by taking the mean of ten estimates of the percentile each based on a sample of 500 observations. Estimates with standard errors of about .002 or less are reported to three decimal places. The estimates reported with two decimal places have standard errors between about .002 and .008.

B.4 Tables of orthogonal polynomials

TABLE B.8. Orthogonal polynomial contrasts

$t = 3$			$t = 4$			$t = 5$		
L	Q	C	L	Q	C	L	Q	C
−1	1		−3	1	−1	−2	2	−1
0	−2		−1	−1	3	−1	−1	2
1	1		1	−1	−3	0	−2	0
			3	1	1	1	−1	−2
						2	2	1
$t = 6$			$t = 7$			$t = 8$		
L	Q	C	L	Q	C	L	Q	C
−5	5	−5	3	5	−1	−7	7	−7
−3	−1	7	2	0	1	−5	1	5
−1	−4	4	1	−3	1	−3	−3	7
1	−4	−4	0	−4	0	−1	−5	3
3	−1	−7	1	−3	−1	1	−5	−3
5	5	5	2	0	−1	3	−3	−7
			3	5	1	5	1	−5
						7	7	7
$t = 9$			$t = 10$			$t = 11$		
L	Q	C	L	Q	C	L	Q	C
−4	28	−14	−9	6	−42	−5	15	−30
−3	7	7	−7	2	14	−4	6	6
−2	−8	13	−5	−1	35	−3	−1	22
−1	−17	9	−3	−3	31	−2	−6	23
0	−20	0	−1	−4	12	−1	−9	14
1	−17	−9	1	−4	−12	0	−10	0
2	−8	−13	3	−3	−31	1	−9	−14
3	7	−7	5	−1	−35	2	−6	−23
4	28	14	7	2	−14	3	−1	−22
			9	6	42	4	6	−6
						5	15	30

L: linear, Q: quadratic, C: cubic

B.5 Tables of the Studentized range

TABLE B.9. $Q(.95, r, dfE)$

dfE	r = 2	3	4	5	6	7	8	9	10	11
1	18.0	27.0	32.8	37.1	40.4	43.1	45.4	47.4	49.1	50.6
2	6.09	8.33	9.80	10.88	11.74	12.44	13.03	13.54	13.99	14.39
3	4.50	5.91	6.83	7.50	8.04	8.48	8.85	9.18	9.46	9.72
4	3.93	5.04	5.76	6.29	6.71	7.05	7.35	7.60	7.83	8.03
5	3.64	4.60	5.22	5.67	6.03	6.33	6.58	6.80	7.00	7.17
6	3.46	4.34	4.90	5.31	5.63	5.90	6.12	6.32	6.49	6.65
7	3.34	4.17	4.68	5.06	5.36	5.61	5.82	6.00	6.16	6.30
8	3.26	4.04	4.53	4.89	5.17	5.40	5.60	5.77	5.92	6.05
9	3.20	3.95	4.42	4.76	5.02	5.24	5.43	5.60	5.74	5.87
10	3.15	3.88	4.33	4.65	4.91	5.12	5.31	5.46	5.60	5.72
11	3.11	3.82	4.26	4.57	4.82	5.03	5.20	5.35	5.49	5.61
12	3.08	3.77	4.20	4.51	4.75	4.95	5.12	5.27	5.40	5.51
13	3.06	3.74	4.15	4.45	4.69	4.89	5.05	5.19	5.32	5.43
14	3.03	3.70	4.11	4.41	4.64	4.83	4.99	5.13	5.25	5.36
15	3.01	3.67	4.08	4.37	4.59	4.78	4.94	5.08	5.20	5.31
16	3.00	3.65	4.05	4.33	4.56	4.74	4.90	5.03	5.15	5.26
17	2.98	3.63	4.02	4.30	4.52	4.71	4.86	4.99	5.11	5.21
18	2.97	3.61	4.00	4.28	4.50	4.67	4.82	4.96	5.07	5.17
19	2.96	3.59	3.98	4.25	4.47	4.65	4.79	4.92	5.04	5.14
20	2.95	3.58	3.96	4.23	4.45	4.62	4.77	4.90	5.01	5.11
24	2.92	3.53	3.90	4.17	4.37	4.54	4.68	4.81	4.92	5.01
30	2.89	3.49	3.85	4.10	4.30	4.46	4.60	4.72	4.82	4.92
40	2.86	3.44	3.79	4.04	4.23	4.39	4.52	4.64	4.74	4.82
60	2.83	3.40	3.74	3.98	4.16	4.31	4.44	4.55	4.65	4.73
120	2.80	3.36	3.69	3.92	4.10	4.24	4.36	4.47	4.56	4.64
∞	2.77	3.31	3.63	3.86	4.03	4.17	4.29	4.39	4.47	4.55

These tables are largely those from May (1952) and are presented with the permission of the Trustees of *Biometrika*. Comparisons with several other tables have been made and the values that appear to be most accurate have been used. In doubtful cases, values have been rounded up.

TABLE B.10. $Q(.95, r, dfE)$

dfE	r								
	12	13	14	15	16	17	18	19	20
1	52.0	53.2	54.3	55.4	56.3	57.2	58.0	58.8	59.6
2	14.75	15.08	15.38	15.65	15.91	16.14	16.37	16.57	16.77
3	9.95	10.15	10.35	10.53	10.69	10.84	10.98	11.11	11.24
4	8.21	8.37	8.53	8.66	8.79	8.91	9.03	9.13	9.23
5	7.32	7.47	7.60	7.72	7.83	7.93	8.03	8.12	8.21
6	6.79	6.92	7.03	7.14	7.24	7.34	7.43	7.51	7.59
7	6.43	6.55	6.66	6.76	6.85	6.94	7.02	7.10	7.17
8	6.18	6.29	6.39	6.48	6.57	6.65	6.73	6.80	6.87
9	5.98	6.09	6.19	6.28	6.36	6.44	6.51	6.58	6.64
10	5.83	5.94	6.03	6.11	6.19	6.27	6.34	6.41	6.47
11	5.71	5.81	5.90	5.98	6.06	6.13	6.20	6.27	6.33
12	5.62	5.71	5.80	5.88	5.95	6.02	6.09	6.15	6.21
13	5.53	5.63	5.71	5.79	5.86	5.93	6.00	6.06	6.11
14	5.46	5.55	5.64	5.71	5.79	5.85	5.92	5.97	6.03
15	5.40	5.49	5.57	5.65	5.72	5.79	5.85	5.90	5.96
16	5.35	5.44	5.52	5.59	5.66	5.73	5.79	5.84	5.90
17	5.31	5.39	5.47	5.54	5.61	5.68	5.73	5.79	5.84
18	5.27	5.35	5.43	5.50	5.57	5.63	5.69	5.74	5.79
19	5.23	5.32	5.39	5.46	5.53	5.59	5.65	5.70	5.75
20	5.20	5.28	5.36	5.43	5.49	5.55	5.61	5.66	5.71
24	5.10	5.18	5.25	5.32	5.38	5.44	5.49	5.55	5.59
30	5.00	5.08	5.15	5.21	5.27	5.33	5.38	5.43	5.48
40	4.90	4.98	5.04	5.11	5.16	5.22	5.27	5.31	5.36
60	4.81	4.88	4.94	5.00	5.06	5.11	5.15	5.20	5.24
120	4.71	4.78	4.84	4.90	4.95	5.00	5.04	5.09	5.13
∞	4.62	4.69	4.74	4.80	4.85	4.89	4.93	4.97	5.01

TABLE B.11. $Q(.99, r, dfE)$

dfE	2	3	4	5	6	7	8	9	10	11
1	90.0	135	164	186	202	216	227	237	246	253
2	14.0	19.0	22.3	24.7	26.6	28.2	29.5	30.7	31.7	32.6
3	8.26	10.6	12.2	13.3	14.2	15.0	15.6	16.2	16.7	17.1
4	6.51	8.12	9.17	9.96	10.6	11.1	11.6	11.9	12.3	12.6
5	5.70	6.98	7.80	8.42	8.91	9.32	9.67	9.97	10.24	10.48
6	5.24	6.33	7.03	7.56	7.97	8.32	8.61	8.87	9.10	9.30
7	4.95	5.92	6.54	7.01	7.37	7.68	7.94	8.17	8.37	8.55
8	4.75	5.64	6.20	6.63	6.96	7.24	7.47	7.68	7.86	8.03
9	4.60	5.43	5.96	6.35	6.66	6.92	7.13	7.33	7.50	7.65
10	4.48	5.27	5.77	6.14	6.43	6.67	6.88	7.06	7.21	7.36
11	4.39	5.15	5.62	5.97	6.25	6.48	6.67	6.84	6.99	7.13
12	4.32	5.05	5.50	5.84	6.10	6.32	6.51	6.67	6.81	6.94
13	4.26	4.96	5.40	5.73	5.98	6.19	6.37	6.53	6.67	6.79
14	4.21	4.90	5.32	5.63	5.88	6.09	6.26	6.41	6.54	6.66
15	4.17	4.84	5.25	5.56	5.80	5.99	6.16	6.31	6.44	6.56
16	4.13	4.79	5.19	5.49	5.72	5.92	6.08	6.22	6.35	6.46
17	4.10	4.74	5.14	5.43	5.66	5.85	6.01	6.15	6.27	6.38
18	4.07	4.70	5.09	5.38	5.60	5.79	5.94	6.08	6.20	6.31
19	4.05	4.67	5.05	5.33	5.55	5.74	5.89	6.02	6.14	6.25
20	4.02	4.64	5.02	5.29	5.51	5.69	5.84	5.97	6.09	6.19
24	3.96	4.55	4.91	5.17	5.37	5.54	5.69	5.81	5.92	6.02
30	3.89	4.46	4.80	5.05	5.24	5.40	5.54	5.65	5.76	5.85
40	3.83	4.37	4.70	4.93	5.11	5.27	5.39	5.50	5.60	5.69
60	3.76	4.28	4.60	4.82	4.99	5.13	5.25	5.36	5.45	5.53
120	3.70	4.20	4.50	4.71	4.87	5.01	5.12	5.21	5.30	5.38
∞	3.64	4.12	4.40	4.60	4.76	4.88	4.99	5.08	5.16	5.23

TABLE B.12. $Q(.99, r, dfE)$

	r								
dfE	12	13	14	15	16	17	18	19	20
1	260	266	272	277	282	286	290	294	298
2	33.4	34.1	34.8	35.4	36.0	36.5	37.0	37.5	38.0
3	17.5	17.9	18.2	18.5	18.8	19.1	19.3	19.6	19.8
4	12.8	13.1	13.3	13.5	13.7	13.9	14.1	14.2	14.4
5	10.70	10.89	11.08	11.24	11.40	11.55	11.68	11.81	11.9
6	9.49	9.65	9.81	9.95	10.08	10.21	10.32	10.43	10.5
7	8.71	8.86	9.00	9.12	9.24	9.35	9.46	9.55	9.65
8	8.18	8.31	8.44	8.55	8.66	8.76	8.85	8.94	9.03
9	7.78	7.91	8.03	8.13	8.23	8.33	8.41	8.50	8.57
10	7.49	7.60	7.71	7.81	7.91	7.99	8.08	8.15	8.23
11	7.25	7.36	7.47	7.56	7.65	7.73	7.81	7.88	7.95
12	7.06	7.17	7.27	7.36	7.44	7.52	7.59	7.67	7.73
13	6.90	7.01	7.10	7.19	7.27	7.35	7.42	7.49	7.55
14	6.77	6.87	6.96	7.05	7.13	7.20	7.27	7.33	7.40
15	6.66	6.76	6.85	6.93	7.00	7.07	7.14	7.20	7.26
16	6.56	6.66	6.74	6.82	6.90	6.97	7.03	7.09	7.15
17	6.48	6.57	6.66	6.73	6.81	6.87	6.94	7.00	7.05
18	6.41	6.50	6.58	6.66	6.73	6.79	6.85	6.91	6.97
19	6.34	6.43	6.51	6.59	6.65	6.72	6.78	6.84	6.89
20	6.29	6.37	6.45	6.52	6.59	6.65	6.71	6.77	6.82
24	6.11	6.19	6.26	6.33	6.39	6.45	6.51	6.56	6.61
30	5.93	6.01	6.08	6.14	6.20	6.26	6.31	6.36	6.41
40	5.76	5.84	5.90	5.96	6.02	6.07	6.12	6.17	6.21
60	5.60	6.67	5.73	5.79	5.84	5.89	5.93	5.97	6.02
120	5.44	5.51	5.56	5.61	5.66	5.71	5.75	5.79	5.83
∞	5.29	5.35	5.40	5.45	5.49	5.54	5.57	5.61	5.65

B.6 The Greek alphabet

TABLE B.13. The Greek alphabet

Capital	Small	Name	Capital	Small	Name
A	α	alpha	N	ν	nu
B	β	beta	Ξ	ξ	xi
Γ	γ	gamma	O	o	omicron
Δ	δ, ∂	delta	Π	π	pi
E	ϵ, ε	epsilon	P	ρ	rho
Z	ζ	zeta	Σ	σ	sigma
H	η	eta	T	τ	tau
Θ	θ	theta	Υ	υ	upsilon
I	ι	iota	Φ	ϕ	phi
K	κ	kappa	X	χ	chi
Λ	λ	lambda	Ψ	ψ	psi
M	μ	mu	Ω	ω	omega

B.7 Tables of the F distribution

TABLE B.14. 90th percentiles of the F distribution

Den.	Numerator degrees of freedom							
df	1	2	3	4	5	6	7	8
1	39.862	49.500	53.593	55.833	57.240	58.204	58.906	59.439
2	8.5263	9.0000	9.1618	9.2434	9.2926	9.3255	9.3491	9.3668
3	5.5383	5.4625	5.3908	5.3427	5.3092	5.2847	5.2662	5.2517
4	4.5449	4.3246	4.1909	4.1072	4.0506	4.0098	3.9790	3.9549
5	4.0604	3.7797	3.6195	3.5202	3.4530	3.4045	3.3679	3.3393
6	3.7760	3.4633	3.2888	3.1808	3.1075	3.0546	3.0145	2.9830
7	3.5895	3.2574	3.0741	2.9605	2.8833	2.8274	2.7849	2.7516
8	3.4579	3.1131	2.9238	2.8065	2.7265	2.6683	2.6241	2.5894
9	3.3603	3.0065	2.8129	2.6927	2.6106	2.5509	2.5053	2.4694
10	3.2850	2.9245	2.7277	2.6054	2.5216	2.4606	2.4140	2.3772
11	3.2252	2.8595	2.6602	2.5362	2.4512	2.3891	2.3416	2.3040
12	3.1765	2.8068	2.6055	2.4801	2.3941	2.3310	2.2828	2.2446
13	3.1362	2.7632	2.5603	2.4337	2.3467	2.2830	2.2341	2.1954
14	3.1022	2.7265	2.5222	2.3947	2.3069	2.2426	2.1931	2.1539
15	3.0732	2.6952	2.4898	2.3614	2.2730	2.2081	2.1582	2.1185
16	3.0481	2.6682	2.4618	2.3328	2.2438	2.1783	2.1280	2.0880
17	3.0263	2.6446	2.4374	2.3078	2.2183	2.1524	2.1017	2.0613
18	3.0070	2.6240	2.4160	2.2858	2.1958	2.1296	2.0785	2.0379
19	2.9899	2.6056	2.3970	2.2663	2.1760	2.1094	2.0580	2.0171
20	2.9747	2.5893	2.3801	2.2489	2.1583	2.0913	2.0397	1.9985
21	2.9610	2.5746	2.3649	2.2334	2.1423	2.0751	2.0233	1.9819
22	2.9486	2.5613	2.3512	2.2193	2.1279	2.0605	2.0084	1.9668
23	2.9374	2.5493	2.3387	2.2065	2.1149	2.0472	1.9949	1.9531
24	2.9271	2.5384	2.3274	2.1949	2.1030	2.0351	1.9826	1.9407
25	2.9177	2.5283	2.3170	2.1842	2.0922	2.0241	1.9714	1.9293
26	2.9091	2.5191	2.3075	2.1745	2.0822	2.0139	1.9610	1.9188
28	2.8939	2.5028	2.2906	2.1571	2.0645	1.9959	1.9427	1.9002
30	2.8807	2.4887	2.2761	2.1422	2.0492	1.9803	1.9269	1.8841
32	2.8693	2.4765	2.2635	2.1293	2.0360	1.9669	1.9132	1.8702
34	2.8592	2.4658	2.2524	2.1179	2.0244	1.9550	1.9012	1.8580
36	2.8504	2.4563	2.2426	2.1079	2.0141	1.9446	1.8905	1.8471
38	2.8424	2.4479	2.2339	2.0990	2.0050	1.9352	1.8810	1.8375
40	2.8354	2.4404	2.2261	2.0909	1.9968	1.9269	1.8725	1.8289
60	2.7911	2.3932	2.1774	2.0410	1.9457	1.8747	1.8194	1.7748
80	2.7693	2.3702	2.1536	2.0165	1.9206	1.8491	1.7933	1.7483
100	2.7564	2.3564	2.1394	2.0019	1.9057	1.8339	1.7778	1.7324
150	2.7393	2.3383	2.1207	1.9827	1.8861	1.8138	1.7572	1.7115
200	2.7308	2.3293	2.1114	1.9732	1.8763	1.8038	1.7470	1.7011
300	2.7224	2.3203	2.1021	1.9637	1.8666	1.7939	1.7369	1.6908
400	2.7182	2.3159	2.0975	1.9590	1.8617	1.7889	1.7319	1.6856
∞	2.7055	2.3026	2.0838	1.9449	1.8473	1.7741	1.7167	1.6702

TABLE B.15. 90th percentiles of the F distribution

Den.	Numerator degrees of freedom							
df	9	10	11	12	13	14	15	16
1	59.858	60.195	60.473	60.705	60.903	61.072	61.220	61.350
2	9.3805	9.3915	9.4005	9.4080	9.4144	9.4198	9.4245	9.4286
3	5.2401	5.2305	5.2226	5.2158	5.2098	5.2047	5.2003	5.1964
4	3.9357	3.9199	3.9067	3.8956	3.8859	3.8776	3.8704	3.8639
5	3.3163	3.2974	3.2816	3.2682	3.2568	3.2468	3.2380	3.2303
6	2.9577	2.9369	2.9195	2.9047	2.8920	2.8809	2.8712	2.8626
7	2.7247	2.7025	2.6839	2.6681	2.6545	2.6426	2.6322	2.6230
8	2.5613	2.5381	2.5186	2.5020	2.4876	2.4752	2.4642	2.4545
9	2.4404	2.4164	2.3961	2.3789	2.3640	2.3511	2.3396	2.3295
10	2.3473	2.3226	2.3018	2.2841	2.2687	2.2553	2.2435	2.2331
11	2.2735	2.2482	2.2269	2.2087	2.1930	2.1792	2.1671	2.1563
12	2.2135	2.1878	2.1660	2.1474	2.1313	2.1173	2.1049	2.0938
13	2.1638	2.1376	2.1155	2.0966	2.0802	2.0659	2.0532	2.0419
14	2.1220	2.0954	2.0730	2.0537	2.0370	2.0224	2.0095	1.9981
15	2.0862	2.0593	2.0366	2.0171	2.0001	1.9853	1.9722	1.9605
16	2.0553	2.0282	2.0051	1.9854	1.9682	1.9532	1.9399	1.9281
17	2.0284	2.0010	1.9777	1.9577	1.9404	1.9252	1.9117	1.8997
18	2.0047	1.9770	1.9535	1.9334	1.9158	1.9004	1.8868	1.8747
19	1.9836	1.9557	1.9321	1.9117	1.8940	1.8785	1.8647	1.8524
20	1.9649	1.9367	1.9129	1.8924	1.8745	1.8588	1.8450	1.8325
21	1.9480	1.9197	1.8957	1.8750	1.8570	1.8412	1.8271	1.8147
22	1.9328	1.9043	1.8801	1.8593	1.8411	1.8252	1.8111	1.7984
23	1.9189	1.8903	1.8659	1.8450	1.8267	1.8107	1.7964	1.7837
24	1.9063	1.8775	1.8530	1.8319	1.8136	1.7974	1.7831	1.7703
25	1.8947	1.8658	1.8412	1.8200	1.8015	1.7853	1.7708	1.7579
26	1.8841	1.8550	1.8303	1.8090	1.7904	1.7741	1.7596	1.7466
28	1.8652	1.8359	1.8110	1.7895	1.7708	1.7542	1.7395	1.7264
30	1.8490	1.8195	1.7944	1.7727	1.7538	1.7371	1.7223	1.7090
32	1.8348	1.8052	1.7799	1.7581	1.7390	1.7222	1.7072	1.6938
34	1.8224	1.7926	1.7672	1.7452	1.7260	1.7091	1.6940	1.6805
36	1.8115	1.7815	1.7559	1.7338	1.7145	1.6974	1.6823	1.6687
38	1.8017	1.7716	1.7459	1.7237	1.7042	1.6871	1.6718	1.6581
40	1.7929	1.7627	1.7369	1.7146	1.6950	1.6778	1.6624	1.6486
60	1.7380	1.7070	1.6805	1.6574	1.6372	1.6193	1.6034	1.5890
80	1.7110	1.6796	1.6526	1.6292	1.6087	1.5904	1.5741	1.5594
100	1.6949	1.6632	1.6360	1.6124	1.5916	1.5731	1.5566	1.5417
150	1.6736	1.6416	1.6140	1.5901	1.5690	1.5502	1.5334	1.5182
200	1.6630	1.6308	1.6031	1.5789	1.5577	1.5388	1.5218	1.5065
300	1.6525	1.6201	1.5922	1.5679	1.5464	1.5273	1.5102	1.4948
400	1.6472	1.6147	1.5868	1.5623	1.5408	1.5217	1.5045	1.4889
∞	1.6315	1.5987	1.5705	1.5458	1.5240	1.5046	1.4871	1.4714

TABLE B.16. 90th percentiles of the F distribution

Den.	Numerator degrees of freedom							
df	18	20	25	30	40	50	100	200
1	61.566	61.740	62.054	62.265	62.528	62.688	63.006	63.163
2	9.4354	9.4408	9.4513	9.4579	9.4662	9.4712	9.4812	9.4861
3	5.1900	5.1846	5.1747	5.1681	5.1597	5.1546	5.1442	5.1389
4	3.8531	3.8444	3.8283	3.8175	3.8037	3.7952	3.7781	3.7695
5	3.2172	3.2067	3.1873	3.1741	3.1573	3.1471	3.1263	3.1157
6	2.8481	2.8363	2.8147	2.8000	2.7812	2.7697	2.7463	2.7343
7	2.6074	2.5947	2.5714	2.5555	2.5351	2.5226	2.4971	2.4841
8	2.4381	2.4247	2.3999	2.3830	2.3614	2.3481	2.3208	2.3068
9	2.3123	2.2983	2.2725	2.2547	2.2320	2.2180	2.1892	2.1743
10	2.2153	2.2008	2.1739	2.1554	2.1317	2.1171	2.0869	2.0713
11	2.1380	2.1231	2.0953	2.0762	2.0516	2.0364	2.0050	1.9888
12	2.0750	2.0597	2.0312	2.0115	1.9861	1.9704	1.9379	1.9210
13	2.0227	2.0070	1.9778	1.9576	1.9315	1.9153	1.8817	1.8642
14	1.9785	1.9625	1.9326	1.9119	1.8852	1.8686	1.8340	1.8159
15	1.9407	1.9243	1.8939	1.8728	1.8454	1.8284	1.7928	1.7743
16	1.9079	1.8913	1.8603	1.8388	1.8108	1.7935	1.7570	1.7380
17	1.8792	1.8624	1.8309	1.8090	1.7805	1.7628	1.7255	1.7059
18	1.8539	1.8368	1.8049	1.7827	1.7537	1.7356	1.6976	1.6775
19	1.8314	1.8142	1.7818	1.7592	1.7298	1.7114	1.6726	1.6521
20	1.8113	1.7938	1.7611	1.7382	1.7083	1.6896	1.6501	1.6292
21	1.7932	1.7756	1.7424	1.7193	1.6890	1.6700	1.6298	1.6085
22	1.7768	1.7590	1.7255	1.7021	1.6714	1.6521	1.6113	1.5896
23	1.7619	1.7439	1.7101	1.6864	1.6554	1.6358	1.5944	1.5723
24	1.7483	1.7302	1.6960	1.6721	1.6407	1.6209	1.5788	1.5564
25	1.7358	1.7175	1.6831	1.6589	1.6272	1.6072	1.5645	1.5417
26	1.7243	1.7059	1.6712	1.6468	1.6147	1.5945	1.5513	1.5281
28	1.7039	1.6852	1.6500	1.6252	1.5925	1.5718	1.5276	1.5037
30	1.6862	1.6673	1.6316	1.6065	1.5732	1.5522	1.5069	1.4824
32	1.6708	1.6517	1.6156	1.5901	1.5564	1.5349	1.4888	1.4637
34	1.6573	1.6380	1.6015	1.5757	1.5415	1.5197	1.4727	1.4470
36	1.6453	1.6258	1.5890	1.5629	1.5282	1.5061	1.4583	1.4321
38	1.6345	1.6149	1.5778	1.5514	1.5163	1.4939	1.4453	1.4186
40	1.6249	1.6052	1.5677	1.5411	1.5056	1.4830	1.4336	1.4064
60	1.5642	1.5435	1.5039	1.4755	1.4373	1.4126	1.3576	1.3264
80	1.5340	1.5128	1.4720	1.4426	1.4027	1.3767	1.3180	1.2839
100	1.5160	1.4944	1.4527	1.4227	1.3817	1.3548	1.2934	1.2571
150	1.4919	1.4698	1.4271	1.3960	1.3534	1.3251	1.2595	1.2193
200	1.4799	1.4575	1.4142	1.3826	1.3390	1.3100	1.2418	1.1991
300	1.4679	1.4452	1.4013	1.3691	1.3246	1.2947	1.2236	1.1779
400	1.4619	1.4391	1.3948	1.3623	1.3173	1.2870	1.2143	1.1667
∞	1.4439	1.4206	1.3753	1.3419	1.2951	1.2633	1.1850	1.1301

TABLE B.17. 95th percentiles of the F distribution

Den.	Numerator degrees of freedom							
df	1	2	3	4	5	6	7	8
1	161.45	199.50	215.71	224.58	230.16	233.99	236.77	238.88
2	18.513	19.000	19.164	19.247	19.296	19.329	19.353	19.371
3	10.128	9.552	9.277	9.117	9.013	8.941	8.887	8.845
4	7.709	6.944	6.591	6.388	6.256	6.163	6.094	6.041
5	6.608	5.786	5.409	5.192	5.050	4.950	4.876	4.818
6	5.987	5.143	4.757	4.534	4.387	4.284	4.207	4.147
7	5.591	4.737	4.347	4.120	3.972	3.866	3.787	3.726
8	5.318	4.459	4.066	3.838	3.687	3.581	3.500	3.438
9	5.117	4.256	3.863	3.633	3.482	3.374	3.293	3.230
10	4.965	4.103	3.708	3.478	3.326	3.217	3.135	3.072
11	4.844	3.982	3.587	3.357	3.204	3.095	3.012	2.948
12	4.747	3.885	3.490	3.259	3.106	2.996	2.913	2.849
13	4.667	3.806	3.411	3.179	3.025	2.915	2.832	2.767
14	4.600	3.739	3.344	3.112	2.958	2.848	2.764	2.699
15	4.543	3.682	3.287	3.056	2.901	2.790	2.707	2.641
16	4.494	3.634	3.239	3.007	2.852	2.741	2.657	2.591
17	4.451	3.592	3.197	2.965	2.810	2.699	2.614	2.548
18	4.414	3.555	3.160	2.928	2.773	2.661	2.577	2.510
19	4.381	3.522	3.127	2.895	2.740	2.628	2.544	2.477
20	4.351	3.493	3.098	2.866	2.711	2.599	2.514	2.447
21	4.325	3.467	3.072	2.840	2.685	2.573	2.488	2.420
22	4.301	3.443	3.049	2.817	2.661	2.549	2.464	2.397
23	4.279	3.422	3.028	2.796	2.640	2.528	2.442	2.375
24	4.260	3.403	3.009	2.776	2.621	2.508	2.423	2.355
25	4.242	3.385	2.991	2.759	2.603	2.490	2.405	2.337
26	4.225	3.369	2.975	2.743	2.587	2.474	2.388	2.321
28	4.196	3.340	2.947	2.714	2.558	2.445	2.359	2.291
30	4.171	3.316	2.922	2.690	2.534	2.421	2.334	2.266
32	4.149	3.295	2.901	2.668	2.512	2.399	2.313	2.244
34	4.130	3.276	2.883	2.650	2.494	2.380	2.294	2.225
36	4.113	3.259	2.866	2.634	2.477	2.364	2.277	2.209
38	4.098	3.245	2.852	2.619	2.463	2.349	2.262	2.194
40	4.085	3.232	2.839	2.606	2.449	2.336	2.249	2.180
60	4.001	3.150	2.758	2.525	2.368	2.254	2.167	2.097
80	3.960	3.111	2.719	2.486	2.329	2.214	2.126	2.056
100	3.936	3.087	2.696	2.463	2.305	2.191	2.103	2.032
150	3.904	3.056	2.665	2.432	2.274	2.160	2.071	2.001
200	3.888	3.041	2.650	2.417	2.259	2.144	2.056	1.985
300	3.873	3.026	2.635	2.402	2.244	2.129	2.040	1.969
400	3.865	3.018	2.627	2.394	2.237	2.121	2.032	1.962
∞	3.841	2.996	2.605	2.372	2.214	2.099	2.010	1.938

TABLE B.18. 95th percentiles of the F distribution

Den.	Numerator degrees of freedom							
df	9	10	11	12	13	14	15	16
1	240.54	241.88	242.98	243.91	244.69	245.36	245.95	246.46
2	19.385	19.396	19.405	19.412	19.419	19.424	19.429	19.433
3	8.812	8.786	8.763	8.745	8.729	8.715	8.703	8.692
4	5.999	5.964	5.936	5.912	5.891	5.873	5.858	5.844
5	4.772	4.735	4.704	4.678	4.655	4.636	4.619	4.604
6	4.099	4.060	4.027	4.000	3.976	3.956	3.938	3.922
7	3.677	3.637	3.603	3.575	3.550	3.529	3.511	3.494
8	3.388	3.347	3.313	3.284	3.259	3.237	3.218	3.202
9	3.179	3.137	3.102	3.073	3.048	3.025	3.006	2.989
10	3.020	2.978	2.943	2.913	2.887	2.865	2.845	2.828
11	2.896	2.854	2.818	2.788	2.761	2.739	2.719	2.701
12	2.796	2.753	2.717	2.687	2.660	2.637	2.617	2.599
13	2.714	2.671	2.635	2.604	2.577	2.554	2.533	2.515
14	2.646	2.602	2.566	2.534	2.507	2.484	2.463	2.445
15	2.588	2.544	2.507	2.475	2.448	2.424	2.403	2.385
16	2.538	2.494	2.456	2.425	2.397	2.373	2.352	2.333
17	2.494	2.450	2.413	2.381	2.353	2.329	2.308	2.289
18	2.456	2.412	2.374	2.342	2.314	2.290	2.269	2.250
19	2.423	2.378	2.340	2.308	2.280	2.256	2.234	2.215
20	2.393	2.348	2.310	2.278	2.250	2.225	2.203	2.184
21	2.366	2.321	2.283	2.250	2.222	2.197	2.176	2.156
22	2.342	2.297	2.259	2.226	2.198	2.173	2.151	2.131
23	2.320	2.275	2.236	2.204	2.175	2.150	2.128	2.109
24	2.300	2.255	2.216	2.183	2.155	2.130	2.108	2.088
25	2.282	2.236	2.198	2.165	2.136	2.111	2.089	2.069
26	2.265	2.220	2.181	2.148	2.119	2.094	2.072	2.052
28	2.236	2.190	2.151	2.118	2.089	2.064	2.041	2.021
30	2.211	2.165	2.126	2.092	2.063	2.037	2.015	1.995
32	2.189	2.142	2.103	2.070	2.040	2.015	1.992	1.972
34	2.170	2.123	2.084	2.050	2.021	1.995	1.972	1.952
36	2.153	2.106	2.067	2.033	2.003	1.977	1.954	1.934
38	2.138	2.091	2.051	2.017	1.988	1.962	1.939	1.918
40	2.124	2.077	2.038	2.003	1.974	1.948	1.924	1.904
60	2.040	1.993	1.952	1.917	1.887	1.860	1.836	1.815
80	1.999	1.951	1.910	1.875	1.845	1.817	1.793	1.772
100	1.975	1.927	1.886	1.850	1.819	1.792	1.768	1.746
150	1.943	1.894	1.853	1.817	1.786	1.758	1.734	1.711
200	1.927	1.878	1.837	1.801	1.769	1.742	1.717	1.694
300	1.911	1.862	1.821	1.785	1.753	1.725	1.700	1.677
400	1.903	1.854	1.813	1.776	1.745	1.717	1.691	1.669
∞	1.880	1.831	1.789	1.752	1.720	1.692	1.666	1.644

TABLE B.19. 95th percentiles of the F distribution

Den.	Numerator degrees of freedom							
df	18	20	25	30	40	50	100	200
1	247.32	248.01	249.26	250.09	251.14	251.77	253.04	253.68
2	19.440	19.446	19.456	19.462	19.470	19.475	19.486	19.491
3	8.675	8.660	8.634	8.617	8.594	8.581	8.554	8.540
4	5.821	5.803	5.769	5.746	5.717	5.699	5.664	5.646
5	4.579	4.558	4.521	4.496	4.464	4.444	4.405	4.385
6	3.896	3.874	3.835	3.808	3.774	3.754	3.712	3.690
7	3.467	3.445	3.404	3.376	3.340	3.319	3.275	3.252
8	3.173	3.150	3.108	3.079	3.043	3.020	2.975	2.951
9	2.960	2.936	2.893	2.864	2.826	2.803	2.756	2.731
10	2.798	2.774	2.730	2.700	2.661	2.637	2.588	2.563
11	2.671	2.646	2.601	2.570	2.531	2.507	2.457	2.431
12	2.568	2.544	2.498	2.466	2.426	2.401	2.350	2.323
13	2.484	2.459	2.412	2.380	2.339	2.314	2.261	2.234
14	2.413	2.388	2.341	2.308	2.266	2.241	2.187	2.159
15	2.353	2.328	2.280	2.247	2.204	2.178	2.123	2.095
16	2.302	2.276	2.227	2.194	2.151	2.124	2.068	2.039
17	2.257	2.230	2.181	2.148	2.104	2.077	2.020	1.991
18	2.217	2.191	2.141	2.107	2.063	2.035	1.978	1.948
19	2.182	2.156	2.106	2.071	2.026	1.999	1.940	1.910
20	2.151	2.124	2.074	2.039	1.994	1.966	1.907	1.875
21	2.123	2.096	2.045	2.010	1.965	1.936	1.876	1.845
22	2.098	2.071	2.020	1.984	1.938	1.909	1.849	1.817
23	2.075	2.048	1.996	1.961	1.914	1.885	1.823	1.791
24	2.054	2.027	1.975	1.939	1.892	1.863	1.800	1.768
25	2.035	2.007	1.955	1.919	1.872	1.842	1.779	1.746
26	2.018	1.990	1.938	1.901	1.853	1.823	1.760	1.726
28	1.987	1.959	1.906	1.869	1.820	1.790	1.725	1.691
30	1.960	1.932	1.878	1.841	1.792	1.761	1.695	1.660
32	1.937	1.908	1.854	1.817	1.767	1.736	1.669	1.633
34	1.917	1.888	1.833	1.795	1.745	1.713	1.645	1.609
36	1.899	1.870	1.815	1.776	1.726	1.694	1.625	1.587
38	1.883	1.853	1.798	1.760	1.708	1.676	1.606	1.568
40	1.868	1.839	1.783	1.744	1.693	1.660	1.589	1.551
60	1.778	1.748	1.690	1.649	1.594	1.559	1.481	1.438
80	1.734	1.703	1.644	1.602	1.545	1.508	1.426	1.379
100	1.708	1.676	1.616	1.573	1.515	1.477	1.392	1.342
150	1.673	1.641	1.580	1.535	1.475	1.436	1.345	1.290
200	1.656	1.623	1.561	1.516	1.455	1.415	1.321	1.263
300	1.638	1.606	1.543	1.497	1.435	1.393	1.296	1.234
400	1.630	1.597	1.534	1.488	1.425	1.383	1.283	1.219
∞	1.604	1.571	1.506	1.459	1.394	1.350	1.243	1.170

TABLE B.20. 99th percentiles of the F distribution

Den.	Numerator degrees of freedom							
df	1	2	3	4	5	6	7	8
1	4052	5000	5403	5625	5764	5859	5928	5981
2	98.50	99.00	99.17	99.25	99.30	99.33	99.36	99.37
3	34.12	30.82	29.46	28.71	28.24	27.91	27.67	27.49
4	21.20	18.00	16.69	15.98	15.52	15.21	14.98	14.80
5	16.26	13.27	12.06	11.39	10.97	10.67	10.46	10.29
6	13.75	10.92	9.78	9.15	8.75	8.47	8.26	8.10
7	12.25	9.55	8.45	7.85	7.46	7.19	6.99	6.84
8	11.26	8.65	7.59	7.01	6.63	6.37	6.18	6.03
9	10.56	8.02	6.99	6.42	6.06	5.80	5.61	5.47
10	10.04	7.56	6.55	5.99	5.64	5.39	5.20	5.06
11	9.65	7.21	6.22	5.67	5.32	5.07	4.89	4.74
12	9.33	6.93	5.95	5.41	5.06	4.82	4.64	4.50
13	9.07	6.70	5.74	5.21	4.86	4.62	4.44	4.30
14	8.86	6.51	5.56	5.04	4.69	4.46	4.28	4.14
15	8.68	6.36	5.42	4.89	4.56	4.32	4.14	4.00
16	8.53	6.23	5.29	4.77	4.44	4.20	4.03	3.89
17	8.40	6.11	5.19	4.67	4.34	4.10	3.93	3.79
18	8.29	6.01	5.09	4.58	4.25	4.01	3.84	3.71
19	8.19	5.93	5.01	4.50	4.17	3.94	3.77	3.63
20	8.10	5.85	4.94	4.43	4.10	3.87	3.70	3.56
21	8.02	5.78	4.87	4.37	4.04	3.81	3.64	3.51
22	7.95	5.72	4.82	4.31	3.99	3.76	3.59	3.45
23	7.88	5.66	4.76	4.26	3.94	3.71	3.54	3.41
24	7.82	5.61	4.72	4.22	3.90	3.67	3.50	3.36
25	7.77	5.57	4.68	4.18	3.85	3.63	3.46	3.32
26	7.72	5.53	4.64	4.14	3.82	3.59	3.42	3.29
28	7.64	5.45	4.57	4.07	3.75	3.53	3.36	3.23
30	7.56	5.39	4.51	4.02	3.70	3.47	3.30	3.17
32	7.50	5.34	4.46	3.97	3.65	3.43	3.26	3.13
34	7.44	5.29	4.42	3.93	3.61	3.39	3.22	3.09
36	7.40	5.25	4.38	3.89	3.57	3.35	3.18	3.05
38	7.35	5.21	4.34	3.86	3.54	3.32	3.15	3.02
40	7.31	5.18	4.31	3.83	3.51	3.29	3.12	2.99
60	7.08	4.98	4.13	3.65	3.34	3.12	2.95	2.82
80	6.96	4.88	4.04	3.56	3.26	3.04	2.87	2.74
100	6.90	4.82	3.98	3.51	3.21	2.99	2.82	2.69
150	6.81	4.75	3.91	3.45	3.14	2.92	2.76	2.63
200	6.76	4.71	3.88	3.41	3.11	2.89	2.73	2.60
300	6.72	4.68	3.85	3.38	3.08	2.86	2.70	2.57
400	6.70	4.66	3.83	3.37	3.06	2.85	2.68	2.56
∞	6.63	4.61	3.78	3.32	3.02	2.80	2.64	2.51

TABLE B.21. 99th percentiles of the F distribution

Den.	Numerator degrees of freedom							
df	9	10	11	12	13	14	15	16
1	6022	6056	6083	6106	6126	6143	6157	6170
2	99.39	99.40	99.41	99.42	99.42	99.43	99.43	99.44
3	27.35	27.23	27.13	27.05	26.98	26.92	26.87	26.83
4	14.66	14.55	14.45	14.37	14.31	14.25	14.20	14.15
5	10.16	10.05	9.96	9.89	9.82	9.77	9.72	9.68
6	7.98	7.87	7.79	7.72	7.66	7.60	7.56	7.52
7	6.72	6.62	6.54	6.47	6.41	6.36	6.31	6.28
8	5.91	5.81	5.73	5.67	5.61	5.56	5.52	5.48
9	5.35	5.26	5.18	5.11	5.05	5.01	4.96	4.92
10	4.94	4.85	4.77	4.71	4.65	4.60	4.56	4.52
11	4.63	4.54	4.46	4.40	4.34	4.29	4.25	4.21
12	4.39	4.30	4.22	4.16	4.10	4.05	4.01	3.97
13	4.19	4.10	4.02	3.96	3.91	3.86	3.82	3.78
14	4.03	3.94	3.86	3.80	3.75	3.70	3.66	3.62
15	3.89	3.80	3.73	3.67	3.61	3.56	3.52	3.49
16	3.78	3.69	3.62	3.55	3.50	3.45	3.41	3.37
17	3.68	3.59	3.52	3.46	3.40	3.35	3.31	3.27
18	3.60	3.51	3.43	3.37	3.32	3.27	3.23	3.19
19	3.52	3.43	3.36	3.30	3.24	3.19	3.15	3.12
20	3.46	3.37	3.29	3.23	3.18	3.13	3.09	3.05
21	3.40	3.31	3.24	3.17	3.12	3.07	3.03	2.99
22	3.35	3.26	3.18	3.12	3.07	3.02	2.98	2.94
23	3.30	3.21	3.14	3.07	3.02	2.97	2.93	2.89
24	3.26	3.17	3.09	3.03	2.98	2.93	2.89	2.85
25	3.22	3.13	3.06	2.99	2.94	2.89	2.85	2.81
26	3.18	3.09	3.02	2.96	2.90	2.86	2.81	2.78
28	3.12	3.03	2.96	2.90	2.84	2.79	2.75	2.72
30	3.07	2.98	2.91	2.84	2.79	2.74	2.70	2.66
32	3.02	2.93	2.86	2.80	2.74	2.70	2.65	2.62
34	2.98	2.89	2.82	2.76	2.70	2.66	2.61	2.58
36	2.95	2.86	2.79	2.72	2.67	2.62	2.58	2.54
38	2.92	2.83	2.75	2.69	2.64	2.59	2.55	2.51
40	2.89	2.80	2.73	2.66	2.61	2.56	2.52	2.48
60	2.72	2.63	2.56	2.50	2.44	2.39	2.35	2.31
80	2.64	2.55	2.48	2.42	2.36	2.31	2.27	2.23
100	2.59	2.50	2.43	2.37	2.31	2.27	2.22	2.19
150	2.53	2.44	2.37	2.31	2.25	2.20	2.16	2.12
200	2.50	2.41	2.34	2.27	2.22	2.17	2.13	2.09
300	2.47	2.38	2.31	2.24	2.19	2.14	2.10	2.06
400	2.45	2.37	2.29	2.23	2.17	2.13	2.08	2.05
∞	2.41	2.32	2.25	2.18	2.13	2.08	2.04	2.00

TABLE B.22. 99th percentiles of the F distribution

Den.	Numerator degrees of freedom							
df	18	20	25	30	40	50	100	200
1	6191	6209	6240	6261	6287	6302	6334	6350
2	99.44	99.45	99.46	99.46	99.47	99.48	99.49	99.49
3	26.75	26.69	26.58	26.50	26.41	26.35	26.24	26.18
4	14.08	14.02	13.91	13.84	13.75	13.69	13.58	13.52
5	9.61	9.55	9.45	9.38	9.29	9.24	9.13	9.08
6	7.45	7.40	7.30	7.23	7.14	7.09	6.99	6.93
7	6.21	6.16	6.06	5.99	5.91	5.86	5.75	5.70
8	5.41	5.36	5.26	5.20	5.12	5.07	4.96	4.91
9	4.86	4.81	4.71	4.65	4.57	4.52	4.41	4.36
10	4.46	4.41	4.31	4.25	4.17	4.12	4.01	3.96
11	4.15	4.10	4.01	3.94	3.86	3.81	3.71	3.66
12	3.91	3.86	3.76	3.70	3.62	3.57	3.47	3.41
13	3.72	3.66	3.57	3.51	3.43	3.38	3.27	3.22
14	3.56	3.51	3.41	3.35	3.27	3.22	3.11	3.06
15	3.42	3.37	3.28	3.21	3.13	3.08	2.98	2.92
16	3.31	3.26	3.16	3.10	3.02	2.97	2.86	2.81
17	3.21	3.16	3.07	3.00	2.92	2.87	2.76	2.71
18	3.13	3.08	2.98	2.92	2.84	2.78	2.68	2.62
19	3.05	3.00	2.91	2.84	2.76	2.71	2.60	2.55
20	2.99	2.94	2.84	2.78	2.69	2.64	2.54	2.48
21	2.93	2.88	2.79	2.72	2.64	2.58	2.48	2.42
22	2.88	2.83	2.73	2.67	2.58	2.53	2.42	2.36
23	2.83	2.78	2.69	2.62	2.54	2.48	2.37	2.32
24	2.79	2.74	2.64	2.58	2.49	2.44	2.33	2.27
25	2.75	2.70	2.60	2.54	2.45	2.40	2.29	2.23
26	2.72	2.66	2.57	2.50	2.42	2.36	2.25	2.19
28	2.65	2.60	2.51	2.44	2.35	2.30	2.19	2.13
30	2.60	2.55	2.45	2.39	2.30	2.25	2.13	2.07
32	2.55	2.50	2.41	2.34	2.25	2.20	2.08	2.02
34	2.51	2.46	2.37	2.30	2.21	2.16	2.04	1.98
36	2.48	2.43	2.33	2.26	2.18	2.12	2.00	1.94
38	2.45	2.40	2.30	2.23	2.14	2.09	1.97	1.90
40	2.42	2.37	2.27	2.20	2.11	2.06	1.94	1.87
60	2.25	2.20	2.10	2.03	1.94	1.88	1.75	1.68
80	2.17	2.12	2.01	1.94	1.85	1.79	1.65	1.58
100	2.12	2.07	1.97	1.89	1.80	1.74	1.60	1.52
150	2.06	2.00	1.90	1.83	1.73	1.66	1.52	1.43
200	2.03	1.97	1.87	1.79	1.69	1.63	1.48	1.39
300	1.99	1.94	1.84	1.76	1.66	1.59	1.44	1.35
400	1.98	1.92	1.82	1.75	1.64	1.58	1.42	1.32
∞	1.93	1.88	1.77	1.70	1.59	1.52	1.36	1.25

TABLE B.23. 99.9th percentiles of the F distribution

Den.	Numerator degrees of freedom							
df	1	2	3	4	5	6	7	8
1	405292	500009	540389	562510	576416	585949	592885	598156
2	998.54	999.01	999.18	999.26	999.31	999.35	999.37	999.39
3	167.03	148.50	141.11	137.10	134.58	132.85	131.59	130.62
4	74.138	61.246	56.178	53.436	51.712	50.526	49.658	48.997
5	47.181	37.123	33.203	31.085	29.753	28.835	28.163	27.650
6	35.508	27.000	23.703	21.924	20.803	20.030	19.463	19.030
7	29.245	21.689	18.772	17.198	16.206	15.521	15.019	14.634
8	25.415	18.494	15.830	14.392	13.485	12.858	12.398	12.046
9	22.857	16.387	13.902	12.560	11.714	11.128	10.698	10.368
10	21.040	14.905	12.553	11.283	10.481	9.926	9.517	9.204
11	19.687	13.812	11.561	10.346	9.578	9.047	8.655	8.355
12	18.643	12.974	10.804	9.633	8.892	8.379	8.001	7.710
13	17.816	12.313	10.209	9.073	8.354	7.856	7.489	7.206
14	17.143	11.779	9.729	8.622	7.922	7.436	7.077	6.802
15	16.587	11.339	9.335	8.253	7.567	7.092	6.741	6.471
16	16.120	10.971	9.006	7.944	7.272	6.805	6.460	6.195
17	15.722	10.658	8.727	7.683	7.022	6.563	6.223	5.962
18	15.379	10.390	8.488	7.459	6.808	6.355	6.021	5.763
19	15.081	10.157	8.280	7.265	6.623	6.175	5.845	5.590
20	14.819	9.953	8.098	7.096	6.461	6.019	5.692	5.440
21	14.587	9.772	7.938	6.947	6.318	5.881	5.557	5.308
22	14.380	9.612	7.796	6.814	6.191	5.758	5.438	5.190
23	14.195	9.469	7.669	6.696	6.078	5.649	5.331	5.085
24	14.028	9.339	7.554	6.589	5.977	5.550	5.235	4.991
25	13.877	9.223	7.451	6.493	5.885	5.462	5.148	4.906
26	13.739	9.116	7.357	6.406	5.802	5.381	5.070	4.829
28	13.498	8.931	7.193	6.253	5.657	5.241	4.933	4.695
30	13.293	8.773	7.054	6.125	5.534	5.122	4.817	4.581
32	13.118	8.639	6.936	6.014	5.429	5.021	4.719	4.485
34	12.965	8.522	6.833	5.919	5.339	4.934	4.633	4.401
36	12.832	8.420	6.744	5.836	5.260	4.857	4.559	4.328
38	12.714	8.331	6.665	5.763	5.190	4.790	4.494	4.264
40	12.609	8.251	6.595	5.698	5.128	4.731	4.436	4.207
60	11.973	7.768	6.171	5.307	4.757	4.372	4.086	3.865
80	11.671	7.540	5.972	5.123	4.582	4.204	3.923	3.705
100	11.495	7.408	5.857	5.017	4.482	4.107	3.829	3.612
150	11.267	7.236	5.707	4.879	4.351	3.981	3.706	3.493
200	11.155	7.152	5.634	4.812	4.287	3.920	3.647	3.434
300	11.044	7.069	5.562	4.746	4.225	3.860	3.588	3.377
400	10.989	7.028	5.527	4.713	4.194	3.830	3.560	3.349
∞	10.828	6.908	5.422	4.617	4.103	3.743	3.475	3.266

TABLE B.24. 99.9th percentiles of the F distribution

Den.	Numerator degrees of freedom							
df	9	10	11	12	13	14	15	16
1	602296	605634	608381	610681	612636	614316	615778	617058
2	999.40	999.41	999.42	999.43	999.44	999.44	999.45	999.45
3	129.86	129.25	128.74	128.32	127.96	127.65	127.38	127.14
4	48.475	48.053	47.705	47.412	47.163	46.948	46.761	46.597
5	27.245	26.917	26.646	26.418	26.224	26.057	25.911	25.783
6	18.688	18.411	18.182	17.989	17.825	17.683	17.559	17.450
7	14.330	14.083	13.879	13.707	13.561	13.434	13.324	13.227
8	11.767	11.540	11.353	11.195	11.060	10.943	10.841	10.752
9	10.107	9.894	9.718	9.570	9.443	9.334	9.238	9.154
10	8.956	8.754	8.587	8.445	8.325	8.220	8.129	8.048
11	8.116	7.922	7.761	7.626	7.510	7.409	7.321	7.244
12	7.480	7.292	7.136	7.005	6.892	6.794	6.709	6.634
13	6.982	6.799	6.647	6.519	6.409	6.314	6.231	6.158
14	6.583	6.404	6.256	6.130	6.023	5.930	5.848	5.776
15	6.256	6.081	5.935	5.812	5.707	5.615	5.535	5.464
16	5.984	5.812	5.668	5.547	5.443	5.353	5.274	5.205
17	5.754	5.584	5.443	5.324	5.221	5.132	5.054	4.986
18	5.558	5.390	5.251	5.132	5.031	4.943	4.866	4.798
19	5.388	5.222	5.084	4.967	4.867	4.780	4.704	4.636
20	5.239	5.075	4.939	4.823	4.724	4.637	4.562	4.495
21	5.109	4.946	4.811	4.696	4.598	4.512	4.437	4.371
22	4.993	4.832	4.697	4.583	4.486	4.401	4.326	4.260
23	4.890	4.730	4.596	4.483	4.386	4.301	4.227	4.162
24	4.797	4.638	4.505	4.393	4.296	4.212	4.139	4.074
25	4.713	4.555	4.423	4.312	4.216	4.132	4.059	3.994
26	4.637	4.480	4.349	4.238	4.142	4.059	3.986	3.921
28	4.505	4.349	4.219	4.109	4.014	3.932	3.859	3.795
30	4.393	4.239	4.110	4.001	3.907	3.825	3.753	3.689
32	4.298	4.145	4.017	3.908	3.815	3.733	3.662	3.598
34	4.215	4.063	3.936	3.828	3.735	3.654	3.583	3.520
36	4.144	3.992	3.866	3.758	3.666	3.585	3.514	3.451
38	4.080	3.930	3.804	3.697	3.605	3.524	3.454	3.391
40	4.024	3.874	3.749	3.642	3.551	3.471	3.400	3.338
60	3.687	3.541	3.419	3.315	3.226	3.147	3.078	3.017
80	3.530	3.386	3.265	3.162	3.074	2.996	2.927	2.867
100	3.439	3.296	3.176	3.074	2.986	2.908	2.840	2.780
150	3.321	3.180	3.061	2.959	2.872	2.795	2.727	2.667
200	3.264	3.123	3.005	2.904	2.816	2.740	2.672	2.612
300	3.207	3.067	2.950	2.849	2.762	2.686	2.618	2.558
400	3.179	3.040	2.922	2.822	2.735	2.659	2.592	2.532
∞	3.097	2.959	2.842	2.742	2.656	2.580	2.513	2.453

TABLE B.25. 99.9th percentiles of the F distribution

Den.	Numerator degrees of freedom							
df	18	20	25	30	40	50	100	200
1	619201	620922	624031	626114	628725	630301	633455	635033
2	999.46	999.46	999.47	999.48	999.49	999.49	999.50	999.50
3	126.74	126.42	125.84	125.45	124.96	124.67	124.07	123.77
4	46.322	46.101	45.699	45.429	45.089	44.884	44.470	44.261
5	25.568	25.395	25.080	24.869	24.602	24.441	24.115	23.951
6	17.267	17.120	16.853	16.673	16.445	16.307	16.028	15.887
7	13.063	12.932	12.692	12.530	12.326	12.202	11.951	11.824
8	10.601	10.480	10.258	10.109	9.919	9.804	9.571	9.453
9	9.012	8.898	8.689	8.548	8.369	8.260	8.039	7.926
10	7.913	7.804	7.604	7.469	7.297	7.193	6.980	6.872
11	7.113	7.008	6.815	6.684	6.518	6.417	6.210	6.105
12	6.507	6.405	6.217	6.090	5.928	5.829	5.627	5.524
13	6.034	5.934	5.751	5.626	5.467	5.370	5.172	5.070
14	5.655	5.557	5.377	5.254	5.098	5.002	4.807	4.706
15	5.345	5.248	5.071	4.950	4.796	4.702	4.508	4.408
16	5.087	4.992	4.817	4.697	4.545	4.451	4.259	4.160
17	4.869	4.775	4.602	4.484	4.332	4.239	4.049	3.950
18	4.683	4.590	4.418	4.301	4.151	4.058	3.868	3.770
19	4.522	4.430	4.259	4.143	3.994	3.902	3.713	3.615
20	4.382	4.290	4.121	4.005	3.856	3.765	3.576	3.478
21	4.258	4.167	3.999	3.884	3.736	3.645	3.456	3.358
22	4.149	4.058	3.891	3.776	3.629	3.538	3.349	3.251
23	4.051	3.961	3.794	3.680	3.533	3.442	3.254	3.156
24	3.963	3.873	3.707	3.593	3.447	3.356	3.168	3.070
25	3.884	3.794	3.629	3.516	3.369	3.279	3.091	2.992
26	3.812	3.723	3.558	3.445	3.299	3.208	3.020	2.921
28	3.687	3.598	3.434	3.321	3.176	3.085	2.897	2.798
30	3.581	3.493	3.330	3.217	3.072	2.981	2.792	2.693
32	3.491	3.403	3.240	3.128	2.983	2.892	2.703	2.603
34	3.413	3.325	3.163	3.051	2.906	2.815	2.625	2.524
36	3.345	3.258	3.096	2.984	2.839	2.748	2.557	2.456
38	3.285	3.198	3.036	2.925	2.779	2.689	2.497	2.395
40	3.232	3.145	2.984	2.872	2.727	2.636	2.444	2.341
60	2.912	2.827	2.667	2.555	2.409	2.316	2.118	2.009
80	2.763	2.677	2.518	2.406	2.258	2.164	1.960	1.846
100	2.676	2.591	2.431	2.319	2.170	2.076	1.867	1.749
150	2.564	2.479	2.319	2.206	2.056	1.959	1.744	1.618
200	2.509	2.424	2.264	2.151	2.000	1.902	1.682	1.552
300	2.456	2.371	2.210	2.097	1.944	1.846	1.620	1.483
400	2.429	2.344	2.184	2.070	1.917	1.817	1.589	1.448
∞	2.351	2.266	2.105	1.990	1.835	1.733	1.495	1.338

References

Aitchison, J. and Dunsmore, I. R. (1975). *Statistical Prediction Analysis*. Cambridge University Press, Cambridge.

Atkinson, A. C. (1973). Testing transformations to normality. *Journal of the Royal Statistical Society, Series B*, **35**, 473–479.

Atkinson, A. C. (1981). Robustness, transformations and two graphical displays for outlying and influential observations in regression. *Biometrika*, **68**, 13–20.

Atkinson, A. C. (1985). *Plots, Transformations, and Regression: An Introduction to Graphical Methods of Diagnostic Regression Analysis*. Oxford University Press, Oxford.

Bailey, D. W. (1953). *The Inheritance of Maternal Influences on the Growth of the Rat*. Ph.D. Thesis, University of California.

Bartlett, M. S. (1937). Properties of sufficiency and statistical tests. *Proceedings of the Royal Society, A*, **160**, 268–282.

Baten, W. D. (1956). An analysis of variance applied to screw machines. *Industrial Quality Control*, **10**, 8–9.

Beineke, L. A. and Suddarth, S. K. (1979). Modeling joints made with light-gauge metal connector plates. *Forest Products Journal*, **29**, 39–44.

Berger, J. O. (1985). *Statistical Decision Theory and Bayesian Analysis*, Second Edition. Springer-Verlag, New York.

Berk, K. N. and Picard, R. R. (1991). Significance tests for saturated orthogonal arrays. *Journal of Quality Technology*, **23**, 79–89.

Bethea, R. M., Duran, B. S., and Boullion, T. L. (1985). *Statistical Methods for Engineers and Scientists*, Second Edition. Marcel Dekker, New York.

Bickel, P. J., Hammel, E. A., and O'Conner, J. W. (1975). Sex bias in graduate admissions: Data from Berkeley. *Science*, **187**, 398–404.

Bliss, C. I. (1947). 2×2 factorial experiments in incomplete groups for use in biological assays. *Biometrics*, **3**, 69–88.

Bliss, C. I. and James, A. T. (1966). Fitting the rectangular hyperbola. *Biometrics*, **22**, 573–602.

Box, G. E. P. (1950). Problems in the analysis of growth and wear curves. *Biometrics*, **6**, 362–389.

Box, G. E. P. and Cox, D. R. (1964). An analysis of transformations. *Journal of the Royal Statistical Society, Series B*, **26**, 211–246.

Box, G. E. P. and Draper, N. R. (1987). *Empirical Model-Building and Response Surfaces*. John Wiley and Sons, New York.

Box, G. E. P., Hunter, W. G., and Hunter, J. S. (1978). *Statistics for Experimenters*, John Wiley and Sons, New York.

Box, G. E. P. and Meyer, R. D. (1986). An analysis for unreplicated fractional factorials. *Technometrics*, **28**, 11–18.

Box, G. E. P. and Tidwell, P. W. (1962). Transformations of the independent variables. *Technometrics*, **4**, 531–550.

Brownlee, K. A. (1960). *Statistical Theory and Methodology in Science and Engineering*. John Wiley and Sons, New York.

Burt, C. (1966). The genetic determination of differences in intelligence: A study of monozygotic twins reared together and apart. *Br. J. Psych.*, **57**, 137–153.

Carroll, R. J. and Ruppert, D. (1984). Power transformations when fitting theoretical models to data. *Journal of the American Statistical Association*, **79**, 321–328.

Cassini, J. (1740). *Eléments d'astronomie*. Imprimerie Royale, Paris.

Christensen, R. (1987). *Plane Answers to Complex Questions: The Theory of Linear Models*. Springer-Verlag, New York.

Christensen, R. (1989). Lack of fit tests based on near or exact replicates. *Annals of Statistics*, **17**, 673–683.

Christensen, R. (1990a). *Linear Models for Multivariate, Time Series, and Spatial Data*. Springer-Verlag, New York.

Christensen, R. (1990b). *Log-Linear Models*. Springer-Verlag, New York.

Christensen, R. (1991). Small sample characterizations of near replicate lack of fit tests. *Journal of the American Statistical Association*, **86**, 752–756.

Christensen, R., and Huzurbazar, A. V. (1996). A note on augmenting resolution III designs. *The American Statistician*, to appear.

Cochran, W. G. (1941). The distribution of the largest of a set of estimated variances as a fraction of their total. *Annals of Eugenics*, **11**, 47–52.

Cochran, W. G. and Cox, G. M. (1957). *Experimental Designs*, Second Edition. John Wiley and Sons, New York.

Coleman, D. E. and Montgomery, D. C. (1993). A systematic approach to planning for a designed industrial experiment (with discussion). *Technometrics*, **35**, 1–27.

Conover, W. J. (1971). *Practical Nonparametric Statistics*. John Wiley and Sons, New York.

Cook, R. D. and Tsai, C.-L. (1985). Residuals in nonlinear regression. *Biometrika*, **72**, 23–29.

Cook, R. D. and Tsai, C.-L. (1990). Diagnostics for assessing the accuracy of normal approximations in exponential family nonlinear models. *Journal of the American Statistical Association*, **85**, 770–777.

Cook, R. D. and Weisberg, S. (1982). *Residuals and Influence in Regression*. Chapman and Hall, New York.

Cornell, J. A. (1988). Analyzing mixture experiments containing process variables. A split plot approach. *Journal of Quality Technology*, **20**, 2–23.

Cox, D. R. (1958). *Planning of Experiments*. John Wiley and Sons, New York.

Cramér, H. (1946). *Mathematical Methods of Statistics*. Princeton University Press, Princeton.

Daniel, C. (1959). Use of half-normal plots in interpreting factorial two-level experiments. *Technometrics*, **1**, 311–341.

Daniel, C. (1976). *Applications of Statistics to Industrial Experimentation*. John Wiley and Sons, New York.

David, H. A. (1988). *The Method of Paired Comparisons*. Methuen, New York.

Day, B. B. and del Priore, F. R. (1953). The statistics in a gear-test program. *Industrial Quality Control*, **7**, 16–20.

Day, N. E. (1966). Fitting curves to longitudinal data. *Biometrics*, **22**, 276–291.

Deming, W. E. (1986). *Out of the Crisis*. MIT Center for Advanced Engineering Study, Cambridge, MA.

Devore, J. L. (1991). *Probability and Statistics for Engineering and the Sciences*, Third Edition. Brooks/Cole, Pacific Grove, CA.

Diggle, P. J. (1990). *Time Series: A Biostatistical Introduction*. Oxford University Press, Oxford.

Dixon, W. J. and Massey, F. J., Jr. (1969). *Introduction to Statistical Analysis*, Third Edition. McGraw-Hill, New York.

Dixon, W. J. and Massey, F. J., Jr. (1983). *Introduction to Statistical Analysis*, Fourth Edition. McGraw-Hill, New York.

Draper, N. and Smith, H. (1966). *Applied Regression Analysis*. John Wiley and Sons, New York.

Emerson, J. D. (1983). Mathematical aspects of transformation. In *Understanding Robust and Exploratory Data Analysis*, edited by D.C. Hoaglin, F. Mosteller, and J.W. Tukey. John Wiley and Sons, New York.

Everitt, B. J. (1977). *The Analysis of Contingency Tables*. Chapman and Hall, London.

Fienberg, S. E. (1980). *The Analysis of Cross-Classified Categorical Data*, Second Edition. MIT Press, Cambridge, MA.

Finney, D. J. (1964). *Statistical Method in Biological Assay*, Second Edition. Hafner Press, New York.

Fisher, R. A. (1925). *Statistical Methods for Research Workers*, 14th Edition, 1970. Hafner Press, New York.

Fisher, R. A. (1935). *The Design of Experiments*, Ninth Edition, 1971. Hafner Press, New York.

Fisher, R. A. (1947). The analysis of covariance method for the relation between a part and the whole. *Biometrics*, **3**, 65–68.

Fisher, R. A. (1956). *Statistical Methods and Scientific Inference*, Second Edition, 1959. Hafner Press, New York.

Forbes, J. D. (1857). Further experiments and remarks on the measurement of heights by the boiling point of water. *Transactions of the Royal Society of Edinburgh*, **21**, 135– 143.

Fuchs, C. and Kenett, R. S. (1987). Multivariate tolerance regions on F-tests. *Journal of Quality Technology*, **19**, 122–131.

Garner, N. R. (1956). Studies in textile testing. *Industrial Quality Control*, **10**, 44–46.

Hader, R. J. and Grandage, A. H. E. (1958). Simple and multiple regression analyses. In *Experimental Designs in Industry*, edited by V. Chew, pp. 108–137. John Wiley and Sons, New York.

Hahn, G. J. and Meeker, W. Q. (1993). Assumptions for statistical inference. *The American Statistician*, **47**, 1–11.

Hare, L. B. (1988). In the soup: A case study to identify contributors to filling variability. *Journal of Quality Technology*, **20**, 36–43.

Hartley, H. O. (1938). Studentization and large sample theory. *Journal of the Royal Statistical Society, Supplement*, **5**, 80–88.

Heyl, P. R. (1930). A determination of the constant of gravitation. *Journal of Research of the National Bureau of Standards*, **5**, 1243–1250.

Hochberg, Y. and Tamhane, A. (1987). *Multiple Comparison Procedures*. John Wiley and Sons, New York.

Holcomb, Z. C. (1992). *Interpreting Basic Statistics: A Guide and Workbook based on Excerpts from Journal Articles*. Pyrczak Publishing, Los Angeles.

Holms, A. G. and Berrettoni, J. N. (1969). Chain-pooling ANOVA for two-level factorial replication-free experiments. *Technometrics*, **11**, 725–746.

Inman, J., Ledolter, J., Lenth, R. V., and Niemi, L. (1992). Two case studies involving an optical emission spectrometer. *Journal of Quality Technology*, **24**, 27–36.

Jaswal, I. S., Mann, R. F., Juusola, J. A., and Downie, J. (1969). The vapour-phase oxidation of benzene over a vanadium pentoxide catalyst. *Canadian Journal of Chemical Engineering*, **47**, No. 3, 284–287.

Jensen, R. J. (1977). Evinrude's computerized quality control productivity. *Quality Progress*, **X, 9**, 12–16.

John, P. W. M. (1961). An application of a balanced incomplete block design. *Technometrics*, **3**, 51–54.

John, P. W. M. (1971). *Statistical Design and Analysis of Experiments*. Macmillan, New York.

Johnson, F. J. (1978). Automated determination of phosphorus in fertilizers: Collaborative study. *Journal of the Association of Official Analytical Chemists*, **61**, 533–536.

Johnson, R. A. and Wichern, D. W. (1988). *Applied Multivariate Statistical Analysis*. Prentice-Hall, Englewood Cliffs, NJ.

Jolicoeur, P. and Mosimann, J. E. (1960). Size and shape variation on the painted turtle: A principal component analysis. *Growth*, **24**, 339–354.

Kempthorne, O. (1952). *Design and Analysis of Experiments*. Krieger, Huntington, NY.

Koopmans, L. H. (1987). *Introduction to Contemporary Statistical Methods*, Second Edition. Duxbury Press, Boston.

Lazerwitz, B. (1961). A comparison of major United States religious groups. *Journal of the American Statistical Association*, **56**, 568–579.

Lehmann, E. L. (1975) *Nonparametrics: Statistical Methods Based on Ranks*. Holden-Day, San Francisco.

Lenth, R. V. (1989). Quick and easy analysis of unreplicated factorials. *Technometrics*, **31**, 469–473.

McCullagh, P. and Nelder, J. A. (1989). *Generalized Linear Models*, Second Edition. Chapman and Hall, London.

McDonald, G. C. and Schwing, R. C. (1973). Instabilities in regression estimates relating air pollution to mortality. *Technometrics*, **15**, 463–481.

Mandel, J. (1972). Repeatability and reproducibility. *Journal of Quality Technology*, **4**, 74–85.

Mandel, J. (1989a). Some thoughts on variable-selection in multiple regression. *Journal of Quality Technology*, **21**, 2–6.

Mandel, J. (1989b). The nature of collinearity. *Journal of Quality Technology*, **21**, 268–276.

Mandel, J. and Lashof, T. W. (1987). The nature of repeatability and reproducibility. *Journal of Quality Technology*, **19**, 29–36.

Marquardt, D. W. (1963). An algorithm for least-squares estimation of nonlinear parameters. *SIAM Journal of Applied Mathematics*, **11**, 431–441.

May, J. M. (1952). Extended and corrected tables of the upper percentage points of the studentized range. *Biometrika*, **39**, 192–193.

Mercer. W. B. and Hall, A. D. (1911). The experimental error of field trials. *Journal of Agricultural Science*, **iv**, 107–132.

Miller, R. G., Jr. (1981). *Simultaneous Statistical Inference*, Second Edition. Springer-Verlag, New York.

Milliken, G. A. and Graybill, F. A. (1970). Extensions of the general linear hypothesis model. *Journal of the American Statistical Association*, **65**, 797–807.

Mosteller, F. and Tukey, J. W. (1977). *Data Analysis and Regression*. Addison-Wesley, Reading, MA.

Mulrow, J. M., Vecchia, D. F., Buonaccorsi, J. P., and Iyer, H. K. (1988). Problems with interval estimation when data are adjusted via calibration. *Journal of Quality Technology*, **20**, 233–247.

Nelson, P. R. (1993). Additional uses for the analysis of means and extended tables of critical values. *Technometrics*, **35**, 61–71.

Ott, E. R. (1949). Variables control charts in production research. *Industrial Quality Control*, **3**, 30–31.

Ott, E. R. (1967). Analysis of means – A graphical procedure. *Industrial Quality Control*, **24**, 101–109.

Ott, E. R. and Schilling, E. G. (1990). *Process Quality Control: Trouble Shooting and Interpretation of Data*, Second Edition. McGraw-Hill, New York.

Patterson, H. D. (1950). The analysis of change-over trials. *Journal of Agricultural Science*, **40**, 375–380.

Pauling, L. (1971). The significance of the evidence about ascorbic acid and the common cold. *Proceedings of the National Academy of Science*, **68**, 2678–2681.

Pritchard, D. J., Downie, J., and Bacon, D. W. (1977). Further consideration of heteroscedasticity in fitting kinetic models. *Technometrics*, **19**, 227–236.

Quetelet, A. (1842). *A Treatise on Man and the Development of His Faculties*. Chambers, Edinburgh.

Rao, C. R. (1965). *Linear Statistical Inference and Its Applications*. John Wiley and Sons, New York.

Rao, C. R. (1973). *Linear Statistical Inference and Its Applications*, Second Edition. John Wiley and Sons, New York.

Reiss, I. L., Banward, A., and Foreman, H. (1975). Premarital contraceptive usage: A study and some theoretical explorations. *Journal of Marriage and the Family*, **37**, 619–630.

Ryan, T. P. (1989). *Statistical Methods for Quality Improvement*. John Wiley and Sons, New York.

St. Laurent, R. T. and Cook, R. D. (1992). Leverage and superleverage in nonlinear regression. *Journal of the American Statistical Association*, **87**, 985–990.

Satterthwaite, F. E. (1946). An approximate distribution of estimates of variance components. *Biometrics*, **2**, 110–114.

Scheffé, H. (1959). *The Analysis of Variance*. John Wiley and Sons, New York.

Seber, G. A. F. and Wild, C. J. (1989). *Nonlinear Regression*. John Wiley and Sons, New York.

Shapiro, S. S. and Francia, R. S. (1972). An approximate analysis of variance test for normality. *Journal of the American Statistical Association*, **67**, 215–216.

Shewhart, W. A. (1931). *Economic Control of Quality*. Van Nostrand, New York.

Shewhart, W. A. (1939). *Statistical Method from the Viewpoint of Quality Control*. Graduate School of the Department of Agriculture, Washington. Reprint (1986), Dover, New York.

Shumway, R. H. (1988). *Applied Statistical Time Series Analysis*. Prentice Hall, Englewood Cliffs, NJ.

Smith, H., Gnanadesikan, R., and Hughes, J. B. (1962). Multivariate analysis of variance (MANOVA). *Biometrics*, **18**, 22–41.

Smith, L. S. (1988). Ethnic differences in knowledge of sexually transmitted diseases in North American Black and Mexican-American migrant farmworkers. *Research in Nursing and Health*, **11**, 51–58.

Snedecor, G. W. (1945a). Query. *Biometrics*, **1**, 25.

Snedecor, G. W. (1945b). Query. *Biometrics*, **1**, 85.

Snedecor, G. W. and Cochran, W. G. (1967). *Statistical Methods*, Sixth Edition. Iowa State University Press, Ames, IA.

Snedecor, G. W. and Cochran, W. G. (1980). *Statistical Methods*, Seventh Edition. Iowa State University Press, Ames, IA.

Snedecor, G. W. and Haber, E. S. (1946). Statistical methods for an incomplete experiment on a perennial crop. *Biometrics*, **2**, 61–69.

Stigler, S. M. (1986). *The History of Statistics*. Harvard University Press, Cambridge, MA.

Sulzberger, P. H. (1953). The effects of temperature on the strength of wood, plywood and glued joints. Aeronautical Research Consultative Committee, Australia, Department of Supply, Report ACA–46.

Tukey, J. W. (1949). One degree of freedom for nonadditivity. *Biometrics*, **5**, 232–242.

Weisberg, S. (1985). *Applied Linear Regression*, Second Edition. John Wiley and Sons, New York

Williams, E. J. (1959). *Regression Analysis*. John Wiley and Sons, New York.

Wilm, H. G. (1945). Notes on analysis of experiments replicated in time. *Biometrics*, **1**, 16–20.

Woodward, G., Lange, S. W., Nelson, K. W., and Calvert, H. O. (1941). The acute oral toxicity of acetic, chloracetic, dichloracetic and trichloracetic acids. *Journal of Industrial Hygiene and Toxicology*, **23**, 78-81.

Yates, F. (1935). Complex experiments. *Journal of the Royal Statistical Society, Supplement*, **1**, 181–223.

Younger, M. S. (1979). *A Handbook for Linear Regression*, Duxbury Press, Belmont, CA.

Author index

Subject index